Earth Science

FIFTH EDITION

David M. Quigley

Rachel Santopietro

bju press

Greenville, South Carolina

NOTE: The fact that materials produced by other publishers may be referred to in this volume does not constitute an endorsement of the content or theological position of materials produced by such publishers. Any references and ancillary materials are listed as an aid to the student or the teacher and in an attempt to maintain the accepted academic standards of the publishing industry.

EARTH SCIENCE
Fifth Edition

David M. Quigley, MEd
Rachel Santopietro, MEd

Consultants
Bradley Batdorf, EdD
Timothy Clarey, PhD
Robert Hill, EdD
Emil Silvestru, PhD
Andrew Snelling, PhD
Donna Bishop, MEd
Donald Congdon, MA
Amy Corey, MS
Kimberly Jackson

Biblical Worldview
Tyler Trometer, MDiv

Academic Oversight
Jeff Heath, EdD

Project Editor
Rick Vasso, MDiv

Concept and Cover Design
Aaron Dickey
Sarah Lompe

Page Layout and Design
Katie Cooper
Jessica Johnson
Sarah Lompe

Permissions
Sharon Belknap
Sylvia Gass
Lily Kielmeyer
Carrie Walker

Illustrators
John Cunningham
Terrance Egolf
Zach Franzen
Sarah Lompe
David Schuppert
Del Thompson

Project Coordinator
Donald Simmons

Fourth Edition by Terrance Egolf and Rachel Santopietro

Photo credits appear on pages 699–705.

All trademarks are the registered and unregistered marks of their respective owners. BJU Press is in no way affiliated with these companies. No rights are granted by BJU Press to use such marks, whether by implication, estoppel, or otherwise.

The mineral on the front cover is multicolored fluorite. On the back cover is a sample of red soil. The spine shows a carbon print fossil of a leaf.

BJU Press extends special thanks to Dr. Jonathan Sarfati, Dr. Andrew Snelling, Dr. Timothy Clarey, Dr. Emil Silvestru, Mr. Michael Oard, and Dr. Ron Samec for providing introductions to each of the six units in this textbook.

© 2018 BJU Press
Greenville, South Carolina 29609

Fourth Edition © 2012 BJU Press
First Edition © 1979 BJU Press

Printed in the United States of America
All rights reserved

ISBN 978-1-62856-283-5

15 14 13 12 11 10 9 8 7 6 5 4 3 2 1

CONTENTS

THIS BOOK IS FOR YOU! — IX

1 INTRODUCTION TO EARTH SCIENCE — XIV

CHAPTER 1
THE WORLD OF EARTH SCIENCE — 2
- **1A** WHY STUDY EARTH SCIENCE? — 3
- **1B** A CHRISTIAN APPROACH TO EARTH SCIENCE — 7
- **1C** EARTH SCIENCE IN ACTION — 14

CHAPTER 2
MATTER, FORCES, AND ENERGY — 23
- **2A** MATTER — 24
- **2B** FORCES AND MATTER — 31
- **2C** ENERGY AND MATTER — 36
- **2D** COMPOSITION OF MATTER — 41

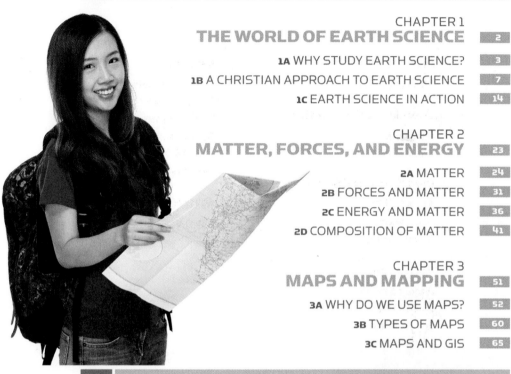

CHAPTER 3
MAPS AND MAPPING — 51
- **3A** WHY DO WE USE MAPS? — 52
- **3B** TYPES OF MAPS — 60
- **3C** MAPS AND GIS — 65

2 THE RESTLESS EARTH — 74

CHAPTER 4
GEOLOGY—THE EARTH SPEAKS — 76
- **4A** THE EARTH, A SPECIAL PLACE — 77
- **4B** GEOLOGY, THE SCIENCE — 85
- **4C** THE EARTH'S STRUCTURE — 90
- **4D** THE EARTH'S NATURAL RESOURCES — 93

CHAPTER 5
THE CHANGING EARTH — 100
- **5A** OBSERVING THE EVIDENCE — 101
- **5B** THE OLD-EARTH STORY — 108
- **5C** THE YOUNG-EARTH STORY — 115

CHAPTER 6
EARTHQUAKES — 128
- **6A** TECTONIC FORCES — 129
- **6B** FAULTS AND JOINTS — 133
- **6C** EARTH WAVES AND SEISMOLOGY — 136
- **6D** EFFECTS OF EARTHQUAKES — 141

CHAPTER 7
MOUNTAINS AND HILLS — 152
- 7A WHAT IS A MOUNTAIN? — 153
- 7B TECTONIC MOUNTAINS — 158
- 7C NON-TECTONIC HILLS AND MOUNTAINS — 165

CHAPTER 8
VOLCANOES AND VOLCANISM — 174
- 8A FIRE MOUNTAINS — 175
- 8B CLASSIFYING VOLCANOES — 184
- 8C INTRUSIVE VOLCANISM — 192

3 EARTH'S ROCKY MATERIALS — 200

CHAPTER 9
MINERALS AND ORES — 202
- 9A DESCRIBING MINERALS — 203
- 9B IDENTIFYING MINERALS — 205
- 9C MINERALS AS RESOURCES — 212

CHAPTER 10
ROCKS — 222
- 10A CLASSIFYING ROCKS — 223
- 10B IGNEOUS ROCKS — 226
- 10C SEDIMENTARY ROCKS — 230
- 10D METAMORPHIC ROCKS — 237
- 10E THE ROCK CYCLE — 242

CHAPTER 11
FOSSILS — 248
- 11A FOSSILIZATION — 249
- 11B PALEONTOLOGY — 258
- 11C FOSSIL FUELS — 265

CHAPTER 12
WEATHERING, EROSION, AND SOILS — 276
- 12A WEATHERING — 277
- 12B EROSION AND DEPOSITION — 281
- 12C SOIL — 293

4 THE WATER WORLD — 302

CHAPTER 13
OCEANS AND SEAS — 304
- 13A OCEAN BASINS — 305
- 13B SEAWATER — 317
- 13C OCEAN ENVIRONMENTS — 323

CHAPTER 14
OCEAN MOTIONS — 330
- 14A TIDES — 331
- 14B CURRENTS — 338
- 14C WAVES — 345

CHAPTER 15
OCEAN EXPLORATION — 355
- 15A THE HISTORY OF OCEAN EXPLORATION — 356
- 15B OCEANOGRAPHY IN ACTION — 360
- 15C ENTERING AN ALIEN WORLD — 368

CHAPTER 16
SURFACE WATERS — 378
- 16A STREAMS — 379
- 16B LAKES AND PONDS — 386

CHAPTER 17
GROUNDWATER — 398
- 17A UNDERGROUND RESERVOIRS — 399
- 17B GROUNDWATER CHEMISTRY — 406
- 17C WATER AS A RESOURCE — 408
- 17D GROUNDWATER LANDFORMS — 414

5 THE ATMOSPHERE — 426

CHAPTER 18
EARTH'S ATMOSPHERE — 428
- 18A WHAT IS THE ATMOSPHERE? — 429
- 18B SPECIAL ZONES IN THE ATMOSPHERE — 439
- 18C ENERGY IN THE ATMOSPHERE — 445

CHAPTER 19
WEATHER — 452
- 19A WHAT IS WEATHER? — 453
- 19B WINDS — 458
- 19C CLOUDS AND PRECIPITATION — 464

CHAPTER 20
STORMS AND WEATHER PREDICTION — 474
- 20A AIR MASSES AND FRONTS — 475
- 20B SEVERE WEATHER — 481
- 20C WEATHER FORECASTS — 495

CHAPTER 21
CLIMATE AND CLIMATE CHANGE — 502
- 21A WHAT IS CLIMATE? — 503
- 21B CLIMATE ZONES — 511
- 21C CLIMATE CHANGE — 515

6 | THE HEAVENS — 528

CHAPTER 22
THE SUN, MOON, AND EARTH SYSTEM — 530
- **22A** THE SUN — 531
- **22B** THE MOON — 539
- **22C** THE SUN, MOON, AND EARTH AS A SYSTEM — 545

CHAPTER 23
OUR SOLAR SYSTEM — 559
- **23A** MODELING THE SOLAR SYSTEM — 560
- **23B** THE PLANETS — 567
- **23C** NON-PLANETARY OBJECTS — 577

CHAPTER 24
STARS, GALAXIES, AND THE UNIVERSE — 587
- **24A** STARS — 588
- **24B** GAS TO GALAXIES — 601
- **24C** THE UNIVERSE AND ITS ORIGIN — 606

CHAPTER 25
SPACE EXPLORATION — 618
- **25A** TELESCOPES — 619
- **25B** ROCKETS, SATELLITES, AND PROBES — 624
- **25C** MANNED SPACE EXPLORATION — 635

BIBLICAL ORIGINS

THE GAP THEORY	7
THE DAY-AGE THEORY	124
PROGRESSIVE CREATIONISM	253
THE ANALOGOUS DAYS THEORY	403
THE FRAMEWORK HYPOTHESIS	509
THEISTIC EVOLUTION	615

CASE STUDIES

MODELING THE SOLAR SYSTEM	17
THE BUILDING BLOCKS OF EVERYTHING	43
NICOLAUS STENO, BISHOP OF GEOLOGY	89
THE ARCTIC QUEST TO UNDERSTAND CLIMATE AND GEOLOGY	107
EARTHQUAKES—THE BIG ONE!	151
BRAD WASHBURN: ADVENTURER, CARTOGRAPHER	157
ROCKS AND THE AGE OF THE EARTH	247
WHAT HAPPENED TO GEORGE WASHINGTON?	280
THE $6,000,000 CLOCK	357
THE INFLUENCE OF RIVERS ON US HISTORY	387
WHITE NOSE SYNDROME	425
SKYDIVING FROM SPACE	431
CO_2 IN THE ATMOSPHERE	522
JOURNEY TO MARS	573
NO SAFE RETURN	638

LIFE CONNECTION

SCIENCE IN MICROGRAVITY	35
GIS AND DISASTER RELIEF	67
AN "OUT-OF-BALANCE" WORLD	99
THE FLOOD, THE ARK, AND SPECIES TODAY	121
QUAKES AND CRITTERS	145
"IN THE ZONE" ON MOUNT KILIMANJARO	170
THEY CAN TAKE THE HEAT	180
MUD PIES FOR MACAWS	219
ROCK SWEET ROCK	224
DEEP, DARK SECRETS	268
LIVING SOIL	294
SWIMMING THROUGH A RAINFOREST	316
OCEAN MIGRATIONS	337
A FARM ON YOUR ARMS	365
BOGS, BAY, AND BLOODTHIRSTY PLANTS	392
CURIOUS CAVE CRITTERS	417
UV LIGHT AND LIFE	448
WINDS AND MIGRATION	463
WHAT HAPPENS TO ANIMALS DURING HURRICANES?	492
ARE POLAR BEARS ON THIN ICE?	525

JUST ADD WATER	570
ASTROBIOLOGY	605
CRITTERNAUTS	645

SERVING GOD AS A(N)...

CARTOGRAPHER	70
SEISMOLOGIST	138
VOLCANOLOGIST	190
SEDIMENTOLOGIST	235
PEDOLOGIST	301
OCEANOGRAPHER	360
SPELEOLOGIST	422
CLIMATOLOGIST	510
ASTROGEOLOGIST	584
AEROSPACE ENGINEER	634

WORLDVIEW SLEUTHING

DRIVERLESS CARS	73
MOUNT ST. HELENS	187
ÖTZI	257
DRINKABLE WATER FROM SEAWATER	320
GREAT PACIFIC GARBAGE PATCH	343
SEVERE WEATHER RESPONSE	495
EXTRATERRESTRIAL INTELLIGENCE	599

APPENDIXES

APPENDIX A: UNDERSTANDING SCIENTIFIC TERMS	649
APPENDIX B: MATH PRINCIPLES AND GRAPHING	651
APPENDIX C: PERIODIC TABLE OF THE ELEMENTS	655
APPENDIX D: TOPOGRAPHIC MAP SYMBOLS	656
APPENDIX E: GEOLOGIC TIME SCALE	659
APPENDIX F: MODIFIED MERCALLI INTENSITY SCALE	660
APPENDIX G: LANDFORM REGIONS OF THE CONTINENTAL UNITED STATES	661
APPENDIX H: RELATIVE HUMIDITY	662
APPENDIX I: TORNADO/HURRICANE CATEGORY SCALES	663
APPENDIX J: SOLAR SYSTEM DATA	664

GLOSSARY 666

INDEX 683

THIS BOOK IS FOR YOU!

Welcome to Earth Science 5th Edition! As you begin this course, here are two people you need to meet.

This guy is a scientist. He enjoys his work and takes it very seriously. He thinks the world and everything in it developed naturally over billions of years. For this reason, he tries to discover how and when things formed. He also thinks that science will eventually provide answers to all the important questions that people have. So he tries to help other people with his work.

This fellow is also a scientist. He loves God and the Bible, which he holds to be the only absolutely reliable source of truth. And because of this, he also loves what he does. He believes that praising God through his discoveries and helping people to live better lives are his highest callings. He is certain that the world is relatively young and a special place because the Bible teaches us these things.

Sometimes these two can agree, and sometimes they can't, especially on answers to scientific questions about events that shaped Earth's history. You will meet them from time to time throughout this textbook. Carefully think about what they say. You'll learn that people's views affect how they see and study the world and the universe.

ABOUT THIS BOOK

Now that you've met our two featured scientists, let's get to know your book. We have arranged the topics in this book so that you can learn about earth and space sciences from the ground up—literally!

Before you begin your journey, you need to pack up some tools for the trip. In Unit 1, you'll learn how to use science to study the earth. You'll learn the nuts and bolts underlying all sciences. And you'll learn how to get around the earth using maps.

When building a house, the foundation is always a good place to start. That's what you will be studying in Unit 2—the foundations of the earth. You'll learn about its history, structures, land features, earthquakes, volcanoes, and plate tectonics—the forces and processes that reshape the earth's surface.

Next, you'll examine what's under your feet. In Unit 3, you'll be looking in some detail at Earth's rocky materials—minerals, ores, rocks, fossils, and soil. These materials have some interesting stories to tell.

Unit 4 finds you up to your neck in oceans, seas, lakes, and rivers. Earth is mainly a water planet. If it has to do with Earth's waters and their importance to us, this is where you will find it.

Next, you'll take a deep breath—and then examine just what goes into that breath! Everything having to do with the atmosphere is the topic of Unit 5. Clean air, weather prediction, storms, and climate are all important issues in our modern world.

In Unit 6, you will take a giant leap into the heavens. You'll see how the sun, moon, and earth work together. The planets, their moons, and all the rest of the space rocks await your exploration. Beyond them, the stars, galaxies, and universe put on a light show like no other. How old is the universe, really?

This is the adventure of earth science. Hopefully, this course will be just the first step you take on a lifelong journey exploring God's good Earth.

HOW TO USE THIS BOOK

Every journey has a path leading to the destination. Along the way, you will need road signs and mile markers to help you stay on the path and measure your progress. Let's discuss features of this textbook that will help you make the most of your trip.

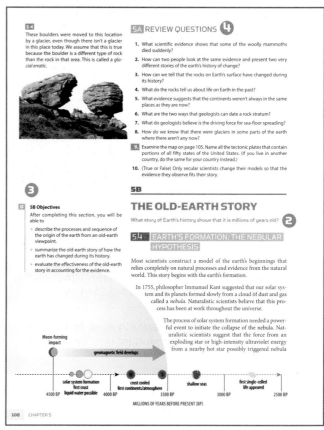

① **Chapter opener**—Each chapter opens with a story that highlights issues and developments in earth science that need to be examined from a biblical worldview.

② **Essential question**—This is the big question that you will learn about in a section.

③ **Objectives**—Marked with an , these are smaller tasks that you can complete along the way through a section to enable you to answer the essential question.

④ **Review questions**—These give you practice in applying what you've learned in a section or chapter. Problem-solving and extra-thought questions are marked with a blue box. For example, a **1.** means that you need to think a little harder to answer Question 1!

⑤ **Find it!**—These boxes provide coordinates for locations related to the text and are marked with a **Fi**. You can enter these locations into Google Earth and fly there.

⑥ **Info boxes**—These boxes offer extra information that pertains to the topic of each chapter, including worldview sleuthing activities, case studies, careers, life connections, and views on biblical origins.

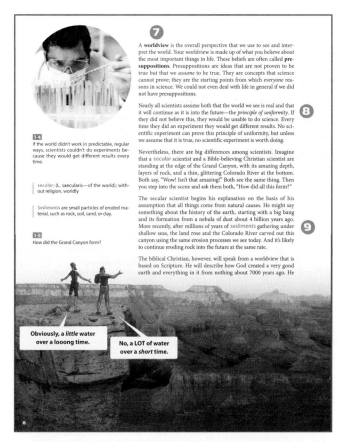

⑦ Bold-faced terms—These are terms that you need to know and are found in the Glossary.

⑧ Italic terms—These are terms that will be defined later in the text or are important in other fields of science.

⑨ Blue terms—These terms have etymologies or extra information in the margin to go along with them.

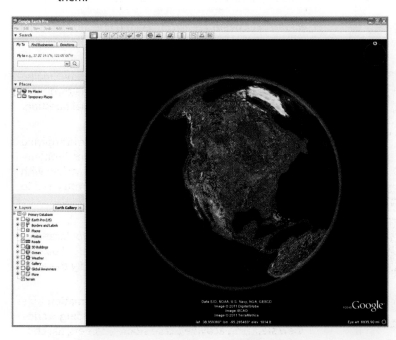

⑩ Chapter summary—This is a handy statement of the big ideas of the chapter, including a vocabulary list.

3D EARTH VIEWER

Perhaps the most valuable thing you will do this year is make virtual visits to the earth and space features that you will study. Imagine being able to fly to the world's tallest volcano, visit a seamount at the bottom of the ocean, or explore the surface of the moon or Mars with the click of a mouse. Today, the tools exist for you to do this! We encourage you to install on your home computer the program Google Earth, freely available on the Internet. Other 3D Earth-viewing programs are also available that are either free or for sale. With these programs, you can visit all the places listed in the **Fi** and examine them in 3D.

The Lab Manual that accompanies this edition of the textbook provides an introductory lesson in using Google Earth. As with any software, Google Earth is being revised and improved all the time, so visit our website for up-to-date instructions at **www.bjupress.com/resources/science/earth-science/**.

1 INTRODUCTION TO EARTH SCIENCE

CHAPTER 1
THE WORLD OF EARTH SCIENCE 2

CHAPTER 2
MATTER, FORCES, AND ENERGY 23

CHAPTER 3
MAPS AND MAPPING 51

DR. JONATHAN SARFATI
CHEMIST AND CREATIONIST

"Christianity makes sense. You shouldn't be scared of logic and reason. They are essential to being a Christian! Without them it is impossible to learn anything from the Bible, the Christian's final authority. This applies to Creation, one of the key teachings of Christianity. Logical reasoning can support the truth of biblical Creation and demonstrate problems in the arguments given by those who hold to evolutionary/old-earth ideas.

Logic is the science of the relationships between propositions, which are statements about the world that we believe are true. Logic can tell us what we can gather from a given proposition. But logic can't tell us whether a proposition is true in the first place. All worldviews rely on starting assumptions—statements called *axioms*—which, by definition, cannot be proved true. It makes sense for us to accept as true the axioms revealed by the infallible God in the 66 books of the Bible. This is how we can love God, not only through faith, but with our whole mind."

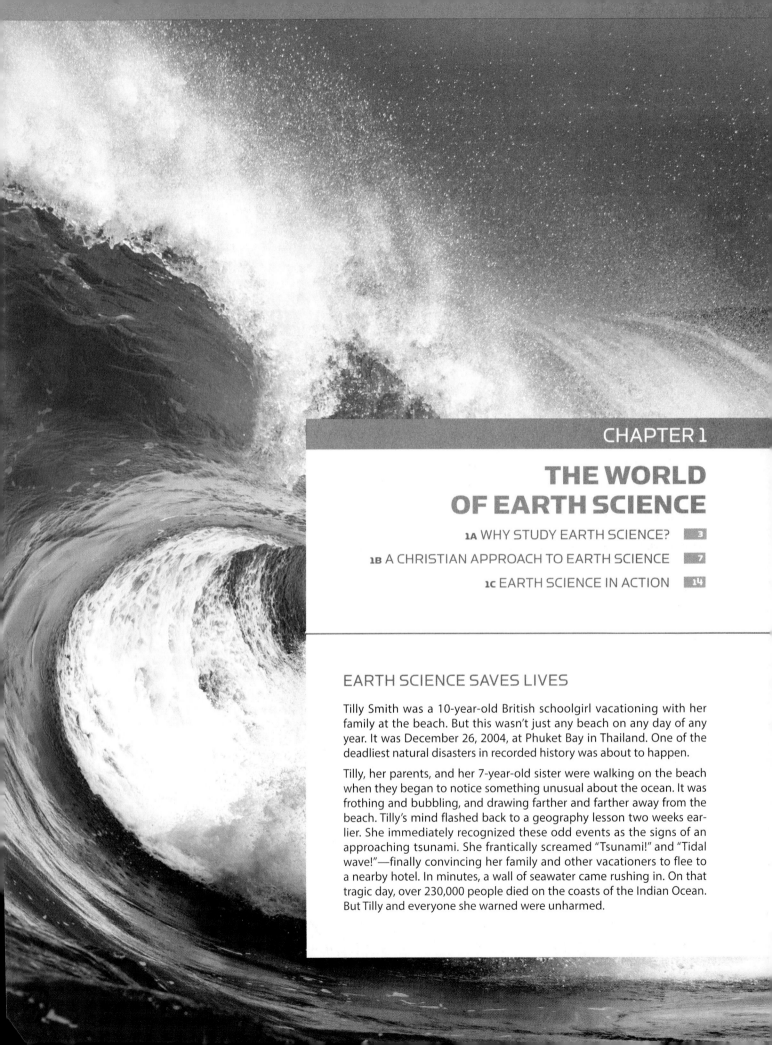

CHAPTER 1

THE WORLD OF EARTH SCIENCE

- **1A** WHY STUDY EARTH SCIENCE? — 3
- **1B** A CHRISTIAN APPROACH TO EARTH SCIENCE — 7
- **1C** EARTH SCIENCE IN ACTION — 14

EARTH SCIENCE SAVES LIVES

Tilly Smith was a 10-year-old British schoolgirl vacationing with her family at the beach. But this wasn't just any beach on any day of any year. It was December 26, 2004, at Phuket Bay in Thailand. One of the deadliest natural disasters in recorded history was about to happen.

Tilly, her parents, and her 7-year-old sister were walking on the beach when they began to notice something unusual about the ocean. It was frothing and bubbling, and drawing farther and farther away from the beach. Tilly's mind flashed back to a geography lesson two weeks earlier. She immediately recognized these odd events as the signs of an approaching tsunami. She frantically screamed "Tsunami!" and "Tidal wave!"—finally convincing her family and other vacationers to flee to a nearby hotel. In minutes, a wall of seawater came rushing in. On that tragic day, over 230,000 people died on the coasts of the Indian Ocean. But Tilly and everyone she warned were unharmed.

1A
WHY STUDY EARTH SCIENCE?

How can we use earth science to fulfill God's commands to the human race?

1.1 HELPING PEOPLE

You may be wondering, "Why do I have to study earth science this year?" You can see one good reason from Tilly Smith's story—earth science is a powerful tool to help others.

Christians, of all people, should want to help others. In Genesis 1, we see that God made people to have dominion over the earth (Gen. 1:26–28). This command, sometimes called the **Creation Mandate**, includes the idea that people should fill the earth and study it to discover the best ways to use it.

As we study and work, we should use the knowledge we get to help other people. We know this is true because Genesis 1 also reveals that we are valuable to God. We are made in God's own image (Gen. 1:26–27). Though we are finite and He is infinite, in our own limited way we are like God. So we need to engage in dominion in a way that helps other people because people are important to God, and they should be to us too.

This is what Jesus was talking about when he gave the two most important commandments. The first states, *Thou shalt love the Lord thy God with all thy heart* (Mark 12:30; see also Deut. 6:5). And the second, Jesus said, was like the first: *Thou shalt love thy neighbour as thyself* (Mark 12:31; see also Lev. 19:18).

1A Objectives
After completing this section, you will be able to
» explain why Christians do science.
» explain how earth science helps Christians declare God's glory.
» defend the idea that earth science can play a role in God's work of redemption.

1-1
Coast guard vessels sail the seas doing search and rescue and maintaining ocean buoys that collect valuable data about the oceans.

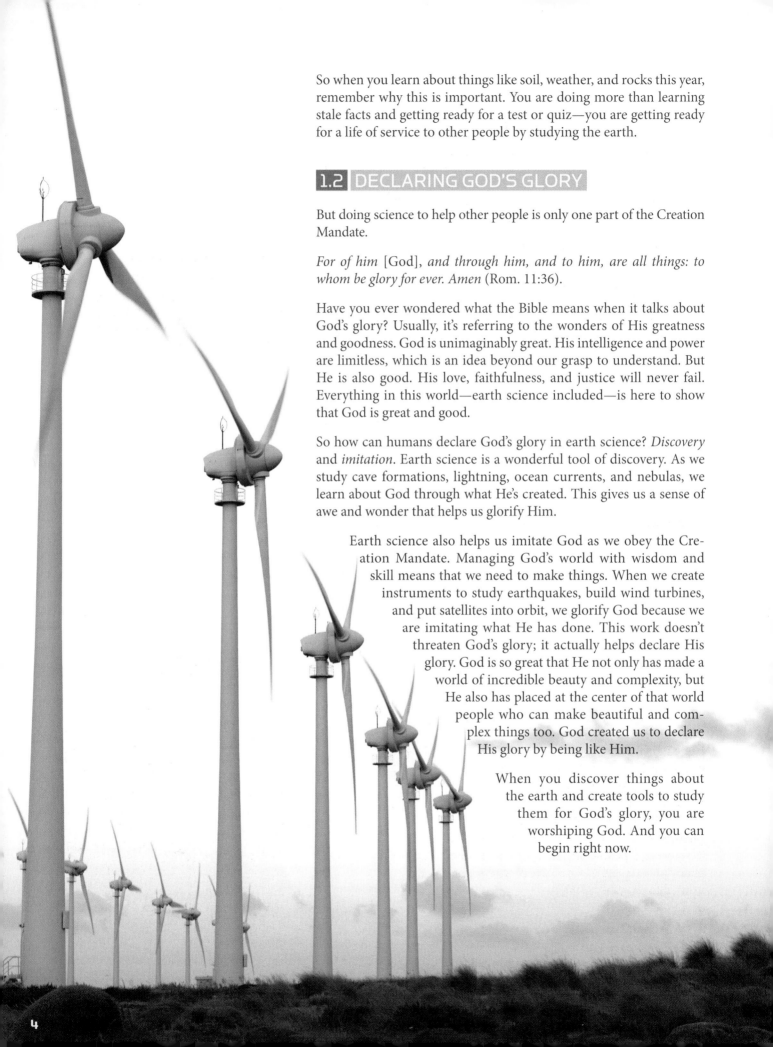

So when you learn about things like soil, weather, and rocks this year, remember why this is important. You are doing more than learning stale facts and getting ready for a test or quiz—you are getting ready for a life of service to other people by studying the earth.

1.2 DECLARING GOD'S GLORY

But doing science to help other people is only one part of the Creation Mandate.

For of him [God], *and through him, and to him, are all things: to whom be glory for ever. Amen* (Rom. 11:36).

Have you ever wondered what the Bible means when it talks about God's glory? Usually, it's referring to the wonders of His greatness and goodness. God is unimaginably great. His intelligence and power are limitless, which is an idea beyond our grasp to understand. But He is also good. His love, faithfulness, and justice will never fail. Everything in this world—earth science included—is here to show that God is great and good.

So how can humans declare God's glory in earth science? *Discovery* and *imitation*. Earth science is a wonderful tool of discovery. As we study cave formations, lightning, ocean currents, and nebulas, we learn about God through what He's created. This gives us a sense of awe and wonder that helps us glorify Him.

Earth science also helps us imitate God as we obey the Creation Mandate. Managing God's world with wisdom and skill means that we need to make things. When we create instruments to study earthquakes, build wind turbines, and put satellites into orbit, we glorify God because we are imitating what He has done. This work doesn't threaten God's glory; it actually helps declare His glory. God is so great that He not only has made a world of incredible beauty and complexity, but He also has placed at the center of that world people who can make beautiful and complex things too. God created us to declare His glory by being like Him.

When you discover things about the earth and create tools to study them for God's glory, you are worshiping God. And you can begin right now.

1.3 LIVING FOR REDEMPTION

But working at dominion isn't easy. The world we live in isn't only created; it's also fallen. It's filled with tragedy, pain, and death. Things are this way because the human race has fallen into sin. When Adam rebelled against God, He changed His very good creation to work against us (Gen. 3:1–19). So everything you will study this year is cursed and broken. In fact, we are under God's curse. On our own we can't see the earth as it should be seen.

Is God doing anything about this? That's where God's redemption comes in. In the Bible, redemption means restoring something to the way it should be by paying a price. Christians often use the word to refer to God's salvation of individual sinners. He saves them from slavery to sin by buying them with the price of the blood of Christ (1 Pet. 1:18–19; 1 Cor. 6:20). That is certainly an accurate view of the term. However, redemption can also refer to God's work in the past, present, and future to restore His fallen creation (Rom. 8:20–23). Redemption can also refer to the way a Christian lives his life. Paul tells believers to be busy "redeeming the time" (Eph. 5:16). Instead of imitating the unwise, sinful choices of unbelievers, Christians should live in this world the way humans were always meant to live. Christians should love others, work hard, be generous, be pure, and keep their promises. They should endeavor to live redemptively in this present evil age.

What does all this mean for earth science? A Christian can use earth science to live redemptively in this day dominated by scientific unbelief. They can use earth science to exercise dominion according to biblical principles. Earth science helps us deal with the problems of a fallen world. It also can help us understand God's world in the light of God's Word. Earth science can help Christians live richly redemptive lives.

Redemption and Earth Science

Respond to the following statement:

"God's work of redemption is His work of saving people from their sins. It has nothing to do with studying earth science or how we use earth science in life."

HEALING HURTS, HEALING HEARTS

When Jesus sent His disciples to preach the **gospel**, He sent them to imitate Him. The disciples not only preached, but they also healed just as Jesus did. Healing people's hurts was important because it played a key role in the gospel they preached. These acts of mercy were proofs that God's Redeemer had indeed come. As Christians multiplied, they spread across the whole earth. They were a people whose hearts had been changed—a people who loved others and wanted to do good to them.

gospel: gos- (O.E.: god—good) + -spel (O.E: spel—spell); i.e., good news

1-2
Children in South Sudan enjoy an ample supply of clean water available to them through wells dug by Samaritan's Purse. Meeting the physical needs of others provides us access to meet their spiritual needs with the gospel.

Today, earth science can be a powerful tool for helping and preventing hurt. Technology can provide clean, fresh water for thirsty people. Weather satellites can predict the motion of hurricanes. Tsunami warning systems can warn of devastating waves hours before they arrive. Wind models and satellites can help direct airlines to avoid clouds of volcanic ash.

As Christians get involved in this kind of work, they can save lives and help people live better lives. If Christians do this with love and concern, they can show others—the people they work with or the people who benefit from their labors—that Christianity is no storybook fable. It is real. Jesus has redeemed their lives, and He wants to redeem the lives of many other people too.

REDEEMING THE MIND

What does God redeem when He saves a person? Certainly He redeems people from a guilty conscience (Heb. 10:22) and a sinful way of life (1 Pet. 1:18). But He also redeems His people from wrong thinking. The human mind is part of who we are, and it is fallen. Because of sin, our thinking is flawed (1 Cor. 2:14; Eph. 4:17–18). For this reason, some of the Bible's most important statements about redemption focus on how God saves His people from their fallen, corrupt minds (Rom. 12:2; Eph. 4:23; Col. 2:3).

So growing in godliness involves more than learning just to act the way Scripture teaches that Christians are to act. A believer also needs to think the way Christ would have him to think (2 Cor. 10:5). This applies, of course, to having pure and wholesome thoughts, and also to the friends we choose and what we do with our free time. But it also applies to earth science. Wrong thinking is easy to spot in earth science. Many people believe that earth science proves that the Bible is not true.

In contrast, Christians examine the earth and its systems from a Christian viewpoint. We have the advantage of building on the strong foundation of biblical assumptions, allowing us to reach sound scientific conclusions. What you learn this year could get you ready to help others see the glory of God in what He's made. In fact, this could better equip you to be a part of God's work of redemption.

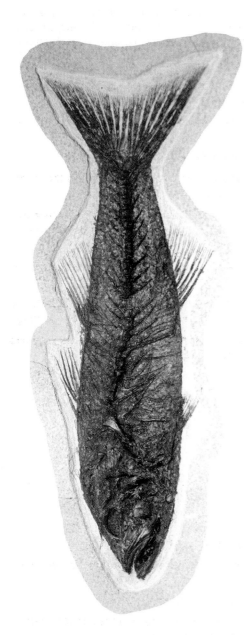

1-3
How old is this fish fossil, and what story does it tell? The answer to this question depends not on the evidence but on your worldview.

1A REVIEW QUESTIONS

1. Why should a Christian want to help others?
2. What is the Creation Mandate?
3. What are the two great commandments?
4. What is the glory of God?
5. Give two ways that earth science helps us declare the glory of God.
6. (True or False) Studying Mars's surface with robotic rovers is not part of exercising good and wise dominion.

BIBLICAL ORIGINS: THE GAP THEORY

The Bible is the Word of God (2 Tim. 3:16). Since God knows everything and does not lie (Ps. 147:5; Titus 1:2), we must trust in all that the Bible says. Certainly, we must believe what the Bible says about God, Jesus, and eternal life. But we also must believe what God says about the origin of the earth and its history. If we take the Bible seriously, we learn that the earth is about 7000 years old. This view is called the *young-earth creationist view*. If you were to ask someone, "How old is the earth?" you could find out right away whether he believed something else.

And yet there are Christians who believe something else. They think that the Bible records the key events of history, but they also believe that its record needs to fit modern scientific theories. Let's look at one common interpretation of the Bible's description of Creation.

THE GAP THEORY

Some Christians believe that after the world was created in Genesis 1:1, it appeared somewhat like it is now. It was full of plants and animals, and populated by humans that lived long before Adam and Eve. They believe that sometime later, it was utterly destroyed, possibly because God banished Lucifer from heaven (Ezek. 28:12–16). This triggered a great spiritual battle that produced sin and corruption in the earth. A global flood completely destroyed the earth in the process. Then God started over, creating the earth again as the story continues in Genesis 1:2 and following.

Most Gap theorists believe millions or billions of years could fit between Genesis 1:1 and 1:2. During this time, the sediments and fossils of the geologic column were deposited. Many generations of people lived and died. This idea ultimately allows Christians to believe in both the Bible and old-earth geology and evolution. The Scofield Bible has popularized this view of origins for generations of Christians.

The Gap theory has serious problems. Can you explain why?

Questions to Consider

1. According to the Bible, what was the reason that death entered the world?
2. If fossils were formed before the world's destruction due to the fall of Lucifer, then what evidence do we have for the Genesis Flood?
3. Read Genesis 1:2. Could the fossils we see today come from before that time in history?
4. Why is it dangerous for a Christian to say that between Genesis 1:1 and 1:2 is a history several billion years long?
5. Does the Gap theory help make standard geology and the Bible agree, as intended?

1B

A CHRISTIAN APPROACH TO EARTH SCIENCE

How does a biblical worldview affect earth science?

1.4 EARTH SCIENCE AND WORLDVIEW

You might be confused about how a Christian should view earth science. You know that there are some earth scientists who say things that are wrong. At least, you think they are wrong. But what is right? The difference is not *what* we look at, like rocks, weather, and stars. The difference is *how* we look at them. Or, to say it another way, it's a difference of worldviews.

1B Objectives

After completing this section, you will be able to

» define *worldview*.
» explain how one's worldview is a key part of doing science.
» compare and contrast the secular and Christian worldviews.
» explain how models are important to science.
» define *science*.

1-4
If the world didn't work in predictable, regular ways, scientists couldn't do experiments because they would get different results every time.

secular: (L. saecularis—of the world); without religion, worldly

Sediments are small particles of eroded material, such as rock, soil, sand, or clay.

1-5
How did the Grand Canyon form?

A **worldview** is the overall perspective that we use to see and interpret the world. Your worldview is made up of what you believe about the most important things in life. These beliefs are often called **presuppositions**. Presuppositions are ideas that are not proven to be true but that we *assume* to be true. They are concepts that science cannot prove; they are the starting points from which *everyone* reasons in science. We could not even deal with life in general if we did not have presuppositions.

Nearly all scientists assume both that the world we see is real and that it will continue as it is into the future—the *principle of uniformity*. If they did not believe this, they would be unable to do science. Every time they did an experiment they would get different results. No scientific experiment can prove this principle of uniformity, but unless we assume that it is true, no scientific experiment is worth doing.

Nevertheless, there are big differences among scientists. Imagine that a secular scientist and a Bible-believing Christian scientist are standing at the edge of the Grand Canyon, with its amazing depth, layers of rock, and a thin, glittering Colorado River at the bottom. Both say, "Wow! Isn't that amazing!" Both see the same thing. Then you step into the scene and ask them both, "How did all this form?"

The secular scientist begins his explanation on the basis of his assumption that all things come from natural causes. He might say something about the history of the earth, starting with a big bang and its formation from a nebula of dust about 4 billion years ago. More recently, after millions of years of sediments gathering under shallow seas, the land rose and the Colorado River carved out this canyon using the same erosion processes we see today. And it's likely to continue eroding rock into the future at the same rate.

The biblical Christian, however, will speak from a worldview that is based on Scripture. He will describe how God created a very good earth and everything in it from nothing about 7000 years ago. He notes that man's sin brought God's judgment on the world through a global flood. That flood probably formed the layers of rock, and the retreating floodwaters gouged out the canyon. And seeing this canyon reminds us that God judges sin.

Which worldview is right? The worldview that accepts the Bible as the Word of God is the right perspective. We cannot understand God's world unless we study it through the lens of God's Word.

1.5 EARTH SCIENCE AND MODELS

We are limited in our ability to observe and reason. To understand complicated things, we sometimes have to simplify the problem and leave some information out. In science, this means that a scientist's explanation is never complete or perfectly accurate. But that doesn't mean it isn't useful.

Scientific **models** have been created for every area of science. Models explain, describe, or represent something in the world. No usable scientific knowledge exists apart from an explanatory model. Models are valuable because they *work*—they are useful for understanding the world. The main purposes of science are to make existing models better and to create new models as new things are discovered.

Scientists interpret and use the models they create within their worldview. Many models are workable within both secular and Christian worldviews. But some secular scientists create and agree with models that describe the world in a way that does not agree with the Bible. They even try to use their models to show that the Bible is not true. Christian scientists, on the other hand, try to create models that accurately model the world but also are faithful to the Bible. Nowhere in science are the models created in the two worldviews more different than in biology and earth science.

Earth science has models for the solar system, the universe, hurricanes, the inside of the earth, and so on. Without a useful model of how tsunamis move through the ocean, Tilly Smith could never have recognized the symptoms of an approaching tsunami and saved lives.

1.6 STRUCTURE OF SCIENCE

Scientific models do not just pop up fully formed from the imagination of a scientist. Most models take time to develop and there is structure to them. Scientific models include *hypotheses*, *theories*, and *laws*, and can be both physical and mathematical. They can include things like scaled versions, mathematical equations, graphs of data, and computer programs that simulate real-world processes. Each kind of model has a specific role in furthering scientific knowledge.

When a scientist has a question to investigate, he first suggests a **hypothesis**. A hypothesis is an initial explanation for the problem. It also provides a direction for testing the hypothesis.

Wrong and Right Thinking

How are the following quotations examples of wrong and right thinking?

"The biblical story of the perfect and finished creation from which human beings fell into sin is pre-Darwinian mythology and post-Darwinian nonsense."

Bishop John Shelby Spong, 1999

"I believe that one day the Darwinian myth will be ranked the greatest deceit in the history of science."

Søren Løvtrup, 1987

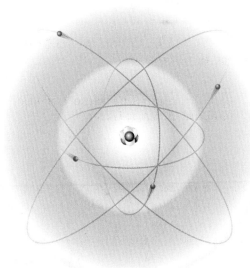

1-6
This model of an atom is useful, but it does not show how a real atom is constructed.

hypothesis, p. -ses (hye PAHTH eh sis, -sees): hypo- (Gk. *hypo*—under) + -thesis (Gk. *thesis*—placing). To suggest a hypothesis for an observation places it under an explanation.

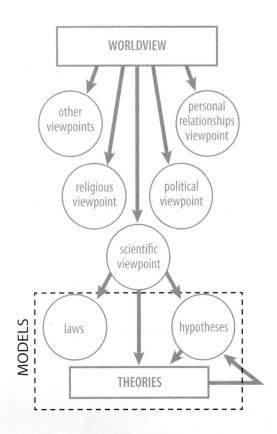

Scientists may have to revise hypotheses. They may even try different ones until they discover one that works well to explain the problem they are working on. After they have extensively tested it, they can use it to explain other observations. Scientists may even use the hypothesis to make predictions about things they haven't observed yet.

As the original hypothesis and related hypotheses become more accepted by scientists, they begin to view them as parts of a **theory**. Theories are models that scientists use as frameworks to explain their observations. Some theories are very broad and include many other theories that address more limited concepts. ***Theories explain***.

Not all scientific models explain. Some describe the way different kinds of measurable physical quantities relate to each other. These models are scientific **laws**. For example, the moon orbits the earth under the influence of *gravitational attraction*. All objects exert a gravitational force on all other objects because of the *matter* they contain. This mathematical equation states the *law of universal gravitation*.

$$F_g = G \frac{Mm}{r^2}$$

In this equation, *symbols* represent measurable quantities. F_g stands for the gravitational force, M is the mass of the earth, sun, or another planet, m is the mass of an object, like a satellite, r is the radius or distance between them, and G is a constant number. This law allows us to describe the gravitational force between two objects at any distance apart. ***Laws describe***.

Many scientific laws like this can be expressed as equations. Notice that there is no explanation given for why these quantities are related this way. It simply describes how they are always associated when considering gravity. The law just works.

Other laws in science are sentences of words rather than equations. They state rules or relationships found in nature. Scientific laws such as these model the ordinances that God created to govern the operation of the universe. They are *not* perfect statements of God's laws, but are our imperfect attempts to describe the laws, known only to God, by which He governs the universe.

1.7 DATA

To be useful, models must be based on observations. **Data** is any information scientists collect by observing nature. When they measure something, they collect **measured data**. They use instruments with scales, like rulers and thermometers, to produce reliable data. Scientists measuring the same quantity using the same instrument should get the same results. They can also be counted, which may not require an instrument. They are numbers that are usually combined with units, like inches or grams.

When scientists observe and write descriptions of their observations, they collect **descriptive data**. Descriptive data normally looks like words, not numbers. A geologist may describe a mineral as having a glassy luster or a rosy color. But another geologist's description of the mineral could be a bit different. So, descriptive data may not be as reliable or repeatable as measured data.

Sometimes, scientists take their data and analyze it to produce *derived* data. You could too. For example, you could create a graph of sea level data taken every hour over a 24-hour period. The graph in Figure 1-7 uses two kinds of data—sea level and time. You could then use this graph to see how fast the sea level changes by finding the *slope* of the graph in feet per hour. (See Appendix B for a discussion of slope of a graph.) You can also easily note low and high tides. These bits of information weren't part of the original data. They were derived by observing relationships among the data points and are definitely useful.

data: (L. *datum*—that which is given). Something known or assumed to be true. In common usage, data may be either singular or plural.

derive (dee RIVE): (L. *derivare*—to turn a river, as from its riverbed); to obtain information from other information

1-7

Changes in sea level at Anchorage, Alaska, over a 24-hour period

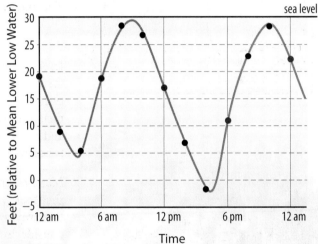

1-8

When the tides change in Alaska, they can produce waves called *tidal bores* that surfers can ride.

Look at Figure 1-7 again. What if the scientist had recorded only one data point? Would that single piece of information have been worth anything? What if he had taken data every six hours, as indicated by the red dots? Would their pattern alone have shown the actual motion and range of the tide? Earth scientists need to be sure that they gather enough data so that they can draw correct conclusions, or *inferences*, from the data. This makes the model work better and predict better. Years of collected data about tsunamis and how they work made it possible for Tilly Smith to save so many lives.

inference: (L. *inferre*—to carry in, show, infer); drawing a conclusion from something known or assumed to be true

1.8 SO, WHAT IS SCIENCE?

Defining science is not so easy. You can see that it includes the idea of carefully observing nature. It involves using these observations to make models that can explain and predict things. It's affected by your worldview. And the reason for it all is to fulfill the Creation Mandate. Consider the following definition:

Science is *the collection of observations, explanations, and models produced through an organized study of nature and the processes found in nature, for the purpose of enabling people to exercise good and wise dominion over God's world.* The word *science* is also used to refer to *the organized methods that produce the observations, explanations, and models.*

This definition of science states both the "what" and "how" behind scientific knowledge. It also declares the "why." Since we will approach scientific studies from a biblical point of view, we need to be sure that we clearly state that science first and foremost is to be used to glorify God and to benefit our fellow human beings. Recall that the Bible says that humans are the bearers of God's image. Helping other humans is a sacred objective of good science.

So **earth science** is simply the knowledge and organized methods of observing the earth and its processes using appropriate tools. It involves explaining these observations and creating models. It means interpreting our observations within the framework of God's Word. And we do it all to glorify God and to help other human beings live safe, healthy, and meaningful lives in a fallen and dangerous world.

1B REVIEW QUESTIONS

1. What is a worldview?
2. How can two different scientists study the same rock formation and give two different explanations for how it formed?
3. Give the major themes in the Bible that frame a Christian worldview.
4. What does it mean that a theory or model is *workable*?
5. Why do scientists make hypotheses?
6. What is the main difference between scientific theories and laws?
7. Contrast measured data and descriptive data.
8. What could happen if a scientist doesn't collect enough data?
9. What should be the features of a good definition of science?
10. (True or False) Presuppositions are what we assume to be true according to our worldview.

> **1C Objectives**
>
> After completing this section, you will be able to
>
> » explain how scientists do science.
> » compare operational and historical science.
> » identify various earth scientists and briefly describe their work.

1C

EARTH SCIENCE IN ACTION

How does earth science work in the real world?

1.9 THE SCIENTIFIC PROCESS

When scientists do their work, they have a plan of attack. They use a **scientific process**. The scientific process separates science from nearly all other human activities. Often the scientific process is taught as a checklist of things that a scientist always does as he works. But this is almost never the way a real scientist works. Other factors—the kind of work he is doing, the types of instruments he is using, and other details—all influence the way he approaches his work. But most scientists follow a similar path when they study something.

❶ FORMING A SCIENTIFIC QUESTION

Scientists usually do not sit around trying to think up a question. They normally are working to solve the many unanswered questions that have already come up from other research. But sometimes a question pops up out of the blue during a conversation, driving down a road, or even from daydreaming! Life presents us with problems and opportunities for exercising biblical dominion.

❷ CONDUCTING INITIAL RESEARCH

Once a scientist has a question to tackle, he has to find out what people already know about the subject. He reads related scientific research papers. He may have to get more education or consult other scientists who know more than he does about the question.

❸ STATING A HYPOTHESIS

If the scientist is an expert in the field, he might be able to suggest a preliminary explanation with little or no research. But after researching the topic, he will be able to form a much more detailed hypothesis that suggests explanations that can be tested. This possible explanation is the most important function of a hypothesis. Without it, the scientist has little direction to his research.

❹ COLLECTING DATA

This is where the scientist can get his hands dirty! Data is needed to test his hypothesis. You probably think of scientific experiments as the most common form of scientific test of a hypothesis. A *controlled experiment* allows the scientist to vary only one factor at a time in order to see how that factor affects the thing he is studying. But you can get data in lots of ways. Scientists can make a collection of samples, draw maps, do field surveys, or even collect data during unexpected events like supernovas or earthquakes.

❺ ANALYZING DATA

During and after data collection, the scientist will create a model. This will organize and explain the data. It is this process that will support or not support his hypothesis. He begins to conclude whether his hypothesis is a good one. Scientists often make minor, or even major, adjustments to their hypotheses as they work through the analysis phase of their research.

❻ DRAWING CONCLUSIONS

After interpreting and analyzing the data, a scientist will decide whether the data supports his original hypothesis. Usually, this is not a yes-or-no result. His conclusions will also include questions that remain unanswered, as well as suggestions for future research.

❼ PUBLISHING RESULTS

No research is complete or useful until it's written down and shared. Scientists need to be able to read about other people's work to help them with their own scientific problems. Scientists also need to be able to check one another's work. Scientific research is usually published in *scientific journals*, such as *Nature*, *Geology*, or *Creation Research Society Quarterly*. Today, there are many specific journals that publish research in nearly every possible scientific and technological field.

1-9 Scientists often have to go where the data is located. A scientist (top) is preparing to take an ice core sample. Scientists in the Antarctic are interested in how the glacier moves; the scientist (middle) is setting up a site for GPS measurements of the glacier (bottom).

THE WORLD OF EARTH SCIENCE   15

1.10 WHAT SCIENTISTS DO

OPERATIONAL SCIENCE

The majority of scientists do the kind of work that we just described in Subsection 1.9. They are trying to find explanations for scientific questions that apply to present-day problems. This kind of science is called operational science because the subjects and results of research operate in the present. An example of this is developing a detailed map of the faults around Los Angeles, California. Such a map could potentially help predict earthquakes in the region.

Operational science investigates events and facts that can be observed in the here and now.

1-10
This is a 3D image of the system of faults (red lines) near the large city of Los Angeles, California.

The results of operational science extend theories and make them more accurate. This work also helps determine the limits of scientific models. So operational science creates more accurate models of the atmosphere. It plots the orbits of newly discovered asteroids. It monitors the changes to Antarctic ice shelves. Most scientists spend their entire lives working within operational science.

HISTORICAL SCIENCE

But how do you scientifically study something that people can't observe today? How can you apply the scientific process to confirm hypotheses that you can't test?

Scientists rely on two key ideas in their work: the **principle of uniformity** and the **principle of cause and effect**. The first assumes that the world operates in a reliable and unvarying way. This means that the same process will always produce the same results. The second principle tells us that for anything that is an observable result of a process—an *effect*—there must be an adequate *cause*. Both of these principles are presuppositions of science—they are beliefs that allow science to work.

1-11
Note this gravity map image of a depression in the crust below the sediments on the floor of the Gulf of Mexico. Is this a meteorite crater formed about 65 million years ago? Or is it something else formed during the Flood about 5500 years ago? Such historical science questions come from one's worldview.

Scientists observe mountains, rivers, oceans, living things, and the sun, moon, and stars today. Figuring out how they got here is **historical science.** As you probably have guessed, historical science relies almost entirely on a scientist's worldview. Did all things come from natural causes? Have all processes we can see *always* operated in the past as they do today? Or was God the cause of all things? And were there times in Earth's history when God made things work very differently than they do today, as the Bible tells us? Earth science has become a major battleground between these two worldviews. You will learn more about historical earth science and the conflict of worldviews in later chapters.

Historical science investigates things that happened in the unobservable past.

CASE STUDY: MODELING THE SOLAR SYSTEM

When you sit on your couch and surf the Internet on your smartphone, you certainly don't feel like you're being flung at 113,301 km/h (70,402 mi/h) though the galaxy. In fact, if you had lived in Europe during the Middle Ages (fifth through fifteenth centuries) you would have thought that anyone telling you such a thing was crazy!

For thousands of years, people thought that the whole sky was a great sphere that revolved around a stationary Earth. Such a model was called the *geocentric theory*.

Sometimes, however, the natural world doesn't behave itself according to our neat little models. As astronomers observed the planets and their moons, a pile of problematic evidence began to build against the geocentric theory. It was time for a new model of the solar system.

Then, in 1543, Nicholas Copernicus's book *The Revolutions* was published (after his death). It made the shocking suggestion that perhaps the earth wasn't stationary. In fact, it wasn't the center of the solar system at all! Instead, the sun was at the center, orbited by other bodies, including the earth. This model is called the *heliocentric theory*. Many people, including some Bible-believing Christians, were very suspicious of these new ideas. This drastic change in the view of the solar system is called the *Copernican Revolution*.

Scientists like Johannes Kepler improved on the heliocentric theory. He collected data on the movement of the planets to develop his *laws of planetary motion*. These laws are mathematical equations that predict the way planets move around the sun. We still use the heliocentric theory today as we send probes into the far reaches of our solar system.

Questions to Consider

1. Suggest why people believed in the geocentric theory for so long.
2. How is the heliocentric theory a model?
3. Why are Kepler's mathematical equations predicting the motions of planets called the laws of planetary motion?
4. How could a person confirm that the sun is indeed the center of the solar system?
5. Many people felt that the Copernican Revolution had an impact on the way people thought about the value of mankind. Suggest why they felt this way.
6. How would you respond to someone who said that modern science has shown that the Bible isn't true because of passages such as Ecclesiastes 1:5, which says that the sun rises and sets?
7. *General* revelation is what God reveals about himself through nature; *specific* revelation is what God reveals about himself through His Word. Is the heliocentric theory part of general revelation? Explain.

SCIENTIFIC EXPERTS

Few people have the experience and knowledge of scientists in their fields. Governments, courts of law, and corporations often consult scientists to help them make decisions. These consultant scientists provide an important service and are paid well for their work.

EDUCATORS

Many research scientists teach or write textbooks. A scientist uses this experience to educate the next generation of scientists. The student scientists who work with these teachers learn the presuppositions of their field and are often the ones who make some of the most important discoveries.

1.11 WORKING IN EARTH SCIENCE

Since earth science covers so much knowledge, nearly all scientists can specialize in one or more areas of earth science. Check out a partial list of earth science specialties below. Maybe you'll find in one of them something you would be interested in pursuing.

GEOLOGISTS

1-12 A petrologist examines mineral crystals under a special microscope.

Geologists study the solid, rocky part of the earth, though their work overlaps with other areas of science. There are lots of different kinds of work that geologists can do:

» *Petrology*—Geologic materials scientists identify rocks and minerals, determining their properties and studying how they formed.

» *Mineralogy*—In this part of petrology, geologic chemists focus on identifying the chemicals in minerals that make up rocks, their crystal structure, and how they formed.

» *Structural Geology*—Structural geologists study the arrangements of rocks, fractures, and faults within rocks and minerals, and the history of their formation.

» *Seismology*—Seismologists study earthquakes, seismic waves, and the earth's interior.

» *Stratigraphy*—Geologists classify and identify strata (layers of rocks) and find underground resources associated with certain rock strata.

HYDROLOGISTS

Hydrologists study surface and underground fresh water. They study the chemical properties, movement, and storage of ground water. *Limnologists* focus on the basin geology, chemistry, biology, and history of lakes.

MARINE SCIENTISTS

A marine scientist studies anything related to the ocean. This would include the ocean's biology, the ocean basin, seawater chemistry, or ocean movements. You can see that marine science overlaps lots of different areas of science. There are marine geologists, marine biologists, marine chemists, marine physicists, and even marine engineers!

METEOROLOGISTS

Meteorologists study weather and model the earth's atmosphere. They study winds, energy movement in the atmosphere, and storms. Recently, some atmospheric scientists have been attempting to model how the atmosphere changes over long periods of time in support of *climatology* studies.

ASTRONOMERS

Astronomy also overlaps many areas of science. Astronomers study everything outside the earth's atmosphere. Space scientists include theoretical astronomers, astrophysicists, observational astronomers, and astrogeologists.

ENGINEERS

Engineers use technology to solve problems. They design and build computers, telescopes, particle accelerators, space probes, microscopes, submarines, communications systems, aircraft, and land vehicles. Engineers design different tools for different areas of earth science. They also take the theoretical results of earth scientists and turn them into useful technology, like power dams and wind turbines.

1-13
Marine scientists sampling ocean water

1-14
Many space science titles begin with the prefix *astro-*, which comes from the Greek word for star. Astronauts study many aspects of earth science. Their view from space gives them a unique perspective.

1-15
Engineers are working on using a huge drill like this to bore a tunnel under the Bering Strait between Eastern Russia and Alaska. (Note the man standing on the drill face for scale.)

THE WORLD OF EARTH SCIENCE

1.12 EARTH SCIENCE AND YOU

So what about you? Will you use earth science to make a difference with your life? Think of Tilly Smith, someone who was younger than you. All it took for her to save lives with earth science was being a good student of a good teacher. But the teacher needed the knowledge obtained from scientists who went out and observed tsunamis and their effects. Someone must do this kind of work. Why not you?

But obtaining knowledge is not the greatest need in science. Tragically, many earth scientists have a secular worldview. True dominion science needs people with a biblical worldview to take up careers in earth science as their calling in life. We need to reclaim earth science for God's glory and for good and wise dominion. Who knows how God can use you in the future?

1-16
Tilly Smith saved many lives by remembering a lesson on tsunamis. What you learn in school *does* matter!

1C REVIEW QUESTIONS

1. How is scientific work different from nearly all other human activities?
2. Why is initial research important to the scientific process?
3. A mineralogist is studying different kinds of beryl, a mineral found all over New Hampshire. How should she collect data?
4. Contrast operational science and historical science.
5. Name the kind of scientist who studies each of the following:
    a. waves generated by earthquakes
    b. the geography of the ocean floor
    c. rocks on Mars
    d. the basins of the Great Lakes
6. (True or False) Data collection relies entirely on scientific experiments.

CHAPTER 1 REVIEW

CHAPTER SUMMARY

» Earth science is a powerful tool for helping others because it is one way to fulfill God's commands.

» We can declare God's glory through discovering His works in nature and imitating His creativity.

» Christ's redemption restores both people and the world.

» Redeeming earth science means restoring us to the work of dominion and helping people in a fallen world.

» A worldview is the perspective and beliefs through which we understand and live in the world.

» Secular scientists view earth science differently than Christians do because of a difference of worldviews.

» A scientific model is a workable explanation or description of something in nature.

» The main purpose of science is to use models to account for how the world works.

» Hypotheses, theories, and laws serve different purposes in science.

» Data is the raw material of scientific models. It can be either measured or descriptive.

» To be useful, there needs to be enough data to make valid inferences.

» A definition of science should include what is studied, how it is studied, and why we need to use science as a tool to exercise good and wise dominion over the earth.

» Science is different from other human activities because it uses a special approach called the *scientific process*—a sequence of activities that scientists follow in order to ensure that their work meets the standards of scientific work.

» The two major types of science are operational science and historical science.

» Earth scientists may serve society in many ways and in many different kinds of occupations in the field of earth science.

REVIEW QUESTIONS

1. Why is helping people important to God?
2. What is our role on God's earth?
3. Give at least two examples of how we can help others with earth science.
4. Why do secular and Christian scientists often arrive at different conclusions about the same geologic data?
5. Contrast the time frame of a secular view of Earth's history with that of a biblical view.
6. Why are scientific models incomplete or not perfectly accurate?

Key Terms

Term	Page
Creation Mandate	3
worldview	8
presupposition	8
model	9
hypothesis	9
theory	10
law	10
data	11
measured data	11
descriptive data	11
science	12
earth science	13
scientific process	14
operational science	16
principle of uniformity	16
principle of cause and effect	16
historical science	17

7. What is the most important quality of a good scientific model?
8. Discuss the difference between a hypothesis and a theory.
9. Compare a theory and a law.
10. What are the three kinds of data discussed in this chapter?
11. Define *earth science* in your own words.
12. List the activities usually associated with the general scientific process.
13. Give one example each of operational science and historical science. Which kind of science can usually produce more accurate models?
14. Identify the earth scientist who does the following work:
    a. maps the cracks and faults in large units of rock
    b. creates models that explain the formation of tornadoes
    c. observes the heavens using various kinds of telescopes
    d. maps undersea mountain ranges
    e. studies the flow of ground water

True or False

15. Using the knowledge we gain from earth science to improve human life, like building dikes to hold back the sea, imitates God's creative works.
16. Using scientific knowledge properly is just one way to worship God.
17. Showing mercy to others through our work in earth science is a very effective way to show the truth of the gospel.
18. When we change the way we view the earth's story so that it agrees with the Bible, we demonstrate that our thinking is being redeemed.
19. A scientist's presuppositions are not as important as the observed data itself when suggesting a hypothesis.
20. A good hypothesis should agree with a scientist's presuppositions and explain all known observations.
21. Scientific theories explain, while laws describe.
22. The fact that a fossil weighs 3.3 lb (1.5 kg) is a piece of descriptive data.
23. Science is done mainly to satisfy human curiosity.
24. God designed the world to operate in an understandable, repeatable way.
25. The conclusions of historical science are not as certain as those of operational science.
26. A scientist who studies the composition and structural strength of reinforced concrete is a petrologist.
27. The only way earth science will be profitable to you is if you become a geologist or some other kind of earth scientist.

CHAPTER 2
MATTER, FORCES, AND ENERGY

2A MATTER — 24
2B FORCES AND MATTER — 31
2C ENERGY AND MATTER — 36
2D COMPOSITION OF MATTER — 41

PARTICLES OF THE UNIVERSE

About 100 meters (300 ft) underground near the Switzerland-France border, the world's largest and most complex scientific instrument operates. This instrument took 14 years to construct at a cost of 7 billion dollars. What is its main purpose? To study the smallest particles of which everything is made. These particles are even smaller than the protons, neutrons, and electrons you already know about.

The instrument studying these particles is a particle accelerator called the *Large Hadron Collider (LHC)*. Particle accelerators take particles and speed them up to over 99% of the speed of light. That is over 184,000 miles per second! Scientists cause these particles to collide with each other. Out of these collisions, they hope to discover new, even smaller particles. They think that these particles are the key to understanding the universe and its origin.

> **2A Objectives**
>
> After completing this section, you will be able to
> - show the impact of worldview on science.
> - describe matter and the different forms it can take.
> - describe how matter changes from one state to another.
> - demonstrate three ways to measure matter.

2A

MATTER

What is matter and why is it important to earth science?

2.1 WORLDVIEW AND DARK MATTER

Why do we spend so much time and money on the LHC? Its size and expense show how eager people are to explain the world around them. The collider will likely reveal the nature of the tiniest particles that make up matter. This is a valid and worthy goal.

However, the real reason researchers keep the LHC going is to explain the structure of the universe on the basis of the current scientific model. This model says that over billions of years galaxies were shaped by gravity, and we know that gravity is caused by matter. However, there are many observations that don't fit with this model of the universe. It seems that the visible matter in the universe doesn't provide enough gravity to shape it.

To solve this problem, scientists have suggested that there must be a lot of matter in space which is not visible. This unseen matter would provide the mass needed to explain these observations. They call this substance *dark matter*. Physicists hope that LHC experiments will reveal the particles that make up dark matter, if it exists.

So the efforts of thousands of physicists from all over the world are aimed at providing a better explanation of the universe. Some scientists expect that the LHC experiments will demonstrate that the universe began and has changed through completely natural processes. Others will view evidence that is based on the Bible's story. These two competing worldviews have scientists looking at the same data through very different perspectives.

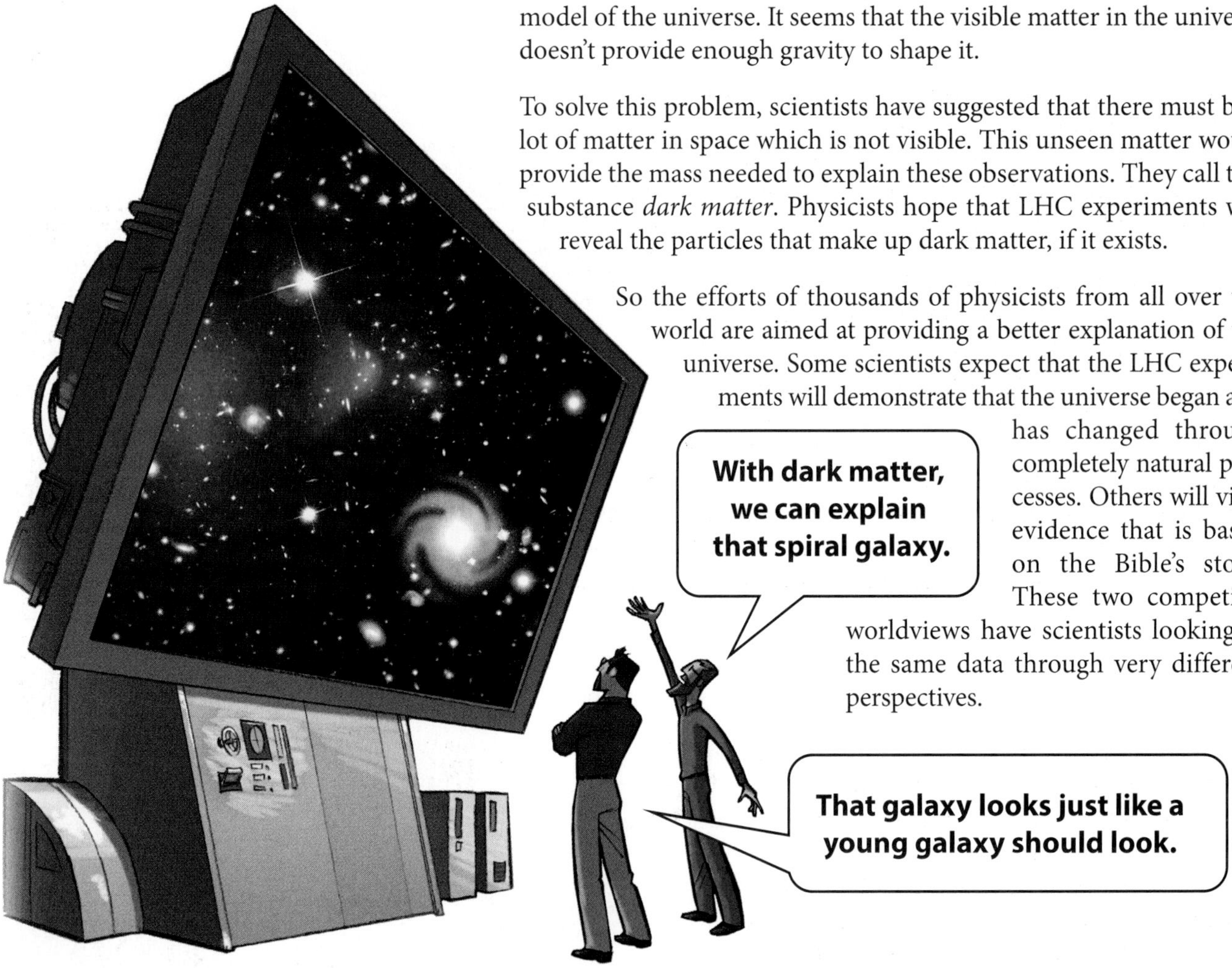

With dark matter, we can explain that spiral galaxy.

That galaxy looks just like a young galaxy should look.

2-1 These astronomers have colored a region of space that they believe is occupied by a significant amount of dark matter.

2.2 WHAT IS MATTER?

Scientists define **matter** as anything that occupies space and has mass. This kind of statement is an *operational definition*. Most definitions use more basic words to describe a concept. But in science, that is often not possible. Defining matter is just such an example. To work around this problem, an operational definition states the tests that the thing defined has to meet.

So anything that is matter must fill a three-dimensional space or *volume*. And it must have *mass*. Mass is a measure of an object's tendency to resist a change in motion. Mass gives things weight.

Looking around, you can see all kinds of matter. There is soft matter, hard matter, liquid matter, and gaseous matter. Some matter is living and some is nonliving. Matter is classified by what it is made of, or its *composition*. It is also classified by how it appears and acts, or its *physical properties*.

An *operational definition* of a scientific term tells what tests something must pass to meet the definition.

Matter that comes from living things is called *organic matter*. Matter that does not come from living things is called *inorganic matter*.

KINDS OF MATTER

Shortly after Creation, humans discovered that the earth held **pure substances** (see Gen. 2:11–12, 4:22). These are materials made of just one kind of particle. Some examples of pure substances are gold, pure water, and pure oxygen.

Mixtures are physical combinations of two or more substances. Parts of a mixture can be separated relatively easily, perhaps by filtering, by using a magnet, by evaporation, or by crushing. Mixtures may be *solutions*, uniform mixtures such as sugar and water, or salt and water. Granite is a nonuniform mixture of three kinds of minerals—dark biotite, light-pinkish feldspar, and white or clear quartz. These substances can be seen as distinct crystals within granite. Even your blood is a mixture of many kinds of blood cells, chemicals, and water. Foods like mayonnaise, gelatin, and whipped cream are also mixtures.

STATES OF MATTER

Scientists in the 1700s and 1800s began to suspect that matter was made of really tiny particles, so small that they couldn't see them with microscopes. Eventually, after much reasoning and experimentation, they concluded that matter must be made of separate particles that are always moving and have some attraction to each other. This model helped to explain the **states of matter**. For example, water can be ice, liquid water, or steam. Most pure substances can be in one of these three states, or *phases*.

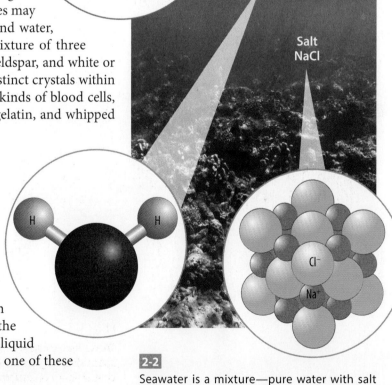

2-2
Seawater is a mixture—pure water with salt and oxygen dissolved in it.

1 Temperature is the hotness or coldness of an object or substance. It is actually the measure of how fast an object's particles are moving, or their average kinetic energy. But we can't measure their energies directly, only how hot or cold the object containing them is.

Scientists measure temperature with thermometers and other instruments sensitive to temperature. Units of temperature are degrees Fahrenheit (°F), degrees Celsius (°C), or kelvins (K).

A substance's properties and **temperature** usually determine its state of matter. Particles in a cool substance move slowly. As the substance warms, its particles move faster. Similar particles also attract each other, so to move farther apart, they must overcome their attraction. In **solids**, the attraction between particles is greater than their ability to move away from each other. Their particles vibrate in a fixed place. Solids hold their shape, have a constant volume (their size doesn't change), and can't be easily squeezed into a smaller volume. The particles in **liquids** can move around but are still close together. Liquids do not have a fixed shape, but their volumes are constant. They can't be squeezed into a smaller volume. The particles in **gases** are far apart and move very fast. They have almost no attraction for each other. Gases don't have a fixed shape or a constant volume. See Table 2-1 for a summary of the properties of the three common states of matter.

But what about the sun? Is it a solid, liquid, or gas? The matter in the sun and all visible stars is in the **plasma** state—the fourth state of matter. Comparing the size of stars to everything else in the universe, you could say that the most common state of matter in the universe is plasma. Plasmas act like gases, but their particles move so fast and collide so hard that they break apart and become electrically charged. The plasma of the surface of the sun is approximately 8700 °F (4800 °C).

2-3
A solar prominence is an arc of plasma from the sun's surface.

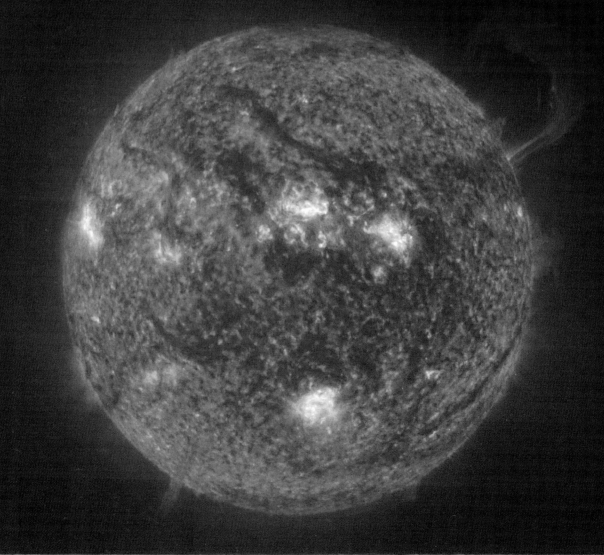

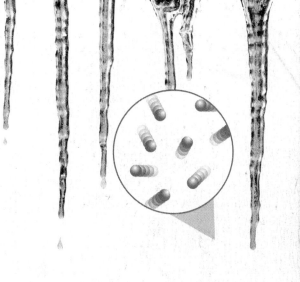

Table 2-1
States of Matter

Property	Solid	Liquid	Gas
Temperature (see note below)	cold to cool	warmer	hot
Shape	fixed	variable	variable
Volume	fixed	fixed	variable
Compressible?	no	not easily	yes
Motion of particles	vibrate in fixed positions	somewhat faster; mobile	fast; not affected by other particles
Distance between particles	close	close	distant

Temperatures in this table are compared to the temperature at which the substance is solid. Most metals and rocks can remain solid up to many hundreds of degrees.

2.3 CHANGES OF STATE

As a substance warms or cools, it can change state. This should not surprise you, since you have seen this happen many times yourself.

MELTING AND FREEZING

The particles in matter are always moving. In a solid they are always vibrating in a fixed position. As a solid warms, the particles vibrate faster. When it gets warm enough, the attraction between the particles is not enough to hold them in their fixed positions. They start to move apart. This process of a solid turning to a liquid is called **melting**.

If a liquid cools, the opposite happens. The particles slow down until their attraction to each other locks them into fixed positions. This is called **freezing**, or *solidification*. The temperature at which a substance melts or freezes is the same; we call it the *melting point* (or *freezing point*).

VAPORIZATION AND CONDENSATION

Have you ever left a glass of water on a window sill for several days? What happened? The water disappeared. What happens if you strongly heat a pan of water on a stove? It bubbles and also reduces gradually. In both cases, the liquid water becomes a gas. The gaseous state of a pure liquid substance is its *vapor*. The change from the liquid state to a vapor is **vaporization**. Let's see how these two forms of vaporization occur.

> **1** The vapor state of a substance is not visible to the eye. What is commonly called vapor, such as clouds or "steam" from a kettle, is tiny drops of liquid.

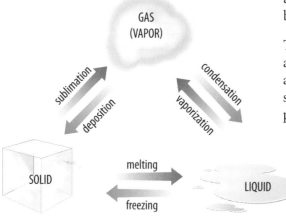

BOILING

Vaporization can happen quickly in a process called **boiling**. As the liquid heats up, its particles move faster and push on each other more to exert pressure. When the pressure becomes sufficiently high, the particles push far enough apart to form a bubble. Ever think about what is in those bubbles at the bottom of a pot of boiling water? They are bubbles of water vapor, not air. The temperature at which these bubbles form is the *boiling point*.

The boiling point depends on the kind of liquid and the air pressure above it. If a liquid has strong particle attraction, higher temperatures are needed to give the particles enough speed to change into the vapor state. As the particles try to form a bubble of vapor, air particles try to push the vapor particles back into the liquid. The higher the air pressure, the harder it is for liquid to vaporize. So a higher temperature is needed to boil the liquid.

Lower air pressure requires lower temperature for boiling. People who live at higher elevations, such as Denver, Colorado, where air pressure is lower, must boil food longer to cook it. Due to the lower air pressure, the boiling point of water is lower than it is at sea level. As a result, they are cooking food at a lower temperature and will have to cook the food longer.

EVAPORATION

Vaporization can also occur at temperatures below the boiling point. This is a slower process called **evaporation**. Particles in a liquid move around, bumping and jostling each other. Sometimes several particles will bump a certain particle at the same time, giving it a strong push. If this particle happens to be at the surface of the liquid, it may be able to escape the attraction of its neighbors and turn into a vapor. Evaporation can occur at any temperature between the melting point and the boiling point of the liquid. The warmer the liquid, the faster it evaporates.

2-4 Hot springs are natural sources of groundwater heated by geothermal processes. Due to the high temperature of the water, evaporation takes place rapidly.

CONDENSATION

The opposite of vaporization is **condensation**. As a vapor cools, the particles slow down and have less energy. When they collide, they no longer have enough energy to overcome their attractive forces. As vapor particles clump together, they begin to form droplets of liquid. These droplets can either float in the air or collect on cool surfaces, such as window panes or grass. Floating droplets form clouds or fog, while droplets on surfaces form dew. In fact, the temperature at which dew forms is called the *dew point*, or *condensation temperature*. The dew point depends on how many vapor particles are present. If there are more vapor particles, the dew point will be higher; if there are fewer particles the dew point is lower.

SUBLIMATION AND DEPOSITION

Though occurring less often, solids can change directly into vapor without first becoming a liquid. This process is called **sublimation**. The solid appears to shrink and disappear. You may notice this as ice cubes in a freezer sublimate over many weeks.

Deposition is the opposite of sublimation, as particles change directly from vapor to solid. Particles in the vapor states can collect on cold surfaces, forming the solid state without becoming a liquid first. Frost is the deposition of water vapor on a sub-freezing surface. Snowflakes form when water deposits on particles in the air.

2-5

The image on the left is of a snowflake taken with a scanning electron microscope. Snowflakes are formed by the deposition of solid water on particles in the air.

2.4 MEASURING MATTER

Your teacher hands you a gold nugget. How much metal do you have? There are several ways to answer this question. First, you can measure its **mass**. Mass is the quantity of matter in the nugget. The mass of an object is the same no matter where you measure it. Mass is measured in grams (g) or kilograms (kg) in the metric system.

You can also measure the gold nugget's **weight**. Weight is the gravitational attraction for an object's mass. This *can* vary with location because *gravity* is not exactly the same from place to place. For example, things weigh more at sea level than at the top of a tall mountain. Units of weight include the *pound* (lb) and the *newton* (N). You will learn more about force and weight in Section 2B.

MATTER, FORCES, AND ENERGY 29

WEIGHT VERSUS MASS

"How much do you weigh?" If you were deep in outer space, you could honestly say that you do not weigh anything. On Earth, you weigh less on the top of a mountain than at the beach. This sounds confusing because in our everyday conversation we don't make a difference between weight and mass.

Weight refers to the gravitational pull exerted on an object by another very large object, such as the earth. The farther you are from the center of the earth, the less gravitational pull you experience. That is why you weigh less at high elevations.

If you live in the United States, you probably report weights in pounds (lb). Scientists report weight in metric units called newtons (N). Force is a push or pull on an object, so weight is just a special name for the gravitational force at a particular location.

The weight of an object can be found by multiplying its *mass* by the *gravitational acceleration* at the location of the object. Let's examine each of these quantities.

Mass is the amount of matter in an object. The metric unit for mass is the kilogram (kg). (Some people mistakenly use the kilogram as a unit of weight because a kilogram always weighs the same at the surface of the earth—about 2.2 lb.) Moving an object from the earth to the moon doesn't change the amount of matter in the object. Therefore an object with a mass of 1 kg on the earth would also have a mass of 1 kg on the moon.

What is gravitational acceleration? You probably know that as an object falls, it speeds up. If you drop a ball from the top of a tall building, after one second, it will be falling 9.8 meters per second (9.8 m/s), or 32 feet per second (32 ft/s). After two seconds, the ball will be dropping 19.6 m/s, or 64 ft/s. After three seconds, it will be falling 29.4 m/s, or 96 ft/s, and so on. Notice that the ball's speed changes 9.8 m/s (32 ft/s) every second. This change in speed as the ball falls is called *acceleration*. Near the earth's surface, the average acceleration caused by gravity is a constant 9.8 m/s per second (9.8 m/s/s, or 9.8 m/s^2). In other words, during one second, a falling object will speed up 9.8 m/s. In the English system, gravitational acceleration is 32 feet per second per second (32 ft/s^2). All freely falling objects drop with this acceleration near the earth's surface unless hindered by air resistance.

To calculate the weight of a 5 kg object, the equation would look like this:

$$5 \text{ kg} \times 9.8 \text{ m/s}^2 = 49 \text{ kg·m/s}^2 = 49 \text{ N}$$

Notice that you multiply units together as well as numbers. The combination of metric units kg·m/s^2 is the same as the newton (N). This equation shows that there is a mathematical relationship between mass and weight determined by gravitational acceleration. On the moon, gravitational acceleration is only 1/6 of that on the earth—about 5.3 ft/s^2, or 1.6 m/s^2. Thus an object's metric weight on the moon would be

$$5 \text{ kg} \times 1.6 \text{ m/s}^2 = 8.0 \text{ N}.$$

Geologists and other earth scientists are mainly concerned about mass in their studies of earth materials. They use mass per volume (density) to identify minerals and rocks. The mass of lava that a volcano ejects determines the size of the eruption. Astronomers measure the effects of gravity on objects in space by determining their masses.

So… what is your mass today?

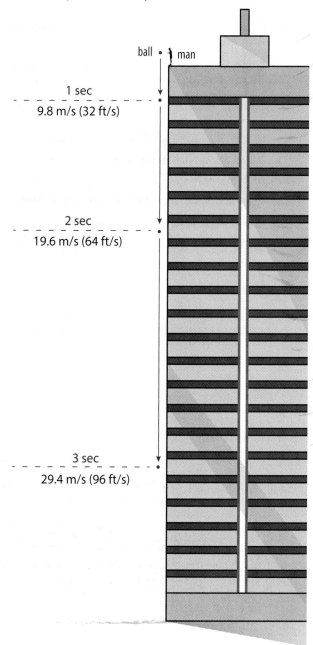

You could also measure the nugget's **volume**, or the space it occupies. Quarts (qt), gallons (gal), cubic inches (cu. in. or in³), liters (L), milliliters (mL), cubic meters (m³), and cubic centimeters (cm³) are all units of volume. Different units of volume may be used for different states of matter.

Measuring is important in science and is a skill you will need for life! You will learn much more about measuring as you explore other areas of earth science and do the lab activities in your Lab Manual.

2-6

A spring scale can be used to weigh an object and is calibrated in units of weight.

2A REVIEW QUESTIONS

1. Define matter. Why is it hard to completely define it?
2. Contrast pure substances and mixtures.
3. What main factor determines a material's state of matter? How does it affect the particles of a substance?
4. What are the differences in the motions and arrangements of particles in solids and liquids? between liquids and vapor?
5. How is the plasma state of a substance different from the gas of the same substance?
6. Name the change of state for each of the following:
    a. Water vapor becomes liquid water droplets.
    b. Molten steel hardens.
    c. On a cold, clear night, ice forms on the windshield of your family car.
    d. A block of dry ice slowly disappears.
    e. A beaker of liquid oxygen bubbles and forms oxygen gas.
7. Name three ways to measure matter.

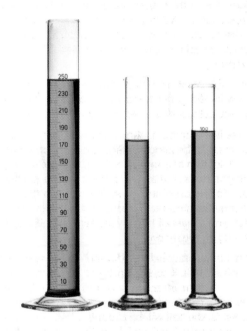

2B

FORCES AND MATTER

How does matter interact to affect the world around us?

2.5 WHAT IS A FORCE?

Did you know that you are using forces right now? Your heart is exerting a force as it pumps blood through your body. You exert a force on your chair, and your chair exerts a force on you.

So what *is* a force? A **force** is a push or a pull. You can see that this is another operational definition. It doesn't really define *force* in more basic terms. It just describes the tests that something must pass to be a force. Besides that, the words *push* and *pull* can't be defined unless you use the word "force."

2B Objectives

After completing this section, you will be able to

» classify forces and identify the various kinds of forces.

» investigate and describe how forces work in the universe.

» discuss the significance of the force of gravity.

When something exerts a force on an object, that object exerts a force back. In fact, all forces act in pairs. When you lean against a wall, you are pushing against the wall, but the wall is also pushing against you. That's why you don't fall through the wall. You can feel the force it exerts.

Fluids like air and water exert forces on objects moving through them. They also can exert lots of force when they flow against fixed objects. Think of the damage done by floodwaters and hurricane winds.

In most situations, scientists are interested in only one of the pair of forces. When manned spacecraft return to Earth, they use the atmosphere to slow down. So when engineers calculate how this will happen, are they interested in the force the vehicle exerts on the atmosphere, or the force the atmosphere exerts on the vehicle? The latter, of course. While the spacecraft does exert a force on the atmosphere, this is of little interest to flight engineers. So, scientists must identify the object that is of interest to them when studying the effects of forces.

2.6 TYPES OF FORCES

CONTACT FORCES

There are two types of forces—contact forces and field forces. *Contact forces* occur when objects touch each other. Here are some important contact forces:

Tension pulls on an object, usually in a straight line. Suspension bridges are held up with tensional forces. Players in a tug-of-war exert tension on a rope. Tension can build up in rocks and release to create an earthquake.

The glacier in this valley is gone, but this U-shaped valley was created by the *friction* between the glacier and the earth. Friction opposes the motion of objects in contact. The valley floor is covered with small rock fragments left by the receding glacier.

Shear causes the particles in an object to move across each other in layers. Shear forces occur when layers of rock move past one another. When you spread a deck of cards in front of you, you exert a shearing force on the deck.

Compression pushes objects closer together. A compression force crushed this car. You use a compression force when you crush a soft drink can.

MATTER, FORCES, AND ENERGY

FIELD FORCES

Field forces act on objects even when separated by distance—they don't have to touch. The object exerting a field force can attract or repel objects in the space around it. This region is called the *field*. Field forces all have a similar property: the force exerted between two objects decreases rapidly as the objects are farther apart. However, the effects of a field force never disappear, even at great distances. Three important field forces are the magnetic force, the electric force, and gravity.

2.7 GRAVITY

Earth scientists are interested in gravity since it affects everything on, above, and within the earth. **Gravity** gives things *weight*. It acts on every particle of a substance. This is why the gas particles in the atmosphere don't float off into space. You stand on the ground because gravity pulls you toward the center of the earth. But the ground is also pushing against your feet. These balanced forces allow you to stand still, neither rising nor sinking. On the other hand, gravity acting on water flowing off a cliff will draw it down because there is no force to hold it back. When the water reaches a pond, a lake, or the ocean, the forces on the water particles are balanced. The water doesn't flow down anymore by gravity alone.

Of all the field forces, gravity's effects are the most noticeable. Gravity is what holds planets, solar systems, and galaxies together. You learned at the beginning of the chapter that many scientists do not believe that there is enough visible matter in the universe to explain the gravitational observations. This is why scientists are looking for dark matter. Since dark matter cannot be directly seen in space, physicists are hoping that the high-energy interactions inside the LHC will produce the super-massive particle to make dark matter here on Earth.

Magnetic force—exerted between magnets and magnetic materials, such as some metals. Magnetism also affects moving electrical charges. The earth and many of the planets and moons in the solar system have extensive magnetic fields.

Electric force—exerted between electrical charges. For example, lightning is caused by the electric force moving charges between a cloud and the ground.

Gravity—exerted by matter on other matter. The earth exerts a gravitational force on everything on its surface and near it in space. For a geologist, this is probably the most important field force that he deals with.

LIFE CONNECTION: SCIENCE IN MICROGRAVITY

Did you know that astronauts in the International Space Station (ISS) experience gravity? Even though they float around, gravity is there. Obviously, the earth's gravity holds the moon in orbit. So if Earth's gravity acts on the moon, then it also acts on anything between here and the moon.

So why do astronauts seem gravity-proof? Being in orbit, astronauts are actually continually falling. Have you ever been on a roller coaster that made you float for a split second? That is similar to being in orbit. Since everything in the ISS is "falling" in orbit around the earth at the same rate, there is no way a surface can push "up" to give the sensation of weight.

This condition where gravity cannot be measured is called a *microgravity environment*. And scientists are using this in the ISS to do some pretty cool experiments. What effect does gravity have on human bones? on the flow of liquids? on plants? on animals? on bacteria that cause disease? Astronauts in the microgravity environment of the ISS are experimenting to get answers that may help us with problems here on Earth.

One of these problems is how to better treat cancer patients. Astronauts on the ISS will conduct an experiment named *CASIS PCG5*. They will be growing crystalline molecules known as *monoclonal antibodies*. These molecules attach to specific molecules in the body to target diseases.

Astronauts will grow the monoclonal antibodies in the microgravity environment on the ISS to allow more in-depth analysis. The researchers expect that results from this experiment will improve drug manufacturing and delivery, as well as actually store the antibodies. We may be able to improve many lives as a direct result of this research.

Astronaut Tim Peake using a microscope to see the fine detail on a microgravity experiment onboard the ISS

2B REVIEW QUESTIONS

1. What is a force? In what two ways can a force relate to an object?
2. What are the two types of forces? How are they different?
3. Name the force responsible for the following:
    a. lightning
    b. sand wearing away rock as water flows over the rock
    c. increasing pressure as you go deeper into the ocean
    d. a chunk of a glacier falling off into the ocean to form an iceberg
    e. a rope holding up a cave explorer
    f. a landslide (two possible answers)
4. What force acts against gravity to keep you from sinking into solid ground when you stand?
5. (True or False) Your feet exert a compressive force on the ground where you stand.
6. Which force is most significant for earth scientists?
7. What is true about the forces acting on a rock sitting on the ground?

2-7
Many forces can act on you at the same time. If they are balanced, then you don't accelerate.

2C Objectives

After completing this section, you will be able to

- » define work and energy.
- » classify different types of energy.
- » discuss the significance of the principle of the conservation of energy.

2C
ENERGY AND MATTER

What is energy, and how does it affect matter?

2.8 WHAT IS ENERGY?

Do you feel like you have energy right now? Could you jump up and run around the block? Or do you just want to sit in front of the TV because you lack energy? What does it really mean to have energy? To a scientist, *energy is the ability to do work.* This statement is yet another operational definition. It really doesn't define energy; it just describes what it does. And what is *work*?

When scientists talk about **work**, they define it as a force that acts on an object as it moves through a distance. When you lift a box from the floor to your shoulder height, you do work on the box. When gravity acts on a boulder falling from a cliff, gravitational work is being done on the boulder.

Doing work on an object changes its **energy**. Energy is added to or taken from an object when work is done on it by a force. When you lift a box, you expend energy lifting it, and the box gains energy as it is lifted. You can see that energy and work are very closely related.

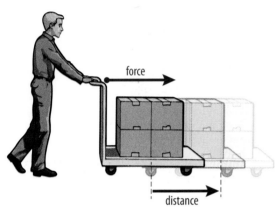

2-8 Mechanical work happens when a force acts on an object that moves through a distance.

2.9 KINDS OF ENERGY

KINETIC ENERGY

Matter can have two forms of energy. A moving object has *kinetic energy*. This form of energy depends on the mass of an object and its speed. The kinetic energy of an object changes as its speed changes, since its mass can't really change. Actually, an object's kinetic energy depends much more on its speed than its mass.

Kinetic energy is important in earth science because so many things move or can be moved. Moving water erodes soil and rocks. Hurricane and tornado winds can destroy buildings. So earth scientists need to understand kinetic energy. The tiny particles in the LHC have very little mass, but they have immense amounts of kinetic energy because of the great speeds they gain as they move around the underground loop.

2-9 A meteorite striking the earth expends its kinetic energy by blasting out a crater and vaporizing earth materials.

POTENTIAL ENERGY

Objects may also have *potential energy*. An object has this kind of energy when a force acting on it gives it the *potential* to move. For example, gravity acts on a boulder teetering at the top of a cliff. The boulder has the potential to fall. Its potential energy depends on its mass and the distance it could fall. The higher it is, the greater its potential energy.

2.10 SOURCES OF ENERGY

Kinetic and potential energy are associated with the physical condition of an object. Scientists identify these as forms of *mechanical energy* because they occur as a result of the action of physical forces on distinct objects, just like the principles by which machines work. There are many other sources of energy earth scientists need to study.

THERMAL ENERGY

You know that all matter contains tiny particles in constant random motion. These particles have different amounts of kinetic energy. They also exert forces on each other, so they create potential energy between particles. No one can measure these mechanical energies in an object, because the particles are too small and too numerous. But the sum total of these energies equals a kind of energy called *thermal energy*. All objects have thermal energy. The hotter an object is, the more thermal energy it has. The flow of thermal energy from a hot place to a cooler place is called *heat*. Any source of heating produces thermal energy. The sun is the largest and most important source of thermal energy in the solar system.

2-10 Mexican Hat Rock in Utah has a large amount of potential energy.

2-11 Glowing molten lava, flowing from Mount Kilauea, Hawaii, contains large amounts of thermal energy.

MATTER, FORCES, AND ENERGY 37

2-12
The compression wave (shock wave) from a T-38 flying at supersonic speeds. Sound waves are compression waves.

SOUND ENERGY

Sound energy also is produced by particle motion. While thermal energy comes from particles vibrating in random directions, sound energy exists when particles in matter vibrate repeatedly back and forth in one direction. The regular motion of particles creates waves in matter called *sound waves*. Any object that vibrates can produce sound energy. Sound waves heard with the unaided ear are called *audible* sounds. Sound vibrations that are too fast to hear form *ultrasonic* sound. If the vibrations are too slow to hear, as in earthquake waves, they are *infrasonic* sounds.

Waves

Energy can move through space and matter using vibrations. The vibrations appear as waves. Waves have properties that relate to the amount of energy they contain. The distance between two similar points on a wave is the *wavelength*. The height of a peak or valley is its *amplitude*. The number of full waveforms that pass a point per second is the wave's *frequency*. Here is a labeled diagram of the parts of a wave.

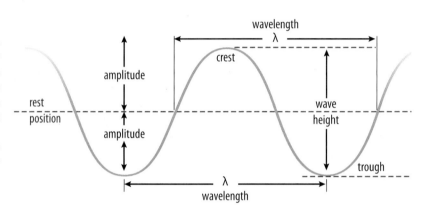

ELECTRICAL ENERGY

Electrical energy is the ability to do work with electrical charges. There are two kinds of electrical charge—positive and negative. Opposite charges attract each other. Like charges repel each other. When an electrical force acts on charges, it produces electric energy. Lightning is an example of electrical energy, though there are many man-made sources too.

MAGNETIC ENERGY

Magnetic energy results from the motion of electrical charges. The moving charges may be inside particles fixed in a magnet. They can be charges moving through wires, or even through space. Magnets have two poles, named the north and south poles. Opposite poles attract and like poles repel as they exert magnetic forces. The energy from magnetic forces results in work on magnetic materials, such as iron. Magnetic forces also cause the movement of charged particles—rotating magnets in electrical generators produce electrical current. Earth itself is a huge magnet, a feature designed by God that you will learn about in later chapters.

2-13 A display of the aurora borealis, or the northern lights, seen here from low-earth orbit, happens when energetic particles become trapped in the earth's magnetic field.

CHEMICAL ENERGY

The particles in matter often form connections to each other. These connections, called *chemical bonds*, produce many of the properties of every kind of matter. Chemical bonds contain stored energy. When these bonds break during chemical changes, they can release the stored *chemical energy*, usually as other forms of energy. Fuels such as coal, oil, and natural gas are important sources of chemical energy.

LIGHT ENERGY

Everything that gives off light produces *light energy*. Visible light is just a small part of light energy. Light energy also includes radio waves, microwaves, infrared light, ultraviolet light, x-rays, and gamma rays. Though we can't see most of these kinds of light, we can use them in many ways in earth science.

NUCLEAR ENERGY

Very strong forces hold the particles in the nuclei of atoms together. When an atom is split, some of the *nuclear energy* that produces these forces is released. Nuclear changes can happen naturally or can be made to happen artificially. In a nuclear reactor, a very small amount of a nuclear fuel, such as uranium, can produce huge amounts of heat that boil water to steam. This steam is used to turn steam turbines connected to electrical generators. In nature, all the energy emitted by the sun starts as nuclear energy.

2-14 Rainbows occur when water droplets break up visible light into the colors present in the light.

MATTER, FORCES, AND ENERGY

2-15

These falling, house-size chunks of ice gain kinetic energy, which is then transferred to the water to form wave energy, heat, and sound. Their original potential energy is conserved in the resulting forms of energy. This photo was taken where Harding Icefield meets Prince William Sound in Alaska.

Thermodynamics is the science of energy and the way it moves between different objects.

2.11 CONSERVATION OF ENERGY

What happens to energy after it is used? Does it just disappear? Is it destroyed? When energy is added to an object as work is done, it is changed into other types of energy. These forms of energy may not be useful in doing more work, but they are still there. So, energy is never lost. It is always saved up. Scientists call this principle the *conservation of energy*. Energy cannot be created or destroyed, but can only be changed from one form to another. This is one way of stating the **first law of thermodynamics**. Of course, Christians believe that all energy (and matter) was created in the beginning by God. This law does not deny God's work. It simply means that in our otherwise orderly universe, no energy is coming into existence out of nothing and no energy is disappearing.

Think of some examples of the conservation of energy. The chemical bonds found in burning wood release thermal energy, sound energy, and light energy. A crashing ocean wave converts its kinetic and potential energy into sound and thermal energy. The concept of conservation of energy will be important to remember when you learn about how the sun warms the atmosphere and the earth.

2C REVIEW QUESTIONS

1. How would a scientist define *work*?
2. Define *energy*. What happens to an object's energy when work is done on it?
3. What are the two kinds of mechanical energy? When a skydiver jumps out of a plane, what factors determine each of these two energies?
4. Identify the main kind of energy released by the objects below.
   a. piano
   b. toaster
   c. a computer monitor screen
   d. static on a sweater
   e. food
5. What happens to energy when it is used? What is this principle called?
6. (True or False) An avalanche of rocks is dangerous because of the huge amount of kinetic energy it develops as the rocks fall.

2D

COMPOSITION OF MATTER

What makes up matter?

2.12 ATOMS AND IONS

ATOMS

You have learned that matter is made of particles. But what are these particles, and why are they important? The smallest building block of matter is the **atom**. This word comes from the Greek word that means "indivisible." Early scientists believed that atoms could not be divided. Since the late 1800s, however, scientists have learned that atoms themselves are made of smaller parts.

According to the atomic model that scientists use today, an atom occupies an extremely tiny spherical space. At its center is the *nucleus*. Two major kinds of particles can be in the nucleus. Protons are always present, and usually there are neutrons. *Protons* have a relatively large mass and a single positive charge. *Neutrons* are slightly more massive than protons but they have no electrical charge. In the space around the nucleus are *electrons*. Each electron carries a single negative charge. You often see diagrams of atoms with electrons moving in circular paths around the nucleus, but their motion is much more complicated than that. Protons and neutrons have about 1840 times more mass than an electron, so you can see that nearly all the mass of the atom is in the nucleus.

2D Objectives

After completing this section, you will be able to

» describe the structure of atoms.
» recognize that protons determine an element's identity.
» compare and contrast ions and atoms.
» distinguish between elements and compounds.
» show how a chemical formula is used to identify the elements in a molecule.
» describe the structure of matter at the atomic level.
» state ways that we can know that different changes of matter have taken place.

2-16
This model of the atom places the nucleus at its center surrounded by electrons.

Atoms are extremely tiny. Consider that 2,900,000 iron atoms could fit across the width of the dot in the exclamation point at the end of this sentence!

Physicists have discovered that the protons and neutrons in atoms are made of even smaller particles. Hundreds of them are known. The LHC was built to discover even more of these truly basic building blocks of matter.

Since a single atom is not normally charged, the number of its electrons must equal the number of protons in its nucleus. The negative charges equal and cancel the positive charges. Naturally occurring atoms can have as many as 92 protons in the nucleus, as is the case with uranium. So neutral uranium atoms must have 92 electrons to balance the 92 protons.

IONS

The negatively charged electrons surrounding the nucleus are held to the atom by the electrical attraction from the nucleus (opposite charges attract). Electrons move freely around the nucleus. Their positions are not fixed as is the case with protons. Sometimes, they can be pulled from their parent atom, or free electrons can be attracted to a neutral atom. When this happens, the positive and negative charges in the atom aren't balanced anymore. If there are more electrons than protons in the atom, it has an overall negative charge. If there are fewer electrons than protons, then the atom has a positive charge. Atoms with more or fewer electrons than protons are called **ions**.

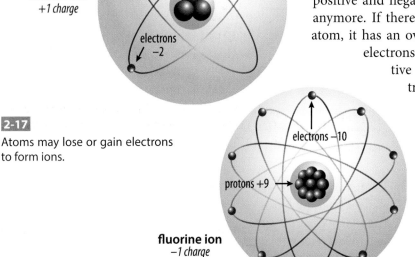

2-17
Atoms may lose or gain electrons to form ions.

Ions exist all over the earth and in space. The oceans are solutions of salt ions. The plasma state of a substance is made entirely of ions. They form during high-energy collisions between hot gas atoms that strip electrons away from the nuclei. Many important earth materials are made of substances containing ions.

2.13 ELEMENTS AND COMPOUNDS

Earlier in Subsection 2.2 you learned that matter can be classified by composition, or the *chemicals* that make up matter. There are two types of pure chemical substances—elements and compounds.

A chemical **element** is a pure substance made of one kind of atom. As of this printing, there are 118 elements. Of these, 92 occur naturally, and the remaining 26 are man-made. The atom of an element is identified by the number of protons in its nucleus. So every atom of hydrogen has 1 proton, every atom of oxygen has 8 protons, and every atom of iron has 26 protons. Elements are identified by their names, and also by the number of protons in their nuclei, called the *atomic number*.

CASE STUDY: THE BUILDING BLOCKS OF EVERYTHING

Of what is everything made? Man has searched for the answer to this question for thousands of years. Aristotle believed that there were only four elements—air, water, earth, and fire. These elements could be cut into infinitely smaller portions. Around 400 BC, Democritus proposed that matter was made of tiny indivisible particles. In fact, the word *atom* means "indivisible" in Greek. Amazingly, our understanding of atomic theory remained unchanged for the next 2000 years!

In 1897, J. J. Thompson discovered what we call the *electron*. Thompson concluded that electrons were a removable part of all atoms. This meant that the model of an indivisible atom was wrong! It was time for a new model. Using his observations, Thompson developed the plum pudding model. Thompson described atoms as solid spheres of positively charged "pudding" with negatively charged "plums" (electrons) spread throughout. You may have never eaten plum pudding, but you can think of this model like a chocolate chip cookie. The cookie is the positive part of the atom, and the chocolate chips are the electrons.

In 1911, Ernest Rutherford made another discovery. He was doing an experiment in which he was firing small, positively charged particles at thin sheets of gold. Most of the particles passed straight through the foil. It was like firing a bullet at a piece of paper and watching it pass through. We would expect that, right? However, 1 out of 20,000 particles bounced off at large angles. That's like the bullet deflecting off the sheet of paper!

Rutherford had to modify the concept of the atom again. He suggested that the atom was made of a very small, very dense, positively charged nucleus with negatively charged electrons travelling around it. This would mean that the atom is mostly open space!

Since that time, many scientists have worked on our understanding of the atom. We discovered the proton in 1911 and the neutron in 1932. Scientists have worked to understand how electrons are arranged and interact. Physicists at the LHC continue to work to discover the particles that make up the protons, neutrons, and electrons, trying to learn how these subatomic particles interact in the world around us.

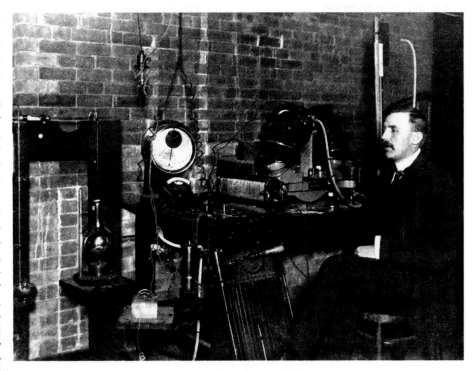

Ernest Rutherford sits in his lab in 1905. Rutherford discovered the positively charged nucleus of the atom. He is also credited with discovering the proton.

Questions to Consider

1. Why is it difficult for us to study the atom?
2. Think back to our discussion of models in Chapter 1. How is our understanding of the atom a model?
3. What effect did J. J. Thompson's discovery of the electron have on the Greek model of the atom? Why?
4. How did Rutherford's experiment demonstrate that the nucleus is very dense?
5. How did Rutherford's experiment demonstrate that the nucleus is very small?
6. According to a biblical worldview, why is the work at the LHC important?

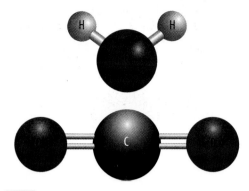

2-18 The compound water always has two hydrogen atoms for every oxygen atom. Carbon dioxide always has two oxygen atoms for every carbon atom. These models show the actual arrangement of the atoms in their molecules.

Pure elements exist in nature. They form most of the earth's atmosphere. The element nitrogen makes up nearly 80% of air, and oxygen makes up about 20%. Our sun and stars are made mainly of a mixture of the elements hydrogen and helium.

The atoms of one element can bond with atoms from other elements to form a **compound**. A compound is a pure substance because it always contains the same elements in the same proportions. The atoms are always bonded in the same way. You've probably heard about a few compounds, such as water, a compound of the elements hydrogen and oxygen, and carbon dioxide, a compound of carbon and oxygen. Most of the matter in the earth's rocks and oceans is made of compounds. You'll explore these compounds later in this textbook. For a visual outline of the classification of matter, study the chart below, which summarizes the kinds of matter and shows how they are related.

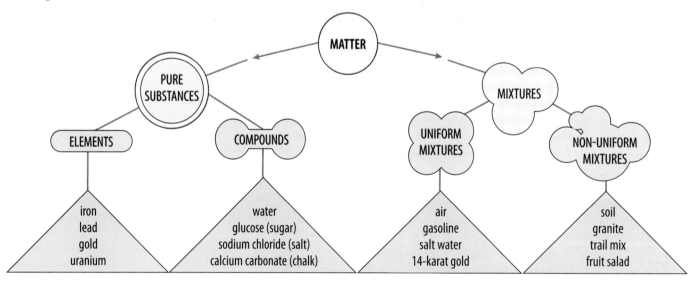

2.14 THE STRUCTURE OF MATTER

God designed atoms to combine in different but useful ways. Some elements exist as single atoms. Under normal conditions, these elements are gases. Examples include the elements helium (He), neon (Ne), and argon (Ar).

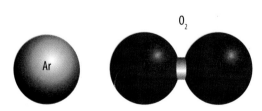

Many gaseous elements consist of two-atom particles. Distinct particles formed by the combination of two or more atoms are called **molecules**. A molecule of oxygen contains two oxygen atoms. The symbol for an oxygen molecule is O_2. Molecules of elements can contain two or more atoms, depending on the element. The element sulfur occurs naturally in 8-atom molecules shaped like a ring (S_8).

Compounds can also occur as molecules. The compound water consists of two atoms of hydrogen and one atom of oxygen (formula: H_2O). Molecular compounds in nature can contain many atoms. The compound methane found in natural gas contains one atom of carbon and four atoms of hydrogen (CH_4). Organic molecules found in living things can have hundreds or even thousands of different atoms.

The most common materials forming the rocky parts of the earth are also compounds. These very rigid materials form when lots of particles join in a repeating pattern called a *crystalline*, or crystal-like, structure. Sometimes they make recognizable *crystals*—solid shapes with regular flat faces—like salt and quartz. Other substances, such as bismuth (right), may have a crystalline particle structure, but they may appear as shapeless masses, like some metals. Atoms and molecules of both elements and compounds can form crystalline structures. Diamond is a crystalline element of only carbon atoms. Sugar crystals result from the bonding of innumerable sugar molecules in a repeating pattern.

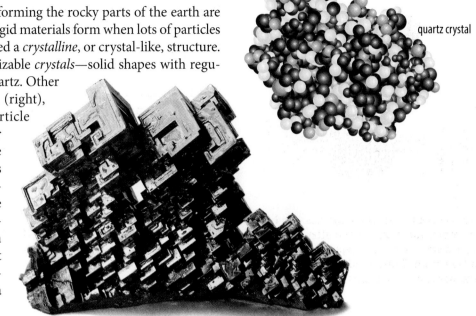

quartz crystal

2.15 CHANGES IN MATTER

You know that matter doesn't always stay the same over time. For instance, you've learned that temperature determines a substance's state of matter. Liquid water, steam, and ice are still water. You have observed metals rusting and wood burning. And what goes on inside a nuclear reactor? What kinds of changes are these?

PHYSICAL CHANGES

When matter changes in a way that doesn't change its chemical identity, it is called a **physical change**. Physical changes result in a change of the appearance of a substance but do not change its chemical composition. Some common physical changes include:

» melting/freezing
» evaporation/boiling/condensing
» crushing
» polishing/grinding
» cutting
» dissolving
» bending

2-19

Diamond and coal are both made of carbon. Pressure and heat cause coal to undergo a physical change to form a diamond.

MATTER, FORCES, AND ENERGY

CHEMICAL CHANGES

When a pure substance changes into another substance, a **chemical change** takes place. There are a few ways this can happen. When two or more pure substances chemically combine, a new substance can form. For example, the gases hydrogen and oxygen combine when they burn to form water. Another way is to heat a compound until it breaks down into elements or other simpler compounds. When you strongly heat the reddish mineral cinnabar, it releases the gaseous element oxygen and the liquid metal mercury. Many other kinds of chemical changes can happen for more complex substances.

Chemical changes, or *chemical reactions*, can be recognized by one or more of the following signs:

- » a release of energy (an explosion of sound or light)
- » a temperature change (either heating or cooling)
- » a permanent color change
- » the appearance of a new or different substance from a mixture of chemicals in solution
- » a gas is produced (not by evaporation)

You will see how chemical reactions form and change earth materials later in this course.

NUCLEAR CHANGES

In unusual circumstances, the very atoms of matter may change. Certain kinds of atoms, especially those with many protons and neutrons in their nuclei, can suddenly and randomly change in one of several ways.

- » Atoms can emit one or more particles.
- » A proton can change to a neutron, or a neutron to a proton.
- » Atoms can emit a gamma ray.

These events are **nuclear changes**. When an atom emits protons, it changes into another element. When an atom emits a neutron, the atom's identity doesn't change, but it has less mass. In the most common change of this sort, a heavy atom will emit a particle made of two protons and two neutrons, called an *alpha particle*. If the atom emits only a gamma ray, there is no permanent change to the atom's nucleus. It just has less energy.

2-20

A solid sulfur/zinc fuel is used to launch this rocket. The sulfur and zinc—both pure substances—react, releasing heat and light and forming a new substance (zinc sulfide).

Sometimes, a heavy nucleus splits into smaller nuclei. This releases lots of energy along with several of its neutrons. This kind of nuclear change is called a *fission*. Fissions are fairly rare in nature, but humans use nuclear fission for electrical power generation.

The sun and stars rely on another kind of nuclear change called *fusion*. The smallest atoms, hydrogen and helium, smash together at high temperatures in the solar plasma to form heavier nuclei. The fusion reactions release huge amounts of energy, even more than fissions. Humans have been able to control this kind of nuclear reaction only in weapons. Scientists are actively researching ways to harness fusion reactions to generate electricity.

CONSERVATION OF MATTER

In this section you have learned how matter can change in certain ways. But when matter changes, it is always still matter. Matter can't be destroyed. This is an important, basic, scientific idea that all scientists believe. It is another part of the first law of thermodynamics. Matter, like energy, can't be created or destroyed. This principle is the *law of the conservation of matter*. In every physical and chemical change, matter is conserved.

Nuclear energy may seem to be an exception, but it isn't. Physicists long ago discovered that matter and energy can be converted from one to the other in nuclear reactions. That is how usable nuclear energy is obtained. Since energy is a property of all matter, scientists have expanded the first law of thermodynamics to cover both matter and energy. So, a complete statement of the first law is, *"Matter and energy can neither be created nor destroyed; they can only change forms."*

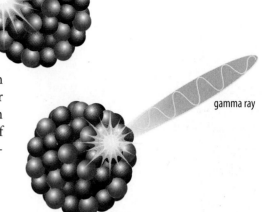
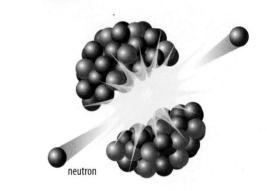

2-21
An atom's nucleus can emit a particle, emit a gamma ray, or split in a fission.

2-22
Some minerals are radioactive. This is a sample of uraninite, or uranium ore.

2D REVIEW QUESTIONS

2-23 Mountains and hills seem immovable, but rockslides represent a significant danger, especially near roadways. A rockslide is an example of a physical change.

1. Discuss the three main subatomic particles. Where are they located within the atom? What are their relative masses? Describe the electrical charge of each.
2. Compare neutral atoms and atomic ions.
3. What is a chemical compound?
4. Using the Periodic Table of the Elements in Appendix C, indicate which element has seven protons.
5. In what three ways can particles in matter be arranged? What kind of particles make up most of the earth's atmosphere?
6. The compound NH_3 is made up of which elements and how many atoms of each element?
7. Describe three ways that matter can change.
8. State the kind of change in matter in each example below:
    a. Ocean waves and sand erode a sea cave in a solid rock cliff.
    b. Fallen snow compacts into glacier ice.
    c. A solid rock decomposes into a different substance (called clay) from exposure to sun and rain.
    d. A sample of radioactive uranium slowly turns into the element lead over a long period of time.
    e. A pond dries up.
    f. Acid rain gradually eats away the features of a limestone statue in New York City.
9. How is the use of nuclear reactions for power generation an example of good and wise dominion science?
10. (True or False) When a log burns, its matter disappears.

CHAPTER 2 REVIEW

CHAPTER SUMMARY

» Without dark matter, scientists have difficulty explaining observation of the universe on the basis of their current models.

» Operational definitions are often needed in science. They set up tests that a defined term has to pass.

» Matter occupies space and has mass.

» There are two main types of matter: pure substances and mixtures.

» Matter can be in one of four states: solid, liquid, gas, and plasma. Each of these states has its own properties.

» The states of matter depend on temperature and a substance's properties.

» Matter can be measured by its mass, weight, and volume.

» A force is a push or pull. There are many types of forces, but the force of gravity is very important in earth science.

» Work on an object results from a force acting on the object as it moves through a distance.

» Energy is the ability to do work. There are many different forms of energy. Energy is never created or destroyed. It only changes form.

» Atoms are made of protons, neutrons, and electrons. When an atom contains unequal numbers of electrons and protons, the atom is a charged particle called an ion.

» There are two types of pure substances—elements and compounds. Elements are made of only one kind of atom. Compounds are made of two or more different elements.

» Matter may undergo physical, chemical, or nuclear changes. Physical changes alter the shape or state of matter but not its composition. Chemical changes alter the chemical identities of substances but not the atoms. Nuclear changes can transform the identities of atoms.

REVIEW QUESTIONS

1. What is dark matter and why are scientists looking for it?
2. How are operational definitions in science examples of the models that scientists create of the world (see Chapter 1)?
3. Draw a diagram that illustrates the relationships between matter, pure substances, mixtures, solutions, elements, and compounds.
4. How do particle motion and distance between particles change as a solid substance becomes a gas?
5. Which kind of vaporization can occur at any temperature between the freezing and boiling points of a substance?
6. If a lunar astronaut wanted to know the quantity of matter in a moon rock that he had collected, would he want to know its mass or weight? Explain.
7. A glass of milk sits on a table. What forces act on the glass of milk?

Key Terms

Term	Page
matter	25
pure substance	25
mixture	25
state of matter	25
temperature	26
solid	26
liquid	26
gas	26
plasma	26
melting	27
freezing	27
vaporization	27
boiling	28
evaporation	28
condensation	29
sublimation	29
deposition	29
mass	29
weight	29
volume	31
force	31
gravity	34
work	36
energy	36
first law of thermodynamics	40
atom	41
ion	42
element	42
compound	44
molecule	44
physical change	45
chemical change	46
nuclear change	46

8. Would you do more work lifting a box from the floor to the top of a table, or from the floor to a shelf above your head? Why?

9. Which will probably break a pane of glass, a small rock sitting on the glass, or the same rock thrown at the glass? Why?

10. State the main kind of energy associated with the following sources.
    a. earbuds
    b. a match before it is struck
    c. the LED on a DVD player display
    d. a battery
    e. a stove burner
    f. a power plant that uses uranium for fuel

11. Does the identity of an uncharged atom change when it becomes an ion? Explain.

12. Why is a compound considered a pure substance?

13. What is the difference between the molecules of an element and the molecules of a compound?

14. Which changes in matter can produce an atom different from the original?

True or False

15. Classifying matter by state is a simple and reliable way to tell mixtures and pure substances apart.

16. The most common state of matter in the universe is the solid state.

17. The more vapor particles in a given volume of air, the higher the condensation temperature, or dew point, will be.

18. Mass, weight, and volume are all ways to measure matter.

19. When you exert a force, two forces are actually at work.

20. Magnetism is the most important field force in geology.

21. Mechanical energy is always related to a force.

22. Sound energy and thermal energy result from different kinds of particle motion.

23. Chemical energy can become other kinds of energy during chemical changes.

24. Some energy is permanently lost when one kind of energy is converted into another.

25. Mechanical work is equal to the product of a force acting on an object and the distance the object moves. If a rock weighing 15 lb falls to the foot of a 55 ft cliff, how much work has gravity done on the rock as it hits the ground?

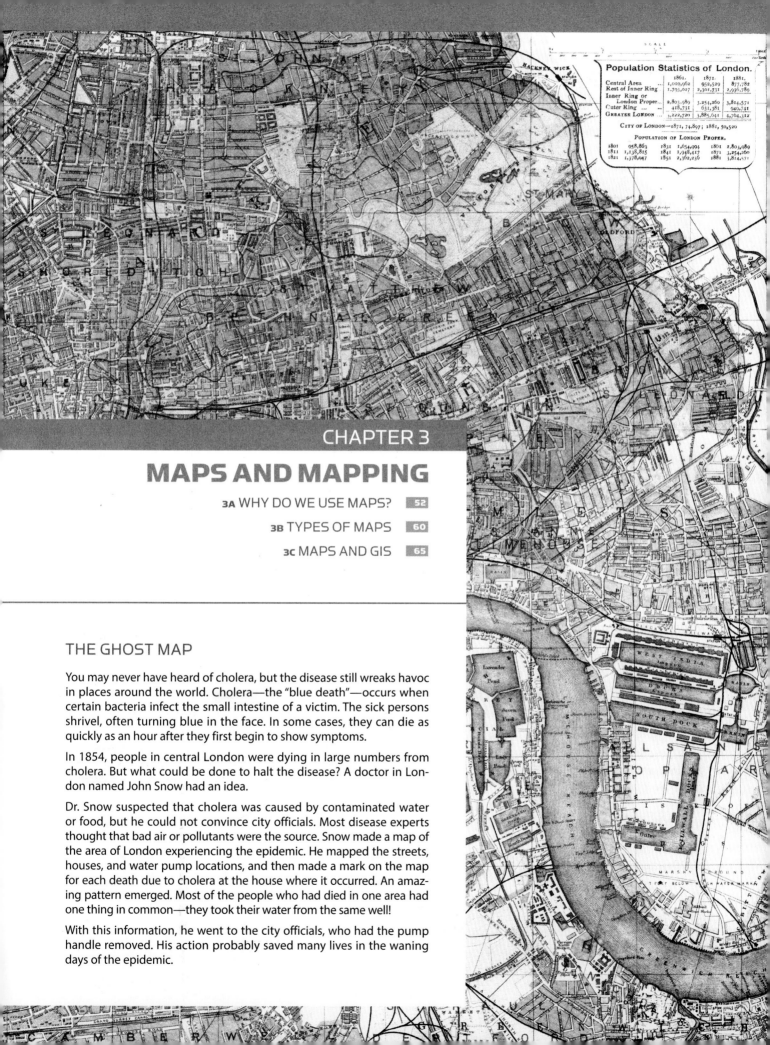

CHAPTER 3
MAPS AND MAPPING

- **3A** WHY DO WE USE MAPS? 52
- **3B** TYPES OF MAPS 60
- **3C** MAPS AND GIS 65

THE GHOST MAP

You may never have heard of cholera, but the disease still wreaks havoc in places around the world. Cholera—the "blue death"—occurs when certain bacteria infect the small intestine of a victim. The sick persons shrivel, often turning blue in the face. In some cases, they can die as quickly as an hour after they first begin to show symptoms.

In 1854, people in central London were dying in large numbers from cholera. But what could be done to halt the disease? A doctor in London named John Snow had an idea.

Dr. Snow suspected that cholera was caused by contaminated water or food, but he could not convince city officials. Most disease experts thought that bad air or pollutants were the source. Snow made a map of the area of London experiencing the epidemic. He mapped the streets, houses, and water pump locations, and then made a mark on the map for each death due to cholera at the house where it occurred. An amazing pattern emerged. Most of the people who had died in one area had one thing in common—they took their water from the same well!

With this information, he went to the city officials, who had the pump handle removed. His action probably saved many lives in the waning days of the epidemic.

3A Objectives

After completing this section, you will be able to
- show why maps are important for life.
- explain how mapmaking is modeling.
- discuss why maps need coordinate systems.
- describe how to find your location on a map.
- summarize standard map features.

3A
WHY DO WE USE MAPS?

What features on most maps help us use them?

3.1 LIFE SAVING MAPS

John Snow's map (see Chapter opener) was an early breakthrough for the science of *epidemiology*—the study of things that affect a population's health. Today government health organizations make similar kinds of maps to locate cities and places within cities that show high rates of certain diseases. Concentrations of cases help officials find the sources of infections or disease so that they can be eliminated. Vaccination programs that target areas with high rates of infections are far more effective and cost less than vaccinating a whole population. These maps have even identified hidden chemical pollution that causes genetic diseases. Maps like John Snow's help us love others by exercising good and wise dominion.

3.2 MAPS AS MODELS

What exactly is a map? A **map** is a simplified image that is a model of part of the earth's surface. As with any model, maps do not perfectly show the earth, but only the information that a user needs. The term *map* usually refers to those used to find your way on land. Aviation maps and maps of the ocean used by sailors are called *charts*.

Early maps tried to show how buildings and natural features would appear from above. This is important because it helps us better measure distances and directions. Showing the earth's surface from a tilted or *oblique view* is a more natural way of viewing a city, as if from a hilltop. But an oblique view makes it harder to measure true distances and tell directions. This is because the scene appears shortened in the direction of the view—a form of distortion.

A *globe* is a special kind of map that shows the earth's surface as it appears from space. Globes show the true shape of continents and oceans on the surface of a sphere.

Today we see maps almost everywhere. But digital maps are becoming much more common. You can find maps of just about any place on Earth on the Internet. These maps can be scaled to any size with just a click of a mouse. You can change their appearance by switching between map drawings and actual satellite images of the earth's surface. And that doesn't even scratch the surface! There are all kinds of portable electronic devices that show your location on a map.

Legend:
- H Ebola treatment center
- L Field laboratory
- L National laboratory
- ○ National capital
- (hatched) No active transmission
- (dark) Areas with confirmed and probable cases
- (gray) Areas reporting suspected cases

3-1 Starting in 2014, West Africa experienced the largest Ebola outbreak in history. This map could help epidemiologists or government workers address the spread and treatment of the disease through the region.

MAP SCALE

Since maps are models, they can't show all the details. That means some information has to be left out. It is impossible to show every feature on the earth's surface on any given map. For example, 1 inch on the map may represent 1 mile in the real world. This ratio, 1 in. = 1 mi, is called the **map scale**. At this scale, a 75-ft house would be a dot about 1/70 inch long! But you might not need a map that shows individual houses. Maps at this scale do better showing roads between towns, entire towns and cities, and lakes.

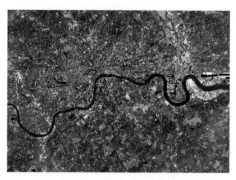

3-2 Both images show London.

Mapmakers, known as **cartographers**, often express the scale of a map as a fraction or ratio of equal units. In our example above, 1 in. = 1 mi gives the ratio 1:63,360. This says that 1 inch on the map shows 63,360 inches on the earth, since there are 63,360 inches in a mile. A scale ratio is useful because you can then apply it to any kind of unit of length. On this map, 1 cm on the map equals 63,360 cm over land, and 1 ft = 63,360 ft. In other words, the size of the map is 1/63,360 of the actual land area that it shows. For convenience, cartographers usually choose round numbers for a map scale, such as 1:10,000 or 1:40,000.

Maps with larger scale ratios show smaller areas, and features appear larger. We call these *large-scale maps*. A cartographer will choose a large scale ratio when he wants to show more details for the area. In a different situation, a cartographer may use a *small-scale map* to show a large area. However, the drawback to small-scale maps is that many of the details are not visible.

Notice the three maps in Figure 3-3. The map showing the United States has a very small scale ratio of 1:170,000,000. You can't see any cities or roads on this map. The map of Florida has a larger scale ratio (1:9,000,000). There you can see cities and some major highways. The third map zooms in on Jacksonville. This map has a scale of 1:1,000,000. You can see major and minor roads, lakes, and rivers on this map.

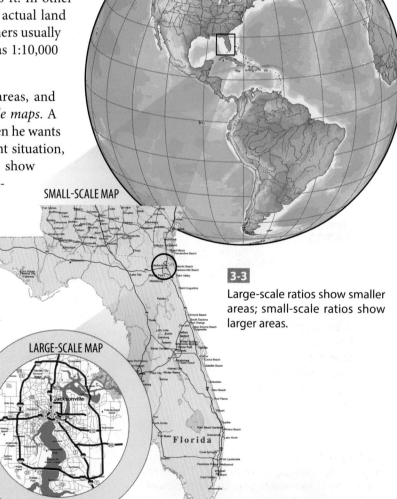

3-3 Large-scale ratios show smaller areas; small-scale ratios show larger areas.

MAPS AND MAPPING 53

geometry: (Gk. *geo*—earth + *metria*—to measure). Geometry is the scientific and mathematical study of the relationships of points, lines, shapes, and solids. A person who does geometry is a geometer.

3-4

The shadow during a lunar eclipse is always curved. A flat earth would produce straight-line shadows at times.

3.3 LANDMARKS

Have you ever been lost? It's not fun! Think of all the challenges people face when they get lost in the wilderness. How will they find food? How will they find water? Where is their family? Where is their home? They want to know where they are because they have to know where they are in order to survive!

But some people have been lost in the wilderness even with a map! Maps do a fine job of modeling the earth, but they do nothing to tell you where you are. To find your position, you need a recognizable landmark that you can find on the map. And then you have to find your position relative to the landmark. In a city where many streets cross each other, this is fairly easy. But what if you are out on open ocean? No landmarks! Cartographers solved this problem by creating an invisible system that takes the place of visible landmarks. This system relies on something familiar—the **cardinal directions**—north, south, east, and west.

Early cartographers focused on the general shape of continents and countries. Later, they realized that it was important to show the shapes and locations accurately, including true distances and directions. The science of geometry was born. Geometers, and later, cartographers, decided that it made sense to set up a north-south, east-west grid to locate positions on land. On a map, a geographic grid is ideally a system of crossing vertical and horizontal lines. Vertical lines run north-south. Horizontal lines run east-west.

THE EARTH AS A SPHERE

Most educated people at least several hundred years before Christ believed the earth was spherical. This is what they observed:

» Ships far out to sea gradually disappear below the horizon.

» Some stars can be seen only from certain parts of the earth at the same time. If the earth were flat, all stars above the earth should be seen from every location on the planet at once.

» During a lunar eclipse, Earth's shadow on the moon is always round.

A system of rectangular grid lines works for small-area maps. But maps of larger areas have problems. As you move farther from the center of a flat map, positions don't line up with the grid. The map is distorted compared with the actual earth's surface. This is because the two-dimensional flat map surface tries to model a small patch of a much larger three-dimensional sphere. Rectangular grid lines don't work on a sphere. Cartographers had to create a grid that worked on a sphere.

LATITUDE LINES

When you draw the largest possible circle around a sphere, it forms a *great circle*. A great circle always has the same diameter as the sphere. Perhaps the most important great circle to a cartographer is the **Equator**. It divides the Northern Hemisphere from the Southern Hemisphere. Earth's rotational axis runs right through the center of this circle and is perpendicular to the plane of the Equator.

Add more circles to the sphere parallel to the Equator. These are called *small circles* because their diameter is less than that of a great circle. Cartographers size these circles so that they mark equal angles between the Equator and the poles. These circles on a globe are called lines of **latitude**, or **parallels**, because they appear parallel to the Equator when the globe is viewed from the side. The Equator is 0 degrees latitude.

The North and South Poles of the earth are 90° from the Equator. They mark where the earth's rotational axis passes through the earth's surface. Latitude angles to the north of the Equator are followed by the letter N. An S follows latitude angles south of the Equator. Washington DC is located at latitude 38.9°N. The geographic North Pole is at latitude 90°N, and the South Pole is at latitude 90°S.

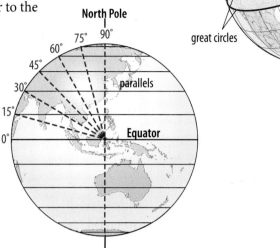

When you are looking at a map, latitude lines look flat.

LONGITUDE LINES

Imagine a half circle on a sphere that directly connects the North and South Poles. This forms a *semicircle*, that is, half of a great circle. If you rotate the sphere a little around its poles, you can draw another semicircle between the poles. These are called **longitude** lines on a globe. Lines of longitude are also known as **meridians**. The term *meridian* comes from the Latin word for "midday." A meridian marks the east-west position of the sun at its highest point in the sky. Timewise, this sun position marks local noon.

3-5

Local noon is the moment when the sun is at its highest point in the sky at the observer's location. It is also the time that the sun crosses the local meridian. Local noon and 12 p.m. on a clock are rarely the same time because of time zones.

Meridians are semicircular arcs of great circles that mark longitude. The 0° longitude line passes through Greenwich, England. Cartographers call this line the **Prime Meridian**. Longitude angles measure from 0° to 180° east and west of the Prime Meridian. The longitude of the Washington Monument is 77.03°W.

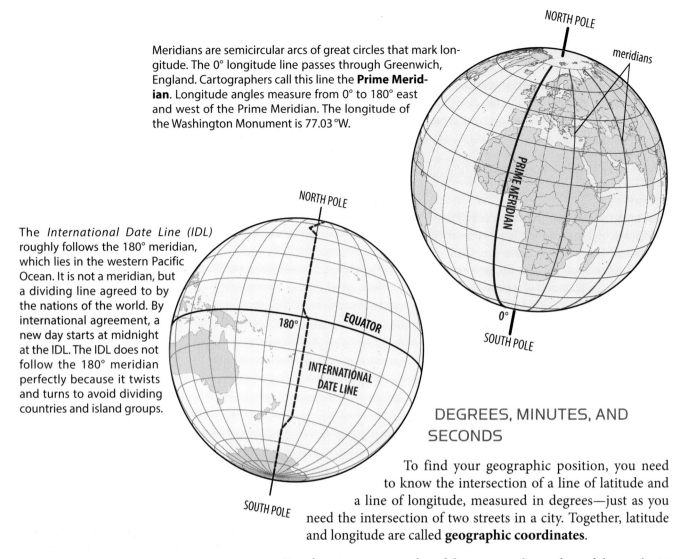

The *International Date Line (IDL)* roughly follows the 180° meridian, which lies in the western Pacific Ocean. It is not a meridian, but a dividing line agreed to by the nations of the world. By international agreement, a new day starts at midnight at the IDL. The IDL does not follow the 180° meridian perfectly because it twists and turns to avoid dividing countries and island groups.

DEGREES, MINUTES, AND SECONDS

To find your geographic position, you need to know the intersection of a line of latitude and a line of longitude, measured in degrees—just as you need the intersection of two streets in a city. Together, latitude and longitude are called **geographic coordinates**.

One degree can cover a lot of distance on the surface of the earth. At the Equator, a degree of latitude or longitude is equal to about 60 **nautical miles** on the earth's surface! For more exact positions, degrees are divided into smaller units, the same way feet are divided into inches. A degree contains 60 minutes, so a *minute* is 1/60 of a degree. An angle of 1 minute is about 1 nautical mile along a meridian.

A **nautical mile** (nm) is about 6076 ft (or *exactly* 1852 m) long, compared to a statute (or land) mile, which is only 5280 ft (or 1609 m) long.

For more exact positions, an angle of 1 minute contains 60 seconds, so an angle of 1 *second* is 1/60 of a minute and equals about 34 yards (31 meters) along a meridian. See the box in the margin for the special symbols used to indicate geographic angles. When timepieces became more accurate several centuries ago, the units for angles were borrowed for timekeeping. Today, people associate minutes and seconds more with time than with angles.

Angular Relationships:
1° (degree) = 60′ (minutes)
1′ = 60″ (seconds)
1° = 60′ = 3600″

By using decimals, you can use these units to write geographic coordinates as exactly as you need. For example, the position of the Washington Monument in Washington DC is written as

38°53′22.084″N latitude 77°2′6.864″W longitude.

These coordinates give the position of the monument to the nearest inch!

Latitude angles are measured along a meridian. There are 180° in a meridian from the North to South Pole (it is a semicircle). Every meridian is practically the same length, so a degree of latitude is the same distance anywhere along a meridian.

Longitude angles are measured along parallels. Latitude circles become smaller closer to the poles. All circles have the same number of degrees (360°), but a smaller circle has a shorter circumference. So degrees of longitude cover shorter distances as you get closer to a pole. At the poles, the meridians come together, so a degree of longitude has no meaning at the poles.

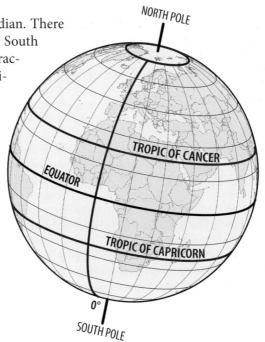

3.4 YOU ARE HERE

So now you know how latitude and longitude lines pinpoint landmarks on a map. But how do you know where you are, even if you have a map with these lines? You must find your latitude and longitude coordinates, and then plot them on the map. This is called *fixing your position*, and your location is a *position fix*. But how can you find your latitude and longitude?

Sailors and explorers long ago looked to the skies for navigation. In ancient times, they could find only their latitude accurately. They measured the angle of the sun above the horizon at local noon using special instruments called *astrolabes*, or *sextants* (see below), to calculate their latitude.

Sometimes, maps used only in the Southern Hemisphere will have south toward the top. This can be very confusing!

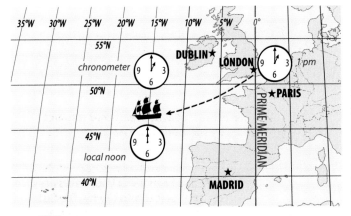

Longitude was much more complicated. Not until the invention of accurate seagoing clocks in 1761 could navigators easily determine longitude. The clock was set using an accurate shore-based clock in their home port just before leaving port. Far at sea, the navigators watched how the time of local noon changed as they traveled. They compared the time of the ship's local noon to the time of local noon at their home port to find the ship's longitude. Sailors navigated this way until electronic navigation systems were invented after World War II.

3-6
Finding longitude by time difference. The one-hour time difference equals 15° of longitude west of the longitude of the home port.

MODERN POSITIONING METHODS

Today, there are many ways to obtain position fixes. Aircraft and ships use various kinds of radio navigation. For you, the most familiar way is probably the Global Positioning System (GPS). It is free to the public and is used by millions of people all over the world. GPS uses the signals from twenty-four Earth satellites. A GPS receiver must be able to detect signals from at least three satellites to get a horizontal position fix. Position accuracy can be within just a few yards with off-the-shelf GPS receivers. Surveyors and military receivers can obtain GPS positions within fractions of an inch. See the box on the next page to learn more about GPS.

3.5 STANDARD MAP FEATURES

Maps may be used for all kinds of purposes, and they can be almost anywhere. If you look for certain features that maps usually have, you will find them easier to use.

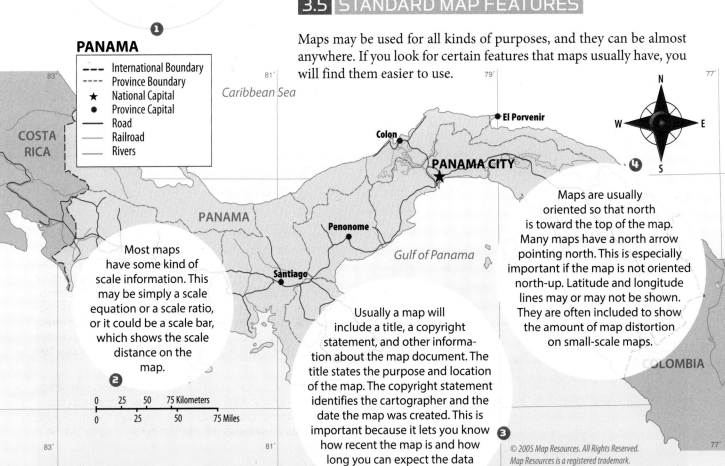

1 For maps that display many different kinds of information, there may be a *map legend*. This is a box somewhere on the map that defines all the symbols and colors used on the map. Study the legend to understand what the map is showing you.

2 Most maps have some kind of scale information. This may be simply a scale equation or a scale ratio, or it could be a scale bar, which shows the scale distance on the map.

3 Usually a map will include a title, a copyright statement, and other information about the map document. The title states the purpose and location of the map. The copyright statement identifies the cartographer and the date the map was created. This is important because it lets you know how recent the map is and how long you can expect the data to be valid.

4 Maps are usually oriented so that north is toward the top of the map. Many maps have a north arrow pointing north. This is especially important if the map is not oriented north-up. Latitude and longitude lines may or may not be shown. They are often included to show the amount of map distortion on small-scale maps.

GPS (GLOBAL POSITIONING SYSTEM)

It was 1991, and the United States was in the middle of its first war with Iraq. Can you imagine being a soldier trying to find your way in a desert with no landmarks? Or trying to destroy military targets without injuring close-by civilians? These and many other urgent military needs depended on precise, real-time location information.

To meet these challenges, the US military began using a new navigation system called the *Global Positioning System (GPS)*. US troops used mobile receivers on the ground that communicated with the GPS satellites. Signals from three satellites could tell a solider his location. With a fourth signal, he would also know his altitude. Ships, missiles, tanks, aircraft, and foot soldiers all used GPS signals to navigate. Missiles were so accurate, they could fly through a window to hit their targets! This positional accuracy made every soldier and every guided weapon more effective. Fewer innocent civilians were injured or killed than in similar wars in the past.

If it takes three satellites to know where you are, then there have to be a lot of GPS satellites "up there"! The whole group of satellites in orbit around the earth is called the GPS "constellation." For GPS to work right, there are at least twenty-four satellites orbiting in six planes tilted about 55° to the Equator at all times. These satellites are evenly spaced around the world. Extra satellites provide backups when others fail or wear out.

A constellation of at least twenty-four satellites serves the Global Positioning System.

The overall standard accuracy of the GPS has been much improved since it was placed into service. The first major improvement used signals from a fixed GPS base station to correct the positions found by a mobile GPS receiver. A computer calculated the difference between the base station's own computed GPS position and its actual position on a map. This position difference was then radioed to the mobile receiver to produce a more accurate receiver position. A more recent improvement includes similar corrections made directly into the entire GPS to improve regional GPS accuracy to within a few meters. GPS positions are now so accurate that commercial airlines can land and take off using GPS-guided autopilots!

At any given moment, there are between four and five satellites above the horizon. Since GPS signals travel by line-of-sight, this helps us find our location anywhere on Earth. All you need is a GPS receiver. Your cell phone, car, laptop, or other portable digital device likely contains a GPS receiver. GPS units that communicate back to a satellite can help police track stolen vehicles or identify the location of an emergency 911 call. Emergency GPS beacons can even report the exact location of a car wreck, a plane crash, or a survivor lost at sea.

Researchers, archaeologists, and many scientists use GPS to give them location information. But one of the most scientifically important uses of GPS is in mapping and measuring the earth's movements resulting from earthquakes and volcanoes.

MAPS AND MAPPING

3A REVIEW QUESTIONS

1. How did drawing a map help Dr. John Snow save lives?
2. How is a map a model of the earth?
3. What kind of map is a globe?
4. How does the detail of a large-scale map compare to that of a small-scale map?
5. What real-world distance does 1 inch equal on a map with a scale of 1:25,000?
6. Do you need visible landmarks to find your position on a map? Explain.
7. Why are maps often distorted?
8. Give the following for both latitude and longitude.
   a. the name of the reference, or 0° line
   b. the range of possible values, including directions
9. The invention of what instrument allowed sailors to accurately determine longitude?
10. What standard map feature explains symbols used for main highways, secondary roads, and unpaved roads?
11. (True or False) The location of any meridian can be adjusted to avoid dividing continents or island groups.

3B Objectives

After completing this section, you will be able to

- identify the three main types of map projections.
- briefly discuss the properties and uses of the common map projections.
- identify three standard types of maps.
- briefly discuss the use of contour lines in topographic maps.
- discuss the concept of a map theme.
- identify thematic maps.

3B

TYPES OF MAPS

What must cartographers consider when designing a map?

3.6 WHAT IS A MAP PROJECTION?

Examine the two world maps below. Consider the size and shape of the continents. Look at the areas near the North and South Poles. Observe the shapes of the maps. What do you notice? These maps are different, even though they both show the same areas of the world.

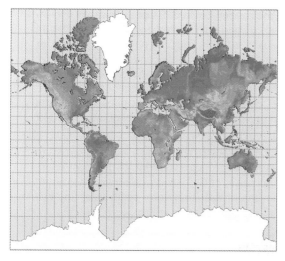
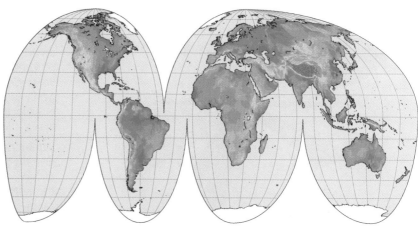

In Section 3A you learned that every two-dimensional map of the earth's surface is at least slightly deformed. You can't flatten a sphere without distorting features on its surface. In maps that show small areas, this distortion isn't noticeable. But maps of larger areas, like our world maps, display significant distortion.

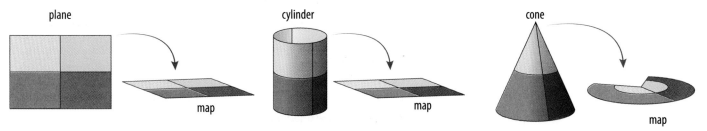

A cartographer's job is to create a map that has some kinds of distortion but removes others. The choice he makes is determined by the purpose of the map. He constructs the map by transferring each point on the earth's spherical surface to a surface that is flat, or one that can be made flat. The surface may be a plane, a cylinder, or a cone. This geometric process is called **map projection**. The resulting map still contains distortions, but map features will keep at least some of the same relationships (such as, shape, area, directions, or distances) as they had on a globe. However, the cartographer cannot keep all of them on the same flat map.

3-7 A plane, cylinder, or cone can be flattened into a plane that can be useful as a map.

3.7 TYPES OF MAP PROJECTIONS

Cartographers classify maps by the shape of the surface on which the projection is made. As you just learned, this could be a plane, a cylinder, or a cone. A projection can further be classified by how the projection surface touches the globe. When it touches or wraps around the Equator, it is an *equatorial projection*. If it touches a pole, it is a *polar projection*. If it touches somewhere between the Equator and a pole, it is an *oblique projection*.

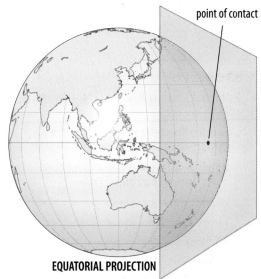

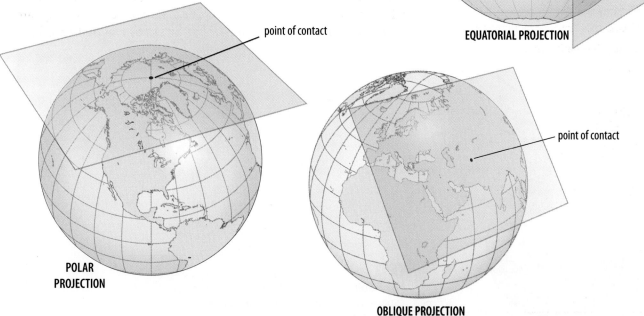

3-8

Notice the lamp at the center of the earth. The light projects through the globe onto the flat surface, forming a gnomonic polar projection.

gnomonic (no MON ick): (Gk. *gnomon*—indicator rod of a sun dial + -ic); in maps, a pattern similar to that of a sun dial

Look at Figure 3-8. The points on a globe's surface are transferred to a map plane along a ray, like a light ray from a bulb. You can get many kinds of projections by changing the position of the light bulb, that is, the source of the rays. We use the word "projection" to describe this process since it is very similar to using a photographic slide or digital projector.

BASIC MAP PROJECTIONS

The simplest kind of projection involves a flat plane touching the globe at a point. When the "light bulb" is at the center of the earth, it creates a *gnomonic projection*. These kinds of maps show the shortest distance between two points, since great circles form straight lines. They are often used for maps of the polar regions.

Imagine taking a sheet of paper and wrapping it around a globe. The sheet of paper forms a cylinder. A projection created this way forms a *cylindrical projection*. The cylinder touches the globe along a great circle (Figure 3-9). When you cut the cylinder vertically and spread it flat, it forms a two-dimensional map.

If a map's projection surface is shaped like a cone, a *conic projection* is created. The projection surface touches the globe along a small circle (see Figure 3-10). Cartographers like to use conic projections for maps of the middle-latitudes of the world.

3-9

An equatorial cylindrical projection contacts the earth along the Equator. Cylindrical projection surfaces may touch the earth along great circles other than the Equator (right).

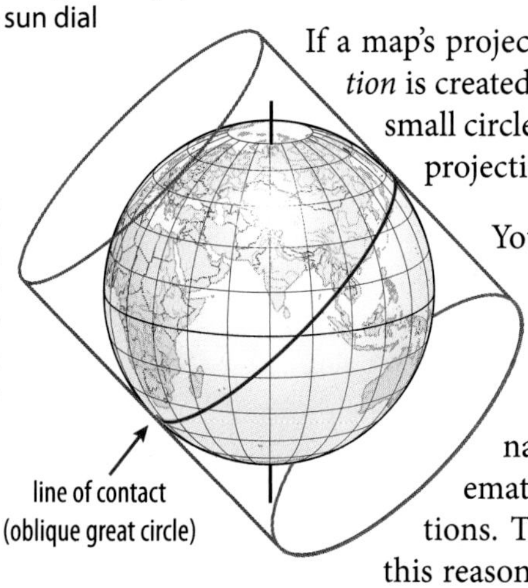

line of contact (oblique great circle)

3-10

A conic projection (below) is often used for middle-latitude maps.

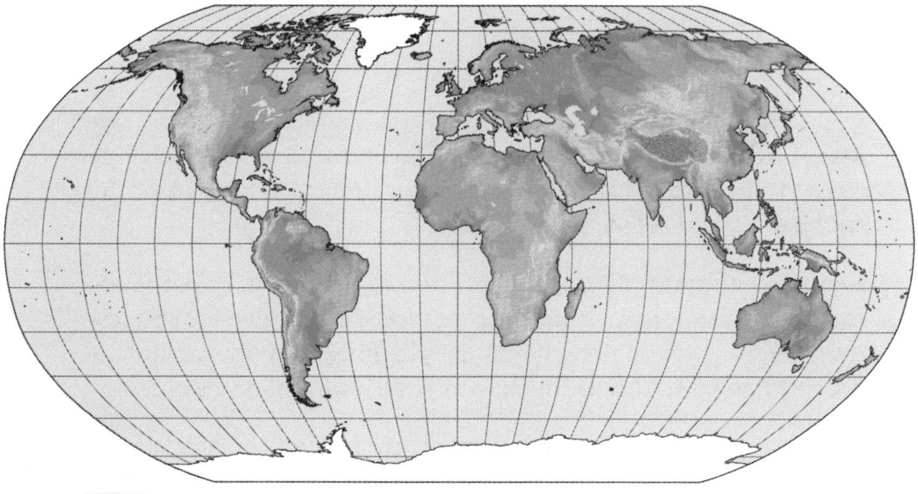

You probably have seen a world map like the one shown in Figure 3-11 in a textbook, or even on a wall map. The maps do not preserve the actual shapes of features, and their areas are usually not proportional, but they show them in a more natural relationship. This is because they are mathematical rather than geometric or "light bulb" projections. They are called "false cylindrical" projections for this reason. The most popular world map of this type is the Robinson, shown below.

3-11

The Robinson world map was designed especially for student textbooks because it looks right.

In every kind of map projection, the projection surface is imagined to touch the earth at a point or line. The cartographer chooses a projection to minimize the distortion in the areas near where the mapping surface touches the globe.

3.8 STANDARD MAPS

Map projections determine how an area looks on a map. But what kind of information can you find on a map? Standard maps may be any of the following types:

Political—shows names, boundaries, and shapes of countries, states, counties, cities, and towns—any feature established by local, state, or national governments

Geographic—focuses on names, symbols, and locations of man-made and natural physical features, such as roads and highways, streams, lakes, coastlines, and oceans

Topographic—displays elevation and landform information, usually indicated by special symbols, called contour lines, or by landform shading

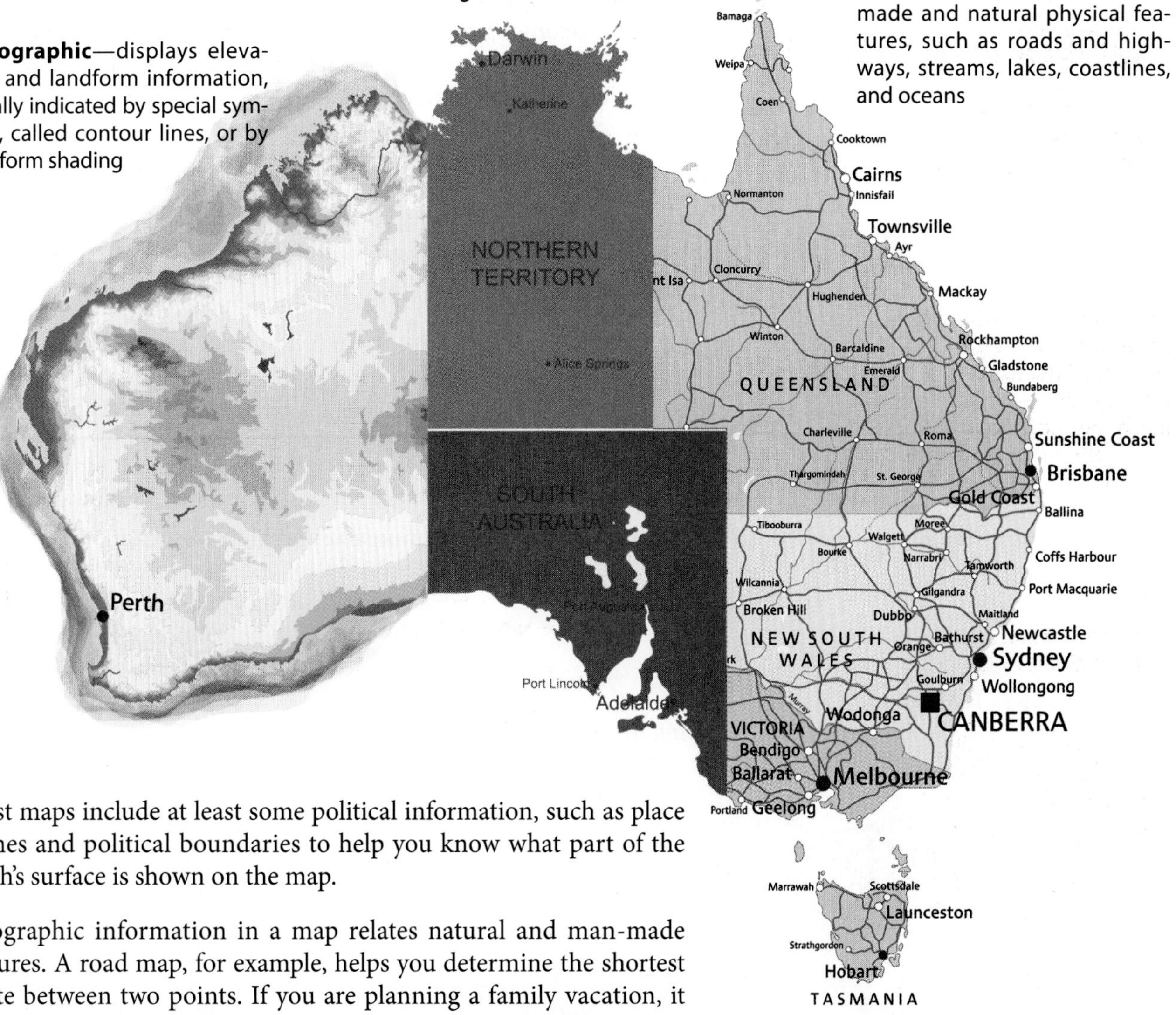

Most maps include at least some political information, such as place names and political boundaries to help you know what part of the earth's surface is shown on the map.

Geographic information in a map relates natural and man-made features. A road map, for example, helps you determine the shortest route between two points. If you are planning a family vacation, it helps to know what towns are near mountains, lakes, or the ocean.

If you want to see how land changes in elevation or height, you should use a *topographic* map. Such maps show hills and mountains and their relationship to other geographic or political features.

topographic (top uh GRAF ick): (Gk. *topo*—place + *graphos*—describe + -ic); a symbolic description of the lay of the land

MAPS AND MAPPING

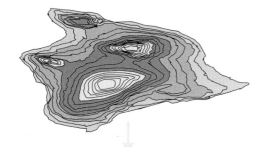

3-12

Compare the topographic map (top) with the relief map (bottom). Notice that contour lines connect points of equal height above sea level. They form closed loops on a topographic map.

spatial (SPAY shul): (L. *spatium*—space); having to do with space, position, or location

Contour lines show elevations on the map above some standard height, such as sea level. These lines connect points of equal elevations that are some standard height apart. For example, a hill appears as a series of rings nested inside one another (see Figure 3-12). Contour lines far apart indicate a gentle slope, while lines close together show a steep slope. If you were hiking a mountain, you would want to know this information! You will learn more about topographic maps in Lab Manual map activities. Appendix D lists many of the symbols found on standard maps.

3.9 THEMATIC MAPS

Standard maps can be used to plot data that by itself is not political, geographic, or topographic. Think back to the beginning of the chapter. John Snow plotted the locations of cholera deaths on a street map of London. A street map is a kind of political and geographic map. The death of a person by itself has no geographic significance. But if you plot the location of that person's death on a map, you give it spatial meaning. When Dr. Snow plotted the locations of the cholera deaths of many people, a pattern emerged that could be seen only on a map. He used a *theme* or unifying idea when plotting his data. From this **thematic map**, he obtained important clues in his investigation to identify the well that was infected. Many people's lives were saved by his insight.

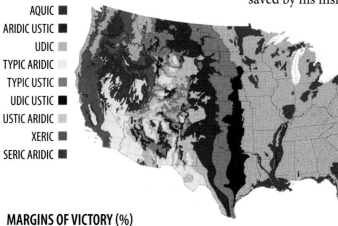

Earth scientists use thematic maps for most of their mapping needs. When they plot the locations of earthquakes on a standard map of the world, they create a seismic thematic map that shows clusters of earthquakes—the places that are more likely to have earthquakes. Geologists who need to know the types of rocks that make up the earth's surface create a thematic bedrock map. While political boundaries may be included to orient the user to the locations on the map, the colors and symbols indicating the different kinds of bedrock provide the most important information. You will learn to use and even create thematic maps later in this course.

3-13

A wide variety of information can be displayed on thematic maps. One example displays different soil types found across the United States (above left). To the left, a thematic map shows voter turnout and vote results from the 2004 presidential election.

3B REVIEW QUESTIONS

1. Why are maps of the earth's surface distorted?
2. Name three kinds of projection surfaces commonly used for mapmaking.
3. What method do cartographers use to minimize distortion in the middle of a map, especially one showing a large part of the earth's surface?
4. What kind of map was created specially for classroom textbooks?
5. Identify Maps A, B, and C (below) as gnomonic, cylindrical, or conic projections. The red dot or line indicates the way that the map projection touches the globe.

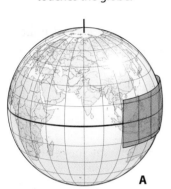
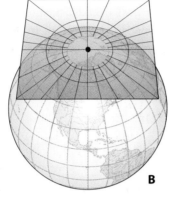
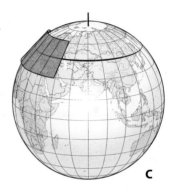

A   B   C

6. What standard type of map would you choose if you were interested mainly in the shapes and boundaries of the countries of the world?
7. Using a topographic map, how can you tell that the road on which you intend to bicycle runs up a steep hill?
8. Why do cartographers and scientists create thematic maps?
9. (True or False) The type of projection used to make a map is based on a cartographer's preference, not on the map's purpose.

3C

MAPS AND GIS

How can maps help people solve problems?

3.10 WHAT IS GIS?

Have you ever used the Internet to get directions or to plan a cross-country trip? Then you have used GIS technology. **GIS** stands for **geographic information system**. It is the fastest-growing use of computerized map data. There are three main parts to a GIS:

» computerized geographic data,
» programs, hardware, and data storage systems that allow mapmakers to display and use this information, and
» trained users who analyze and plot the data to make maps.

A GIS produces thematic maps that help people solve problems.

3C Objectives

After completing this section, you will be able to

» define a geographic information system (GIS).
» state the main uses for a GIS.
» identify sources of GIS data.
» explain how GIS maps are used to help people.

You can find uses of GIS almost anywhere. GIS programs run on computers, smartphones, or tablets. On the Internet, GIS helps you plan trips, get directions, look at property boundaries, and even virtually "walk" down a street in a neighborhood.

Local governments rely on GIS for community planning, emergency response, and power and water system maintenance. Within the past decade, businesses have begun using GIS to plan where to build stores to best serve customers and maximize profitability. News and weather programs have been using GIS technology for many years now.

The US military has very complex GIS applications. In the battlefield or at sea, GIS technology helps commanders keep track of friendly and enemy units in real time using satellites. It also helps them to know the range of weapon systems in relation to the battle picture and to predict the geographic effects of attacks. GIS technology has virtually unlimited potential to help us use geographic information.

3-14
A GIS consists of digital geographic data—computers and software to use and display the data on maps—and trained people who collect data, create the maps, and make decisions using the maps.

GIS maps are created from computer data. Individual GIS data can be stored only as points, lines, or areas bounded by straight line segments called *polygons*. These standard *data types* can be stored in digital computer memory. A GIS can also store digital pictures. The *point data type* identifies specific locations, like the position of a well. The *line data type* represents features with length, such as rivers. The *polygon data type* shows features that cover an area, like a lake.

But what makes GIS special is its ability to organize data into *layers* of information. The layers of data overlap or stack on top of each other (see Figure 3-15). For example, in a thematic map of a city, the city limits—a "polygon"—is on one layer. Its streets—line features—are on a second layer. Fire station locations are points on a third layer. Another layer could show both underground water main pipes (line features) and the locations of fire hydrants (points) connected to them. Yet another layer could show the city parks (polygons).

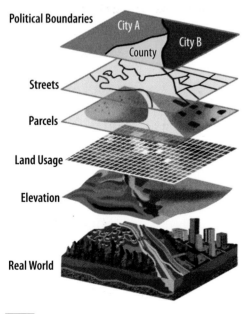

A user can display or turn off each layer as needed. The ability to create thematic layers and display them together on a GIS map helps people making decisions to better understand the relationships between different kinds of data. If John Snow had been able to use a GIS program, he could have created a street map of London on one layer, the limits of the epidemic on a second, the locations of wells on a third, and the locations of cholera deaths on a fourth. As it was, he used GIS map layer principles on a single map to save lives during an epidemic.

3-15
A GIS allows organizing different kinds of data into layers that are displayed on the same map.

3.11 GIS DATA COLLECTION METHODS

All of this GIS data in a computerized database needs to come from somewhere. How is GIS data collected? How can cartographers locate features accurately in relation to the geographic coordinate grid?

SURVEYING

Until about the middle of the last century, people gathered geographic information only by *surveying*. Most of these methods are still used today. Engineers called *surveyors* start from a position with known geographic coordinates. From this point, they use many tools to find the distance and direction from that point to another point or object to be surveyed. Surveying a small piece of land is quick and easy, but what about an extensive river system or a mountain range? Surveys of large natural features can take months and aren't very accurate. Modern surveyors use GPS. This helps surveyors quickly and easily gather data on even large-area surveys of ground features.

To be used in a GIS, survey positions must be entered into a database table. Other descriptive information about each position and feature must also be entered. Modern GIS software programs can create these tables automatically from special GPS survey instruments.

3-16

Modern surveyors use GPS for precise positions, as well as laser transits to determine directions and distances.

Benchmarks

Known geographic locations used for surveying are called *benchmarks*. These are often bronze disks permanently fastened to rock or small concrete monuments.

LIFE CONNECTION: GIS AND DISASTER RELIEF

On January 12, 2010, a magnitude 7.0 earthquake shook the island country of Haiti. Tens of thousands of people were buried under the rubble, and initial reports indicated that more than a million were left homeless or starving. The actual numbers may never be known for certain. How could other nations help? And how could scientists understand the aftershocks? That's where GIS came in.

Scientists used seismic data to plot the earthquake and its aftershocks. After making ground shake calculations, geologists plotted the information on a GIS map of Haiti to spot all cities and villages that were within the area of concern.

Various satellite imaging companies immediately made pre-earthquake satellite photos of the area available on the Internet. Satellites took post-earthquake photos of the same areas. GIS analysts created maps with both sets of photos in layers. They created damage estimates and located areas that were able to receive ships and aircraft. This helped relief workers worldwide find places to offload supplies.

As weeks passed, cartographers received full information about the amount of damage to buildings, water supplies, and power lines. This information was layered with population information on GIS maps. Relief workers used this data to best place centers for medical aid and food distribution. Responders used the maps to plan urgent repairs to help the most people. Repair efforts were ranked according to need on the basis of maps showing hospitals, schools, and areas where the most people lived.

This disaster shows how GIS technology is an essential tool for saving lives and helping workers to provide rapid and effective aid to victims. Planners can also use GIS information to make important decisions and guide rebuilding efforts. GIS is truly a redemptive use of technology!

This image is a composite map showing the location of water sources (diamonds and triangles) available to the displaced resident centers (triangles) in Port-au-Prince, Haiti, after the devastating earthquake. The base layer is an aerial photo of the city. Streets and place names are on a separate GIS map layer.

REMOTE SENSING

Most GIS data is not collected by surveying today. Rather, it is done using instruments located at some distance from the feature being measured. *Remote sensing* instruments are usually mounted in aircraft or satellites so that they can rapidly and accurately map large areas of the earth's surface. Sensors include cameras that can take pictures in visible light, infrared, and radar. (See Chapter 2 to review these kinds of light energy.) Laser sensors on aircraft can also be used to map landforms using a remote sensing system called LIDAR. Satellites or even manned spacecraft are used when mapping very large portions of the earth's surface. For instance, during one mission NASA's space shuttle mapped most of the earth's surface with a special height-measuring radar.

Radar is probably a familiar term. It was coined during World War II and stands for <u>Ra</u>dio <u>D</u>etection <u>a</u>nd <u>R</u>anging. It uses radio waves to find distance and directions to objects. LIDAR is similar, but uses laser light instead of radio waves. The name stands for <u>L</u>ight <u>D</u>etection <u>a</u>nd <u>R</u>anging.

Most remote sensory data are stored as large digital pictures. Just like an image on a computer screen, these pictures are made of thousands or millions of tiny points of data called *pixels*. Each pixel has its own digital geographic address. A pixel contains sensor data, such as a temperature, an elevation, or another property of the surface in the image.

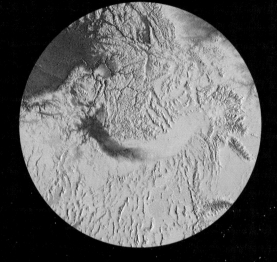

This remote sensing image of Washington State was created from thousands of pixels, each associated with an elevation value.

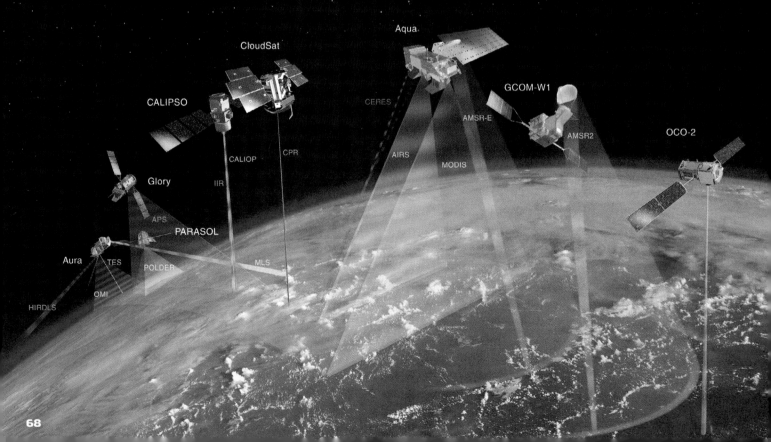

There are a variety of remote sensing platforms. Scientists use these to collect vast amounts of data over a wide area.

ACOUSTIC SENSING

Surveying and remote sensing work for surface features on land, but what about the ocean floor? Three-fourths of the earth's surface is under the oceans of the world. Mapping the ocean bottoms is important for scientific, economic, and military reasons. However, the technology for mapping large land surfaces does not work well for ocean bottoms. Light energy, which includes radar waves and laser light, doesn't penetrate very far into the water before it is absorbed. How then can we map the sea floor?

Since before the start of the last century, scientists have known that sound travels great distances underwater. Many countries used sound or acoustic sensors to detect enemy submarines in World War I. During World War II, this underwater submarine-hunting technology became known as *sonar*. At the same time, scientists discovered that they could also use sonar to create sea-floor maps. With the help of computers, scientists can even make 3D images of the sea floor. This helps us better understand the connections between water movement, undersea geography, and living things in the oceans.

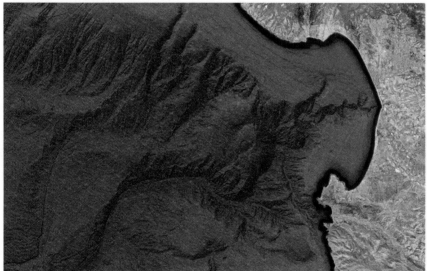

© Google Earth Pro/Data MBARI/Data SIO, NOAA, U.S. Navy, NGA, GEBCO/©2010 Google/Image AMBAG

3-17
This 3D digital map of the sea floor of Monterey Bay, California, is an example of submarine GIS applications.

3.12 DOMINION USING MAPS

Can you imagine living without maps? Maps help to make our daily lives possible, from predicting weather and planning cities to analyzing soil that farmers use to grow our food. GIS technology can be used to respond to natural disasters (see the Life Connection box on page 67), identify high crime areas for law enforcement officials, or even plan the location of a new fire station. Scientists use GIS and maps to identify pollution and to control insect pests and forest fires. In all these cases, we are modeling our world through maps using technology in order to love our neighbors and exercise good and wise dominion.

3C REVIEW QUESTIONS

1. Describe in your own words what a GIS is.
2. Give two examples that show how GIS information can help people.
3. Which of the three standard GIS data types is used to show the location of a well on a map?
4. What feature of GIS maps sets them apart from standard maps?
5. What is the best way to quickly collect geographic data over large areas?
6. What is the main geographic data collection method for the ocean bottom?
7. Discuss the usefulness of GIS technology from a dominion science perspective.
8. (True or False) GIS principles have been in use only since the development of computer mapping.

SERVING GOD AS A CARTOGRAPHER

JOB DESCRIPTION

How do you think people made maps in the time of Christopher Columbus? They couldn't exactly look at a satellite photo! Ancient maps were hand-drawn, often relying on explorers' descriptions. Of course, this contributed to the errors of early maps. Now we have impressive technology to help with accurate mapmaking.

But good information is not the only thing you need for a good map. Mapmaking is as much art as science. First, a cartographer must know the purpose of the map. Who is going to use it? What part of the world will it show? How much of the world? Second, he must decide what information to include on the map, and then find it. How old is the information? How accurate is it? Sometimes, the most recent data is not always the best. Finally, he must design the map so that it will look good and work as intended, considering colors, symbols, and labels.

EDUCATION

Because making maps involves art, geography, and math, cartographers may have earned a degree in any of these areas. Sometimes cartographers work as apprentices or assistants before getting their own projects. Mapmakers who use GIS programs need extra training. These mapmakers are GIS specialists or GIS analysts. Many colleges and universities offer bachelor's degrees in GIS cartography.

POSSIBLE WORKPLACES

Cartographers can work for businesses or for local, state, or federal government agencies. Organizations such as the National Geographic Society, the United States Geological Survey (USGS), the National Oceanographic and Atmospheric Administration (NOAA), and the National Geospatial-Intelligence Agency (NGA) all create and use maps. In addition, many universities and nonprofit organizations have GIS specialists on their staffs. These employees assist with planning changes to campus facilities and studying the effectiveness of programs affecting a geographic area.

DOMINION OPPORTUNITIES

Cartographers create models of the world's surface. These models are tools to help us exercise good and wise dominion. Maps can also help us respond to natural disasters, such as hurricanes and earthquakes. By creating maps from satellite photos on the spot, people delivering aid can find places to land airplanes and helicopters, pilot boats, or drive trucks delivering supplies.

Thematic maps help us efficiently monitor and use natural resources. They help us manage animal and plant species and plan communities to best conserve our resources. More than ever before in history, cartographers help us to best use our world.

CHAPTER 3 REVIEW

CHAPTER SUMMARY

» Maps are models of the earth's surface.

» We make maps to find where we are, to know the positions of other things on the face of the earth, and to understand how they relate to each other.

» The scale of a map determines what features it can show. Large-scale maps are useful for showing small geographic features, while small-scale maps can show only large features.

» Maps use artificial landmarks—a geographic grid—to locate features.

» Flat maps show a distorted view of the earth's spherical surface.

» On a sphere, a geographic grid consists of latitude lines, which are small circles parallel to the Equator, and longitude lines, which are great circles passing through the North and South Poles.

» Geographic positions are combinations of latitude and longitude coordinates, expressed in degrees, minutes, and seconds.

» Latitude measures your position north or south of the Equator, while longitude measures your position west or east of the Prime Meridian.

» Maps often display many standard features, including a map scale, a symbol legend, and a north arrow.

» Map projections are ways to accurately transfer details on the earth's surface to a flat mapping surface while minimizing certain kinds of distortion.

» The three main geometric map projections are gnomonic, cylindrical, and conic.

» Map projections attempt to minimize distortion of earth features in the area of interest. Gnomonic polar projections are used for polar maps of the poles, conic projections for middle-latitude maps, and cylindrical equatorial projections for maps along the Equator or other great circles.

» Standard maps can be political, geographic, topographic, or combinations of these.

» Thematic maps display information organized with a single concept or theme plotted on a standard map. Often thematic information is not related to its physical location on the earth's surface, but when plotted on a map, it gains geographic significance.

» Geographic information system (GIS) technology greatly improves the ability of maps to reveal unsuspected relationships among different kinds of data.

» GIS maps display information on thematic layers, which can be turned off or on as desired.

» Modern mapmaking receives information from many sources, including standard surveying, remote sensing sources, and ocean-floor mapping.

» Maps and mapping are important tools for humans to exercise wise dominion and benefit other people.

Key Terms

map	52
map scale	53
cartographer	53
cardinal direction	54
Equator	55
latitude	55
parallel	55
longitude	55
meridian	55
Prime Meridian	56
geographic coordinates	56
map projection	61
contour line	64
thematic map	64
geographic information system (GIS)	65

3-18

Scientists use computers to generate 3D maps of Earth features. This is a 3D map of mountain landforms.

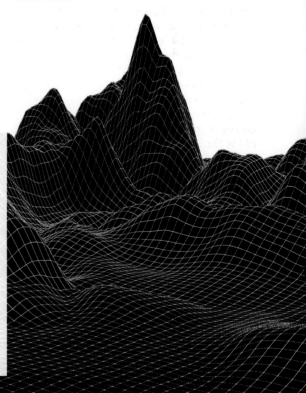

REVIEW QUESTIONS

1. How was John Snow's "Ghost Map" an application of GIS principles?
2. Explain why a map is simply a model of part of the earth's surface.
3. Discuss the differences between large-scale and small-scale maps.
4. How are map grid lines like landmarks? How are they different?
5. Which grid lines on a standard map are the vertical lines? Which are the horizontal lines?
6. Before modern navigation methods, such as GPS, how did sailors find their longitude?
7. List three features on a standard map that help you use it better.
8. Why is a map projection needed?
9. What are the advantages of plane, cylindrical, and conic projection surfaces in cartography?
10. What kind of projection would show the continent of Antarctica with the least amount of distortion?
11. What are three types of standard maps? Which would likely show glaciers and lakes?
12. Why are thematic maps useful?
13. Name the three main parts of a geographic information system (GIS).
14. Give an example of an earth feature that could be represented by each of the GIS data types.
15. What is the most efficient way to collect GIS data over large areas?

True or False

16. Maps generally use an oblique view to show as much of the earth's surface as possible.
17. A map scale of 1:50,000 is larger than a scale of 1:300,000.
18. North, south, east, and west are always the same, no matter where you are on Earth.
19. Great circles on a globe do not have to pass through the North and South Poles.
20. A position fix is a pair of geographic coordinates.
21. If the north arrow on a map points to the left, you would expect longitude lines to be vertical and latitude lines to be horizontal.
22. All map projections can show the true shape of countries and their true sizes.
23. The terms *gnomonic*, *cylindrical*, and *conic* refer to the shapes of the projection surfaces for each kind of projection.
24. The steepness of a slope on a topographic map is indicated by the angle that a contour makes with the horizontal.
25. A map that shows the location of all the geysers in Yellowstone National Park is a thematic map.
26. Geographic information system (GIS) applications can be used only by specially trained persons.
27. GIS mapping is more effective for complex analysis than standard maps because layers of thematic data can be displayed as needed.

3-19
Engineers have designed highly precise 3D maps of roadways to control driverless cars.

28. A degree of angle contains 360 angular seconds.

Map Exercises

29. Find a bird's-eye view of your house using an online mapping program. Make a screenshot of it and print out the image to show to your class.

30. Sketch your neighborhood for several blocks around your house (or obtain a map from an online mapping service such as Google Maps™ and print it out). Locate all the fire hydrants and mark their locations on the map. What kind of map is this?

WORLDVIEW SLEUTHING: DRIVERLESS CARS

We live in a society where driving a car is a common and necessary part of life. But think about the 5 million car accidents that happen in the United States each year. Driver errors cause more than 90% of them. And another thing: fleets of taxis in New York City and other large cities around the world take a lot of money to man. One way to alleviate these problems and completely rethink driving may be simple—eliminate the driver.

INTRODUCTION

You are an engineer working at Google with a team developing self-driving cars. Your goal is to build cars that can safely travel without a human driver and accurately arrive at the desired destination.

TASK

Your job is to plan the driverless cars project. You will create a presentation using presentation software to outline the challenges of developing driverless cars to your team of "engineers" (your class or family). You may suggest some solutions to these challenges, especially considering what you've learned about mapping in this chapter.

PROCEDURE

1. Research tools and technology that you can build into your car to keep it from colliding with other cars or pedestrians by doing a keyword search on "driverless car sensors."

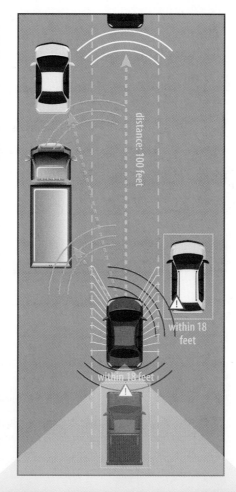

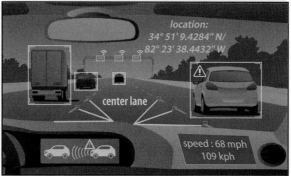

2. Research tools and technology that you can use to help the car know its location and the location of the road by doing a keyword search for "autonomous vehicle GPS."

3. Plan your presentation and collect any photos or videos you need, being careful to give proper credit.

4. Create your presentation and show it to another person for feedback.

5. Deliver your presentation to your class or family.

CONCLUSION

Companies such as Audi, Google, and Bosch have been working on driverless cars since 2009. Google's most recent driverless car can be summoned with an app on a smartphone. Driverless cars have driven over a million miles in road tests in California, Washington, and Arizona. While there are still improvements to be made, driverless car technology has the ability to save and improve lives. Exploring this kind of technology is important because God has made people to work, and travel is often a necessary part of work. People's lives are important to protect because they are made in God's image.

MAPS AND MAPPING

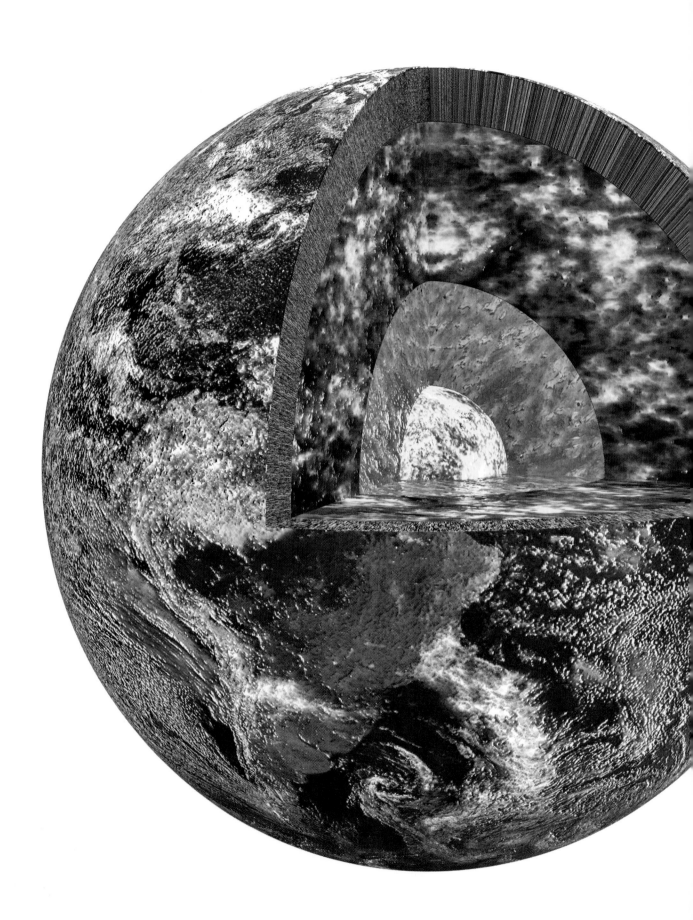

2 THE RESTLESS EARTH

CHAPTER 4
GEOLOGY—THE EARTH SPEAKS 76

CHAPTER 5
THE CHANGING EARTH 100

CHAPTER 6
EARTHQUAKES 128

CHAPTER 7
MOUNTAINS AND HILLS 152

CHAPTER 8
VOLCANOES AND VOLCANISM 174

DR. ANDREW SNELLING
GEOLOGIST

"A visit to some copper mines when I was 9 years old stimulated my interest in rocks and minerals. By the time I was in junior high, geology had become my passion. God used this passion to lead me to train in geology at two Australian universities, eventually receiving my PhD. After working four and a half years in northern Australia in mineral exploration and geologic research, I have researched full-time for more than twenty years in Creation and Flood geology, studying extensively in the Grand Canyon.

I have been excited to see the fruits of this research. Creationist scientists have found overwhelming evidences of a recent, global, cataclysmic, biblical Flood. Global patterns of geologic features, such as fossils, rock deposition, large-scale geologic structures, and indicators of strong water currents all point to the Flood. The scientific case for a young earth is increasingly irrefutable as radioactive methods for dating rocks are being shown to be totally unreliable.

We need more people to research, and to spread and teach what we find. I warmly encourage you to consider whether God might be calling you to an exciting life of creation science."

CHAPTER 4

GEOLOGY—
THE EARTH SPEAKS

4A THE EARTH, A SPECIAL PLACE   77
4B GEOLOGY, THE SCIENCE   85
4C THE EARTH'S STRUCTURE   90
4D THE EARTH'S NATURAL RESOURCES   93

MESSAGE FROM THE MOON

It was 1968, and the world was torn by conflict. The Vietnam War, riots, struggles against social injustice, and the Cold War had taken their toll on the lives and morale of Americans and even the world.

There was also the hotly contested space race between the Soviet Union and the United States. In the latest leg of this race, *Apollo 8* launched on December 21, 1968, with three astronauts on board—Frank Borman, Jim Lovell, and Will Anders. They were going to orbit the moon without landing on its surface to prepare for future moon missions. After three days of travel, they entered lunar orbit, traveling around the moon more than ten times in twenty hours.

It was Christmas Eve, and it was time to address the world in a spirit of unity and human achievement. NASA had planned to have a broadcast from space that night. The astronauts had the largest television audience in history, and they decided to read the opening verses of Genesis 1 to the world . . . from the moon.

"For all the people on Earth the crew of *Apollo 8* has a message we would like to send you. '*In the beginning God created the heaven and the earth. . . .*'"

4A

THE EARTH, A SPECIAL PLACE

What scientific evidence confirms that Earth was designed for life?

4A Objectives

After completing this section, you will be able to

» explain why Earth is well-suited for life.

» show how Earth is unique by comparing it to other planets.

» explain how Earth's design helps humans explore the heavens.

4.1 THE GOOD EARTH

For the first time ever, during the *Apollo 8* mission, humans viewed their planet from the remote, alien, desolate, hostile environment of the moon. These astronauts had come all this way to observe the moon, yet they couldn't pull their gaze away from their home.

And so, after seeing scenes of the life-filled, colorful earth setting behind the desolate horizon of the moon, they concluded their TV broadcast:

"'And God called the dry land Earth; and the gathering together of the waters called he Seas: and God saw that it was good.' And from the crew of *Apollo 8*, we close with good night, good luck, a Merry Christmas, and God bless all of you—all of you on the good Earth."

4.2 EVIDENCES FOR DESIGN

The conclusion to this TV broadcast shocked many in the scientific community. For the past two centuries, scientists in general had discarded a Christian view of the earth. After many years of scientific observation, they had concluded that our earth, sun, and even our galaxy were average. To them, there was no reason to think that the earth and its life are unique. On the other hand, the Bible's presentation of the earth as a unique place seemed antiquated, unscientific, and even silly.

The *Apollo 8* astronauts saw things differently. While speeding through the heavens far from Earth, they found that the biblical story had a remarkable pull on their thoughts and emotions. Far from being a silly account from long ago, the Bible provided a narrative that seemed to fit what they were experiencing.

Christians should attempt to look at everything from a biblical worldview. As they do this, they find that the Bible fits with the world they see. The Bible reveals that God designed the earth to sustain the life that He planned for it: "*He* [God] *hath established it* [the earth], *he created it not in vain, he formed it to be inhabited*" (Isa. 45:18b).

Have you ever thought about how special your home planet is? It is true that because of the Fall and God's curse on mankind's sin, Earth's atmosphere is broken, its surface is scarred, and parts are not friendly to life. But we can still see evidences of God's good design if we look at Earth from the Bible's perspective.

4-1 In December 1968, kids gather around the television as *Apollo 8* astronauts broadcast a Christmas message.

4-2
What is so special about where you are?

Every aspect of Earth's physical structure, and its relationship with its moon, the sun, and the rest of the universe, is linked to God's design of living things. The immensity and precision of this grand system reminds us that God is great. Its remarkable beauty and life-sustaining power remind us that God is good.

EARTH'S MASS AND STRUCTURE

The strength of Earth's gravity is designed for life. The earth's 6 billion trillion metric tons of mass produces just the right amount of gravity *at its surface*. Too little gravity affects living organisms by producing weaker bones and muscles. You've probably heard about how astronauts have to exercise in the "weightless" environment of space to keep this from happening. But too much gravity would rapidly tire muscles and require heavier bones to avoid breakage. Organisms would have to use much more energy to move against gravity. Flight of birds and other animals might not even be possible in a stronger gravity. Earth's gravity is strong enough to hold the kind of atmosphere we have, which is essential to life on Earth. The size and rate of the moon's orbit also depends on Earth's gravity. See the box on page 83 to see how the moon affects life on Earth.

The structure of the earth is designed for life. Only half the planets in our solar system have a solid surface. The other planets are balls of toxic gases with no accessible solid surfaces. The earth is also constructed in layers. The outer layer, or *crust*, is made of distinct pieces called *tectonic plates*. The separation of these plates has influenced the rise of diverse human cultures and has resulted in the development of a vast variety of living things from the original created kinds. The inner parts of the earth also help protect life in special ways; we will see that in the next few pages.

4-3
Strong gravity would be a problem for flying creatures.

EARTH'S TILT AND ROTATION

Earth's tilt and rotation create environments that are designed for life. As is the case with most of the planets, the earth's orbit is a nearly circular ellipse and lies in a plane. The sun is near the orbit's center. The earth itself spins on its own axis. If the earth's spin axis were perpendicular to its orbital plane, the sun's rays would always shine straight down near the Equator and at a steep slant near the poles. As a result, it would be much hotter near the Equator and much colder near the poles year-round. There would be no seasons. Living things would be crowded into smaller habitable areas between the poles and the Equator.

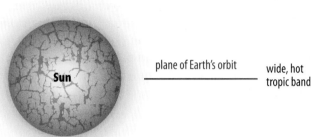

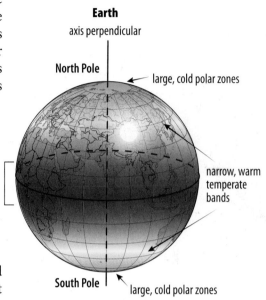

But the earth's axis is *not* perpendicular to its orbital plane. It is angled 23½° from the vertical. No one spot on Earth is heated or chilled at the same rate through the year. To a ground observer, the sun's highest position in the sky continually changes from day to day as the earth revolves around the sun. This creates broad areas of moderate temperatures between the Equator and the poles. Thus, living things can live almost anywhere on the planet. Earth's tilted axis makes much more land and sea suitable for living things. For this reason, we also enjoy a larger variety of creatures.

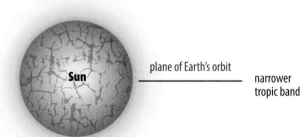

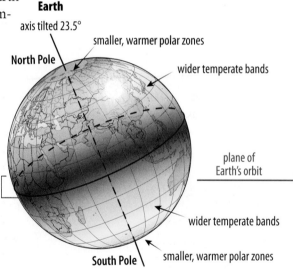

4-4
With a tilted axis, the earth's surface is heated more evenly by the sun.

The speed of the earth's rotation is designed for life. It is perfectly matched to other characteristics of the earth. If the earth turned too slowly or not at all, very few organisms would be able to survive under the sun through the long or constant day. Temperatures would soar and the ground would scorch. Plants that get their energy through photosynthesis would not be able to live with long periods of darkness in the cold.

Too fast a rotation would also be harmful to life. The rapid day and night cycles would limit hunting or food gathering, and sleep periods would be too short. Many familiar plants could not survive the rapid cycle of day and night. The entire nature of the earth's wind and ocean circulation patterns would be very different, and weather would probably be more violent.

GEOLOGY—THE EARTH SPEAKS

EARTH'S LIQUID WATER

The earth has liquid water, which is necessary for life. No other planet has liquid water on its surface. And the earth doesn't just have a little liquid water—nearly three-fourths of the earth's surface is covered with it! Water is essential to life, but it is not the only thing life needs. Some scientists believe that if a planet has water, then life is possible, but frozen water can't support life. And only the simplest organisms live in near-boiling water.

Water has many special properties that make it essential for life. It dissolves and transports nutrients that living things need to function. It has a much greater ability to absorb heat than similar liquids. Animals use watery body fluids to regulate their internal temperature. The vast quantities of water on the earth's surface help limit the swings in temperatures around the world. Ocean currents transport warm water to cool areas of the world, and cool water to warm areas. This makes even more of the world habitable. Frozen water floats, making it possible for animals to survive in ponds during the winter or even flourish under year-round polar ice.

4-5
The high heat capacity of water helps to warm cool areas of the earth and cool warm areas.

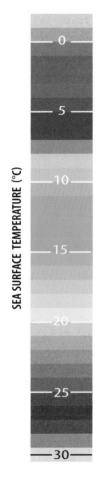
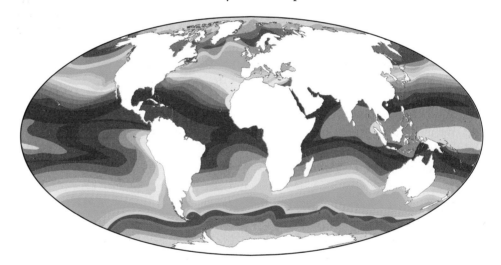

EARTH'S ATMOSPHERE

The earth has an atmosphere that is designed for life. The mere presence of an atmosphere around a planetary body is rare in our solar system. Of the eight main planets and hundreds of moons, only seven have significant atmospheres. And of these, only Earth's is transparent because it is made of gases that are colorless and clear. Other planets have dense, murky atmospheres made of compounds that filter out light. It is important for life that light from the sun can reach the earth's surface. Humans, and all large and complex forms of life, ultimately depend on plant photosynthesis for food.

Earth's atmosphere also offers protection to its inhabitants. The atmosphere blocks dangerous wavelengths of light from the sun from reaching living things on the surface. Only a very narrow band of light energy wavelengths necessary for warmth and photosynthesis gets through the atmosphere.

The atmosphere is also a physical shield against bombs from space—*meteors*. Of the tens of thousands of meteors that enter the atmosphere each day, less than a hundred actually reach the surface, and most of these are the size of small pebbles.

Our atmosphere also contains just the right amount of oxygen necessary for life. Nitrogen dilutes the oxygen to just the right amount for living things to use. No other body in the solar system has an atmosphere with this feature. Furthermore, all other atmospheres contain poisonous gases.

EARTH'S MAGNETIC FIELD

Earth has a strong magnetic field, making it safer for life. Most of the planets and even some moons in the solar system also have strong magnetic fields. Of the planets, only Venus and Mars lack a magnetic field, as does our own moon. Most earth scientists believe that a planet's magnetic field comes from the movement of liquid metals in its center. While this may or may not be true, Earth's magnetic field is vital to life. If it were not for this magnetic field, dangerous energetic radiation particles from space could reach Earth's surface, harming or killing living things. The field also deflects or traps the continual stream of high-energy, charged particles coming from the sun, called the *solar wind*. These particles could greatly change the composition of the atmosphere if they were able to reach it.

EARTH'S SUN

The properties of the sun make it designed for life. The sun's mass and its composition are essential for life on Earth. These factors determine the sun's surface temperature, which controls the wavelengths of visible light that it radiates. These wavelengths have just the right energy to drive photosynthesis on Earth.

The sun's size and temperature determine the location and size of a region around it called its *habitability zone*. Within this zone, water can exist as a liquid, and carbon dioxide remains a gas. The earth orbits toward the inner edge of the sun's habitability zone.

4-6
Our atmosphere is in layers, each of which benefits those living on the earth.

Nitrogen and Design
Interestingly, the plentiful nitrogen in the atmosphere is fixed in the soil by certain plants and bacteria. Nitrogen compounds are essential as natural fertilizer for plant growth. Humans use a chemical process to create artificial fertilizers. Without nitrogen compounds, the earth could not feed the amount of life that it has.

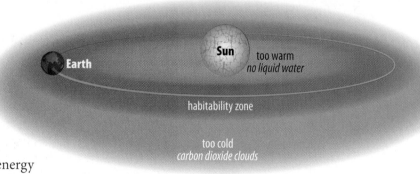

4-7
The sun's habitability zone within which the earth orbits

4-8 Our solar system is located well away from the galaxy core and the dusty spiral arms.

A light-year (ly) is the distance that light travels in one year, about 6 trillion miles, or 9.5 trillion kilometers.

4-9 If our solar system were located within the dusty regions of our galaxy, we would not be able to see the immensity of the universe, which reflects the immensity of our God.

If the earth were much closer, it would be too hot for liquid water. If it were too far out, carbon dioxide would form ice clouds high in the atmosphere. This would block sunlight, making Earth's surface a cold, barren desert like Mars.

Earth orbits the sun, which is unusually stable. Scientists have studied many stars, looking for any comparable to our sun, hoping to find one that might support life. They have found a few that are similar to the sun, but there are crucial differences. The energy output of those stars varies twenty to thirty times more than our sun. These variations would produce significant temperature variations on planets around these stars. By comparison, our sun is very stable, providing a friendly neighborhood for the earth.

EARTH'S PLACE IN THE SOLAR SYSTEM

Earth's place in the solar system is designed for life. You might wonder why the rest of the solar system would be important to life here. Earth is tucked in relatively close to the sun. Five planets, including the four massive "gas giants," Jupiter, Saturn, Uranus, and Neptune, lie much farther out.

These huge gas giants are vital for deflecting comets and similar objects that enter the solar system away from the inner planets. Without these planets and their strong gravitational fields, hundreds of comets from a region much farther out (called the *Kuiper Belt*) of comet-like objects might frequently swing through the inner solar system. This would create a real possibility of a devastating collision with Earth. In a similar way, Jupiter's immense gravity is especially important for guiding the minor planets, or *asteroids*. Most of these rocky objects orbit between Mars and Jupiter. Jupiter's gravity helps keep them away from Earth.

EARTH'S PLACE IN THE MILKY WAY

Earth's place in the Milky Way galaxy is designed for life. Even the sun's location in its galaxy is important for preserving life. The Milky Way is an immense collection of stars, other objects, and dust more than 100,000 **light-years** across. Like many galaxies, the Milky Way is shaped like a flattened disk, with a bulging center and dusty spiral arms. Astronomers believe that the center of the galaxy contains many densely packed stars. At its very center, there may be at least one huge black hole, holding the galaxy together with its intense gravity. The galaxy center would be a very dangerous place for Earth, with lots of exploding stars giving off high-energy radiation.

The spiral arms of the galaxy are not so dangerous, though they still have many more stars that could explode than the region we find ourselves in. The earth's sun and its family of planets are located about halfway to the rim of the galaxy from its center, well away from the deadly activity there. We are also between two main spiral arms of the galaxy; there are relatively few stars nearby that could be a danger to us if they exploded.

OUR MOON: UNIQUE IN THE SOLAR SYSTEM

Astronomers are amazed by our moon. It has properties that make it distinct among all the moons in the solar system. Consider its size. It is 100 times larger than the average moon in the solar system. And when compared to the size of its planet, the ratio of the moon's diameter to Earth's is 5 times greater than any other moon/planet ratio. Secular scientists cannot account for such a large moon except due to random chance, but the Bible does. Genesis 1 indicates that God made the moon in a direct act of creation. God suggests the reason for its size in verse 16, where He calls it the "lesser light to rule the night." Our moon is large enough to light up the entire night sky. The earth is the only planet with a moon that can do this.

The moon's mass is so large compared to Earth's that some astronomers consider the Earth-moon system a double planet. The moon's gravitational pull causes tides, which benefit life along the world's coasts. Its gravity also stabilizes the tilt of the earth's axis to keep the earth from wobbling like a top.

One of the most unusual aspects about the moon is that its apparent size in the sky is the same as the sun's. This means that when the sun, moon, and earth are aligned properly, the moon barely blocks out the sun in a total solar eclipse. Though the moon is 400 times smaller than the sun, it is 400 times closer to the earth. Total solar eclipses are predictable and are awe-inspiring events. But even more amazing is the fact that when the moon just covers the sun's disk, astronomers can study the sun's atmosphere, which is usually obscured by the glare of the sun. Before special satellites, this would not have been possible without the providential arrangement of the sun, earth, and moon.

The Bible also says that God specially created the moon "for signs, and for seasons, and for days and years" (Gen. 1:14). Most cultures on Earth now use, or at some point in the past used, a calendar based upon a lunar month—the number of days from new moon to new moon. We can also use predictable eclipses to date events of the past. The stable moon, "a faithful witness in heaven," reminds us that God is faithful to keep His promises and to provide for our needs (Ps. 89:37).

4.3 EARTH, THE PLACE FOR BIBLICAL DOMINION

As we peer through our transparent atmosphere into the vastness of space or study the depths of the earth with earthquake waves, we are reminded that we are small pieces in a very big world. But as we look at the earth and the heavens from the perspective of Scripture, we see that we play an important, central role. God has made us in His image. So we have the ability and desire to study the world God has made. God has also called us to have dominion over the earth. The earth wasn't just created for life; the earth was created for us—intelligent, spiritual, human life. As we study the earth and wisely use its systems to improve our lives, we are not selfishly abusing it, not if we intend to declare God's glory through good and wise dominion. We are fulfilling our God-given calling.

Some people think of the universe as a cold, dark place and Earth as a lonely blue marble speeding through space. The Bible teaches us that the universe is a huge cathedral filled with the praises of God and Earth is the home of His beloved image bearers. We do not know why God has chosen to honor us as He has. But we do know that it is our task to love Him and praise Him for His greatness and goodness.

When I consider thy heavens, the work of thy fingers, the moon and the stars, which thou hast ordained; What is man, that thou art mindful of him? … Thou madest him to have dominion over the works of thy hands: thou has put all things under his feet … O LORD our Lord, how excellent is thy name in all the earth! (Ps. 8)

4A REVIEW QUESTIONS

1. What is the relationship between the design of the earth, its moon, and the sun, with the living things that inhabit the earth?
2. Give two reasons why the earth's mass is designed for life.
3. Explain how the tilt of the earth's rotational axis is designed for life on Earth.
4. What special properties of water on the earth are designed for life?
5. List three properties of the earth's atmosphere that make it important for life.
6. Why is Earth's magnetic field critical for life on Earth?
7. Give two reasons why the sun's mass is designed for life on Earth.
8. How does Earth's position in the solar system protect life?
9. How does Earth's position in the Milky Way galaxy help protect life?
10. How does Earth's design help us learn about God by observing the vastness of what He's created?
11. (True or False) Earth's special features and position in the universe give strong evidence that our planet was specially created.

4B

GEOLOGY, THE SCIENCE

How does our worldview affect our understanding of geology?

4.4 HISTORY OF GEOLOGY

4B Objectives

After completing this section, you will be able to

» summarize the history of geology.

» explain the dangers of viewing the earth as very old and as the product of natural processes.

» describe how geology is used.

geology: (Gk. *geo*—earth + *-logia*—to speak of)

One of the ways that we can exercise dominion on God's specially created earth is through **geology**—the study of the earth. People have been studying the earth for a long time, probably since right after the Creation and Fall (see Gen. 4:22). After the Genesis Flood, people needed geology to identify new sources of metals and building materials.

Written records of humans studying the earth are very old. Most of the ancient centers of learning included the subject of geology, and these writings led to the rediscovery of geology in Europe following the Middle Ages. Most European scientists during the 1500–1700s had no problem with a biblical history of the earth. They believed from the ancient writings and their own observations that most rocks were formed under a great flood.

But things began to change. Scottish physician and geologist James Hutton and his fellow scientists believed that the earth's features were mainly shaped over time by the slow geologic processes that we see today. He concluded in 1795 that the earth had to be much older than the Bible seems to show. A few thousand years weren't long enough for sediments to form the rocks that he could see or for mountains to rise at the rates they were growing in his day. This was the beginning of **uniformitarianism**—the belief that earth-forming processes are natural, have always been the same, and have always happened at the same gradual rate. The opposing view to uniformitarianism was *catastrophism*, or the belief that one or only a few catastrophic events in the past shaped the earth's surface, followed by long periods of no change.

In 1830, English geologist Charles Lyell published his book *Principles of Geology*. This book defined many of the terms and methods that geologists use today. Lyell also strengthened Hutton's theories. He summed up uniformitarianism with the statement, "The present is the key to the past." His book greatly influenced Charles Darwin, who extended these ideas to the origin of life. For more than 150 years, geologists viewed uniformitarianism as the most important and basic principle of geology.

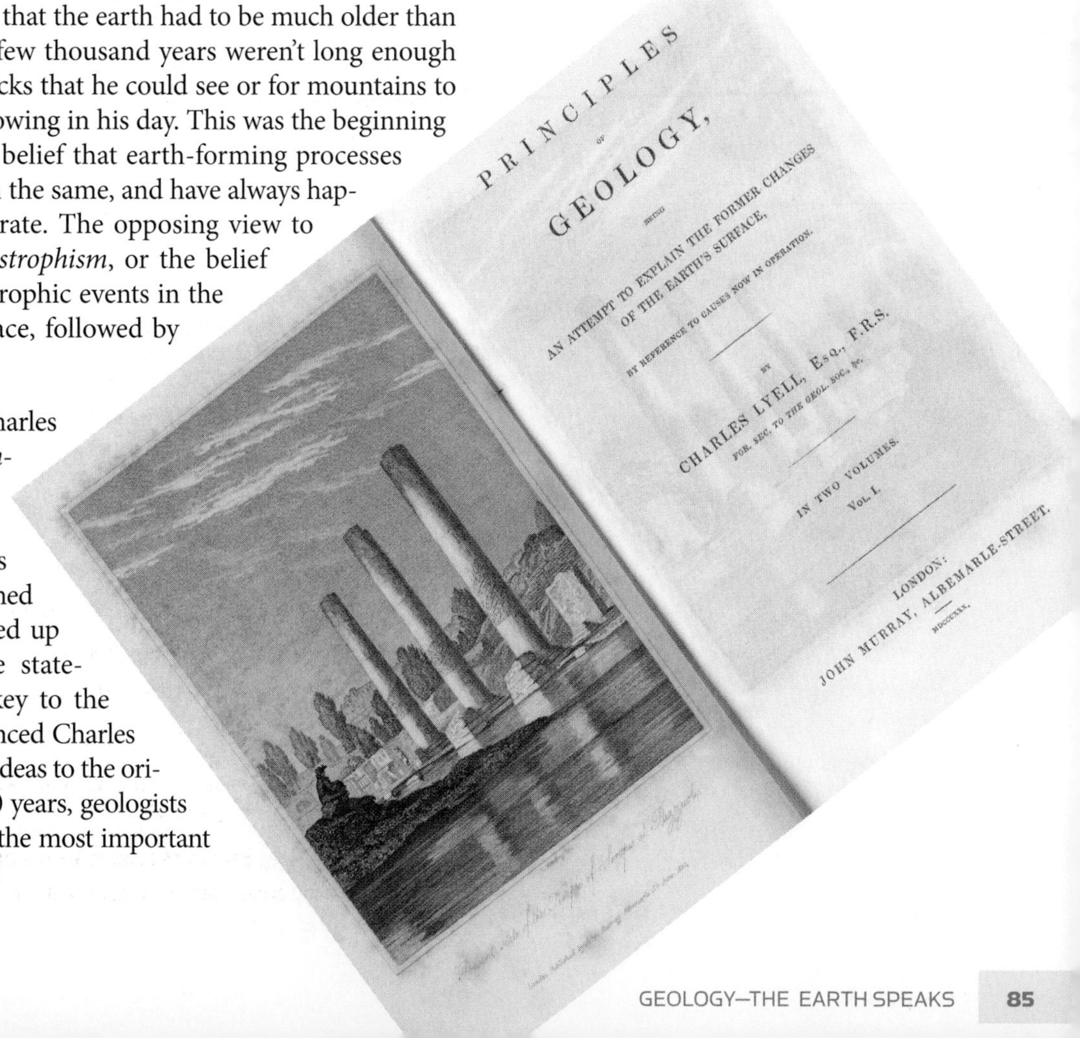

Discoveries in geology in the 1800s and 1900s seemed to strengthen the idea of an ancient earth, a perspective known as the **deep-time view**. Most scientists today believe that the earth is around 4.5 billion years old! On the other hand, most biblical creationists hold to a young-earth view. They believe that the earth is about 6000–7000 years old.

Geologists focused on defining periods of the earth's history and their ages. Discoveries showed that there were times in the earth's past when great changes happened. Christian pastors and theologians began to feel the need to reinterpret the Bible to make it fit with these findings. They looked for places in the Creation narrative to insert millions and billions of years to make room for the theories of geology.

In 1912, the German geologist Alfred Wegener suggested that the continents were once one big land mass that broke up into pieces over time and drifted apart. In the 1960s, this idea became the plate tectonics theory. So the old-earth, gradualistic view had to be modified to include this evidence.

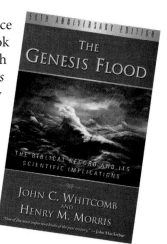

A young-earth view of geology and of science revived in the 1960s. Just as Lyell's 1830 book influenced many to believe in an old earth and gradual change over time, *The Genesis Flood*, written by John Whitcomb and Henry Morris in 1961, encouraged people to re-examine the scientific data from a Christian worldview. Bible-believing geologists and even secular scientists have shown that the scientific evidence actually fits the Bible's story of a recent creation event and the Flood better than the old-earth view. Even a plate tectonics theory fits into the young-earth time frame.

You can see the battle of worldviews in the field of geology. All geologists have a shared purpose—understand the earth's structure and processes better. Even though they all look at the same data and evidence, they come to radically different conclusions because of their presuppositions. Geologists working from a Christian worldview study and work to provide the tools needed for good and wise dominion.

4.5 OPERATIONAL GEOLOGY

So what do geologists do today? Geologists can learn all kinds of things about the earth through observation. Observation is a key aspect of operational science. Geologists identify rocks and minerals by their makeup and properties. They draw maps showing where to find different kinds of rocks and the vertical order of rock layers. They study cracks in the rocks to look for movement where building foundations will go. Soil geologists identify different layers of soil and learn how underlying rocks make these layers. This information is helpful to farmers.

4-10

Geologists may work in labs but many also do field work. This geologist is collecting a rock sample—an example of operational geology.

You can see how *operational geology* is really important for obeying the Creation Mandate given to us in Genesis 1:28. We need geology to use the earth's materials wisely and to protect people from natural disasters. At the same time, geology reveals the wonderful designs and patterns within the earth that point to the Creator of the earth.

4.6 HISTORICAL GEOLOGY

How does the history of geology as a science affect geologic observations today? As you learned in Chapter 1, the observations that scientists make are affected by the presuppositions of their worldviews. Both the questions that geologists ask and the answers that they find are affected by their worldviews. The most widespread worldview in geology is secular, where natural processes similar to today's acted at more or less the same rate through time. But the earth's history included catastrophic events, such as widespread volcanic activity, ice ages, and meteorite impacts. Secular geologists believe that these kinds of rare events dramatically changed the earth and seriously affected all living things.

4-11
How long ago was the last meal for this fish? This question may be answered by historical geology.

All scientists believe that the world is regular. The principles of uniformity and cause and effect state that nature is regular and predictable. God created this world, and He is faithful and unchangeable. The Bible also says that He continues to regulate His creation so that it is a stable place (Gen. 8:22). So the principle of uniformity is an important part of operational science.

But there's a problem when scientists extend uniformity into the unobservable past. Some of their conclusions contradict clear statements in Scripture. Scientists assume that the past has always been as regular as things seem today. But they ignore written evidence of special events like the Creation of the world and global catastrophes like the Flood. This is why deep-time gradualism is not reasonable and leads to wrong conclusions.

What's in a Name?

Modern secular geologists call their approach to geology *neocatastrophism*, rather than uniformitarianism. This is because they have realized that changes during Earth's history really haven't been uniformly slow and due to uniform processes. The name separates them from the *catastrophists* of the 1700s and 1800s, most of whom were young-earth creationists. Thus, modern geologists are "new catastrophists."

So secular scientists and Bible-believing scientists interpret data and answer questions much differently because they use different models of Earth's history. Where did the mountains come from? When did dinosaurs live and die? How did fossils form? How long did it take minerals to form? These are questions that one must answer within his scientific model.

Similarly, scientists cannot predict with certainty what will happen in the future. Assuming that earth processes continue as they do in the present because of uniformity, scientists can make certain predictions about the future. But our models of long-term earth processes, like the changing global climate, are incomplete at best. These predictions may or may not prove to be accurate. As with studying the past, future predictions are based on worldview assumptions. Secular geologists believe that the earth will continue to change slowly over time and likely experience unpredictable catastrophes along the way. But the Bible shows us that specific devastating events await the earth sometime in the future (2 Pet. 3:7–13).

4B REVIEW QUESTIONS

1. What was uniformitarianism? Who first published this idea as a basic approach to geologic science?
2. Why did uniformitarianism require that the earth be old?
3. What is the name for the viewpoint that the earth is billions of years old?
4. Give some examples of direct scientific observation in geology.
5. What two kinds of information does a geologist use when explaining how a rock formed in the past?
6. What basic principles of operational science do most secular scientists *not* properly apply to historical science? Name two biblical events this problem causes most scientists to ignore.
7. How does Genesis 8:22 relate to the principle of uniformity?
8. (True or False) Scientists who do not believe the Bible accept the principle of uniformity.

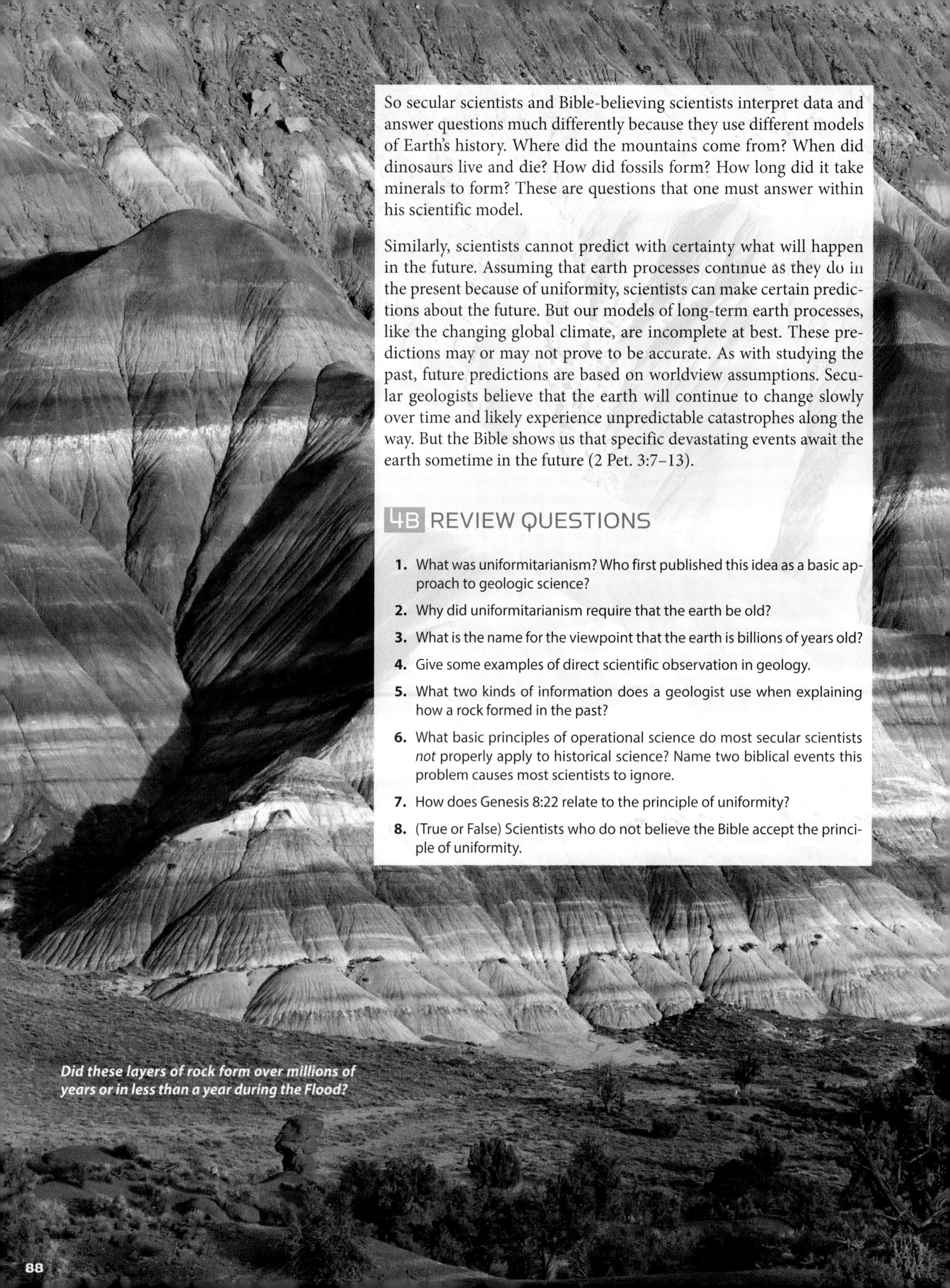

Did these layers of rock form over millions of years or in less than a year during the Flood?

CASE STUDY: NICOLAUS STENO, BISHOP OF GEOLOGY

Some scientists think that there is a conflict between geologic evidence and the Bible. However, the conflict is actually between worldviews and is new to the field of geology. Years before Charles Lyell was called the "Father of Geology," another scientist earned this title. And he happened to be a firm believer in the Bible.

Born in 1638, Nicolaus Steno trained as a doctor. His study of biology led him to geology and paleontology. He focused mostly on stratigraphy—the study of rock layers and how they form. He developed four principles in geology that have become the bedrock of modern geology.

1. Lower layers of rock formed before the layers above them—the principle of superposition.
2. Rock layers always form in horizontal layers.
3. Rock layers are continuous across the world unless interrupted by other objects.
4. Any interruption in rock layers occurred after the layer formed.

In addition to being a scientist, Steno was also committed to God's Word. The Bible provided the framework for Steno's scientific work and guided how he interpreted his observations.

According to Steno's fourth principle we know that the faults interrupting the rock layers occurred after the rock layers formed.

He explained the formation of rock layers on the basis of Creation and the Genesis Flood. Steno was convinced that the Flood had occurred 4000 years before his time. His writings clearly show that evidence in nature and Scripture agree.

Later geologists built their theories on the work of Nicolaus Steno, including James Hutton. Even though Steno based his work on Scripture, Hutton came to completely naturalistic conclusions. Hutton's work led Charles Lyell to propose long periods for geologic processes. Lyell's *Principles of Geology* framed geology for a hundred years.

Questions to Consider

1. As scientists look at rock layers, where do scientists expect the oldest layer to be? Explain how they assess the age of the layers.
2. Geologists often find layers that are not horizontal. Can you think of why a layer may no longer be horizontal?
3. How did Steno and Hutton come to significantly different conclusions when looking at the same data?

According to the principle of superposition, the lowest layers were laid down first. Therefore the bottom layers are the oldest and newer layers are above older layers.

4C Objectives

After completing this section, you will be able to

- explain how scientists study the interior of the earth.
- describe the different layers of the earth and their properties.
- sketch the earth's interior, labeling its regions and layers.

seismic (SIZE mick): (Gk. *seismos*—shaking, earthquake + -ic); having to do with earthquakes and earthquake waves moving through the earth

4C

THE EARTH'S STRUCTURE

What is inside the earth and how do we know?

4.7 STUDYING EARTH'S STRUCTURE

What do you think the inside of the earth looks like? Is it a uniform ball of rock, or a hollow shell, or something else? And how do we know? We can't exactly dig very far or look into it with telescopes!

Geologists learn about the earth's interior by studying how earthquake waves travel through it. Earthquake waves are waves of energy that form when part of the earth's *crust* moves suddenly. They're also called **seismic** waves. You will learn more about earthquakes in Chapter 6.

Geologists use these waves to create computer models of the earth's interior. They use a process very similar to medical x-ray CAT scans. The earth has three main layers: the *crust*, the *mantle*, and the *core*. These layers contain different kinds of materials. We know this because seismic waves move through the layers differently.

Natural seismic waves (earthquakes) are not the only way that geologists can learn about the earth's interior. They can create their own seismic waves using explosives. These less powerful waves allow scientists to locate mineral and oil deposits up to several miles below the earth's surface. Photography and ground-penetrating radar images from space can tell us about earth structures just a few meters underground. Geologists can also fly over a region in special aircraft that record tiny changes in the earth's magnetic or gravitational fields that identify deeper structures. But these technologies can indicate conditions less than a few hundred miles below the earth's surface at best. They can't show us the structures deep within the earth. Only seismic waves can do that.

4-12

Scientists gain much of their knowledge of Earth's structure from studying earthquakes. The map above was automatically generated by data from a seismic sensor network.

4.8 THE CRUST

The earth's outer layer, the **crust**, is solid, relatively low-density rock. Recall that density is the amount of matter in a given volume. This rock's thickness averages 3–6 mi (5–10 km) under the oceans to 12–56 mi (20–90 km) under the mountain chains on continents. The crust averages six times as thick under the continents as under the ocean basins.

Rock becomes more dense farther into the crust. We know this because seismic waves gradually move faster as they pass deeper into the crust. In 1909, **Andrija Mohorovičić** noticed that earthquake wave speed suddenly begins to slow at a certain depth far below the earth's surface. Such a sudden change is called a *discontinuity*. Scientists know that wave speed depends on the material through which it travels, so this change in speed told Mohoravičić that the earth material had changed. He concluded that this must be the boundary between two layers of the earth. The boundary between the crust and the mantle is called the *Mohorovičić discontinuity*, or more commonly, **Moho**, in his honor. The Moho marks the bottom of the crust.

> Andrija Mohorovičić (AHN-dree-ya MORE-OH-vih-chick) (1857–1936) was a Croatian seismologist (earthquake scientist) who was far ahead of his time in studying earthquakes and the damage they caused to buildings.

4.9 THE MANTLE

The **mantle** stretches from the Moho to a depth of about 1800 mi (2900 km) from the surface. The mantle takes up 84% of the earth's volume. Seismic waves travel differently through certain parts of the mantle. Scientists use this data to divide the mantle into sections. Directly below the Moho is the *upper mantle*, which is solid, but is made of rock different from the crust. The upper mantle is about 40 mi (70 km) thick beneath the ocean basin crust and 80–160 mi (125–250 km) thick under the continents. This section of the upper mantle tends to move with the tectonic plates. Geologists call the combination of this segment of the upper mantle and the crust the **lithosphere**.

Seismic waves normally move faster at greater distances into the earth. But below the lithosphere, the rate of change of seismic wave speed in the upper mantle is much lower than it is deeper in the earth. Geologists call this layer the *asthenosphere*. It extends from the bottom of the lithosphere with a thickness of about 125 mi (200 km). (It is thicker under continents.) Geologists think that the higher pressures and temperatures make this rock closer to melting. This would explain why seismic waves slow down here.

The upper mantle continues down to about 420 mi (670 km), the deepest point from where earthquakes can start. Below this is the *lower mantle*, which extends to 1800 mi (2900 km) from the surface. This is about 2200 mi (3500 km) from the center of the earth.

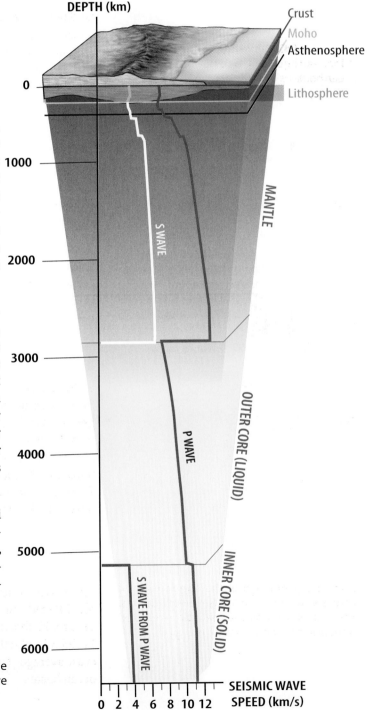

4-13 Seismic waves reveal the complicated structure of the earth's crust and upper mantle. Note how seismic wave speed changes with depth into the earth.

GEOLOGY—THE EARTH SPEAKS

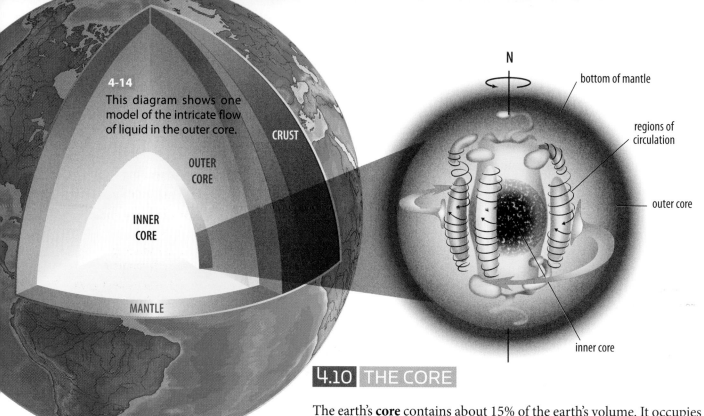

4-14 This diagram shows one model of the intricate flow of liquid in the outer core.

Some geologists estimate that the outer core has a temperature of 7200–9000 °F (4000–5000 °C) and a density of about 10 g/mL. The inner core may have a temperature of 9000–11,000 °F (5000–6000 °C) and a density of 13 g/mL.

The density of normal iron at room temperature is about 7.9 g/mL.

4.10 THE CORE

The earth's **core** contains about 15% of the earth's volume. It occupies the center of the earth below the mantle. Many geologists don't even try to guess what the core material is like because it is so different from anything we know. The core must be extremely hot and dense, possibly more than twice as dense as ordinary iron. We cannot imagine the pressures and temperatures in the core.

The core has two parts. The outer core is about 1400 mi (2260 km) thick. Geologists believe it is a liquid or has liquid properties because certain kinds of seismic waves, which do not pass through liquids, do not pass through the outer core either. The inner core at the very center is about 1500 mi (2440 km) across. Geologists believe it is completely solid.

It's possible that the outer core at least partially creates Earth's magnetic field. Creationist physicists, though, believe that the field could also come from other portions of the earth's interior. Geologists believe that the core is probably 85% iron, 4–5% nickel, and the rest, lighter elements such as oxygen and sulfur.

4C REVIEW QUESTIONS

1. Describe a seismic wave.
2. Why can't radar from aircraft or satellites show us structures deep within the earth?
3. Draw a diagram of the earth's interior. Label the crust, mantle, upper mantle, lower mantle, lithosphere, asthenosphere, inner core, and outer core. Include the distances given in these sections.
4. The earth's crust and the solid, rocky upper part of the mantle together form what feature of the earth?
5. How would you describe the rock in the lower mantle?
6. Contrast the outer and inner core.

4D

THE EARTH'S NATURAL RESOURCES

How can we wisely use Earth's resources to best glorify God and help others?

4D Objectives

After completing this section, you will be able to

- » identify natural resources.
- » explain how to manage natural resources.
- » list factors that affect environmental quality.
- » explain why Christians should be concerned about the environment.
- » analyze the relationship between Earth's resources and population.

4.11 WHAT ARE NATURAL RESOURCES?

What kind of substances did you use to get ready for school? What are you wearing? What did you eat today? What does your car use for fuel?

Did you ever stop to think about where the materials for these things came from? The plastic for your toothbrush, electricity for your hair dryer, ceramic for your dishes, and gasoline for your car all came directly or indirectly from the earth. Some of these materials are highly processed, like the plastic case of your cell phone. You can't find plastic in nature—people rework raw petroleum to make it. Other materials we use are in their natural state, like the water you drink, the air you breathe, and the apple you eat at lunch. Any raw material that we use from our environment is a **natural resource**.

People classify natural resources into *biological* and *nonbiological* natural resources. Forests and ocean fisheries are two examples of biological or living resources. Trees and fish grow wild. People go into their natural environments and harvest them. Some people consider agriculture a natural resource. Nonbiological resources include soil, groundwater, clean air, sunlight, and minerals and ores in the earth.

Scientists may also classify a resource by availability. If it is being used right now, then it is an *actual resource*. A coalfield being mined is an actual resource. If prospectors or geologists find a vein of gold or an oil reservoir, these materials are *potential resources*. When workers begin to remove these materials for use, they become actual resources.

4-15

Alternating coal and sandstone layers. Coal is a nonliving, nonrenewable natural resource.

4.12 RENEWABLE AND NONRENEWABLE RESOURCES

You've probably seen gas prices change dramatically. If there's a hurricane that shuts down refineries or oil rigs, prices go up. If there's a lot of gas available, they go down. What makes prices change? They change because the fuel's availability changes. That's the principle of "supply and demand."

GEOLOGY—THE EARTH SPEAKS 93

Natural resources are like that too. The availability of some resources is affected by how fast people use them and how fast they can be produced. **Renewable resources** have an unlimited supply or are easily replenished. Water, sunlight, wind, and even air are examples of renewable resources.

Some resources take longer to produce or replenish than others. A lake that supplies a power dam may drop in level during a dry season, reducing electricity output. It may fill up again during a rainy season. Food crops require a growing season to restore the food supply. Forests may take ten to twenty years to replace trees cut for lumber. The soil fertility of a patch of land may take hundreds of years to recover naturally.

Many natural resources just get used up and can't be replenished. These are called **nonrenewable resources**. Materials such as silver, copper, and uranium exist in fixed amounts in the ground. This is probably true for oil, coal, and natural gas too. There's no known natural process that makes more of these resources at the present.

4.13 ACCESSING NATURAL RESOURCES

As the *Apollo 8* astronauts looked back at Earth, they saw a place with enormous supplies of air, water, and food. Their tiny space capsule held only enough of these things to get them through their mission and home again. These precious and essential resources were very limited, but they were just where they were needed.

The value of a natural resource depends partly on where it is found. One of the top needs in the world is a supply of clean drinking water. Many people around the world die every year of thirst or of diseases they catch from polluted water. Does clean drinking water exist in the world in large quantities? Absolutely! But it is not accessible to these people. So a large, clean lake of drinkable water is more valuable in Africa than in remote northern Canada. The existence of such resources affects where people build cities and villages. It can affect their culture too.

Building materials and energy sources are also valued by how easily they can be obtained. Lumber in a treeless area like the western prairies of the United States is quite expensive. So is gasoline in Hawaii. The cost of limited resources is directly tied to the expense and effort of transporting the resource to where it will be used.

4-16
Hydropower uses water stored behind the dam for energy. The water and energy are renewable resources.

4-17 Not much lumber here!

4-18 These forests are being managed to maximize the amount of lumber that can be harvested over time.

4.14 RESOURCES AND DOMINION

God created a world full of natural resources. What are humans to do with them? We are to manage them wisely, and at the same time we are to increase our numbers (Gen. 1:28). Though the Fall makes these goals more difficult, God never revoked these two commandments.

Today, we know much about the earth and its resources. We have amazing tools to locate, measure, and observe the things that we obtain from the earth. We monitor land use, water quality, and animal populations. We keep careful track of what we harvest and place limits on hunting and fishing certain kinds of animals. The effort to oversee and control the use of natural resources is called **resource management**.

Resource management is simply good and wise dominion. Managing renewable resources focuses on the most economical way to use them. But sometimes people who need a resource, such as energy, don't want an economical, readily available supply of that resource to be produced in their community. Landowners who favor the idea of wind-generated electricity will often fight against having a wind farm built near their properties because they believe wind mills are noisy or unsightly. So proper resource management can often be political too.

To manage resources that recover slowly, we need to know how long the resource takes to be replenished. Think about what you would need to know if you were in charge of operating a power dam on a lake, logging an area of national land, or managing the hunting of white-tailed deer or black bears in a national forest. What kind of information would you need? What decisions would you make? What are the possible consequences of those decisions? We need to use resources at a rate that allows them to replenish. This is called maintaining a *sustainable yield*. It takes hard work to obtain accurate information about a resource and to enforce the laws to use it properly.

Unwisely managing resources can create problems like endangered species, irrecoverable land erosion, pollution, and loss of a resource.

GEOLOGY—THE EARTH SPEAKS 95

Alternatives to Nonrenewable Resources

Nonrenewable resources can sometimes be replaced by an artificial substitute. For example, oil is expected to become scarce in the centuries to come. Gasoline, which comes from oil, could become very expensive and rationed. Scientists today are working to invent electric fuel cells that run on water. They hope that fuel cells will eventually completely replace gasoline engines and eliminate the need for gasoline.

But what about resources that can't be replenished or those that may not always be available in the quantities that are needed? Minerals, ores, and probably fossil fuels fall into this category. We need to conserve these materials by using them wisely and not wasting them. One method is *recycling*. Resources like scrap metals can be recycled. Even renewable resources like wastewater are reprocessed where their availability is limited. Resources that are directly consumed by humans, animals, and plants must also be protected from pollutants. You will learn more about how we control *pollution* of our most important resources later in the textbook.

Exercising biblical dominion of the earth's resources does *not* mean using them any way we like. Resource management is especially important because of the other part of God's command to Adam in Genesis 1:28—to populate the earth. Humans are the most important part of God's plan for the earth. Humans were created in the image of God, and they declare His glory by being like Him. Some people say that the earth's biggest problem is that there are too many people! If we highly value God, we cannot believe that. God wants our planet to be full of people to wisely enjoy what He has provided for us on this good earth.

4D REVIEW QUESTIONS

1. What is a natural resource?
2. Classify the following as a biological or nonbiological resource.
   - a. iron ore
   - b. fresh water
   - c. ocean tuna fish
   - d. pine trees for lumber
   - e. petroleum
   - f. soil
3. What makes a resource renewable?
4. What do you think makes a resource valuable? Why?
5. Why should a Christian care about the environment?
6. What are two ways to get the most out of nonrenewable resources?
7. (True or False) Wind power is always accepted by the community.
8. Imagine that you are a forest ranger for a piece of federal land in Colorado. Hunting, fishing, camping, and logging are allowed in this wilderness. Explain how you would manage the resources on this piece of land, especially focusing on how you would know there was a problem and what you would do to solve it.

4-19 Recovering and recycling nonrenewable metals shows good stewardship of our natural resources.

CHAPTER 4 REVIEW

CHAPTER SUMMARY

» Though humans are physically small, and our world and sun are seemingly insignificant, as God's image bearers, we are the main focus for the design of all things.

» The earth's mass, rotation rate, liquid water, atmosphere, magnetic field, location in the universe, and life itself are all designed to work harmoniously together.

» God created the earth as a good place for us to live, to wisely manage His creation, and to observe the works of His hands.

» Geology began within a biblical framework of Earth's history, but scientists gradually left this worldview to embrace the belief system called uniformitarianism.

» Geologists today can make observations that are useful in many practical ways.

» The principle of uniformity is supported by Scripture (Gen. 8:22), and it enables us to do observational science.

» Secular geologists assume that uniformity of processes, including the occasional global catastrophe in nature, extends into the unobservable past and applies even where it contradicts clear statements in the Bible.

» Geologists answer questions differently according to their worldviews.

» Earth's interior contains different layers of materials. The major layers are the crust, the mantle, and the core.

» The upper mantle and the crust form the lithosphere, which is made up of tectonic plates.

» Geologists believe that the earth's center is a liquid outer core and a dense, solid inner core.

» Natural resources are raw materials we use that come directly or indirectly from the earth.

» Resources are renewable if they can be replenished in some way. If their amount is fixed, then they are nonrenewable resources.

» Part of what makes a resource valuable is how available and accessible it is.

» Resource management is crucial to the wise use of both renewable and nonrenewable resources. Efforts to conserve and recycle are part of resource management, which is good and wise dominion.

Key Terms

geology	85
uniformitarianism	85
deep-time view	86
crust	90
Moho	91
mantle	91
lithosphere	91
core	92
natural resource	93
renewable resource	94
nonrenewable resource	94
resource management	95

REVIEW QUESTIONS

1. Even though God cursed the earth because of man's sin, and parts are unfriendly to life, how can we tell that God designed the earth to sustain life?
2. What four things do most kinds of life need?
3. How does the tilt of the earth's axis improve conditions for life on Earth?
4. How does water help distribute heat more evenly around the world?
5. Name three features of the atmosphere that support life on Earth.
6. What might happen to the earth if it did not have a strong magnetic field?
7. Explain what scientists mean by the habitability zone.
8. If the earth orbited a smaller sun at its current distance, what would our planet be like?
9. What could happen if the earth were in the center of the galaxy or in one of its spiral arms?
10. If God had created the earth with an opaque atmosphere, would Genesis 1:14 make any sense? Explain.
11. Why were James Hutton and Charles Lyell wrong when they applied the principle of uniformity in order to explain events in the distant past?
12. What term identifies a viewpoint that is based on the belief in a very old Earth?
13. Why is it possible to even *make* predictions in science?
14. Medical x-rays are to doctors as _____ are to geologists.
15. How was the Moho discontinuity discovered?
16. What special protective feature do geologists believe originates deep within the earth, most likely within the core?
17. Why are coal, petroleum, and natural gas considered nonrenewable resources?
18. What do we call all efforts to identify, obtain, and use natural resources wisely?
19. What can happen to natural resources needed by living things that makes them unfit for use?

True or False

20. The main point of the *Apollo 8* Christmas message was to celebrate how similar the earth and the moon were.
21. The features and properties of Earth are so suitable for life that there is an extremely low possibility that they came about coincidentally.
22. Of the eight planets, Earth has the largest moon in proportion to its size.
23. The earth appears to be designed not only for life, but for *intelligent* life.
24. Geologic evidence collected since Charles Lyell published his book in the 1800s has proved that the earth is very old.
25. The work of mineralogists to find and identify metal ores is observational science.

4-20

Crew of *Apollo 8* (left to right): Command Module Pilot James Lovell Jr., Lunar Module Pilot William Anders, and Commander Frank Borman II

LIFE CONNECTION: AN "OUT-OF-BALANCE" WORLD

Have you ever heard people talk about the "balance of nature"? They use this phrase to describe lots of different things in the world, like how chemical elements move through the environment and the balance of predator and prey populations. But our fallen world has lost the perfect balance that God intended. Without question, God's gracious care preserves conditions suitable for life. But there are no guarantees that a particular species of animal or plant will always be with us.

In the time it takes you to read a lesson, some secular biologists believe it is likely that at least one animal or plant species somewhere in the world will become extinct. This is nothing new. The fossil record contains clear evidence that many species have vanished forever from our planet. God's promise to Noah that he would never again destroy the earth with a flood contained no guarantees that every kind of living thing would be preserved forever. Extinctions continue to this day.

As stewards of this planet and its diverse life, should it be our goal to save every species at any cost? What if protecting a species prevents humans from using a necessary natural resource? Many environmentalists feel that all species of plants and animals have the same right as humans to survive. According to them, all living things evolved from the same beginnings and are equal in value. Because of their worldview, they miss the fact that most present-day species on Earth are simply variations of other biblical kinds. They are not the unique products of millions of years of evolution.

Even so, we are to practice wise dominion over God's creation. Though extinction is one result of the Fall, we should manage the world's resources for man's benefit while not causing unnecessary extinctions. In the end, we can look forward to a future day when God will restore perfect balance and order to His Earth, with no death and no extinctions (Isa. 11:6–9).

---

26. As you go deeper into the earth, seismic wave speeds gradually and smoothly increase.

27. Earth's crust is thickest under mountains and thinnest under the oceans.

28. Back in the 1960s, ocean scientists discovered huge fields of mineral-rich pebbles and stones littering the bottom of the ocean, but they were too deep to mine. These nodules would be considered an *actual* mineral resource.

29. Biblical dominion allows us to use natural resources as much and as quickly as we want to support population growth.

4-21
A sample of uranophane

GEOLOGY—THE EARTH SPEAKS

CHAPTER 5
THE CHANGING EARTH

- **5A** OBSERVING THE EVIDENCE — 101
- **5B** THE OLD-EARTH STORY — 108
- **5C** THE YOUNG-EARTH STORY — 115

MYSTERY IN THE ICE

A very old mystery lies locked in the frozen ground of Siberia. Scientists think that the bones of thousands or even millions of woolly mammoths are buried there, along with woolly rhinoceroses, tigers, horses, and antelopes. But how did they get there?

As scientists learn more, this mystery deepens. They find mammoth fossils buried in permanently frozen ground. Some are carcasses with skin, muscle, and internal organs. Many of these mammoth remains still have grass from fertile grassland in their stomachs and even buttercup flowers in their mouths! These facts mean that Siberia was a much different place long ago, and whatever happened to the mammoths was very sudden. A few of them were buried as if they were standing up. Several show signs of suffocation. They seem to have died at different times of the year, many showing signs of decay. And all of this is found on top of land that has not been scraped by glaciers. How can scientists piece together a story from this evidence?

5A

OBSERVING THE EVIDENCE

How does what we see tell us where Earth came from and how it has changed?

5A Objectives

After completing this section, you will be able to

» explain why creating a story of Earth's history depends on your worldview.

» analyze the evidence that the solar system gives us of its history of change.

» analyze the evidence that the earth gives us of its history of change.

5.1 UNRAVELING THE MYSTERY

Mysteries of science hide all around us. The story of the mammoths is one of those mysteries, and getting to the bottom of this story depends a lot on your worldview.

Geologists try to tell the story of earth structures that we see today on the basis of models that they create to support their story. When observations from the natural world don't fit their model, they modify it. However, they are reluctant to change their story because it relates to their worldview. Both creationist and secular scientists do this. The best story would account for all the evidence we observe.

However, there can be only one true story! How do we know what actually happened? No human was there to observe the earth's beginnings. In fact, our oldest writings go back only several thousand years. When we observe things like earthquakes, volcanoes, canyons, deep gouges on rocks, meteorite craters, and river valleys, we can't help but wonder how much the earth has changed over time. Let's look at the evidence that we observe in the natural world that hints at the earth's history of change.

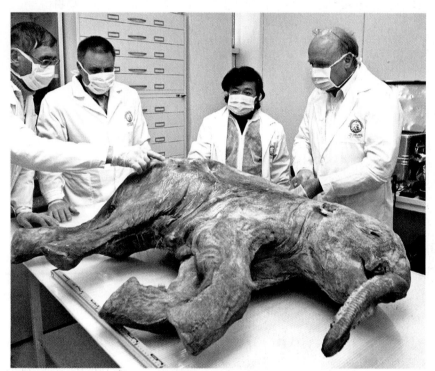

5-1
A baby woolly mammoth recovered from the frozen ground in northern Russia

5.2 OBSERVING THE SOLAR SYSTEM

When we look at our solar system, a few patterns emerge. The sun exerts a gravitational pull on its orbiting bodies, but it doesn't seem that gravity is enough to create what we currently observe. The sun itself is a much smaller star than the central stars in other solar systems.

There are patterns in the composition and movement of the planets. The planets closest to the sun have rocky bodies, and those farthest from the sun are made mostly of gases. Half of the rocky planets have magnetic fields, indicating that they might have a liquid core. The rocky planets have fewer moons than the gaseous planets farther out.

All the planets rotate in the same direction except two: Venus rotates in the direction opposite that of the other planets, and Uranus rotates at an almost 90° angle to the other planets.

5.3 OBSERVING THE EARTH

The earth itself gives us the biggest clues of beginning times as we piece together its history of change. Geologists believe that the rocks are a record of Earth's history. Since no one observed what happened in the beginning, geologists try to build a history from the rock record. They look for different kinds of rocks, changes to the rocks, fossils in the rocks, and the order of the rock layers that we see today.

THE GEOLOGIC COLUMN

A rock layer is called a **stratum**. Multiple layers of rocks are *strata*. By comparing rock strata from many locations all over the world, geologists have constructed a model of rock strata that documents the history of the earth. This model is called the **geologic column**.

Using the geologic column, geologists assign a *fixed*, or *absolute*, age of a rock by measuring the relative amounts of certain elements found in it. They do this by measuring the nuclear changes within its substances—a process called *radiometric dating*. Recall from Chapter 2 that when a nuclear change occurs in an atom, it can become an atom of a different element. These changes normally happen at predictable rates. Geologists can measure the amounts of the original elements and the new ones produced by nuclear changes that occur over long periods. Making certain assumptions, they believe that the relative amounts of each element indicate the age of the rock, as though these elements act like stopwatches.

Geologists also estimate the age of a rock layer by comparing its position to nearby layers and then assigning a *relative age*. Geologists assume that rock strata normally formed in a vertical sequence, like stacking books one at a time on a table. The oldest stratum would be at the bottom, and younger layers placed in order above it. This view is the **principle of superposition**. Geologists rely on this natural law when assigning relative ages to rock strata.

Fossils are the remains of living things preserved by natural processes. Geologists often assign an age to a rock layer on the basis of the absolute or assumed ages of nearby strata and the presence of certain key fossils of accepted ages. They believe that if they know the ages of strata above and below a given stratum, then they can estimate the age of the stratum between them. However, there are a few difficulties with the geologic column. Nowhere in the world do we find rocks from *all* the layers stacked on top of each other in one place. In some locations, the layers seem out of order. See Appendix E for a more detailed description of the geologic column.

1. recent erosion
2. sediment
3. lava or basalt sill
4. erosion surface
5. tilted strata
6. basement rock

5-2
Ammonites are extinct mollusks that we find as fossils. When geologists find them embedded in rock, they use them to date that rock. When fossils are used this way, we refer to them as *index fossils*.

TECTONICS

Geologists look at rock layers in a particular location to learn about its history. But what about the continents? As early as the 1600s, scholars noticed that the shape of the African, South American, and North American continents could fit together like puzzle pieces. They thought that maybe these continents used to be connected but somehow broke apart and moved. These ideas were the beginning of the **Continental Drift theory**, that is, that continents have slowly drifted to their current positions over long periods of time. German meteorologist and geophysicist Alfred Wegener is best known for first describing these ideas (see the box on page 107). Wegener's hypothesis was initially rejected because it was hard to imagine how such large pieces of the earth's crust could have moved such great distances.

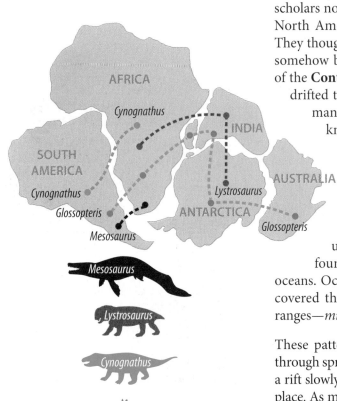

5-3 Scientists find similar fossils on opposite sides of oceans, suggesting that continents may have been joined at these places in the past.

But data supporting the idea of moving continents continued to pile up. During World War II, ships discovered underwater mountain ranges and deep-sea trenches. They found that continents extended underwater far out into the oceans. Ocean drilling uncovered another surprise. Geologists discovered that the rock on opposite sides of underwater mountain ranges—*mid-ocean ridges*—contained the same magnetic patterns.

These patterns suggest that hot molten rock—**magma**—oozes up through spreading cracks in the ocean basin, called *rifts*. As the sides of a rift slowly pull away, more magma pushes up from below to take its place. As molten rocks cool, crystals containing magnetic compounds line up with the earth's magnetic field like tiny magnetic compass needles. The cooling rock on both sides of the rift records the earth's magnetic field at the time it is formed and creates new ocean crust.

Geologists have also observed magnetic properties that suggest that the continents have moved. Geologists believe that a supercontinent

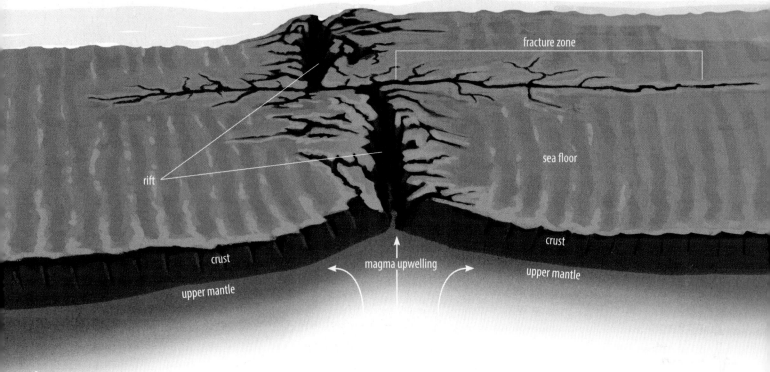

would have a consistent magnetic alignment. On the basis of this assumption, scientists believe that magnetic patterns show that continents may have rotated when the supercontinent broke apart.

The formation of new ocean crust at the mid-ocean ridge is called **sea-floor spreading**. Geologists believe that sea-floor spreading is driven from below Earth's surface. Blobs of hot, lower-density magma in the mantle rise like hot air balloons until they hit the bottom of the earth's lithosphere. The plumes spread out in opposite directions, dragging the lithosphere with them. This drag causes the rift to spread.

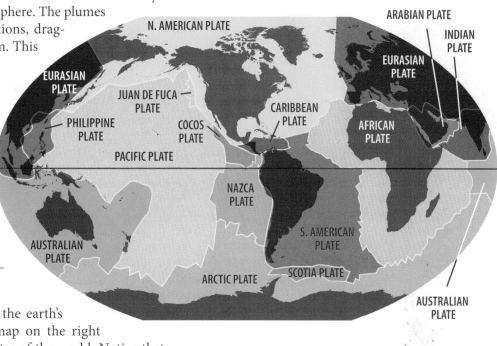

Consider all the matching patterns—fossils, minerals, and magnetic clues—preserved in rocks from different continents! The study of earth-shaping processes and the structures they form is called **tectonics** and involves learning about things like earthquakes, volcanoes, mountains, and bedrock.

Geologists call the pieces of the earth's crust **tectonic plates**. The map on the right shows the fifteen tectonic plates of the world. Notice that most of them include both continental and oceanic crust. The jagged boundaries between plates that lie near the middle of the ocean basins are the mid-ocean ridges.

So if the sea floor spreads, what happens to the extra ocean crust? The earth isn't getting any bigger, so it needs to go somewhere. Geologists have noticed many deep trenches along the edges of some ocean basins. They also exist along strings of islands called *island arcs*. It seems as though some oceanic plates are sliding under continental plates. Geologists call this process **subduction**. As the ocean crust bends and drags on the continental plate, it creates a trench.

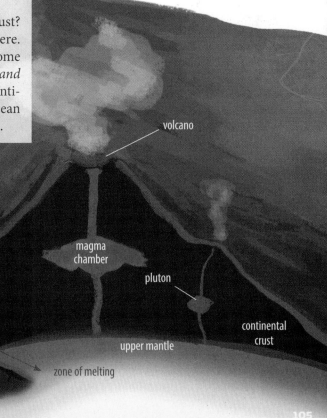

GLACIERS

Land around the world looks like it has changed in drastic ways also because of *glaciers*. A glacier is a mass of ice produced when snow builds up and compacts. Glaciers move downhill through gravity, like frozen, flowing rivers, changing Earth's geography in major ways. In some parts of the world we see evidence of glaciers that no longer exist—glacial ridges, U-shaped valleys, and other land features bulldozed by glaciers in the past.

The earth is telling us a story. The planets, rocks, and oceans contain data that hints at the story of their origin, but it takes something more to piece together this story. As you'll see in the next section, people's assumptions about the world affect how they interpret the evidence in the natural world.

CASE STUDY: THE ARCTIC QUEST TO UNDERSTAND CLIMATE AND GEOLOGY

Looking ahead, all Alfred Wegener and his two partners could see was snow and ice. No route markers directed them to their destination. It was late fall and they were one-third of the way across Greenland. Wegener, a meteorologist and climatologist, was fearful for his team's survival. He was trying to reach a camp to supply his team so that they could survive the winter on the ice.

Why was Wegener in Greenland? His love of meteorology and climatology had led him to this arctic world. He had gone on his first expedition in 1906 with the dual goal of mapping the Greenland coast and studying arctic weather patterns. He would return in 1913, 1929, and now this time, in 1930. Alfred Wegener saved his team but he himself didn't survive this expedition, dying before he could make it back to base.

However, despite dying tragically, Wegener would prove to be instrumental in our understanding of plate tectonics. He was the first to propose an in-depth theory of *continental drift*, the theory that stated that the continents are slowly moving relative to one another. He suggested that all the modern continents had started as a single large continent called *Pangaea*. He linked the movement of the continents with the formation of mountain ranges. He also noted similar fossil groups at the edges of the continents where they would fit together. This indicated to him that the continents were joined in the past. His interests in climatology and geology helped scientists piece together the story of Earth's history of change.

Alfred Wegener and Rasmus Villumsen in Greenland, 1930

Questions to Consider

1. Why is it important to study glaciers and meteorology?
2. What is Alfred Wegener primarily known for?
3. How does Wegener's life show the importance of scientific study?

5-4

These boulders were moved to this location by a glacier, even though there isn't a glacier in this place today. We assume that this is true because the boulder is a different type of rock than the rock in that area. This is called a *glacial erratic*.

5A REVIEW QUESTIONS

1. What scientific evidence shows that some of the woolly mammoths died suddenly?
2. How can two people look at the same evidence and present two very different stories of the earth's history of change?
3. How can we tell that the rocks on Earth's surface have changed during its history?
4. What do the rocks tell us about life on Earth in the past?
5. What evidence suggests that the continents weren't always in the same places as they are now?
6. What are the two ways that geologists can date a rock stratum?
7. What do geologists believe is the driving force for sea-floor spreading?
8. How do we know that there were glaciers in some parts of the earth where there aren't any now?
9. Examine the map on page 105. Name all the tectonic plates that contain portions of all fifty states of the United States. (If you live in another country, do the same for your country instead.)
10. (True or False) Only secular scientists change their models so that the evidence they observe fits their story.

5B

THE OLD-EARTH STORY

What story of Earth's history shows that it is millions of years old?

5B Objectives

After completing this section, you will be able to

» describe the processes and sequence of the origin of the earth from an old-earth viewpoint.

» summarize the old-earth story of how the earth has changed during its history.

» evaluate the effectiveness of the old-earth story in accounting for the evidence.

5.4 EARTH'S FORMATION: THE NEBULAR HYPOTHESIS

Most scientists construct a model of the earth's beginnings that relies completely on natural processes and evidence from the natural world. This story begins with the earth's formation.

In 1755, philosopher Immanuel Kant suggested that our solar system and its planets formed slowly from a cloud of dust and gas called a *nebula*. Naturalistic scientists believe that this process has been at work throughout the universe.

The process of solar system formation needed a powerful event to initiate the collapse of the nebula. Naturalistic scientists suggest that the force from an exploding star or high-intensity ultraviolet energy from a nearby hot star possibly triggered nebula

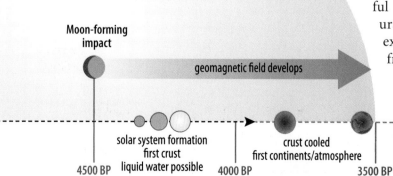

MILLIONS OF YEARS BEFORE PRESENT (BP)

collapse. Once this happened, the force of gravity continued pulling the matter together into a *stellar disk*.

At the center of the stellar disk, the dust and gas became very dense. This produced immense pressure that eventually formed a star. According to secular scientists, Earth began forming approximately 4.5 billion years ago. Clumps of dust formed *planetesimals*—small bodies of rock, dust, and ice—which gravity pulled together to form planets like Earth. Others of these small bodies were captured by planets as moons. Some planets like Earth collected gases from the nebula to form an atmosphere.

The above explanation for a solar system's origin is called the **nebular hypothesis**. This story predicts that everything formed from the nebula spins or orbits in the same direction as the original nebula. This includes the planets, their moons, and the star itself. Dense, rocky planets should form near the star, and less dense, gaseous planets form farther out. This model also predicts that rocky planets should have a dense core of heavy metals and have lighter materials near the surface.

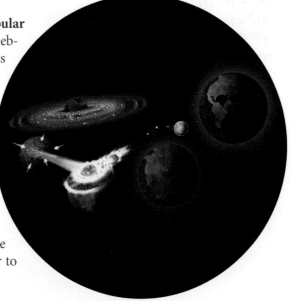

Scientists believe that within the first 30–50 million years after Earth's surface had cooled, a smaller rogue planet about the size of Mars crashed into then-smaller Earth. This catastrophic event remelted the young planet and added dense matter to the earth's core. Some of the debris that was flung from the crash went into orbit around the earth and later came together to form the moon.

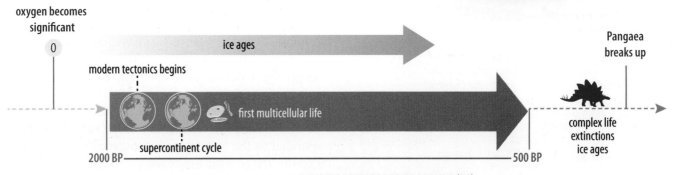

THE CHANGING EARTH 109

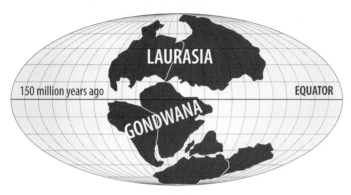
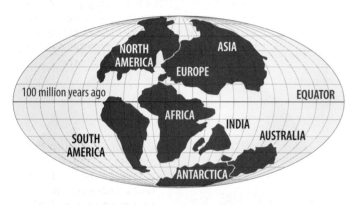
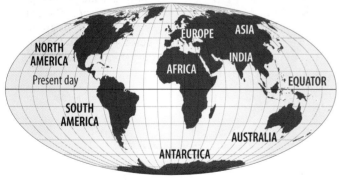

5-5 This series of maps shows the way that secular scientists believe the continents drifted from the supercontinent of Pangaea to our current continents. Notice the long periods of time assigned by geologists for the continents to have moved.

As the earth cooled from this collision, geologists believe that the oldest existing rocks formed, about 4 billion years ago. Gases collected to form an atmosphere. However, that atmosphere was poisonous and lacked oxygen. Water would not have existed on the newly formed planet. Some secular scientists believe that ice from comets that collided with the earth melted to form the oceans. Other scientists hypothesize that water may have been trapped with the magma and released. However the oceans formed, as they rose, the first continents formed.

5.5 THE CHANGING EARTH

Secular geologists have also formed models about how the earth has changed since its formation through the nebular hypothesis. Wegener's ideas about continental drift led scientists to believe that after Earth's formation, a global ocean surrounded the continents. Over the last 4 billion years, the continents grew and then broke apart. The separate parts drifted to different places, and later rejoined like the pieces of a huge jigsaw puzzle. We know that the continents drift very slowly today; geologists believe that this slow movement has occurred throughout Earth's history. It has only occasionally been interrupted by cataclysms.

Geologists believe that the continents have repeatedly crashed together to form **supercontinents** and then have broken up. The supercontinent called **Pangaea** was the last one to form, about 300 million years ago. A global sea—Panthalassa—surrounded this supercontinent. Geologists have constructed a complex history of how the tectonic plates moved from place to place.

At some point, Pangaea broke apart and eventually formed the continents that we see today. Using satellite technology, geologists today know that the continents currently move a few centimeters per year. Assuming this constant rate, scientists calculated how long it would have taken to form the Atlantic Ocean and concluded that Pangaea must have broken apart 180 million years ago.

This assumption of slow, naturalistic processes formed the foundation for science from Wegener's time forward. Around the turn of the eighteenth century, German geologist Abraham Werner studied Earth's strata using Nicholas Steno's principles of geology (see box on page 89). He developed a timescale covering the supposed 4.5-billion-year history of the earth that corresponded to the layers that he observed in rocks. He called this the *geologic time scale*.

Geologists know that earthquakes, glaciers, weathering, and tectonic forces all change rocks. Most of these changes are slow processes, and secular geologists assume that they have always been as slow. You can see how their worldview can lead someone to see great age in the geologic column.

Geologists who study the evolution of life, called *paleontologists*, assume that animals and plants were very simple at the beginning and became larger and more complex with time. Therefore, they expect that the fossil record would show an evolutionary progress from simple to more complex. They also expect to find the simplest fossils in the oldest strata at the bottom of the geologic column as well as fossils of extinct animals that transition to modern animals.

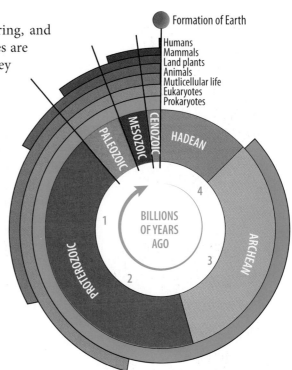

5-6

Old-earth geologists assign long periods of time to the different sections of the geologic column. Evolutionists expect to find specific kinds of fossils in layers of the geologic column that correspond to these periods of time.

5.6 ICE AGES

According to old-earth geologists, at least five major ice ages happened during the earth's history. What is an ice age? It is a period, also called a *glacial period*, when deep continental glaciers covered a large part of the earth's surface.

Geologists and climatologists believe that ice ages have occurred due to gradual cooling of the earth. They have hypothesized that there could have been many different things to cause this cooling, from cycles in the sun's heat output, to a reduction of gases that trap heat within the earth's atmosphere, to a repeating wobble in the earth's orbit.

Many geologists believe that the first major ice age started around 2.1 billion years ago. After that, a series of three ice ages started around 800 million years ago. These ice ages were so severe that many geologists believe that they produced ice sheets that reached from the poles all the way to the Equator. They

5-7

The graph below shows the way that many scientists believe global temperatures changed in the past, producing multiple glacial periods. The data comes from deep ice cores (left) extracted from glaciers and ice caps.

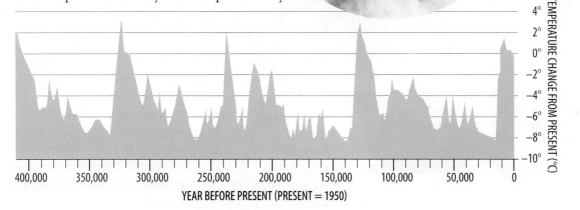

THE CHANGING EARTH 111

call these periods of time "Snowball Earths" because ice covered nearly all the earth's land. They also think that many living things became extinct during these periods.

Naturalistic geologists say that the latest ice age began about 2.8 million years ago and ended about 10,000 years ago. They believe that most of the woolly mammoths died out between 8000 and 12,000 years ago. Today, the only continental glaciers left are on Greenland, Antarctica, and a few smaller islands in the Arctic.

5.7 FITTING THE EVIDENCE INTO THE OLD-EARTH STORY

So how does the old-earth story fit the evidence in the natural world? When we look at our solar system, it seems that there isn't enough gravity to have formed the sun and planets from nebular dust. This means that we would have needed a supernova or some kind of high-energy-star emission to have kick-started the process. But if it's true that we needed a star to form a star, where did the first star come from?

We can also look at solar systems outside of our own. Other stars have planets that orbit them; we call them *exoplanets*. How do these fit with the nebular hypothesis? Many exoplanets have properties that are inconsistent with the nebular hypothesis.

The story of the shifting continents and the formation and breakup of Pangaea makes sense when we consider the evidence of magnetic patterns in rocks and fossils in the geologic column. However, the main issue with the old-earth view of the geologic column is that scientists assume that things have typically happened at the same rate as they do now.

There are two problems with this assumption. As layers were forming over millions of years, the rock would have become weathered and eroded. Geologic force would have tilted, cracked, or folded these rock layers. So then, apart from a catastrophe such as a local flood, we wouldn't expect to see thick, horizontal layers of rock. But that is *exactly* what we observe in many places on Earth.

The second problem is the way that naturalistic scientists assign dates to the layers in the geologic column. We don't know whether atoms are reliable stopwatches. Have radioactive atoms always decayed as they do now? Were the amounts of radioactive atoms in the atmosphere the same in the past as they are now? The assumptions of radioactive dating are not safe for dating rocks, and fossils aren't reliable either.

We find fossils buried in rock layers across the globe. However, fossilization requires rapid burial. This is a problem for secular scientists who believe that fossils formed over long periods of time. Recently, soft tissue has been discovered within fossils said to be 65 million years old. Secular paleontologists are having a difficult time explaining why the tissue remains soft after so many years.

5-8
Secular scientists once thought that this fossil—*Archaeopteryx*—was a transitional form between reptiles and birds. This creature was probably a bird. Paleontologists no longer believe that present-day birds descended from *Archaeopteryx*.

In general, we find simple fossils in lower strata and more complex fossils in higher strata. However, no fossils show one kind of organism evolving into another. Additionally, old-earth scientists look at the fossil record as a record of life on the earth, when in reality it is a record of death on the earth. Finally, why are there so many invertebrate fossils, very few fossils from mammals, and even fewer human fossils?

And how does the nebular hypothesis work with the view of multiple ice ages? Think about really cold winter mornings—they are usually cloud-free because the air tends to be very dry. Glaciers need lots of snow to form, and snow can fall only if there is enough moisture in the air. If the earth cools slowly, the oceans will cool also. This means that there wouldn't be enough water vapor in the air to produce the huge snowfall needed to form glaciers and continent-sized ice sheets several miles thick.

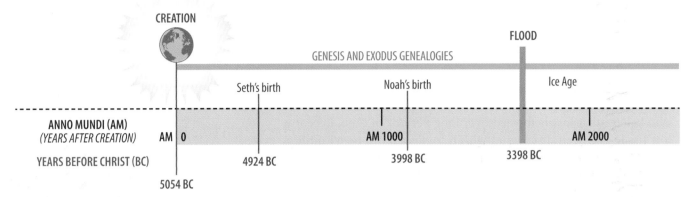

We've seen some evidence that is problematic to the old-earth story. What should old-earth scientists do? Well, maybe they should change their story, or at least their supporting models. However, though scientists often change their models, they are often reluctant to change their story. This means that they must live with evidence that contradicts their story. But is there another story that fits our observations of the universe better? There is indeed.

5B REVIEW QUESTIONS

1. Where do naturalistic scientists believe that rocky planets would form in a developing solar system?
2. How do secular scientists think the earth's oceans formed?
3. Why is Venus a problem for the nebular hypothesis?
4. How do you think a secular scientist would explain the reason that Uranus rotates on its side using his model?
5. According to secular geologists, what kind of landmass formed repeatedly as continental plates came together and separated throughout Earth's history?
6. How many major ice ages have occurred in the earth's past according to secular geologists? How severe do they believe they were?
7. Why is the presence of fossils in rocks all over the world a problem for the old-earth story?
8. Suggest why people cling to their beliefs about the story of Earth's history of change in spite of evidence that contradicts this story.
9. (True or False) Evolutionists find proof for the natural origin of life from simple to complex organisms with transitional forms as observed in the geologic column.

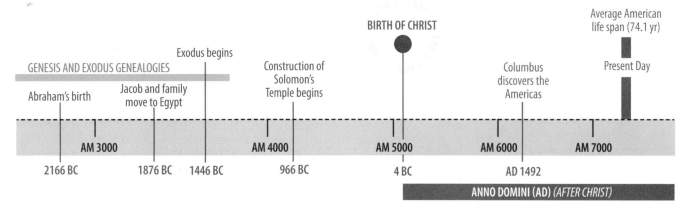

5C

THE YOUNG-EARTH STORY

What story of Earth's history shows that it is thousands of years old?

5.8 EARTH'S CREATION

There is another story that we can tell to understand the evidence that we observe. However, this story is informed not just by God's revelation in the natural world but also by His revelation in His Word. Unlike the old-earth model, the Bible's account of creation is simple. It tells us in Genesis 1 and 2 that God created all things by the power of His Word, including the natural processes that we see today. A straightforward reading of Scripture leads us to believe that it took six normal days to create everything—He spoke and things appeared.

God created the world in a specific order. On Day 1, He created light and separated it from darkness. On Day 2, He separated the waters under the expanse from the waters above. God created dry land and filled it with plants on Day 3. God filled the sky with the sun, moon, and stars on Day 4. He created all the flying and swimming creatures on Day 5. On Day 6, He created all the land creatures and crowned His work with the special creation of mankind.

This order of creation is not how life would have evolved, even if it were possible. No evolution was necessary. Everything God created was fully functional and just right for its purpose. *And God saw every thing that he had made, and, behold, it was very good. And the evening and the morning were the sixth day* (Gen. 1:31).

5C Objectives

After completing this section, you will be able to

» describe the processes and sequence of the origin of the earth from a young-earth viewpoint.

» summarize the young-earth story of how the earth has changed during its history.

» evaluate the effectiveness of the Bible's story in accounting for the evidence.

THE CHANGING EARTH

5.9 CHANGING EARTH: THE FLOOD

Sadly, God's good Earth didn't stay good. In Genesis 3 humans sinned, corrupting God's "very good" creation. Adam's sin brought upon us the judgment of death. God's judgment affected not only people and animals, but also the physical earth (Gen. 3:17–18; Rom. 8:22). What had been a pleasant life with good and meaningful work in the Garden of Eden became a struggle to survive in a world of weeds, thorns, and increasingly dangerous animals. The effects of sin began to accumulate until the time when, nine generations after Adam, God acted. When Noah was 480 years old, God announced that judgment was coming (Gen. 6:13). After 120 years, during which Noah built the vessel, God commanded Noah and his family to load the animals onto the vessel and move aboard. Let's see how the Flood story unfolds.

Day 0
Noah boards the Ark with animals, provisions, and his closest family. (Gen. 7:1–9)

Day 7
God closes the door, the fountains of the deep burst forth, and torrential rain falls. (Gen. 7:10–12, 16)

Day 47
The rain stops, and the mountains are covered by at least 15 cubits (22–27 ft of water). All air-breathing animals outside the Ark are now dead. (Gen. 7:12–22)

Day 157
God makes a wind to pass over the earth. Waters begin to recede from the continents. The Ark is grounded on the mountains of Ararat. (Gen. 8:1–4)

Day 227
The tops of mountains near the Ark emerge from the floodwaters. (Gen. 8:5)

Imagine what Noah and his family saw when they left the Ark! A flood of the magnitude just described would have completely altered the earth's surface. Their world was a very different environment from the one they had known before. Innumerable volcanoes filled the sky with clouds and dusty haze. They probably recoiled from the sulfurous taint. Plants may have just begun to make the earth green again, but there were no mature trees. In the distant valley far below, they could see a river full, fast-flowing, and brown with eroded sediments. It is likely that there were frequent small earthquakes as the earth's crust continued to readjust after the Flood. Noah and his family began life anew, after God reminded them of the Creation Mandate (Gen. 9:1–7), refreshed them with a new covenant (Gen. 9:8–15), and even provided a reminder of that covenant, the rainbow (Gen. 9:16–17).

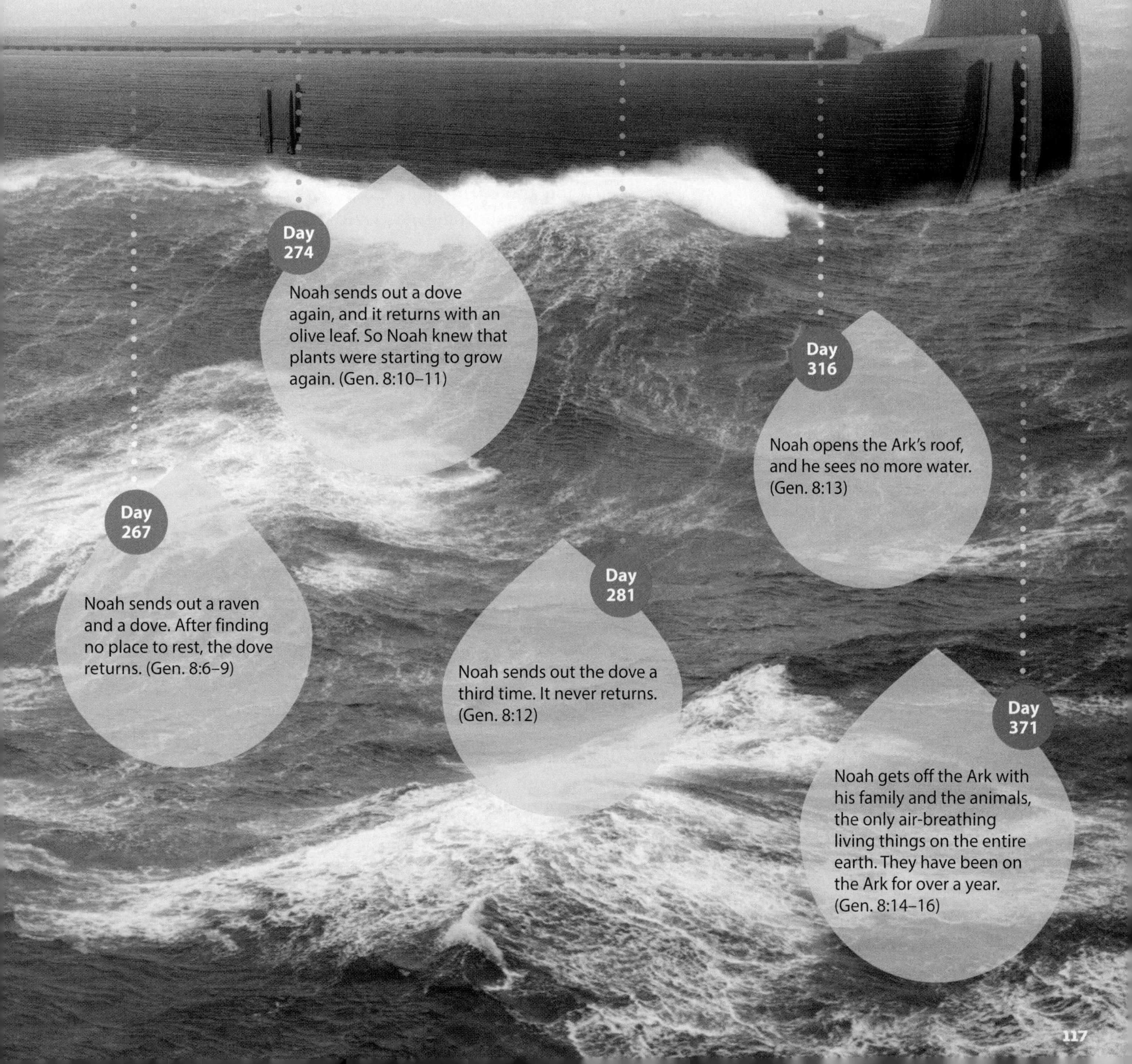

Day 267
Noah sends out a raven and a dove. After finding no place to rest, the dove returns. (Gen. 8:6–9)

Day 274
Noah sends out a dove again, and it returns with an olive leaf. So Noah knew that plants were starting to grow again. (Gen. 8:10–11)

Day 281
Noah sends out the dove a third time. It never returns. (Gen. 8:12)

Day 316
Noah opens the Ark's roof, and he sees no more water. (Gen. 8:13)

Day 371
Noah gets off the Ark with his family and the animals, the only air-breathing living things on the entire earth. They have been on the Ark for over a year. (Gen. 8:14–16)

THE FLOOD AND THE GEOLOGIC COLUMN

The Genesis Flood is the most significant physical event in the earth's history after Creation. As it is, the Flood marks the dividing line between the originally created world and the world we live in today. It provides us a lens to look clearly at the geologic column and the fossils that it contains.

Creationists have developed a model of the earth's history to interpret these same rock layers. We know that when God created the earth, there was at least one continent (Gen. 1:9) where everything lived. Creation geologists think that the continent foundation was probably the rock we call *granite*, which makes up the deepest rocks today. We also know from the Bible that plants grew before the Flood. This means that soil covered the basement rocks.

Now let's consider the fossil record, which provides evidence of when and how things died. The order of fossils in the geologic column relates to an organism's ability to flee or survive the rising floodwaters. The sediments moved by floodwaters would have trapped small, simple organisms quickly, so we are not surprised to find these fossils in the lowest layers of rock laid down by the Flood. Upper layers of rock contain organisms that are more complex; these organisms, being larger and stronger, survived longer during the Flood.

Creation geologist Tasman Walker has rethought the geologic time scale from, well, the ground up! At the lowest levels of the geologic column are strata that date back to Creation. These rocks have no fossils since there was no death at that time. Above this are strata from the time between Creation and the Flood. These layers have few to no fossils and only minor erosion and account for about 1700

5-9 If the Flood had simply buried organisms as the waters rose, this is the burial sequence we would expect to see.

years of geologic history. Most of the layers of the geologic column correspond to those formed during the Flood. They are rich in the fossils of animals killed during that event. The topmost and most recent levels are recent sediments formed by erosional processes at work today. You'll learn more about these processes in Chapter 12. Walker called this the **diluvial** (Flood) geologic time scale.

THE FLOOD AND TECTONICS

When reading the old-earth story of plate tectonics, you may have thought that it seemed logical. And it is. Young-earth geologists agree that there is much evidence for tectonic plates, plate subduction, and the features they create.

But *when* these events occurred and *how long* they took are key differences between the two stories. Young-earth geologists believe that a global catastrophe, the Genesis Flood, created the dramatic features that we see today just thousands of years ago in a short amount of time.

The young-earth, Bible-believing geophysicist John Baumgardner has proposed a version of plate tectonics that fits well with the Bible's Flood account. He calls it **catastrophic plate tectonics**. He modeled this process using powerful supercomputers, even verifying some of the model's assumptions with laboratory experiments on rocks similar to those in the lithosphere.

In his model, Baumgardner proposed that God created the continents in the beginning as a single large landmass, similar to Wegener's Pangaea. At the start of the Flood, this supercontinent shattered into large plates with the breaking up of the "fountains of the great deep" described in Genesis 7:11. Many of the continental plates were carried along as the oceanic plates rapidly slid under adjacent continental plates at speeds of many miles per hour, not just a few centimeters per year. Baumgardner called this process *runaway subduction*.

In the newly formed Atlantic Ocean and in other places, rapid sea-floor spreading between continental plates built new ocean crust. The plate edges at mid-ocean ridges were the sources of new crust needed to fill the gaps that had been created by continental plates moving apart.

5-10 This computer simulation of the catastrophic plate tectonics model shows that folds in the mantle could be remnants of subducted oceanic crust. A model like this doesn't prove anything, but it does provide an explanation of a part of the story of Earth's history of change.

But how could this have happened so fast? Where oceanic plates began to slide under continental plates, the rocks in the oceanic plates heated up. This greatly weakened the rocks, making the subduction zone even more slippery. With faster motion, the rocks became even hotter, and the plates slipped faster. These conditions provide a simple yet effective explanation for the evidence of plate tectonics within a young-earth time frame.

5.10 THE ICE AGE

The effects of the Flood weren't over after the waters dried up. All the changes brought on by the Flood greatly affected Earth's weather. Though the Bible doesn't explicitly mention an ice age, the book of Job contains more references to ice, snow, and freezing conditions than any other book of the Bible. Most Creation geologists see clear evidence for a glacial period after the Flood, and it's possible that the events recorded in Job took place at that time.

Creation scientists suggest that as the crust broke apart, continents moved and millions of underwater volcanoes formed around the world. The molten rock, or *lava*, heated the ocean water. Even after the continents had emerged from the world ocean, volcanoes probably continued sending lots of ash high into the atmosphere.

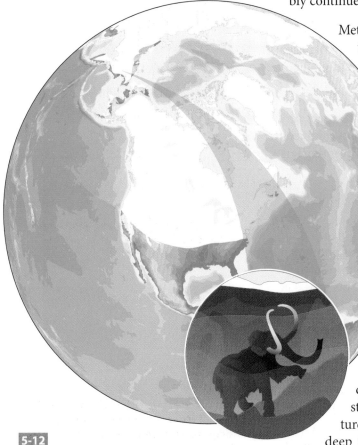

5-11
This is how the earth may have appeared without cloud cover at the height of the Ice Age.

5-12
Huge dust storms quickly buried a few woolly mammoths alive and suffocated them near the end of the Ice Age.

Meteorologist and creationist Michael Oard believes that all this activity set the stage for the Ice Age. He suggests that water evaporating from the warm oceans and dust from the volcanoes blocked sunlight. This allowed the continents to cool quickly. Depending on the season, heavy rains or snows fell from the water-filled clouds. Eventually, huge continental glaciers formed, 3–4 km (2–2.5 mi) thick at the peak of the Ice Age, 500–800 years after the Flood (see Figure 5-11).

Toward the end of the Ice Age, the earth began to warm up overall. The question then arises, "Why do we find Siberian mammoths in frozen ground?" Most of the mammoths likely died from exposure to the Arctic cold and lack of food. As the glaciers melted, rising sea levels trapped many mammoths on Siberian islands. These died and decomposed. Wind-blown sediments covered their bones. In the few cases where carcasses were quickly preserved, it seems that sudden, heavy dust storms buried them alive. Shortly after, frigid temperatures would have quickly preserved them in a permanent deep freeze before they could decay. The mystery of the mammoths comes to light with the Bible's story.

LIFE CONNECTION: THE FLOOD, THE ARK, AND SPECIES TODAY

What was the world like before the Flood? No one knows for sure, but we see fossilized tropical and temperate plants all over the world, including in the rocks of Antarctica. So we can imagine that pre-Flood weather involved seasons that were mild to tropical. This makes sense in a "very good" world.

Conditions have changed quite a bit since then. The earth now has frozen ice caps and broiling deserts, with surface temperatures ranging from nearly −100 °F to 130 °F! Such different climates would have threatened the survival of some animal kinds that came off Noah's Ark. In His wisdom, God created mechanisms in the original kinds of animals and plants so that they or their offspring could change in order to live in different conditions. These changes produced the species we see today within the same kinds of living things.

So, with all these changes, how did we end up with the huge variety of living things in the present world? Some creation scientists believe that there could have been more than 10,000 animals on the Ark. (Of course, no one knows how many there actually were because the Bible doesn't tell us.) If that number is correct, there could have been nearly 5000 or more "kinds" of animals. But there are lots more animal species in the world today than this. How do we explain the great number of species?

How did the Ark's few members of the "frog kind" lead to the more than 6000 species of toads and frogs seen today? Think about it—there is a big difference between a bright golden spray toad (¼ oz and ¾ in. long) and a cane toad (2.9 lb and 9.4 in. long)! Evolutionists claim that the variety of animals alive today came about through many hundreds of thousands of years of slow genetic changes. Young-earth creationists, on the other hand, believe that this variety developed since the Flood as the animals reproduced and repopulated the earth.

From studying animals and plants, we know that God designed great variability into the genetic materials of their cells. We can see these variations in humans, plants, and animals. No two individuals are exactly alike. Breeders and plant scientists have known for centuries that animals and plants can be artificially bred to produce certain desirable characteristics. After the Flood, animals, plants, and humans spread out from the early centers of civilization. Variations best suited to survive in the new harsh conditions produced new species. This process, called *speciation*, happened many times in different locations in the world. It allowed the relatively few kinds of animals that left the Ark to become the wide variety of species that we see today—not more kinds, just more species. The ability of organisms to thrive after the Flood shows the amazing wisdom of God in His care for His creation.

Both the tiny spray toad (top) and the large cane toad (bottom) may have come from a few members of the frog kind aboard the Ark.

5.11 FITTING THE EVIDENCE INTO THE YOUNG-EARTH MODEL

Those who believe the Creator's Word are in the best position to understand His world. The evidence in the solar system that seems so problematic for the nebular hypothesis isn't a problem for the Bible's story. It isn't necessary to try to give a natural explanation for these things if you believe that everything ultimately came from God.

Rock Formation

Recent geological events, such as the 1980 eruption of Mount St. Helens in the state of Washington, have demonstrated that rock can form from sediments and ash in just a few days.

Looking at the rock strata today from a biblical viewpoint, we can see clear evidence for the key geological phases of the earth. The Flood transported immense quantities of soil, sediment, and rock and redeposited them quickly in different locations all over the world. These layers hardened into new rock strata. In some places, water contained dissolved chemicals that settled out to make rock strata. These rock strata contain fossils in patterns that the Bible's story can explain. And a single Ice Age offers a simple explanation for the evidence of glaciation in the world today.

The rock strata that biblical and naturalistic geologists observe are the same, but their perspectives are completely different. The Bible provides a framework through which we can logically understand the evidence on our changing Earth. And it does more than just that. The Bible assures us that the same God who created the earth has a purpose for its future—to redeem His creation from sin's dominion.

BIBLICAL ORIGINS: THE DAY-AGE THEORY

Some Christians assume that the language of Genesis 1 is symbolic. They suppose that the word "day" used in Genesis 1 means a period and not a literal, sunset-to-sunset day. After all, they claim, 2 Peter 3:8 states *"one day is with the Lord as a thousand years, and a thousand years as one day."* Therefore, a day could be of any length—one second, or a billion years.

Those who hold to the Day-Age theory believe that the days of Creation were long periods, or ages, during which God created the universe and all its contents. They try to fit the standard geologic sequence of events to the sequence of events of the Creation week.

For example, secular scientists claim that there was the big bang, then star formation. Day-Age theorists would say that the creation of light (Day 1) corresponds to the time of the big bang, and that Days 2–4 roughly correspond to the eons that old-earth scientists require for star and planet formation. Days 5–6 correspond to the billions of years for birds, fish, and animals to evolve into the species we see today.

The Day-Age theory has serious problems. Can you explain what they are?

Questions to Consider

1. What evidence given in Genesis 1 indicates that the days of creation were literal, sunset-to-sunset days?
2. Read Exodus 20:10–11. What does this passage suggest about the days of Genesis 1?
3. What event on Day 3 would have presented a problem if the sun didn't appear until a long time into the future?
4. Does the Day-Age theory help make the secular history of the earth and the Bible agree, as intended?

5C REVIEW QUESTIONS

1. Suggest one way that the order of Creation events is different than how the solar system and life are believed to have evolved.
2. State three ways in which the biblical Flood affected the world and people.
3. How does the presence of fossils in rocks all over the earth fit with the Bible's story?
4. Identify the four main periods of Earth's history as represented by layers in the geologic column according to the Bible's story.
5. According to the catastrophic plate tectonics model, how long ago did the tectonic plates spread apart? How long did this process take?
6. What two factors associated with the Flood may have contributed to a single ice age following the Flood?
7. How could a young-earth creationist explain the evidence in the solar system, such as tilted planets and weird magnetic fields? Comment on the validity of such an argument.
8. (True or False) The Genesis Flood lasted forty days and forty nights.

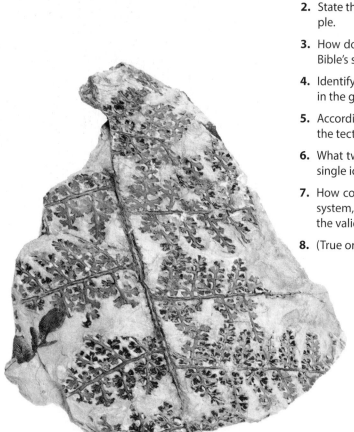

CHAPTER 5 REVIEW

CHAPTER SUMMARY

» A history of the earth must include how it began and how it changed over time.

» The solar system shows patterns that scientists study as they try to understand its formation. Rocky planets are near the sun and gas giants farther away. Gas giants tend to have more moons than rocky planets. Almost all the bodies in the solar system revolve and rotate in the same direction.

» A geologic column is a record of the history of changes in the earth's surface.

» Visible patterns on the earth demonstrate that portions of the earth's crust can move.

» Geologists use the plate tectonics theory to explain geologic structures and processes in the past and present.

» There is evidence on the earth of extensive glacial activity in the past.

» The nebular hypothesis assumes that the solar system formed out of a huge cloud of dust and gas.

» According to the nebular hypothesis, the earth has a complicated history that began more than 4.5 billion years ago.

» Old-earth geologists believe that the earth's surface changed through billions of years as continental plates broke apart and collided. The last major reshaping began about 200 million years ago and is still in progress.

» The old-earth geologic column classifies strata of rock according to specific periods over its existence.

» Old-earth scientists believe that at least five major ice ages have occurred during the earth's history. The most recent ended around 10,000 years ago.

» The nebular hypothesis doesn't accurately predict all that we observe about the solar system.

» Biblical creationists believe that God created the universe, the earth, and everything in it in six days.

» Biblical creationists believe that God judged the fallen world with a global flood that completely rearranged Earth's surface after Creation.

» Though it is not specifically mentioned in the Bible, many creation scientists believe that conditions after the Flood triggered a single ice age that ended around 800 years after the Flood.

» Creation geologists believe that most of the strata in the geologic column were laid down during the Flood.

» Many creation geologists believe that the Flood events explain the evidence for sea-floor spreading and continental drift.

» Significant atmospheric changes at the end of the ice age trapped, killed, and preserved woolly mammoths and other Arctic animals.

Key Terms

stratum	102
geologic column	102
principle of superposition	103
Continental Drift theory	104
magma	104
sea-floor spreading	105
tectonics	105
tectonic plate	105
subduction	105
nebular hypothesis	109
supercontinent	110
Pangaea	110
diluvial	119
catastrophic plate tectonics	119

THE CHANGING EARTH 125

REVIEW QUESTIONS

1. Name one of the patterns that we notice in our solar system.
2. Explain the principle of superposition.
3. Explain absolute age and relative age in reference to rock layers.
4. Name two pieces of evidence showing that the continents have moved over time.
5. What is one piece of evidence that glaciers were more widespread in the past?
6. What are two of the hypothetical triggers for the collapse of a nebula into our solar system, according to the nebular hypothesis?
7. On what did secular scientists base their calculations to determine that Pangaea broke apart 180 million years ago?
8. According to secular scientists, how do ice ages start?
9. What is the primary problem with the secular view of ice ages?
10. Outline the biblical account of Creation.
11. How do creation scientists explain the fossil record in the geologic column?
12. How do biblical scientists view the geologic column?
13. What conditions do young-earth scientists believe led to the Ice Age? What caused these conditions?
14. How can a Bible believer explain the wide variety of dogs that we have today?

In another example of catastrophic change, a strike-slip fault is shown below. The two sides of the fault slide past each other. Notice how the layers of rock no longer line up.

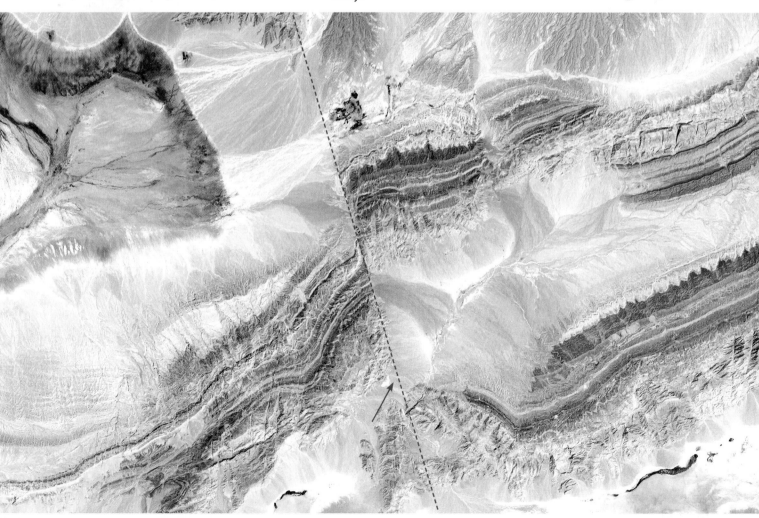

True or False

15. Scientists are more likely to change their story than they are to change their models.
16. Earth deviates from the pattern in our solar system in that it is a rocky planet that is close to the sun.
17. Scientists determine the absolute age of rocks by comparing the relative amounts of elements contained in the rocks.
18. There is no credible evidence for the movement of continents.
19. Glaciers form by snowfall that accumulates and compacts over many years.
20. According to the nebular hypothesis, a star is needed to form a new star.
21. Supercontinents are part of only the secular view of the earth.
22. Secular scientists all agree that changes in the sun's energy output caused ice ages.
23. Slow burial rates are a problem for the secular model when dealing with fossils.
24. According to the Creation account, God created the sun on Day 1 when He created the light.
25. Young-earth and old-earth scientists agree on the evidence for the processes of tectonics.
26. The geologic column used by creation scientists is completely different from the one used by secular scientists.
27. The Bible discusses the Ice Age in the book of Genesis.

CHAPTER 6
EARTHQUAKES

6A TECTONIC FORCES `129`
6B FAULTS AND JOINTS `133`
6C EARTH WAVES AND SEISMOLOGY `136`
6D EFFECTS OF EARTHQUAKES `141`

A FEW SECONDS

What would you do if you were on an elevator when an earthquake struck? What if you were on a train? Or what if you were undergoing surgery? What if you were driving a car? A couple of seconds of warning could make all the difference. Seismologists can't predict earthquakes, but perhaps technology could buy a few precious seconds.

6A

TECTONIC FORCES

What causes earthquakes?

6.1 EARLY EARTHQUAKE WARNINGS

Countries like Japan, Turkey, and Mexico already have earthquake early warning systems in place. Here's how they work. A system of **seismometers** detects earthquake waves. The seismometer sends a message to a central government office. In Japan, the Japan Meteorological Service coordinates these messages. Then this agency sends out a warning message on radios, TVs, cell phones, and public intercoms. The farther someone is from the center of the earthquake, the more time he has to respond. At most, someone may have a full minute after the warning before large tremors reach him. But an early warning system isn't the same thing as predicting when an earthquake will happen. An earthquake has to occur to activate the system.

In earthquake-prone California, researchers are urgently working on developing a similar earthquake early warning system. Scientists from leading universities are pooling their knowledge to come up with a solution. One of the most important things to avoid is false alarms. The key to doing that is to have lots of sources of earthquake data.

But having large numbers of seismometers is expensive. To solve this problem, researchers are using regular people—nonscientists—to collect data. University of California, Riverside, and Stanford University are working to create the Quake Catcher Network (QCN). They are creating a network of sensors using personal computers (which now have sensors that can detect movement) and apps on smartphones and tablets. People use these hand-held devices every day, all around the world, so why not use them to monitor earthquakes? By connecting to a central computer, we can have a worldwide earthquake detection network.

For this network, scientists have developed a computer program that filters out all but the heaviest shakes sensed by the accelerometer. (Users can purchase inexpensive plug-in accelerometers for computers that don't have one.) The program minimizes false alarms from users bumping the computer or from the vibrations of heavy road traffic. It accepts data only when it detects similar shocks from other computers or smart devices in the same area.

This relatively low-tech network has several thousand participants and is growing. Their computers are linked to a database by the Internet. Anyone with properly equipped devices can join the network voluntarily and free of charge. The system could eventually help warn people of earthquakes over a large area. Just think—average people like you and your family could have a part in saving lives through real-time earthquake alerts!

6A Objectives

After completing this section, you will be able to

» summarize how tectonic forces trigger earthquakes.

» show how certain kinds of tectonic processes are most likely the cause of earthquakes.

» identify the material properties of rocks that help cause earthquakes.

seismometer (size MAHM ih ter); an instrument that measures earthquake waves

6-1

What difference could a few seconds have made?

6.2 WHAT IS AN EARTHQUAKE?

People like to think that the ground they stand on is secure and solid. But earthquakes demonstrate that the earth's surface can be quite unstable.

An **earthquake** is any kind of shaking of the ground measurable by seismic instruments. Most earthquakes result from faults slipping, but measurable ground shaking can also come from volcanoes and the movement of molten rock deep underground, from landslides and avalanches, and even from meteorite impacts. Man-made quakes come from such things as nuclear and other kinds of explosions, mine cave-ins, and aircraft crashes. Earthquakes produce a series of low-frequency waves, somewhat like sound waves, that travel through the earth. Some waves from earthquakes travel along the earth's surface, but most move through its interior. Most earthquakes are too slight for a person to feel. Still, a few earthquakes each year shake the ground enough to frighten people nearby. Occasionally a strong earthquake near a city causes great destruction.

6.3 FORCES IN THE EARTH

Do you remember what a force is? It is a push or a pull on something. Immense forces underground cause earthquakes. But where do these *tectonic forces* come from? Tectonic plate motion builds up stresses at plate boundaries and within the plates themselves.

A **stress** is a force exerted inside a material. There are three main kinds of stress. If you carefully step on an empty soft drink can placed upright on the floor, your foot exerts a **compression** stress on the can. Place more weight on the can and you can crush it. When you pull on a rope in a tug-of-war, you exert a **tension** stress on the rope. Compression and tension forces are quite common in earth materials. But the most important stress in producing earthquakes is **shear** stress. Shear exists when two forces acting in opposite directions attempt to slide parts of the object past each other. If you press sideways on a block of gelatin (see Figure 6-2), you produce a shear stress in the middle of the gelatin.

> A **stress** is a measure of force exerted within an object.

6-2
This block of gelatin is experiencing a shear stress.

You learned in Chapters 4 and 5 that the mantle rock flows very slowly. It acts like a piece of stiff plastic that slowly bends or stretches when you pull on it. Geologists believe that the mantle's temperature rises as you go deeper into the earth. Hotter rock is less dense than cooler rock, so it floats in denser rock. **Buoyant force** is an upward force on a less dense object in a denser material. Gravity pulls the cooler, denser rock downward while the buoyant force lifts hotter, less dense rock. The less dense rock moves upward in slow-moving currents or plumes. Gravity and the buoyant force drive this process.

In Chapter 5 you also learned that geologists believe that sea-floor spreading happens today mainly because rising mantle plumes push against the rocky lithosphere and then spread out, dragging the lithosphere with them (see Figure 6-4). These spreading zones along some plate margins are called **divergent boundaries**. The plate sections are moving apart, or *diverging*.

In other locations, tectonic plates move toward each other. They may collide and crumple up, forming mountains. Or a thinner oceanic plate may be sliding under a much thicker continental plate. These zones of collision or subduction are called **convergent boundaries**. The plates are moving toward each other, or *converging*. Great compression and shear forces build up along a convergent boundary.

Places where plates slide past each other in opposite directions along long cracks in the crust are **transform boundaries**. Transform boundaries usually appear in ocean basins and are connected with mid-ocean ridges. Shear forces are the most important forces in these zones.

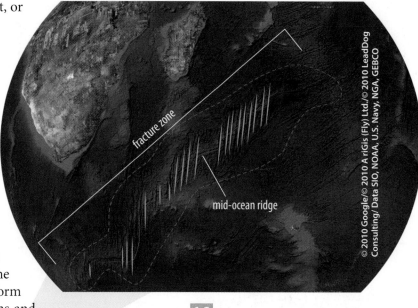

6-3
Transform tectonic boundaries (blue lines) are common where sea-floor spreading occurs.

6-4
Geologists believe that hot mantle rock slowly pushes up against sections of the oceanic plates, spreading them apart.

6.4 STRAIN AND FRACTURE

How do the rocks in the earth's crust react to stresses at these boundaries? Materials respond to shear stress in different ways. Gases and liquids simply flow in different directions. But solids are rigid. Soft or plastic-like solids can absorb shear stresses up to a point. They slowly deform or change shape but don't break until greatly stressed.

Any change in shape of a solid under stress is called **strain**. The amount of strain that a material can endure without breaking is related to its property of *ductility*. Soft metals like copper, gold, and silver are very ductile. Metalworkers use the ductility of these metals to draw them into thin wires without breaking.

Other solids are very rigid. With just a little shear stress, they shatter or break along stress fracture planes, called *joints*. If the two parts of such materials move along the fracture, a *fault* forms. Hard, easily fractured materials are *brittle*. Glass, ice, diamonds, and quartz crystals are examples of brittle materials.

You might not think of a rock being stretchy. But under the immense forces involved in earth movements, even rocks can bend, stretch, or compress a little. A rock's *elasticity* or stretchiness most significantly influences earthquakes. Just like a rubber band or a spring, many rocky materials show strain by stretching, compressing, or bending very slightly. If they don't break, they store up potential energy under stress. Beyond a certain amount of stress, the materials break. All that stored up energy is suddenly released, as when a broken rubber band snaps back.

These properties of solids often determine whether an earthquake occurs and how strong it is. Tectonic forces build up stresses deep in the earth. If the rocks are weak and brittle, they fracture easily and do not store up energy. Soft, ductile rocks simply change shape under stress. Elastic rocks, or strong but brittle rocks, can fail catastrophically, releasing lots of energy suddenly. In the rest of the chapter, we'll see what it looks like when this happens.

Strain is the distortion, bending, stretching, or compression of a material under stress.

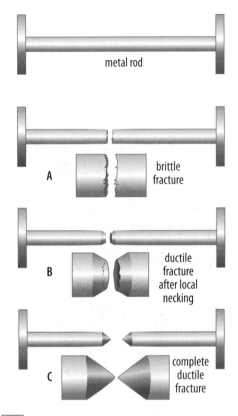

6-5
Ductile materials, such as metals and plastics, stretch somewhat before breaking. Stretching is an example of strain. A material under tension may fracture (break) without any strain (A); it might narrow (strain) a little prior to fracturing (B); or it might deform significantly (C) before breaking.

6A REVIEW QUESTIONS

1. How might students like you help provide earthquake warnings?
2. What is an earthquake?
3. What kind of physical quantity is a stress? Which kind of stress is most likely to cause an earthquake?
4. What is happening at convergent tectonic boundaries?
5. What are two ways that stress in a rock can be relieved?
6. What properties of rocks could contribute to a serious earthquake?
7. (True or False) Many geologists believe that it is magnetism that ultimately drives sea-floor spreading.

6B

FAULTS AND JOINTS

How do we classify faults?

6.5 JOINTS

6B Objectives

After completing this section, you will be able to

» explain how joints, faults, and earthquakes are related.

» summarize how an earthquake happens.

Have you ever seen a crack in a concrete sidewalk? Cracks very similar to these occur in rocks. You learned in Section 6A that rocks under stress form cracks when they begin to fail. These cracks are called **joints**. There is no visible movement of the rock pieces along the surface of a joint. Joints can vary in size from microscopic to many meters long.

Joints in rocks can also vary in appearance. They can be single joints, but they usually occur in groups. In rocks that are uniform, joints form at an angle to the stress. In certain kinds of rocks, joints can form along weak directions in rock materials or between different rock strata.

Joints can result from many geologic processes. When molten rocks cool and contract, they can form joints, sometimes in hexagonal patterns. When solid rock strata fold under stress, the layers on the outside of the fold crack. Tectonic activity can create numerous joints in rocks. But when rocks break as they form joints, they produce only very weak seismic waves because there is no motion at the joint.

6.6 FAULTS

But what happens if the rock moves as it cracks? A **fault** occurs when the sections of rock on opposite sides of the crack move relative to each other. Some faults are only a few centimeters long; others are hundreds of kilometers. Large faults are associated with earthquakes and with certain kinds of mountains. They also exist at the boundaries between tectonic plates. In every case, tectonic activity forms faults. It's no surprise then that faults are important in geology.

DESCRIBING FAULTS

Let's consider a model fault for a moment. Ideally, a rock breaks, making a flat surface between the two pieces. This *fault plane* can be oriented in any direction and at any angle. Geologists define the direction of a fault by its **strike**. If you could draw a level, or horizontal,

6-6

Rock joints can occur in many forms. Volcanic basalts often form in geometric joint patterns when they cool (top right). Slate rocks form useful joint patterns that make them easy to split (left). Joints in rocks form patterns that relate to the direction of the stresses that caused them (bottom right).

6-7
The strike of a fault is the compass direction of a horizontal line on the face of a fault. The dip is the slope of the face of the fault measured from the horizontal.

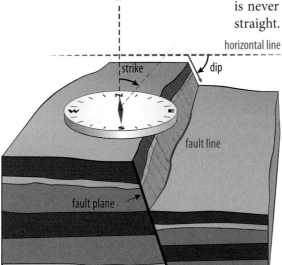

line on one face of the fault, the compass direction of this line would be the strike (see Figure 6-7).

Most fault planes tilt, so geologists measure the angle of the fault face downward from the horizontal. This angle is the fault's **dip**. When a field geologist surveys a fault, he will measure both the strike and dip of a fault at many places along the fault. The face of a real fault is never flat. It is wavy or has bumps on its surface. It is also never straight. So the fault's strike and dip can change along its length.

MAPS AND FAULTS

Maps are used to show faults in different ways. If you can see a fault on the earth's surface, or its location is known for certain, it is shown as a solid line. When a fault runs below ground and out of sight, geologists estimate its position on a map with a dashed line. Figure 6-8 shows how the strike and dip of a fault appear on a map. Figure 6-9 is a map of the continental United States, showing the patterns of prominent faults. Most major faults in the United States are in *mountain ranges*.

6-8
Faults as shown on a geologic map. The strike and dip of the fault to the left are given at several locations.

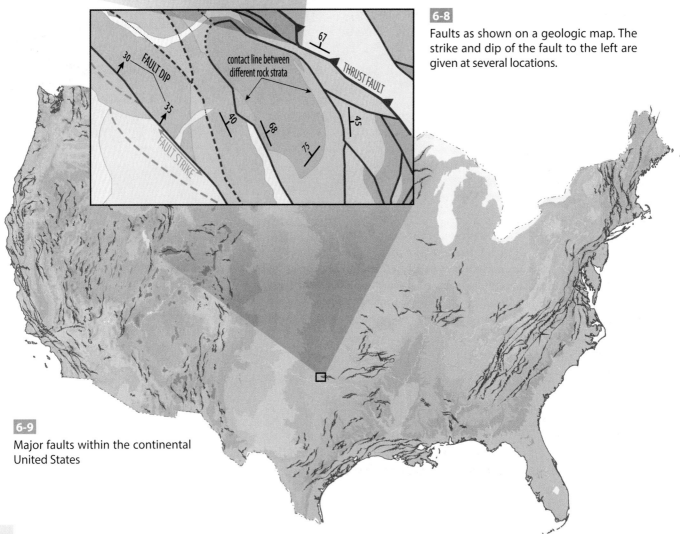

6-9
Major faults within the continental United States

CLASSIFYING FAULTS

Geologists classify faults by the way the two blocks of rock moved to form the fault. If a fault's motion is parallel to its dip (up and down), then it is called a **dip-slip fault**. There are two kinds of dip-slip faults. When the block above the fault surface dropped in relation to the block under the fault, then a **normal fault** resulted (see Figure 6-10, top). If the upper block rose in relation to the lower block, then the fault is a **reverse fault** (Figure 6-10, middle). This term usually applies if the fault's dip is steeper than 45° from the horizontal. If the reverse fault's dip is less than 45°, it is called a *thrust fault*. See the box on pages 142–43 for a discussion of thrust faults.

If the only motion along the fault is horizontal, then it is a **strike-slip fault** (Figure 6-10, bottom). Long strike-slip faults between sections of the same or different tectonic plates are also called *transform faults*. While such faults may be visible on the surfaces of continents, they are most numerous deep in the oceans where they form transform boundaries at mid-ocean ridge mountain systems (see Figure 6-4 on page 131). The San Andreas Fault in the western portion of southern California is a strike-slip fault. This fault separates the Pacific and North American tectonic plates (see page 140). All transform boundaries are strike-slip faults.

Of course, real faults may move in a direction that is a combination of both strike- and dip-slip faults. Geologists try to classify a fault by the direction of its greatest motion.

6.7 FAULTS AND EARTHQUAKES

Most of the earthquakes that we can feel happen at depths less than 70 km (45 mi) below the earth's surface. Earthquakes deeper than this usually cannot be felt, though we can measure them with seismic instruments. Geologists believe that these deep earthquakes happen where one plate subducts beneath another. The stresses of the colliding plates slipping past the other cause faults. As you learned earlier, if the blocks of rock on both sides of the fault do not move, or they move easily and continuously, then no significant ground shaking occurs. If there is no jerky movement along the fault, there are no earthquakes.

A fault's resistance to movement depends on how smooth and "slippery" the fault surface is. Just like oil acts as a lubricant to help the metal parts in a car's engine slide past each other, certain kinds of minerals and even water lubricate faults. Well-lubricated faults don't create serious earthquakes. But if the fault face is bumpy, or the minerals are sticky, or water is squeezed from the fault, it can stick.

A *locked fault* builds up shear stresses as the adjacent plates continue to move slowly. Strong, elastic rocks will flex slightly, but as long as the fault doesn't slip, it remains locked. When the stress becomes too great, rocks along the fault surface may suddenly fracture. The fault slips with a jerk, releasing immense amounts of elastic energy into

6-10

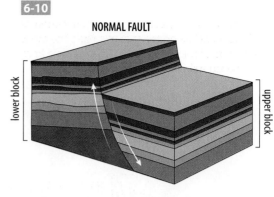
NORMAL FAULT

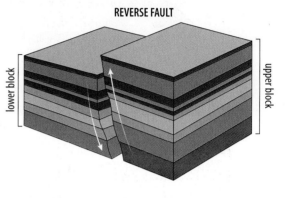
REVERSE FAULT

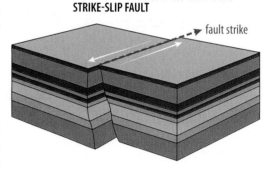
STRIKE-SLIP FAULT

the ground. Geologists believe that these stressed faults cause most earthquakes around the world. The fault blocks may move only a few centimeters. But in very large earthquakes, the slippage may be 10 m (33 ft) or more along the fault.

The released elastic potential energy causes seismic waves that radiate from the fault slip. Often, slippage in one location of the fault increases the stress on other locked portions of the fault, and then those sections slip sometime later. These secondary earthquakes are *aftershocks*. Some earthquakes have hundreds of aftershocks that can continue for days, weeks, or months after the main earthquake, creating even more devastation.

6B REVIEW QUESTIONS

1. What distinguishes a joint from a fault?
2. Name the two angular measurements that describe the orientation of a fault. What does each measure?
3. How do geologists classify faults?
4. What kind of fault is the San Andreas Fault?
5. Geologists believe that most earthquakes come from what kind of earth feature?
6. (True or False) Pumping groundwater out of a rock formation containing an active fault could result in a more severe earthquake.

6C

EARTH WAVES AND SEISMOLOGY

How do scientists collect earthquake data?

6.8 EARTH WAVES

An earthquake sends out seismic waves in all directions from its source. Seismometers installed at *seismic stations* detect these earthquake waves. *Seismographs* include both a seismometer and a way to record or graph the changing earthquake waves. Early seismographs

6-11
Aerial view of the San Andreas Fault in the Carrizo Plain area of central California

6C Objectives

After completing this section, you will be able to

» describe how seismologists collect earthquake wave data.
» compare and contrast the types of seismic waves.
» explain how to find an earthquake's epicenter.

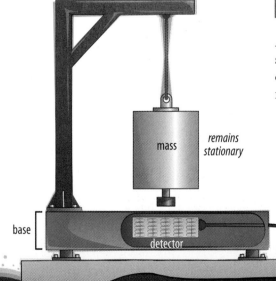

6-12
A *seismometer* consists of just the support frame, stationary mass, and position-sensing circuits, such as the instrument shown in this diagram. A *seismograph* includes the recording system.

consisted of a large mass supported by a frame mounted in bedrock. The frames of seismometers had to be attached to a rigid foundation so that they would move with the earth. Professional units were usually mounted directly on bedrock in deep dry wells to isolate them from man-made vibrations. A mechanical linkage connected the mass to the recording device. A pen or scribe traced a line on a rotating drum covered by paper. Later systems used a thin light beam that projected onto photographic paper mounted on the drum. Modern seismometers still use large masses and a stationary frame, but the instruments sense the motions of the earth using electronic and digital components.

Under normal conditions, a seismograph records a nearly straight line. When the earth shakes, the seismograph's mass does not move because of its inertia. But the mass is connected to the recording system, which moves with the earth. The recording system traces a zigzag line instead of a straight one. The height, or amplitude, of the "zigs" and "zags" in the graph of the earth waves shows how strong they were. A seismographic recording is called a *seismogram*.

As with all waves, seismic waves have amplitude and frequency (see Chapter 2). Seismometers measure both the amplitude and frequency of earth waves. Recall that wavelength is related to frequency, so wavelengths can be computed too. These properties are used to describe the earthquake waves.

Seismometer data is transmitted by radio and over the Internet to earthquake analysis centers. Earthquakes are usually displayed on large monitors in these centers and even on personal computers. Locations of earthquakes can be checked in near-real time at many websites using seismic GIS applications. By comparing seismic data with other stations around the world, seismologists can determine the location of the earthquake, the kinds of seismic waves it created, and the strength of the earthquake.

6-13
Most seismic stations are unmanned, remotely monitored instrument sites, like this one located on the slopes of Mount St. Helens in the state of Washington.

TYPES OF SEISMIC WAVES

Most earthquake waves pass right through the earth. These are called *body waves*. The fastest body waves are called **P waves**. The *P* stands for the Latin word *primus*, meaning "first." P waves have relatively small amplitudes and short wavelengths. They travel through both solid and liquid rock in the earth's interior.

SERVING GOD AS A SEISMOLOGIST

JOB DESCRIPTION

A day in the life of a seismologist can be very exciting! Seismologists are geologists who study the earth's movements and the structure of its interior. A seismologist may travel around the world, visiting geologic faults or volcanoes and earthquake or tsunami zones. He may do lots of fieldwork collecting data. He reads instruments, makes measurements, and simply observes geologic processes.

Seismologists also analyze data and, using logical reasoning, model earthquake processes. This leads to understanding earthquakes, possibly allowing us to predict what could happen in the future. Some seismologists look for geologic structures deep underground containing natural resources like oil and gas.

EDUCATION

You can prepare for a career in seismology even before you graduate from high school. Take earth science, chemistry, physics, and math classes. The most advanced classes can be the most profitable. Visit museums, caves, and geologically interesting places. Build your own seismometer using plans available on the Internet or at a library.

When you enter college, choose a physics, geophysics, seismology, or earth science major. Advanced degrees can further your understanding and opportunities in the field.

POSSIBLE WORKPLACES

The most likely jobs for seismologists are with companies that search for resources deep underground, such as oil companies. Companies exploring for geothermal steam sources need these scientists too. Government agencies such as the United States Geological Survey (USGS) employ many seismologists in their earthquake monitoring network. Seismologists also work for engineering consulting companies that specialize in geology studies for bridge, dam, and nuclear power plant construction.

DOMINION OPPORTUNITIES

Seismologists have special opportunities to exercise biblical dominion and serve their fellow man. Technology developed for earthquake and tsunami warning systems has the potential to save lives. Understanding the physics of earth movements helps engineers and architects design buildings and bridges that are more earthquake resistant. Finding natural resources like fossil fuels or mineral ores and developing wise plans to use them can make people's lives better.

S waves are the second (L. *secundus*) type of waves to arrive at a seismic station. S waves have much larger amplitudes compared with P waves. They are both stronger and slower than P waves. Like P waves, they travel through the earth's interior, so they are also body waves. But geologists don't observe S waves traveling through the core. S waves are a type that can't pass through liquids. This makes geologists suspect that at least the outer portion of the earth's core is liquid.

The last waves to reach a seismic station are **surface waves**. They travel along the earth's surface rather than through its interior. Surface waves are *not* S waves. Surface waves have two basic wave forms—Rayleigh waves and Love waves. These waves move at about the same speed as S waves, but they are far more destructive because they affect human structures as they disturb the earth's surface.

6.9 LOCATING EARTHQUAKES

Although an earthquake is normally most severe at its epicenter, a strong earthquake may cause ground shakes hundreds of kilometers from the epicenter. Deep rock formations and certain kinds of earth materials at the surface can sometimes concentrate and amplify seismic waves to produce greater shaking in certain areas.

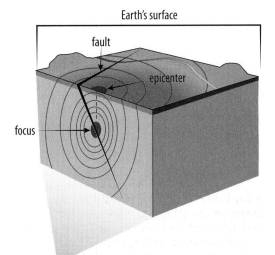

It's important to geologists to find where an earthquake has happened. The center of an earthquake's activity is called its **focus**, located deep underground. This may be as much as 670 km (420 mi) below the earth's surface. Measurable earthquakes do not seem to occur deeper than this, and most are much shallower. An earthquake's **epicenter** is the spot on the earth's surface directly above the focus. The closer the focus is to the surface, the more the ground shakes at the epicenter.

Seismologists can find the distance from a seismic station to the epicenter of an earthquake by examining the time interval between the arrival of P and S waves at the station. The farther the station is from the earthquake focus, the longer the time interval (see Figure 6-15 on the next page).

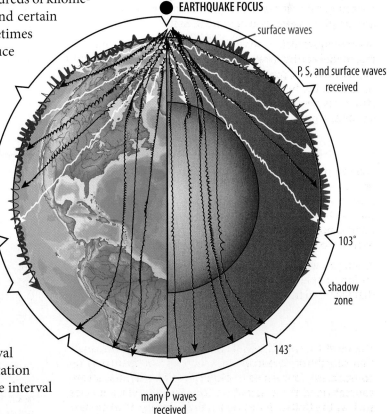

6-14
Earthquake waves and the direct paths they travel. From 103° to 143° are "shadow zones"—locations that do not receive any S or P waves by direct transmission. In addition, stations beyond 143° of the earthquake focus don't receive S waves by direction transmission.

A single station can tell the distance to the earthquake's epicenter but not its location. This creates a circle of possible locations. Two stations can show two possible locations for the epicenter (where the circles intersect). Three stations can usually pinpoint its epicenter to one location (three intersecting circles). By studying the arrival times of P waves, S waves, and surface waves, seismologists can calculate the depth of the earthquake's focus below its epicenter.

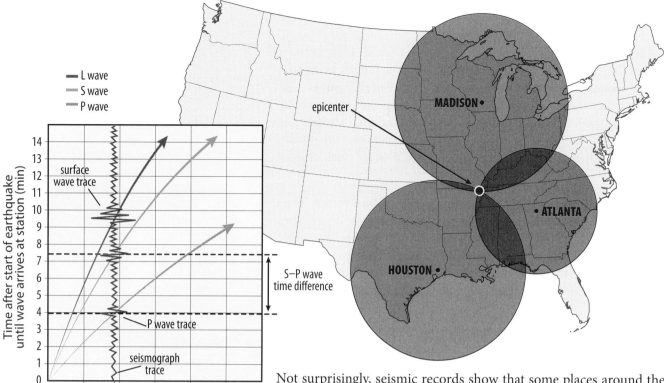

6-15
The epicenter of an earthquake is determined by using observations from at least three seismic stations. The length of time between the arrival of S and P waves is used to calculate the distance between a seismic station and the epicenter.

6-16
Earthquakes are most common along convergent, transform, and divergent tectonic plate boundaries.

Not surprisingly, seismic records show that some places around the world have been particularly prone to earthquakes in the past. Most earthquakes happen in faults along tectonic plate boundaries. Figure 6-16 shows Earth's tectonic plates, convergent and divergent boundaries, and the density of earthquakes (areas shaded in red). Notice that the boundaries of the Pacific Plate is seismically very active.

Earthquakes may also occur at other places where faults exist, far from tectonic boundaries. One famous example in the United States was the series of earthquakes near New Madrid, Missouri, in 1811–12. The last quake was so powerful that it caused the Mississippi River to briefly flow backward. Ground shocks were felt as far away as Norfolk, Virginia. The New Madrid earthquakes originated within a fault zone in the middle of the North American tectonic plate.

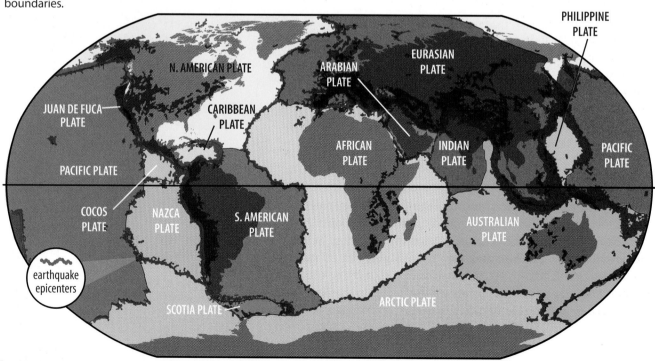

6C REVIEW QUESTIONS

1. What kind of instrument detects and transmits earth wave data to a central analysis station?
2. What is the most basic part of a seismic instrument?
3. List the types of earth waves that an earthquake produces.
4. Through which structure inside the earth can S waves not pass? What is the significance of this fact?
5. Compare an earthquake's focus with its epicenter.
6. Why are there lots of earthquakes around the rim of the Pacific Ocean?
7. (True or False) Because of the movement of seismic waves, scientists believe that the core of the earth may be at least partly liquid.

6D

EFFECTS OF EARTHQUAKES

What are the effects of earthquakes on living things?

6D Objectives

After completing this section, you will be able to

» describe how scientists rate earthquakes.
» explain why earthquakes can be so dangerous.
» evaluate the difficulty and benefits of predicting earthquakes.

6.10 EARTHQUAKE MAGNITUDE

The strengths of earthquakes are reported using the **Richter scale**. This scale, given in units of **magnitude**, indicates the energy released by the earth movement. Most measuring scales thay you may be familiar with are related directly to what they measure. For example, if you graphed the number of inches in a measurement to the number of feet in a measurement, you would obtain a straight line. Mathematicians call this a *linear relationship*.

But the Richter scale is *not* a linear relationship. An earthquake of magnitude 2 does not have twice the energy of an earthquake of magnitude 1. Instead, each additional 1 Richter unit stands for about 31.6 times more energy released. So an earthquake of magnitude 2 has 31.6 times as much energy as a magnitude 1 earthquake. An earthquake of magnitude 3 has 31.6 times as much energy as a magnitude 2 earthquake, and it has 31.6 × 31.6, or almost 1000, times as much energy as a magnitude 1 earthquake.

Richter designed his scale for the local, moderate earthquakes that occur in southern California. The scale does not work well for earthquakes stronger than about magnitude 7. Also, detections of earthquakes farther than about 700 km (435 mi) away from the seismic station cannot be accurately measured on the original Richter scale. Seismologists use other magnitude scales, such as the *moment magnitude* scale, that can also measure stronger and more distant earthquakes. However, the public is more familiar with earthquake magnitudes reported in Richter units. For that reason, seismologists convert all public earthquake reports to equivalent magnitudes on the Richter scale.

EARTHQUAKE MAGNITUDE

> 9.0 massive earthquake regional destruction

8.0–8.9 great earthquake extensive destruction near epicenter

7.0–7.9 major earthquake; serious damage

damage to well-built buildings	6.0–6.9
damage to poorly built buildings	5.0–5.9
felt by most people	4.0–4.9
felt by some people	3.0–3.9
possibly felt by people	2.0–2.9
recorded; generally not felt	< 2.0

magnitude less than 7.0

6-17
A tsunami wave (see page 147) crashing over the seawall in Miyako City on the northeastern shore of Honshu Island, Japan, after a magnitude 9.0 earthquake struck just off the coast in March 2011

The earthquake magnitude scale covers a wide range of earthquakes. The most destructive quakes usually have had magnitudes between 8 and 9. Calculations show that if an earthquake with a magnitude of 10 or more occurred anywhere on the earth, people everywhere would feel the shaking. It is unlikely that a fault slip could produce such a ground shake. But a large asteroid impact could. Note that there is no upper limit to earthquake magnitude.

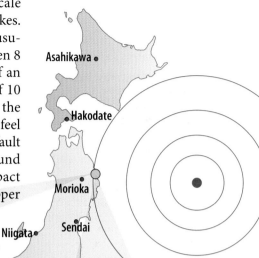

UPSIDE DOWN MOUNTAIN

In Glacier National Park is a mountain that appears to be upside down. At Chief Mountain, geologists have found what they believe to be older rock layers above rock layers that are younger! They think this because complex fossils are in strata below simpler fossils. Geologists assume that more complex fossils are found in newer rock layers and simple fossils are found in older rock layers, a view that fits with both the young- and old-earth stories. This unexpected ordering of rock layers directly contradicts the principle of superposition. But what could have caused this to happen?

A fault develops in sedimentary strata (A). Strata on the left rides up over the right side (B). Erosion removes much of the overlapping material (C).

It's obvious that only some unusual process could have altered the typical rock strata sequence. Somehow, the older rocks moved above the younger rocks. Both naturalistic and creationist geologists believe that the movement is explained by an overthrust, which happens along a thrust fault, a reverse fault that dips less than 45°. In a reverse fault, the upper block of rock strata slides back along the fault plane, and then up and over nearby younger rocks.

After an overthrust occurs, geologists believe that the upper, younger strata of the overthrusted rock erode, leaving the older rocks in the lower part of the upper fault block exposed. (See the sketches below to understand this idea.) This leaves older layers in place above younger strata. Geologists

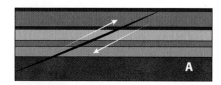
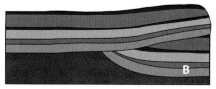
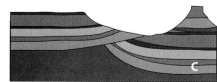

UPSIDE DOWN MOUNTAIN

have identified cases of overthrust on a small, local scale. In these places, the surface between the upper and lower fault blocks is eroded. There is a rough, uneven contact surface and lots of gravel and broken rock rubble from one block dragging over the other. These overthrusts, though, show less than a few hundred meters of motion.

Around the globe, there are mountains that exhibit this same reverse fossil profile. However, moving this much rock around is not local, and it's not on a small scale. For example, Heart Mountain near Yellowstone National Park also shows a reversed strata sequence. Geologists believe that Heart Mountain formed when a block of limestone 500 m thick, covering an area of 1100 km^2, slid over the ground for about 60 km! Even if a large enough force had been available to move these trillions of tons of rock, the rock would have shattered long before it had traveled that distance. This is especially true if it had occurred slowly.

Both secular and creation geologists agree that there is evidence that this overthrust occurred rapidly. Between the overthrust layer and the stationary layer, there is a thin layer of very fine rock material. Analysis of this layer shows that it formed quickly and at high temperatures. What could provide the conditions for the overthrust to happen? This is where worldview comes in. Secular scientists continue to search for a natural mechanism for the overthrust, but creation scientists look to the Genesis Flood.

Creation geologists believe that an explosion (volcanic, perhaps) during the recessionary period of the Flood could have triggered overthrusts like the one at Heart Mountain. Once triggered, the slab of rock could have easily moved along a layer of high-pressure fluid with minor erosion because of reduced friction. The conditions during the Flood explain the features of Heart Mountain and of other upside down mountains all over the world.

Chief Mountain is an example of a hypothetical overthrust.

The forces needed to form Chief Mountain must have been immense!

Yes, you could even say they were catastrophic!

On April 11, 2011, a magnitude 9 earthquake struck off the coast of Japan causing significant damage. The subsequent tsunami added to the devastation, including causing the meltdown of the Fukushima Daiichi nuclear power plant.

6.11 EARTHQUAKE INTENSITY

Earthquakes can threaten people's lives and welfare. To measure the actual damage that an earthquake causes, seismologists need a different scale. An earthquake's **intensity** is a measure of how much damage actually results from the earthquake. Earthquake intensity depends on lots of things. Its Richter magnitude is a big factor. But almost as important are the depth of its focus, the distance of its epicenter from populated places, the quality of building construction, the kinds of soil and underground rock formations, and especially the earthquake's duration.

Let's look at earthquake duration. Earthquakes can last anywhere from a few seconds to many minutes, but most earthquakes last less than a minute. Some buildings can withstand a few seconds of violent shaking but fall apart with longer exposures. The main shock of the 2004 Indian Ocean Sumatra-Andaman earthquake lasted nearly 10 minutes. The 2010 Haiti earthquake and a flurry of immediate aftershocks lasted more than half an hour.

Scientists call the earthquake destructiveness scale the *Modified Mercalli Intensity (MMI) scale*. Roman numerals from I to XII identify the levels of destruction (see Appendix F). An earthquake that leaves no visible damage receives a rating of I, while XII equates to total destruction to all buildings and services and a change of the appearance of the land. The 1994 Northridge, California, earthquake, which was rated only 6.7 on the Richter scale, was a ten (X) on the MMI scale. Because it lasted for 20–30 seconds, it did much more damage than many earthquakes with higher magnitude ratings.

LIFE CONNECTION: QUAKES AND CRITTERS

Presently, we can't predict earthquakes. But it's possible that we could get some help from a very unlikely source—animals. People have noticed for a long time that livestock, pets, wild animals, and even insects act very strangely before an earthquake strikes. Even hibernating snakes have left their burrows in winter and have frozen in the snow! Because of this, some scientists suggest that certain animals are trustworthy predictors of earthquakes.

Though some scientists have their doubts, let's suppose that some animals can be aware of a coming earthquake. Just what kinds of changes might they be sensing? A fault may send out weak, low-frequency seismic waves not readily detectable by instruments. This might trigger some strange animal behaviors. Many stories, however, describe animals acting oddly for hours or even days before an event. Some theorize that these animals might detect a slight tilting of the land, increasing groundwater seepage, or even the effects of radioactive radon gas released from the ground as earth stresses build. Others suggest that animals are responding to electrical or magnetic changes. The cause might be due to several of these factors.

These frogs massed just before a devastating earthquake in China in 2008.

While we may never know for sure whether animals can detect earthquakes or how they might be doing it, there is one thing we can know with certainty. The same Creator who said that a sparrow could not fall to the ground without His knowledge (Matt. 10:29) could have designed creatures with the ability to sense dangers like earthquakes.

6.12 EARTHQUAKE HAZARDS

A *hazard* is a source of danger; a *risk* is a possibility of injury or death to people and damage to property. The most dangerous and damaging earthquakes happen near cities. More people suffer in cities from an earthquake's effects. But ground shaking alone seldom causes deaths. Scientists and engineers cannot change earthquake hazards, but they thoroughly understand the risks. We can do much to reduce or even eliminate the risks.

Landslides

Earthquakes can cause avalanches and mudslides. These are particularly dangerous in the mountains, where sliding rock and debris can gain great speeds on steep slopes and then plow through villages.

The hazards and risks associated with earthquakes can also create great loss of life and property. But Christians recognize that such events are under the direct control of our sovereign God. Though His purposes are unknown, Christians can certainly trust that He is acting according to His holy and gracious character. One purpose that God often reveals is the increased opportunity to spread the Gospel. For example, following the 2010 earthquake in Haiti, an annual pagan Vodou celebration scheduled in advance was replaced by a Christian evangelistic meeting that led to the salvation of many Haitians.

Bi Wealth and Domain

Shortly after the devastating earthquake in Bam, Iran, in 2003, American writer Josh Chavetz made the following observation: "In the long term, we can help the people of Iran to live under a more responsive government, and we can help them build a twenty-first century economy—the kind of economy that produces enough wealth to allow them to build safer buildings." Should a Christian agree or disagree with this statement? Why? (Quotation from World Magazine, January 10, 2004, p. 11)

Building Collapse

Most earthquake-related deaths occur when man-made things fail, such as when buildings collapse or when fires start from broken gas mains or electrical wires. So if an earthquake strikes a city, especially in a poor country where buildings may not be built well, the death toll is high. In contrast, if an earthquake strikes a modern city that enforces strict building codes and has well-designed earthquake-proof buildings, there are fewer deaths and less property damage. The magnitude 7.0 Haitian earthquake near the capital Port-au-Prince in 2010 clearly showed this. It was initially reported that over 220,000 people died and more than a million became homeless. Though the actual numbers may never be known for sure, what is certain is that the high number of casualties was mostly due to poor building construction.

Architects know how to design buildings that can resist collapsing in earthquakes up to magnitude 8.0. For older buildings, they have learned how to reinforce structures that were not built to the highest earthquake-resistant standards.

Tsunamis

Earthquakes can also trigger devastating waves called **tsunamis**. A tsunami forms when an earthquake displaces a large volume of water. This jolt adds energy to the water, forming several deep waves that can travel thousands of miles. When these waves approach shore, the shallower bottom lifts the waves into huge swells of water as they crash ashore. The Great Chilean earthquake of 1960 produced waves as high as 25 m (82 ft). A tsunami wave can carry large ships anchored in a harbor hundreds of meters inland and topple large buildings.

Tsunamis often claim more lives than the earthquakes that cause them. They also spread devastation far from the epicenter of the earthquake. For example, the 2004 Indian Ocean tsunamis killed about 230,000 people in fourteen countries. If a tsunami warning network had been installed in the Indian Ocean, like the one built earlier in the Pacific, loss of life would have been a tiny fraction of this number.

6.13 EARTHQUAKE PREDICTION

Right now, even with all our technology, we can't predict earthquakes. But seismologists are working on this. A few days' or hours' notice can't save buildings, but they can save lives. Seismic maps using GIS technology show the locations of most of the faults in earthquake-prone areas. Seismologists collect data all over the world and analyze it for patterns. Every earthquake is an opportunity to learn more. But reliable earthquake forecasting is probably still years or even decades away because we can't recognize for sure the signs of an earthquake that is about to happen.

To give crude predictions of earthquakes, seismologists study the history of areas with faults. For example, they have discovered that a major earthquake strikes San Francisco about every 150 years. The city's last major earthquake was in 1906 and had an estimated magnitude of 7.9–8.3. So does that mean that we can expect another major quake in San Francisco within 50 years? This kind of prediction is imprecise and unreliable. And these kinds of predictions don't make evacuation possible.

Another way of predicting earthquakes is to notice what areas have *not* had an earthquake recently. If major earthquakes have occurred all along a fault except in one area—called a *seismic gap*—that area may be due for an earthquake. However, the seismic gap may simply be less likely to have earthquakes than other sections of the fault. So again, prediction is unreliable.

Fire

Another significant earthquake hazard in populated areas is fire. Earthquake shaking breaks electrical wires, creating sparks that can easily ignite flammable materials. The most significant fire hazard comes from broken natural gas supply pipes. Leaking gas and sparking electrical wires cause explosions. Broken water supply pipes make the fire problem even more difficult. A lack of water pressure can prevent fire fighters from quickly dousing fires. To reduce the risk of these problems, engineers can design safety features. Sensors can automatically shut off gas if a rupture in a gas line occurs. Similarly, automatic switches can turn off electrical supplies if they detect electrical problems.

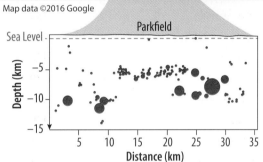

6-18

The thematic map above shows the location of aftershocks of the 2004 Parkfield, CA, earthquake. The upper view shows the distribution of shock foci (red circles) along the San Andreas Fault under Parkfield. The graph shows the depths of the foci.

But we are making progress. As you learned at the beginning of the chapter, Japan and other earthquake-prone locations are trying to identify the early warning signs of an impending earthquake. For example, seismologists successfully predicted a large earthquake in Liaoning Province of northeast China in 1975. A combination of an increased number of minor quakes, changing *water table* levels, and unusual animal behavior (see the Life Connection box on page 145) signaled that an earthquake was a possibility. Officials ordered an evacuation. Several days later, a 7.3 magnitude earthquake struck. Very few people died or were injured. An estimated 150,000 casualties might have resulted without the warning. However, this success is the rare exception rather than the rule, even today.

In the meantime, when we must build in seismically active areas, we need to carefully design structures that will minimize the potential for seismic damage. We need to develop earthquake prediction tools and early warning communication systems. Through these and many other ways, we use the wisdom God has given us to save lives. In this way, we exercise dominion through science to lessen the dangers of living in a fallen world.

6D REVIEW QUESTIONS

1. How much does an earthquake's energy change with each number on the Richter scale?
2. What is the highest magnitude on the Richter scale?
3. Besides the magnitude of an earthquake, what other major factor can increase seismic damage in even well-built areas?
4. What is the maximum value on the intensity scale and what would be the observed result?
5. Why is exercising dominion more about minimizing the risks associated with an earthquake than minimizing earthquake hazards? Give an example of minimizing earthquake risks.
6. Why is it difficult to predict earthquakes?
7. How do seismologists determine which locations have the greatest likelihood of earthquakes?
8. (True or False) An earthquake with an intensity of VI is 31.6 times more damaging than an intensity V earthquake.
9. How much more energy is released in a magnitude 9 earthquake than in a magnitude 2 earthquake (which can just be felt)?

CHAPTER 6 REVIEW

CHAPTER SUMMARY

» Early earthquake warnings can save many lives.

» Earthquakes make the ground shake as earthquake waves pass through an area of the earth's crust.

» Tectonic forces cause stresses in the earth's crust.

» Convergent, divergent, and transform tectonic plate boundaries are the locations of the greatest stresses in the crust.

» Rocks can flex or fail under stress. When failure occurs, they crack to form joints and faults.

» Faults form when rocks on opposite sides of a crack move relative to each other.

» Geologists describe faults by their strike (direction) and their dip (tilt).

» A fault is classified by how the blocks of rock on opposite sides of the fault plane move relative to each other.

» Faults may be normal, reverse, strike-slip, or a combination of these.

» Earthquakes happen due to the sudden, jerking motion of previously existing or newly formed faults.

» Geologists use seismometers to detect seismic waves.

» There are three kinds of seismic waves: P waves, S waves, and surface waves. Surface waves are the most destructive.

» An earthquake actually occurs at its focus, which can be anywhere from a point on the earth's surface to deep underground.

» The point on the earth's surface directly above the earthquake's focus is its epicenter.

» The Richter scale rates an earthquake on the basis of the energy of its seismic waves.

» The Modified Mercalli Intensity scale rates the destructiveness of the earthquake.

» Hazards from earthquakes include falling buildings, electrical and gas fires, lack of firefighting water, tsunamis, and landslides.

» Risks of damage or injury from earthquakes depend on building designs and preparations that we make to avoid injuries during earthquakes.

» Geologists are working to develop the ability to predict earthquakes accurately.

Key Terms

Term	Page
seismometer	129
earthquake	130
stress	130
compression	130
tension	130
shear	130
buoyant force	131
divergent boundary	131
convergent boundary	131
transform boundary	131
strain	132
joint	133
fault	133
strike	133
dip	134
dip-slip fault	135
normal fault	135
reverse fault	135
strike-slip fault	135
P wave	138
S wave	139
surface wave	139
focus	139
epicenter	139
Richter scale	141
magnitude	141
intensity	145
tsunami	147

6-19
The Himalaya Mountains, which contain nine of the ten highest peaks in the world, are located along the borders between Bhutan, China, India, Nepal, and Pakistan in a subduction zone.

REVIEW QUESTIONS

1. What is the difference between warning of an earthquake and predicting an earthquake?
2. What main geologic processes are associated with earthquakes?
3. What is the main source of stresses in the rocks of the earth's crust?
4. The Himalaya Mountains seemed to have formed from the collision of the plate containing the India subcontinent with the plate containing Asia. What kind of tectonic boundary is this?
5. Rocks that are _____ tend to break easily and not store up stresses.
6. Examine the maps in Figure 6-9 and Appendix G. With what landforms are most of the faults associated in the United States?
7. What kind of motion along a fault produces a measurable earthquake?
8. What are two main differences between P and S seismic waves?
9. At a minimum, how many seismic stations are needed to identify the epicenter of an earthquake?
10. How much more energy does a magnitude 3.5 earthquake have than a magnitude 2.5 earthquake. (*Hint*: How many magnitudes different are these two earthquakes?)
11. For developing countries, what is usually the main factor that results in deaths during an earthquake? (Ignore secondary effects such as tsunamis and landslides.)
12. Why is being aware of the approximate time between major earthquakes for San Francisco not good enough to provide an earthquake prediction?
13. When speaking of earthquakes, what is one important purpose of dominion through science?

True or False

14. The best earthquake warning systems can give up to a one-hour notice in advance of an earthquake.
15. Any change of shape of an object due to stress is called strain.
16. Rocks that are elastic and store energy are more seismically dangerous than brittle rocks.
17. A fault may have many strikes and dips along its length.
18. Faults are classified mainly by their geologic ages.
19. Seismic gaps might be caused by locked faults.
20. Earthquakes may occur anywhere from the surface of the earth to its core.
21. The Richter scale is based on how an earthquake feels and the amount of damage that it can cause.
22. The destructiveness of an earthquake is related only to its magnitude.
23. The wealthier a nation, the better able it is to protect its citizens from natural disasters.

CASE STUDY: EARTHQUAKES—THE BIG ONE!

It was April 11, 2011, and a group of seismologists was meeting to learn more about earthquakes. The conference was taking place in Japan, a country known for its frequent earthquakes, so no one was surprised when an earthquake interrupted the event. What they did not expect was how long the quake lasted. It continued for about six minutes, so everyone knew that Japan had just had a magnitude 9 earthquake. Knowing that quakes of this size form under the ocean, they knew what was coming next … tsunami!

Seismologists know that there is a relationship between an earthquake's duration and its magnitude (or strength). Most quakes are small and last less than a minute. A magnitude 7 or 8 earthquake may shake the ground for one to three minutes. When a quake continues beyond that, scientists know that they are dealing with a huge earthquake.

People who live in earthquake prone areas are always concerned about the "Big One." And people in the United States always think of the San Andreas Fault. But surprise! The "Big One" will not happen along that fault, but instead will happen farther north. The San Andreas Fault can produce large earthquakes, but they have an upper limit around 8.2 because the San Andreas is a strike-slip fault. The real threat is from a recently discovered fault called the *Cascadia Subduction Zone*.

This zone is located under the Pacific Ocean off the coast of the Pacific Northwest—northern California, Oregon, Washington, and British Columbia. This 1000 km long fault is where the Juan de Fuca oceanic plate is sliding under the North American plate. However, this sliding doesn't occur smoothly—the plates are actually stuck. When the stress finally releases, seismologists predict that the North American plate will suddenly move more than 3 m west and drop 2 m.

This will produce an earthquake of magnitude 9+ in the region. Since the fault is underwater, the quake will also produce a tsunami, similar to the tsunami caused by the earthquake in Japan in April 2011.

Geologists studying this region can't forecast when the next big one will occur, but they know that it will happen eventually. Seismologists attempt to predict the impact of such a large earthquake and tsunami, and the predictions are grim. They expect the shaking to last four to five minutes, damaging buildings and roads and severing electric and gas lines. Once the shaking stops, residents along the coast will have fifteen to twenty-five minutes to get to higher ground before the arrival of the tsunami. While there is a tsunami warning system, this short amount of time requires that people are prepared to evacuate. Such a quake will have a severe impact for a long time. Seismologists are working now to help others prepare for the "Big One."

Questions to Consider

1. Which types of earthquake waves are the first to arrive? Which tend to do the most damage?
2. What is the most recognizable fault in California?
3. Why is the San Andreas Fault likely to produce earthquakes of magnitudes no greater than 8.2?
4. What type of boundary is in the Cascadia Subduction Zone?
5. Why do scientists expect a tsunami to occur along with an earthquake along the Cascadia Subduction Zone?
6. How can seismologists work to prepare people in the Pacific Northwest for the next big earthquake?

CHAPTER 7
MOUNTAINS AND HILLS

- **7A** WHAT IS A MOUNTAIN? `153`
- **7B** TECTONIC MOUNTAINS `158`
- **7C** NON-TECTONIC HILLS AND MOUNTAINS `165`

TOP OF THE WORLD

In 1953, a beekeeper from New Zealand named Sir Edmund Hillary and a Sherpa named Tenzing Norgay reached the summit of Mount Everest. They had made history. But how did they know it was the highest mountain on Earth?

In 1856, England surveyed the Himalaya Mountains. Many of the earth's highest peaks are located there. Scientists used geometry to measure the soaring heights of these mountains. They calculated that Mount Everest (then called Peak XV) was 29,002 feet above sea level.

In 1999, one of the world's leading mountain cartographers, Brad Washburn, installed a solar-powered GPS unit on Everest's summit. He made a few interesting discoveries. He found that the exact height of the mountain was 29,035 ft. Scientists were even more surprised to find that the Himalayas appear to be rising 2.4 inches per year! How can we explain this observation?

7A

WHAT IS A MOUNTAIN?

What factors determine the height of a mountain?

7.1 GOING UP?!

The British first measured the elevation of Mount Everest in 1856. Later measurements showed that the peak was growing taller. Why was the elevation of Mount Everest changing? We've explored how dramatically the earth's surface has changed over time. Change keeps occurring today but not at the catastrophic rates of the past. Mount Everest and the mountains near it are in a zone of the earth's crust that usually rises. Geologists believe that it rises due to the Indian tectonic plate colliding with the Eurasian plate. Today its elevation is 8848 m (29,029 ft), 2 m lower than its elevation in 2012. Why has it suddenly lost elevation? Geologists believe that the plates settled during recent large earthquakes in Nepal resulting in a slightly smaller world's highest peak.

7.2 TOPOGRAPHY AND ISOSTASY

The thematic map in Figure 7-2 below uses colors to indicate the height of land above and below sea level. For land, green zones are closer to sea level, while beige and tan zones are higher. In the oceans, lighter blue means shallower depths, and darker blue is deeper. Notice that every continent and ocean basin has a variety of elevations arranged in different ways. These factors determine the **topography** (tuh PAH gruh fee) of an area.

Have you ever thought about why the rocks in the earth's crust are on top? The crust, or lithosphere, "floats" on the mantle, something like a boat in water. In Chapters 4 and 5 you learned that oceanic crust is fairly thin and dense. It sinks into the mantle, just as a heavier boat floats lower in water. The oceanic crust thus forms low spots on the earth's surface. This is why the waters of the oceans fill in these *basins*.

7A Objectives

After completing this section, you will be able to

» discuss the relation of topography to the principle of isostasy.

» summarize processes that contribute to orogeny.

» identify and describe various kinds of mountain and hill landforms.

» differentiate between elevation and actual height.

7-1
Mountaineer and cartographer Brad Washburn (1910–2007), shown here with his wife, made the study of mountains his life's work.

7-2
This topographic map of the world uses colors to indicate land elevation and ocean depth.

Ocean depth	Land elevation
-1000– 0 meters	>4800 meters
-2000– -1000 m	3000–4800 m
-3000– -2000 m	1800–3000 m
-4000– -3000 m	1200–1800 m
-5000– -4000 m	600–1200 m
-6000– -5000 m	300–600 m
-7000– -6000 m	150–300 m
-8000– -7000 m	0–150 m
< -8000 m	

MOUNTAINS AND HILLS

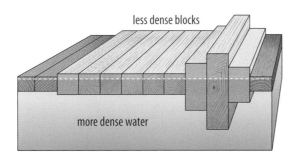

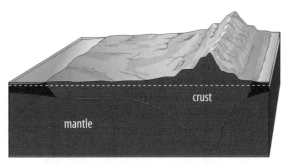

7-3

Isostasy accounts for the extra thickness of the lithosphere needed to support higher land features.

> **isostasy** (eye SAHS teh see): iso- (Gk. *isos*—equal) + -stasy (Gk. *stasis*—to stand still); neither rising nor falling; balanced

On the other hand, the continental crust is made of lighter, less-dense rocks, and it is generally much thicker, especially under mountains. Thus, the continental lithosphere extends more deeply into the mantle, and it rises higher above the earth's average surface height, which is about sea level. Mountains create higher topography and add more weight to the crust. To support these heavier regions, the lithosphere must extend even more deeply into the mantle to hold up the extra weight.

Geologists use the term **isostasy** to describe the balance of the weights of rock, water, and ice with the upward force of the mantle. The elevation of the earth's surface at a given location is due to the total thickness, as well as the density, of the materials forming the crust of the earth there. Believe it or not, isostasy determines the topography of the land you walk on.

7.3 LANDFORMS AND OROGENY

A big part of geology is identifying and naming the shapes of the structures on the earth's surface, or its **landforms**. When you think about it, there are all kinds of landforms to classify. There are mountains, valleys, beaches, plains, deserts, plateaus, lakes, and many others that you will learn about in this book. Look at the landform maps in Appendix G to appreciate the many landform zones in the United States.

Mountains and hills are probably the most obvious kinds of landforms. Mountains make up huge portions of the continents and even of the underwater topography. From a Flood geology viewpoint, most of the high mountains that we see today likely formed during or after the Flood. Tectonic processes that geologists believe resulted in the making of mountains fall under the geologic term **orogeny**. So let's learn what could cause orogeny and identify some of the important kinds of mountains.

> **orogeny** (or AH jin ee): oro- (Gk. *oros*—mountain) + -geny (Gk. *-geneia*—the way to make); having to do with the making of mountains through tectonic processes

7.4 ELEVATION AND HEIGHT

What is a mountain? We can define a **mountain** as "a natural elevation of the earth's surface rising more or less abruptly to a **summit**." But this definition applies to a **hill** too. The difference between mountains and hills is mostly height, but it is also related to local customs in an area.

> The **summit** is the highest point on a hill or mountain.

DENALI (MOUNT MCKINLEY)
6190 m (20,310 feet) above sea level

MOUNT WHITNEY
4421 m (14,505 feet) above sea level

From what point do we measure the height? For example, the highest mountain on Earth is Mount Everest, at 8848 m. Is that

» 8848 m above its base or
» 8848 m above the lowest land on Earth or
» 8848 m above sea level or
» 8848 m above the lowest point on the ocean bottom?

The heights of Mount Everest and all other mountains, as well as any location on land, are usually given in meters or feet above *average* or *mean sea level* (*MSL*). (See Subsection 13.4 for more information on this.) This measurement is also called the mountain's **elevation**. The **actual height** of a mountain is the height of its summit above the lowest elevation of the surrounding terrain, or its base.

Although Mount Everest has the highest elevation on Earth, it's not especially impressive in its setting. Nearby mountains are nearly as high, and its base is about 5200 m above sea level. So Everest's actual height is only about 3600 m.

Other mountains have greater actual heights. Mauna Kea (MAW nah KAY yah), which forms a large part of the island of Hawaii, has the greatest actual height of any mountain on Earth. Its summit is over 10,000 m from its submerged base. That is more than two-and-a-half times that of Everest! Mauna Kea's elevation above sea level, however, is only 4207 m, which is less than half of Everest's. Most of Mauna Kea's actual height is below the ocean's surface.

A mountain's actual height is related to how mountainous or hilly a region is. This is the **relief** of the region, the difference in height between the highest and lowest elevations of the terrain. For example, the highest mountain in California, Mount Whitney, is 4421 m above sea level. The lowest point, in Death Valley, is 86 m *below* sea level. So, the relief of the state of California is about 4500 m. Mountainous areas have *high relief*, while plains and plateaus have *low relief*.

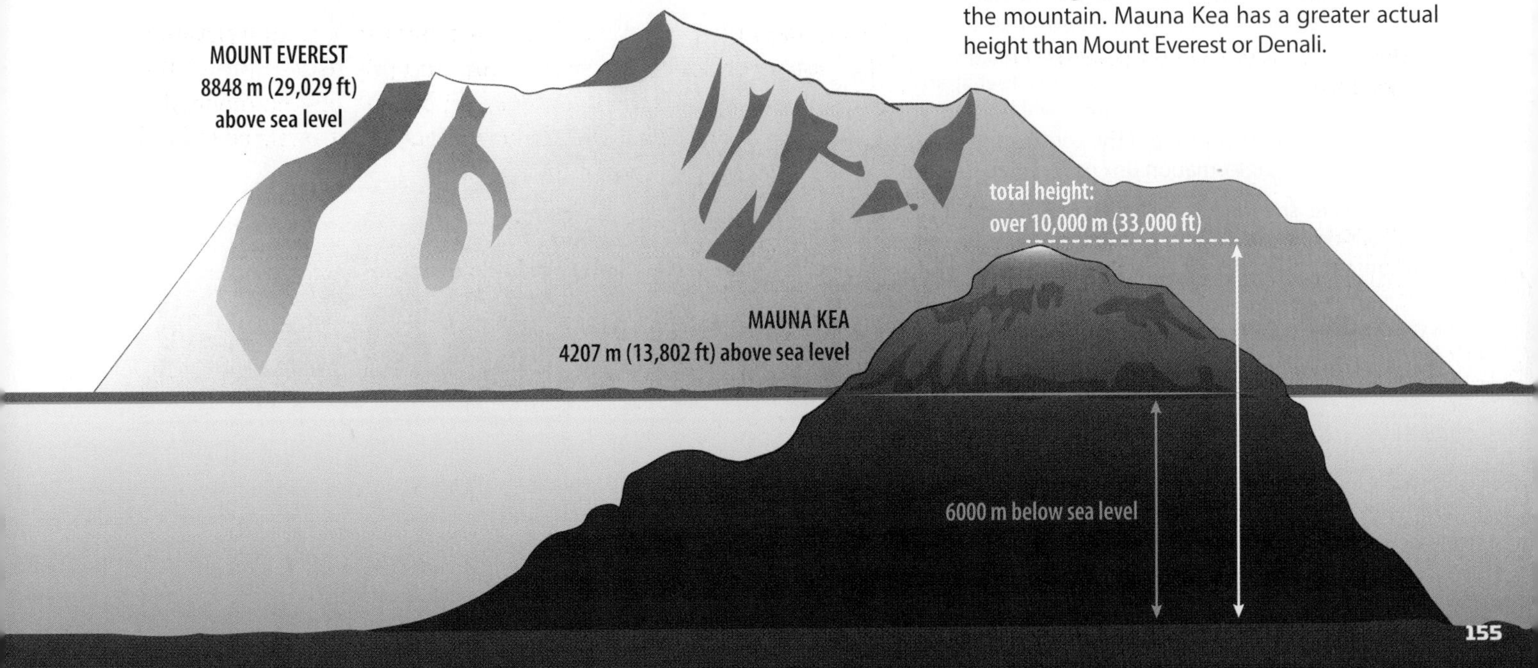

terrain (tehr AIN): terrain (L. *terranum*—earth; of the earth); the nature of the surrounding land; topography

A mountain's **base** is the lowest elevation of nearby terrain from which the mountain rises to a peak.

7-4

Elevation is measured from sea level, whereas actual height is measured from the base of the mountain. Mauna Kea has a greater actual height than Mount Everest or Denali.

MOUNT EVEREST
8848 m (29,029 ft) above sea level

total height: over 10,000 m (33,000 ft)

MAUNA KEA
4207 m (13,802 ft) above sea level

6000 m below sea level

7.5 GROUPS OF MOUNTAINS

Mountains exist either by themselves or in groups. A series of mountain peaks in the same geographical area is called a **mountain range**. The Cascade Range in northwestern United States extends from Oregon northward through Washington State and into British Columbia.

A group of mountain ranges connected over a much larger area is a **mountain system**. The highest mountain system is the Himalaya-Karakoram, located north of Pakistan and India. This system includes 96 of the 109 peaks in the world that are higher than 7300 m (24,000 ft).

But do you know where the longest mountain system is? It's under the Atlantic Ocean! The Mid-Atlantic Ridge extends for 16,000 km (10,000 mi) down the middle of the ocean. It is only one segment of a much longer system of mid-ocean ridges that wraps around the world like the stitching on a baseball. The total length of this mountain system is about 80,000 km (50,000 mi)!

You learned in Chapter 5 that most geologists believe that sea-floor spreading created the Mid-Atlantic Ridge when the Atlantic Ocean was forming. A few of its peaks are high enough to appear above the surface of the water. The highest of these is Mount Pico in the Azores with an elevation of 2351 m (7713 ft). Its actual height is about 4600 m (15,000 ft) above the Atlantic Ocean floor.

Fi Mount Pico: 38.47°N 28.4°E

RELIEF MAPS

A map that shows topography using thematic colors, shading, contour lines, or some other method is a *relief map*. The most useful type of relief map uses contour lines, which join points of equal elevation.

The elevation of each contour line is either labeled or can be determined by nearby lines. Heavier *index contour lines* are usually labeled with the elevation. The number of lighter lines between the index lines is determined by the *contour interval*, or the number of feet or meters between lines. For example, if the index contour lines are 10 feet apart, and the contour interval is 2 feet, then there will be four unlabeled lines between index contours. (The contour interval is usually stated in the map information box on the map itself.)

Contour lines showing a mountain or hill form closed loops of increasing height. As you go higher and closer to the peak, the contour lines enclose smaller areas. Closely spaced contour lines show a steep slope, while widely spaced lines show a gentle slope.

CASE STUDY: BRAD WASHBURN: ADVENTURER, CARTOGRAPHER

Bradford Washburn was a mountaineer, cartographer, photographer, and scientist. This picture is of him surveying Denali. Bradford was the first to make a map of the mountain. His wife was the first woman to summit Denali in 1947.

What do you want to be when you are older? Brad Washburn answered this question when he was about your age. Washburn began climbing mountains in his early teens and reached the summits of Mont Blanc and the Matterhorn in the Alps by the age of 16. He is still known today as a mountaineer, mapmaker, photographer, and scientist.

However, it was Washburn's first failure in mountain climbing that showed him a direction for his life. At the age of 19, Brad attempted to climb Mount Fairweather in Alaska. He considered the expedition a failure because he never reached the summit. But this experience led him to recognize the need for better maps of the area. Gathering support from the National Geographic Society, universities, and government agencies, Washburn funded expeditions to study and map the mountains of Alaska.

From 1932 through 1947, Washburn led many expeditions to the interior of Alaska. He used creative ways to produce better maps of the area. He was the first to use large format and aerial photography to aid mapping and scientific study. His greatest feat was Operation White Tower, a quest to climb and map Denali. Washburn extended his work to many regions of the world. He continued to innovate, mapping the Grand Canyon with lasers and using satellites to map Mount Everest. His efforts allowed him to create detailed maps of these remote regions.

Washburn recognized early in his life that his purpose was to further cartography through his interest in and love of mountaineering, photography, science, and aviation. Many consider Washburn a pioneer in his field who has benefited people around the world by his devotion to his purpose.

Questions to Consider

1. What experience from Brad Washburn's life demonstrates how failure can have a positive influence?
2. What are some examples of Washburn using unique approaches to mapmaking and scientific research?
3. Why do you think Washburn chose more creative ways to map Mount Everest and the Grand Canyon?
4. Can someone serve God by doing work similar to the work that Brad Washburn did?

7A REVIEW QUESTIONS

1. What is an area's topography?
2. Relate topography, the thickness of the lithosphere, and isostasy.
3. What geologic term describes the tectonic origins of mountains and the processes that build them?
4. What is the difference between a mountain and a hill?
5. How do we measure the elevation of a mountain's summit?
6. Compare mountain elevation and actual height.

(continued)

7. Describe the relief of
   a. the Himalaya Mountains.
   b. the Great Plains in the United States Midwest.
8. Where is the most extensive mountain system in the world located?
9. (True or False) The relief and elevation of the small volcanic island in the adjacent photo are the same thing.

7B

TECTONIC MOUNTAINS

How do mountains form?

7.6 TECTONIC FORCES AND LANDFORMS

7B Objectives

After completing this section, you will be able to

- relate tectonic forces to orogeny.
- identify various convergent, divergent, and volcanic mountain landforms.
- identify various uplift and subsidence landforms.

As with most things in science, we humans are eager to classify what we see. When Adam named the creatures in Genesis 2, he was classifying things by giving them names. This is part of fulfilling the Creation Mandate. And this applies to mountains too. Geologists classify mountains on the basis of the way that they think the mountains formed. In this section, you will learn about mountains shaped by tectonic forces.

In Chapter 5 you learned how tectonic plates move. In many places in the world, these plates converge, creating huge forces in the crust. When the forces are stronger than the rock, the crust crumples like two cars in a crash. The wrinkles in the earth's crust are tectonic mountains. So we expect to find tectonic mountains near convergent tectonic plate boundaries.

But tectonic mountains also form at divergent zones, where plates are moving apart. The mid-ocean ridge system seems to exist because tectonic forces are slowly pulling rocks apart. Most geologists agree that liquid mantle rock forces its way upward, lifting the rift zone above the surrounding crust. As the plates spread slowly apart, the cooled rock cracks and faults. The blocks eventually sink to lower levels as they move farther away from the rift.

While we find most divergent-zone mountains underwater, there are several notable ones on land. For example, sections of the African continental plate are pulling apart in East Africa. Using Google Earth, you can view convergent and divergent tectonic boundaries around the world. You can also study them using the map on page 140.

However, not all tectonic forces are horizontal. Sometimes forces act up and down. These forces create *uplift landforms* or *subsidence landforms*. You will learn a little later in this chapter about forces that create these kinds of landforms.

So tectonic mountains formed when the pieces of the crust crunched together, pulled apart, rose up, or sank downward. If these processes happen at all today, they happen very slowly. But for the earth to be as young as the Bible suggests, these processes must have been much faster in the past. What kinds of landforms do tectonic forces create?

CONVERGENT LANDFORMS

7.7 FOLD MOUNTAINS

When water deposits sediment, it forms a flat or nearly flat layer. In the past, sediments became *sedimentary* rock strata. When we find slanted or upturned strata, we know that something else happened! Some kind of force in the crust tilted or bent the rock. These forces can act in any direction.

Layers of softer, unsolidified materials, such as clays and moist sand, easily bend under a force. Many folded strata show no breaks at all or only minor joints caused by the rock drying and hardening. This appearance suggests to Flood geologists that the soft sediments laid down by the floodwaters were shaped by tectonic forces during the active portion of the Flood. Solid, brittle rock strata, strongly folded (as seen in the center image on the next page), would have cracked and crumbled if forces had attempted to fold them.

Most major mountain ranges contain some folded rocks. The Rockies, the Appalachians, the Alps, and the Himalayas all include folded strata. Mountains that form from folded rock strata are called **fold mountains**.

> **Sediment** is a deposit of eroded earth materials. It collects in low spots, usually in streambeds or under bodies of water. Sediments can also be carried and deposited by ice or wind.

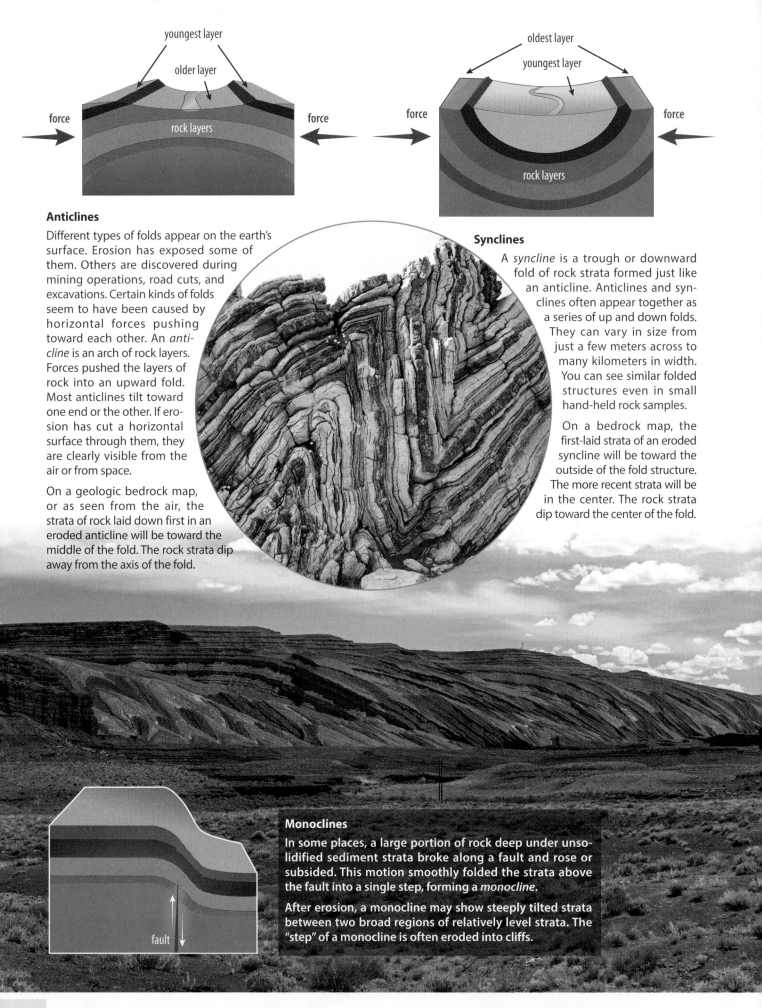

Anticlines

Different types of folds appear on the earth's surface. Erosion has exposed some of them. Others are discovered during mining operations, road cuts, and excavations. Certain kinds of folds seem to have been caused by horizontal forces pushing toward each other. An *anticline* is an arch of rock layers. Forces pushed the layers of rock into an upward fold. Most anticlines tilt toward one end or the other. If erosion has cut a horizontal surface through them, they are clearly visible from the air or from space.

On a geologic bedrock map, or as seen from the air, the strata of rock laid down first in an eroded anticline will be toward the middle of the fold. The rock strata dip away from the axis of the fold.

Synclines

A *syncline* is a trough or downward fold of rock strata formed just like an anticline. Anticlines and synclines often appear together as a series of up and down folds. They can vary in size from just a few meters across to many kilometers in width. You can see similar folded structures even in small hand-held rock samples.

On a bedrock map, the first-laid strata of an eroded syncline will be toward the outside of the fold structure. The more recent strata will be in the center. The rock strata dip toward the center of the fold.

Monoclines

In some places, a large portion of rock deep under unsolidified sediment strata broke along a fault and rose or subsided. This motion smoothly folded the strata above the fault into a single step, forming a *monocline*.

After erosion, a monocline may show steeply tilted strata between two broad regions of relatively level strata. The "step" of a monocline is often eroded into cliffs.

160 CHAPTER 7

Complex Tectonic Mountains

Think about what happens when a snowplow pushes against deep snow. Snow piles up and forms a large heap with wrinkles and folds. Similarly, as two tectonic plates converged, they produced piles of mountains with complex folds and faults. Geologists recognize that most of these rocks were already hard and brittle when they were broken up. Mount Everest and the entire Himalayan range continue to rise because of this process.

In this image, you can see folded mountains in West MacDonnell National Park, Northern Territory, Australia.

DIVERGENT LANDFORMS

7.8 RIFT VALLEYS

Most divergent zones lie between oceanic tectonic plates, mainly along mid-ocean ridges. One of the dramatic features of these zones is a rift. A *rift* is a series of steep-walled valleys lined with high cliffs rising from faults parallel to the valley. In the center of the rift are low areas where rocks cracked and dropped downward. These are called *grabens*. The long, faulted cliffs parallel to the grabens are prominent features on the ocean bottom.

Though most rift zones are underwater, tension forces have also created a long series of grabens in the East African Rift zone. These run for hundreds of kilometers from the southern end of the Red Sea to the western and eastern margins of the depression that holds Lake Victoria, a large, shallow lake in eastern Africa. Part of the system of rift valleys forms the basin for Lake Tanganyika (Figure 7-5).

graben (GRAH ben): graben (Ger. *graben*—ditch); a steep-sided valley framed by normal faults parallel to the length of the valley

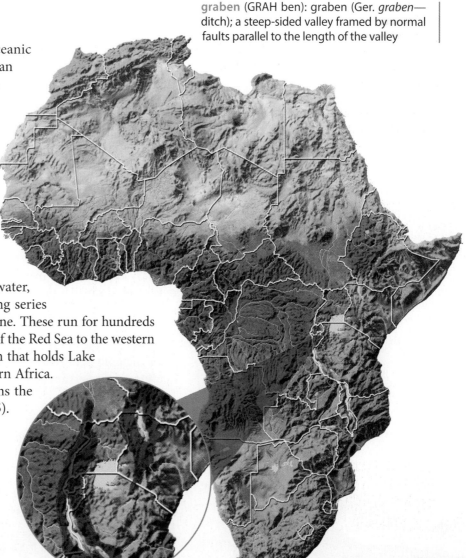

7-5
The system of rift valleys in eastern Africa forms the basins of many of the African Great Lakes.

MOUNTAINS AND HILLS

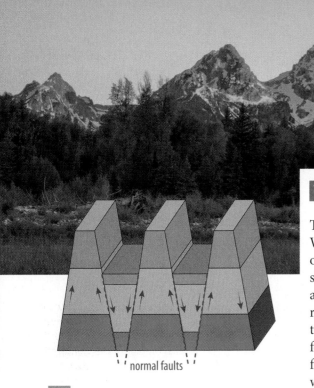

7-6
Fault-block mountains are bounded by at least one fault. A landmass on one side of the fault rises above the landmass on the other side. The Grand Tetons (above) are fault-block mountains.

> A geologic **province** is a region with characteristics that separate it from surrounding terrain.

7.9 FAULT-BLOCK MOUNTAINS

The mountains in the image above are called *fault-block mountains*. What do you notice about them? These are mountains set apart by one or more normal faults. The Sierra Nevada range in California is a classic example of this type. The rocks on the west side of each fault rose and tilted above the level of the rocks to the east. The area of these raised blocks is enormous, measuring about 120 km (75 mi) from west to east and about 640 km (400 mi) from north to south. It appears that forces pushed the eastern end upward about 3 km (2 mi). Other forces from different directions broke the blocks into separate mountains with gentle western slopes and steep eastern cliffs. California's highest peak, Mount Whitney, is part of this range.

The Great Basin and Range **Province** of Nevada and Utah has many fault-block mountains (see Appendix G). A complex system of faults has cut the earth's crust into thousands of blocks, forming many valley basins and mountain ranges. The Wasatch Range that borders the Colorado Plateau is well known in this area.

UPLIFT AND SUBSIDENCE LANDFORMS

7.10 LANDFORMS FROM UPLIFTING

PLATEAUS

Plateaus (pla TOHS) are broad regions of relatively undisturbed sedimentary deposits lifted by some tectonic process. Most old-earth geologists would say that plateaus began by slowly forming under shallow seas over millions of years as sediments collected. Rock strata in a plateau are usually level or gently tilted in one direction. Erosion at its edges reveals the often beautifully striped patterns of rock strata. Plateaus themselves or their erosional remnants often became mountain landforms (see Section 7C).

7-7
Mountain Castildetierra in Bardenas Reales Nature Park, Navarra, Spain

Flood geologists think that many plateaus are remnants of rapid sedimentation during the Flood. Since sediment-laden waters completely covered every continent, it isn't surprising to find sedimentary plateaus all over the world.

Vast, thick lava deposits formed some plateaus. In these locations, so much lava spewed from cracks in the earth that it flooded hundreds of square kilometers. When the lava cooled and hardened into a layer many meters thick, it formed an erosion-resistant plateau.

DOMES

A *dome* is a landform with sedimentary strata that looks like an upside-down bowl. A dome usually forms when molten magma deep underground pushes up on a relatively small area of overlying rock strata. On a geologic map, an eroded dome is a roughly circular or elliptical pattern formed by exposed strata. The rock strata tend to dip away from the middle of the dome. Because of its structure, the inner exposed strata on a map must have been laid down before the outer rings. In South Dakota, the Black Hills seem to be the remnants of a large geologic dome.

7-8

Domes usually form as magma collecting in a chamber deep underground pushes upward (below). Rock in the center of the dome is older than surrounding rock. The Grenville Dome formation in Wyoming is shown on the left.

Location 41.76°N 107.1°W

7.11 LANDFORMS FROM SUBSIDENCE

BASINS

A basin is the opposite of a dome. Most basins seem to have formed when a magma chamber deep underground emptied. Without any support, the overlying rock strata sagged into a bowl-shaped structure. On bedrock maps, eroded basins have outer rings of exposed strata that were placed before the strata that form the inner rings. A basin's strata tend to dip toward the middle of the structure.

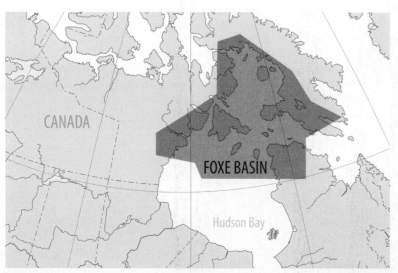

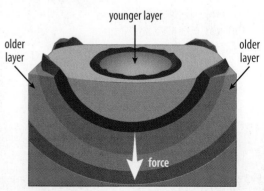

7-9

Basins form as the landform above an empty magma chamber sinks (below). The Foxe Basin (left) covers much of northeastern Canada.

MOUNTAINS AND HILLS 163

VOLCANIC LANDFORMS

7.12 VOLCANOES AND SEAMOUNTS

Some mountain landforms are volcanoes. These mountains form by the eruption of molten rock on the earth's surface from deep tectonic zones all over the world. We will study volcanoes in Chapter 8.

*Seamount*s are extinct volcanoes in ocean basins. There are perhaps millions of these features scattered throughout the ocean basins of the world. You will learn more about seamounts in Chapters 8 and 13.

7B REVIEW QUESTIONS

1. How are tectonic mountains created?
2. What kinds of folds appear in fold mountains?
3. Where are most rift valleys located? Give an example of an exception.
4. How do tectonic forces create fault-block mountains?
5. How are plateaus related to other mountain landforms?
6. Compare how domes and basins look on a geologic bedrock map. Be sure to mention the arrangement and ages of the rock strata.
7. Why are volcanoes and seamounts considered tectonic mountains?
8. (True or False) Through Earth's history, rock strata have piled up and become highly folded along mid-ocean ridges.

7-10
Volcanos are depositional landforms often caused by tectonic action. The volcanoes shown below sit beside Laguna Verde in the Atacama Desert, Bolivia.

7C

NON-TECTONIC HILLS AND MOUNTAINS

What other ways can mountains form?

7.13 EROSION AND DEPOSITION

7C Objectives

After completing this section, you will be able to

» explain in basic terms the processes of erosion and deposition.
» describe the various processes that produce erosional mountains.
» describe related tectonic processes that contribute to some residual landforms.
» discuss the major processes that created depositional mountains.

Land changes. Whether it's the Grand Canyon or your own backyard, land shows signs of wear. Even rocks show signs of wear. Wind, water, ice, and even plants can break down rocks into gravel, sand, and fine silt.

Gravity, flowing water, or wind carries away these particles. This is called *erosion*. Erosion happens all the time, but how fast it happens depends on all kinds of things. Erosion affects mountains, valleys, shorelines, and every other landform. And the particles of rock and soil have to go somewhere.

The ability of wind and water to carry sediment is directly related to the speed of flow. The faster the movement, the more sediment they can carry and the larger the pieces. Fast-flowing water can carry boulders, while slow-moving water can carry only silt.

Under certain conditions, when wind or water carries eroded materials, they drop to the underlying surface. That surface may be the ground, a river or lake bottom, or the ocean floor. These sediments build up over time in a process that geologists call *deposition* or *sedimentation*.

Today, sedimentation is fairly slow. For example, marine geologists estimate that the sedimentation rates for the deep ocean floors are just a few centimeters in 1000 years. Though deposition can quickly happen today in some places, it's usually much slower than it appears to have been in the past. Flood geologists believe that erosion and deposition must have been much faster earlier in the earth's history to produce the landforms that we see today.

Though erosion and deposition are not usually related to tectonic processes, they act on geologic features produced by tectonic forces. Let's look at some other kinds of mountain landforms associated with erosion and deposition.

7-11
Erosional mountains are seen in the mesas of Monument Valley, Arizona/Utah (above). Severe erosion often leaves just pinnacles, as seen in Bryce Canyon, Utah (shown at the bottom of the page).

Bi | Stones of Witness
Since God has given us the record of the Flood and its consequences, how should we respond to the strange beauty of the mountains of Monument Valley and Bryce Canyon?

7.14 EROSIONAL HILLS AND MOUNTAINS

Erosional mountains, or *residual mountains*, are mountains that were carved out by extensive erosion, usually from a plateau. As erosion cut apart, or *dissected* a plateau, some parts of the plateau remained intact. These parts often contained erosion-resistant materials, such as hardened lava. In other cases, the residual parts simply weren't washed away with the rest of the plateau. The Catskill Mountains of southeastern New York and the mountains of the Allegheny Plateau of West Virginia are erosional mountains.

MESAS AND BUTTES

In the western United States, broad, flat-topped hills remaining from the erosion process are called *mesas* (MAY suh). Smaller flat-topped hills are *buttes* (BYOOTS). Both mesas and buttes have steep, cliff-like sides. Sloped piles of broken rock called *talus* (TAY lus) collect at the base of the cliffs. As with mountains and hills, the difference between mesas and buttes is not well defined.

PINNACLES

Pinnacles are interesting erosional remnants. They are tall spires of rock that are all that remain of an eroded plateau. They can be found alone, in groups, or connected together in bladelike formations called *fins*. They typically are capped with an erosion-resistant rock that prevents the pinnacle from being eroded from the top down. Bryce Canyon National Park in Utah is famous for its collection of freestanding pinnacles and fins of orange-red sandstone and mudstone.

Old-earth geologists believe that pinnacles, mesas, and buttes formed long ago during a wetter climate. The land was slowly pushed upward as rivers gradually cut downward into the rock. This model has some problems. First, it assumes that climate didn't change for tens of millions of years to allow river erosion to take place. It also fails to explain why pinnacles, buttes, and mesas exist at all. They should have eroded away over time with the rest of the landscape.

MONADNOCKS

Certain kinds of erosional mountains are all that are left of an ancient magma chamber. Long ago, molten rock pushed its way upward into sedimentary rocks and formed domed chambers. Later, the rock cooled to form very hard, erosion-resistant masses. Then the overlying softer sedimentary rocks eroded away. The solidified magma is all that remains, forming domed rock "islands" standing out of a flat plain surrounding them. These distinctive lone mountains are *monadnocks* (muh NAD nahks). Mount Monadnock in New Hampshire gives the name to these kinds of mountains in the United States. Stone Mountain in Georgia and the Spitzkoppe in Namibia, Africa, are other famous monadnocks.

According to the Flood model, the existence of erosional hills and mountains makes a lot of sense. The geologic evidence from over a large area of the American West points to rapid erosion on a continental scale. Places where the floodwaters deposited thick rock strata early in the Flood probably experienced lots of erosion near the end of the Flood and shortly after it. These sedimentary rocks had not hardened fully, so running water would easily erode them. And there was no shortage of water to do the work!

OTHER REMNANTS OF EROSION

Flood geologists believe that the Flood created many other interesting erosional landforms as it receded. Stone arches, natural bridges, solution caves, and volcanic necks are just a few.

Stone arches are often found along with fins and pinnacles. Geologists believe that water weakened a section of rock at the base of a fin, and some of the fin collapsed, leaving the arch. It is hard to imagine how these could have formed gradually! Fins and arches are clear evidence for rapid erosion.

Volcanic necks or plugs are all that remain of ancient volcanoes. In some ways, they are like monadnocks. As a volcano erupts and builds a mountain, the magma tube to the crater at the top grows longer. If the volcano goes extinct, the magma hardens in the tube. Over time, the rocks that make up the mountain erode away, leaving behind the hardened magma in the neck. It can stand like a tower above the surrounding landscape. In Chapter 8, you'll explore more about volcanoes and their features.

7-12
Towering over the surrounding plains is the granite form of Spitzkoppe—a monadnock—located in Namibia.

Mount Monadnock: 42.861 °N 72.108 °W
Stone Mountain: 33.805 °N 84.145 °W
Spitzkoppe: 21.826 °S 15.194 °E

7-13
Rapidly flowing water forms stone arches by eroding soft rock. While scientists understand the fact that something happened, they can't always explain how it happened.

I can't explain how a river could have eroded this arch.

You know ... I can't explain how the Flood could have either.

7-14
Khongoryn Els dunes in the southern Gobi Desert of Mongolia (below, left). They are known as the Singing Sand Dunes for the sounds they create in high winds.

7-15
These underwater sand dunes near the Bahamas (above, right) are similar to those formed by wind.

7.15 DEPOSITIONAL HILLS AND MOUNTAINS

When sediments build up on land, they can form **depositional mountains** and other landforms. These sediments can be carried by wind or glaciers or deposited by volcanoes.

WIND-FORMED

You've probably seen sand dunes at the beach or any other sandy place. *Sand dunes* are wind-deposited hills of sand. Sand and dust settle out when the wind slows down. Though sand dunes at the beach are small, others can be huge! Dunes in the Sahara Desert can be as high as 430 m (1400 ft) and may cover several square kilometers!

Some geologists believe that in the past wind may have deposited vast stretches of sand that later hardened into sandstone. However, these dunes often include large ripples, which means that they may have formed under flowing water. Oceanographers can see underwater sand dunes today. It is often very difficult to tell the difference between the results of the two processes. If this interpretation is correct, then they could be depositional features of receding floodwaters.

GLACIER-FORMED

Glaciers can produce several types of hills. Remember from Chapter 5 that a glacier is a mass of ice produced when snow builds up and compacts. The compressed ice can flow slowly downhill under the influence of gravity. Wherever a glacier touches the underlying bedrock, it breaks out chunks of rock and grinds them into pebbles, gravel, sand, and fine silt. This crushed material is *glacial till*.

Glacial till is carried along inside a glacier and pushed out to its sides and in front. When a glacier melts, the till drops to the ground, forming distinctive landforms. *Moraines* (moh RAYNS) are long, relatively low ridges of glacial till that form mostly at the front and sides of glaciers. *Eskers* are winding deposits of till that followed meltwater streams as they flowed beneath the glacier. *Kames* are steep-sided piles of till. The material first collected in depressions on top of the glacier before it melted. Finally, *drumlins* are long, low hills that may have been formed as glaciers dragged over piles of till.

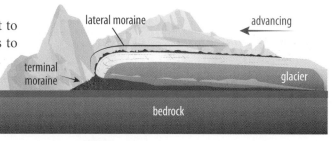

7-16
A terminal moraine is a ridge of rock debris that a glacier piled up at its front.

VOLCANIC

Volcanoes are important depositional mountains found in active tectonic zones. So volcanoes are both tectonic and depositional landforms. Geologists estimate that there are more than a million extinct and active volcanoes all over the world. They grow as various kinds of lava collect on their slopes. In Chapter 8 you'll discover how volcanoes form and shape the land around them.

7C REVIEW QUESTIONS

1. What process affects all rocks after they have formed? What kind of earth materials does this process create?
2. How are sediments moved about?
3. Assume that sediments collect on the deep ocean floor at a rate of 3 cm per 1000 years. How old might the ocean floor be if sediments are 90 cm thick? Is this a reasonable answer?
4. Describe and compare mesas and buttes.
5. How are monadnocks different from most other erosional mountains?
6. What kind of depositional mountains or hills is formed by wind?
7. What is glacial till? Give two ways that glaciers can move till.
8. Why are volcanoes considered depositional mountains?
9. (True or False) Some fossilized sand dunes seen in sandstone formations may have actually formed underwater.

7-17
Eskers and drumlins are deposition landforms. The esker above is located in Wisconsin. The drumlins below are located in Transylvania, Romania.

LIFE CONNECTION: "IN THE ZONE" ON MOUNT KILIMANJARO

Welcome to Kilimanjaro, the highest mountain in Africa and also the highest volcano on Earth. Kilimanjaro is in the country of Tanzania, near its border with Kenya. You have joined a trekking team on the Umbwe route to the Uhuru peak. This is the highest point on the mountain, located on Kibo, one of three volcanic cones. You will be on the mountain for about a week and pass through six different zones, or biomes. (A biome is an area that has a specific type of weather and is home to a specific group of plants and animals.) So pack your down jacket, sunscreen, tent, and crampons. Here we go!

You'll begin at the foot of the mountain, in the cultivated fields of the Chagga people, where the locals farm. They've cleared the forest land, and grasses thrive here in the rich, volcanic soil. The snowy summit of Kilimanjaro almost looks like a mirage from here.

You begin to ascend the mountain and venture into the tropical rain forest. Blue monkeys and colobus monkeys chatter in the treetops. Palms heavy with dripping moss absorb moisture like gigantic sponges. Streams and rivers contribute to the moisture-laden air. Wildflowers are everywhere, and an occasional lion, leopard, giraffe, or buffalo saunters through. Everywhere there is life.

As you continue to climb, you encounter the heath. Temperature and moisture drop, and towering trees give way to grasses and wildflowers. Mice scamper about, collecting seeds and fruit. Elephants wander here to graze.

Heath gives way to moorland, and the vegetation becomes bizarre. Cabbage-like senecio trees and flowering lobeliuses are everywhere here, yet are found nowhere else on Earth. A brilliant red and green bird, the scarlet-tufted malachite sunbird, drinks nectar from the lobeliuses. Rainfall is scarce, and temperatures continue to drop.

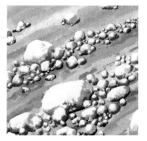

Then you reach the alpine desert. Temperatures fall to freezing at night yet can reach 100 °F during the day. Very little grows here except a few hardy flowers and tussock grasses. This is the country of buzzards, eagles, and vultures. It's getting harder to breathe, as you are pretty high in elevation.

You're almost there! You reach the ice cap, the famous year-round snows of Kilimanjaro. These are special because this mountain is so close to the Equator. Time to break out the down jacket! Lichens are the only living thing here, and if you were to go to the nearby Kibo crater, you could smell the sulphur of a volcano in hibernation.

Congratulations! You've made it to the top of Africa, and you've seen six different biomes along the way.

CHAPTER 7 REVIEW

CHAPTER SUMMARY

» Topography is a description of a region's geologic features.

» Isostasy accounts for the thickness of the lithosphere anywhere on Earth.

» Landforms are features on the earth's surface. Geologists classify the things that they study on the earth's surface into different types of landforms.

» Orogeny is the tectonic process that formed mountains.

» Mountains and hills are both landforms that rise above the surrounding terrain.

» Elevation is measured from sea level. Actual height is measured from the base of the mountain, which may be above or below sea level.

» Mountains can be found by themselves, but they usually are part of a mountain range. Groups of mountain ranges form a mountain system.

» Convergent tectonic zones produced fold mountains.

» Divergent tectonic zones can produce rift valleys walled in by steep, rugged cliffs. Fault-block mountains also result from tension in the crust.

» In the past, tectonic uplifting formed thick plateaus of nearly horizontal sedimentary rock strata.

» Domes are roughly circular or elliptical structures where the rock strata were uplifted in the middle, like an inverted bowl. After erosion, the oldest strata are in the middle, and strata dip away from the center of the formation.

» Basins are roughly circular or elliptical structures where sedimentary rock sagged in the middle. After erosion, the most recent strata are in the center, and strata dip toward the center of the formation.

» Erosion and deposition are normal and important processes that shape and create mountain landforms.

» Plateaus have eroded, creating mesas, buttes, pinnacles, fins, and arches.

» Massive erosion, combined with tectonic and volcanic processes, can create monadnocks and volcanic necks.

» Depositional landforms include sand dunes and glacial features such as moraines, eskers, kames, and drumlins.

» Volcanic mountains are both tectonic and depositional landforms.

Key Terms

Term	Page
topography	153
isostasy	154
landform	154
orogeny	154
mountain	154
hill	154
elevation	155
actual height	155
relief	155
mountain range	156
mountain system	156
fold mountain	159
erosional mountain	166
depositional mountain	168

The Matterhorn is one of the highest peaks in Europe. It is one of the first significant mountains that Brad Washburn climbed.

REVIEW QUESTIONS

1. What do geologists think is causing the Himalaya Mountains to rise?
2. What main principle of geology do geologists use to account for the topography at any given location on Earth?
3. What is the geologic term for tectonic mountain building?
4. Compare mountains and hills.
5. Compare elevation and actual height.
6. Which contain more mountains, mountain systems or mountain ranges? Explain.
7. How do geologists explain the presence of many mountain landforms near the boundaries of tectonic plates?
8. Give an explanation for how rock strata could have been smoothly folded.
9. Compare anticlines, synclines, and monoclines.
10. What is another name for the valley-like formations found in rift systems?
11. Why do geologists associate normal faults with fault-block mountains? Why did these faults form this way?
12. Compare domes and basins.
13. Explain why volcanoes are so different from other mountains.
14. What can erode rock?
15. What agents are mainly responsible for depositing sediment?
16. Mesas are to buttes as mountains are to _____.
17. What seems to be the best explanation for pinnacles and stone arches? Why?
18. What kind of material are nearly all glacial depositional landforms made of?

True or False

19. Topography expresses the differences of elevations of mountains and the differences of depths in a body of water.
20. Mount Everest has the greatest actual height of any mountain in the world.
21. The earth's most extensive mountain system is underwater.
22. Movement of molten magma within the crust can create tectonic forces.
23. A rift valley forms when blocks of rock strata between parallel faults sag, creating cliff-like valley walls.
24. Creationist geologists believe that all plateaus originally formed from Flood sediments.
25. The ages of the oceans can be calculated using the sedimentation rates and the thickness of the sediments.
26. Monadnocks are large masses of hardened magma that first formed under sedimentary rock but were then exposed by erosion.
27. Geologists believe that all "fossilized sand dunes" in sandstone are evidence for wind deposition.

28. Reread the Top of the World story on page 152. Let's compare the 1856 and 1999 heights of Mount Everest and see whether the current rate of rise can be supported by observations over long periods of time.

Make a table on a piece of paper with two columns. In the first column, put "Date," and in the second column, put "Elevation." Consider the 1999 data on Mount Everest to be the most accurate. So put 1999 in the first column and put 29,035 feet in the second column. Assume that Mount Everest gets 0.2 ft (2.4 in.) higher every year.

   a. Let's check this. Put 1856 in the date column. Use the rate of 0.2 ft/y to calculate the elevation of Mount Everest in that year. How does your result compare with what the nineteenth-century English scientists calculated with geometry?

   b. Let's assume that Mount Everest has been rising at the same rate since the Himalayan orogeny began. Using a growth rate of 0.2 ft/y, calculate how long it would have taken for the mountain to grow from sea level to its 1999 height. Is this acceptable within a young-earth time frame? Why or why not?

   c. At this growth rate, calculate how long it would take before Everest would reach the top of the weather layer of the atmosphere, at 6.8 miles, or 35,900 ft. Is this result reasonable?

   d. To answer these last two questions, you had to make an assumption about the rate of Everest's growth over time. This was a hypothetical model. What do your results indicate about the nature of models, especially historical science models?

Mount Etna on the northeast coast of Sicily is the most active volcano on the earth.

CHAPTER 8
VOLCANOES AND VOLCANISM

8A FIRE MOUNTAINS 175
8B CLASSIFYING VOLCANOES 184
8C INTRUSIVE VOLCANISM 192

FIRE AND ICE

In the spring of 2010, fire and ice met. A volcano under the Eyjafjallajökull (AY a fyat la YO kutl) icecap in southern Iceland blew in a series of eruptions. These eruptions released lava flows, floods of meltwater, and an impressive ash cloud that triggered magnificent lightning. When the cold meltwater met the hot lava, the molten rock instantly solidified and shattered. The very fine glass particles moved with the wind in a huge ash cloud over all Europe. This delayed and grounded European air traffic for weeks in what was the most significant disruption of air travel since World War II.

8A

FIRE MOUNTAINS

Why should we study volcanoes?

8.1 WHY STUDY VOLCANOES?

Evacuations, livestock confined to buildings, waterways muddy with volcanic ash, and airports packed with stranded people—these were some of the consequences of Eyjafjallajökull's eruptions in 2010. Thankfully, no one died.

Why are we so fascinated by volcanoes? Maybe it's because they are so dangerous. Volcanoes create devastation, disrupt our lives, and kill people. This is why we need to study them—knowing about volcanoes can save lives. There were no deaths in the Eyjafjallajökull eruptions because local scientists had carefully monitored the area and made accurate predictions. Airlines avoided damage to aircraft engines and possible crashes because of atmospheric models that showed where the ash cloud would drift.

8.2 WHAT IS A VOLCANO?

A **volcano** is a depositional mountain landform built from solidified magma from deep in the earth. Our English word *volcano* comes from Vulcano Island (*Isola Vulcano*), a now-dormant volcano in the Mediterranean Sea. The ancient Romans named this island for Vulcan, their god of fire. They believed that the volcano was the chimney to his workshop and that the ash accumulated as he cleaned his forge from time to time.

Volcanoes are made of igneous rocks. The word **igneous** in geology refers to anything having to do with molten earth materials, including the rocks and minerals that come from them. Because volcanoes form on top of the earth's crust, they are a type of **extrusive volcanism**. Processes that form igneous structures related to volcanoes inside the crust are *intrusive volcanism*. You will learn about intrusive volcanism later in this chapter.

8A Objectives

After completing this section, you will be able to

» explain how earth science helps reduce the risks of natural hazards like volcanoes.

» describe the structure of a volcano.

» list and describe the main kinds of volcanic emissions.

» associate the locations of extrusive igneous features around the world with tectonically active regions.

» infer from the global distribution of volcanoes the amount of volcano activity during the Genesis Flood.

Eyjafjallajökull (AY a fyat la YO kutl): eyja- (Ic. *Eyja*—island) + -fjalla- (Ic. *Fjälla*—mountain) + -jökull (Ic. *jökull*—glacier); island-mountain glacier

Vulcano Island: 38.40°N 14.96°E

extrusive (from extrude—to thrust out)

8-1
The glacier on top of the Eyjafjallajökull volcano melted as it erupted, sending floods of water and debris into the valleys below.

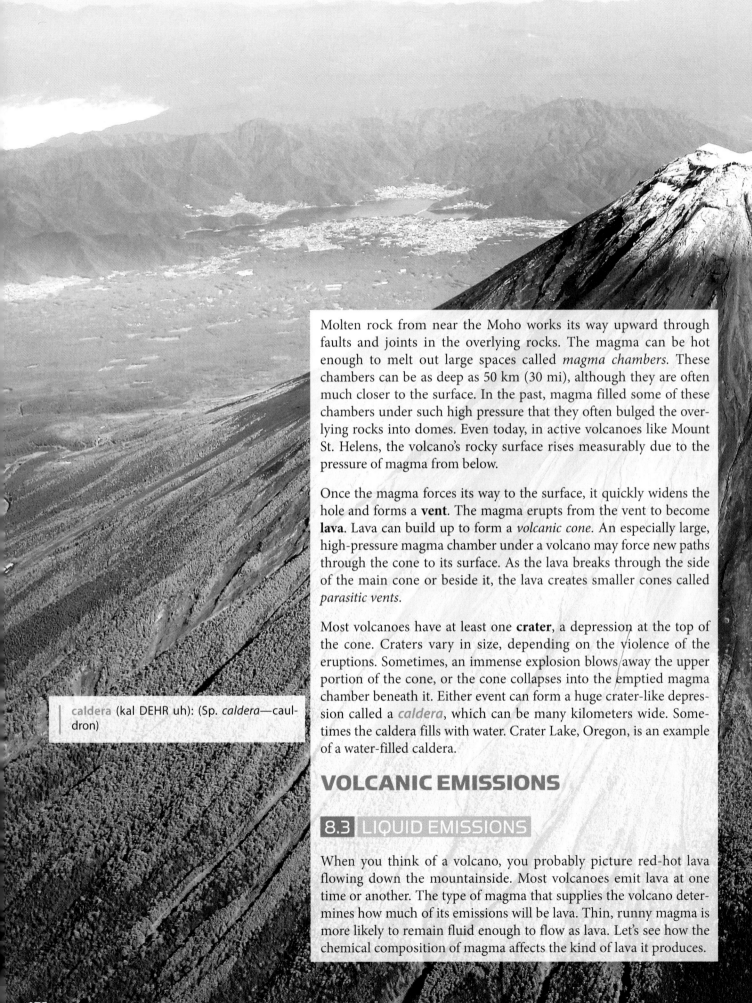

Molten rock from near the Moho works its way upward through faults and joints in the overlying rocks. The magma can be hot enough to melt out large spaces called *magma chambers*. These chambers can be as deep as 50 km (30 mi), although they are often much closer to the surface. In the past, magma filled some of these chambers under such high pressure that they often bulged the overlying rocks into domes. Even today, in active volcanoes like Mount St. Helens, the volcano's rocky surface rises measurably due to the pressure of magma from below.

Once the magma forces its way to the surface, it quickly widens the hole and forms a **vent**. The magma erupts from the vent to become **lava**. Lava can build up to form a *volcanic cone*. An especially large, high-pressure magma chamber under a volcano may force new paths through the cone to its surface. As the lava breaks through the side of the main cone or beside it, the lava creates smaller cones called *parasitic vents*.

Most volcanoes have at least one **crater**, a depression at the top of the cone. Craters vary in size, depending on the violence of the eruptions. Sometimes, an immense explosion blows away the upper portion of the cone, or the cone collapses into the emptied magma chamber beneath it. Either event can form a huge crater-like depression called a *caldera*, which can be many kilometers wide. Sometimes the caldera fills with water. Crater Lake, Oregon, is an example of a water-filled caldera.

VOLCANIC EMISSIONS

8.3 LIQUID EMISSIONS

When you think of a volcano, you probably picture red-hot lava flowing down the mountainside. Most volcanoes emit lava at one time or another. The type of magma that supplies the volcano determines how much of its emissions will be lava. Thin, runny magma is more likely to remain fluid enough to flow as lava. Let's see how the chemical composition of magma affects the kind of lava it produces.

caldera (kal DEHR uh): (Sp. *caldera*—cauldron)

There are three main types of magmas. All contain an important compound called *silica*. Silica, or silicon dioxide, is made of the elements silicon and oxygen. Geologists classify magmas mainly by how much silica they contain. Generally, magma with a high percentage of silica is thicker and flows less easily. In other words, it is more **viscous**.

A **viscous** liquid resists flowing. Molasses has a higher viscosity than water.

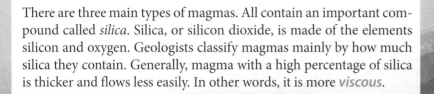

Rhyolitic magma contains more than 63% silica. It forms the hard *rhyolitic* and *dacite* rocks. They are lighter in color and may contain many light-colored mineral crystals. They fall under the felsic chemical category of rocks. Rhyolitic lava is liquid at temperatures as low as 650–750 °C (1200–1380 °F), making it the most viscous type of lava.

Andesitic magma contains 52–63% silica. It forms *andesitic* rock when cool. This rock is usually dark in appearance, with a fine-to-coarse texture. Andesitic lava is liquid at temperatures lower than in basaltic lava, 750–950 °C (1380–1740 °F), so it is less viscous.

Basaltic magma contains less than 52% silica. When cool, it forms *basalt*, a common volcanic rock. Geologists also call it *mafic magma*, which is a chemical classification. Basaltic lava is runny at high temperatures, 950–1200 °C (1740–2200 °F), but it is still more viscous than water.

177

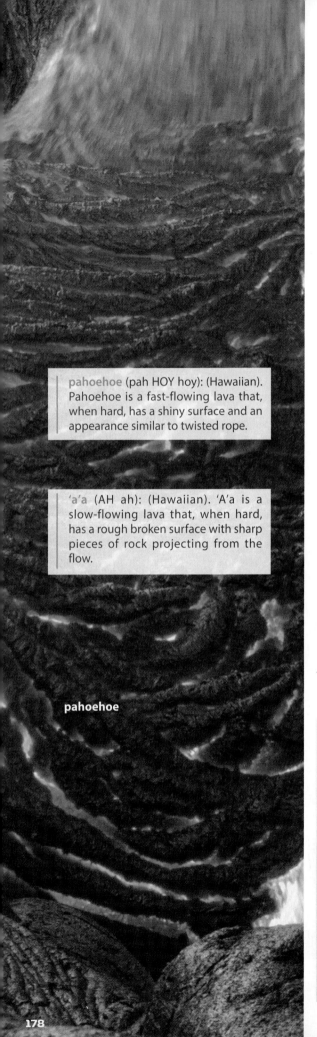

pahoehoe (pah HOY hoy): (Hawaiian). Pahoehoe is a fast-flowing lava that, when hard, has a shiny surface and an appearance similar to twisted rope.

'a'a (AH ah): (Hawaiian). 'A'a is a slow-flowing lava that, when hard, has a rough broken surface with sharp pieces of rock projecting from the flow.

Although lava looks dangerous, it causes relatively few deaths. Lava usually doesn't flow faster than a person can walk, so it's easy to avoid. Lava is far more damaging to fixed structures like buildings and roads. Volcanic lava flows have destroyed entire cities, even in the recent past.

Rhyolitic lavas tend to be lighter in color and very thick. They flow like cold molasses or drying concrete, with speeds measured in meters per day.

How lava solidifies depends on what it is made of and how it cools. Runny basaltic lavas flow great distances at speeds up to 30 km/h (19 mi/h) near the vent. As this kind of lava cools farther from the vent, it forms a skin that wrinkles while the still-molten lava inside continues to flow. The "skin" greatly slows the lava movement down to walking speed. Lava with a smoothly wrinkled, ropey surface is called *pahoehoe*.

Another kind of basaltic lava flow moves more slowly, forming a splintered, sharply fractured surface. This kind of lava is called *'a'a*. Gases released from 'a'a lava often form large, bubble-like voids and sharp edges.

If lava erupts underwater, the outside hardens immediately to form a shell. Then the melted lava inside bursts through the shell to form a new shell. This process repeats, forming what looks like a stack of pillows, earning for itself the name *pillow lava*. Because pillow lava forms only underwater, geologists use this kind of lava formation to identify where volcanoes erupted underwater in the past.

8.4 GASEOUS EMISSIONS

But volcanoes produce other substances besides lava. Gases and solids also come from within the earth with volcanic eruptions.

Magma contains hot gases dissolved in it. These gases separate from the magma as it approaches the volcano's vent, just like a bottle of a soft drink fizzes when you take the cap off. High-pressure gas bubbles form. Gases exploding from the vent are often the first sign that a volcano is going to blow.

Sometimes a volcano erupts a mixture of hot solid particles suspended in water vapor or other gases. This hot mixture, called a *glowing avalanche*, is so dense that it flows down the volcano's slope instead of rising into the air. At night, the entire flow glows a dull red. Also called a *pyroclastic flow*, these emissions can move at speeds up to 150 km/h (90 mi/h) and contain gases at temperatures between 100 and 800 °C (212–500 °F). They suffocate and burn everything in their path. This type of eruption from Mount Vesuvius (vuh SOO vee us) destroyed the Roman cities of Herculaneum and Pompeii in AD 79. In 1997, 20 people died in a glowing avalanche from the Soufriere Hills volcano on the Caribbean island of Montserrat (MOH seh AH).

Volcanoes emit large quantities of steam and water vapor. Although these emissions are usually harmless, they can have tragic consequences on a snow-capped volcano. The 1985 eruption of Nevado del Ruiz in Colombia was mostly steam, but it melted some of its heavy snow cover. The melted snow caused a volcanic mudslide, called a *lahar*, that buried about 23,000 people. A lahar is like a flow of concrete and can contain large boulders and rocks. Heavy rains or the collapsing wall of a crater lake can also cause lahars.

Fumaroles are vents in the ground where steam and other gases from volcanic activity escape. Carbon dioxide leaking from fumaroles can be dangerous. Since it is about 1.5 times denser than air, it collects in depressions in the ground and displaces oxygen. Death Gulch in Yellowstone National Park is such a place. On calm days, grizzly bears and other animals have suffocated there.

8-2
A pyroclastic flow produced by superheated gases carrying volcanic debris downhill

lahar (lah HAR): (Jav. *lahar*—a mudflow of volcanic ash mixed with water)

fumarole (FYOO muh ROHL): (L. *fumariolum*—smoke chamber)

8-3
Fumaroles on Fourpeaked Volcano in Alaska

VOLCANOES AND VOLCANISM **179**

8.5 SOLID EMISSIONS

When volcanoes erupt, they can also emit ash, cinders, bombs, and sometimes large chunks of rock. This solidified lava is called **pyroclastic material**. It ranges in size from microscopic dust to chunks weighing many tons.

pyroclastic (PIE row CLASS tic): pyro- (Gk. *pyr*—fire) + -clastic (Gk. *klastikos*—broken); referring to small, sharp particles emitted from a volcano

When gas-laden lava explodes in the volcano's vent, the fine mist of solid lava falls as *ash*. Volcanic ash is made of tiny angular glassy fragments and resembles the dusty ash that comes from burning wood. These particles are between microscopic dust and about 2 mm (0.1 in.) in size. Clouds of ash can travel all the way around the world and endanger aircraft that fly through them (see Chapter opener).

LIFE CONNECTION: THEY CAN TAKE THE HEAT

A black smoker. Notice the abundant life found in this inhospitable place.

Have you ever scalded your foot in a bathtub? Plumbers warn that hot water heaters should be set at or below 50 °C (120 °F) to avoid painful burns. Now, imagine a creature who lives in water that is 80 °C (180 °F)! The water that this animal lives in is not in a bathtub but in the inky blackness of the Pacific Ocean, more than 2 km (1.2 mi) below the surface.

The 10 cm (4 in.) Pompeii worm secretes a papery tube from which it extends its head and tentacles to feed and breathe. The worms live in colonies attached to black smokers. No, we're not talking about the fancy grill in your neighbor's yard!

Black smokers are deep-ocean hydrothermal vents where scalding water shoots up through rifts in the crust in mid-ocean ridges. When the hot, mineral-rich water mixes with the surrounding cold seawater, dissolved minerals precipitate out of the solution. Over time, these deposits form the chimney-like smokers.

The habitat surrounding a black smoker is one of the most bizarre known to man. Since absolutely no sunlight penetrates to this depth, plant photosynthesis can't happen. Instead, special bacteria absorb and process minerals from the water. These bacteria are called *extremophiles* (ex TREEM oh files), since they survive in such extreme conditions. Extremophiles form the base of the food chain that supports clams, snails, and even worms like the Pompeii worm.

Although not discovered until the early 1980s, the Pompeii worm has been the subject of several studies. Remarkably, the worm's head temperature is much cooler (22 °C, or 72 °F) than its body. Some scientists believe that the key to this fragile worm's survival is a unique relationship with bacteria. A shaggy gray coat of bacteria covers the worm's body. It seems that the bacteria feed on mucus secreted by the worm. The bacteria appear to insulate the worm against the scalding blasts of water that would otherwise cook it in its own home. Further studies suggest that the bacteria may also be a food source for the worms.

Cinders are like ash but are larger, about 2 to 64 mm (0.1 to 2.5 in.) in diameter. Larger blobs of lava fall as blocks. Blocks range from baseball-size to the size of a house. *Bombs* are lava blobs and blocks larger than 64 mm that solidify in a streamlined shape as they fly through the air. They usually fall within 3 km (2 mi) of the vent, but the other kinds of pyroclastics can travel much farther.

Tephra is a layer of loose pyroclastic materials and ash covering the ground. Most of the people who died in the 1980 eruption of Mount St. Helens suffocated in volcanic ash. Heavy tephra is a common cause of death from volcanoes and of damage to buildings and vehicles.

VOLCANO GEOGRAPHY

8.6 VOLCANO LOCATIONS

Scientists think that there are about 1500 active and dormant volcanoes on land. Counting the extinct volcanoes, including those on the ocean floor, brings the total to many millions! Most active volcanoes and many extinct ones are located in two volcano belts, the Alpine-Himalayan and the Circum-Pacific belts.

The *Alpine-Himalayan volcano belt* lies above the collision boundary of the African, Indo-Australian, and southern Eurasian plates. Figure 8-4 below shows the locations of volcanoes in this belt.

The *Circum-Pacific volcano belt* surrounds most of the Pacific Ocean. This is where the tectonic plates that make up the Pacific basin are subducting beneath adjacent continental plates. Recall from Chapter 5 that geologists believe that subduction produces great amounts of magma as the subducted plate melts in the mantle. Chains of volcanoes along the margins of continents and volcanic island arcs are the result of these processes. For this reason, the Circum-Pacific belt is sometimes called the "Ring of Fire." See Figure 8-4 below.

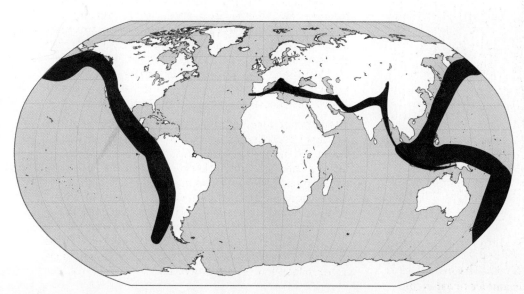

8-4
This world map shows the Alpine-Himalayan volcano belt (blue), which extends along the southern edge of the Eurasian tectonic plate, and the Circum-Pacific volcano belt (red).

Active volcanoes result from tectonic activity in other places too. There are lots of submarine volcanoes and volcanic islands in the mid-ocean rift zones where the sea floor is spreading. Iceland and the Azores are volcanic islands of the Mid-Atlantic Ridge mountain system. Volcanic Easter Island in the eastern Pacific Ocean sits near the mid-ocean ridge system there.

Some volcanic islands rise from the middle of tectonic plates rather than at their edges. Most geologists think that hot spots welling up in the mantle melted the overlying crust. The resulting magma created volcanoes as the thin oceanic plates moved over the hot spots (or the hot spots moved under the plates). This process resulted in lines of volcanic islands. The Hawaiian Islands and the Emperor Seamounts are a long string of volcanoes that may have formed this way. They are located in the northwestern portion of the Pacific Plate.

Flood geologists agree that nearly all volcano landforms probably came into existence during the Genesis Flood or shortly after. The great number of these features confirms the unimaginable devastation of the Flood event. It wasn't just heavy rain and wind. It was the end of the world. Just one smallish volcano like Eyjafjallajökull affected most of Western Europe for several weeks. Imagine conditions on Earth with thousands of large volcanoes erupting at the same time for many years!

Fi Easter Island: 27.1°S 109.33°W

Bi **Stones of Witness**

State a hypothesis that is based on a biblical worldview of geologic history that could explain why there are so many more extinct volcanoes than active ones.

8-5

Easter Island—a volcanic island—is famous for the stone statues that encircle the island. In the background, you can see Puakatike Volcano.

8.7 VOLCANO MAPS

You can easily spot volcanoes on maps, especially topographical maps. Since most volcanoes are conical, the elevation contours often appear regularly spaced and encircle the cone. A deep crater or a caldera really stands out because its interior is lower than the rim. Cartographers indicate this with hatch marks along the contour. See an example map in Figure 8-6.

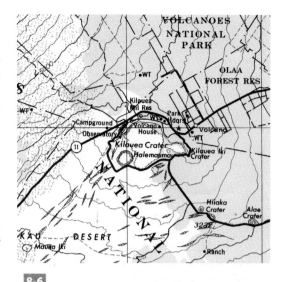

8-6
Notice the hatched contour lines outlining the cone and indicating the depression of the Kilauea Crater in Hawaii.

8.8 FLOOD BASALT ZONES

Thick layers of igneous rocks cover many places in the world, including the ocean floor. These regions are called **flood basalts**, or *basalt traps*. Flood basalts can be thousands of meters thick.

But how did they get there? Flood basalts are not just lava fields from individual volcanoes. Old-earth geologists believe that many of these basalt traps formed as the ancient supercontinent, Pangaea, broke apart about 180 million years ago. As the crust cracked, huge volumes of magma from the mantle flowed upward through the cracks and covered the surface of the crust. The largest known continental flood basalt zone is the Mackenzie Large Igneous Province. This zone extends from northern Canada southward to Lake Superior. It includes portions of the Canadian Shield (see Figure 8-7) and exceeds 2.7 million square kilometers (1 million square miles)!

Flood geologists tend to agree about the processes that formed the basalt traps but suggest that they formed recently and rapidly during the Flood as part of the tectonic activity that you learned about in Chapter 5. The size and number of these formations all over the globe indicate, again, the extreme conditions that probably existed on Earth during and after the Flood. The Flood was God's judgment on man's sin. The extent and completeness of the Flood's devastation by water, volcanism, and tectonics reveals how completely God hates sin. God often waits for a long time before judging His image bearers, giving them opportunity for repentance. But He *will* judge them. And His judgment is thorough and complete.

The Canadian Shield is a very broad region of surface igneous rocks and granite bedrock that covers Eastern and Central Canada and the North-Central United States. Old-earth geologists believe that it contains some of the oldest rocks on Earth.

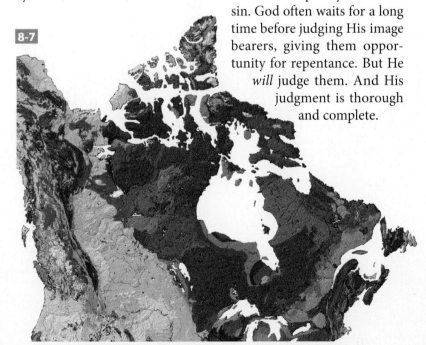

8-7

8A REVIEW QUESTIONS

1. How did earth scientists help prevent deaths from the 2010 Eyjafjallajökull volcano eruptions? Why is this important?
2. Which kind of igneous structure is a volcano?
3. Describe the general features of a volcano.
4. Name the three main types of volcanic emissions. Briefly describe or list the most common forms of each type.
5. Why do we find volcanoes near active tectonic zones?
6. What are the main volcano belts in the world? Briefly describe their geographic extent.
7. What do the number and extent of volcanoes and flood basalt zones indicate about conditions during the Flood?
8. (True or False) A glowing avalanche is a downhill rush of hot volcanic gases carrying volcanic rocks and cinders.

8B

8B Objectives

After completing this section, you will be able to

» identify and categorize volcanoes by their shape and composition.
» infer the activity of a volcano on the basis of its eruption history and seismic activity.
» analyze the definition of volcanic activity from both young-earth and old-earth viewpoints.
» classify the destructiveness of a volcanic eruption on the basis of the Volcanic Explosivity Index.

CLASSIFYING VOLCANOES

What are the different types of volcanoes?

If you were a volcanologist, how would you classify volcanoes? Would you group them by their shape and structure? Or maybe how often they erupted? Or perhaps how explosive they were? Real volcanologists do all of these things. They classify volcanoes by *structure*, *activity*, and *explosivity*.

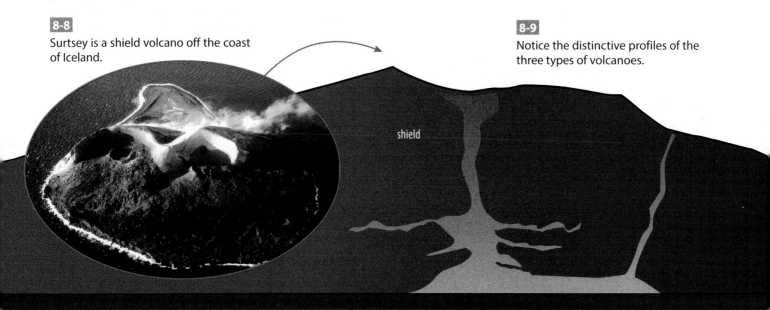

8-8 Surtsey is a shield volcano off the coast of Iceland.

8-9 Notice the distinctive profiles of the three types of volcanoes.

Distance (150 km)

8.9 STRUCTURE

A volcano's emissions determine its shape and composition. Some volcanoes emit mostly lava, while others give off mostly ash and cinders. Some release lava and ash in alternating eruptions.

Quiet eruptions of mostly runny, basaltic lavas form **shield volcanoes**. Like a broad shield, they are a flat, dome shape with gently sloping sides. The Hawaiian Islands are excellent examples of shield volcanoes. Mauna Loa has a greater area within its base on the sea floor than any other volcano in the world. Nearby Mauna Kea holds the world's record for actual height. From its underwater base, it stands over 10,000 m (33,000 ft) high. Surtsey, a shield volcano located south of Iceland, formed a new island in 1963. Today, the island supports plant life, and animals have migrated to the island. This demonstrates that islands don't have to take a long time to form and to become populated with life.

Mauna Loa: 19.47°N 155.6°W
Surtsey (SURT say): 63.302°N 20.603°W
Parícutin: 19.493°N 102.251°W

Eruptions of bursts of gases that eject mostly pyroclastic materials (ashes, cinders, and bombs) form **cinder cones**. These cones consist primarily of gravel-size bits of tephra. Most cones are small and have relatively steep slopes because the cinders interlock with each other. This prevents landslides of volcanic material. One short eruption is enough to produce a cinder cone. Parícutin is a famous Mexican cinder cone that first erupted on February 20, 1943. It started as a small smoking crack in a farmer's cornfield as the farmer and his son were plowing. It quickly increased in size and people could see it from a great distance within a month. After about a year, it had reached a height of nearly 340 m (1120 ft). The volcano erupted off and on for nine years. It has been quiet since 1952.

8-10
Mount Rainier (above) is an example of a composite volcano. Below is Eves Cone, a cinder cone in Mount Edziza Plateau, British Columbia, Canada.

Eruptions that alternately eject lava and pyroclastic materials form *composite cones* or **stratovolcanoes**. These names identify the layered structure of the cones. Quiet eruptions produce the lava layers. Gaseous, explosive eruptions produce the layers of tephra. Types of eruptions often, but do not always, alternate with each other. Stratovolcano slopes are generally steeper than shield volcanoes but gentler than cinder cones. Usually much larger than cinder cones, they also have the symmetrical cone shape that people associate with volcanoes. Most of Earth's volcanoes are stratovolcanoes. Mount Fuji in Japan, Mount Etna in Sicily, Mount Vesuvius in Italy, Mounts Rainier and St. Helens in Washington, and Mount Mayon in the Philippines are well-known composite volcanoes.

Mount Stromboli is one of three active volcanoes in Italy.

Fi Mount Stromboli: 38.79°N 15.21°E

Historical times can vary from place to place. Not all places have been inhabited or even explored for the same amount of time. The Mediterranean has probably been known for more than 5000 years. But the US Pacific Northwest has been inhabited by people who could have kept records of volcanic activity for less than 300 years.

8-11

Oregon's Crater Lake is set in the caldera of an extinct volcano, Mount Mazama.

Location: 42.94°N 122.10°W

8.10 HISTORICAL ACTIVITY

A volcano's activity is difficult to define. Even volcanologists have different ideas about how to classify volcanoes depending on when they have erupted. The United States Geological Survey (USGS) defines volcanic activity as follows:

» **Active volcano**—currently erupting or has been observed to erupt in historical times. Even without observed eruptions, the USGS considers related seismic and volcanic activity as significant. Earthquakes under the volcano, nearby fumaroles and hot springs, and swelling of the mountain are all taken as signs that a volcano is still active and could erupt again.

» **Dormant volcano**—an inactive but uneroded volcano. Some occasional deep earthquakes may occur, indicating a future eruption might be possible.

» **Extinct volcano**—a volcano that shows no signs of seismic or volcanic activity. It is often heavily eroded, even to the point of leaving only a volcanic stack or neck. Volcanologists are confident that these volcanoes will never erupt again, though they can never be sure.

Active volcanoes include Mauna Loa in Hawaii, Mount Vesuvius in Italy, and Mount St. Helens in Washington. Mount Stromboli in the Aeolian Islands of Italy has been erupting as often as several times each hour for at least 2000 years. Such volcanoes are called *continuous volcanoes*.

Mount Shasta in California and Mount Rainier in Washington are both dormant. Geologists classify Mauna Kea in Hawaii as dormant, and astronomers are very confident that Mauna Kea is dormant, or actually extinct. They have even built some of the world's largest and most expensive telescopes on its summit.

Although the terms *active*, *dormant*, and *extinct* help scientists classify volcanoes, these can be misleading. The uninformed may believe that dormant or extinct volcanoes are not dangerous. However, a volcano formerly considered dormant or extinct *can* erupt disastrously. **Mount Vesuvius** had been dormant for nearly 300 years when it violently erupted in AD 79. The eruption wiped out two cities built near its base because their citizens assumed that the volcano posed no danger. Similarly, Mount St. Helens had not erupted for over a hundred years before its 1980 eruption. Nearly any volcano can erupt at any time. Active volcanoes are simply more likely to erupt—their recent history suggests it. A volcano is not truly extinct until its magma chamber has been exposed by erosion.

Mount Vesuvius: 40.82°N 14.43°E

(continued on page 189)

WORLDVIEW SLEUTHING: MOUNT ST. HELENS

Many in the scientific community hold to a worldview that is based on long geologic ages. The reason for this is that most processes are slow and take long periods to produce the results we see. The difficulty is that we rarely get the opportunity to watch cataclysms in action. That changed on May 18, 1980, as many geologists got a front row seat for a volcanic eruption and its aftermath.

INTRODUCTION

You are a college student studying geology. A friend has asked you for an article for the school paper related to your field of study. As a Christian, you want the opportunity to share geology from a Biblical worldview.

TASK

You need to produce an article for the school paper, with the length determined by your teacher. The focus for your article should be the eruption itself, the relation of the eruption to Creation, and how the area has recovered since the eruption. The topic—what we learned from the Mount St. Helens eruption on May 18, 1980.

PROCEDURE

1. Research the eruption by doing an internet keyword search for "Mount St. Helens Eruption."
2. Go to the Answers in Genesis website. Research the relation of the eruption to Creation by doing a keyword search on their website for "Mount St. Helens Origins."
3. Go to the Answers in Genesis website. Research lessons learned from the eruption by doing a keyword search on their website for "Mount St. Helens lessons learned."
4. Complete your article, being careful to give proper credit to your sources.

CONCLUSION

Mount St. Helens had been silent for over 120 years, which had convinced many that the mountain was not a threat. The volcano awoke in March of 1980, so looking back, geologists had two months before the actual eruption to observe and study its behavior. The eruption itself was devastating, blasting 1300 feet off the top of the mountain, and gave geologists an opportunity to see firsthand the destructive force of a volcano. With this knowledge, geologists will be better able to help others through forecasting volcanoes.

These photos were taken before and after the May 1980 eruption of Mount St. Helens. In the left image, you can see a bulge (arrow), which created great concern for scientists. The right photo shows what remained in the aftermath of the eruption. Notice the large caldera from where the top of the mountain was blown away.

MOUNT TAMBORA: THE MOST POWERFUL ERUPTION IN HISTORY

About two hundred years ago, 140,000 native people and European colonists lived on the island of Sumbawa in what is now Indonesia. On a peninsula of this island there was a 4000 m (13,000 ft) volcano named Mount Tambora. Mount Tambora had not erupted in about 1100 years, according to carbon-14 dating. The 12,000 people living at the foot of this huge mountain felt secure. In the fifty years prior to this time, three volcanoes on nearby islands had erupted very explosively. The people knew what devastation an erupting volcano could bring, but Mount Tambora remained quiet. However, in the spring of 1814, Tambora awoke. Nearby residents noticed a few mild bursts of volcanic ash. Then all was quiet for a few months.

In February 1815, Mount Tambora sent out several more spurts of volcanic ash. Then, on April 5, a new vent opened on the north side at the summit, spewing out a huge spray of ash and cinders over the Flores Sea. A series of earthquakes sent shocks throughout the group of islands. About 320 km (200 mi) to the north, the captain of a British East India Company warship, the *Benares*, heard the commotion and thought it was a naval battle, perhaps between pirates and British shipping vessels just over the horizon. He was ordered to go investigate. The ship found nothing, of course.

Over the next several days, ash in the sky reduced the daylight to a gray overcast. The Europeans on Java to the west and in a settlement on the east coast of Sumbawa became concerned that a major volcanic event was taking place. However, they believed that it was one of the other volcanoes that had recently erupted.

Starting on the evening of April 10 and continuing through the early morning hours of April 12, a 4000 km (2500 mi) length of the East Indian island chain rocked with heavy, repetitive shocks. Mount Tambora erupted in a series of mighty blasts that produced immense pyroclastic flows, destroying the villages on its flanks. A column of ash and cinders rose over 45 km (28 mi) and covered the islands to the west. Volcanic bombs ranging in size from walnuts to double fists rained down on the village of Sangir, less than 40 km (25 mi) from the volcano and in sight of the summit. As the peak collapsed, it produced ocean surges that swamped Sangir with walls of water more than 4 m (13 ft) high. Later, blasts of air leveled trees and buildings in the battered village.

By the end of the second day, the few survivors witnessed a sky that was dark as night, caused by the ash that filled the sky. The volcano continued to produce huge explosions that finally subsided during the night of April 11. The next day, floating pumice (sponge-like volcanic rock) up to two feet thick choked the ocean for miles to the west of the island, making sea travel difficult.

Volcanologists estimated that Mount Tambora lost more than 1000 m (3300 ft) of height through explosions and collapse. A caldera 6 km (4 mi) in diameter formed where the summit had been. Tephra up to two feet thick settled on buildings for miles around, crushing them and their occupants. By the morning of April 12, some 10,000 people had died as a direct result of the eruption. Only fifty who stayed on the peninsula, or who fled by canoe after the eruptions began, survived. In the weeks and months that followed, with crops buried, famine took thousands of lives on Sumbawa, and an estimated 37,000 lives on Lombok Island to the west of Sumbawa. In total, Mount Tambora's eruption killed more than 87,000 people.

Wind currents distributed the ash around the world, reducing the amount of sunlight that reached the ground in the Northern Hemisphere. As a result, the winter of 1815–16 was bitterly cold. The 1816 summer growing season throughout the Northern Hemisphere was very short, or there was none

MOUNT TAMBORA

at all. Food supplies failed, and thousands of additional people died of disease and starvation.

Terrible winter-like storms raged all year long. One particular storm especially impressed novelist Mary Godwin (soon to become Mary Shelley) at Lake Geneva, Switzerland. Her description became the "dark and stormy night" setting for her novel *Frankenstein*, or *The Modern Prometheus*.

Mount Tambora's 1815 eruption was rated at VEI 7. It is one of only four eruptions in recorded history of this intensity. Other eruptions such as the 1883 explosion of Mount Krakatau or the 1980 eruption of Mount St. Helens are more famous. None, however, matches or exceeds Mount Tambora for sheer suddenness, shortness of duration, and destruction. Yet, as awe-inspiring as that event was, it cannot compare to the cataclysms that God unleashed with the Flood.

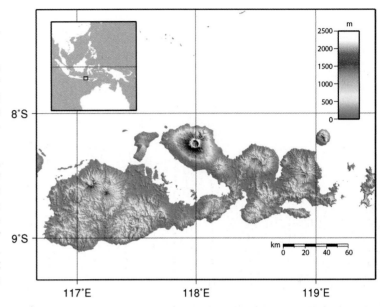

Young-earth geologists look at volcano activity differently than those who hold old-earth views. Most if not all of the volcanoes we see today probably erupted during and after the Flood. We believe that the Flood happened about 5500 years ago. By the USGS definition, virtually all volcanoes could be considered active, since humans have always existed since Creation. It is probably safer to classify volcanoes by seismic activity and other signs of volcanism than by when they were last observed to erupt.

8.11 EXPLOSIVITY

Volcanoes are exciting because they explode. But volcanic explosions are also much more dangerous than lava flows because they happen faster and are less predictable. Explosions often destroy the volcano. Mount St. Helens lost 400 m (1300 ft) of its height in its 1980 explosion. Volcanic islands often disappear beneath the ocean after explosive eruptions. Volcanic explosions can be very loud. People heard the 1883 explosion of Krakatau, Indonesia, nearly 4800 km (3000 mi) away. Instruments detected the explosion's pressure wave on the opposite side of the globe. That wave circled the globe repeatedly for five days!

Just as with earthquakes, scientists classify volcanoes by the intensity and destructiveness of their explosions. They look at how much energy an eruption released and how it has affected human populations. Probably the best indicator of a volcano's destructiveness is its **Volcanic Explosivity Index (VEI)**. This index, an open-ended scale starting at 0, is an estimate of the explosive power in a volcanic eruption. Explosivity is based on the amount of tephra that a volcano produces, the height of the eruption cloud, and a descriptive estimate of the explosion.

Warnings Unheeded

The AD 79 eruption of Mount Vesuvius and the 1980 eruption of Mount St. Helens were both of the same magnitude. But while the number of deaths due to Vesuvius was likely many thousands, only fifty-seven people are known to have died during the Mount St. Helens eruption.

However, in both cases, people failed to heed the warning signs of impending eruption. People in the towns near Vesuvius had many days of seismic activity and even small eruptions but tried to flee only when the main eruption began. Many of the St. Helens victims ignored or disobeyed the USGS restrictions and warnings. Thankfully, in the latter case, modern technology provided information that likely saved hundreds of lives.

If an eruption produces less than 10,000 m³ of tephra, it's rated a 0 on the VEI. A VEI 1 eruption blows off from 10,000 to 1,000,000 m³ of debris. This is equal to 6 to 15 times the volume of the Lincoln Memorial building in Washington DC. A VEI 2 eruption is 100 times more explosive than a VEI 1. Beginning at VEI 3, higher explosivities are 10 times more powerful than the next smaller number on the scale. So, for example, a VEI 3 eruption is 10 times stronger than a VEI 2. The most powerful volcanic explosions ever observed by people were VEI 7. Such eruptions produce more than *100 km³* of tephra! Geologists believe that there were more than forty eruptions in prehistoric times that rated a VEI 8 or greater. These probably occurred during or shortly after the Flood and added to the catastrophe. They were truly devastating events with global effects.

8-12
The Yellowstone Caldera is believed to be the remnant of one of the largest volcanic explosions ever.

SERVING GOD AS A VOLCANOLOGIST

Can you imagine visiting an erupting volcano? The crunch of sharp tephra under heat-resistant boots, the sulfur stench in the air, the waves of tremendous heat, and the dance of fountains of lava fascinate the scientists who study them. They are volcanologists.

Maurice (top) and Katia (bottom) Krafft were a French couple with a passion for volcanology. They began visiting volcanoes at their own expense. Then they started filming them, capturing the imagination of the scientific world. They were often the first volcanologists to arrive at an erupting volcano. Sadly, they were among the forty-three people killed in 1991 during a pyroclastic flow at Mount Unzen in Japan.

JOB DESCRIPTION

Volcanologists are like detectives. They visit a volcano, gather clues, and link events to understand volcanoes better. Volcanologists observe both active volcanoes and ones that haven't erupted in decades or centuries. Hiking up and down mountains, collecting samples, lugging heavy equipment, and working in dangerous conditions are all part of the job. These men and women have to develop a sense for what is safe.

EDUCATION

So how can you enter this "hot" career? There are lots of possibilities. Courses in math, physics, chemistry, geography, oceanography, and even biology can help. Probably the most important subjects to master are geology and geophysics. The more you learn, the more opportunities you'll have.

POSSIBLE WORKPLACES

Volcanologists find employment in many places. Most professionals work for government agencies, directly or as consultants. Most cities located near volcanoes maintain full-time volcano observation staffs. Some university geology or physics faculty members specialize in volcanology and conduct research programs. And then there are those like the Kraffts who are publicly or privately funded to travel the world and develop educational and scientific videos.

DOMINION OPPORTUNITIES

What's the reward for this dangerous work? Volcanology saves lives and property. One of the films the Kraffts made helped the Philippine government decide to evacuate during the eruption of Mount Pinatubo in 1991, saving many lives.

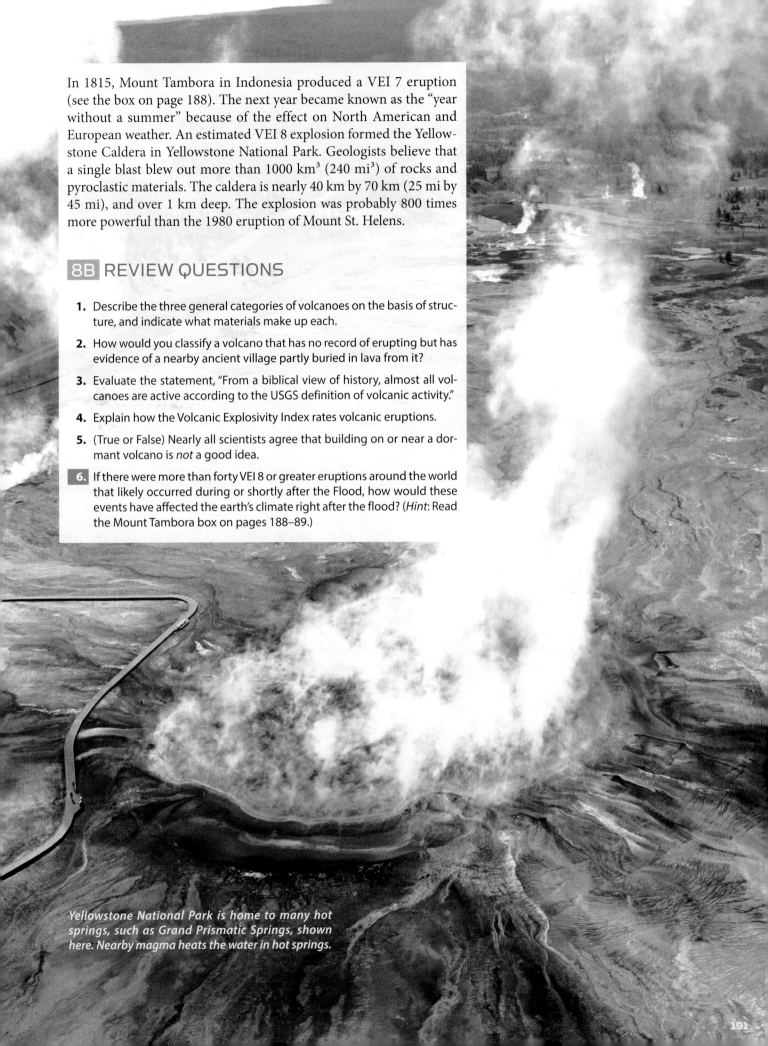

In 1815, Mount Tambora in Indonesia produced a VEI 7 eruption (see the box on page 188). The next year became known as the "year without a summer" because of the effect on North American and European weather. An estimated VEI 8 explosion formed the Yellowstone Caldera in Yellowstone National Park. Geologists believe that a single blast blew out more than 1000 km³ (240 mi³) of rocks and pyroclastic materials. The caldera is nearly 40 km by 70 km (25 mi by 45 mi), and over 1 km deep. The explosion was probably 800 times more powerful than the 1980 eruption of Mount St. Helens.

8B REVIEW QUESTIONS

1. Describe the three general categories of volcanoes on the basis of structure, and indicate what materials make up each.
2. How would you classify a volcano that has no record of erupting but has evidence of a nearby ancient village partly buried in lava from it?
3. Evaluate the statement, "From a biblical view of history, almost all volcanoes are active according to the USGS definition of volcanic activity."
4. Explain how the Volcanic Explosivity Index rates volcanic eruptions.
5. (True or False) Nearly all scientists agree that building on or near a dormant volcano is *not* a good idea.
6. If there were more than forty VEI 8 or greater eruptions around the world that likely occurred during or shortly after the Flood, how would these events have affected the earth's climate right after the flood? (*Hint*: Read the Mount Tambora box on pages 188–89.)

Yellowstone National Park is home to many hot springs, such as Grand Prismatic Springs, shown here. Nearby magma heats the water in hot springs.

8C

INTRUSIVE VOLCANISM

How does intrusive volcanism differ from extrusive volcanism?

In Section 8A, you learned that volcanoes are a form of extrusive volcanism. They occur on top of the earth's crust. So can you figure out where intrusive volcanism happens?

8.12 SILLS AND DIKES

Intrusive volcanism occurs when magma works its way into cracks and spaces inside the crust near volcanic vents. **Intrusions** cool into igneous rock. In many places around the world, you can see where magma has wedged its way between sedimentary rock strata, spreading them apart. Horizontal, shelflike igneous formations are called *sills*. In many places, after the surrounding rocks erode, large sills form an igneous plateau. It seems that sills often act as protective caps. During the latter stages of the Flood, intrusive volcanism reduced the rate of erosion of the underlying sedimentary rocks.

If intruding lava works its way into vertical joints in the rock, the solidified remnants form a wall-like *dike*. Dikes, unlike sills, always *cross* rock strata. Volcanoes often form dikes extending out from the lava conduit into the surrounding cone. After the softer tephra erodes away from large dikes, they appear as freestanding walls of dark igneous rocks. Both sills and dikes can vary in size from just a few centimeters thick to many meters.

8C Objectives

After completing this section, you will be able to

» compare intrusive volcanism with extrusive volcanism.

» describe various intrusive igneous formations and how they formed.

» define the geothermal gradient and describe how it varies with depth into the earth.

» discuss hydrothermal processes and identify volcanic features associated with heated groundwater.

» describe how energy can be extracted from geothermal sources.

An *intrusion* is an unexpected or unwanted entry of something into a place. In geology, an intrusion is the process of injecting a different kind of rock type into the existing rocks. It can also apply to the intruding rock itself.

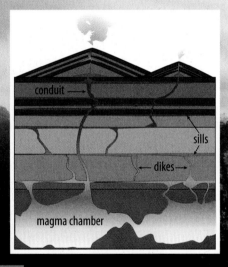

8-13
Sills are horizontal, shelflike igneous intrusions. Dikes are vertical, wall-like intrusions often associated with volcanoes.

Fi **Half Dome, Yosemite National Park, CA:** 37.745°N 119.535°W

8.13 PLUTONS

Probably the most important intrusive igneous features are **plutons**. These formed when large amounts of magma collected in huge chambers within the crust and then cooled and hardened.

The largest plutons are called *batholiths*. They are often made up of many smaller plutons. The upper surface of a batholith has an area larger than 100 km² (40 mi²). Geologists believe that they originally formed at deep or medium depths within the crust or below sedimentary rocks. Erosion has exposed some batholiths, which appear as broad, lumpy formations of igneous rocks.

Smaller plutons that seem to have formed from blobs of magma collected at the tops of deep dikes. They cooled to form dome-shaped intrusions called *laccoliths*. Laccoliths are often discovered under sedimentary rock dome formations.

Stocks are very small (less than a few square kilometers wide) vertical igneous intrusions. They may have supplied volcano vents in the past.

pluton (PLOO tahn): pluton (L. *Pluton*—the god Pluto, ruler of the underworld); something formed deep underground

batholith (BATH oh LITH): batho- (Gk. *bathos*—deep or depth) + -lith (Gk. *lythos*—rock); a deep structure formed of igneous rocks

laccolith (LACK oh LITH): lacco- (Gk. *lakkos*—reservoir) + -lith; a small storage chamber for igneous rocks

This rocky exposure in the Sierra Nevada Mountains, called Half Dome, *is part of a batholith that was exposed by erosion.*

geothermal: geo- (Gk. *geo*—earth) + -thermal (Gk. *therme*—heat)

gradient (GRAY dee ent): gradient (L. *gradientem*—to walk or step); referring to gradual or step-by-step changes in a quantity

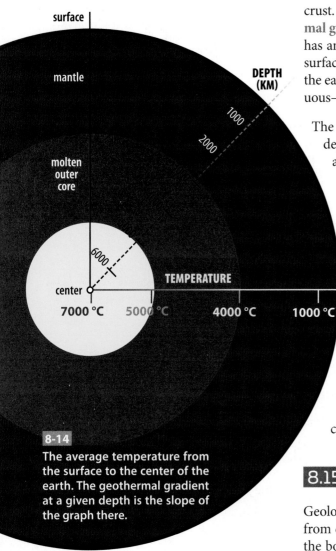

8-14 The average temperature from the surface to the center of the earth. The geothermal gradient at a given depth is the slope of the graph there.

Can the Earth Boil Water?

Use the average value for the geothermal gradient to determine the temperature at a depth of 5.0 km (3.1 mi) into the earth's interior. Is this hot enough to boil water?

Answer:
Calculate: 5.0 km × 25 °C/km = 125 °C (257 °F). Added to the average ground surface temperature, this temperature would be more than hot enough to boil water if the water were brought to the earth's surface.

8.14 THE GEOTHERMAL GRADIENT

Long ago, miners working in deep mines found that the earth's temperature naturally increases with depth. Deep drilling confirms that rock temperature rises continuously as the drill bites deeper into the crust. The rate of temperature change with depth is called the **geothermal gradient**. The temperature gradient exists because the earth's core has an estimated temperature of 5000 °C (9000 °F), while the earth's surface temperature averages about 30 °C (86 °F). It makes sense that the earth's interior temperature must increase in a more or less continuous—although not constant—rate as you go deeper into the earth.

The geothermal gradient within the crust varies with location. At depths above 120 m (400 ft), the climate at the earth's surface affects crust temperature. Below this depth, toward the middle of continental tectonic plates and away from intrusive igneous activity, the gradient is about 25 °C/km (72 °F/mi).

Secular geologists, with their old-earth view of history, believe that there are two main sources of heat in the earth. They think that about 20% of the heat is left over from the earth's formation, when planetesimals collided and melted to build up the early earth (see Chapter 5). The remaining 80% of the heat comes from radioactive decay—similar to how nuclear reactors produce heat.

Young-earth geologists believe that the earth's high internal temperatures are simply part of God's original design. They don't deny that radioactive decay could contribute heat as well. But a young-earth history doesn't depend on a natural or catastrophic origin for the earth's interior heat.

8.15 HYDROTHERMAL FEATURES

Geologists were surprised when they first began studying rock cores from deep drill holes. They had assumed that the great pressures at the bottom of these holes would squeeze all water out of the rocks. But they were wrong. Some estimate that deep igneous rocks could contain as much as 50% of their weight in water. This water can be chemically combined with minerals, or as liquid water it can fill tiny spaces within the rocks.

Hot water deep underground flows relatively easily, while solid rocks in the crust are immobile. Because most rocks contain cracks or *pores*, heated groundwater can work its way through rock and along faults and joints to the earth's surface. The flow of water usually starts near the surface. Cooler groundwater can sink by gravity downward into the geothermally heated rocks, displacing the hotter, less dense water there. The hot water rises nearer to the surface, where it loses heat to the cooler rocks. The water can then sink again, repeating the cycle. Geologists call this continuous, geothermally driven flow of water a *hydrothermal circuit*. Hydrothermal features are common near volcanoes because the circuit is much shorter near igneous intrusions connected to volcanoes.

hydrothermal: hydro- (Gk. *hydro*—water) + -thermal (Gk. *thermos*—hot or heat); hot water

HOT SPRINGS

A *hot spring* is a place where heated water rises to the earth's surface as a liquid. The spring at Warm Springs, Georgia, has a temperature of 31 °C (88 °F). The thermal gradient heats surface waters after they descend about 1 km (0.6 mi) into the crust. Shallow igneous intrusions heat the hot springs in the western United States.

Hot water typically dissolves chemicals better than cool water, so water in hot springs usually contains lots of dissolved minerals. Minerals, which settle or precipitate out when the water cools, often produce stepped rock formations called *terraces*. Algae and bacteria, which grow in the heated water, help form the deposits. They also color the terraces red, blue, and brown. Mammoth Hot Springs in Yellowstone National Park, Wyoming, is famous for its beautiful terraces.

Hot springs in places where there are thick deposits of fine volcanic tephra form muddy, bubbling springs called *mud pots*. Steam bubbles break at the surface, sending splatters of mud into the air.

8-15

Terraces of carbonate minerals left by flowing water from local hot springs in Pamukkale, Turkey

Location: 37.924 °N 29.123 °E

GEYSERS

A **geyser** (GY zur) is a hot spring that forcefully ejects its water from the ground at regular intervals. Scientists believe that geysers happen when an igneous intrusion heats a long, twisting chamber filled with water.

A delicate balance between temperature and pressure determines when a geyser erupts. Groundwater seeps into the water-filled chamber, or pipe, and is heated by the rocks. Pressure determines the boiling point of a liquid. The pressure of the tall column of water in the pipe keeps the water from boiling. So the water gets hotter and hotter. Eventually, the water at the top of the pipe is hot enough to boil. As this water boils away, the pressure at the bottom of the pipe drops. When pressure at the bottom drops enough, the superheated water there flashes into steam, violently pushing the remaining water out of the pipe. The pipe refills and the process repeats.

Geysers often create incredible mineral deposits near their vents. These formations are made of a material called *geyserite*, a mineral chemically similar to quartz. Probably the most famous geyser in the world is the Old Faithful geyser in Yellowstone National Park, Wyoming. The geyserite cone from which it erupts is about 2 m high.

Fi Old Faithful geyser: 44.46 °N 110.828 °W

8-16 Mineral deposits can be seen around the Pohutu and Prince of Wales Geysers on the North Island of New Zealand.

8.16 HARNESSING GEOTHERMAL ENERGY

People have found ways to harness the heat produced by magma intrusions using **geothermal energy**. As early as 863 BC, the Celts in England enjoyed the healing and cleansing effect of hot mineral springs. About AD 43, the Romans built extensive public bath facilities in the town that became Bath, England. Several of these ornate structures and their hot springs remain today. Modern resorts around the world use hot springs for bathing and swimming.

But heated groundwater can do much more. Hundreds of projects around the world use geothermal water and steam for all kinds of things. Geothermal energy can be used for heating homes and businesses, supplying fish and algae farms (*aquaculture*), feeding and watering livestock, heating greenhouses, supplying industrial chemical plants, and, most significantly, generating electricity. Readily available geothermal water or steam economically heats more than 87% of the homes in Iceland.

8-17 Geothermal plants use heated groundwater to produce electricity.

Today, the United States produces more electricity from geothermal energy than any other country, over 15 billion kilowatt-hours every year. This is enough to supply 1.4 million typical American households continuously! But compared with the country's total energy supplies, geothermal power represents just 0.4%. All the geothermal plants in the United States make just 2% of the energy generated by all of its nuclear plants. However, geothermal energy is an attractive option for those who believe that pollutants from the generation of electricity contribute to climate change. And once a plant is built, the electricity it produces can cost even less than that produced by other kinds of power plants. We can expect to see more geothermal energy projects in the near future. All industrialized nations in tectonically active regions of the world are building geothermal power plants.

Projects that tap geothermal energy are showing real promise. Still, only a few places have intrusive heat sources close enough to the surface to be usable. It's also expensive and risky to find good places that offer geothermal energy sources. These factors make developers very cautious about investing in such projects. For now at least, geothermal energy is only a minor source of energy. Maybe in the future we'll learn how to more effectively use this God-given resource. Maybe you could be the scientist who makes these breakthroughs!

8C REVIEW QUESTIONS

1. What kind of igneous formations are sills, dikes, and plutons?
2. How are sills and dikes arranged in relation to the surrounding rocks?
3. Name three kinds of plutons from smallest to largest.
4. As you get closer to the center of the earth, how does the temperature of rocks change?
5. Give the two different views about where the earth's internal heat comes from.
6. Why are hydrothermal features often found near volcanoes?
7. Contrast hot springs and geysers.
8. (True or False) The temperature of rocks rises at a constant 25 °C/km with depth to the center of the earth.

Krakatoa putting on a great show as lava flows and pyroclastic material flies through the air. Scientists rate Krakatoa's 1883 eruption at a VEI 6 event.

Key Terms

volcano	175
igneous	175
extrusive volcanism	175
vent	176
lava	176
crater	176
pyroclastic material	180
tephra	181
flood basalt	183
shield volcano	185
cinder cone	185
stratovolcano	185
active volcano	186
dormant volcano	186
extinct volcano	186
Volcanic Explosivity Index (VEI)	189
intrusive volcanism	192
pluton	193
geothermal gradient	194
geyser	196
geothermal energy	196

CHAPTER 8 REVIEW

CHAPTER SUMMARY

» Studying volcanoes helps scientists predict eruptions and reduce the risks associated with their hazards.

» Volcanoes are extrusive igneous depositional mountains.

» A volcano builds up from lava coming out of a vent to form a cone with a crater at the top. It can also form parasitic craters nearby.

» Collapsed volcanoes may form huge craters called calderas that sometimes fill with water.

» Volcanic liquid emissions differ in chemical composition and viscosity. The more silica in the liquid, the thicker the lava is.

» Volcanoes emit many kinds of heated gases, including steam and carbon dioxide. Violent flows of heated gases can carry hot cinders and ash in what are called glowing avalanches.

» Solid materials ejected by volcanoes are hardened lava. They can vary in size from fine ash to huge blocks.

» Volcanoes occur in tectonically active areas, such as convergent and divergent zones between tectonic plates. Some geologists believe that volcanoes also developed above hot spots in the mantle in the past, producing strings of volcanic islands.

» Flood basalt zones are broad plains of thick lava rock that flowed from deep cracks in the crust. Some zones cover very large areas of land.

» Volcanoes exist on all other earthlike planets and on many of the larger moons in the solar system.

» Geologists classify volcanoes by their structures, their historical activity, and their explosivity.

» Many intrusive features and processes are linked to volcanic activity. These include sills, dikes, and different kinds of plutons.

» The geothermal gradient is a measure of how the temperature of rock changes with depth into the earth's crust.

» Rocks deep underground can heat water to produce hot springs and geysers on the surface.

» Geothermal energy from hot water and steam is a relatively clean and economical energy source. Many countries are developing geothermal energy to supplement other kinds of energy.

REVIEW QUESTIONS

1. Why were there no deaths from the Eyjafjallajökull volcano eruption in 2010?

2. Why is a volcano an extrusive igneous feature rather than an intrusive one?

3. How does the amount of silica in lava affect its viscosity and the temperature at which it melts?

4. Explain why gases escape from lava as it rises in the neck of a volcano.

5. What is the main difference between the types of solid volcanic emissions?
6. What is the Pacific Ring of Fire?
7. What is the most common type of volcano? Describe its structure.
8. From a young-earth view of Earth's history, what is the difference between an active volcano and a dormant one? What may be a safer way of classifying a volcano's activity?
9. In general, how is the energy of a Volcanic Explosivity Index (VEI) number related to the energy of the next lower value?
10. Sills, dikes, and batholiths are examples of what kind of volcanism?
11. Explain what causes the geothermal gradient.
12. Compare the source(s) of Earth's interior heat as suggested by old-earth and young-earth geologists.
13. Evaluate the basis for old-earth and young-earth geologists' theories for the origin of the earth's interior heat.
14. Why does heated groundwater easily create special mineral deposits like travertine and geyserite?
15. Why is geothermal energy not more widely used?

True or False

16. Most volcanoes erupt from a magma chamber relatively near the earth's surface.
17. Flowing lava is the most dangerous hazard from volcanoes.
18. Pillow lava forms from the action of pyroclastic flows near the crater of a volcano.
19. The term *pyroclastic* refers to all the different types of solid volcanic emissions.
20. Hawaii, in the middle of the Pacific Ocean, is an example of a volcano that rises from a mid-ocean ridge system.
21. On a contour map, hatched contour lines indicate lower elevations completely enclosed by higher elevations.
22. The number of active and extinct volcanoes around the world and the extent of flood basalts suggest that the Flood event was far more horrific than just a global flood with heavy rain and tsunamis.
23. Cinder cones are among the largest volcanoes in the world.
24. Even though no VEI 8 eruption has ever been observed, we can see geologic evidence for past volcanic explosions that could have been rated with a VEI of 8 or more.
25. The geothermal gradient is often higher near tectonic plate borders than at the interior of the plates.
26. Geothermal energy has only recently been discovered and put to use.

3 EARTH'S ROCKY MINERALS

CHAPTER 9
MINERALS AND ORES 202

CHAPTER 10
ROCKS 222

CHAPTER 11
FOSSILS 248

CHAPTER 12
WEATHERING, EROSION, AND SOILS 276

DR. TIMOTHY CLAREY
GEOLOGIST

> "My interest in science began at an early age while collecting rocks that contained fossil seashells in the fields near my home in Michigan. Throughout my high-school years, my love of science and history was fueled by reading *The Genesis Flood*, by John Whitcomb and Henry Morris. This book explained how the great Flood of Noah was responsible for the fossils I was finding. My excitement for science again peaked as I took my first geology class in college. Finally, I found a science that combined my love for fossils, rocks, and Earth history.
>
> I went on to earn a BS and MS in geology and worked successfully as an oil exploration geologist for nearly a decade. God had other plans for me, however, and as oil prices dropped, I found myself unemployed. God never fails. He provided this as an opportunity for me to obtain a PhD in geology, even securing funding for my degree.
>
> After teaching at a secular college for 17 years, I took a full-time position as research geologist for the Institute for Creation Research. All of my experiences have better prepared me for what I am doing now. My oil industry training is allowing me to map out the Flood deposits across entire continents. The world's rocks and fossils truly declare that there was a global flood only thousands of years ago. Many new discoveries of original dinosaur soft tissue provide further evidence of this recent event. There has never been a more exciting time to be a creation geologist!"

CHAPTER 9
MINERALS AND ORES

- **9A** DESCRIBING MINERALS — 203
- **9B** IDENTIFYING MINERALS — 205
- **9C** MINERALS AS RESOURCES — 212

GOLD, PLATINUM, AND… ALUMINUM?

During the reign of Napoleon III of France in the mid-1800s, honored guests ate off aluminum plates, while the rest made do merely with gold plates. In those days, aluminum was more expensive than either gold or platinum. A few decades later, aluminum was a little cheaper—one ounce cost only an entire day's wage! Today, you probably throw away aluminum soda cans without thinking about it. What's changed?

9A
DESCRIBING MINERALS

How do I know whether a substance is a mineral?

9.1 MINERALS, ORES, AND DOMINION

Aluminum is the metal of our modern society. It's special because it is light, fairly strong, and rust-resistant. Without aluminum, transportation, building construction, and electronics would all be a lot less economical, if not impossible. So where does aluminum come from?

The Greeks and Romans used natural compounds called *alumen* for centuries. Unknown to them, these substances contained the element aluminum (symbol: Al). They used alumen for medicines and dyes. In the early 1800s, chemists discovered an aluminum-bearing rock called *bauxite*. But people found it very difficult to get pure aluminum from bauxite. That was why it cost so much. In the 1880s, two scientists discovered a practical way to extract aluminum from bauxite in a process called *electroplating*. In electroplating, electric current passing through a water solution carries metal atoms from a source of the metal and coats, that is, *plates*, it onto another piece of metal. The cost of aluminum dropped more than 90% after this breakthrough. Today, pure aluminum costs less than 8¢ per ounce.

God originally created the earth with many kinds of *minerals* and *ores* for us to use and enjoy. Minerals and ores have played a significant role in human history and culture. But understanding and using these resources wisely, as God intended, involves science.

9A Objectives

After completing this section, you will be able to

» relate the study and use of minerals to exercising biblical dominion.

» determine whether a substance is a mineral.

» classify natural materials as either native or compound minerals, or mixtures of these.

Scientists gave elements symbols as a shorthand way of writing chemical *formulas*, which represent compounds. The Periodic Table of the Elements (Appendix C) lists all the known elements and some of their properties.

9.2 DEFINING MINERALS AND ORES

What exactly is a mineral? A **mineral** is a naturally occurring, inorganic, crystalline solid that has a definite chemical composition. To help you understand what a mineral *is*, let's see what it is *not*. Minerals can't be substances such as steel and artificial gems because those are man-made. They can't be pearls or coal because those are organic, or come from living or once-living organisms. They can't be air or liquid water because those are not solids (though ice *is* considered a mineral). A mineral can't be a natural glass like igneous obsidian, because natural glasses don't have regularly arranged atoms like crystals do. Minerals *do* include substances such as natural gold, natural gems, asbestos, and quartz.

Ores are the rocks that contain one or more minerals that are the source of elements or compounds important to industry or agriculture. You probably have heard of iron ore, from which iron is extracted in a blast furnace. You will learn more about ores later in this chapter and in Chapter 10.

9-1
Fluorite is a mineral made of calcium fluoride.

MINERALS AND ORES 203

9.3 ORIGIN OF MINERALS

Where did minerals come from? It's simple enough and true to say that God created them. But most minerals probably do not appear in the same forms as they did at Creation, or in the same locations. Of all the elements and compounds created in the beginning, only gold, bdellium, onyx, copper (brass), and iron are named in the first six chapters of Genesis. But it's likely that many other minerals existed at Creation too.

After the Flood, the Bible mentions at least thirty-six more rocks and minerals. The Flood completely transformed the earth's surface. The catastrophic processes of the Flood broke apart the original materials that God had made during the Creation week, both physically and chemically. New minerals and ores formed in different places. Though the Flood is a testimony of God's judgment in the devastation of the earth, it is also a testimony of His grace in providing new minerals for man's use.

9.4 MINERALS AND MATTER

Minerals, like all other matter, have mass and take up space (have volume). They are also made of atoms. As you learned in Chapter 2, chemical compounds are pure substances made of the atoms of two or more different elements. Recall that the different atoms of a compound always combine in the same ratio.

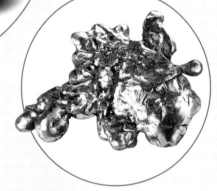

Compound Minerals

Most of the known minerals are compounds, which are pure substances. For example, the mineral quartz is a compound of silicon (Si) and oxygen (O). Quartz (formula: SiO_2) always contains twice as many oxygen atoms as silicon atoms. If the ratio is different, then the compound is something else, but not quartz. Most **compound minerals** are more complicated than simple chemical compounds. For this reason, *mineralogists* normally work with *families of minerals* that are chemically similar.

Native Minerals

A few minerals are made of atoms of just one element. These are called **native minerals**. Such minerals were important in early history, even before the Flood, because they were the first sources of pure elements. They include gold (Au), silver (Ag), copper (Cu), and sulfur (S).

Mixtures of Minerals

Do you remember what a mixture is? It is a combination of two or more pure substances physically combined. The substances are not chemically bonded in a mixture. You can separate a mixture into its pure substances by physical processes, such as sifting, melting, crushing, or evaporating. Minerals are never mixtures because minerals have a definite composition. We can change the composition of a mixture but not a mineral. In contrast, most *rocks* are a mixture of two or more minerals. This is how you can tell most rocks and minerals apart. We'll talk more about rocks in Chapter 10.

9A REVIEW QUESTIONS

1. Why was aluminum so precious in the mid-1800s?
2. List the key properties of a mineral.
3. How are the minerals we have today evidence of God's grace?
4. Why are minerals considered matter?
5. Of the three kinds of substances that make up matter in nature, which one forms most minerals?
6. A mineral sample is nothing but the element sulfur. What category of mineral is it?
7. Explain why quartz is *not* a mixture of silicon and oxygen.
8. (True or False) All the atoms that make up the minerals we see today appeared at Creation.
9. Metalsmiths call pure gold *24-karat gold* because it is a ratio of 24 parts gold to 24 parts of the sample (24/24 gold). One common gold and silver alloy is called "18-karat gold" because it is 18 parts gold to 24 parts of the alloy (18/24 gold). What is the ratio of gold to alloy in a piece of 12-karat gold?

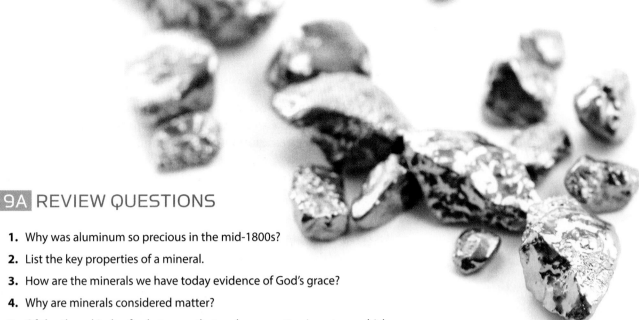

9-2
This pure gold is uncombined with any other element, making it a native mineral. Most gold that we purchase is not pure gold.

9B

IDENTIFYING MINERALS

How do mineralogists identify minerals?

9.5 MINERALOGY

Imagine you are a geologist. A mining company wants you to identify the minerals in an *outcrop* or rocky area that they recently explored. What would you do? You'd begin by getting lots of samples. You would look at the rocks in which you found the samples. Then you'd do tests on those samples. You would closely examine their color and texture. You might weigh them, or test them with certain chemicals. In fact, you will actually do some of these things in your lab activities for this chapter.

This kind of work is called *mineralogy*, and geologists who do these things are called **mineralogists**. Mineralogists look at the physical and chemical properties of small samples of a mineral taken from the location of study. Such tests help geologists identify, classify, and analyze minerals and better understand the geologic setting that they come from.

9B Objectives

After completing this section, you will be able to

» define mineralogy and explain what mineralogists do.

» describe characteristics used for mineral identification.

9.6 PHYSICAL PROPERTIES

To be confident that they've identified a mineral correctly, mineralogists use more than a single test. Although some minerals have properties that make recognition easy, many minerals can appear to be like quite different minerals, or they can contain impurities that make identification difficult. Following are some important properties that can help identify all minerals.

COLOR

Mineralogists begin by observing a sample's color. This is not as easy as it sounds. Many minerals have similar colors. And some minerals change color when they are exposed to air because their surfaces chemically react with water vapor, oxygen, or some other chemical in the air. To make things even more interesting, impurities can give samples of the same mineral different colors. Quartz, for example, can be green, pink, blue, white, or dark gray. Amethyst is quartz with manganese (Mn) impurities that make it violet. Another example is the mineral corundum. Usually it is colorless, but if a little of the element chromium (Cr) is present, the mineral turns a shade of red and forms a ruby. If a little iron (Fe) and titanium (Ti) are present, quartz becomes a sapphire. You can see that though impurities can make it harder to identify a mineral, they can also affect its value!

9-3
Small amounts of impurities give this quartz its rose color.

9-4
Smoky quartz contains small amounts of free silicon.

STREAK

Mineralogists also test a mineral for the color of its powder. Even though the color of a **specimen** can vary because of impurities, its **streak** is the same color in any case. Light doesn't reflect from the tiny particles of the powder in the same way that it does from the mineral's crystal structure. For most minerals, you produce a streak of powder by rubbing a surface with a sample. This is appropriately called a mineral's *streak test*.

| Sometimes mineralogists call a small piece of a mineral a **specimen**.

9-5
Even though impurities change the color of a mineral, its streak remains the same.

Mineralogists normally test a mineral's streak by using a *streak plate*, which is a piece of unglazed porcelain. Not all minerals produce a streak. Some minerals are harder than a streak plate and they will just scratch it. In this case, the mineralogist rasps the mineral with a file or other tool to produce powder of the mineral. To accurately identify the streak or powder color, you need a contrasting white or black streak plate.

LUSTER

A mineral's **luster** is the amount and quality of the light that it reflects from its surface. Materials can have either metallic luster or nonmetallic luster. Gold, silver, and copper are metals and therefore have metallic luster. Scientists further categorize nonmetallic luster on the basis of the appearance of the sample. Because of this, determining nonmetallic luster is more challenging. Quartz reflects light like glass. Mineralogists say that it has a *glassy* or *vitreous luster*. The luster of gypsum is *pearly* because it has the soft glow of an oyster pearl. The brilliant flare of a diamond is an *adamantine* (ad ah MAN teen) *luster*. Asbestos, with its fibrous texture, has a *silky luster*. Other recognized lusters include *greasy*, *waxy*, *earthy* (or *dull*), and *resinous*.

9-6 Asbestos (above) has a silky luster, and muscovite (left) has a pearly luster.

CRYSTAL SHAPE AND GROWTH

The Roman philosopher Pliny, who lived shortly after the time of Christ, was one of the first people to describe mineral crystals. He noted that mineral crystals come in many fascinating shapes. Crystal sizes range from the microscopic crystals of kaolin (KAY uh lin; clay) to the giant crystals of beryl or feldspar that can sometimes weigh several tons.

When they have room, crystals build a shape based on the arrangement of atoms in the mineral. They grow larger by *accretion* (uh KREE shun), where atoms or ions add one-by-one to the existing crystal structure. This can happen in water with lots of dissolved pure substances. It can also happen in cooling magma or as hot gases of pure substances cool near a volcano vent. Each mineral tends to form distinctive crystals with characteristic faces, called *facets*, arranged at certain angles. For example, quartz crystals are hexagonal (six-sided) rods. Halite crystals (table salt) are cubes. Mineralogists have identified about thirty-two classes of mineral crystals. See the box on page 211 for more information on some common kinds of crystals.

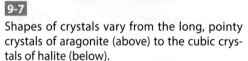

9-7 Shapes of crystals vary from the long, pointy crystals of aragonite (above) to the cubic crystals of halite (below).

CLEAVAGE AND FRACTURE

The ways that minerals break help mineralogists identify them. Minerals break in two ways—cleavage and fracture. Some minerals can split into flat sheets or along certain planes, creating new facets. Mineralogists call this property **cleavage**. A mineral's cleavage is related both to how its atoms are arranged in the crystal and to the strength of the chemical bonds between them. Mineralogists have five ratings for mineral cleavage: *perfect*, *good*, *fair*, *poor*, and *none*.

At one end of the scale are minerals, such as mica, which separate easily into thin sheets. These have perfect cleavage. At the other end of the scale are minerals, such as quartz, which shatter and do not

MINERALS AND ORES

9-8 Mica is quickly identified by its easy cleavage into thin sheets.

show cleavage at all. Cleavage also involves the number of directions in which a mineral breaks. Mica cleaves in only one direction, halite cleaves in three directions, and sphalerite (SFAL er ite) cleaves in six directions.

Minerals without cleavage planes can still break in ways that help identify them. Their **fracture** property may be *uneven* (rough surface), *hackly* (fine points), *fibrous*, or conchoidal (kahn KOID el; glassy, clamshell-like chips).

HARDNESS

Mineralogists also identify minerals by their **hardness**. Some minerals are so soft that you can scratch them with your fingernail. Others are so hard that almost nothing can scratch them. In 1822, German mineralogist Friedrich Mohs created a scale to classify and identify minerals by their relative hardness. On the **Mohs scale**, minerals range from 1 (very soft) to 10 (very hard). Mohs assigned talc a hardness of 1 and diamond a hardness of 10.

It's fairly simple to test a mineral's hardness. For example, if a mineral scratches gypsum (hardness 2) but is scratched by calcite (hardness 3), its hardness is between 2 and 3. Table 9-1 gives the relative hardness of some common materials that mineralogists use to determine the hardness of minerals in the field.

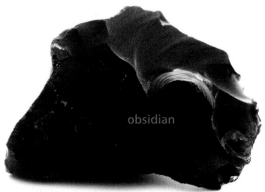

9-9 Obsidian's property of breaking along smooth, curved surfaces is called *conchoidal fracture*.

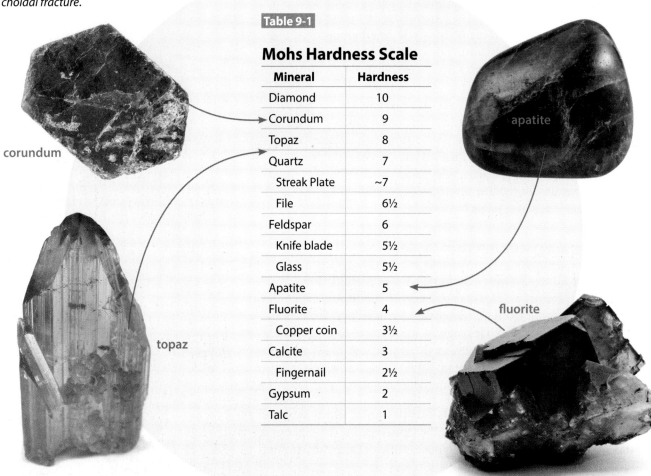

Table 9-1

Mohs Hardness Scale

Mineral	Hardness
Diamond	10
Corundum	9
Topaz	8
Quartz	7
Streak Plate	~7
File	6½
Feldspar	6
Knife blade	5½
Glass	5½
Apatite	5
Fluorite	4
Copper coin	3½
Calcite	3
Fingernail	2½
Gypsum	2
Talc	1

The numbers of the Mohs scale show only their relative order of hardness, not their actual hardness as measured by special instruments. The actual hardness of diamond, for instance, is four times that of corundum, but corundum is only twice as hard as topaz.

SPECIFIC GRAVITY

The **specific gravity** (**sp gr**) of a mineral is the ratio of its density to the density of water at 4 °C (39 °F), that is, 1 g/mL. Some minerals, such as gypsum, have low specific gravities. Others, such as gold, are dense and have high specific gravities. If a sample has a density of 5 g/mL then the mineral's specific gravity is as follows:

$$\text{sp gr} = \frac{\text{density of mineral}}{\text{density of water}} = \frac{5 \frac{g}{mL}}{1 \frac{g}{mL}} = 5$$

Sometimes a mineralogist can estimate the specific gravity of a mineral by handling a sample of each of two different minerals. If the specimens are about the same size, he can compare their weights to estimate their specific gravities. With practice, a person can become good at estimating whether a mineral's specific gravity is low, average, or high.

9.7 SPECIAL PROPERTIES

Mineralogists identify most minerals by their color, streak, luster, crystal shape, cleavage, fracture, hardness, and specific gravity. A few minerals, however, have special properties that make them easier to identify.

FLAME TEST

Some minerals produce characteristic colors of flame when they burn. Mineralogists use a *flame test* to identify these minerals. For example, when you use a platinum wire loop to pick up a few grains of table salt and hold the loop in a clear flame, the flame turns yellow. The element sodium (Na) in table salt (sodium chloride, or NaCl) gives the flame this color. Potassium (K) produces a violet flame, and calcium (Ca) gives an orange-red flame.

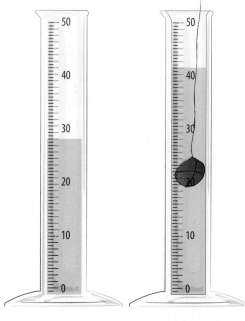

9-10
A mineral's density is its mass divided by its volume. The volume of small samples can be measured by water displacement using a graduated cylinder.

The **specific gravity** of a substance is simply the number value of its density.

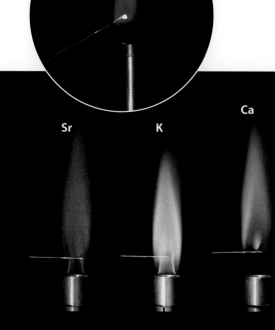

9-11 The reaction of some minerals with acid is another method of identification.

9-12 Magnetite is a natural magnet. It will pick up small iron objects just as other magnets do.

ACID TEST

Mineralogists can use the **acid test** to identify minerals in families such as the carbonates, sulfides, and sulfites. If you place a drop of hydrochloric acid (HCl) onto any of these minerals, they fizz, giving off bubbles of gas. This happens because the acid reacts with the mineral, just as baking soda fizzes when you mix it with vinegar, which is also an acid. The minerals calcite, dolomite, and galena all react with acids. Dolomite reacts only with warm acid, or with cool acid in powder form, because the chemical bonds are stronger in this mineral.

MAGNETISM

Some minerals are magnetic. Mineralogists simply test to see whether a magnet will attract a mineral or its powder. The best example of a magnetic mineral is magnetite. Most magnetic materials like magnetite contain iron. Minerals that contain other metals like copper or aluminum are nonmagnetic. Some are even antimagnetic: they weaken a magnetic field when placed in one.

OTHER TESTS

Geologists use three additional tests on minerals, though not as often: radioactivity, luminescence, and refraction.

9B REVIEW QUESTIONS

1. Why do geologists study minerals?
2. What is the most reliable way to identify a mineral using color? Why is this test more reliable?
3. What familiar mineral has an adamantine luster?
4. What kind of luster should native silver have?
5. By what process do crystals grow? What are two conditions under which this can happen?
6. Which mineral property describes how well a crystal breaks along certain flat surfaces?
7. Study Table 9-1. What are five common objects that can be used to estimate a mineral's hardness?
8. You are following a compass course as you hike through some rough terrain filled with dark-colored, broken rocks. Suddenly, as you near an outcrop, your compass needle veers 20° to point toward it. What might be causing this?
9. You are trying to identify a sample of a mineral. It is brassy yellow with metallic luster, has conchoidal fracture, a hardness between 6 and 6.5, and a specific gravity of 5.08. What is it? (If helpful, use Lab Manual Appendix B or do an Internet keyword search for "mineral identification chart.")
10. (True or False) The Mohs hardness scale gives the relative hardness of minerals in comparison to each other.

MINERALS AND THEIR CRYSTALS

Minerals can be classified by their crystal shapes, and their shapes are determined by the shape and positions of the faces of the crystals. A crystal face is called a *facet*. The position of the facets around the center of the crystal determines the symmetry of the crystal. *Symmetry* is a geometry term that describes how one part of a geometric figure is related to its other parts. If you connect the center point of a facet to the center of the crystal with an imaginary line, you create an *axis of symmetry*. There are seven basic mineral systems. Study the small ideal crystal system diagram and see how the real mineral crystal relates to this ideal.

9C Objectives

After completing this section, you will be able to

- » evaluate the dominion uses of minerals in view of their practicality and beauty.
- » discuss where native minerals are found.
- » describe the key identifying properties of native minerals.
- » give specific examples of the usefulness of minerals.
- » differentiate between native and compound minerals.
- » weigh the benefits and adverse effects of mining for minerals.

9C

MINERALS AS RESOURCES

How do we use minerals?

9.8 MINERALS, BEAUTY, AND THE IMAGE OF GOD

As we study some of the most important minerals that humans use, we learn something about our God. He gave this earth not only structure but also beauty because He is a God of order and beauty. This idea is important to us because God made us in His image. If we are to carry out dominion in a way that declares His glory, we must learn how to appreciate the beauty of minerals as well as their usefulness. We show by how we use minerals that dominion is not just science but also art. The results of dominion are not just practical but also pleasing to the eyes and other senses.

9.9 NATIVE MINERALS

METALS

When you think of beautiful minerals, gemstones probably come to mind first. But remember that precious metals like gold, silver, copper, and platinum are also minerals. You can find most of these metal elements in their pure states in nature as native minerals. They may fill cracks in rocks called *veins* or can be found as chunks and nuggets. When the rocks containing these native metals erode, their particles wash into nearby streams and onto beaches. These mineral sediments are called *placer deposits*, which can be rich sources of precious metals. Most native deposits of the industrial metals like copper and iron have been used up. They are now obtained only from ores.

Native metals, like most metal elements, are much denser than water. They have bright metallic lusters and their colors are distinctive. Gold is a bright yellow. Silver is usually dark gray because it tarnishes, but a fresh surface is bright and mirror-like.

9-13
The International System standard kilogram is made of a platinum alloy. Platinum was chosen for this purpose because of its resistance to corrosion.

9-14
An astronaut's visor is coated with a film of gold for protection.

Native copper has a dull red color, and platinum is steel gray. Just like other natural resources, the value of native metals depends on how common they are, how easy it is to obtain them, and how they are used.

You are probably most familiar with the use of these metals as jewelry. But did you know that these beautiful minerals are also very useful for other things? Gold, silver, platinum, and copper are all good *conductors*—they easily allow both heat and electricity to pass through them.

Gold has long been a precious metal. It has been used as an international money standard and still is one kind of financial investment. It is also used for fine wires, and tarnish-proof coatings in electronics, as well as for jewelry, utensils, and objects of art.

Silver is a historically precious metal used for coins and plating metal utensils. It is the best conductor of electricity. It also is used in quality photo printing and for the reflective surfaces of quality mirrors. Its main drawback is that it tarnishes when exposed to air.

When copper tarnishes, a tough surface layer forms. This layer seals out oxygen and resists further corrosion. Because of this, copper is used for plumbing pipes and for corrosion-proof architectural coverings, like roofs and dome plating.

Platinum reacts chemically with hardly anything, so it is valuable for products where corrosion must be avoided. It is the active material in automobile catalytic converters, which reduce air pollution in car exhausts. It will likely be important for the future of fuel-cell energy technology.

NONMETALS

Native minerals can also be elements that are nonmetals. *Diamond* is a form of pure carbon. Carbon is a nonmetallic element that is important in living cells. However, diamond is still a mineral because its carbon doesn't come from living things. Prospectors may find diamonds in placer deposits, but usually they are mined in *diamond pipes*. These are underground remnants of the magma necks of extinct volcanoes. Diamonds formed under great heat and pressure as igneous minerals. They are usually pale yellow or colorless, but red, orange, green, blue, and brown diamonds exist. Diamonds with these rare colors are quite valuable.

9-15
The familiar green color of the Statue of Liberty, which is made of copper, is a result of its copper skin tarnishing. The tarnish now actually protects the statue from further corrosion. The torch is covered with a thin layer of gold.

MINERALS AND ORES

9-16

Diamonds in the rough are of various sizes and colors (below). The accompanying rock is kimberlite. The Hope Diamond (right) is the largest blue diamond known. It was cut from a 112-carat rough-cut gemstone.

We know diamonds mostly for their beauty as gems. The value of gem diamonds depends on their color, clarity, size, and the skill with which they are cut. The size of a stone is measured in a unit of mass called the *carat*. A two-carat stone is worth three to four times as much as a one-carat stone of equal color, clarity, and cut. This is because larger stones are rarer.

Diamond's hardness makes it really useful beyond jewelry. Natural diamonds that are of little value as gems because of flaws are often ground into powder and used as industrial abrasives. Drill bits and saw blades that are used to cut very hard materials often have cutting edges coated with tiny diamond fragments. Artificial diamonds also serve this purpose.

The element sulfur can occur as a native mineral in bright yellow deposits. It occurs in its pure form or in compounds near the rims of volcanoes and in sedimentary beds deep underground. Sulfur has a hardness of 1.5–2 and can occur in several crystalline forms. It melts into a deep red liquid at around 113 °C (235 °F) and it burns with a distinctive blue flame. The rotten-egg smell from some spoiled foods is due to a toxic sulfur compound.

9-17

The crystalline form of the mineral sulfur. Sulfur deposits can be found around the world.

Sulfur is one of the most important elements to modern industrial nations because it is the main component of sulfuric acid. This acid is an essential compound for many industrial processes. Most sulfur produced in the United States comes from the refining and processing of petroleum and coal.

9.10 COMPOUND MINERALS

While some minerals can occur naturally as just one kind of element, most minerals are chemical compounds of two or more elements. Mineralogists group minerals with similar combinations of elements together into families. And each of these families has its own special properties and uses. Let's look at a few of these important mineral families.

SILICATES

Silicates make up about 25% of all known minerals and about 40% of the common ores. More than 90% of the earth's crust is made of silicates. This class of minerals contains mainly silicon and oxygen. Feldspar is the most common silicate mineral in the crust. Quartz, the second most common, is a pure silicate compound called silica (SiO_2). Important silicate minerals include chalcedony, opal, mica, hornblende, olivine, garnet, talc, and even kaolin. Silicates are the major rock-forming class of minerals. But they can also be used directly for all kinds of important products, such as glass, jewelry, and ceramics. In addition, they are the raw materials in computer chip manufacturing.

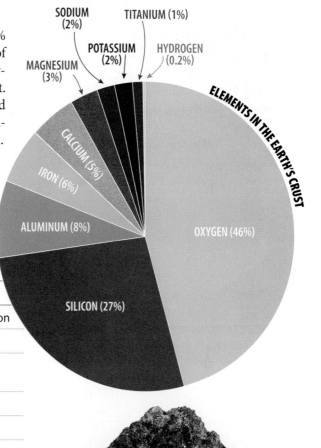

9-18
Opal is a form of quartz that is often used in jewelry.

ELEMENTS IN THE EARTH'S CRUST
- OXYGEN (46%)
- SILICON (27%)
- ALUMINUM (8%)
- IRON (6%)
- CALCIUM (5%)
- MAGNESIUM (3%)
- SODIUM (2%)
- POTASSIUM (2%)
- TITANIUM (1%)
- HYDROGEN (0.2%)

Table 9-2
Selected Silicate Materials

Mineral Class	Name	Uses
olivines	forsterite/fayalite	gemstone, steel production
pyroxenes	augite	mineral collections
pyroxenes	beryl	gemstones (emeralds, aquamarine, and others)
amphibole	hornblende	found in granite
micas	biotite	radiometric dating
micas	muscovite	manufacturing and construction materials
feldspars	orthoclase	gemstones and porcelain
feldspars	plagioclase	ceramics and clays
quartz	quartz	gemstones, semiconductors, and electronics

OXIDES

Oxides are minerals made of oxygen and some other element, usually a metal. These are economically important because they are the chief sources of major metals used in industry. At the beginning of the chapter, we mentioned that aluminum is extracted from bauxite, a member of the oxide family. Other common metal oxides include hematite and magnetite (iron oxides), from which we obtain iron, and cassiterite, a tin oxide.

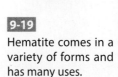

9-19
Hematite comes in a variety of forms and has many uses.

SULFIDES

Sulfides are minerals made of one or more metals and sulfur. Like oxides, these minerals contain metals economically important to us. Many sulfides are opaque. Yet they have characteristic colors and often have colored streaks. Galena (lead sulfide), chalcocite (copper sulfide), cinnabar (mercury sulfide), realgar (arsenic sulfide), stibnite (antimony sulfide), and pyrite (iron sulfide) are all important metal sulfide ores.

CARBONATES

Carbonates are minerals containing one or more metal ions and the carbonate ion (CO_3^{-2}). This ion particle is made of one carbon and three oxygen atoms. Carbonates are another of the major rock-forming minerals. They form vast beds of sedimentary rocks. Many solution caves are dissolved out of carbonate rocks. The strange and beautiful formations found in these caves come from the accretion of carbonate minerals. Calcite (calcium carbonate), dolomite (calcium and magnesium carbonate), rhodochrosite (manganese carbonate), and malachite and azurite (copper carbonates) are some of the more common minerals in this class. Carbonates are used in manufacturing many different products like cement, paper, and steel.

9-20
Galena, or lead sulfide, a lustrous, blue-gray mineral that usually crystallizes in cubes, is the principal source of lead.

9-21
A sample containing both azurite and malachite showing the typical blue and green colors of a copper compound. These two minerals are examples of carbonate materials.

The world's most productive diamond mine is the Argyle in Australia.

HALIDES

Halides are mineral salt compounds. These form when atoms such as sodium, potassium, magnesium, or calcium combine with fluorine, chlorine, bromine, or iodine.

The most familiar halide mineral is halite. Halite is a compound of sodium and chlorine (NaCl). You probably use some on your food as table salt. Another important halide is a salt used on icy roads called calcium chloride ($CaCl_2$). It is less damaging to cars and plant life than regular table salt or halite, but it still melts ice. Calcium chloride is also used as a *desiccant* (a drying agent) because it easily absorbs water from the air. Other halides are used in certain fertilizers, in metal production, and in photographic chemicals.

9.11 MINING AND DOMINION

Before and after the Flood, God gave us minerals and ores to supply man's needs (Gen. 2:11–12; Job 28:1–11). He even declared to the Israelites that minerals in the hills of Canaan would be one of the gifts they could look forward to when they possessed the land (Deut. 8:9). But we have to work for them! For most minerals, we have to dig into the ground or use some method to separate the mineral from other earth materials. The most common method is mining, or opening a hole in the ground to get at the mineral. The harder it is to obtain a mineral, the more precious it is.

Many minerals can be found deep underground, buried by flood sediments or formed there by the tectonic processes during the Flood. To reach these resources, *deep mining* is necessary. This usually involves digging a vertical mine shaft and then tunneling sideways to follow the mineral layers or veins. Many coal mines are like this. The deepest mine in the world today is the Tau Tona gold mine in South Africa, which is about 3.9 km (2.4 mi) deep. The deepest *diamond* mine in the world, named the Kimberley, was in South Africa. Mining stopped at a depth of about 1100 m (3600 ft) because the diamond pipe became too narrow.

Some minerals lie exposed at the earth's surface or are just under the surface in rocks covered by soil. These minerals are far easier to obtain by *surface mining*. Companies usually just strip off the overlying soil, and then remove the mineral in *strip mines*. If the minerals are not at the surface but are close enough that mine shafts are unnecessary, miners blast and remove the overlying rock and then dig huge open pits in the earth's surface. Gold, copper, and iron can be mined this way.

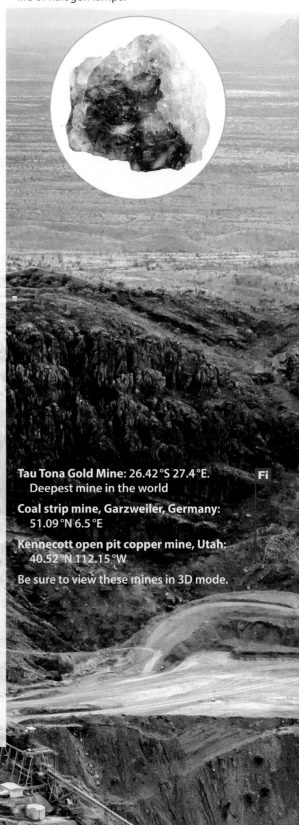

9-22
Halite mineral crystals usually are cubic. Halides are essential to the brightness and long life of halogen lamps.

Tau Tona Gold Mine: 26.42 °S 27.4 °E.
Deepest mine in the world

Coal strip mine, Garzweiler, Germany: 51.09 °N 6.5 °E

Kennecott open pit copper mine, Utah: 40.52 °N 112.15 °W

Be sure to view these mines in 3D mode.

Sluicing, dredging, and panning for gold in streams are other examples of surface mining. Miners scoop up gold and other minerals found in gravel beds and then wash the sediments with lots of water to separate them from other rock materials.

You can see that mining operations, which are necessary for providing raw materials, can also be very destructive to the environment, and even dangerous to miners. Strip mines and open-pit mines drastically change the earth's surface in their location. These operations eliminate forests and watersheds. The byproducts of mining operations can displace or even poison wildlife and people nearby. Many mines leave huge piles of rubble called *tailings* that create unsightly landscapes. Mine cave-ins have endangered miners for centuries. Mine rescues are some of the riskiest as well as most rewarding examples of people showing love for other human beings.

These problems point out how even beneficial efforts to exercise dominion have unintended negative consequences. We must learn to use the earth's resources to the best of our ability while minimizing the hazards, especially to humans and living creatures. This is one reason why mining requires a thorough knowledge of earth science.

TRAPPED UNDERGROUND

Every day mine workers enter dark shafts to bring treasures to the sunlit surface. We need mines to provide us with resources to produce heat, building materials, and even works of art. Most of us don't think about the benefits and risks of mining. That all changed on August 5, 2010, when part of the San José copper and gold mine in Copiapó, Chile, collapsed.

Trapped! The thought of that terrifies people. Imagine being trapped 700 m (2300 ft) underground. This is what happened to thirty-three miners in the San José mine. For the next seventeen days, they had no communication with the surface. Rescuers didn't know whether the miners were alive, and the miners didn't know whether a rescue was in progress.

The mining company and the Chilean government worked around the clock to rescue the miners. Rescuers brought in experts from NASA and the Chilean navy. Teams from the United States and Canada brought in drilling rigs to attempt to reach the miners. Crews worked the drills day and night while the world watched. Days turned into weeks and weeks into months.

Finally, on October 12, 2010, they were successful in removing the miners. For twenty-two hours, crews lowered and retrieved a rescue capsule and raised, one at a time, all thirty-three miners out of the earth. After sixty-nine days underground, all the miners returned safely to their families.

Questions to Consider

1. The San José mine was a copper and gold mine. What types of minerals are copper and gold?

2. What do we call the naturally occurring solid material from which we obtain minerals?

3. This story relates to the Dominion Mandate in two ways. Name two ways that it demonstrates good and wise dominion.

LIFE CONNECTION: MUD PIES FOR MACAWS

When you were young, did you ever make mud pies? Did you ever eat them? Yuck! No way! But certain insects, reptiles, birds, and even mammals and people eat earth materials as a regular part of their diet.

For many years, people have been watching a colorful, raucous, and fascinating spectacle on certain exposed banks of clay (kaolin) in the jungles of Peru called *clay licks*. On a regular basis, hundreds of macaws descend and perch on banks along waterways just to lick and eat clay. And they're picky too! Some jealously protect their favorite spot from intruders. They use their feet and beaks to dig out tootsie-roll-sized lumps of slippery earth. Some swallow the soil on the spot. Parents may fly back to their nests to stuff their waiting chicks with the not-so-appetizing (to us) treat.

Why would a bird want to eat clay when the jungle is loaded with fresh fruits, nuts, and seeds? That's a question that scientists are working to answer. They do know that some foods in the parrots' diet contain very bitter chemicals and even poisons. Because of this, scientists first assumed that the clay was able to absorb or neutralize these harmful chemicals. Chemical analysis shows that this probably is not true. Also, many birds eat sand or small gravel to help grind food in their digestive tract as it passes through. But macaws have such powerful beaks that they don't need to do this. And the clay particles are much too small to be used for that anyway.

Now some researchers suggest that the clay contains certain minerals that the birds lack in their natural diet. Sodium, an element essential for life, is common in many types of clay. Because sodium is rarely found in the macaw's typical food, it may be something they crave. This is especially true in rain forests, where frequent and heavy rains wash nutrients like sodium out of the soil. Clay is resistant to water, so sodium and other ions stay put. To further support this hypothesis, scientists note that these clay licks are found only inland, not near the seacoasts where sodium is much more available from sea salt.

9C REVIEW QUESTIONS

1. Where are placer deposits of gold and other metals found?
2. Name the native mineral that is the best conductor of electricity. What is one disadvantage of this precious metal?
3. What properties make copper such a useful native mineral?
4. What geologic feature seems to be a common location for finding diamonds?
5. Diamonds are composed of what element(s)? Why are diamonds considered minerals as defined in this chapter?
6. With what unit is the size or mass of gems measured?
7. Name the most abundant family of minerals in the earth's crust. What are the two most common minerals in this family?
8. Oxides, sulfides, and carbonates are good sources of what kind of materials?
9. What problem often comes with removing minerals from the earth? What consideration should always be a part of wise dominion in a fallen world?
10. (True or False) When exercising biblical dominion, people should not use the raw materials of the earth for decoration or beauty's sake. This is wasteful because God intended for us to use these materials only in practical ways for man's benefit.

Key Terms

Term	Page
mineral	203
ore	203
compound mineral	204
native mineral	204
mineralogist	205
streak (test)	206
luster	207
cleavage	207
fracture	208
hardness	208
Mohs scale	208
specific gravity (sp gr)	209
acid test	210
silicate	215
oxide	215
sulfide	216
carbonate	216
halide	217

9-23 If you looked only at the top half of this photo you might think that you were looking through a microscope. However, you can see a scientist studying these giant crystals—selenite crystals that formed in water and are saturated with gypsum.

CHAPTER 9 REVIEW

CHAPTER SUMMARY

» A mineral is a naturally occurring, inorganic, crystalline solid with a definite chemical composition.

» An ore is a rocky material from which a valuable mineral or a metal can be extracted.

» Minerals and ores were originally created by God. New ones probably formed as the result of events and processes after Creation, especially during the Flood.

» Minerals can be compounds or pure elements (native minerals). They can be found in mixtures with other minerals to form rocks.

» Mineralogists are professionals whose job is to identify minerals.

» All minerals have distinguishing physical properties of color, streak, luster, crystal shape, cleavage, fracture, hardness, and specific gravity.

» Some minerals may be identified by using special tests such as the flame or acid tests, or by identifying special properties such as magnetism, radioactivity, luminescence, and refractivity.

» God provided minerals for our use, not only for practical purposes, but also for our appreciation of their beauty, reflecting God's love of beauty.

» Native minerals are pure elements, such as gold, silver, copper, platinum, diamond, or sulfur.

» Minerals are most often found as compounds. Major compound mineral groups are the silicates, oxides, sulfides, carbonates, and halides.

» Nearly all minerals occur in special ores, such as the aluminum ore bauxite and iron ores.

» Mining minerals can result in devastation of the land. Proper dominion must minimize these consequences while making the best use of natural resources.

REVIEW QUESTIONS

1. Explain what makes one mineral so precious or costly compared to another.
2. Why is a crystal of pure calcium carbonate made in a lab not a sample of the mineral calcite?
3. What is the likely relationship between mineral deposits that exist today and those that were created before the Flood? Explain your answer.
4. Compare compound minerals, native minerals, and rocks.
5. What is the main reason that a mineralogist identifies minerals?
6. Explain why a mineral's color is not a totally reliable way to identify it. Which test related to color is more reliable?
7. Under what conditions do geologists believe most mineral crystals formed?
8. Compare mineral cleavage and fracture.
9. The Mohs hardness scale shows the relative hardness of a mineral, not its actual hardness. Explain.

10. Why do you suppose that water was used for comparing the densities of other substances in order to compute specific gravity?

11. Why are diamonds often found within ancient solidified volcanic necks or pipes?

12. Compared to other elements, how abundant is aluminum in the earth's crust?

13. What is the difference between an oxide mineral and a mineral that contains the element oxygen?

14. Why are the sulfide minerals economically important?

15. Where could you find beautiful rock formations formed of carbonate minerals?

16. What kind of minerals are the halides?

True or False

17. Aluminum is a very important metal to a modern society.

18. Coal, a kind of rock made from fossilized plant matter that we can burn for energy, is a mineral.

19. Most minerals that we find today probably formed after the Flood.

20. Minerals may be pure elements or compounds.

21. Taste is an important way to classify minerals.

22. If a mineral is harder than the streak plate ceramic, then you won't be able to create a mineral powdered streak. You will just scratch the plate surface.

23. Making jewelry out of gold, silver, and platinum is a selfish and vain misuse of God's resources that He has given us for the purpose of exercising good and wise dominion.

24. Copper is cheaper than gold and platinum because it is more plentiful.

25. Most of the sulfur produced in the United States is from sulfur mines.

26. The only important minerals are silicates, made of silicon, oxygen, and other elements.

The Jefferson Memorial in Washington DC is made of marble, which is formed from carbonate minerals.

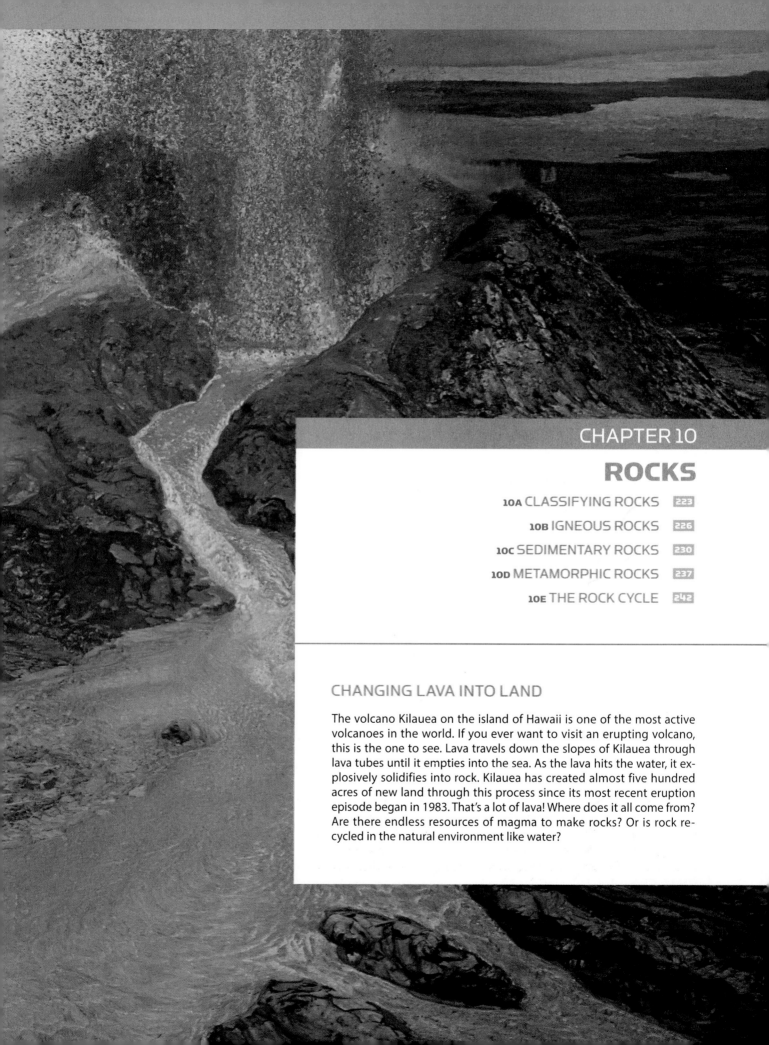

CHAPTER 10
ROCKS

- 10A CLASSIFYING ROCKS — 223
- 10B IGNEOUS ROCKS — 226
- 10C SEDIMENTARY ROCKS — 230
- 10D METAMORPHIC ROCKS — 237
- 10E THE ROCK CYCLE — 242

CHANGING LAVA INTO LAND

The volcano Kilauea on the island of Hawaii is one of the most active volcanoes in the world. If you ever want to visit an erupting volcano, this is the one to see. Lava travels down the slopes of Kilauea through lava tubes until it empties into the sea. As the lava hits the water, it explosively solidifies into rock. Kilauea has created almost five hundred acres of new land through this process since its most recent eruption episode began in 1983. That's a lot of lava! Where does it all come from? Are there endless resources of magma to make rocks? Or is rock recycled in the natural environment like water?

10A

CLASSIFYING ROCKS

How do the types of rock differ?

10A Objectives

After completing this section, you will be able to

» define *rocks*.

» classify types of rocks on the basis of how they formed.

» identify properties of rocks used to classify them.

10.1 RECYCLING ROCK?

If you think about it, the earth has only a certain amount of matter. Remember that the law of conservation of matter says that matter can't be created or destroyed. So that means that the earth's matter can change in form only, not in mass. Now if volcanoes spew lava from the interior of the earth onto its surface, that lava has to come from somewhere. And it seems reasonable that something has to take its place. Otherwise, huge hollow spaces would grow beneath an ever-thicker crust.

A person's view of Earth's history affects how he addresses this problem. Old-earth geologists think that after a long time, most rocks that form from lava at the surface eventually return to deep within the earth. It's as if they were on some sort of very slow geologic conveyor belt.

But is this theory correct or even reasonable? We'll look at how different people answer these questions. First, though, you need to learn about the different kinds of rocks that play a part in this story.

10.2 WHAT ARE ROCKS?

Rock, the solid material in the earth's crust, is a natural combination of minerals or other materials. Most rocks are made of two or more minerals, though a few rocks contain only one mineral. *Petrology* is the study of rocks, their classification, and their history. Geologists who specialize in identifying rocks are called *petrologists*.

When a petrologist studies an unknown rock sample, the first thing he tries to do is identify its general type. He asks two questions: "How did this rock get here?" and "How has it changed since it formed?" This helps a petrologist to classify a rock as *igneous*, *sedimentary*, or *metamorphic*.

You learned in Chapter 8 that igneous rocks are hardened lava or magma. Compared to the past, not many of these rocks are forming now. You also learned that sedimentary rocks form from small particles of eroded rocks or from dissolved chemicals. Large amounts of these kinds of rocks don't seem to be forming today either. Also in Chapter 8, we briefly mentioned how heat and pressure can change igneous and sedimentary rocks into different kinds of rocks in a process called *metamorphism*. Any rocks that result from changes to existing rocks are metamorphic rocks. Geologists assume that some metamorphism still occurs today near geologically active areas.

petrology (pe TRAH luh jee): petro- (Gk. *petra*—rock) + -logy (study of)

metamorphic (met ah MORF ick): meta- (Gk. *meta*—change of) + morph- (Gk. *morphos*—having the form of) + -ic; showing or causing a changed form

10.3 OBSERVING ROCKS

The minerals in rocks help us to classify them. Since rocks are mixtures of minerals, streak tests and hardness tests don't really work on a rock sample as a whole. But using acids or testing for magnetism, radiation, and luminescence can help identify rocks by the minerals they contain. For example, some minerals are dark. If the rock is made mostly of these minerals, it will look dark in color, like igneous basalt. Other rocks may contain a mixture of light and dark minerals, or colored minerals, like pink feldspar or reddish-brown iron oxide. Depending on the size and arrangement of the mineral crystals, the rock may appear speckled, like granite or gabbro, with an overall pinkish appearance.

LIFE CONNECTION: ROCK SWEET ROCK

If through some crazy scientific experiment gone haywire you were shrunk to the size of an ant, what would be your first response? Okay—after you raided the picnic area—where would you go? If you were wise, you would look for some shelter. The world is a dangerous, hostile place when you are that small. Chances are good that you would seek shelter in the space beneath a rock, since they are so common in many habitats.

In your new "digs" you would have some protection from larger creatures who might want to eat you. At the least, the rock would keep you out of their sight. Those predators that are too large to come in after you but too small to overturn the rock would have to search elsewhere for a meal. Another plus is that lots of good food can be found beneath a rock. If you're not a picky eater, you might be able to dine without ever leaving home.

Your *real* home is probably heated and air-conditioned to avoid temperature and humidity extremes. Would you believe that rocks do much the same thing? By blocking the sun's rays, a rock moderates the temperature beneath it, keeping you from getting overheated. The sun's heat is released more slowly over a period of hours as the rock warms. After the sun sets and the air cools, your rock will continue to give off warmth to keep you from getting too cool. Worried about drying out? A rock also helps to maintain a proper humidity level between it and the ground. Even in a desert, there is some life-sustaining moisture beneath most rocks.

Life under a rock is so attractive that you might have to put up with some unusual roommates. Slugs, worms, and a host of arthropods from pill bugs to spiders commonly find shelter under rocks. Larger creatures like salamanders, lizards, and snakes also take advantage of rocks, especially for their temperature- and humidity-control benefits. Many rodents, ranging from mice to marmots, live on, among, and under rocks. The rocks not only provide protection from aerial predators, but also offer dark, cool places to stash food to be eaten later.

But even if you are able to make yourself comfortable and safe in your new home, you might not want to get too relaxed. I think I hear the footsteps of a student trying to find one more specimen for his rock collection!

Did you feel an earthquake?

Petrologists also classify rocks by how they look and feel—that is, by their **texture**. The size, shape, and appearance of the individual rock particles, mineral crystals, and even fossils in a rock determine its texture. These pieces and particles are called **grains**.

The size, shape, and pattern of the grains in a rock help geologists to classify it. Some grains are visible to the unaided eye and can be very large. Petrologists say that these rocks have a *coarse-grained* texture. Grains can be so small that you need a microscope or hand lens to see them. These are *fine-grained* rocks. Grains may be sharp and angular or smooth and rounded. They may be randomly arranged or they may appear in bands of different sizes or colors of grains. In some rocks, grains form wavy or swirly patterns. Some rocks have no grains at all. These rocks often look smooth, glassy, or transparent.

Grains can provide clues to how a rock formed. Tiny, sand-like mineral grains or larger particles of broken rock may indicate that the rock formed from sediments. Closely fitted mineral grains suggest that the mineral crystals grew in place to form the rock. Long, thin grains could be the result of heat or pressure squashing or stretching the rock.

Each of the rock types has textures that are based on the size, shape, and pattern of its grains. These grains are the key to classifying rocks and understanding how they formed. In the following sections, we'll look at each of these types of rocks in detail.

10A REVIEW QUESTIONS

1. State the law of science that establishes the permanence of matter in the universe.
2. Suggest one reason why the law in Question 1 might support the idea of recycling rock in the earth.
3. List the three kinds of rocks classified according to how they formed. Do any of these rocks still seem to be forming in large amounts today, if at all?
4. Relate the source of metamorphic rocks to the other two types of rocks.
5. Why can't you use mineral hardness tests or streak tests to identify most rocks?
6. What are grains in a rock? How do they assist in classifying rocks?
7. What properties of rock grains affect a rock's texture?
8. (True or False) According to a young-earth view of geologic history, it is very likely that metamorphic rocks were part of the original created world.

> **10B Objectives**
>
> After completing this section, you will be able to
>
> » evaluate the theories of origin of igneous rocks.
> » classify igneous rocks by texture and magma types.
> » identify common intrusive and extrusive igneous rocks.
> » explain why igneous rocks have been used by humans throughout history.

10B

IGNEOUS ROCKS

How do intrusive and extrusive igneous rocks differ?

10.4 ORIGIN OF IGNEOUS ROCKS

You know from Chapter 8 that new **igneous rocks** come from volcanoes. But where did the first igneous rocks come from? Old-earth geologists believe that the first igneous rocks made up the earth's crust as it began to cool and harden four billion years ago. They assert that new intrusive and extrusive igneous rocks continue to form as they did in the past.

Young-earth geologists believe that God created the original igneous rocks during the first three days of creation. He probably finished this process on the third day, when He separated the dry land from the water. Modern research shows that the underlying rock forming the continents is mostly granite. So we believe that granite was probably the original basement rock under the created land. If the Flood was the starting point of all significant tectonic activity on Earth, then both intrusive and extrusive igneous rock formations happened only in the past 5500 years since the Flood.

10.5 TEXTURE AND IGNEOUS ROCKS

All intrusive and extrusive igneous rocks contain silicate minerals. These two groups of rocks from the same magmas are usually chemically similar. Until recently, geologists thought that the only real way to tell the difference between intrusive and extrusive rocks lying on the surface was by their average grain size.

Petrologists believed that the slower a rock cooled, the larger the crystals, because there was more time for the atoms to move and arrange themselves into crystals. The faster a rock cooled, the smaller its crystals. They decided that deep intrusive rocks should have large crystals because they were insulated from the cooler surface temperatures, and so they cooled more slowly. Extrusive rocks should have small crystals because lava lacks the insulation of surrounding rocks and cools far faster. Some textbooks still teach this model today.

10-1
A coarse-textured igneous rock like granite (above) contains large, distinct crystals of different minerals. A fine-textured rock like basalt (below) has tiny, even microscopic, crystals.

granite

basalt

10-2
Granite is considered by most geologists to be the original foundation rock forming the continents.

But geologists have discovered that things are not so clear. They have found some extrusive lava flows with large mineral grains, even though they cooled quickly. On the other hand, they've also found underground intrusive formations with very fine grain textures in rocks that should have cooled slowly. It seems now that mineral texture is *not* a sure way to tell where an igneous rock originally formed.

So what *does* affect grain size, and how does grain size help identify the type of igneous rock? It all goes back to the original magma and the surrounding conditions. The magma's composition, temperature, high pressure, the existence of *seed crystals*, and the rate of crystal formation all affected grain size. Groundwater often flowed around and through intrusive magma, cooling it rapidly. But these conditions deep underground actually aided large crystal formation. For extrusive rocks, less favorable crystal-forming conditions, not just fast cooling, usually resulted in small crystals or none at all.

10-3

As you can see, the pegmatite rock shown above is made of very large crystals.

COARSE-GRAINED IGNEOUS ROCKS

An igneous rock with large mineral grains is a *coarse-grained igneous rock*. Many of these rocks are intrusive and can be of the basaltic, andesitic, or rhyolitic chemical families. By far, most intrusive rocks are rhyolitic granites.

Petrologists may classify a coarse-grained igneous rock as a *pegmatite*, a *porphyry*, or a *phanerite*. These textures indicate to a petrologist how fast the crystals in these rocks formed and what kind of magma they came from. See the infographic on the next page for a comparison of the different textures.

Mount Rushmore, Keystone, South Dakota: 43.879°N 103.459°W

FINE-GRAINED IGNEOUS ROCKS

An igneous rock with small mineral grains is a *fine-grained igneous rock*. Geologists believe that extrusive volcanism produces fine-grained rocks because they form at atmospheric pressure within turbulent lava flows or as ejected materials. These conditions do not allow crystals to grow very large. The extrusive rock families are basalt (basaltic magmas), andesite (andesitic magmas), and rhyolite (rhyolitic or felsic magmas). Basalts are dark gray or greenish, andesites are lighter gray, and rhyolites may be nearly white to pink.

Extrusive igneous rocks are often very fine-grained. This means that you can't see the crystals with your eyes, and often, not even with a microscope. They also may be porphyritic, having large, widely separated individual crystals immersed in a fine-grained rock. These crystals began forming in the neck of the volcano just before the lava erupted.

> Geologists estimate that more than 90% of all intrusive igneous rocks are granite and 90% of all extrusive igneous rocks are basalt. Compare these two. It is clear from their different compositions that the source of the surface lavas is quite different from the rocks that make up the continental basements.

- INTRUSIVE ROCKS
- EXTRUSIVE ROCKS

leopardite • gabbro • basalt • granite • phanerite • rhyolite • andesite • diorite

Volcanic lava can produce some very unusual rocks. Basaltic lava may froth and bubble as it hardens, producing a sharp-edged, sponge-like rock called scoria (right, top). Similarly, if dissolved gas in felsic lava foams as it hardens, pumice forms (right, bottom). This rock has so many gas bubbles that it floats like a cork! On the other hand, felsic lava contains no dissolved gases; it can form a glassy rock called obsidian when it hardens. In fact, obsidian is called *volcanic glass* (see bottommost image on page 225).

10.6 USING IGNEOUS ROCKS

People throughout history have used igneous rocks as raw materials for buildings and monuments because of their strength and resistance to weathering. Of course, these properties also make them hard to work with. Stone workers need special tools to cut and shape them.

Common igneous rocks such as granite and basalt are useful because of their beauty and the fact that we can carve them into intricate shapes. The statues, idols, and buildings of ancient civilizations made from granite and basalt have lasted thousands of years. Archaeologists have found that most of the buildings of the city of Capernaum in Jesus' day were built of basalt blocks. Today, people use granite to make gravestones, statues, and monuments. Many modern state and government buildings are also made of granite. Quarries in the United States, including the world's largest in Vermont, supply high-quality granite for such uses.

10B REVIEW QUESTIONS

1. Compare the origin theories of the first igneous rocks. What were these rocks and where were they located?
2. When did virtually all intrusive and extrusive igneous rocks begin to form?
3. What minerals make up granite?
4. How do petrologists use grain size to tell the difference between intrusive and extrusive igneous rocks?
5. Describe the three general types of coarse-grained igneous rocks.
6. Which type of magma formed most intrusive igneous formations? Which type forms most extrusive rocks? What can you conclude about the source of extrusive magmas?
7. Compare the textures of intrusive and extrusive igneous rocks.
8. Pumice and obsidian are chemically similar. Why do they look so different?
9. (True or False) Igneous rocks are generally ground up for gravel in concrete because they are not very durable.

10C Objectives

After completing this section, you will be able to

» evaluate the theories of origin of sedimentary rocks.

» describe the processes by which clastic and nonclastic sedimentary rocks formed.

» correctly classify sedimentary rocks.

» discuss common uses of sedimentary rocks.

10C

SEDIMENTARY ROCKS

How does eroded material become sedimentary rock?

10.7 ORIGIN OF SEDIMENTARY ROCKS

Erosion shapes the earth's surface, both tearing down the earth and rebuilding it again from sediments. **Sedimentary rocks** are made of these sediments. Sediments accumulate on land, in lakes, and in oceans, relatively slowly today because erosion doesn't happen very fast. But it doesn't seem to have always been this way.

Old-earth geologists assume that erosion and sedimentation in the past mostly happened as slowly as they do now. In concluding this, these scientists misapply the principle of uniformity. Most geologists agree that there were catastrophes in the past that could have temporarily increased the rates of erosion or sedimentation in small areas. But they believe that sedimentary rocks that are miles thick must have taken millions of years to accumulate.

When a young-earth geologist sees these same rocks, he thinks, "Flood!" We've tried to help you picture the geologic upheaval during this event. Unimaginable tectonic forces generated great currents of deep water that flowed across the continents. The currents ripped apart rocks, grinding and pulverizing boulders into pebbles, sand, and silt. They scoured the continents down into the bedrock, and the resulting sediment came to rest in thick layers over three-quarters of the earth's land surface. So how did these sediments form sedimentary rock?

EROSION AND SOLUTION

The first step to becoming sedimentary rock involved erosion or water dissolving chemicals. During the Flood, *erosion* occurred at a scale that we cannot imagine. Even today, flowing water in streambeds carries sand particles and pebbles along. These strike against the grains in the surrounding rocks and break them apart. These new particles add to the sediments in the bed of the stream. Ocean waves do the same kind of work on coastal rocks (Chapter 14). During the Flood, these processes were much more violent, widespread, and rapid.

As you learned in Chapter 2, water is an excellent solvent. It can dissolve many natural substances, making *solutions*. Some materials can dissolve in water more easily than others. The more substance dissolved in a solution, the more *concentrated* the solution becomes. The amount of a substance that can dissolve depends on the kind of substance. The temperature of a solution is especially important and probably was an important rock-forming factor. Warmer solutions can typically dissolve more solid materials.

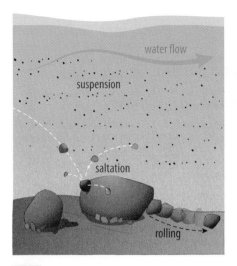

10-4
Eroded sediment can be carried or dragged along by water. Slow-moving or deep water deposits sediments in horizontal layers. Fast-moving water forms structures called *cross bedding*. Both are illustrated below and evident on the facing page.

TRANSPORTATION AND DEPOSITION

Solid particles carried along by water eventually settled out. Layers or strata of sediment under slow-flowing or deep water usually were horizontal. Sediments settled out according to particle size, density, or shape. So it isn't unusual to see a stratum of mainly one kind of mineral or particle size above a stratum of denser or larger particles. Similar pieces of rock and other matter would have all dropped to the bottom about the same time as water speed dropped below a certain value. But dissolved materials could be carried great distances no matter what the water speed. In Chapter 12, you will learn more about water-sediment transportation.

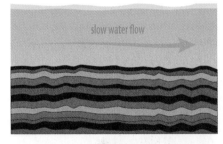

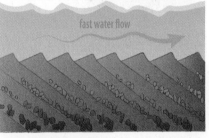

COMPACTION

After deposition into thick layers, sediment particles were pressed more closely together by the weight of sediments above. The spaces or pores between the particles became smaller as the sediments settled. During *compaction*, the water between the particles had to go somewhere. The expelled water flowed upward through the sediments. Some geologists believe that the water flow could have been involved with forming caves in the still-soft rock (see Chapter 17).

10-5
Sediment grains have lots of spaces between them when they first settle out (top). After compaction, there is far less space between grains (bottom).

CEMENTATION

The final stage of becoming sedimentary rock was *cementation*—gluing the particles together. Some minerals like silicates and carbonates are natural cements. These dissolved minerals collected on particles of similar rocks. As the cementing compounds became thicker, the spaces between sediment particles became smaller, forcing out the water and gluing the particles together. In many places, the cements filled the pores completely, making a solid, nonporous rock.

10.8 CLASSIFYING SEDIMENTARY ROCKS

CLASTIC SEDIMENTARY ROCKS

The process of sedimentation formed rocks from pieces of other rocks, called **clasts**. You've seen this word used to describe flows of hot volcanic ash and heated gases, called *pyroclastic* flows. Clasts may be pebbles, sand grains, and microscopic clay particles. Rocks made of eroded fragments are **clastic sedimentary rocks**. Some rocks have only similarly sized clasts. Others may have clasts that differ from very coarse to very fine. Four important types of clastic sedimentary rocks are conglomerates, sandstones, siltstones, and shales.

A **conglomerate** is the coarsest clastic sedimentary rock. It may contain chunks of gravel, pebbles, brick-sized stones, and even large boulders. The spaces between are filled with sand cemented by calcite or silica. Most conglomerates contain rounded, eroded particles greater than 2 mm in diameter. *Breccia* is similar to conglomerate except that its clasts are sharp and angular. They didn't have time to be worn smooth before deposition. The large clasts of conglomerate and breccia rocks suggest that their sediments were carried by fast-flowing water.

The clasts in sandstones are smaller than in conglomerates, between 0.0625 and 2 mm in diameter. Beds of quartz sand grains became strata of sandstone. Silica, calcite, or iron oxide (rust) cemented the grains of sand together. The different colors of sandstone come from the different cementing compounds. These sediments likely settled out in moderate water currents or were formed by wind, glacier, or wave action. Under a microscope, the sand grains often appear rounded by erosion.

Siltstones and shales have very tiny clasts. Particles in these rocks average less than 0.0625 mm in size. You can't see them without a powerful microscope. Siltstone feels gritty and tends to break along surfaces at an angle to its original sediment layers. It does not split into thin sheets at all. Shale is softer and has a finer texture than siltstone. It splits easily into thin sheets parallel to the strata in the rock.

Siltstones and shales are usually gray because of their high clay content, but they also may be white, brown, red, green, or black. These rocks also often contain fossils. Fossil hunters really like shales because they easily split into sheets that reveal fossils of organisms that were trapped as the sediments accumulated.

10-6

Conglomerate and breccia are both composed of larger pieces of rock cemented together. Conglomerate (top) has incorporated rounded fragments; breccia (bottom) includes angular fragments.

clast: (Gk. *klastikos*—broken); rock fragment

conglomerate (kun GLAHM ur it): conglomerate (L. *conglomerare*—to lump together). Geologists often call conglomerate *nature's concrete*.

breccia (BRECH ee uh): breccia (Ger. *brecha*—breaking)

NONCLASTIC SEDIMENTARY ROCKS

Nonclastic or **chemical sedimentary rocks** formed from sediments that weren't clasts of rock. These kinds of rocks formed two different ways: by *chemical precipitation* or as sediments built up from the actions of living things.

INORGANIC NONCLASTIC SEDIMENTARY ROCKS

You know from putting sugar in your iced tea or coffee that it dissolves faster in hot liquids than in cold. Warmer liquids also make it possible to dissolve more sugar, but only up to a point. When no more sugar can dissolve, the solution is *saturated*. If you put more in, it just settles to the bottom. This is true for any saturated solution of a solid compound mixed in water.

Water can dissolve only so much solid at a given temperature and in a given volume. If the temperature of a saturated solution lowers or some of the water evaporates, the solution can't hold all the solid in a dissolved state. Some of the solid appears, making the mixture appear milky. The solid settles out as sediment at the bottom of the solution. This process is chemical **precipitation** (see Figure 10-7).

In the past, chemical precipitation could have easily produced rocks. During the Flood, the world's oceans were probably very warm because of volcanic and other tectonic activity. Such warm water could have dissolved a lot of minerals from the particles suspended in the floodwaters. As the oceans cooled after the Flood and the warm water evaporated, minerals precipitated and formed sediments. Today cooling and evaporation form mineral deposits near geysers and in caves.

Two examples of sedimentary rocks produced this way are limestone and halite. Limestone can precipitate from seawater. It is mainly calcite, a form of calcium carbonate ($CaCO_3$). Broad, deep beds of nonclastic limestone have formed in many places around the world and lie under some very unusual topography. Similarly, geologists have discovered broad beds of halite (natural table salt, NaCl) ranging from 1 m (3 ft) thick to over 60 m (200 ft) thick covered by other sedimentary strata. Some halite deposits have been found deep underground shaped like vertical cylindrical masses called *salt domes*. These deposits seem to have squeezed upward through overlying rock strata under great pressure.

10-7

A beaker containing a saturated solution is shown on the left. Notice the same beaker at a cooler temperature (right). The excess chemical has precipitated.

10-8 Limestone takes many different appearances—a coarsely fossiliferous rock (top) or fine non-fossiliferous precipitate (bottom) are two examples.

ORGANIC NONCLASTIC SEDIMENTARY ROCKS

Perhaps the most interesting nonclastic sedimentary rocks are made of the shells or skeletons of fossils. Because they contain the remains of living organisms, they are organic sedimentary rocks. Some limestones are nothing but compacted masses of clam, snail, and oyster shells of all sizes. These rocks also include fossils of extinct animals like crinoids, brachiopods, and squid-like ammonites. Petrologists call some shell-rich limestones coquina (left).

Chalk is another type of organic sedimentary rock. You may have heard of the White Cliffs of Dover along the north coast of the English Channel. These are beds of chalk nearly 110 m (360 ft) high! Natural chalk is made almost entirely of the tiny shells of single-celled animals called *foraminifera* and plants called *coccoliths*. Chalk beds like this exist all over the world.

10.9 USING SEDIMENTARY ROCKS

People have been using sedimentary rocks for construction for a long time. Many Egyptian pyramids were built of locally quarried limestone. So were important government buildings, statues, monuments, and churches, such as the Washington National Cathedral. Limestone is attractive and relatively easy to work with compared with igneous rocks.

Processed limestone is also an important part of cement. It holds the mixture of sand and gravel (often crushed limestone) together to form concrete. Concrete is probably one of the most common and important construction materials in the world. It was widely used even by the Romans. Brick mortar is similar to cement and includes processed limestone. Crushed limestone, or gravel, is used for construction site preparation beneath poured foundations and roadbeds.

(continued on page 236)

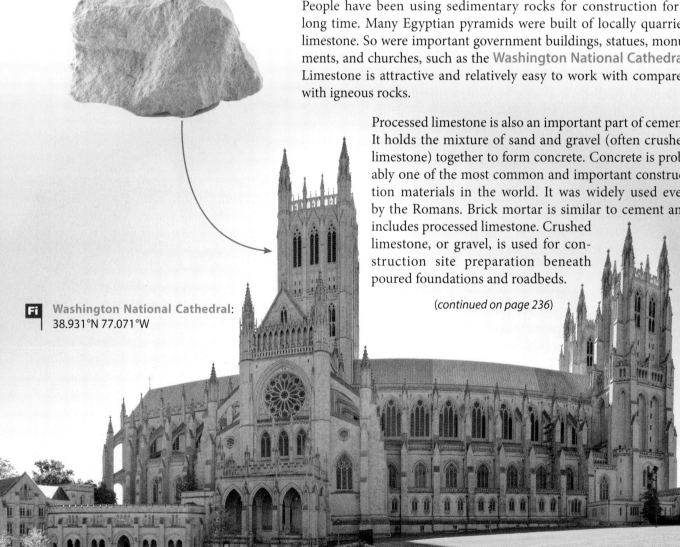

Fi Washington National Cathedral: 38.931°N 77.071°W

SERVING GOD AS A SEDIMENTOLOGIST

JOB DESCRIPTION

Sedimentologists are geologists who study—you guessed it—the materials and processes that formed sedimentary rocks. Though this might sound boring, remember that almost all fossils appear in sedimentary rocks. And fascinating places like canyons and caves are carved out of sedimentary rocks. Now it seems a little more exciting!

People have been doing sedimentology since the 1600s. It's one of the oldest branches of geology. Sedimentologists often call themselves "soft-rock" geologists because most sedimentary rocks are quite soft compared with igneous and metamorphic rocks. This name distinguishes them from igneous petrologists and volcanologists, who call themselves "hard-rock" geologists.

Sedimentologists examine, classify, and identify sedimentary rock samples. This helps them determine the history of the formation from which the sample was taken. They also try to identify the source of the original rocks from which the sediments were eroded. This work can even help us find new sources of minerals, ores, and fossil fuels. Some sedimentologists focus on how certain weathering processes affect the environment.

EDUCATION

As with so many other careers, you can begin to prepare for a career in sedimentology right now. Visit mines and caves, and volunteer in places where sedimentologists are doing research. You can do this by speaking to faculty at university geology departments. Sometimes the best way to see whether a career is right for you is to just get started in any way you can.

A college degree in geology is the starting point in your sedimentology education. Advanced degrees in geology, chemistry, hydrology, and fluid physics, as well as sedimentology, open up more opportunities for work and research.

POSSIBLE WORKPLACES

Sedimentologists get jobs at universities as professors. They work for private companies and with the government. They do research for construction, landfill waste management, and conservation projects. They can work for mining, fossil fuel, and even engineering companies. After they've developed some experience, some set up their own businesses as consultants.

DOMINION OPPORTUNITIES

Keeping our environment healthy, looking for new geologic resources, and ensuring safe building sites are all ways that sedimentologists exercise good and wise dominion over the earth.

But limestone isn't the only useful sedimentary rock. In a few locations in the Middle East, people live in caves carved out of tuff, a form of breccia made of volcanic ash deposits. Halite, or rock salt, is used for salting roads in the winter. Many chemical industry processes also use halite as a raw material. Farm animals lick halite blocks to get the salt that they need in their diets. Halite is also part of many medicines. Likewise, chalk has many uses. You probably think of chalk as a blackboard or sidewalk marker, but athletes also use it to improve their grip. Chalk can be used as a mild abrasive in polishing creams and toothpaste and as a source of calcium carbonate in chemical reactions.

10C REVIEW QUESTIONS

1. When would young-earth geologists say that most sedimentary rocks formed?
2. Explain the process by which most clastic sedimentary rocks probably formed.
3. List four important types of clastic sedimentary rocks. Which of these has the largest variation in clast size within a rock sample?
4. Compare sandstone, siltstone, and shale rocks.
5. How are clastic sediments different from precipitate sediments?
6. What conditions helped the formation of chemical sedimentary rocks during the Flood?
7. What common rock could have formed as both a chemical and an organic nonclastic sedimentary rock at the same time?
8. What is one disadvantage to building with limestone, especially in areas where acid rain falls?
9. (True or False) Fine chalk forms from a precipitate of calcium carbonate.

10D
METAMORPHIC ROCKS

How can rocks change?

10.10 ORIGIN OF METAMORPHIC ROCKS

Rocks probably didn't start out as **metamorphic rocks**. Metamorphic rocks were igneous, sedimentary, or even other metamorphic rocks that have chemically and physically changed over time. Geologists call these original, unaltered rocks *source rocks*. Temperature, pressure, and the presence of hot, watery, chemical solutions called *hydrothermal fluids* make rocks change into metamorphic rocks. These physical and chemical factors are *agents* of metamorphism. They make metamorphism happen.

Metamorphism affected minerals in source rocks in several ways. The agents of metamorphism could have changed the mineral compounds into other minerals, or they could have forced a change in the shape and arrangement of mineral crystals. Often, metamorphism involved all these changes. Geologists recognize three important kinds of metamorphism according to the way that the agents of metamorphism acted: *regional*, *contact*, and *dynamic*. There are also other less common types of metamorphism.

REGIONAL METAMORPHISM

Great forces within tectonic zones crushed rocks together and bent strata to form mountains like the Appalachians, the Alps, and the Himalayas. These tectonic forces also acted along subducting plate margins. Because such zones can extend for hundreds of kilometers, the source rocks changed by these forces display **regional metamorphism**. Regional metamorphism also occurred over large areas simply from the weight of a thickness of rock of overlying strata. When the rocks above them eroded away, it became possible for us to examine them.

(continued on page 239)

10D Objectives

After completing this section, you will be able to

» differentiate between metamorphic rocks and other kinds of source rocks.
» identify the important agents of metamorphism.
» describe important processes of metamorphism.
» correctly classify metamorphic rocks.
» explain why metamorphic rocks have been used throughout history.

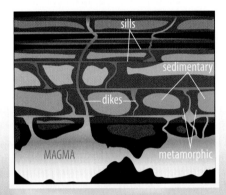

10-9
Pressure and heat from magma will change sedimentary rock to metamorphic rock (shaded dark brown).

10-10
Appearing like great wrinkles in the earth, the Wasatch Mountains in Utah are folded mountains. The processes that created folded mountains are also associated with regional metamorphism.

HYDROTHERMAL FLUIDS

Imagine what Noah and his family experienced in the shelter of the Ark. They could feel and hear the thudding of volcanic explosions and the continual shocks of great earthquakes through the hull. Overhead, along with the incessant roar of the rain, they may have heard bangs and rattles as volcanic bombs showered down on the Ark's roof.

And if this weren't enough to terrify Noah's family, they probably had to endure hot, humid breezes and the sulfurous stench from thousands of active volcanoes. No one born since that time has ever experienced such conditions and lived to tell about it.

As the volcanically heated floodwaters receded, the thick deposits of sediments compacted, forcing out the water from between the sedimentary grains. Since hot water dissolves solids more easily than cold water, the escaping water probably carried lots of dissolved minerals. At some places in the rock, the flow of water rising to the surface through the rock was strong enough to carve out caves.

As the flowing water cooled, it couldn't keep minerals dissolved in solution. So they started to crystallize, forming veins of deposits that filled cracks in rocks, even forming beautiful crystals in caves. In Mexico, Cave of the Crystals, or *Cueva de los Cristales*, holds huge gypsum crystals (see below). These crystals formed underwater, but now they are maintained by hot and humid conditions far underground.

Hydrothermal deposits of gold in quartz are common in some places.

Even today, thousands of years after the Flood, we can still observe magma heating groundwater to form *hydrothermal fluids*. So how did this hot water affect the earth's rocks? And do we see evidence of how hot water affected the earth's rocks?

Hydrothermal fluids during and after the Flood dissolved minerals such as calcium and magnesium carbonate, zinc, lead, copper, silver, and even gold as they flowed through rocks containing these substances. Water solutions move easily through most rocks. So as the water cooled, these minerals precipitated to form the ores we mine today. Hydrothermal fluids were also responsible for chemically changing existing rocks into other kinds of rocks. This process is called *chemical metamorphism*.

These crystals are huge! Notice in comparison the size of the cave explorer.

CONTACT METAMORPHISM

Intrusive or extrusive magma can heat and compress surrounding rocks, altering their crystal structure and chemical composition. We see the results of this kind of metamorphism near dikes and sills, beneath flood basalts, and near volcanic necks and pipes. Since lava forms relatively thin flows on the earth's surface, pressure is not as great a factor. High temperature is the most important agent of metamorphism in nearby source rocks. These changes are the result of **contact metamorphism**. Contact metamorphism is generally limited to small amounts of rock touching igneous features, and it still occurs today near active volcanoes.

DYNAMIC METAMORPHISM

Heat and pressure on relatively stationary rock produce regional and contact metamorphism. But **dynamic metamorphism** seems to have happened when conditions rapidly changed within rocks. For example, rocks along a fault rub and grind against each other as the fault moves. The geologic friction creates enormous amounts of heat and pressure during earthquakes. Mineral crystals in rocks along the fault deform and realign parallel to the fault. This kind of metamorphism is probably very limited today and occurs only near very active faults.

10.11 CLASSIFYING METAMORPHIC ROCKS

Petrologists classify metamorphic rocks as *foliated* or *nonfoliated*. The alterations to these rocks depended mainly on the direction of the pressure that acted on the source rock. Foliated rocks formed when compression forces acted mainly in one direction, flattening the crystal grains. For nonfoliated rocks, forces pressed in evenly from all sides.

FOLIATED ROCKS

Foliated metamorphic rocks have flattened mineral crystals arranged in parallel layers, often making them look banded. They can easily break along these layers. Sometimes the layers peel or flake off. Slate, schist, and gneiss are common foliated metamorphic rocks.

Slate is the metamorphic form of the sedimentary source rock shale. The plate-like clay particles in shale are compacted more evenly in parallel alignment during metamorphism. This makes slate split very easily into broad, thin sheets.

Geologists believe that schist formed when slate rocks became deeply buried and exposed to even greater pressures, temperatures, and hydrothermal fluids. This produced schist's distinct silky, flaky layering. Schist has many thin layers, often highly folded. The layers in schist are *not* normally in the same direction as the original sedimentary strata. Their direction depends on the direction of the

10-11

Schist (shown below) has a high mica content, which makes it split easily into thin flakes.

10-12
This sample of gneiss has the typical banded appearance of foliated metamorphic rock. The folds in the layers show evidence of dynamic metamorphism.

forces that caused the metamorphism. In mica schists, the chemical components of the clay minerals separated into minerals that form at higher temperatures, such as quartz and mica. Using a fingernail, you can easily separate schist into thin flakes of mica.

Under even greater metamorphic stresses, the same minerals found in slates as well as granites formed dense, foliated, coarsely textured rocks with distinct bands of different minerals called gneiss (NICE). The bands include light-colored minerals, such as quartz and feldspar, alternating with darker materials, such as biotite mica. A wide variety of sedimentary or igneous rocks could have transformed into this metamorphic rock. For this reason, the source rock is usually included in the full name of the metamorphic rock, like "granite gneiss."

NONFOLIATED ROCKS

Nonfoliated metamorphic rocks do not have bands or flaky layers. They tend to break into sharp, angular pieces. Common examples include marble and quartzite. *Marble* is metamorphic limestone. In many specimens, the crystals are so small that you can't see them without a microscope, while in others the crystals may be coarse and show cracks along cleavage planes.

Since limestone often contains fossils, marble can too. But the fossils may have been distorted or even completely destroyed by metamorphism. Marble, like limestone, will fizz with the acid test. Pure marble is white, but impurities give it a wide range of colors such as red-brown, green, or black. These colors often flow together and apart, forming the classic marbled texture.

10-13
Michelangelo's famous *Pietà* (pyay TA) is made of marble. The zoom-out shows marble's interesting texture.

Quartzite is metamorphic quartz sandstone. Intense heat and pressure compressed the quartz sand grains, squeezing out spaces and pores until the grains interlocked. Silica deposited by hydrothermal fluids filled in the rest of the space between the crystals. The resulting rock is very hard and nonporous.

10.12 USING METAMORPHIC ROCKS

Just as with sedimentary and igneous rocks, metamorphic rocks have long been used for buildings and decorative stonework. Metamorphic processes often create complex and beautiful patterns of minerals within the rocks. Most metamorphic rocks can be used for essentially the same purposes as their source rocks, but they are often cut and polished to emphasize their beauty. Marble is the main example of a metamorphic rock used this way. Because of cost, architects usually construct a building with another cheaper rock or artificial material and cover the building with a marble *veneer*. Marble countertops are also very popular. Slate is another familiar construction material. It is very much in demand for roofing tiles or interior floor coverings.

But metamorphic rocks are also used for things besides building materials. Quartzite was used in the past for millstones because it is so hard. Today, quartzite is mainly a decorative rock similar to marble. It can be crushed for railway track beds, gravel driveways, and decorative coverings for flower beds. Very pure quartzite is one kind of raw material for silicon, which is used for making integrated circuits.

10-14
Quartzite was used for millstones due to its weight and hardness.

10D REVIEW QUESTIONS

1. What makes the origin of metamorphic rocks different from the other kinds of rocks?
2. What are the three main agents of metamorphism?
3. What kind of metamorphism occurs when intrusive magma forms a sill between two sedimentary rock strata?
4. Name the two main kinds of metamorphic rocks. Explain how geologists believe they formed.
5. For each of the following common metamorphic rocks, identify the source rock and identify whether it was an igneous (I), sedimentary (S), or metamorphic (M) rock.
   a. quartzite
   b. gneiss
   c. slate
   d. marble
   e. schist
6. How do we use metamorphic rocks today?
7. (True or False) Certain metamorphic rocks can form from other metamorphic rocks.

10E Objectives

After completing this section, you will be able to

» explain the key features of the old-earth rock cycle hypothesis.

» evaluate the feasibility of the rock cycle from within a young-earth view of Earth's history.

» refute the assertion that the earth was created to reuse rock natural resources.

10E

THE ROCK CYCLE

Does the rock cycle naturally recycle rocks?

Now that you know about the three different types of rocks, let's revisit the question that we asked at the beginning of the chapter. Do rocks go through a sequence of rock types again and again?

10.13 THE OLD-EARTH ROCK CYCLE

Most natural processes on Earth are cyclic. They repeat a predictable series of steps that allows materials to be reused. God created the earth this way so that living things never run out of what they need to survive.

Not surprisingly, old-earth geologists believe that rock is recycled just like nearly every other natural resource. They call this process the **rock cycle**. According to their deep-time view of Earth's history, they believe that igneous rocks were the first to form billions of years ago as the earth's crust cooled and hardened. Then these rocks eroded over millions of years, and their sediments washed into the seas. Sediments built up into thick layers deep in the ocean basins. They gradually turned into sedimentary rock after more time had passed. This basic process changes igneous to sedimentary rocks. Old-earth geologists assume that this has been happening continuously for more than 4 billion years.

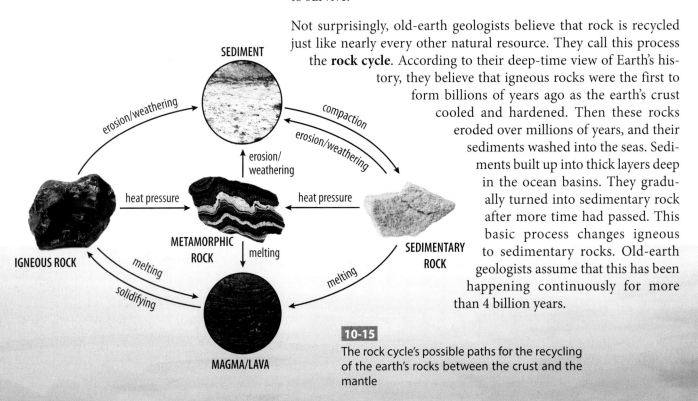

10-15
The rock cycle's possible paths for the recycling of the earth's rocks between the crust and the mantle

But the rock cycle theory assumes that sedimentary rocks can become magma again. Most geologists believe that oceanic plates continually slide under continental plates in subduction zones (see Figure 10-16 below). Sedimentary rocks move along with the crust deep into the earth. There they melt because of the high temperatures in the mantle. Eventually their magma can work its way back up into the crust, where it may form intrusive igneous features. It may even erupt on the surface from a volcano. Thus, the rock cycle ends and begins again after perhaps many millions of years.

In the rock cycle, some rocks can get sidetracked and follow a different path. Because igneous rocks are often close to their source of heat, they can melt again into magma. These rocks never enter the erosion part of the cycle.

In some places, geologists believe that sedimentary rocks rose above sea level as tectonic forces lifted and folded the earth's crust. Erosion and weathering wore down these sedimentary rocks into sediments again. This kept them from returning directly to a molten state in the mantle. There are many places on Earth where sedimentary rocks are made up of clasts from other sedimentary rocks (conglomerates and sandstones, for instance). But old-earth geologists are convinced that most rocks will eventually return to the mantle and melt, given enough time.

Metamorphism is another path that all rocks can take in the rock cycle. After metamorphism, these rocks can follow either of the two paths we just described—to be eroded or to be melted into magma. But has there really been enough time for all this to happen?

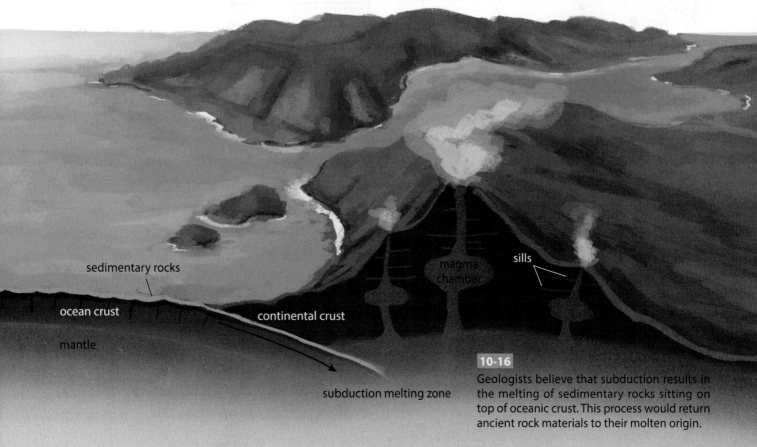

10-16
Geologists believe that subduction results in the melting of sedimentary rocks sitting on top of oceanic crust. This process would return ancient rock materials to their molten origin.

10.14 THE YOUNG-EARTH VIEW

Young-earth geologists agree that certain parts of the rock cycle seem to be happening today. We can observe volcanic eruptions, and we can measure the rates of rock erosion and sedimentation. We observe some kinds of metamorphism. Oceanic plates do slide under continental plates, which would melt some igneous and sedimentary rocks in the mantle.

However, Earth's history as understood by the young-earth view is much shorter than that of the old-earth model. There is not enough time for any given atom within a rock to complete even a tiny fraction of one rock cycle, except under catastrophic conditions. In fact, young-earth geologists conclude that the earth was not originally designed to recycle rocky earth materials at all!

What does this all mean? Most magma and volcanic lava today probably comes directly from the mantle or from deep in the crust. The chemical makeup of most extrusive rocks supports this view. When there is an eruption like the one at Kilauea, the earth's crust gets a little thicker, and it settles a bit to displace the volume of magma that erupted. The law of conservation of matter is satisfied. Only along subduction zones is it possible that some sedimentary rocks could be melted into magma. These zones are probably left over from the tectonic events of the Flood.

10-17
More like a one-way trip . . . (2 Pet. 3:12)

 Cycles and Life
Why do you think that God created water to be recycled in the earth but not rock?

10E REVIEW QUESTIONS

1. Why do scientists think that rock is recycled?
2. According to the old-earth rock cycle, describe the most direct route for the atoms in a rock to be recycled.
3. What are some detours that atoms in a rock could take in the old-earth rock cycle?
4. From a young-earth view of Earth's history, what are the most important conclusions we can come to about the so-called rock cycle?
5. (True or False) No part of the rock cycle is occurring.

CHAPTER 10 REVIEW

CHAPTER SUMMARY

- Rock is a natural, solid material made of one or more minerals.
- There are three kinds of rocks formed from earth materials: igneous, sedimentary, and metamorphic.
- Rock chemical composition is quite variable. It determines the color, type, and shape of minerals that make up the rock.
- Rock texture is the look and feel of a rock sample, determined by the nature of the grains that it contains.
- Igneous rocks are cooled magma or lava. They are classified by their grain size and composition.
- Intrusive igneous rocks tend to have a coarse-grained texture, while extrusive rocks have a fine or glassy texture.
- Most sedimentary rocks probably formed during the Genesis Flood.
- Clastic sedimentary rocks formed by a process of erosion, deposition, compaction, and cementation.
- Clastic sedimentary rocks are classified according to the size of the grains or clasts in the rocks.
- Nonclastic or chemical sedimentary rocks probably formed mainly from chemical precipitation. Some resulted from biological activity.
- Metamorphic rocks are changed igneous, sedimentary, and earlier metamorphic rocks.
- Temperature, pressure, and hydrothermal fluids are the main agents of metamorphism.
- Metamorphism can be regional, contact, or dynamic.
- Metamorphic rocks are classified as foliated and nonfoliated on the basis of the presence or absence of leaf-like layers or color banding.
- Rocks can be used for buildings, sculptures, monuments, protective coverings, decorations, industrial raw materials, and many more purposes.
- The old-earth rock cycle assumes that all rocks began as igneous rocks. They changed into sedimentary rocks or metamorphic rocks and eventually returned to the mantle to be remelted. This cycle can repeat after many millions or even billions of years.
- Young-earth creationists do not believe that the earth was designed to recycle rock. In accordance with the Bible's view of history, there is not enough time to complete even a small part of the rock cycle.

Key Terms

rock	223
texture	225
grain (rock)	225
igneous rock	226
sedimentary rock	230
clast	232
clastic sedimentary rock	232
nonclastic sedimentary rock	233
chemical sedimentary rock	233
precipitation (chemical)	233
metamorphic rock	237
regional metamorphism	237
contact metamorphism	239
dynamic metamorphism	239
foliated metamorphic rock	239
nonfoliated metamorphic rock	240
rock cycle	242

Zuma Rock in Nigeria is a 725 m (2400 ft) pluton.

REVIEW QUESTIONS

1. What physical law assures us that the amount of material in the earth doesn't change?
2. What is a rock?
3. Classify each rock as igneous, sedimentary, or metamorphic from the following descriptions:
    a. a dense rock containing distorted fossils
    b. a hard rock with large, interlocking mineral crystals
    c. a somewhat soft rock with flaky layers or bands of minerals
    d. a rock made almost entirely of fossil shells loosely cemented together
    e. a rock containing sharp, broken pieces of other rocks surrounded by cemented sand grains
4. An igneous rock has large, distinct mineral crystals embedded in fine crystal grains. What kind of texture does it have?
5. Compare the colors of basaltic and felsic (or rhyolitic) rocks.
6. What kind of igneous rock looks like black, broken glass?
7. About how old are the oldest sedimentary rocks according to a young-earth view of Earth's history?
8. Which of the four steps to form sedimentary rocks do *not* apply to a chemical sedimentary rock?
9. How may thick limestone deposits have formed all over the world because of the Flood?
10. Where in the world would you expect to find the results of regional metamorphism?
11. Why could meteorite impact metamorphism be considered a type of dynamic metamorphism?
12. What are two properties that help classify metamorphic rocks?
13. List all of the changes that can happen to rock according to the old-earth rock cycle.
14. On the basis of the earth's design and history, which kind of resource does not seem to be recycled?

True or False

15. The classification and identification of rocks is called *mineralogy*.
16. A key step in classifying a rock is determining how old it is.
17. Color is not a reliable way to classify rocks.
18. A coarse texture applies more to the size and shape of the grains in a rock than to how it feels to the touch.
19. Larger igneous mineral crystals mean that the rock cooled more slowly and probably formed deeper in the earth.
20. Gabbro and basalt are chemically similar.
21. A clast is a fragment of rock or other material cemented into a rock.
22. The main classification difference between siltstone and shale is the way that they feel (siltstone is rougher).
23. Halite probably precipitated from very salty seawater.
24. Length of time is *not* an important agent of metamorphism.

10-18 Notice all the fossils in this limestone.

25. The Himalayas are probably a zone of regional metamorphism occurring today.

26. The bands of colors in foliated metamorphic rocks are normally parallel to the sedimentary strata of the source rock.

27. The rock cycle is an example of a process that God created to reuse natural resources so that we don't run out.

Map Exercise

28. Examine the map to the right and answer the following questions.

    a. What is the theme of the map?

    b. The gabbro rock is an igneous sill that intruded into native limestone. What does the narrow band of altered limestone rock represent? What is this kind of rock called?

    c. Assuming north is up, in which direction is the fault strike?

    d. Did the fault form before or after the magma intrusion?

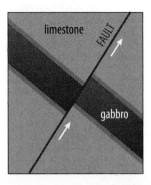

CASE STUDY: ROCKS AND THE AGE OF THE EARTH

Have you ever heard the saying "as old as the hills"? How old are the hills anyway? Rocks give us clues about the age of the earth. Old-earth geologists will tell you that mountains are millions of years old. Young-earth geologists would agree that they are old, but only about 5500 years old since the Flood formed most mountains. The discussion of mountains is a key point of debate between secular geologists and young-earth geologists.

What do old-earth geologists say about rock formation? As you learned in this chapter, igneous rocks can be either coarse-grained or fine-grained. Secular geologists believe that all evidence proves that rocks need millions of years to form. They know that the slower crystals form, the larger they will be. Their model assumes that magma cooled to form large crystal igneous rock only by transferring heat through the earth. This would have taken millions of years. Additionally, geologists see evidence for slow rock formation in sedimentary rocks. While these rocks are forming, living things can disturb the sediments in a process known as *bioturbation*. Old-earth scientists assume that bioturbation occurs slowly. Therefore, if these trace fossils are missing, old-earth scientists assume that some catastrophic event occurred that put down layers before bioturbations could occur. But do both rock cooling and bioturbation have to take a long time?

Creation geologists see the evidence another way. They look at the evidence for coarse-grained igneous rocks and find that they can form quickly. Their model involves large volumes of water removing thermal energy rapidly. Young-earth scientists have demonstrated that the data is consistent with cooling by convection currents in water, which fits well with the Flood. Recent experiments show that bioturbation in sediments can indeed occur rapidly.

Questions to Consider

1. What is the general rule for grain size versus cooling time for igneous rock?

2. Given what we know about movement within the crust, would we expect to see large intrusive igneous rock formations (Half Dome, page 193, or Zuma Rock, page 245) if cooling took millions of years?

3. If rock layers form over long periods, in how many rock layers would secular geologists expect to see bioturbations?

4. Why do you think that we don't see very many layers with bioturbations?

5. Which model seems to fit the evidence better, especially when considered in light of the Bible's story?

CHAPTER 11

FOSSILS

11A FOSSILIZATION 249

11B PALEONTOLOGY 258

11C FOSSIL FUELS 265

TOMB OF TAR

There was a time when the rush-hour traffic in downtown Los Angeles consisted of saber-toothed tigers and giant sloths instead of cars, buses, and trucks. Sound crazy?

The La Brea Tar Pits are right in the middle of present-day downtown Los Angeles. Usually fossils are really hard to find, but in La Brea a whole community of plants and animals is preserved in naturally occurring asphalt that seeps from the ground. There are millions of fossils there. Thousands of saber-toothed tiger bones lie buried in the goo, as well as plants and microfossils, such as freshwater shells, pollen grains, and mice teeth. Many of the fossils are very similar to animals we see today. How did all those fossils get there?

11A

FOSSILIZATION

Where do fossils come from?

11.1 WHAT CAN FOSSILS TELL US?

Scientists believe that the animals and plants whose remains are in the La Brea pits lived during the last glacial period. The asphalt seeps seem to have been there a long time, trapping animals and plants. When a large animal got trapped in the sticky black glop, it died and attracted scavengers. This is one of the reasons why we see so many different kinds of animals there.

The really interesting thing about the bones at La Brea is that most of the plants and animals found there (that are not extinct) are very similar to what we find in coastal California today. They weren't washed into the pits from far away as seems to be the case with other fossil traps. This suggests to scientists that conditions during the last glacial period were pretty much the same as they are now near Los Angeles (except for the freeways). Many scientists believe that fossils can tell us a lot about climates in the past. But is this always true? Let's check this out by first learning about fossils.

11.2 WHAT ARE FOSSILS?

Fossils of large or strange creatures help us imagine what life on Earth was like long ago. Imagine trying to avoid a toothy, 7.5-ton, 45-foot-long dinosaur roaming the countryside! Fossils are simply cool—and they make us think of an Earth that no longer exists. People have collected and studied fossils for thousands of years. But what exactly are fossils?

A **fossil** is any remains or trace of a formerly living thing preserved by natural processes. There are fossils of both plants and animals, ranging in size from bacteria microfossils to huge dinosaurs. Ninety-five percent of all the fossils that scientists find are of **marine organisms**.

11A Objectives

After completing this section, you will be able to

» explain what fossils are and how they form.

» evaluate whether an object is a fossil, a trace fossil, or a non-fossil.

» analyze the origins of fossils that we find today.

fossil (L. *fossilis*—to dig up)

The adjective *marine* refers to anything having to do with the ocean. **Marine organisms** are plants and animals that live in the sea.

11-1
The La Brea Tar Pits became a record of all kinds of animals and plants.

KINDS OF FOSSILS

Casts and Molds. After a dead organism that was trapped in sediment decayed away, the open space it left in the sediment may have filled in with mineral-rich water. This probably happened during the Flood. Eventually, minerals precipitated, filling in the space. Very detailed three-dimensional *casts* of an organism can be the result. This process is called *permineralization*. On the other hand, an empty *mold* could form if minerals encased the organism shortly after it died and then it completely decayed.

Trace Fossils. These fossils are not remains of the organisms themselves, but they are signs that they lived. They include tracks, burrows, eggs and eggshells, castoff exoskeletons, nests, and fossilized waste droppings, called *coprolites*.

Original Materials. Probably the simplest way that fossils formed was by the compacting and cementing of shells and bones into sedimentary rock. Remember from Chapter 10 that certain kinds of limestone are made almost entirely from the shells of marine animals. In more recent, loose sediments, we can find fossized bones and shells from just a few thousand years ago.

Petrified Fossils. In another process called *replacement*, dissolved minerals often replaced the materials in the hard parts of animals and plants. These fossils are not the actual remains of the organisms. Instead, they are rocks that are mineral replicas of the organisms, and they can be very detailed. Petrified trees and mineralized trilobites are examples of this type of fossil.

Microfossils. We find more microscopic fossils than all other kinds of fossils. These fossils are usually less than a millimeter in size. They include tiny mammal teeth, plankton shells, and plant spores.

Carbon Prints. Many fossils are finely detailed carbon prints found pressed between thin layers of shale or sandstone. These *compression fossils* seem to have formed when the crushing weight of overlying sediments squashed the organisms flat. Over time, the original organic tissues chemically changed into thin carbon films. Scientists find many fossils like this of leaves, fish, and even small dinosaurs.

A FOSSIL SAMPLER

Ichthyosaurus
The *ichthyosaurus* is an extinct air-breathing marine reptile. It was very similar in shape and life habit to the modern porpoise, which is a mammal. It sported a long, toothy mouth, large eyes and eardrums, and paddle-shaped legs. Its tail was vertical rather than horizontal like a dolphin's. Paleontologists believe that they gave birth to live young.

Ammonite
Ammonites are extinct squid-like animals that grew within coiled shells. The shells were chambered as the animals grew in a way similar to the modern chambered nautilus, though they are more similar to the octopus than the nautilus. Ammonites were probably free-swimming animals and successful hunters.

Brittle Star
These still-living, bottom-dwelling marine animals are very similar to starfish and sea urchins. Brittle stars are different from standard starfish in that they have five long, flexible snake-like arms, which help them move quickly. Most live in dark, deep water.

Giant Club Moss
These extinct plants are similar to the modern quillwort. The fossil forms of these plants reveal that they were huge, growing to 30 m or more, with a trunk of nearly a meter in diameter. They were vascular plants, but they did not have branches. The leaves grew from the sides of the trunks and left diamond-shaped scars.

Crinoid
Crinoids (below) are interesting marine animals that look like plants. They still live in the oceans from shallow water to depths of 3 mi or more. Crinoids are made of calcium carbonate segments and plates. The mouth is on top of the calyx, or body, of the animal and is surrounded by feathery arms. The calyx is attached to the bottom by a stalk formed of segments connected together like a beaded necklace.

Paddlefish
A paddlefish (right) is a cartilaginous fish similar in some ways to sharks and rays. Modern paddlefish have a long, paddle-shaped projection above their mouth called a *rostrum* that gives the fish its name. Modern paddlefish can grow quite large and are among the largest freshwater fish in North America.

11.3 HOW DID FOSSILS FORM?

DEATH AND DECAY

Fossils came from dead things, but not all dead things ended up as fossils. Under most circumstances, dead things decay. Their cells break down when exposed to bacteria, oxygen, water, and warm temperatures. *Scavengers* also eat the dead tissues. In just a few weeks, all that remains of a dead animal is skin, hair, and bones. A few more weeks, and only bones are left.

Scavengers are animals that eat dead animals for food. *Predators* are animals that hunt and eat live animals.

In some places, decay doesn't happen as fast because bacteria and fungi can't live or grow as well. This helps preserve the remains. Animals that die in dry caves or in the desert simply dry out without decaying. These *mummified remains* can last for many years, or even centuries, without decay. Other organisms come to rest in places that have no oxygen. They can last for many months or years before fully decomposing. Yet others are frozen in place, like a few of the woolly mammoths in the Siberian tundra. Their tissues remain unchanged as long as they are frozen.

But these situations are rather rare. Almost all animals and plants either are quickly eaten or decompose. Except in very rare situations, nothing remains to be fossilized in rocks by processes we see today.

RAPID BURIAL

Fossils, especially large ones, are rare because they form only under special conditions. In every case, the fossils were originally encased in sediments or became sediments themselves. In Chapter 5, you learned that in the deep ocean basins, sediments collect at a rate of just a few centimeters per 1000 years. This is much too slow to cover an organism and keep it from decaying before it can be fossilized.

Ocean and lake geologists find many places where sediments accumulate much faster. And geologists studying ancient sediments believe that sedimentation rates must have been very high in some places. But there are thick sedimentary rock beds covering hundreds of square kilometers that are full of fossils. Old-earth geologists believe that these formed under shallow inland seas over hundreds of thousands of years. This idea suggests that sedimentation rates

BIBLICAL ORIGINS: PROGRESSIVE CREATIONISM

Many Christians hold old-earth creationism ideas like the Gap theory or Framework hypothesis. They believe that scientific evidence for an ancient Earth is convincing. Some also reject biological evolution because it contradicts the clear creation story of the Bible. But they can't conceive any way that the Creation story could have occurred. These Christians hold to a belief known as Progressive Creationism. Dr. Hugh Ross is one of the most vocal supporters of this idea.

Progressive Creationists believe that God created the earth and universe billions of years ago beginning with the big bang. Later, He created the original simple kinds of animals and plants. Through time, He created more complex kinds of organisms, including man. Some of these organisms were preserved as fossils through standard geologic processes, not in a single catastrophic flood. This view considers the fossil record to accurately describe the order of creation from simple to complex organisms.

Questions to Consider

1. According to Progressive Creationism, when did living things begin to die related to the time of Adam's creation?
2. If fossils have formed throughout Earth's history, how significant was the Flood geologically?
3. Ultimately, what is the most reliable source for a Progressive Creationist?
4. Give some examples of Bible passages that deny the validity of Progressive Creationism.

---

must have been very low to keep conditions in the sea basins about the same for such long periods. Otherwise, the seas would have filled in with sediments very quickly. So, with slow sedimentation there shouldn't have been many fossils formed. The organisms wouldn't have lasted long enough to be buried.

Some land animal fossils in China were buried so fast that they left finely detailed casts of soft tissues and even of hair and feathery coverings in the sediments. A dinosaur fossil found in the Badlands of North Dakota had such delicate details preserved in its skin that biologists could even view fossilized impressions of cells! These would not have survived even a few days under normal decay. The only reasonable explanation is that they were rapidly buried under catastrophic conditions such as existed during the Flood.

Reasoning Check

Discuss why the logic (that evolutionists use to explain fossils in extremely thick sedimentary rock strata) breaks down when considering their starting assumptions.

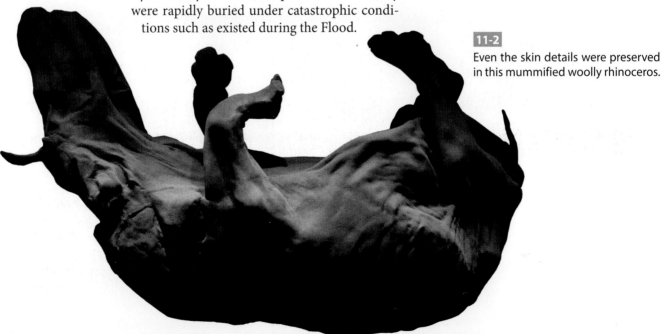

11-2
Even the skin details were preserved in this mummified woolly rhinoceros.

FOSSILS

11-3

Fossils are typically found in sedimentary rock, but can be found in other types of rock. This ammonite fossil is found in limestone—a sedimentary rock.

11-4

This dog-shaped plaster casting was made from a mold resulting from the animal's burial during a pyroclastic flow from the eruption of Mount Vesuvius in AD 79.

11.4 WHERE ARE FOSSILS FOUND?

FOSSILS IN ROCKS

Where should you look for fossils? Geologists and fossil hunters find nearly all fossils in sedimentary rocks. The best-preserved fossils are found in fine-grained rocks like siltstone, shale, and fine sandstone. Coarse-grained sedimentary rocks, like coarse sandstone and conglomerates, rarely contain whole fossil organisms. These kinds of sediments settled out under fast-flowing waters. Animals and plants would have been tumbled by the currents, torn apart, and ground to pieces by the moving sediments.

But other rocks can contain fossils too. Metamorphic sedimentary rocks may contain fossils if the rocks weren't exposed to great temperature or pressure. In some places, beautiful black marble contains many fossils of shelled creatures where the original materials have become calcite. Usually, however, the pressures and temperatures of metamorphism smear or completely destroy all evidence of fossils.

Igneous rocks rarely contain fossils. As you can imagine, organisms covered by molten lava would quickly burn to ash. Scientists have found charred fossil trees embedded in some ancient lava flows that were later buried in sediments. Volcanic ash and tephra often buried organisms and left empty molds behind after the organic material inside decayed. The AD 79 eruption of Mount Vesuvius created many human and animal fossil molds in pyroclastic tephra (Figure 11-4). Plaster castings made from these molds are displayed today in the excavated areas of Pompeii in Italy. Except for these, fossils in igneous rocks are rare.

FOSSILS IN LOOSE SEDIMENTS

Some recent sediments did not turn to rock. Local floods, glacial melting, or volcanic ash falls produced these sediments. Animals and plants were buried quickly, but the thickness of the sediments was not enough to form rocks. For example, below the bottom sediments of Lake Champlain there is a layer of mud that probably was deposited after the Ice Age. Fossils of saltwater creatures have been found, including the skeleton of a small whale called a beluga. Microscopic marine fossils and ocean clamshells are also found in this layer. Geologists call the saltwater body that deposited

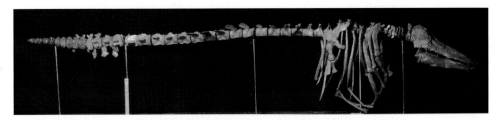

11-5 This skeleton of a saltwater beluga whale was found in loose sediments near present-day freshwater Lake Champlain.

Location: 44.5 °N 73.3 °W

these sediments the Champlain Sea. Below these ocean sediments are different sediments containing freshwater fossils. So a large freshwater lake that geologists call Lake Vermont must have existed before being scooped out by advancing glaciers. None of these sediments are rock, but they do contain fossils.

FOSSIL TRAPS

Amber. Some fossils formed when organisms were trapped in places other than in flood or volcanic sediments. Amazingly lifelike fossil organisms can be found in fossil amber. Amber is a mineralized form of tree sap. For years before the Flood, certain trees produced thick drips of sap that oozed down their trunks. These drips trapped insects, pollen, leaves, and air bubbles before they hardened, sealing them off from the air. Even small vertebrates like lizards were trapped. During the Flood, the sediments buried the hardened sap drips and they became amber. The original tissues remain trapped in these organic stones. Scientists have even recovered DNA from some specimens!

Today, pieces of amber samples are found in many kinds of sedimentary rocks. In some places, there are so many of them that people mine and sell them as jewelry. How about wearing a fossilized insect on your neck!

Ice. As you learned in Chapter 5, ice can trap organisms too. In Canada and Siberia, many bones (and a few whole carcasses) of large animals froze in the permafrost. Even humans are preserved in arctic and glacial deep-freezes. One famous mummified human fossil was nicknamed Ötzi by researchers (see box on page 257). He died and was frozen into a mountain glacier in the Alps near Austria and Italy about 5300 years ago, according to carbon-14 dating.

11-6 This insect was trapped in a drip of resin prior to the Flood, then fossilized during that event.

Asphalt Traps. And then there are asphalt traps like the La Brea pits. These ancient Ice Age time capsules have provided a wealth of information about climate and the wide assortment of animals and plants that lived together at the same time. As dead organisms settled into the asphalt, soil carried by wind and water filled in the pits and surrounded the fossils. Here, modern fossil hunters find tarred bones, leaves, and microfossils. Unlike groups of fossils in most sedimentary rocks, the asphalt traps provide a reliable record of animal and plant communities living in one place.

> **Bi** The book of Genesis indicates that the world is only about 7000 years old. The Flood, which probably produced most fossils, occured about 5500 years ago.

11.5 HOW OLD ARE FOSSILS?

Perhaps some of the controversy about fossils would settle down if they came with date tags. But they don't! Evolutionists assign ages to fossils by the geologic ages of the rocks that contain them. Recall that rocks are dated by their position in the geologic column, by radiometric dating, or by the presence of special fossils called *index fossils*.

As you have learned in previous chapters, evolutionists think that the fossil record spans hundreds of millions of years. Because of such vast periods of time, scientists expect that fossils could not possibly contain original organic materials, except possibly shells or thick bones. These are mostly minerals already. Soft organic tissues simply could not hold together for such vast lengths of time. Until recently, evolutionary scientists simply accepted this as a working assumption.

Even so, scientists have been able to identify many organisms in amber that are the same as living ones. They have even obtained small amounts of DNA from fossil tissues. This is pretty amazing, since the oldest amber fossils are around 140 million years old according to the geologic time scale!

In 2005, a biologist reported that she had found flexible stretchy tissues and remnants of red blood cells inside the broken fossil thigh bone of a 68-million-year-old *Tyrannosaurus rex*. This discovery completely surprised most scientists. Many didn't believe it was possible. Soft tissue simply could not last for millions of years. At about the same time, researchers made similar findings in other fossilized animals.

How do you suppose evolutionists explained these unexpected discoveries? They concluded that *soft tissues can last far longer than anyone could have imagined!* This was the only explanation that fit within their assumptions of an old earth.

For people who don't share their worldview, however, the conclusion is quite different. To address this issue, scientists from the Creation Research Society are working on the iDINO project. In the first phase, they looked at the issue of soft tissue preservation in fossils. They found soft tissue remaining in the horn of a *Triceratops*. These scientists were surprised that the soft tissue had remained since the Flood. This realization led them to the second phase of the project. In this part, the scientists will do an in-depth study of the process of fossilization. They are trying to determine a mechanism that would preserve soft tissue for thousands of years.

These fossils show *recent* burial and fossilization. For soft tissues to survive even the thousands of years since the Flood is amazing. To say that they could be preserved for tens of millions of years is simply incredible. This just shows how tightly people hold to their presuppositions. To say that the dinosaurs lived mere thousands of years ago would unravel the entire evolutionary story of Earth's biological history.

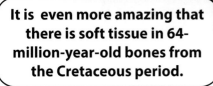

It is amazing that there is soft tissue in 5400-year-old dinosaur bones from the time of the Flood.

It is even more amazing that there is soft tissue in 64-million-year-old bones from the Cretaceous period.

11-7
Stretchy connective tissue (indicated by the arrow above) is probably the remnant of red blood cells found inside a fossilized dinosaur bone.

Images republished with permission of the American Association for the Advancement of Science, from "Soft-Tissue Vessels and Cellular Preservation in Tyrannosaurus rex," Science, 25 March 2005, vol. 307, Mary H. Schweitzer, et al. Copyright © 2005. Permission conveyed through Copyright Clearance Center, Inc.

WORLDVIEW SLEUTHING: ÖTZI

Scientists learn many things by studying fossils. What they *can* learn depends on the type of fossil found, where they discovered it, and other items found with the fossil. The discovery of Ötzi, a mummified man, in the Alps in 1991 has provided a great opportunity to study a human fossil.

INTRODUCTION

You are a reporter for the local TV news program. The manager has assigned you to produce a video report about the discovery of the iceman Ötzi.

TASK

Research, plan, and produce a four-minute video about the discovery of Ötzi. Include in your report the discovery itself, the scientific studies done, and the lessons learned from the discovery.

PROCEDURE

1. Research the discovery of Ötzi by doing a keyword search for "Ötzi discovery." Pay attention to the conditions that allowed for the preservation of the remains.
2. Research the variety of tests done on the fossil. Do a keyword search for "Ötzi scientific testing."
3. Research the different conclusions that scientists have come up with regarding Ötzi by doing a keyword search for "Ötzi conclusions."
4. Plan your video production and collect any photos or videos you need, being careful to give proper credit.
5. Produce the video and show it to another person for feedback.

CONCLUSION

The Bible provides much information about life in the Middle East before and after the Flood. However, we don't have much information about people who moved to other regions after the Tower of Babel. Ötzi provides us with a fascinating view of daily life in present-day Europe within a few centuries of the Flood.

11A REVIEW QUESTIONS

1. Why do the La Brea Tar Pit fossils represent a local community of plants and animals that lived in Southern California during the Ice Age?
2. Is a chicken bone found in a waste dump a fossil? Explain your answer.
3. Which kind of fossilization process leaves behind a three-dimensional shape of an organism made out of minerals?
4. What happens to the remains of nearly all living things that die? What normally *does not* happen to these organisms?
5. What fossil evidence supports rapid and recent burial?
6. Where can we find fossils?
7. Suggest a situation that formed a fossil and preserved its DNA or soft tissue.
8. Why would evolutionists have problems explaining the existence of soft tissues in a mineralized fossil?
9. (True or False) A trace fossil is an impression of an organism that remains after the original material decays or dissolves away.

11B Objectives

After completing this section, you will be able to

» summarize how to classify and name living and extinct organisms.

» identify the factors that lead to the extinction of an organism.

» evaluate efforts to interpret the fossil record in light of one's worldview.

» analyze and evaluate the explanations for the evidence of mass extinctions in the fossil record.

11B

PALEONTOLOGY

What can we learn from fossils?

11.6 CLASSIFYING FOSSILS

Let's go on a dinosaur hunt. The best place to search is probably in the Shandong Province of western China. Here you can dig for dinosaur bones in a fossil graveyard, a place where lots of fossils are found all jumbled together. There are more dinosaur bones here than anywhere else in the world. You painstakingly pick away rock, brush away grit, map the locations of bones, and sort and protect your finds in plaster. You and your team have uncovered the skull of a dinosaur with three horns. What dinosaur is it? Is it similar to anything living today?

Paleontologists excavate Tyrannosaurus rex *fossils in Shandong Province, China.*

The area of science that studies and classifies fossils is called **paleontology**. *Paleontologists* identify fossils just as biologists do living organisms. Both scientists use the classification system that is based on the one developed by Carolus Linnaeus in 1735. This system groups and classifies organisms on the basis of similar characteristics. Each level in the naming system has a category name, such as *kingdom*, *phylum*, *class*, and so on. Biologists assign all living things unique scientific names to set them apart from other similar organisms. These names consist of a genus and species spelled out using the rules of Latin.

Fossil organisms are also given scientific names. For example, the three-horned dinosaur you found on your hunt is classified in the genus *Triceratops* and the species *horridus*. Its scientific name *Triceratops horridus* means "frightful three-horned face" in Latin.

Because most biologists have an evolutionary worldview, they classify living organisms today not only by their appearance but also by genetic similarities. They assume that living things with similar genetic information (DNA and proteins) are related. This is why completely different organisms like whales and hippos are thought to be related by evolution. But this doesn't work for fossils, because most fossils are rock or minerals, and rock has no DNA. So paleontologists classify fossils by the similarity of their appearances to living organisms. When fossils don't seem to fit the system, they sometimes make up new categories, especially for organisms that don't exist today.

In a scientific name, the genus name is always capitalized and both names are italicized.

11-8

The *Triceratops* may actually be a young version of the full-grown *Torosaurus*, a three-horned dinosaur.

Recently, the evolutionary worldview has greatly affected the classification system. Evolutionists are classifying organisms according to their supposed evolutionary relationships. This introduces more complexity into the system. More subgroups had to be introduced to make this work.

11.7 FOSSILS AND EVOLUTION

Even before Charles Darwin published his book *On the Origin of Species* in 1859, some biologists believed that similar animals and plants were somehow related to each other. Darwin suggested that populations of living things changed over time as a result of conditions in their environment. He called this adaptation process *natural selection*. The most successful organisms survived to reproduce. Darwin's theory explained that all similar organisms ultimately descended from less specialized, less complex, common ancestors. Therefore, evolutionists conclude that all life must have evolved from a single kind of organism far in the past.

This idea seems to work in the field of paleontology. In general, relatively simple fossils are found in lower rock strata and more complex fossils are found higher up. According to the principle of superposition, the oldest layers of rock are the lowest in the geologic column. So the simpler fossils must be older than the more complex ones that are higher up, and thus are more recent.

Evolutionary paleontologists classify and arrange fossils using these presuppositions. This isn't easy. Paleontologists work with sedimentary geologists to develop an evolutionary sequence. But often, fossils show up out of order in the geologic column in different places—either earlier or later than the estimated age of the layer of rock in which they were found! Then the scientists have to decide what needs to be corrected—the geologic time scale for a given location, or the time when the fossil lived.

11-9
This is one hypothetical evolutionary sequence showing how fish supposedly evolved into birds.

11-10
This is an artist's conception of a *Panderichthys*. Evolutionary scientists consider this a possible transitional form between fish and tetrapods (see next page).

Fossils that seem to appear reliably in their proper sequence across the world are called **index fossils.** Paleontologists identify index fossils by looking at the geologic column. They look for fossils regularly found only in a narrow range of a local geologic column. Then they look for the same fossils in the same narrow range in other local columns. If they find such a fossil, they call it an index fossil. Index fossils can identify layers in a geologic column and can tell us relative information regarding when a layer formed. We have to beware of implying absolute age on the basis of these index fossils.

So what does the fossil record show? Secular scientists hope that it will clearly show evolution. They anticipate finding a record of one animal slowly changing into a different animal. They search for fossils that show this change—**transitional forms** or missing links. An example is the fossil *Archaeopteryx*, which scientists long thought was a transitional form between dinosaurs and birds. Recent research has shown that this is not a transitional form at all, but just a fossilized bird.

Today, paleontologists are searching for the missing link between fish and tetrapods (TET ruh PODS)—land animals. They think that fish fins evolved into limbs for walking. There are fish with limb buds supporting fins. However, there is no evidence that these buds evolved into legs.

The coelacanth (SEE la KANTH) is a fish that has fins that look like the predecessors of limbs. Scientists had thought that it became extinct 66 million years ago—until one was captured from a South African River in 1938! Now, scientists have turned their attention to *Panderichthys* and *Elpistostege* as possible transitional forms between fish and tetrapods.

So what does the fossil record show? It shows the sudden appearance of species and then remarkably few changes. Organisms appear suddenly in the geologic column and then remain relatively unchanged. Clams remain clams, fish remain fish, and dinosaurs remain dinosaurs (Gen. 1:12, 22, 28). The fossil record is powerful evidence against evolution.

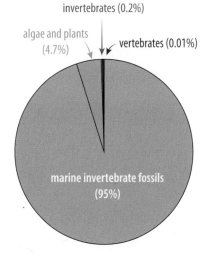

11-11

One challenge to finding answers in the fossil record is the proportion of fossils found. Most fossils are marine invertebrates.

11.8 EXTINCTIONS

Some of the plants and animals that we see as fossils are no longer alive. They became *extinct*. A species is headed for **extinction** when more animals or plants die than are born in a generation. In other words, the population of a species dramatically shrinks over time. This can happen for many reasons, such as disease, predators or parasites, natural catastrophes, or the lack of ability to compete with other species for resources. Paleontologists believe that they can see several great extinctions in the fossil record. If certain fossils exist in earlier, lower strata but not in later, higher strata, then they assume that the fossil species became extinct.

According to the evolutionary interpretation of the fossil record, there have been five major *extinction events* during the past 500 million years. These extinctions did not affect just a few kinds of organisms. Rather, large groups of animals and plants seemed to disappear from the fossil record all at once. The largest of these events was the Permian-Triassic, which is thought to have caused the extinction of 95% of marine animals and 70% of land animals. Paleontologists try to give reasons for these extinctions. These include widespread volcanism, global warming, ice ages, and even nearby exploding stars (supernovas). Geologists believe that they can see evidence for most of these events in the geologic record.

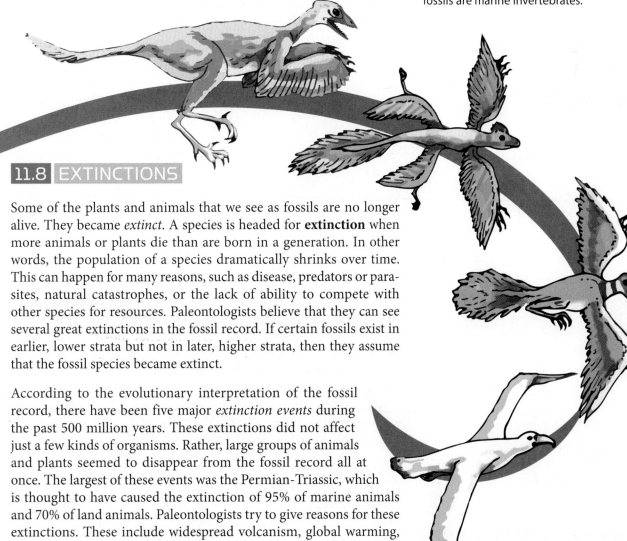

The most famous extinction event is the disappearance of the great dinosaurs from the fossil record around 65 million years ago at the beginning of the Cenozoic era. Recently, old-earth scientists from all over the world have agreed that this change in the fossil record can be blamed on a huge meteorite impact. The debris that it kicked up into the atmosphere supposedly caused global cooling for many years (see the box on the facing page). Some believe that the dinosaurs became extinct because they couldn't adapt to the cold or because of increased competition from mammals in the cooler climate.

11.9 WORLDVIEWS AND PALEONTOLOGY

How should a young-earth Christian view paleontology? The goal of paleontology—the study and classification of fossils—is a legitimate activity of science to understand God's world. Through the study of fossils, we can better appreciate the wondrous abilities of our Creator.

But the dominance of biological evolution and old-earth geology has imprisoned old-earth paleontologists. They *must* interpret fossils within an evolutionary worldview. They *have* to look for evolutionary relationships in the fossil record. They *have* to assume gradual change over time separated by a few catastrophic extinctions. Their worldview requires these things. They also assume that all organisms found in the same stratum probably were part of the same community, like the La Brea Tar Pit fossils.

However, young-earth paleontologists can offer several theories for the variations in fossils at the different levels of the geologic column. We assume that most fossil-bearing sedimentary rocks came from the Flood. Changes in the number and kinds of fossil groups result from the way that they were carried in the moving water, where they were buried, and the order in which they were buried.

> **Bi** **Faith to the Test**
> So what happens if you encounter some scientific evidence that seems to contradict what the Bible says? What should you do?

Many paleontologists believe that a large asteroid impact led to the extinctions of the dinosaurs. But is that so?

WHAT HAPPENED TO THE DINOSAURS?

Imagine how William Foulke felt when he uncovered one of the first documented skeletons of a dinosaur in the sandy soil of Haddonfield, New Jersey, in 1858. What did the living creature look like? How did it die? These huge beasts fascinate us, and we wonder why they disappeared.

The most popular evolutionary explanation for the disappearance of dinosaurs is the meteorite impact theory. This theory suggests that about 65 million years ago, the dinosaurs were wiped out by an asteroid or comet collision with the earth. This would have created a devastating blast hundreds of miles around the impact. The theory also suggests that the impact kicked up debris into the atmosphere that blocked much of the sun's rays for many years. This created year-round winter-like conditions that limited plant growth. Animals needed plants for food, and some of them, used to subtropical climates, couldn't adapt to the cold. The drastic environmental changes are thought to have ultimately killed the dinosaurs.

This theory had its beginning in 1978, when a gravity map (see image to the right) displayed a roughly circular geologic formation about 180 km in diameter, deep under the sediments of the Yucatan peninsula near the town of *Chicxulub* (CHEEK soo loob) (loc.: 21.136°N 89.516°W).

A few years later, scientists began suggesting that a large meteorite may have hit the earth. They found traces of the element iridium (commonly found in meteorites) in greater amounts than is normal at many places in sedimentary rocks around the earth.

Eventually, as geologists compared data, they began to declare that the Chicxulub crater was left by a huge meteorite 10 km in size. They guessed that this impact killed off not only the dinosaurs, but nearly 75% of all the animals on Earth at that time. Evolutionists call this the great Cretaceous-Paleogene extinction. In March 2010, after collecting data for decades, many experts concluded that the impact at Chicxulub caused this extinction. This is now the accepted explanation according to most scientists.

But not all scientists agree with this decision. There are still many who find the evidence presented unconvincing since some data actually contradicts this finding. They believe instead that the dinosaurs could have become extinct from a series of comet or meteorite impacts over many hundreds of thousands of years. But all old-earth scientists agree that all dinosaurs were extinct long before humans evolved.

Young-earth scientists do not see great extinctions in the fossil record. They see the disappearance of organisms in the fossil record as evidence that they were no longer buried and fossilized during various stages of the Genesis Flood.

Because the Bible says that every kind of air-breathing animal went onto the Ark prior to the Flood, we believe that dinosaurs survived the Flood as passengers on the Ark. Many kinds of evidence support this view. The book of Job describes two animals that appear to have been large dinosaurs. Behemoth (Job 40:15–24) was probably a large land dinosaur like *Brachiosaurus*. This dinosaur was up to 26 m long and weighed around 30 tons! Some creationary scientists believe that leviathan (Job 41:1–34) was possibly a huge marine crocodile similar to a recently discovered fossil named *Sarcosuchus imperator*. This beast may have been over 12 m long and weighed 8 tons! No one knows exactly when the events of Job occurred. But evidence from the book indicates that it may have been late in the Ice Age, 600–1000 years after the Flood.

Virtually all cultures have dragon myths. Culture scientists will argue that such myths are fictional, but most scholars agree that myths are often highly distorted versions of some historical truth.

Ancient rock paintings (*petroglyphs*) from many places around the world seem to show large dinosaur-like beasts, including winged pterosaurs. A detailed carving in the wall of an 800-year-old Cambodian temple looks very much like a *Stegosaurus* (a plant-eating dinosaur with large plates of bone along its back).

The Roman historian Cassius Dio reported the destruction of a great beast more than 37 m long during the First Punic War of the Romans against the Carthaginians about 255 BC! And a Medieval painting in a palace illustrates a beast on a leash that one paleontologist believes is very similar to a small plant-eating dinosaur.

These and many other interesting reports from history indicate that dinosaurs may have lived until recent times, long after the Flood. But because people have explored nearly all the earth's surface, and no living dinosaurs have been scientifically observed in the past several hundred years, it is likely that they are now extinct. They couldn't successfully compete with humans and other animals for food and resources. This story is similar to that of many animals that have become extinct since the Flood. However, it is still possible that in some remote location in the world, there may be a few individuals of a dinosaur species left. Living fossil species continue to be discovered all the time.

No one alive today has ever witnessed a natural catastrophe like the Flood. Global volcanism, several hundred huge meteorite impacts, and what seem to be massive extinctions in the fossil record are all probably related to the Flood. Young-earth scientists suggest that the less mobile animals like bottom-dwelling sea creatures were the first to be buried by both water-borne sediments and volcanic ash. More mobile sea creatures were buried many days later. As sea levels rose, the smaller, slower land animals drowned and were buried. Finally, the larger, faster-moving land animals and birds were buried. This agrees in general with what we see in the fossil record.

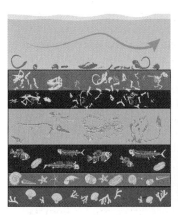 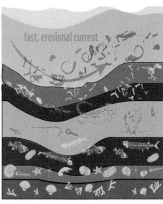 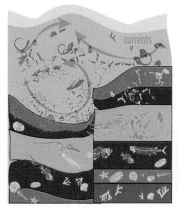 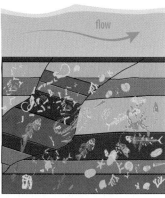

11-12
The great tectonic events that seem to have occurred during the Flood could have moved great quantities of fossils back and forth, creating the stratified fossil beds that we see today.

Though there are exceptions, the Flood is a starting point to explain fossil sequence and extinctions. But if this model is correct, we cannot say for sure that all plants and animals found together in a certain rock stratum actually lived together. Truly, the presuppositions you bring to paleontology determine how you see the fossil record.

11B REVIEW QUESTIONS

1. How do paleontologists classify fossils? How does this compare to the way that biologists classify living animals?
2. How do paleontologists name fossils?
3. With which modern animal category shown below would you classify the fossil in the photograph in the margin?

A   B   C

4. Where did Darwin say all living things came from?
5. Why does the fossil record seem to support evolution?
6. What is the biggest problem evolutionists have when trying to use the fossil record to support biological evolution?
7. Describe the conditions in a species population that will lead to its extinction.
8. According to the evolutionary history of extinctions, which extinction was the largest? What kinds of organisms did it affect?
9. List three kinds of conditions old-earth paleontologists blame for the extinctions that they see in the fossil record.
10. (True or False) DNA studies are a valuable tool for classifying fossils.

TRILOBITES

When you think of fossils, you probably imagine the big ones—dinosaurs, woolly mammoths, and saber-toothed tigers. But did you know that only 0.0125% of discovered fossils are vertebrates (animals with backbones)? About 95% of the fossils paleontologists find are shellfish and other marine invertebrates (sea animals without backbones). Can you think of any reasons why that would be true?

One of the most interesting and common kinds of marine fossils is a trilobite. Trilobites were segmented animals with external skeletons like crabs and pill bugs (roly-poly bugs). The trilobite is named for the way that its exoskeleton was divided into three distinct lobes along its length—a central lobe and a lobe on either side. Its body had a head, called a *cephalon*, a thorax made of from 2 to 16 segments (or more), to which walking legs were attached, and a tail section called the *pygidium*. The legs also included gill-like structures for breathing, like modern lobsters. Most trilobites had complex compound eyes. Some trilobites swam, but most probably crawled along the ocean floor and scavenged for food. Some may have eaten other ocean animals.

Similar to crabs, shrimp, and some insects, they shed their exoskeletons as they outgrew them. In fact, many trilobite fossils are actually cast-off exoskeletons. And there were lots of trilobites—over 17,000 named species! The trilobite fossils we find range from 1 mm to 720 mm (over 2 ft!) long, though the average fossil is about an inch long.

Their exoskeletons had a wide variety of shapes and decorations. Some had interesting bumps and some had pretty dramatic spines. Many varieties even had eyestalks and lacy decorations. We can learn many things about these little creatures from the fossils we find. Sadly, it seems that they are all extinct, since none have been found alive in modern times.

11C

FOSSIL FUELS

Where do fossil fuels come from?

11.10 WHAT ARE FOSSIL FUELS?

No one knows for sure when man first started using substances from the earth for fuel. Records show that people first used *coal* for smelting copper ores in China around 1000 BC. Oily *petroleum* seeping from rocks on the earth's surface was probably discovered nearly as long ago. Some historians believe that early people may have discovered that petroleum could be used as a fuel when they noticed it burning after a lightning strike in an oily bog. *Natural gas* is often linked to petroleum. Its first usage as a fuel was reported by the Chinese philosopher Confucius nearly 2700 years ago.

These natural solid, liquid, and gas materials are good and useful sources of thermal energy, or *fuels*. For a long time, geologists have been certain that coal is fossilized plant matter. They've developed similar theories for the origin of petroleum and natural gas over the past century. So we call these materials **fossil fuels**. Let's take a look at these resources and what makes them so important to our modern world.

11C Objectives

After completing this section, you will be able to

» describe fossil fuels and how we use them.

» evaluate different origin theories for coal, petroleum, and natural gas.

» evaluate the risks and benefits of using fossil fuels.

11.11 COAL

Coal is classified as an organic sedimentary rock. It is not a mineral according to our definition because it is organic—it comes from fossilized plants. There are three types of coal: lignite, bituminous coal, and anthracite, listed in order by increasing density, hardness, and carbon content. Anthracite is considered a metamorphic rock because it has been changed from its original materials by heat and pressure.

11-13 Locations of coal deposits in the United States

bituminous coal

Lignite—brown coal—is dark brown and relatively soft and crumbly. It contains about 30% carbon and a lot of moisture. Lignite burns with a smoky yellow flame and produces lots of soot. Fibrous fossilized plant matter can often be seen in a sample of lignite.

Bituminous (bih TOO mih nus) **coal**, called *soft coal*, is denser and harder than lignite. It is usually black but can be dark brown. It often shows bands of alternating black and brown, left over from its formation. It has a rough fracture that often follows the layering of the compacted fossilized matter. It contains about 70% carbon and includes water and small amounts of oily and tarry organic substances.

Anthracite (AN thruh SITE) is the hardest form of coal. It is very black, fractures with a rough surface, and has a submetallic luster. Anthracite contains more than 90% carbon. It has very little water or nonsolid organic substances. It burns with a very hot, blue, nearly smokeless flame.

Coal can be removed from the earth in standard deep mines, with vertical shafts that provide access to broad seams of coal deep underground. But vast open-pit surface mines also produce much of the world's coal.

Coal is an important resource as a fuel and as a high-temperature heat source for certain industries. Lignite is used mainly for running electric power plants. Nearly half the electrical energy in the United States is from coal-fired plants. Bituminous coal is mainly used for the steel industry, both for heating the furnaces and as **coking coal**, a material used in the iron smelting process. Anthracite mining has mostly stopped worldwide because there is far less of it. Because it burns cleanly, it is still used in power plants where air pollution is a concern.

anthracite

In the steel industry, coke is a smelting fuel and a chemical agent for producing molten carbon steel. **Coking coal** is prepared by heating bituminous coal to a very high temperature to drive off water and tar-like materials, leaving behind gray, porous coke.

Besides its usefulness as an energy source, coal can be processed into other things. Coal and its byproducts can be turned into paints, perfumes, rubber and plastic items, and can even provide ingredients for some edible products.

lignite

LIFE CONNECTION: DEEP, DARK SECRETS

Twelve-year-old Todd Domboski was walking across his grandmother's backyard on Valentine's Day when something caught his attention. Small wisps of smoke were coming from beneath some damp, brown leaves. As he crouched and lifted the leaves to investigate, the ground collapsed beneath him. In seconds, he had sunk to his knees in the moist and noticeably warm soil. Struggling to escape, he only sank more, this time to his waist, and before long, to his chin. In a matter of only a few seconds, he had disappeared from view into a hole slightly wider than his chest.

Todd tried to shove his feet out and even arched his back to stop his descent, but only sank deeper. Clawing at the slippery sides of the hole, his fingers clutched a tree root. Sulfurous fumes choked his nostrils and steam rushed past his face as he began to scream for help. His cousin Erik came running from where he was working on a motorcycle a few yards away. Although Todd's head was now 6 feet below the surface and barely visible through the steam, they were able to stretch their arms and lock hands. Erik, lying on his stomach, gave a giant heave and snatched Todd from what must have seemed like the ultimate Old Testament judgment.

This true story happened in 1981 in the small town of Centralia, Pennsylvania. In the heart of the anthracite coal region, Centralia had by the late 1800s no fewer than fourteen coal mines operating nearby. Changing technology and the Great Depression of the twentieth century had put an end to an industry that had been the region's lifeblood. Although the coal mines had long since closed, abandoned mines and veins of coal formed a spider web of fuel around the quiet mountain town. In 1962, a trash fire on the edge of town ignited a coal vein that extended to the surface. Despite numerous attempts to extinguish the fire, it smoldered underground. Townspeople were aware of the fire, but few were concerned. Todd's near-death experience changed all that.

Thousands of underground mine fires are burning right now around the globe. China has the most, while India has the highest density of these silent but deadly hazards. Dangerous cave-ins and the forest fires spawned by this underground burning coal are only a few of the threats to humans and habitats. Toxic levels of carbon monoxide, sulfur dioxide, and methane are continuously released. Elevated surface temperatures may also affect the plants and animals that can survive in an area.

Underground mine fires may be ignited by careless humans, lightning, or even all by themselves in the presence of oxygen. They may burn for years, or in some cases, centuries before they have consumed the coal or are no longer able to get enough oxygen. In Todd's case, the fire had reduced the underground coal seam to ash. This soft material compacts easily and collapses to form sinkholes, especially after heavy rains. These collapses often open up air shafts, allowing more oxygen to feed the fire.

Todd's brush with death and the carbon monoxide poisoning of several Centralians in their own homes triggered some drastic action. The Federal government offered to buy their homes and relocate their families to safer towns nearby. More than a thousand residents jumped at the offer, although a few refused to go. Today, only a few structures remain, as the government bulldozed the houses as they emptied. Wildflowers bloom where neat row houses once stood. A portion of a state highway had to be rerouted when the underground fire cracked and then buckled the asphalt.

As with many underground mine fires, the Centralia fire is considered inextinguishable. Geologists believe that it is smoldering as deep as 300 feet beneath the surface and may not burn itself out for another 250 years.

11.12 PETROLEUM AND GAS

Petroleum and natural gas are fluid fossil fuels. **Petroleum**, or *crude oil*, is a mixture of many different kinds of organic molecules. These molecules contain mainly the elements carbon and hydrogen. This is why we call them *hydrocarbons*. Hydrocarbon molecules differ by the number of carbon atoms in each, and the number of hydrogen atoms bonded to them. Crude oil is a thick, oily liquid straight from the earth. It usually is black or dark brown, but it can be light amber or other colors.

Natural gas is almost always found with crude oil. Natural gas is mostly the organic gas methane, which is the simplest hydrocarbon. Scientists believe that the different kinds of hydrocarbon gases (methane, ethane, propane, etc.) are simply parts of the complex mix of hydrocarbons that make up petroleum. They are probably products of the same process that formed crude oil. Some hydrocarbon reservoirs are more than 50% natural gas, or one of the liquefied natural gases (LNG), for example, propane or butane.

Oil is pumped out of the ground by wells drilled down through rock layers until they reach an oil reservoir trapped in certain kinds of geologic formations. In many places, these reservoirs are 1.5–5 km (1–3 mi) underground. Because oil and gas are less dense than either water or rock, they tend to flow upward through rock. They collect in *oil traps*, which are usually porous rock strata beneath an impermeable stratum of rock, like shale or slate. In other places, oil can be found in highly fractured igneous rocks. Oil reservoirs may be located beneath land surfaces or under the ocean bottom.

The oil wells themselves may be drilled using *oil derrick* structures built on land or on shallow coastal seabeds. In recent years, deep-ocean drilling has become common since most accessible oil reservoirs on land and in shallow seabeds have already been tapped. These rigs use floating drilling platforms positioned above the point where the well pipe enters the seabed. The sea bottom can be more than 1.5 km (1 mi) below the drilling platform.

Chemical formulas for natural gases:

methane: CH_4
ethane: C_2H_6
propane: C_3H_8
butane: C_4H_{10}

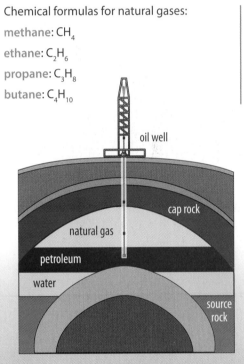

11-14

A typical oil well, showing the location of the gas and oil reservoirs within the geologic dome

impermeable: cannot be penetrated

11.13 ORIGIN OF FOSSIL FUELS

Almost everyone considers coal, petroleum, and natural gas to have come from fossilized organisms. This is what geologists have told us for at least two hundred years. And there seems to be some evidence to support this view.

ORIGIN OF COAL

Old-earth geologists think that coal formed from different fossilization processes than did oil and natural gas. The standard geologic explanation for the origin of coal begins with plants living and dying in a boggy environment between 100 and 300 million years ago. The trees, bushes, and ferns fell into a swamp or shallow sea and became loosely buried with sediment. As the plant matter built up, it compressed into fibrous matter called *peat*. Over a long time, the peat was covered by thick sediments that turned to rock. The extra weight and heat altered the peat, leaving mainly the organic carbon behind. Lignite formed first. With more heat, pressure, and time, the lignite became bituminous coal, and finally, anthracite.

Flood geologists would agree in general with these ideas. However, they interpret geologic history in a much shorter time frame. Since the Flood rapidly buried these materials, the whole process happened quickly. They point out that carbon-14 has been detected in coal. This would not be possible unless the coal were only a few thousands, not millions, of years old. Also, rocks and large boulders have been found in some coal deposits. Large rootless tree trunks exist in coal seams buried at odd angles. This indicates that the plant matter that became coal built up quickly in deep, swift-moving currents, not in shallow, calm waters. Recent experiments in which chemists have produced artificial coal show that little time is needed for coal to form. At high enough temperatures and pressures, coal-like materials can form in less than a day, so long periods of time are not necessary for coal formation.

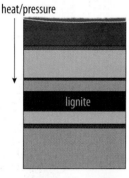
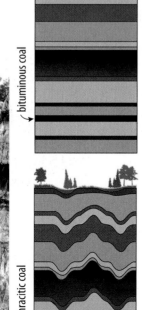

11-15
Old-earth geologists believe that coal formed slowly over great periods of time. Plant matter and sediments first accumulated as peat. The peat turned into lignite over time. Under more metamorphism, lignite became bituminous coal, and sometimes eventually became anthracite.

11-16
An unusual rock specimen that contains both coal and petrified wood

ORIGIN OF PETROLEUM AND NATURAL GAS

Most scientists agree that petroleum and natural gas came from fossilized marine plants called *phytoplankton*. According to accepted old-earth theories, these organisms died, settled to the sea bottom, and mixed in with sediments. Again, heat and pressure converted the organic matter into a waxy substance. With more metamorphism, this material chemically changed into petroleum and natural gas. The oil and gas then moved through porous rocks or along cracks until it was trapped under impermeable rocks. Natural gas dissolved into the oil or collected in spaces above the liquid petroleum deep underground.

For nearly a century, most people have accepted the fossil origin of petroleum and natural gas without question. However, oil geologists have observed some troubling things.

» Some petroleum sources are found in deep igneous bedrock, not sedimentary rock where fossils are located.
» Petroleum and gas occur at all depths in the lithosphere.
» Natural crude oil has a different mix of organic molecules than what would be expected from decomposed marine organisms.
» Inorganic gases, such as helium, hydrogen, and nitrogen, are often found in oil wells.
» Geologists find metals in crude oil, such as mercury and vanadium, that are poisonous to life.
» According to current theories, petroleum formation was a complex chemical process that would have had to follow many steps in a specific order. These steps also had to occur in different types of rock. It is unlikely that these conditions could have occurred repeatedly through geologic history and in many different places around the world.

The theory of petroleum and natural gas forming from fossils is just an assumption that is based on a naturalistic view of Earth's history. People think: "Fossilized plant matter produced coal, so petroleum must have formed under similar (but somewhat different) conditions."

These many problems with biogenic oil formation have led scientists to consider other possibilities. A few secular scientists have considered abiogenic—nonliving—sources of oil. They think that carbon monoxide and carbon dioxide deep within igneous rock may have combined with water to form methane. The methane then moved toward the surface and formed other hydrocarbons. Some creation geologists have considered that oil could have been part of the original creation. They reason that God created many useful resources, such as minerals, so why be surprised by God creating oil for our use?

phytoplankton (FYE toe PLANK tuhn): phyto- (Gk. *phyton*—plant) + plankton (Gk. *planktos*—wandering); any microscopic marine organism capable of photosynthesis

11-17

Geologists believe that most petroleum formed from ocean organisms called phytoplankton.

11-18

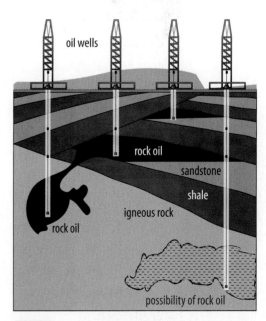

Finding oil in marine sedimentary rocks or in shallow places in bedrock is expected according to accepted theories of oil formation. Finding oil deep in granite bedrocks is not.

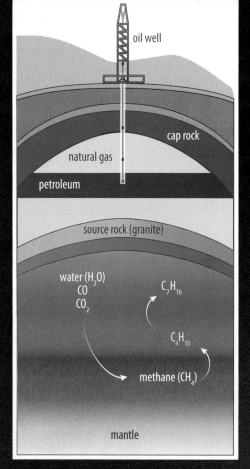

11-19
Some scientists believe that petroleum and natural gas are currently forming deep within the earth from simple carbon compounds.

Carbon dioxide and water vapor are called **greenhouse gases** because they make the atmosphere act like a glass greenhouse, trapping heat close to the earth's surface. But are carbon dioxide emissions a problem? We'll talk about this more in Chapter 21 where we'll study climate change.

However, while most scientists believe that sediments buried plant material that became oil, they do differ on the mechanism and time frame. Secular scientists believe that oil formed by natural processes over hundreds of thousands of years. Creation scientists look to the biblical Flood for both the mechanism and timing of oil formation. For now, most scientists consider oil a fossil fuel.

11.14 FOSSIL FUELS AND DOMINION

Think about all the ways you use fossil fuels. Your car, your clothes, your medicine, the paint on your house, the power plant that generates your electricity, the plastic container that holds your leftovers—all use fossil fuels or products obtained from them. Without them, modern life wouldn't be possible. You would have to live far from civilization in a log cabin or cave, clothed in animal skins or homespun fabrics, and living off the land to avoid using a product of fossil fuels!

What does this do to the environment? Many people believe that fossil fuels should be avoided. Using fossil fuels creates pollution, toxic chemicals, and **greenhouse gases** like carbon dioxide that harm the environment. So should we continue to use them?

This is where good and wise dominion comes in. Fossil fuels are a resource that we need to use responsibly. Energy production is a very important dominion issue. Fuels provide energy for heating and cooling. Vehicles make it easier to find work and move products, and thus help create wealth. Greater wealth in societies makes advanced technology possible. And technology saves lives and reduces pollution. We should use these fuels in a responsible way so that more people can enjoy the benefits of healthier, more productive lives.

11C REVIEW QUESTIONS

1. Why are coal, petroleum, and natural gas called fossil fuels?
2. List the three kinds of coal in order of increasing carbon content.
3. What are coal's two main uses?
4. How are petroleum and natural gas related?
5. Identify the three possible origins of petroleum and natural gas.
6. What chemical differences exist between natural petroleum and the known decomposition products of biological matter?
7. In several paragraphs, evaluate the theories of fossil fuel formation and their reliance on certain presuppositions.
8. (True or False) It is possible that petroleum and natural gas are not really fossil fuels at all.

Key Terms

fossil	249
paleontology	259
index fossil	260
transitional form	260
extinction	261
fossil fuel	265
coal	266
lignite	266
bituminous coal	266
anthracite	267
petroleum	269
natural gas	269

CHAPTER 11 REVIEW

CHAPTER SUMMARY

» Fossils can tell us something of what life was like in the past.

» A fossil is any remains or trace of a formerly living organism that has been preserved by natural processes.

» Fossils are rare because they formed only under special and unusual conditions.

» Fossils may be in the form of original parts of organisms, mineralized tissues, molds and casts, or carbon impressions. They can also indirectly reveal the existence of ancient organisms.

» When things die, they normally decay or are eaten. Most fossils required highly unusual conditions to form.

» Most fossils were buried under sediments. However, sedimentation had to be much faster in the past than what we observe today.

» Geologists assume that fossils in rocks formed hundreds of thousands or millions of years ago. Recent discovery of soft tissues in mineralized dinosaur bones suggests that fossilization was much more recent.

» Paleontology is the science of classifying and studying ancient life.

» Evolutionary paleontologists believe that the fossil record shows that life evolved over the last 500 million years.

» Paleontologists interpret the disappearance of many or most organisms at different times in the fossil record as extinction events.

» A scientist's presuppositions determine how he interprets the fossil record.

» Most geologists consider coal, petroleum, and natural gas to be fossil fuels, or energy sources derived from fossilized organisms.

» There are three types of coal: lignite, bituminous coal, and anthracite.

» Nearly all people agree that coal is a result of geologic heat and pressure acting on masses of dead plant matter long ago.

» Petroleum and natural gas are organic fluids made from hydrocarbon compounds. They are obtained by drilling into underground reservoirs.

» Most people believe that petroleum and natural gas are fossil remnants of ancient marine organisms.

» We must wisely obtain and use fossil fuels to make people's lives better while minimizing air and water pollution.

REVIEW QUESTIONS

1. When were the animals in the La Brea Tar Pits likely trapped and fossilized according to geologic history that is based on the Bible?
2. Why are fossils interesting to most people today?
3. Mineral replacement produces what kind of fossils?
4. What is the main difference between a fossil and a microfossil?

5. List five things that can happen to a dead animal in most places around the world today. Which of these is the most common?
6. What is a good indication that certain fossils were buried rapidly?
7. In what kind of materials are most fossils found?
8. Besides the location in Question 7, name at least two other kinds of materials in which you might find fossils.
9. How do old-earth scientists explain soft tissues inside a supposedly 68-million-year-old dinosaur fossil bone?
10. What makes up a fossil's scientific name?
11. Why does the fossil record *not* show evolution?
12. Why do paleontologists believe that extinctions of nearly all life on Earth have happened several times in the past?
13. How do Flood geologists account for the sequence of fossils?
14. Why do some industries in cities where air pollution is a problem use anthracite, though it is hard to find and more expensive?
15. Why are most petroleum and natural gas reservoirs found trapped beneath a layer of impermeable rock?
16. How do we obtain petroleum and natural gas from the earth?
17. What scientific evidence suggests that the plants that formed coal were *not* buried in quiet, marshy waters?
18. If petroleum came from the decomposition of fossil marine organisms trapped in sedimentary rocks, why are geologists surprised to find petroleum in deep igneous rocks far below sedimentary rocks?
19. If scientists believe that excess carbon dioxide emissions from burning carbon materials as fuel are really the cause of warming climates, why might it be better to help poor countries obtain more advanced power production methods than to restrict the use of fossil fuels in wealthy countries?

True or False

20. According to the Flood model of geology, paleontologists can logically assume that fossilized plants and animals found in the same strata of rock lived in the same time and place.
21. Fossils don't really help us wisely use God's Earth.
22. Fossil organisms found in various kinds of natural traps probably formed before or after the Flood, not during the Flood.
23. It is not logical to date index fossils using the age of the rock that they appear in, and then date other rocks using index fossils.
24. Mineralogists would classify coal as a mineral.
25. The first step in an old-earth view of coal formation is dead plants (buried under sediments) turning into lignite.
26. Since old-earth geologists believe that coal is millions of years old, they would not even try to use the carbon-14 method to determine its age.

CHAPTER 12
WEATHERING, EROSION, AND SOILS

- 12A WEATHERING — 277
- 12B EROSION AND DEPOSITION — 281
- 12C SOIL — 293

HEALING BY EROSION?

The quiet and beauty of Mount St. Helens, near the southern border of Washington State, hid the danger building within. On May 18, 1980, a minor magnitude 5.1 earthquake shook the stillness, and then the mountain literally blew out its northern side. Pyroclastic flows, mudflows, landslides, and a huge ash cloud devastated 230 square mies of forest.

Immediately after the eruption, melting snow and rainwater eroded through dense tephra. Still-living plants were exposed as water cut gullies to the original ground surface. Burrowing animals that survived the blast dug their way to sunlight, opening the ground and letting in air. Returning elk broke up the ash, allowing seeds to take root. Broken and buried trees decayed, returning vital nutrients to the alien, barren soil. Flowing water distributed this organic matter over the landscape.

Grasses, lupines, fireweed, and other wildflowers took root in the volcanic ash almost immediately. They helped anchor the soil and then returned nutrients to it. Today, the forest is growing again and animals have returned. Just thirty years later, life has come back. All this healing started with—believe it or not—erosion.

12A

WEATHERING

What makes rocks break down?

12A Objectives

After completing this section, you will be able to

» explain how rock weathers.
» recognize the effects of weathering.
» analyze what determines the rate of weathering.

12.1 CHANGING EARTH'S MATERIALS

Without *soil*, no complex land biome could survive for long. After the 1980 Mount St. Helens eruption, erosion was key to the beginning of the recovery of the soil there. Without it, the volcanic debris would have remained lifeless far longer, and recovery of the land as a habitat for life would have been much slower. Erosion normally removes soil instead of helping it form. Since this chapter is mainly about how soil forms, let's first look at factors that help produce soil.

In Chapter 10, you learned that today most rocks gradually become sediment and eventually wash into a lake or an ocean—it's a one-way trip. Some of these eroded earth materials end up, often temporarily, as soils. Rocks break down to begin forming soils through three main processes: *weathering, erosion,* and *deposition*. **Weathering** happens when some environmental factor, such as frost or chemicals, breaks rocks down into smaller and smaller pieces. Since most such factors are due to the climate, we call it "weathering," even if plants or animals are involved. **Erosion** picks up and moves these particles of rocks. **Deposition** places them somewhere else as sediments. In this section, you will learn about two types of weathering—*chemical* and *mechanical*. Weathering is *not* the result of only chemical *or* only mechanical factors. It is usually a complex combination of both. In Section 12B, you will see how erosion and deposition work together to build soils.

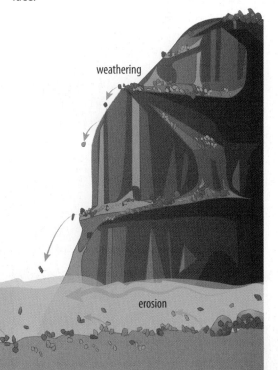

12-1

Weathering, erosion, and deposition are all normal processes that change the earth's surface.

12.2 CHEMICAL WEATHERING

Chemical weathering breaks down rocks through chemical changes. Do you remember how limestone fizzes when you test it with an acid? That is one kind of chemical change. Natural acids are important agents of chemical weathering. Two of these weak acids are carbonic acid and humic acid. Carbonic acid forms when carbon dioxide in the atmosphere dissolves in water. Humic acid forms when water dissolves decaying plant matter. Live lichens can produce acids too. All these acids can dissolve certain minerals in rocks and the cements that hold mineral grains together. But because these acids are weak, such chemical changes usually take a long time.

WEATHERING, EROSION, AND SOILS

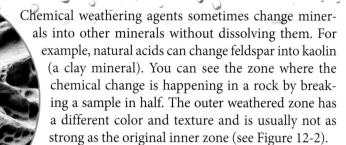

12-2 Chemical weathering dissolves material, producing pits formed by naturally acidified water (right). Note the difference in color and texture between the outer weathered zone and the inner original rock (below).

Chemical weathering agents sometimes change minerals into other minerals without dissolving them. For example, natural acids can change feldspar into kaolin (a clay mineral). You can see the zone where the chemical change is happening in a rock by breaking a sample in half. The outer weathered zone has a different color and texture and is usually not as strong as the original inner zone (see Figure 12-2).

Several things affect chemical weathering. The mineral itself is the most important of these. Calcite or limestone, both made of calcium carbonate, can weather quickly. But quartz, made of silica, is remarkably resistant to weathering under the same conditions.

Climate is also important. Warmth and moisture aid chemical weathering by speeding up the reactions between natural acids and minerals.

Topography and vegetation are factors as well. Places with gentle slopes and abundant vegetation hold rainwater longer—again helping chemical weathering to take place. Steep, bare rocks weather more slowly from chemical changes because they can't hold on to moisture as well.

A final significant factor affecting chemical weathering is the amount of rock surface exposed to the weathering agents. Chemical weathering works directly on the surface of rock and rock particles. The greater the area for a certain mass of rock, the more chemical weathering takes place. So as large rocks break down into smaller pieces, these together have much more surface area than the original rock. Thus, the rate of chemical weathering increases as particles get smaller.

12.3 MECHANICAL WEATHERING

You just learned that chemical weathering speeds up as rocks break down into smaller pieces. So any process that uses mechanical forces to make big rocks into little ones aids the weathering process. This is what happens in **mechanical weathering**. Unlike chemical weathering, mechanical weathering doesn't change the composition of the rocks. It changes only the size and arrangement of rock particles.

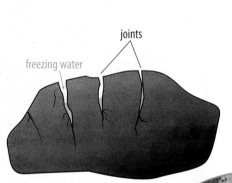

12-3 Frost wedging results from freezing and thawing of water trapped in cracks within rocks.

Water can exert tremendous mechanical forces. When water freezes, it expands about 9%. So as it expands in pores and cracks, it forces materials apart. This weathering process is **frost wedging**. Each time the ice melts and refreezes, it wedges the rock farther apart. Eventually, the rock cracks and small fragments splinter off. Cracks extend farther into the rock so that even larger chunks can break off.

A similar process, *frost heaving*, works to bring underground rocks to the ground surface where they weather faster. In northern climates, frost can reach several feet into the ground during winter. Frozen groundwater below buried rocks expands and lifts them upward a little each winter. Over many seasons, rocks appear at the ground surface in the spring, much to the frustration of farmers! People often also have to deal with broken sidewalks and potholes in roads caused by frost heaving.

Sometimes physical processes make rocks separate into thin layers or slabs. This is called **exfoliation** (eks FOH lee AY shun). When a great amount of weight is removed from above a section of rock, the rock expands and breaks into slabs. This can happen after significant erosion or when a glacier melts, removing the weight of ice on the rock underneath. In hot climates, surface rock exposed to the sun heats up and expands. Thin sections break off when the stresses caused by heating and cooling exceed the strength of the rock. Exfoliation often produces large rounded outcroppings or boulders. It can weather even durable granite.

When mechanical weathering affects exposed rocks on steep inclines, like cliffs, the pieces of broken rock slide or fall. As they pile up at the base of the cliff, a sloped *talus* surface forms.

12-4

Some exposed rocks experience mechanical weathering in which thin sheets separate from the rock—exfoliation (top photo). This rock accumulates at the base of the mountain—talus—as seen at the base of Crowfoot Mountain (bottom of page).

CASE STUDY: WHAT HAPPENED TO GEORGE WASHINGTON?

Think of different statues that you have seen. Do you recall the materials used on these statues? Metal and rock are some of the most commonly used. Why do artists choose these materials? Typically, these materials are chosen for their beauty and durability.

Notice the two statues. The first image shows a 500-year-old marble statue of Moses by Michelangelo, located in Rome, Italy. The second is of a 100-year-old statue of George Washington in New York City. Both statues were made of marble. What do you notice about their conditions?

Questions to Consider

1. What is the term for the changes to rock material by environmental factors, like the rock in these statues?
2. What do you think may have caused the damage to the statue of George Washington?
3. Suggest two reasons that Michelangelo's statue doesn't show damage even though it is five times older.

Biological weathering results from the mechanical action of plants and animals on rocks. Plants are seemingly soft organisms, but they can do great damage to sidewalks and buildings, as well as to rocks. As their roots and stems grow in a crack, their cells exert pressure outward against the walls of the crack. A single cell can't exert much pressure. But when thousands of cells work together, they can force cracks apart and fracture even the strongest rocks. Burrowing animals can open passages deep into soil, which allows water and natural acids to penetrate to bedrock. This process aids chemical weathering.

12A REVIEW QUESTIONS

1. In one sentence, describe how weathering, rocks, and soil are related.
2. Compare the two main types of weathering.
3. State three factors that can affect the rate of chemical weathering.
4. How is water an agent of mechanical weathering?
5. How would you recognize exfoliated rock?
6. (True or False) Plants are stronger than rocks.

12-5
Plants can easily reduce even the strongest rock to rubble as their roots wedge open cracks in rocks.

12B

EROSION AND DEPOSITION

What forces act in the processes of erosion and deposition?

12.4 AGENTS OF EROSION AND DEPOSITION

Weathering is related to erosion because both involve the breakdown and reorganization of earth materials. But erosion implies movement of the materials away from their original location. Let's focus now on the agents that move **eroded** materials.

Eventually, all eroded materials come to rest as sediment. This means that deposition follows erosion. Often, the agent of movement determines the method of deposition. Important agents of erosion and deposition are *gravity, water, wind,* and *glaciers*.

GRAVITY

12.5 GRAVITY AND EROSION

Gravity is the most important force that a geologist deals with. It controls crust topography and sea level. It drives tectonic processes. Gravity can also influence weathering and erosion. Think rockslides! Avalanches! Mudflows! While weathering processes turn big rocks into little ones, gravity moves them to new locations.

Sometimes rocks and soil move by gravity alone, a process that geologists call **mass wasting**. Such action happens only on sloped land, and the slope angle does not have to be very great. Mass wasting can be fast or slow. One type of slow mass wasting is *creep*, the barely visible downhill motion of soil. When a section of loose earth suddenly breaks away from a hillside, but slips only a short distance downhill, it is a *slump*. Plants and grass often hold the slumped ground together, and it leaves a curved scar on the hillside where the sod pulled away.

12B Objectives

After completing this section, you will be able to

» relate stream erosion and deposition to stream speed.
» explain the process of wind erosion and deposition.
» describe the main processes that result in glacial erosion and deposition.
» recognize the effects of erosion.
» identify erosion and depositional features on maps.

erode: e- (L. *ex*—off) + -rode (L. *rodere*—to gnaw)

12-6

This image shows wrinkles in the soil associated with soil creep—a slow process of rock and soil moving over a lower rock layer. Leaning trees, telephone poles, and fences are often the most noticeable indications of creep.

12-7

This picture shows a mass of soil and rock that has rapidly slid downhill. This is a slump—a sudden movement of rock and soil down a slope.

Rapid mass wasting of earth materials downhill is a **landslide**. We name landslides by the kind of debris moving under the force of gravity. *Rockslides* are mainly the fall of broken bedrock. *Mudflows* are soupy mixtures of loose earth debris and water. *Avalanches* are massive downhill slides of ice and snow containing rocks and soil.

Landslides need loose earth debris, a sufficiently steep slope, something that reduces the friction holding the materials in place, and a trigger to get materials moving. The most common triggers are earthquakes and too much water in the soil. People can make landslides more likely if they divert surface water or create road or building excavations in unstable geologic formations.

WATER

12.6 WATER AND EROSION

When you think of erosion, probably the first thing you think about is the action of running water. The energy and force of flowing water are important factors in wearing down the earth's surface. Most flowing water begins its trek as *precipitation* (rain or snow). Then it moves downhill by many paths through the force of gravity. Ultimately, most water on land ends up in the ocean.

When raindrops strike soil, they loosen and kick particles into the air. Rain can either soak in or run off, depending on what the soil is like and how heavy the rain is. Too much rainwater flows like sheets over the ground surface into low spots. Low spots channel the runoff water downhill in trickles. The moving water plucks up soil and rock particles and carries them along.

Erosion increases as trickles feed into rivulets, rivulets into brooks, brooks into creeks, and creeks into rivers. All of these are *streams*. Streams are narrow bodies of water that flow in a *channel* or *streambed*. Rough water, or *turbulence*, does the work of plucking up particles of rock and soil and lifting them into the stream. The more turbulent the water flow, the more sediment the water can carry. If turbulence is strong enough, fast streams can move even large rocks. These rocks do even more work breaking up earth materials in the streambed, further increasing erosion and sediment transport.

Streams move sediment in three ways. The majority of a stream's load—the total mass of sediment that it transports—is *suspended sediments*. These are made of tiny particles of silt, clay, and fine sand that simply flow along with the water. Larger sediment particles hop repeatedly over the stream bottom as they are carried along by the flow. Scientists call this motion *saltation*. Saltating pebbles can break apart and produce sand when they strike other rocks and pebbles. Saltation rounds sediment grains during stream erosion and breaks particles free to become more sediment. Larger rocks can move only by rolling along the bottom. In strong streams, fast-moving rocks crush smaller stones into finer pieces.

Stream erosion moves sediments from the continents and deposits them in lakes and oceans. The amount of material transported by this process is unknown but estimates are staggering. In the United States, about 1.7 billion metric tons of soil are lost to erosion each year.

Erosional landforms around Lake Powell on the Colorado River between Utah and Arizona

12.7 WATER AND DEPOSITION

Streams of water eventually slow down. This happens when the angle of the streambed becomes flatter, the stream becomes wider and shallower, or it enters a larger body of water. As speed drops off, so does flow turbulence. As turbulence lessens, sediment particles of different sizes settle out in an orderly way. The largest rocks and stones stop rolling. Farther along, smaller sediment particles drop to the bottom. As stream velocity continues to slow, smaller and smaller particles settle out. The finest suspended particles—silt—are all that remain after the larger particles have settled. This pattern of sedimentation is called **sorting** because sediment particles become sorted by size. Streams that slow down gradually create *well-sorted sediments*. This means that at any spot in the streambed, most of the sediment clasts are about the same size.

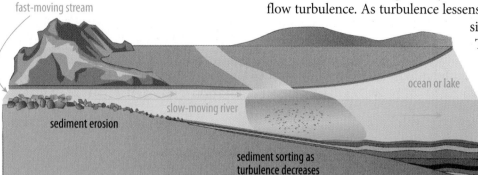

12-8
Stream sorting. As a stream slows, sediments settle out according to their weights and sizes.

You can see sediment **sorting** at the beach where waves wash onto the sand. Larger pebbles and grains are farther out in the water where wave speed and turbulence begin to drop off. Finer grains are closer to the beach where wave speed drops to zero.

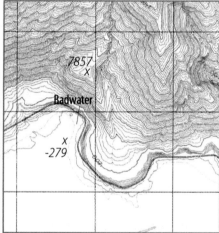

12-9
The Badwater alluvial fan in Death Valley, CA (above). A contour map of the area (right) shows the fan-like distribution of the stream sediment.

Location: 36.225°N 116.772°W

Many large sediment-laden rivers have *deltas* at their mouths. A delta forms where a stream enters a larger body of water, like an ocean or large lake, and the water's speed slows to a near standstill. Deltas build up fertile land out into the body of water. A river delta is named for the shape of the Greek letter delta (Δ), which is equivalent to the English capital *D*. A delta-like deposit may also form where intermittent streams in dry, mountainous areas empty into a flat desert. These are called *alluvial fans*—loose sediments of sand and stones deposited by moving water (see Figure 12-9).

Floods are important ways that sediments can return to the land, at least temporarily. Muddy river water overflows the flat land next to the streambed, called a *floodplain*. Then the water recedes, leaving behind rich, sorted sediment that can quickly turn to soil. Valleys with flooding rivers, such as the Mississippi, the Nile in Egypt, and the Yangtze in China, are known worldwide for their rich farmlands.

Bi — Erosion and Dominion
Would it be a good thing for a Christian to devote his life to managing soil erosion along the Mississippi River? Why or why not?

WIND

12.8 WIND AND EROSION

Wind is second only to water for soil removal by erosion from the world's continents. Wind is similar to water in that it picks up sediments and transports them to other places. But most winds can't carry heavy stones and rocks. Wind can easily erode where sparse vegetation can't hold loose surface sediments in place. Deserts and places experiencing a drought are especially prone to wind erosion. These have low moisture and little vegetation, resulting in loose, dusty soil, and few *windbreaks*. Wind erosion also occurs on sand beaches where some of these same conditions exist.

12-10
Desert pavement in Death Valley, CA

In many deserts, wind may carry away loose surface materials, leaving an empty basin-shaped area called a *blowout*. When the sand and other loose materials have blown away from a larger area, only pebbles and cobbles remain. Rocks left behind can completely cover the desert surface with a *desert pavement*. **Deflation** (dee FLAY shun) is the process of forming desert pavements by wind erosion of loose sediments from flat, dry terrain.

Wind-driven sand erodes exposed rocks, especially softer sedimentary rocks. This action, called **abrasion**, is like sandblasting. Winds in dry places quickly wear away exposed rock formations such as pinnacles. The particles of abraded rock become windblown sediment. In the western deserts of the United States, wind sculpts rocks into very unusual shapes called *ventifacts*—stone formations left after erosion by windblown sand.

12.9 WIND AND DEPOSITION

Sometimes, wind causes *sandstorms* and *dust storms*. These storms can carry particles high into the atmosphere. Astronauts can easily see large dust storms from earth orbit.

Wind transports sediment just like water does. Fine materials rise suspended in the wind, larger particles saltate near the ground, and very strong winds can dislodge and roll stones. Violent winds from tornadoes and hurricanes can hurl rocks and move even small boulders short distances.

12-11
This ventifact in Eduardo Avaroa Andean Fauna National Reserve, Bolivia, was sandblasted into this shape by wind erosion.

Location: 22.052°S 67.882°W (turn on photo layer)

Lots of wind-driven sand can form **dunes**. Dunes develop when something near the ground slows wind enough to cause it to drop the sediments. It may be plants, large rocks, hills, or even existing dunes. Sand dunes are gently sloped upwind, ending at a crest, and have a steep slope on the downwind or *lee* side. Wind easily blows sand up the gentle slope and over the crest. Some of the sand falls and collects in the low-turbulence zone on the lee side. Over time, the dune moves in this direction as the wind erodes and deposits the sand (see Figure 12-12). Dunes can look like irregular waves moving over the ground because of the fluid nature of wind.

> **Bi What do you think?**
> What part do you think that weathering and erosion had in God's original creation? How might they have changed after the Fall?

12-12
A sand dune near Dubai (below) is in motion as sand is eroded from the upwind side of the dune by wind and is deposited on the lee slope. From which direction do the winds tend to blow?

As with water, wind carries the finest particles farthest. Most windborne sediments are clay and silt particles. In the past, there must have been large areas of the world covered in fine dust. This may have happened during the latter stage of the Ice Age because of its tundra conditions. Brief but intense periods of strong global winds seemed to have eroded unprotected soils. This produced thick, fine-grained strata called **loess** deposits. Loess can be many tens of meters thick and cover hundreds of square kilometers. Because it is a fine mixture of silt, sand, and clay, it makes a great soil. Loess is a good base for fertile farmland because plants can easily get nutrients from it. We'll speak more about soils in Section 12C.

GLACIERS

12.10 WHAT ARE GLACIERS?

Water in another form erodes and sculpts land as it flows. A **glacier** is a large mass of dense ice that flows under the influence of gravity. Glaciers form in areas where winter snows fail to melt completely in the summer, year after year. Near the Equator, this happens only on high mountains. In places like Norway, Alaska, Greenland, and Antarctica, glaciers can exist at sea level.

Glacial ice is not like the ice in your soda! Snow accumulates and compacts to form dense opaque ice. It is colored white to blue-gray. But glacial ice color often varies because of particles of eroded rock trapped in the ice called *glacial* or *rock flour*.

There are many types of glaciers. Glaciers in valleys high in the mountains are *valley glaciers*. These are the most common kinds of glaciers. They are also called *mountain* glaciers or *alpine* glaciers. When several valley glaciers merge into a broad mass of ice at the base of a mountain range, a *piedmont glacier* forms.

Langjökull Glacier in Iceland
Location: 64.6 °N 20.2 °W

Ice sheets cover land surfaces that extend far beyond valleys. *Continental glaciers* are broad ice sheets that cover most of a continent. Instead of following a valley, these spread out in all directions, covering everything but the highest mountain peaks. The ice flows slowly from higher, thinner areas of ice cover down into lower, thicker sections.

The period of maximum glaciation during the Ice Age produced continental ice sheets on all the northernmost and southernmost continents. Only two continental ice sheets remain today—in Greenland and Antarctica. The thickness of these ice sheets varies from only a few meters up to several kilometers. In Antarctica, geologists have determined that the weight of ice has depressed the bedrock as much as 2400 m (8000 ft) *below* sea level!

Ice caps are smaller permanent ice covers. Most are on Arctic islands such as Iceland and Ellesmere Island in the Canadian far north. But others also exist in Norway (Svalbard Islands) and in the southern Patagonian ice fields of Argentina.

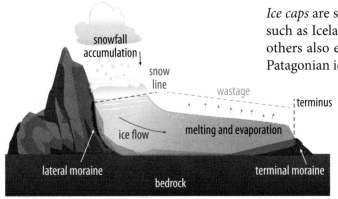

12-13
A typical glacier showing the accumulation zone above the snow line and the wastage zone below the snow line

Glaciers don't just sit where they are—they move. Snow accumulates above the snow line and compacts into dense ice. Surprisingly, ice can slowly flow if it is thicker than 50 m (160 ft). This is why glacial ice flows downhill and fills valleys from side to side. Average speeds are a few centimeters to a few meters per day. As the glacier moves below the snow line, warmer temperatures begin to melt it, and its thickness shrinks. The ice continues to move under its weight and is pushed by thicker ice from higher up. But at the point where the ice melts as fast as it is replaced, the glacier has reached its *terminus*.

If a glacier accumulates more ice than it loses, its terminus moves downhill. If it loses more than it accumulates, the ice continues to move downhill, but its terminus slowly moves *uphill*. To the untrained eye, it looks like it is moving backward, which, of course, is not possible.

GLACIERS AND CLIMATE CHANGE

Glaciologists (glacier scientists) and environmentalists are concerned that the Greenland glaciers are rapidly receding. Knowing whether a glacier is advancing or retreating can give glaciologists information about the local climate, and possibly about global climates as well. If the glacier is retreating, the climate may be warming (increased melting) or becoming drier (less snow accumulation). If the glacier is advancing, the climate may be cooling (decreased melting) or becoming more humid (more snow accumulation). Glaciologists have noted that many of the world's glaciers are retreating, though some are advancing. Many scientists believe that this is caused directly by a slow rise in global temperatures observed since the 1800s, although some glaciers were retreating even before that time.

There is also strong consensus that the rising temperatures relate to CO_2 levels in the atmosphere. Scientists have different models related to this issue, some that blame man, and others that don't. These models make different predictions of what global temperatures will do in the future. While worldview determines which model we accept, Christians must recognize their role regarding this issue. According to the Dominion Mandate, believers should wisely use the resources that God has given them—and that includes not doing unnecessary harm.

Many factors determine how fast a glacier moves—the slope of the bedrock, the rate of snowfall accumulation, the relative lengths of winter and summer, the average temperatures during these seasons, the existence of pools of meltwater on top of the glacier, and the amount of dark glacial debris mixed in with the ice. (Dark materials absorb energy from sunlight, melting the ice.) Glacier movement can literally transform a landscape.

Sometimes glaciers move in rapid spurts of speed. When this happens, it is called a *surge*. The fastest glacier movement ever recorded was in 1953 for a glacier in northern Pakistan called the *Kutiah Glacier*. It moved an average of 112 m (370 ft) per day!

12.11 GLACIERS AND EROSION

Glaciers change the shape of valleys and level almost everything in their path. A mountain stream valley is usually V-shaped. But after a glacier has moved through it, it is more U-shaped.

As they move along, glaciers pick up rocks and other debris. A glacier can even break up bedrock by a process called *plucking*. As the glacier moves, some of its ice melts from friction or pressure, and enters the cracks and pores in the bedrock. Then, as the water freezes again, it enlarges the cracks and loosens large chunks of rock as in frost wedging. Embedded in ice, the chunks of rock get carried away with the glacier.

Glaciated mountains are scarred by glacial erosion. Over time, valley glaciers high in the mountains carve out huge bowl-like depressions nestled in among the peaks. After the glaciers melt, *cirques* (SURKS) are often filled with frigid mountain water, forming small lakes called *tarns*. In heavily glaciated mountain ranges, plucking has eaten away at mountain peaks from several sides. Slender spires called *horns* are all that remain of these peaks. The Matterhorn (loc.: 45.976°N 7.66°E) in Switzerland is a famous horn. Two valley glaciers next to each other can remove the rock from the ridge between them until only a slender blade of rock remains. These narrow, steep-sided ridges are called *arêtes* (uh RATES).

Sometimes coastal glaciers erode their valley floors below sea level. When they retreat, the sea enters the glaciated valley, now called a *fjord* (FYORD). Some fjords extend inland for more than 100 km (60 mi). Fjords are prominent features of the coastal geography of Norway, Iceland, Alaska, and New Zealand's South Island. There are many freshwater fjords around the world as well. Water-filled inland glacial valleys are elongated lakes, such as the Finger Lakes of New York.

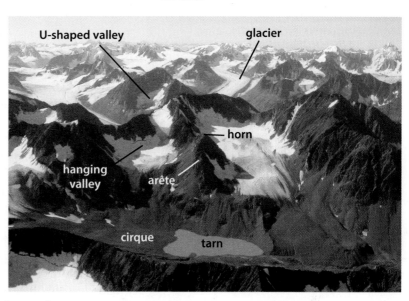

12-14
Glaciers leave unmistakable erosional features on land.

12-15
A fjord is a glaciated valley that reaches the sea. This is a view from Glacier Bay National Park, part of Alaska's Inside Passage.

WEATHERING, EROSION, AND SOILS

12.12 GLACIERS AND DEPOSITION

Glaciers bulldoze a landscape, but they don't leave it flat. They shape some of their eroded materials by their motion. They also deposit their debris in two ways as they melt—either the debris simply drops to the underlying bedrock as the ice melts, or streams of meltwater deposit the debris. All deposits of sediments from glaciers are called **glacial drift**.

As a glacier melts and retreats, all drift that drops directly from the ice forms **till**. Glacial till is unsorted. In other words, it is a well-mixed combination of rocks, pebbles, sand, and silt. Glaciers often deposit ridges of till called **moraines**. There are several types of moraines. A *terminal moraine* is a pile of till that the glacier's terminus pushed in front until the glacier stopped advancing. Terminal moraines mark the glacier's farthest advance.

The pieces of bedrock plucked from the sides of a glacier's valley are carried along in the ice as dark stripes of rock. Only the upper surface of these bands can be seen from the air. When the glacier retreats, the debris drops to the sides of the valley floor as long narrow piles of till called *lateral moraines*. In a similar way, when two valley glaciers merge into a single glacier, the rock debris from their inside walls joins to form a ribbon of rubble in the center of the glacier flow. When the glacier retreats, this forms a *medial moraine* (see facing page). When a glacier melts and retreats, any till from the main body of the glacier simply drops to the underlying bedrock. It forms a blanket of drift called a *ground moraine*. Ground moraines smooth out the terrain, creating gently rolling hills.

Drumlins are long, streamlined hills rising from ground moraines composed of glacial till. From the air, drumlins appear roughly parallel to one another and probably align with the direction that the glacier moved. The thicker part of the hill probably points "upstream," against the flow of the glacier. Not all glaciated regions have drumlins. But they occur in the moraines of North America, northern Europe, Asia, and the southern continents, including Antarctica.

Some glacial drift features result from deposition by flowing glacial water. Meltwater streams flowing over the ground away from a broad glacial terminus carry sand and gravel far from the glacier. As these streams weave back and forth, they produce an *outwash plain*. These deposits are generally well sorted. Larger rocks are in sediments closer to the glacier, and smaller rocks, sand, and silt appear farther away as the streams lose speed.

Sometimes outwash plains and ground moraines contain large potholes called *kettles*. These holes form when chunks of ice separate from the glacier, are buried, and then melt, leaving depressions. A kettle may fill with water and become a *kettle lake*. *Eskers* are long, winding ridges of well-sorted, stratified sediments apparently deposited by meltwater streams flowing under a glacier. *Kames* are steep-sided hills, much like eskers but shorter. Geologists believe that kames form when depressions on top of or inside the glacier collect sorted sediments as meltwater flows through the depressions. When the glacier retreats, it drops the stratified sediment in an irregular pile.

12-16 A drumlin in the Palouse area of eastern Washington

12-17 Glaciers formed this esker near Whitefish Lake, Northwest Territories, Canada (above) and the Kame of Hoy, Orkney Islands, United Kingdom (below).

Note the dark bands of rock that streak this valley glacier. If it melts and retreats, these bands will become medial and lateral moraines.

12B REVIEW QUESTIONS

1. Explain the connection between weathering, erosion, and deposition.
2. Describe the difference between slow and fast mass wasting of earth materials. Give an example of each.
3. What property of flowing water determines how much sediment it can carry?
4. What are the three ways that streams move sediments?
5. Describe the orderly deposition of sediments as water turbulence gradually lessens.
6. What kind of formation is illustrated in the photo to the left? How did it form?
7. What places are especially prone to wind erosion?
8. What are two types of formations resulting from wind erosion?
9. Name two important land features resulting from wind deposition. Which type does not seem to be forming today?
10. How do glaciers form?
11. What factors determine how fast a glacier moves?
12. How do glaciers break up bedrock?
13. What kinds of landforms result from glacial plucking?
14. What kinds of moraines result from the long stripes of glacial till after the glaciers melt and retreat?
15. (True or False) Deflation occurs when a glacier shrinks and retreats.

VARVES

Lakes and other quiet bodies of water that receive streams of glacial meltwater accumulate thin layers of sediment called *varves* (Sw. *varv*—layer). Each varve is a layer with coarser particles at the bottom and finer particles at the top. A varve may be anywhere from a few millimeters to a few centimeters thick.

Most old-earth geologists believe that glacial streams in the past deposited exactly one varve each year. The rapidly flowing water in the summers carried the coarser particles, while the slower flows later in the year could carry only the finer particles. They believe that this characteristic, like tree rings, is useful for dating glacial deposits. However, scientists have discovered that more than one varve may form each year. Storms, floods, and other events can add varves to outwash deposits. Thus, scientists now recognize varve counting to be unreliable as a dating method.

12C

SOIL

How does soil form?

12.13 WHAT IS SOIL?

You've learned how rock breaks down into sediments through weathering and erosion. Sediments can build up in one place or they can be transported to a different place. When you see fine sediments covering the ground, you might think, *soil*! But is it? For most earth materials, this is only the beginning of the makings of a good soil.

Pedologists—scientists who study soils, soil formation, and erosion—define **soil** as a layered formation at the surface of the earth made of inorganic earth materials combined with organic nutrients. Soil is also porous, allowing water and air to move through it.

So where did soils come from? On the third day of Creation, God created soil best suited to support and nourish the plants that He would create later the same day (Gen. 1:9–13). This original soil was the result of a supernatural creative act. It's possible it didn't even look like the soils we see today since today's soils come from weathering and erosion.

As biblical creationists, we can be fairly sure that there aren't any original soils remaining today as they were at Creation. After the Fall and God's curse on Adam's sin, all things changed, even soils. The new difficulty Adam had in farming likely resulted from soil quality and fertility (Gen. 3:7–19). People probably had to start using conservation methods to reduce soil erosion and replenish nutrients in the soil. Then the Flood *really* changed everything. In the upheaval, the floodwaters mixed up the original soil materials with other sediments. But God in His providence and mercy provided for His creation. Soils all over the world were quickly restored through weathering and erosion. But they were probably quite different from the original soil.

12C Objectives

After completing this section, you will be able to

» describe how soil forms, including its horizons.

» analyze how different factors affect soil.

» evaluate ways for using and conserving soil.

pedologist (pe DAHL uh jist): ped- (Gk. *pedon*—ground, soil) + -ology (study of); soil scientist

According to the biblical model of the earth's history, all the soils of the world that exist today must have formed since the Flood, in much less than 5500 years.

12-18
We don't know what the soil composition at Creation looked like, but it is likely that soil was not made up of the products of erosion and decomposition since these processes had not occurred yet.

12.14 RATE OF SOIL FORMATION

How fast does soil form? That's a complicated question. It can happen very quickly or very slowly. The rate at which soil forms depends on the chemical composition of the weathered rocks, temperature, amount of precipitation, seasonal changes, ground slope, and other factors. On some steep, exposed rock surfaces, soil has not formed at all. In dry or cool places, an ideal soil structure seems to have taken thousands of years to form. Yet in other places, weathering and soil formation are more rapid, as in the tropics. Abundant rain and high temperatures speed up chemical weathering of the underlying bedrock and layers within the soils. These processes produce layers of fine particles many meters thick that can make soil in just a few years.

Volcanic debris often can support some plant growth in just a few years after an eruption. For example, the volcano Krakatau formed about 50 cm (20 in.) of soil from volcanic ash within forty-five years after its eruption. Soil can also develop rapidly along the edges of retreating glaciers. About 35 cm (14 in.) of soil formed in Glacier Bay National Park in less than 250 years following retreat of the glaciers there. After only thirty years, soil is beginning to support life again in the pyroclastic zone of Mount St. Helens. Thus, under favorable conditions, soil can form quickly.

12-19 Retreating glaciers can provide large amounts of rock material, which can produce soil rapidly.

LIFE CONNECTION: LIVING SOIL

Benjamin and his family arrived at Arches National Park in Moab, Utah, on a summer morning, eager to explore the fantastic landscape. At their first stop, they hurried out on a trail to a fantastic arch towering into the clear blue sky. On either side of the narrow path, Benjamin noticed large signs warning them to stay on the trail to protect the "biological soil crust" on either side. He knew that "biological" meant that the soil was alive, but to him it just looked like knobby desert sand with a little turquoise, black, and dark brown color splashed on it. Where's the life here?

A short time later, Benjamin learned about biological soil crusts at an information kiosk. In desert places like this, visible plants are few and far between. A mat of living organisms within the soil covers much of the open space. This may extend as much as 10 cm (4 in.) beneath the surface, although it is usually much thinner. Occasionally small mosses and lichens are visible on the top of the soil.

The life found in this crust is mostly specialized bacteria and algae, fungi, lichens, and some mosses. Many of these produce microscopic filaments that bind the rock and soil particles together, preventing erosion and soil loss when the rains finally come. All of these are made of living cells that need water to thrive but can survive long periods of drought.

These living soil crusts lay the foundation for the growth of larger plants by adding nutrients to the soil. Some are able

to capture nitrogen from the air and convert it to forms that plants need. These soil organisms and their fibers also help the soil trap and hold water. Further, the web of fibers prevents the growth of some weeds.

It takes many years for such a crust to develop, so even a careless footprint can reverse years of growth and harm the habitat.

After leaving the park and heading south on the highway, Benjamin noticed a group of ATV riders racing across the desert. Thinking about what he had learned, he hoped they stayed on the marked trails.

12.15 SOIL STRUCTURE

Ideally, a fully developed or *mature* soil has three layers, which pedologists call **horizons**. A horizon in pedology is a horizontal stratum of soil that is distinct in color or texture from other strata of soil. The lowest layer, the *C-horizon*, rests on top of the bedrock underlying the soil. The C-horizon consists of broken and weathered pieces of bedrock. Finer particles in this layer come from more advanced weathering of the bedrock fragments.

Above the C-horizon is the *B-horizon*, or *subsoil*. It contains thoroughly weathered minerals from the C-horizon. For weathered granite, this would be several kinds of clay, reddish iron oxide, and quartz grains. It also contains minerals and other materials that filter down from the upper layer of soil.

The topmost layer is the *A-horizon*, or *topsoil*. It is the most fertile part of the soil. It contains **humus** (HYOO mus)—decayed organic material that provides many nutrients for plants and helps soil hold water. Air also penetrates this layer. Rainwater carries some of these nutrients and finer particles down into the B-horizon.

12.16 SOIL COMPOSITION

Different places in the world have different kinds of soil. What makes the difference? *Native soils* appear to come from the weathering of local bedrocks. The original rock composition affects the texture and color of the soil. *Transported soils* were deposited on top of native bedrocks by water or wind and so they may not be chemically related to them. These soils may not have three standard soil horizons. Most topsoils contain varying amounts of sand, silt, and clay. **Loam**, an especially fertile topsoil, contains about equal parts of sand and silt and about half as much clay. To be fertile for the widest variety of plants, however, it must also contain humus.

Soil scientists use soil classification graphs to identify soil type according to composition. Notice the graph in Figure 12-21. Let's say we have a soil sample that is 55% silt and 25% sand and we want to determine the type of soil. Along the right edge of the graph, we find 55% silt, and along the bottom edge, we find 25% sand. We move down along the guideline from the 55% silt and move up along the guideline from 25% sand until the lines intersect. The portion of the graph now indicates the type of soil—silt loam. If you move directly left along the third guideline to the left edge, you will find that your sample is 20% clay.

The climate in a region can greatly influence a soil's composition. Soils that get plenty of rain differ from desert soils. Enough rain helps plants grow, and this produces humus. Too much rain can dissolve and remove most of the nutrients from the topsoil. Excessive water can also flood the pores in the soil, reducing the amount of oxygen available to the roots. This makes it hard for most plants

12-20

Changes in the soil's coloration mark the soil horizons in this soil profile of an excavation site.

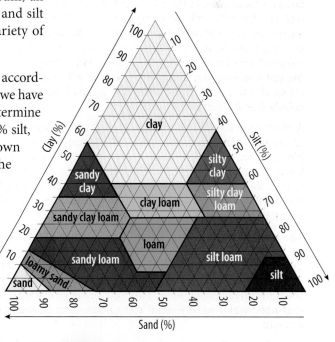

12-21

Soil classification graph

12-22 Prairie dogs bring needed nutrients to the surface, helping fertilize the soil.

12-23 The 1980 eruption of Mount St. Helens devastated hundreds of square miles of forest (top photo). Thirty-two years later, you can see a thriving forest in the same location (bottom photo).

Fi Mount St. Helens, WA: 46.2°N 122.2°W

with shallow roots to grow and stay healthy. Soils in rainy climates are also acidic, like they are in the eastern United States. Roses and pine trees grow well in acidic soils, but most food crops don't. Farmers in rainy climates add lime to their soil to neutralize acidic soil. They call this "sweetening" the soil.

Soils in dry climates, on the other hand, have plenty of nutrients in the topsoil. They often need just water to make them productive farmland. Farmers in southern California take advantage of the fertile desert soil by extensive irrigation. Much of the western United States has similarly dry, but otherwise fertile, soil.

Living things can affect a region's soil too. Dead plants and trees decompose to form humus, something that all fertile soils need. Bacteria and fungi help this happen. Some plants also return needed nutrients to the soil through their roots. Animals such as prairie dogs, moles, earthworms, and ants tunnel through the soil, letting in air and water. Remember that the elk living near Mount St. Helens helped begin the soil restoration process within the devastated forest by breaking up the volcanic tephra with their tracks. They also fertilize soil with their droppings.

Human activities greatly affect soils. Regions covered by natural vegetation undisturbed by people can build up a deep layer of humus as vegetation decays. But when human activities strip land of vegetation, this can't happen. In farmed areas, some farmers may harvest their crops and completely remove the remaining plant matter as animal feed. Then only a little dead plant matter returns to the soil. If this happens for more than a few growing seasons, the ground becomes *infertile*. To avoid this all-too-common problem, farmers fertilize their fields with either chemical fertilizers, or animal or green manures. Many farmers grow cover crops that fix certain nutrients in their tissues, and then plow the crop into the soil. Other soils benefit from just letting them lie *fallow* (rest from cultivation).

12.17 SOIL CONSERVATION

Americans haven't always used soil wisely. During the pioneering years in the United States, people were looking for the fastest and easiest ways to establish their lives and produce food. They didn't consider the long-term consequences of these actions because simple survival was their highest priority. Many overgrazed their pastures. Many didn't rotate crops. They logged trees off huge tracts of land. We cannot blame people for doing these things then as our young country was developing. But these actions couldn't go on for generations without consequences.

By the 1930s, during the Great Depression, topsoil in the Central Plains states had been destroyed by years of misuse and erosion. High winds and droughts completed the devastation in areas with little vegetation coverage in what became known as the Dust Bowl. The personal misery and the loss of farm productivity made the Great Depression even worse.

But things began to change. To give men work during the Depression, President Franklin D. Roosevelt's administration formed the Civilian Conservation Corps (CCC). The goal of this program was to conserve natural resources on public and private lands, as well as put people back to work. One of its priorities was **soil conservation**. Men plowed fields to follow elevation contours. They terraced farm fields. They implemented strip-cropping and planted ground cover. Rows of trees were planted as windbreaks to reduce the wind speed near the ground. In particular, flood control dams were built in many states. All these actions drastically reduced soil erosion by wind and water. As you drive across the Great Plains of the United States today, you can still see the windbreaks that these men planted. Their efforts rescued and revolutionized American agriculture.

Soil is a resource that God has given man for good and wise dominion. It is just as important for life as air and water. This means that we need to study soils so that we can support a growing world population.

12-24

A combination of drought, high winds, and poor management of farm soil created the Dust Bowl of the 1930s.

> Allowing fields to lie fallow was a principle of good agriculture commanded by God to the Israelites (Exod. 23:11). It not only extended the principle of the Sabbath to farming as a sign of faith in God's provision; it also had the effect of helping the land restore some of its fertility.

12-25

Examples of methods for controlling soil erosion: contour plowing (top) and strip cropping (bottom)

12C REVIEW QUESTIONS

1. What is another name for a soil scientist? What does he do?
2. What are the general properties of a soil according to its definition?
3. When did the earth's first soils appear? What happened to them?
4. How do we know that soils don't need thousands of years to form?
5. Describe the layers of a mature soil.
6. What is the composition of ideal, fertile topsoil?
7. Suppose you inherited a 100-acre farm in the mountains. How would you use its soil best, considering what you've learned in this section?
8. Discuss three ways that people can conserve soils.
9. (True or False) There are soils that do not have three horizons in many places around the world.

CONTROLLING EROSION

Flowing water can erode soil (top). Windbreaks (left) and terracing (above) can help control erosion.

Unless you've spent time gardening, soil is probably something you take for granted. But it doesn't just appear; soil usually takes a long time to form! Pedologists believe that before the settlement of Jamestown, Virginia, in 1607, there was an average depth of about 20 cm (8 in.) of topsoil in the United States. Today, topsoil depth averages only 15 cm (6 in.)! What happened? People don't always use soil in the best way. Also, erosion can carry away topsoil. But the nutrients in topsoil are vital to growing crops that sustain our population. Since making more topsoil cannot be done economically, we need to conserve what we have.

There are lots of ways to help control soil erosion. A simple way is to plant vegetation in areas likely to erode. Plant roots hold soil in place well. Sometimes, though, the soil doesn't even have enough nutrients for healthy plants to grow. That is when it becomes important to restore nutrients to the soil, to make sure that it has the minerals plants need to grow. Rotating the crops that are grown in a certain area can help restore those nutrients and build up the soil. Trees can also help keep the soil from blowing away.

Gravity acts directly on water flowing downhill to cause soil erosion. Through contour plowing and planting, farmers arrange crop rows at right angles to the direction the water flows. This arrangement forms many small dams that slow down the flow of water. Strip-cropping, the practice of interspersing strips of row crops such as corn with strips of cover crops such as grass and alfalfa also helps to prevent erosion. Cover crops are plants that have many small roots close to the surface of the soil and dense leaf patterns to break the fall of raindrops. To further slow erosion, property owners may terrace a slope by grading it into nearly level areas with walls around the steep sides.

Recently, farmers have begun to experiment with no-till farming. In this practice, farmers don't plow under the remaining stalks from last year's crop. They break this cover only enough to plant seeds for the new crop. This keeps the ground covered with plant remains all year, preventing soil erosion.

Flooding is a major problem in some parts of the United States. Not only do floods cause the loss of life and property, but they also rapidly displace large amounts of valuable soil. Although this soil may settle downstream in a useful location, it is more likely to settle in the bottom of a river or lake where it can no longer grow crops.

Artificial levees, or walls, on either side of a river are one solution to this problem. These work reasonably well in some areas but prove inadequate in others. Their flaw is that they confine the sediment that normally spreads out over the river's entire floodplain to the river channel. As sediment is deposited in the channel, the stream becomes shallower, causing the water to rise higher the next year. In order to contain floods, the levees must be built progressively higher or moved out farther from the river, or both. Though dredging the sediment inside the levees is possible, it is costly.

Another approach to the problem of flooding is to build temporary reservoirs along the course of a river. At flood stage, these reservoirs take much of the overflow and store it until the river recedes. The reservoirs then slowly release the water back into the river.

Soil is one of our most valuable resources. We need soil to raise crops to feed ourselves and our livestock, and to grow trees that we can use for fruit, lumber, and paper products. Soil is just one more essential component in the complex world God created, without which life would not be possible. In order to properly fulfill the Creation Mandate (Gen. 1:28), a good citizen will do his part to help his fellow humans in the effort to conserve soil.

CHAPTER 12 REVIEW

CHAPTER SUMMARY

» The earth's surface is transformed by weathering, erosion, and deposition.

» Weathering by environmental factors reduces massive rocks to fine particles that can make soil.

» Weathering breaks down rocks through mechanical and/or chemical processes. Living organisms can also aid weathering.

» Gravity, water, wind, and glaciers are all agents of erosion.

» Mass wasting is the downhill slips, slides, or falls of large quantities of soil and/or rocks under gravity's influence.

» A stream is any narrow body of water that flows continuously or seasonally on the earth's surface or underground. Streams are a main source of water erosion.

» A stream moves sediments by suspension, by saltation, and by rolling rocks along its bed.

» A stream's sediment load depends on how fast it is flowing. When it slows down enough, it deposits sediments in an orderly way through sorting.

» Stream erosion removes immense quantities of topsoil annually from the United States.

» Wind can seriously erode arid lands by deflation.

» Wind deposition features include dunes and immense loess deposits.

» Glaciers are masses of ice that move downhill by gravity. They are particularly important agents of erosion and deposition.

» Glaciers form in snowfields above the snow line where winter snow survives the summer and becomes compacted into dense ice crystals called glacial ice.

» Glaciers are classified by their location and size.

» Glaciers change landscapes by erosion and deposition, creating many unique glacial formations.

» Soil is a mix of weathered materials and sediments. It should contain organic materials such as humus to be fertile.

» Mature native soils have three layers or horizons lying on top of bedrock.

» Different areas in the world vary in their soil type because of having different underlying bedrock and also because of the influence of transported soils.

» Soil composition depends on many factors, including sediment type, climate, and biological activity.

» Soils can develop slowly or quickly. They can be lost through erosion even faster.

» People need to conserve soil as a vital resource to be wisely used.

Key Terms

weathering	277
erosion	277
deposition	277
chemical weathering	277
mechanical weathering	278
frost wedging	278
exfoliation	279
biological weathering	280
mass wasting	281
landslide	282
sorting (sediment)	284
deflation	285
abrasion	285
dune	286
loess	286
glacier	287
glacial drift	290
till	290
moraine	290
soil	293
horizon	295
humus	295
loam	295
soil conservation	297

Rice has been traditionally grown using erosion control methods such as contour plowing and terrace farming.

Erosion along the coast can create interesting landforms, such as The Twelve Apostles in Australia.
Location: 38.65 °S 143.1 °E

REVIEW QUESTIONS

1. How do plants help to weather rocks?
2. How can you tell by inspection alone that a piece of rock has been weathered?
3. How does weathering by exfoliation happen?
4. What conditions can help cause a landslide?
5. What property of flowing water makes it such an effective agent of erosion? What is this property related to?
6. What kind of sediment motion in a stream can create finer sediment clasts?
7. Explain how streams deposit sediments as they slow down.
8. What are two ways that wind erodes?
9. How is wind erosion similar to stream erosion?
10. Describe how sediment transport and deposition cause sand dunes to move about.
11. Why are there more glaciers near the poles?
12. Compare continental glaciers and ice caps.
13. What determines whether a glacier will advance or retreat?
14. Where does glacial drift come from?
15. List four types of moraines. How are they similar?
16. Which soil horizon is most fertile? Why?
17. What is the most important part of all fertile soils?
18. How do living things affect soil?
19. From what you've read in this chapter, what causes most land erosion in the United States today?

True or False

20. Chemical weathering can change both the composition and shape of a rock.
21. Living things can break up the strongest rocks.
22. The main features of all mass wasting are their devastation and the high speed of debris fall.
23. Well-sorted sediment results when a stream carries only sediment clasts of the same size.
24. As sediments are eroded in a blowout, only clasts larger than a certain size are left behind.
25. Dunes are products of both wind erosion and wind deposition.
26. Glaciers are pure ice.
27. All glacial deposition features are made of unsorted till.
28. God preserved the originally very good soils for our use today.
29. Soils can form relatively quickly (in a matter of months or several years).
30. All soils have three horizons.
31. Soils used for agriculture must have their nutrients replenished even though plant matter grows on them year after year.

Map Exercises

The following questions assume that you have Google Earth or a similar 3D Earth-viewing program installed on your computer.

32. Find Rice Lake in Canada (loc.: 44.22°N 78.11°W). Activate the Google Earth Daylight tool on the toolbar. Slowly slide the time slider until shadows form on the surrounding landscape. What kind of glacial feature occurs many times around Rice Lake?

33. Identify the erosion or deposition features at the following coordinates.
    a. 47.8065°N 119.34°W
    b. 42.03°N 85.15°E
    c. 43.82°S 172.92°E
    d. 22.46°N 90.59°E

SERVING GOD AS A PEDOLOGIST

JOB DESCRIPTION

Soil is everywhere, so it's not very valuable, right? Wrong! Do you have any idea how precious soil is? Think about how long it takes most soils to form. What would happen if a farmer didn't try to understand or improve his soil? A *pedologist* is a soil scientist, someone who helps people make the most of their soil.

Pedology, or soil science, involves biology, math, physics, hydrology, geology, chemistry, geography, and even technology. The idea is to use Earth's land, soil, and water resources in a way that we can provide all the food we need right now and in the future. Being a pedologist means that you are involved in agriculture, forestry, and environmental science.

EDUCATION

If you are interested in pedology, you could attend an agricultural or nonagricultural university. Studies in agriculture, chemistry, geography and GIS, environmental science, geology, and hydrology all come in handy. You will learn what makes up soil, the weathering processes that produce soil, and how water and air flow through soil. Most importantly, you will learn how to use and protect soil, which is a vital resource for people everywhere.

POSSIBLE WORKPLACES

Being a pedologist may mean that you have to get your hands dirty! You could be using a shovel one day and a laptop the next. Pedologists usually spend time both in the field and in the office.

Pedologists can work for the government or for private companies such as fertilizer companies and laboratories that analyze soils. University extension services and even farm

cooperatives employ pedologists. Nonprofit organizations that offer services for agricultural development in emerging countries also may employ soil scientists.

DOMINION OPPORTUNITIES

What was the first work that God gave us? Adam was to tend and keep the Garden of Eden. And to do that, he probably needed to learn something about the soil. Pedology is one of those sciences that are essential to good and proper dominion of the earth that God gave us. We can use this science to glorify God, improve the lives of His image bearers, and be better stewards of the earth for future generations.

4 THE WATER WORLD

CHAPTER 13
OCEANS AND SEAS 304

CHAPTER 14
OCEAN MOTIONS 330

CHAPTER 15
OCEAN EXPLORATION 355

CHAPTER 16
SURFACE WATERS 378

CHAPTER 17
GROUNDWATER 398

DR. EMIL SILVESTRU
SPELEOLOGIST

> "The present is not the key to the past! More than 75% of all known rocks are sedimentary, most of which are believed to have been formed in the oceans. Over 90% of the rocks formed in the oceans were laid in shallow oceans. But only 10% of present-day oceans are shallow. The sediments the oceans hold are insignificant compared to the millions of tons of sedimentary rock. These facts clearly show that present processes of sedimentary rock formation are not the key to rock formation in the past!
>
> My personal passion is speleology (the science of caves and their geology). I have concluded that most caves formed toward the end of the Flood, created by hot and very aggressive fluids called hydrothermal solutions that ate away the limestone in a matter of months. Modern caves have been reshaped by carbon-dioxide-rich waters that infiltrate cracks and joints in the limestone to form unique and beautiful formations.
>
> God flooded the world with a global ocean which covered and eroded the continents. This is by far the most logical source of rock-forming and rock-eroding processes. Deep canyons cut miles below the ocean surface by forces not present today are scars bearing witness to the monumental forces present during the Flood. This is why studying the oceans is crucial if you want to understand geologic history and appreciate the extent and uniqueness of God's judgment on the fallen world."

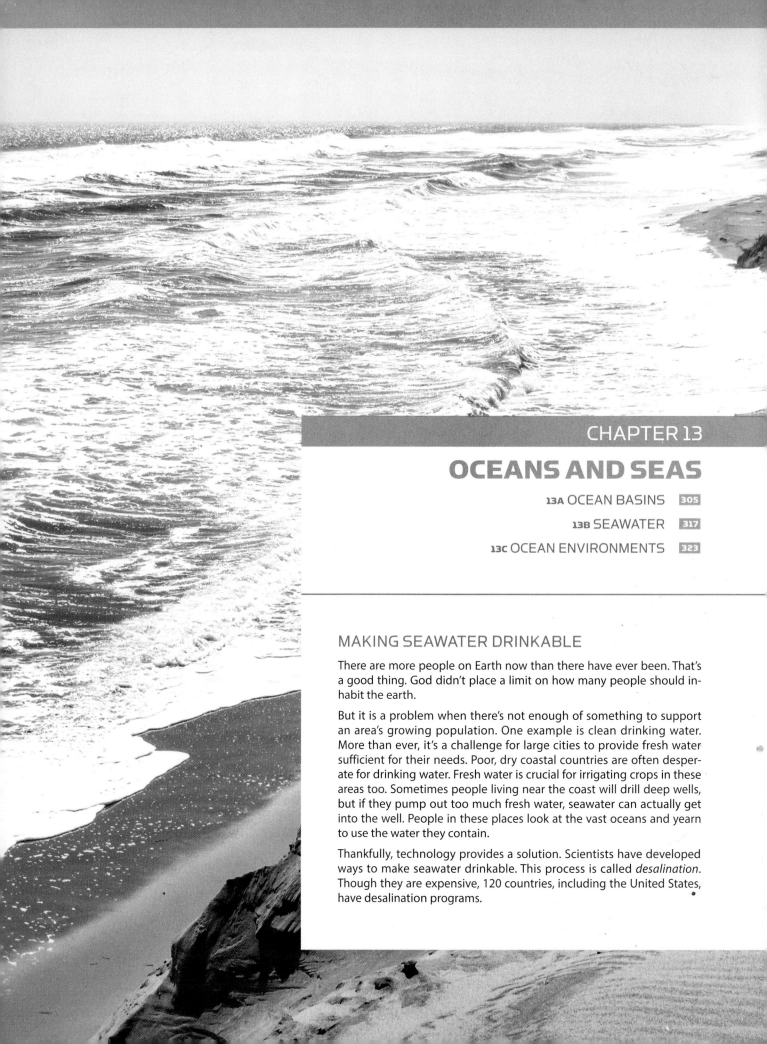

CHAPTER 13
OCEANS AND SEAS

13A OCEAN BASINS 305
13B SEAWATER 317
13C OCEAN ENVIRONMENTS 323

MAKING SEAWATER DRINKABLE

There are more people on Earth now than there have ever been. That's a good thing. God didn't place a limit on how many people should inhabit the earth.

But it is a problem when there's not enough of something to support an area's growing population. One example is clean drinking water. More than ever, it's a challenge for large cities to provide fresh water sufficient for their needs. Poor, dry coastal countries are often desperate for drinking water. Fresh water is crucial for irrigating crops in these areas too. Sometimes people living near the coast will drill deep wells, but if they pump out too much fresh water, seawater can actually get into the well. People in these places look at the vast oceans and yearn to use the water they contain.

Thankfully, technology provides a solution. Scientists have developed ways to make seawater drinkable. This process is called *desalination*. Though they are expensive, 120 countries, including the United States, have desalination programs.

13A
OCEAN BASINS

What does the topography of the oceans look like?

13.1 OCEANS FOR MAN'S USE

The **oceans** are vast bodies of salt water that separate the continents. They cover about 71% of the earth's surface and hold over 97% of all the water on the planet. They absorb most of the sun's energy, which influences the earth's weather and makes its climate more uniform compared with that of other planets. They are home to an abundance of animal and plant life. Perhaps most importantly, most of the oxygen in our atmosphere comes from photosynthetic organisms living in the oceans. Without the oceans, there would be little or no life on Earth. Besides these "big picture" aspects, humans depend on the oceans for many things, not the least of which is a source of fresh water. This is why good and wise dominion of the oceans is important.

To pursue dominion most effectively, we need people who specialize in studying the ocean. *Oceanographers* are marine scientists and engineers from almost every field who work to better understand and use the oceans. *Marine biologists* investigate the plant and animal life in every ocean environment. *Meteorologists* study the influence of the oceans on weather and climate. *Chemists* analyze the distribution of various elements and compounds in the oceans and the ways that those chemicals affect ocean processes. *Physicists* model the distribution of energy in the oceans and the motion of their great currents. They also research ways to use their physical properties to meet our needs. *Marine engineers* develop methods and vehicles to investigate and use the oceans. They design economical ways to produce potable water from seawater as well as electricity from its motions.
As we understand the oceans better, we understand the earth better too.

13A Objectives
After completing this section, you will be able to

» explain the reasons that the oceans are essential to life and some of the ways that we use them.

» evaluate theories that account for the origin of the oceans.

» list the factors that determine mean sea level and describe how sea level varies around the globe.

» describe the general ocean basin topography from the shore to the abyssal plains.

» describe various kinds of coral reefs and atolls, and their origin, geologic significance, and impact on aquatic life.

The word *marine* used as an adjective always refers to anything having to do with the oceans.

13-1
Oceanographers come from many different backgrounds. Engineers, biologists, meteorologists, chemists, and many other scientists can all work in the field of oceanography. Some work in labs while others work on the surface of the ocean or even underwater.

---

potable (POE tuh bul): pota- (L. *potare*—to drink) + -ble; suitable for drinking

13-2
Major ocean basins of the earth, covering the indicated percentage of the earth's surface. Is the earth's surface mostly land or mostly water?

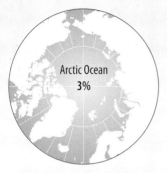

13.2 ARRANGEMENT OF OCEAN BASINS

Compared with land, the ocean surface appears flat, unremarkable, and boring. But below the surface is a very different story. Modern scientists have found that the ocean floor is far from featureless. You just have to look a little deeper. There are endless opportunities for study.

The oceans surround the continents. Because they are all connected, oceanographers say that the earth has only one ocean. However, for convenience, people usually divide the world ocean into five major *basins*—the Arctic, Atlantic, Indian, Pacific, and Southern Oceans. A large section of ocean mostly surrounded by land or islands is called a **sea**. For example, the Mediterranean and Caribbean Seas are separate extensions of the Atlantic Ocean.

It was important for traders and explorers to know about oceans and seas during the age of sailing ships. We have records that explorers studied and mapped the ocean just like they mapped the land. We know the most about the surface of the ocean and its boundaries. But we know less about the sea bottom that forms the ocean basins than we do about the surface of the moon or Mars. There's still a lot to learn.

13.3 ORIGIN OF THE OCEANS

Since the oceans surround the continents, their formation must be closely tied to the history of the continents. Old-earth geologists believe that the oceans have been here from very early in Earth's history. Old-earth geologists hold to one of two theories for the source of the earth's water. The first is that the earth retained water as it formed from the nebular cloud. This water, held within the mantle, slowly moved to the surface through volcanic activity. The other theory claims that icy bodies in space deposited water on Earth during frequent collisions shortly after the earth formed. Old-earth geologists believe that this water filled the oceans and seas that formed as continents appeared and broke apart throughout Earth's history. They believe that the ocean basins that we see today are just the result of an unending process of plate tectonics over billions of years. In this story, the current Atlantic, Indian, and Arctic Ocean basins began forming 300 million years ago. The Pacific Ocean basin is all that remains from an earlier global sea called Panthalassa (see Chapter 5).

Young-earth geologists believe that the earth began as a water planet. God created what was probably a single supercontinent on the third day of Creation (Gen. 1:9–10) around 7000 years ago. That supercontinent was wrenched apart during a one-year flood and its catastrophic aftermath around 5500 years ago. The shapes and sizes of the present-day ocean basins are the result of that single event in Earth's history.

13.4 SEA LEVEL

If you've ever been to the ocean, you know that it's not like the water in your bathtub! So what is **sea level**? Does it depend on the ocean tide? Does it change with each wave? Or is it a fixed height from which all elevations and depths are measured? Sea level is continuously affected by tides, density, salinity, and weather.

There are actually several terms that people may use for sea level, but they have slightly different meanings. *Local sea level* is an always-changing height of the ocean surface at a given location. Oceanographers measure local sea level with *tide gauges* mounted along shorelines or from satellites far above the ocean using radar. The zero point for the tide scale gauge is positioned at the imaginary surface of a mathematical model of the earth. The model is called the *geoid*, or "earthlike" shape. Local sea level is important to people who work on the ocean near the coast. But local sea level is not very useful as a way to determine one's elevation.

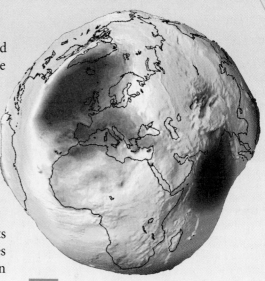

13-3

A mathematical model of the surface of the earth. Differences between a perfect ellipsoid and the mathematical model (the geoid) are indicated by color. The earth looks lumpy in this image because differences are multiplied by 15,000 to make them more noticeable.

Floodwaters From Magma?

A recent Flood geology study suggests that global volcanism and basalt eruptions during the Genesis Flood were significant. They could have liberated enough water vapor from magma into the atmosphere to produce more than a half meter of rainfall per day for the forty days of the active period of the Flood. This water alone could have raised sea level 21 m!

13-4
This diagram shows how mean sea level (MSL) differs from the smooth mathematical model of the earth's surface. The figure uses both a color theme as well as a 3D surface to show relative heights. Greens and blues indicate MSL below the geoid, while yellows and reds show MSL above the geoid surface.

Mean (or average) sea level—what most people call simply "sea level"—is the computed average height of high and low tides at a location. **Mean sea level (MSL)** is the zero height that cartographers, mariners, and aircraft pilots use to measure elevation, depth, and altitude. Since the heights of tides continually vary from day to day, oceanographers measure tide levels over many years and then average them to determine mean sea level.

MSL also changes over time. Right now, it is rising about 2.5 mm each year worldwide. This change is supported by satellite measurements of MSL since 1993. Some locations had even greater increases, and a few reported decreases in mean sea level. Scientists estimate that MSL has risen at an average rate of 1.8 mm per year over the past 100 years, so this recent change suggests that something new must be affecting MSL.

Most scientists think that global warming is causing the increase in MSL. Increasing global temperatures would warm up the oceans and melt ice caps. Water expands as it gets warmer, so that would make sea level higher. Melting glaciers also add water to the oceans. You will learn more about climate change in Chapter 21.

13.5 BASIN TOPOGRAPHY

Let's take a trip into an ocean basin. Imagine climbing inside a self-contained, deep-sea mechanical walking vehicle with a week's worth of food, water, and air. You are standing on the shore, and soon you will begin your trek to the sea, walking on the bottom and describing the topography as you go.

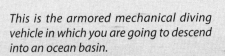

This is the armored mechanical diving vehicle in which you are going to descend into an ocean basin.

THE SHORE

A **shore** is a strip of land that separates the *coastal region* from the ocean. You can make out the shore by the lowest and highest reaches of sea level. You note that the highest limit of the shore results from erosion by the largest storm waves; the lowest limit results from wave erosion at the lowest of low tides (see Chapter 14). The width of a shore can vary from just a few meters to several hundred meters.

The *coastal region* of land is an indistinct zone extending from the shore as far inland as we can recognize land features affected by the ocean. These might include coastal sand dunes, coastal plants, and sandy sediment deposited by past wave or wind action.

BEACHES

The **beach** is an ever-changing place that extends underwater beyond the shore. There are some important terms associated with beaches (see Figure 13-5). Notice that a lot of these terms refer to the *shoreline*, which is the edge of the water at any given time.

The *berm* is the area where only the highest tides or storm waves can reach. It is usually soft, coarse sand, or large cobbles left by the largest storm waves.

The zone between the high- and low-tide shorelines is the *beach face*. It may be wet, compact sand or fine pebbles. Walking over the beach face in your heavy vehicle is easier than on the berm.

The shallow bottom of the beach beyond the low-tide shoreline often contains a *longshore sandbar* where large waves first break and drop any sand that they have moved from deeper water. The breaking action of these waves can also dig a *longshore trough* on the shore side of the longshore bar. The beach extends from the outer longshore bar all the way in to the top of the berm.

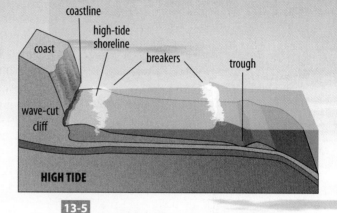

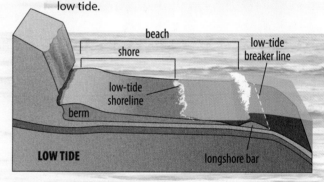

13-5
The beach extends from the highest point of erosion on the shore out to the breaker line at low tide.

The word *longshore* is a combination of the words "along" and "shore." It describes anything that exists or moves parallel to the shoreline.

CONTINENTAL SHELF

Beyond the beach zone, the bottom slopes out gently from the continent, forming a **continental shelf**. As you walk out onto the continental shelf in your mechanical walker, you gradually submerge and move beyond the beach zone. Continental shelves are the submerged edges of continental plates. The topography and geologic formations of the continental shelf are often similar to those on the adjacent land.

A typical continental shelf extends 70 km from the seacoast to an average depth of 135 m. In some coastlines with rugged terrain, such as the West Coast of the United States, the continental shelves are very narrow or nonexistent. In other locations, like the relatively flat US East Coast, the offshore continental shelf is broad, extending up to 1500 km from the shore. The average slope of a continental shelf is about a tenth of a degree. Well before you reach the outer edge of the continental shelf, the sea looks black to you. Only a little sunlight reaches these depths. You have to turn on your vehicle's lights to see where you are going.

CONTINENTAL SLOPE

As you continue walking toward deep water, the angle of the bottom becomes steeper. The **continental slope** begins here. Its upper edge is the beginning of the deep ocean basin. As you pause at this position, you may be 1 to 5 km above the flatter bottom of the ocean basin. The angle of continental slopes varies between 1° and 25°. Steeper continental slopes occur if the edge of the continental plate is a convergent tectonic boundary.

As you carefully pick your way down the continental slope, you come across an immense **submarine canyon** carved into the slope. It might have a fan of clastic sediments and mud at its lower end. These canyons are common around the world. A few of them appear to be continuations of large rivers that we can see on land today. But most submarine canyons do not seem to be extensions of river channels. Young-earth oceanographers believe that these canyons were carved by currents of water full of sediment flowing off the continents late in the Flood. That much water and sediment moving fast could easily have carved out these underwater scars.

CONTINENTAL RISE

The submarine canyon was fascinating, but you must hurry on before your air and other supplies run out! The bottom slope becomes less steep, but you begin to bog down in deep ooze (that's why your vehicle has lifter propellers). Welcome to the **continental rise**! A continental rise is a smooth transition from the continental slope to the deep, relatively flat ocean floor. We find the thickest rises at the mouths of submarine canyons. These locations indicate that the rises formed from materials that slid off the continental shelf over time in downhill currents that flowed through the canyons. These slides are something like underwater landslides called *turbidity currents*. In other places, the rises are either not very thick or not even there.

13-6 A digital model of the continental shelf and slope off the California coast showing the Monterey Submarine Canyon

13-7 A turbidity current flows down an underwater slope in this test tank, carrying sediment and water with it. Such currents can scour out channels and canyons in the continental slope.

Generalized topography of the ocean floor. Vertical scale is exaggerated.

coast
beach
shelf
slope
seamount
rise
canyon
abyssal plain

As you continue down the continental rise, the sloping landscape levels off and meets the ocean floor. The average depth here is about 3.8 km. Better check your control cabin for leaks. The outside water pressure is around 380 times normal atmospheric pressure!

ABYSSAL PLAIN

The relatively flat, deep sea floor is the **abyssal plain**. Sediments cover the ocean floors with varying thickness. Oceanographers study the makeup and history of the sediments by removing long cylindrical samples called *sediment cores*. They get these samples by dropping a sediment corer vertically into the bottom. When the corer is pulled up, a cylinder of mud comes with it. Drilled cores are needed for taking longer core samples of very thick sediments and for sampling the ocean crust bedrock underneath.

abyssal (uh BIS ul): abyss- (Gk. *abussos*—bottomless, deep) + -al; having to do with the deepest ocean environments

Even where deep sediments cover the abyssal plain, it is not featureless. Ridges, valleys, fracture zones, and volcanoes give the ocean floor at least as much topography as the continents. But you cannot visit them now. Your air is running low and you have just enough to get to the surface. You inflate small, tough flotation balloons, which gently lift you off the sea floor. Several hours later, you break the ocean surface and bob like a cork. Your support ship is just a few hundred meters away. Just in time for lunch!

13.6 TECTONIC FEATURES

Though the ocean floor has features created by erosion and deposition, tectonic activity also shapes the ocean floor, just as it does the land.

MID-OCEAN RIDGES

Every ocean basin has one or more **mid-ocean ridges**. As you learned earlier, these ridges are submerged mountain ranges at the margins of diverging tectonic plates. They formed when the mantle pushed up on long, faulted sections of sea-floor crust that line both sides of the plate margins. Many transform faults cross these ridges. They relieved the stresses in the crust as it spread out from the plate margin. Most mid-ocean ridges lie within these *fracture zones* (see Figure 13-8).

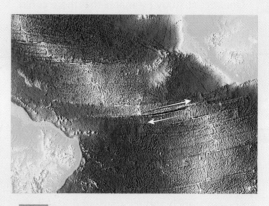

13-8 A fracture zone in the Atlantic Ocean basin

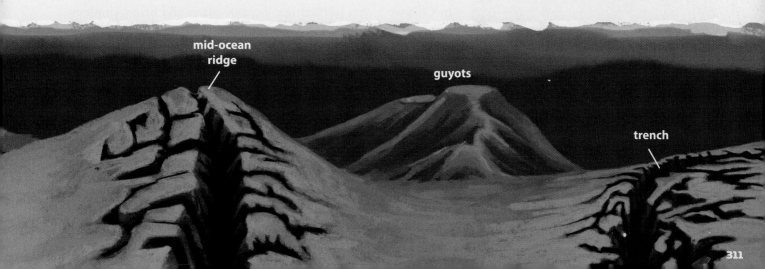

SEAMOUNTS

Oceanographers estimate that more than 100,000 submerged volcanoes and hills, or **seamounts**, taller than 1 km dot the abyssal plains. Most seamounts are volcanic. Some seamounts have flat tops and are called *guyots*. Most geologists think that guyots were once volcanoes that grew into islands above a much lower sea level, perhaps during a glaciation period. Waves eroded the still-soft tephra flat and then sea levels rose. The tops of guyots are now an average of 1.5 km below sea level. However, a Flood model suggests that the volcanoes formed over a short period of time when the sea basins weren't as deep during the Flood. After forming islands, the strong wave action eroded the volcanoes level with the ocean surface. Then sea level rose as the basins sank into the mantle toward the end of the Flood.

What could have produced such widespread volcanism? The old-earth model of plate tectonics suggests that seamounts and guyots were volcanoes that originally formed at the mid-ocean ridges. The spreading sea floor carried them away from the crest of the ridge into deeper water. Young-earth geologists agree that this process could have occurred—but much faster and much more recently—during the Flood. Another explanation is possible. If the Flood model of tectonics is accurate, the crust moved rapidly over the mantle during the early stages of the Flood. This would have created broad regions of molten rock beneath new, very thin oceanic crust. The magma could have easily erupted through the crust in many places, forming the volcanoes near where we find them today. Remember that we just don't know how God caused the Flood. We cannot tell in the present the difference between the results of God's miraculous works and those of natural processes during the Flood.

Many seamounts occur in distinct chains in the middle of an oceanic plate. These may have formed as the thin ocean plate moved over a hot spot in the mantle. Such a chain of volcanoes forms the Hawaiian Islands and **Emperor Seamounts**. Old-earth geologists believe that this process took millions of years, while young-earth geologists think that these features give evidence to the devastating effects of the Flood during a relatively short period.

TRENCHES

In Chapter 5, you learned that oceanic plates slid below continental plates in convergent subduction zones during the Flood. In some places, the plunging oceanic crust dragged against the continental crust, bending its edge downward and creating a deep notch in the ocean floor. These **oceanic trenches** usually have a relatively steep slope on the side toward the continental plate and a gentle slope on the seaward side. Trenches seem to be present only where subduction occurred. Such trenches surround much of the east, west, and northern margins of the Pacific basin. The deepest point in any ocean was found in one of these trenches—the **Challenger Deep**—10,971 m. It is located in the Mariana Trench in the western Pacific.

Guyots (GEE ohs) were named after the Swiss geographer Arnold Guyot by their discoverer, Harry Hess, in 1946.

13-9 This is how a guyot would appear from underwater.

© 2011 Google/© 2011 Europa Technologies/Data SIO, NOAA, U.S. Navy, NGA, GEBCO/Image USDA Farm Service Agency

One recent study estimates that there could be as many as 25 *million* seamounts taller than 100 m scattered across the sea floor.

Emperor Seamounts (middle of chain): 42°N 170°E

13-10 An ocean trench and its associated island arc seem to be related to a subducting oceanic plate.

Challenger Deep: 11.33°N 142.2°E

13-11
Volcanoes along oceanic trenches form island arcs, such as the Aleutian Islands of Alaska.

ISLAND ARCS

The Mariana Trench is only one of many trenches along plate margins far to the east of the Asian and Australian coastlines. As is true near all subduction zones around the world, many volcanoes formed where subducting oceanic crust melted after sliding into the mantle. The hot, low-density magma rose upward through igneous dikes in the overlying crust, creating volcanoes. When these submarine volcanoes broke the ocean surface, they formed long, curved strings of volcanic islands called island arcs. They line the lip of the trench on the side opposite the subducting plate.

Positions are near the middle of each island arc.

Aleutians: 54.5°N 164.5°W
Kurils: 45.4°N 149°E
Ryukyus: 26.6°N 128.0°E
Philippines: 12.7°N 122.7°E
Solomon Islands: 9.6°S 160.1°E

reef: (Ger. *Riff*, poss. from Scan. *rib*—sand bank, reef); a hazard to navigation due to shallow water

13.7 CORAL REEFS

Coral reefs are important marine geologic features that form from the activities of animals. Corals are tiny animals that usually live in warm ocean waters. They live in groups called *colonies*. Most coral colonies secrete an external, rock-like calcium carbonate support structure or skeleton. Many coral formations are massive *coral heads*. Others build delicate, branched skeletons. Over time, skeletal materials have accumulated, forming coral reefs—underwater ridges. In some places, coral reefs emerged above the ocean surface and became islands when the sea level changed.

Many reef-forming corals need light to live because they depend on photosynthetic microorganisms for nutrients. Thus, coral reefs often form in shallow water near land. Other corals can catch their own food, so they can grow in much deeper water. Coral reefs that grow right up to the beach along a coastline are *fringing reefs*. These occur along the coasts of Florida and Bermuda as well as around many Pacific islands.

Barrier reefs are farther from the land. They form a *lagoon* between the reef and the land. Lagoons open to the sea through passages in the reef. The largest barrier reef in the world is the Great Barrier Reef off the northeastern coast of Australia. It extends about 2000 km and covers more area than some countries.

An **atoll** (AT OL or AY TOL) is a ring of low coral islands and reefs surrounding a central lagoon. Geologists believe that most atolls began as barrier or fringing reefs around volcanic islands. There are two hypotheses for the formation of atolls. First, the central volcano could have collapsed, forming a caldera below sea level, leaving the fringing ring of coral islands behind. Second, the sea level could have slowly risen (or the sea bottom could have dropped), drowning the central island. As sea level changed, the coral reefs continued to grow upward, leaving the ring of coral islands and reefs alone at the ocean surface. Several Pacific islands are atolls, including Wake, Midway, Bikini, and Eniwetok (EN ih WEE TAHK). These and many other atolls were sites of important World War II battles in the western Pacific. Many people live on these atolls today.

Great Barrier Reef (middle of reef): 18.8 °S 147.8 °E

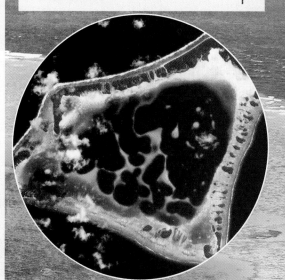

13-12
Atafu Atoll and its lagoon in Tokelau, New Zealand, in the Pacific Ocean (loc.: 8.56 °S 172.5 °W)

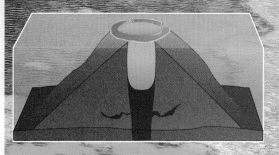

Bikini Atoll: 11.6 °N 165.4 °E

Bikini was the location of American atomic bomb tests during the Cold War.

13A REVIEW QUESTIONS

1. What is the vitally important natural resource contained in the oceans of the world? Why is this resource difficult to obtain from the oceans?
2. List the main ocean basins of the world. Which is the largest?
3. What are the two main differences between the young-earth and old-earth theories for the origin of the ocean basins?
4. People often use mean sea level (MSL) as a reference point for measuring what?
5. List the typical geographic features of ocean basin topography, starting at the coast and moving into the basin.
6. What are three important tectonic features found in oceanic basins?
7. What is a reef? What makes a coral reef?
8. With what kind of tectonic feature do most atolls seem to be associated?
9. (True or False) The Challenger Deep is deeper than the highest mountain in the world is tall.

LIFE CONNECTION: SWIMMING THROUGH A RAINFOREST

You are snorkeling off the coast of Australia through a rainbow of sea life hovering around what appears to be an underwater rock outcrop. As you watch, a yellow-bellied sea snake gobbles up an eel for lunch. Brilliant orange clownfish dart in and out of sea anemones. Blue angelfish graze on sponges and algae. Coming up for air, you hear the screeches of the many sea birds that call this area home.

You are exploring a coral reef, so-called the "rainforests of the sea" because of the diverse forms of life found here. Coral reefs occupy less than 0.1% of the world's ocean floor, yet they are home to 25% of all marine species! Some animals that live here don't live anywhere else in the world.

So what is a coral reef? It's not just rock. Coral reefs actually house thousands of tiny creatures called *coral polyps*. Corals are animals classified in the same phylum as the jellyfish. Unlike jellyfish, they surround themselves with a hard substance containing calcium carbonate, which is chemically the same as limestone. This "suit of armor" acts like an outside skeleton. It provides support and protection for each individual coral polyp. When the organism dies, it leaves behind its skeleton, helping build the reef. Coral reefs appear mostly in the shallow, tropical waters of the Pacific, Atlantic, and Indian Oceans.

Coral reefs support a vast variety of organisms. These living things depend on each other in a tightly knit community. Sea urchins, sea slugs, crabs, prawns, and other invertebrates, as well as many fish, feed on the abundant seaweed growing around coral reefs. Some fish prey on smaller fish or invertebrates, while others help each other survive. Certain small fish eat parasites and other materials off larger fish. Marine biologists call these smaller fish "cleaner fish." Some cleaner fish even act like dental floss, cleaning food out of the jaws of much larger fish that could easily eat them!

Fish also are a source of food for other organisms. Sea snakes, like the one mentioned earlier, feed on fish or fish eggs. Above water, many tropical birds also prey upon the fish of coral reefs. During a visit to a coral reef, you might see pelicans, herons, or even a huge albatross, which has the largest wingspan of any bird.

There are also many animals, such as sponges, clams, and worms that actually live *inside* coral heads. These organisms, called *cryptofauna*, or "hidden animals," bore into the calcium carbonate skeletons of the coral, creating a well-protected hideaway.

Coral reefs have a beauty that mirrors the beauty of the God who made them. But they are also useful. They protect shorelines and are important areas for the fishing industry. Around 6 million tons of fish are harvested in the vicinity of coral reefs each year! But many coral reefs around the world are dying, and we need to figure out why. Some scientists blame global warming and increasing ocean temperatures. But there are many other factors that could cause reef death, including disease or changes in water chemistry. It's important for us to understand coral reefs so that we can protect them and use them wisely. God has commanded that we care for the earth and its creatures. In a fallen world, we cannot do this completely, but God still wants us to do what we can.

13B

SEAWATER

How does seawater differ from fresh water?

Have you ever gone swimming in the ocean? How did it compare to swimming in a freshwater lake? Remember the salty taste? Yuck! The burning eyes? How about the way you floated? Seawater is wet like fresh water, but that is just about where the similarities end!

13.8 ORIGIN OF SALTY OCEANS

So how did the oceans get so salty? According to a young-earth view of the past, everything about our planet is the product of a supernatural creation and a global flood. We cannot know for sure how salty the original ocean was when God created it. We can use the Flood model to predict that the great amount of erosion during and following the Flood *could* have added lots of minerals to seawater. This *could* have made the oceans very salty, at least in some places.

We also know that marine plants and animals that live in salt water cannot survive if they are plunged into fresh water, while most freshwater organisms cannot live if transferred to salt water. Although we cannot know for sure, we can logically deduce that all the *kinds* of aquatic organisms that exist today had to survive for some time in the floodwaters that covered the earth. Therefore, it seems reasonable that the original oceans must have been at least somewhat salty. We assume that as the floodwaters receded, landlocked salty lakes were left behind. Over time, they became freshwater lakes as rivers diluted them. The sea creatures and plants that were trapped in those bodies of salt water at the end of the Flood either gradually became used to living in fresh water because their genetic makeup gave them that ability, or they perished because they couldn't.

Even if the oceans were essentially fresh water following the Flood, the model of a young-earth history can still support the development of the salty oceans that we have today. We know that the oceans gain and lose minerals every day. The streams that flow into the ocean carry large amounts of dissolved minerals with them. Much of the ocean's salts (minerals) could have built up in just a few thousand years from streams. However, if the earth were as old as old-earth geologists say it is, the oceans would probably be far saltier.

13.9 CHEMICAL COMPOSITION

Seawater is different from fresh water because of the chemical compounds dissolved in it, most of which are salts. The ocean's saltiness is its **salinity**. Oceanographers originally measured salinity in grams of salt per kilogram of seawater. The average seawater salinity is about 35 g salt per kilogram of seawater—we can also say this as 3.5% salt or 35 parts per thousand (ppt) salt. The rest of the seawater solution, 96.5%,

13B Objectives

After completing this section, you will be able to

» evaluate different Flood theories that could account for the saltiness of the oceans.

» identify the main chemicals that contribute to ocean salinity.

» list the factors affecting salinity.

» explain how salinity affects important physical properties of seawater.

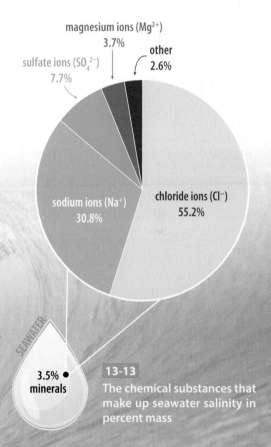

13-13 The chemical substances that make up seawater salinity in percent mass

Table 13-1		
Dissolved Gases (average)		
	Atmosphere	Seawater
nitrogen	780.8 ppt	0.0125 ppt
oxygen	209.5 ppt	0.0070 ppt
argon	9.3 ppt	0.0004 ppt
carbon dioxide	0.4 ppt	0.0900 ppt

> The average surface salinity of the Mediterranean Sea is 10% greater than that of the open ocean.

13-14

Many variables affect salinity. The Dead Sea has an inlet for water but no outlet. Inflowing water continually carries dissolved minerals into the Dead Sea. As the water evaporates, leaving the minerals behind, the salinity increases. The salt crystals below formed along the shore of the Dead Sea.

Location: 31.5°N 35.5°E

is pure water. Today, oceanographers measure salinity on the *practical salinity scale (PSS)*. This method compares the electrical conductivity of seawater to that of a standard salt solution. PSS values have no units.

The most important chemicals in seawater salts are sodium (Na^+) and chlorine (Cl^-) ions. Together they make sodium chloride—table salt (NaCl). Sulfate, magnesium, calcium, potassium, and small amounts of other dissolved minerals form salts too (see Figure 13-13).

Seawater also contains dissolved gases, mostly nitrogen, oxygen, and carbon dioxide. These gases come from the atmosphere. Surface water is usually richer in oxygen than deeper water because it is in contact with the atmosphere. Also, marine plants near the surface, where there is more light, produce oxygen. Carbon dioxide is absorbed in the uppermost layers of the ocean for the same reason. Since most of the dissolved gases enter water from the atmosphere, we would expect that the amounts of gases should be relatively consistent. However, seawater has a higher concentration of carbon dioxide compared with other gases. This is true for a couple reasons. First, carbon dioxide tends to be more soluble than other gases. In addition, carbon dioxide is found in compounds—carbonic acid and carbonates—in seawater. Argon, on the other hand, has a much lower concentration than expected. It is much less soluble and is inert—nonreactive—so it is not present in any compounds found in seawater. See Table 13-1 for a comparison of the main gases in the atmosphere and their solubility in seawater.

13.10 FACTORS AFFECTING SALINITY

If fresh water from rivers constantly flows into the ocean, why is it still salty? On the other hand, you know that rivers carry dissolved minerals from erosion. So should the oceans be growing saltier all the time? For many years, most oceanographers believed that the salinity of seawater around the world seemed to be about the same on average, although it can vary from place to place. Recent research by both young-earth and old-earth scientists suggests that ocean salinity is slowly increasing. Why is this happening?

The ocean gains minerals and water from rivers daily. Other sources of minerals are soil and volcanic ash carried by winds, and hydrothermal water from the earth's interior. But the ocean loses water mainly to evaporation. If minerals continuously built up because of evaporation, then the ocean would only get saltier. This *does* happen in some bodies of water. Lakes with no outlets, such as the Dead Sea in Israel and the Great Salt Lake in Utah, are very salty. Rapid evaporation and higher salinity also occurs in the oceans near the Equator or in warmer seas.

So how do minerals leave seawater? There are several important ways. First, living things use them. For example, the animals that form seashells, corals, and microscopic diatoms remove chemicals from seawater when they build their skeletons. Also, mineral ions cause fine, suspended sediments to clump together. When the clumps become heavy enough, the sediment masses settle to the sea bottom, removing the ions from the seawater. In a third way, chemicals sometimes become so concentrated that they precipitate and form sediments. Creationist scientists believe that this was an important process during the Flood, but it can still happen today. These processes and many others remove chemicals from solution in seawater.

Local conditions can greatly change the seawater salinity. For example, the salinity near the mouths of rivers is much lower than normal. Low-salinity seawater is *brackish*. As ocean tides rise, seawater can flow upriver, especially in large slow-moving rivers like the Mississippi and the Nile. Some large bays are almost completely cut off from the ocean, so the rivers that flow into them dilute the salts, making their waters brackish. The Chesapeake Bay is a good example of this (see Figure 13-16).

The presence of ocean ice can greatly change seawater salinity. As seawater freezes, the dissolved minerals are forced out of the ice. The much saltier *brine* water sinks into the depths under freezing sea ice. In the opposite process, melting sea ice greatly dilutes the salts in water around the ice, reducing salinity. As you will see, salinity can affect many physical properties of seawater.

13-16

This is a thematic map showing the salinity of Chesapeake Bay. It is essentially fresh water at its north end, brackish in the middle, and seawater at its southern mouth.

13-15

Observe the sediment-rich water meeting the cleaner water. This same situation happens when fresh and salt water meet.

Seawater diluted with fresh water is called brackish water. Saltier, more concentrated seawater is called brine.

WORLDVIEW SLEUTHING: DRINKABLE WATER FROM SEAWATER

Seven hundred eighty-three million people around the world do not have access to adequate drinking water, while water covers 71% of the earth. These two facts stand in stark contrast to each other. How can we live on a planet with so much water, yet one out of every ten people on the earth can't get drinking water? For example, California residents have been living with drought conditions and water usage restrictions for several years. An obvious solution would be to figure out a way to make drinking water out of all that seawater.

INTRODUCTION

You are working for the Public Policy Institute of California, where you work in the water management department. Your team is looking into the challenge of providing sufficient drinking water for the state's residents. Public interest in your work is high due to recent severe drought conditions.

TASK

You will produce a presentation for representatives of the California Legislature. The goal is to get the legislature to fund three new desalination plants within the state.

PROCEDURE

1. Research the challenge of providing drinking water by doing a keyword search for "drinking water around the world."
2. Research the process of desalination by doing a keyword search for "desalination."
3. Research desalination plants by doing a keyword search for "desalination plant," "desalination plant design," and "desalination plant cost."
4. Prepare a presentation that includes the need for clean drinking water, the desalination process, and desalination plants.

CONCLUSION

God commands us to have dominion over His creation. Part of this is caring for each other. When 10% of the world population goes without adequate drinking water, there is an opportunity to meet that need and to share the love of God.

A desalination project in Hamburg, Germany

13.11 PHYSICAL PROPERTIES

We've been talking about some of the chemical properties of seawater. Let's look at its physical properties and some factors that affect these properties.

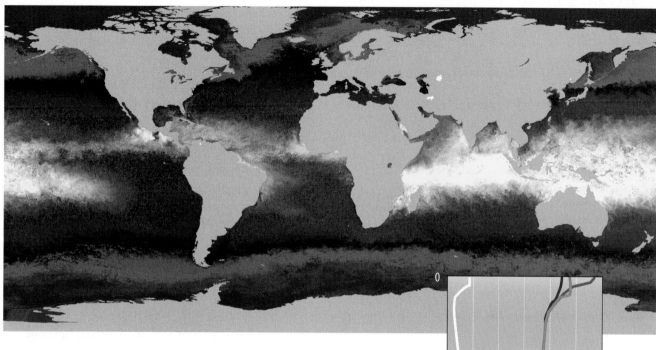

13-17
Thematic world map of surface seawater temperature. The temperature varies from a warm 35 °C (yellow) in the tropics, through red, blue, purple, and green to a freezing −2 °C in the polar regions.

TEMPERATURE

At the ocean surface, seawater temperature depends mostly on the amount of sunlight it absorbs and the temperature of the atmosphere above it. As you might expect, surface temperature varies over the surface of the globe, being warmest near the Equator and coldest in polar regions. Ocean currents move seawater around the ocean basins. This movement helps keep the planet from getting too hot or too cold.

Seawater temperature also changes with depth. Oceanographers measure temperature at many depths with special instruments. While surface temperature is quite variable, as you go deeper in the ocean, the temperature usually gets colder and becomes nearly the same over large areas of the world. The temperature change with depth is nearly constant, especially in deep water. Water temperatures in the deep abysses hover around 0 °C.

Seasons affect seawater temperature too. In the mid-latitudes during summers, wave action mixes the ocean near the surface to create a layer of relatively warm water on top of much colder water. In the spring and fall, when air temperature is colder and the warming by the sun is less, the surface temperature is lower and its change with depth is similar to that in deeper water. In winter, especially in high latitudes, the surface temperature can approach freezing. As you go deeper, the water may actually grow warmer before decreasing as usual. Compare temperature profiles in Figure 13-18.

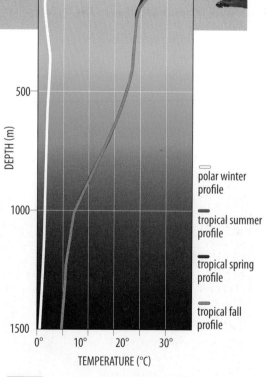

13-18
Ocean seasonal temperature profiles

OCEANS AND SEAS **321**

MELTING/FREEZING POINT

Pure water freezes at 0.0 °C. But seawater doesn't freeze until it is somewhat colder. The water molecules in salt water have a hard time shoving the salt ions out of the way before locking into a crystal lattice. So seawater has to be colder to freeze. Seawater with 3.5% salinity freezes at about −1.8 °C. The saltier the seawater, the colder it has to be to freeze.

PRESSURE

In ocean exploration, the pressure of seawater is one of the most difficult things to deal with. Remember that pressure is the amount of force exerted over a given area. Sea pressure at the ocean surface is the same as atmospheric pressure. As you go deeper, the weight of the mass of water above bears down, increasing pressure. Sea pressure increases at about 1 **atmosphere** (1 atm) every 10 m. To calculate the approximate sea pressure, divide the depth (in meters) by 10 to find the water pressure and add 1 atm for atmospheric pressure. So at 30 m, sea pressure is about 4 atm: 30 m divided by 10 m/atm = 3 atm; adding 1 atm for atmospheric pressure brings the sea pressure calculation to 4 atm. At the bottom of the Challenger Deep (about 11,000 m), sea pressure is about 1100 atm!

> When **atmosphere** is used as a unit, it is abbreviated *atm*. The average sea level atmospheric pressure is 1 atm, or 14.7 psi.

DENSITY

For pure substances, density is an identifying property that we can easily measure. Recall that density is the amount of mass in a given volume. The density of pure water is 1.000 g/mL at 4.0 °C and 1 atm of pressure. But seawater's density at the same conditions is about 1.028 g/mL. So seawater is about 3% denser than pure water because of the dissolved salts that it contains. The density of seawater increases with greater salinity.

> ! Seawater density *increases* with:
> » increasing depth or pressure
> » increasing salinity
> » decreasing temperature

SPEED OF SOUND

The density of seawater is the main factor that determines how fast sound travels through it. This means that depth, salinity, and temperature affect sound speed in seawater. Since these things change with position and time, the speed of sound in seawater is not constant either. The average speed of sound in seawater is about 4.5 times faster than in air, or 1500 m/s.

The speed of sound in seawater may seem like an odd thing to be concerned about. But this property is important for several reasons. First, God created many animals that use sound to find food, communicate, and navigate the ocean. Whales and porpoises, for instance, make loud clicks, chirps, or other sounds, and then listen for an *echo*. The elapsed time to hear the echo indicates the distance to the object that reflected the sound. Marine animals can tell the direction of an echo just like you can—probably better. Using *echolocation*, animals learn from experience how to gauge distance, and can even correct for temperature and salinity. Echolocation depends on the speed of sound in water. Distance to an object is directly proportional to the time it takes to hear an echo from it.

People use echolocation too. Ships use depth sounders to measure water depth. This may be for safety of navigation or for mapping. Fishermen use echolocation to find fish and other marine life. Naval vessels first used this acoustic location technique in *sonar* to find enemy submarines around 100 years ago. You will learn more about how we use artificial echolocation in Chapter 15.

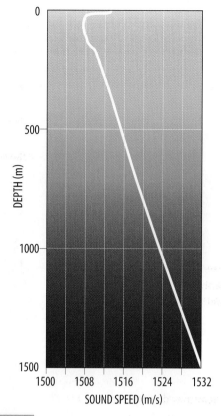

13-19

A typical *sound velocity profile (SVP)* shows how sound speed changes with depth in seawater. Seawater sound speed is highly variable since it depends on anything that can affect water density.

13B REVIEW QUESTIONS

1. Why can we suspect that the original created ocean was probably at least a little salty?
2. Which two chemical elements make up most of the salt in seawater? When combined, what common compound do they produce?
3. Name two sources that add minerals to the oceans. Name two processes that remove minerals.
4. How does salinity affect the freezing point and density of seawater?
5. How does a sperm whale locate a tasty squid in the black depths of the ocean if it can't see it?
6. (True or False) Sea ice is just as salty as the surrounding seawater.
7. What is sea pressure at a depth of 100 m? Give your answer in atmospheres (atm) and in pounds per square inch (psi).

13C

OCEAN ENVIRONMENTS

How does the ocean environment vary from place to place?

13.12 ZONES OF LIFE

In Chapter 4 you learned that Earth is the well-designed home to all the known life in the universe. Land animals and plants live in all kinds of environments—forests, deserts, arctic tundra, mountains, and prairies, to mention just a few. So where do marine organisms live, and what kinds of surroundings exist in the oceans? Water depth, light, and how fast environmental conditions change determine where things live in the ocean. Oceanographers divide the ocean environment into about five life zones, though there are many other unique environments.

13C Objectives

After completing this section, you will be able to

» compare the different biological zones in the ocean.
» summarize the marine carbon and nitrogen cycles.

13-20

The intertidal life zone is a hard place to live. Where animals and plants live here depends on their ability to stand up to wave action and on the wetting and drying of the tide cycle.

INTERTIDAL ZONE

Plants and animals that live on a beach between the low and high tide limits occupy the *intertidal zone*. This zone is very different from the other ocean zones. Every day it is underwater for part of the time and exposed to air for the rest as ocean tides come in and go out. Along exposed seacoasts, only the toughest organisms live in this zone because it is where breaking ocean waves pound anything that lives there. To survive, marine organisms must be very mobile to avoid injury or have a sturdy design and a good grip on underlying surfaces. Marine biologists generally subdivide this zone by how long and how often organisms are underwater (see Figure 13-20).

LITTORAL ZONE

Marine organisms in the *littoral* or *neritic zones* live between the low-tide mark at the shore and the edge of the continental shelf. Conditions are more stable in the littoral zone than within the intertidal zone. There is no wet-and-dry cycle here. This zone is also warmer than deeper water, is well lighted, and has lots of nutrients. However, many factors can still affect living things in this zone—changing salinity and suspended sediments near the mouths of rivers, major changes of temperature from season to season, and storm waves that can stir up the bottom down to 100 m or more.

Beginning in the littoral zone, there are two kinds of lifestyles for marine life in each zone. Bottom-dwelling animals and plants are called **benthic organisms**. Either they are attached to the sea bottom, like kelp, corals, or mussels, or they crawl on the bottom, like crabs and snails. Things that don't live on the bottom are **pelagic** (puh LAY jick) **organisms**. These include phytoplankton, fish, squids, jellyfish, and whales. Pelagic organisms may simply drift with the ocean currents, or they may be active swimmers.

THE OPEN OCEAN

Beyond the continental shelves, oceanographers divide life zones by depth and location. There are three pelagic life zones. They go by various names, but we will call them the *photic*, *aphotic*, and *abyssal zones*. Where the ocean bottom lies within these depth zones, there are corresponding benthic zones, but we will refer to them by just their pelagic zone names.

Photic Zone. The photic zone extends down to 100–200 m in the open ocean, far from continental coastlines. The conditions of light and temperature are very similar to the littoral zone, but salinity and water clarity are much more constant. Because of the way it is defined as an open-ocean environment, the photic zone contains only pelagic organisms. Benthic organisms in this depth zone live closer to shore in the littoral zones of continents, islands, and atoll reefs.

Aphotic Zone. The aphotic zone extends from about a depth of 200 m down to 3000–4000 m. No light penetrates into this zone. Green plants and phytoplankton cannot survive here. All animals that live in this zone are able to live in complete darkness. Most depend on senses other than sight to hunt for food and to avoid predators. Some can create their own light, called *bioluminescence*. This cold biological lighting is similar to that of fireflies. Animals use this for attraction and as a warning signal.

Abyssal Zone. The abyssal zone extends to the floor of the abyssal plane. This zone is perpetually dark and cold. Not the place for an extended holiday, unless you're a jellyfish! Dense, cold, salty water from the polar regions flows along the bottom of the abyssal plains. The only nutrients available to animals fall down from above as *marine snow*—a continuous, gentle shower of fine organic matter. Some oceanographers place the deep trenches in a separate biologic zone called the *hadal zone*. This zone extends from about 6000 m to the bottom of the deepest trenches. Water pressure at these depths is more than a thousand times atmospheric pressure!

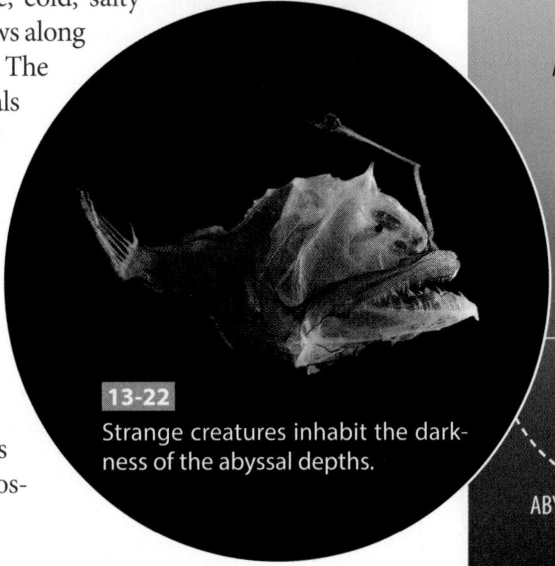

13-21
In the aphotic zone, sight is less important to animals than other senses. However, some animals use bioluminescence to warn, lure, and display in this black realm of the sea. Notice the bioluminescent jellyfish.

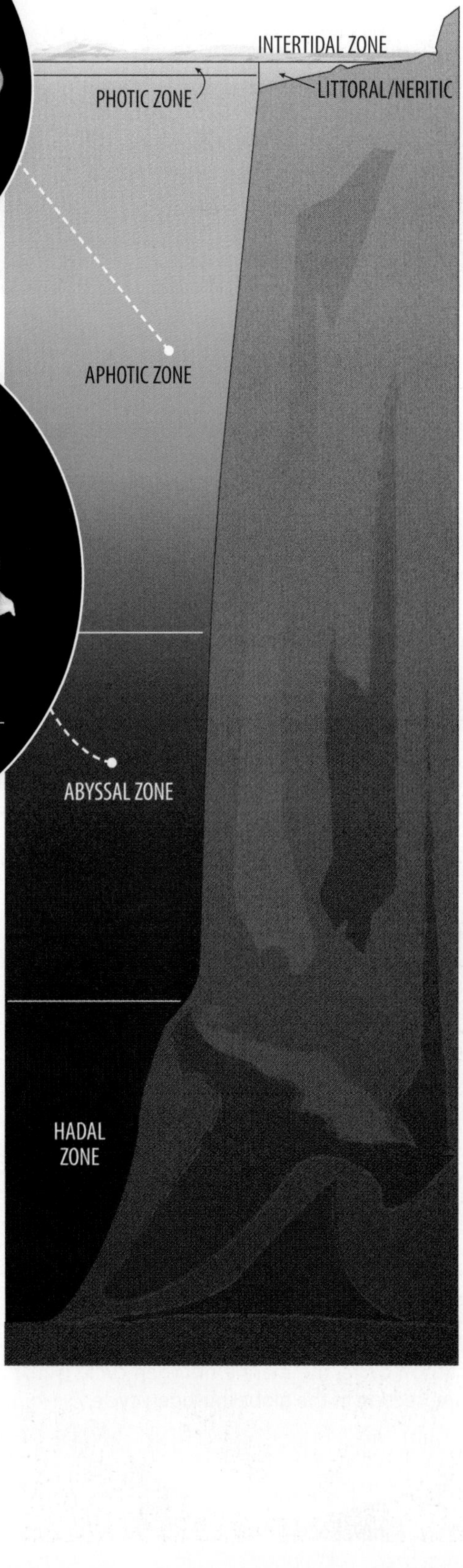

13-22 Strange creatures inhabit the darkness of the abyssal depths.

13.13 MARINE CARBON CYCLE

The chemistry of seawater is all about life. Everything in the sea, from the tiniest one-celled organism to the giant blue whale, needs chemicals in the sea in order to live and grow. One of the most important elements in the food they eat is *carbon*—the main chemical building block of all living things. Though it is readily available in the environment, living things would quickly run out of this essential ingredient if it were not continually replaced by natural processes.

The **carbon cycle** is a vast process operating at the surface of the earth. It includes huge reservoirs of carbon compounds that extend beyond the ocean itself. The marine carbon cycle is just a part of the carbon cycle worldwide. Carbon moves between reservoirs by many biological and nonbiological processes. The main reservoirs of carbon are the *atmosphere*, the *ocean* and its organisms, the *land* and its organisms, *rocks* and *sediments* in the upper lithosphere, and the earth's *deep interior*. Study Figure 13-23 for an overview of the marine carbon cycle.

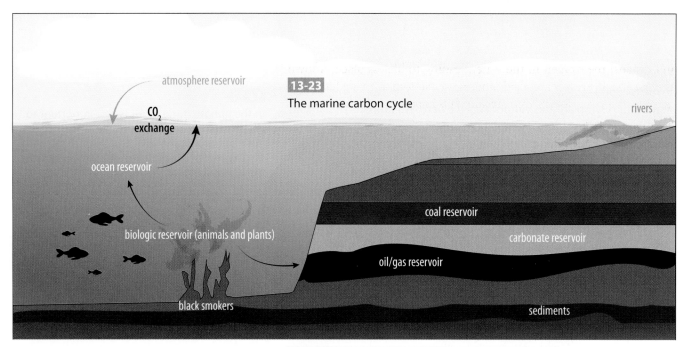

13-23 The marine carbon cycle

13.14 MARINE NITROGEN CYCLE

Like carbon and water, *nitrogen* is a key chemical for life. And like these other resources, it has to be reused. If nitrogen in organisms were not recycled after they died, it would be quickly used up, and life on Earth would soon perish. The **nitrogen cycle** ensures that there is a continuous supply of nitrogen.

Both plant and animal life use nitrogen in their life processes and as building blocks for cells. Air is 78% nitrogen, so you might think that there's plenty to go around. But there's a problem. The nitrogen in the atmosphere is made from two nitrogen atoms bonded together (N_2)—molecular nitrogen. This form of nitrogen is inert. It can't take part in most chemical reactions, including those that build food molecules. So then, how does this essential chemical get into the food that all animals eat?

Certain bacteria turn atmospheric molecular nitrogen into useful *organic nitrogen*. Biologists call this process *nitrogen fixing*. In the shallow photic zone of the ocean, the marine versions of these bacteria are called *cyanobacteria* (blue-green algae). The organic nitrogen compounds are waste products of the bacteria's life processes. Lightning is also an agent of nitrogen fixing. It is so energetic that it breaks apart molecular nitrogen in the atmosphere. The resulting atomic nitrogen reacts with oxygen to form NO_2^-. Organic nitrogen exists as three different chemical compounds containing nitrogen—NO_2^-, NO_3^-, and NH_4^+. Study Figure 13-24 to learn about these forms of fixed nitrogen. These compounds are the starting point for the nitrogen cycle.

Man-made chemical fertilizers are also a source of fixed nitrogen. Most fertilizers not used by crops reach the ocean in rainwater run-off. Ever since the 1920s, fixed nitrogen from fertilizer production has accounted for a significant percentage of the organic nitrogen in the environment.

If too much man-made nitrogen fertilizer enters the ocean, explosions of one-celled organisms called *algal blooms* can occur. The algae die

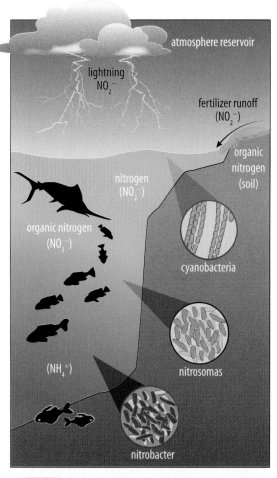

13-24
The marine nitrogen cycle. As with the marine carbon cycle, the marine nitrogen cycle is just one section of the global nitrogen cycle.

and exhaust the oxygen in the water, leading to lifeless places known as dead zones. Preventing algal blooms by using man-made fertilizers carefully is an example of wise stewardship in practicing dominion.

The oceans are a vast part of our world that we are only now beginning to understand. As we continue to study them and the life they contain, we learn again and again that our God is amazingly great. He alone has the power and wisdom to control the mighty deep. As the psalmist said long ago, such a God deserves our praise.

O Lord, how manifold are thy works! in wisdom hast thou made them all: the earth is full of thy riches. So is this great and wide sea, wherein are things creeping innumerable, both small and great beasts. There go the ships: there is that leviathan, whom thou hast made to play therein. These wait all upon thee; that thou mayest give them their meat in due season. That thou givest them they gather: thou openest thine hand, they are filled with good. (Ps. 104:24–28)

13C REVIEW QUESTIONS

1. What are the three factors that determine the kinds of marine life and where they can live in the oceans?
2. How are the littoral zone and the open-ocean photic zone similar? How are they different?
3. Evaluate the statement, "There is no light in the deepest parts of the ocean."
4. What are the four main carbon reservoirs that communicate with the ocean carbon reservoir? See Figure 13-23.
5. Explain why organisms cannot directly use atmospheric nitrogen to build organic compounds.
6. Discuss the importance of the great natural cycles, including the water, carbon, and nitrogen cycles.
7. According to Psalm 104, what does the ocean reveal to us about God?
8. (True or False) Because of the way *benthic* is defined, it is accurate to say that no benthic organisms live in the open-ocean photic zone.

MARINE SNOW

Early in the 1990s, oceanographers discovered that a fine, continuous rain of particles called "marine snow" drifts down from near the ocean's surface to deeper depths and ultimately to the bottom. This "snow" consists of tiny single-celled animals and plants, colonies of cellular organisms, sand, dust (including meteoric dust), chemical precipitates, excrement, and microscopic particles, often bonded together by bacterial colonies. The marine snow can be so dense that it severely limits visibility, even in the lighted zone of the ocean. This gentle rain of particles forms the main source of food for many bottom-feeding and drifting organisms.

Key Terms

ocean	305
sea	306
sea level	307
mean sea level (MSL)	308
shore	309
beach	309
continental shelf	309
continental slope	310
submarine canyon	310
continental rise	310
abyssal plain	311
mid-ocean ridge	311
seamount	312
oceanic trench	312
island arc	313
coral reef	314
atoll	315
salinity	317
benthic organism	324
pelagic organism	324
carbon cycle	325
nitrogen cycle	326

CHAPTER 13 REVIEW

CHAPTER SUMMARY

» The oceans are the main reservoirs of water on Earth. They are essential for life and strongly influence weather and global climate.

» There are five main ocean basins, but they all form a single global ocean.

» Other than the Pacific basin, geologists believe that all other ocean basins are the result of global tectonic activity. Young-earth and old-earth scientists differ on their age and the length of time it took to form them.

» Local sea level is the height of the ocean surface relative to a standard reference surface at a given location. Mean sea level is the long-term average of local high and low tides.

» Mean sea level (MSL) provides us a reference for determining land elevation, ocean depth, and aircraft altitude.

» Most ocean basins have certain topographical features. These include coastal and continental edge features, as well as tectonic features found in the ocean abyss.

» Coral reefs are rock-like features formed by the action of marine organisms. They provide a living environment for many kinds of ocean creatures.

» The oceans are salty, but we do not know how they came to be that way. It is likely that the original seawater was at least somewhat salty.

» The salinity of seawater is measured using the practical salinity scale (PSS).

» Ocean salinity is about the same around the world, but it can differ greatly from place to place.

» Seawater has many measurable physical properties. Salinity strongly affects many of its properties.

» Conditions such as water depth, amount of light, and the rate at which environmental conditions change define five biological zones in the oceans.

» Most marine organisms have one of two different lifestyles: *benthic* (living on the bottom) and *pelagic* (not living on the bottom).

» The marine carbon and nitrogen cycles are essential natural processes that allow reuse of these elements by living organisms.

REVIEW QUESTIONS

1. How is oceanography different from the standard individual sciences, like biology, geology, or chemistry?
2. What are the differences between seas and oceans?
3. Why is it important to be able to predict local sea levels in a harbor from day to day?
4. Give one possible cause for an increase in mean sea level.
5. You are visiting an ocean beach. What determines the width of the shore?
6. Is the continental rise part of the continental plate? Explain.
7. What is the average depth of the abyssal plain?
8. Where are the deepest spots in the ocean located? What may have caused these to form?

9. Give one explanation for how an atoll may have formed.
10. What do we mean by *salinity*? What is the average salinity of seawater?
11. If the average salinity of the ocean is not changing, what must be true about the rates of addition and removal of minerals from seawater?
12. What happens to seawater temperature as you go deeper into the ocean?
13. Explain why sea pressure makes it difficult to explore the ocean depths. Include an example of sea pressure at a depth of your choice.
14. Why is studying the speed of sound important to marine biologists?
15. State two properties of the intertidal zone that make it an especially difficult environment in which to live.
16. What are the two main sources of food for marine animals living in the aphotic zone of the open ocean?
17. What is the main way that carbon becomes food for animals?
18. How have human activities affected the nitrogen cycle since the last century?

True or False

19. The oceans are the source of most of Earth's atmospheric oxygen.
20. There is only one ocean.
21. The ocean's surface makes the earth into a smooth sphere.
22. The beach is broader than the shore at a given location.
23. Geologists recently estimated that there could be around 25 million extinct volcanic seamounts taller than 100 m. This large number suggests that underwater volcanism must have been significant in the past.
24. According to young-earth geologic history, existing coral reefs probably developed after the Flood.
25. The fish kinds found in landlocked lakes today must have originally been ocean fish that survived the Flood.
26. Seawater is 3.5% sodium chloride and 96.5% water.
27. The Dead Sea has a high salinity because the water entering from the Jordan River is even saltier.
28. The freezing point of seawater decreases as salinity increases.
29. A sound velocity profile displays how speed changes with temperature.
30. Marine organisms in the littoral zone are *not* protected from occasional significant changes to their environment.
31. Earth's carbon and nitrogen cycles take place entirely within the oceans.

Map Exercises

Using Google Earth or a similar Earth-viewing program that can display the ocean bottom topography, identify the ocean features located at the following geographic coordinates.

32. 56°N 4°E (general area)
33. from 39.5°N 72.3°W
34. 30°N 168°W (general area)
35. 34.7°S 54.6°E (general area)
36. 25.948°N 20.344°W (multiple features)
37. 9.436°N 138.046°E

CHAPTER 14
OCEAN MOTIONS

14A TIDES 331
14B CURRENTS 338
14C WAVES 345

THE GREAT PACIFIC GARBAGE PATCH

Look around you and name all the things you see that are made of plastic. We use plastic because it is strong and durable. But it lasts for years before it finally breaks down in waste dumps around the world. Do you know where the biggest dump is? It's in the Pacific Ocean! Scientists call it the Great Pacific Garbage Patch. Some think it could be larger than Texas!

Ocean currents gather trash from surrounding lands into the Pacific Gyre, a great rotating current in the northern half of the Pacific Ocean. Though there are larger pieces of trash scattered throughout this area, most of the plastic has been broken down into smaller pieces by the sun's energy. These flecks of plastic also sink into the depths of the oceans, affecting more than just the sea life near the surface.

Decomposed plastic produces harmful chemicals. Sea animals eat this chemical waste and are poisoned. Larger pieces like plastic bags entangle jellyfish and other animals. Smaller pieces fill the stomachs of sea turtles and albatross chicks whose parents mistakenly feed them trash for food. When plastic waste products enter the food chain, they can be a problem for people who eat fish.

14A

TIDES

What causes tides and how can we use them?

14.1 DOMINION AND OCEAN MOTION

The oceans are always on the move. Even when they appear calm, they move in currents, or rise and fall with the tides. On top of these larger motions are ocean waves, which transfer energy over great distances and help shape shorelines through erosion. Understanding all this is necessary in order for us to exercise good and wise dominion over the earth.

After analyzing such motions, oceanographers survey areas where trash collects, like the Great Pacific Garbage Patch. They ask questions like, "How did this move here?" "What effect is this having?" "What can we do to solve the problem?" Reducing man-made marine pollution is clearly a worthy priority for us.

But solving the problem of marine pollution isn't the only way that we can exercise dominion through use of the oceans. We use its currents to aid ship transportation. Knowledge of its motions helps us understand their effects on fish and other wildlife. This allows us to monitor migrations and the health of animal populations and help fishermen improve their catch. Understanding global ocean currents is key to understanding the world's weather. At the same time, changing global climate seems to be affecting sea level, so the height of tides in some places is becoming a concern. Marine physicists are even working on ways to harness tide energy to power our homes. Yet, while working to gather all this practical knowledge, we must not lose the wonder and awe that the oceans inspire in us for the God who created all things. Let's begin this pursuit by looking at the tides.

14.2 WHAT ARE TIDES?

Ocean **tides** are daily or twice-daily changes in local sea level. Tides aren't just oceans sloshing back and forth in their basins. The cause of tides isn't even here on Earth! They are created by the gravitational pulls of the moon and sun.

As you learned in Chapter 13, high and low tides mark the upper and lower limits of a beach. The highest sea level at a given location is high tide; the lowest is low tide. Scientists use *tide gauges* along coastlines to measure tides. Sea surface heights also change with the tides far out to sea. Special satellites are used to measure tides far from coastlines.

14A Objectives

After completing this section, you will be able to

» compare tides with other ocean motions.

» analyze the forces that create and affect tides.

» evaluate the best uses of tides for generating electricity.

The Man Behind the Mission

The Great Pacific Garbage Patch was discovered by a sailor named Captain Charles Moore on his way back from a yacht race to Hawaii in 1997. The experience motivated him to get involved in marine environmentalism. He now captains a specially designed research vessel named the *Alguita*.

14-1

A *tide gauge* is used to measure the height of local sea level. Most are just ruled sticks mounted vertically in the water. Observers read and record sea level by hand. More advanced models include electronic detectors with a flotation sensor, and they transmit their data by radio or the Internet to a central oceanographic data center.

OCEAN MOTIONS 331

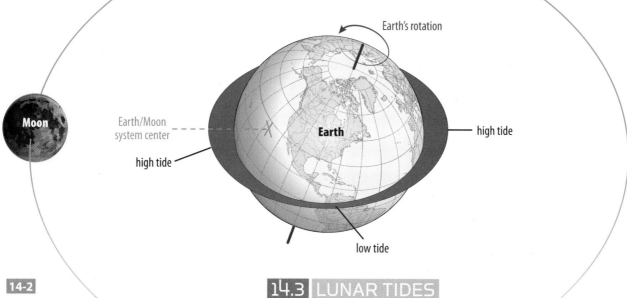

14-2
High and low tides are caused on different sides of the earth because of the gravity of the moon and the rotation of the earth-moon system around a common point.

14.3 LUNAR TIDES

The gravity of the moon affects the oceans more than the sun does. That means that most tides are **lunar tides**. Since ocean waters are liquid, they can flow. On the side of the earth facing the moon, they continually try to flow toward the moon because of its gravitational pull. The water builds up into a heap or tidal bulge. This flow is practically unobservable in the open ocean, but near shore, and especially in waterways, the flow of water can create strong tidal currents.

On the side of the earth opposite the moon, ocean water is affected far less by the moon's gravity than by the earth's gravity. But a second ocean bulge forms there too. Most people view the moon as revolving around the center of the earth. But that is not correct. The earth and moon both revolve around a point between them, like two children holding hands and spinning around each other. Because the earth's mass is so much greater than the moon's, the center of their rotation is much closer to the earth's center. It actually lies deep within the earth, about 1707 km below its surface (see Figure 14-2).

14-3
High tides occur at a given location when that location coincides with the tidal bulges.

So as the moon revolves around the earth, the earth revolves around the moon too. This means that, compared with the center of rotation, the far side of the earth from the moon is moving faster than the near side. This extra speed causes the ocean water on the far side to be thrown into a heap, just like mud gets flung off a spinning wheel. The tidal bulge caused by the spinning of the earth is somewhat smaller than the bulge caused by the moon's gravity.

Why then does sea level rise and fall with the tides? The tidal bulges of ocean water on the near and far sides of the earth stay put in relation to the moon, but the earth rotates beneath them. Because the tidal bulges are higher than the average sea level, the water that supplies the bulges has to come from the oceans between the tidal bulges. Thus, the sea level in these regions is lower than average.

When you are standing at a seacoast, a high tide occurs when your location moves through a tidal bulge as the earth turns. It is more noticeable here than out in the middle of the ocean because changes in sea level can be compared to the height of the shoreline. Similarly, when you move through one of the regions between tidal bulges, you see a low tide.

Along most coasts there are two high tides and two low tides each day. It takes 24 hours and 50 minutes for a point on the earth to experience two complete high-low tide cycles. The high tides are not exactly 12 hours apart because as the earth rotates, the moon has moved forward in its orbit in the same direction. It takes 50 more minutes each day for a location on the earth to catch up with its previous position relative to the moon. Because of the changing tide schedule, tide tables are published so that people can plan their work around the tides.

14-4
A tide schedule

Understand that the diagrams in Figures 14-3 and 14-5 exaggerate the bulges to illustrate the principle. Real tidal bulges are only a few meters high.

14-5
The tide changes as the earth rotates beneath the tidal bulges. The same location experiences low tides when the earth is between the tidal bulges.

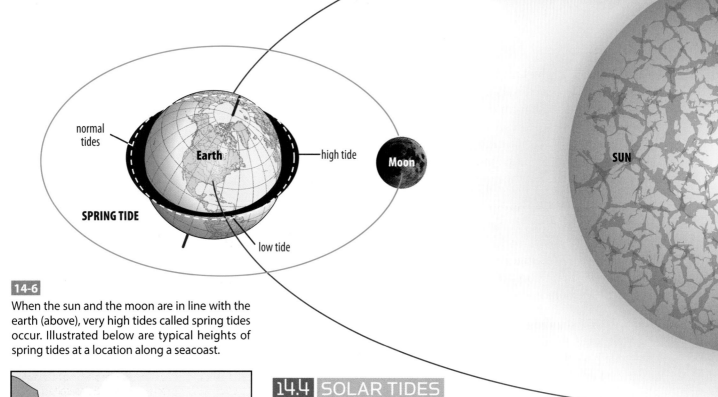

14-6
When the sun and the moon are in line with the earth (above), very high tides called spring tides occur. Illustrated below are typical heights of spring tides at a location along a seacoast.

> It takes the earth about an extra 4 minutes per day to complete one full rotation relative to the sun.

Bay of Fundy: 45°N 66°W

14.4 SOLAR TIDES

The sun's gravity causes tides too, but in most places, they are much smaller than lunar tides. A solar tidal bulge exists only on the side of the earth facing the sun. The earth rotates under this smaller bulge every twenty-four hours plus a few minutes. Since the earth is revolving around the sun, it takes a few minutes longer each day for a given point to directly face the sun.

Each month, the solar tides and the lunar tides align in various ways. Twice a month, at the new moon and full moon, the sun, moon, and earth lie on a straight line (see Figure 14-6). At these times, the sun's gravity works with the moon's to form a higher-than-usual tide called a **spring tide**, which occurs throughout every month and every season of the year, whenever there is a new or full moon. The term *spring tide* relates not to the season but instead refers to springing forth—the tide springs farther than is normal. When the sun, earth, and moon form a right angle and the moon appears half lighted (see Chapter 22 for a full description of moon phases), the sun's gravity works against the moon's to form a lower-than-usual tide called a **neap tide**. Neap tides also occur twice each month. Figures 14-6 and 14-7 show the relationships between the various tides and the sea levels they can produce.

14.5 UNUSUAL TIDES

Lots of things affect the height and time of tides. Some of these include the shapes of the coastlines, the width of the continental shelf, the size and depth of the ocean basin, and the position of the moon north or south of the Equator. In the Mediterranean Sea, the tide seldom rises more than 30 cm. But at the head of the **Bay of Fundy** in eastern Canada, the tide may rise over 16 m! This bay is a V-shaped inlet with its narrowest parts farthest from the sea. It acts as a funnel to bring a large amount of seawater into several narrow bays. During high tide, some of the bay's rivers run backward, creating dangerous currents.

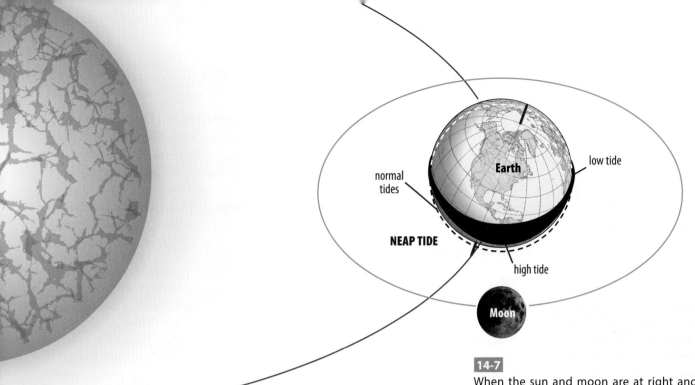

14-7
When the sun and moon are at right angles to each other (above), lower tides called neap tides occur. Note the typical heights of neap tides at a location along a seacoast (below).

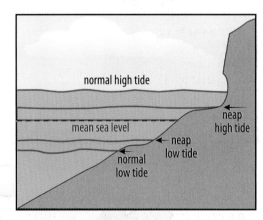

Some bodies of water have only a single high and low tide each day. Much of the Gulf of Mexico experiences only one significant high and low tide per day because most of its shallow basin is enclosed by land, which limits the movement of the tidal bulge into and out of the Gulf.

Unusually high tides can occur with an approaching hurricane. If the hurricane's arrival is at the same time as a spring tide along a coast, extremely high tides result, with sea levels many meters above normal. These conditions can completely devastate a coastal region.

14-8
Notice the drastic change in tides at Hopewell Rocks in the Bay of Fundy from low tide (left) to high tide (right). Scientists use these unusual tides, which range from 6 m to 16 m, to generate electrical energy.

OCEAN MOTIONS

14-9
A tidal power resource map for North and South America shows possible locations for future tidal power generation plants.

- ■ good potential regions
- ■ fair potential regions

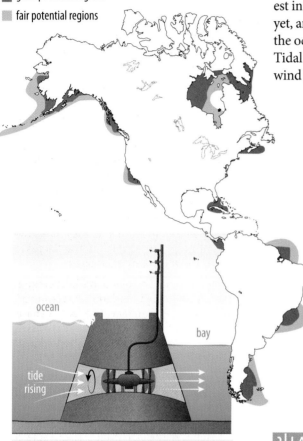

14-10
Power generation using the tides depends on placing a bladed turbine in the tidal stream. Ideally, the turbine works both when the tide comes in and when it goes out.

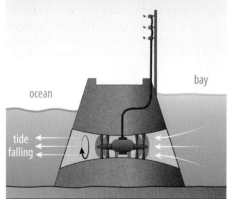

Fi Annapolis Tidal Power Plant, Nova Scotia, Canada: 44.753 °N 65.512 °W

14.6 TIDAL POWER GENERATION

As world population increases, so does the demand for electricity. At the same time, scientists are concerned that air pollution from fossil fuel power plants contributes to climate change. So there is great interest in finding other sources of electricity that do not pollute and, better yet, are renewable energy sources. Some believe that one such source is the ocean's tides. Electricity can be generated as sea level rises and falls. Tidal energy is nonpolluting, renewable, and much more reliable than wind power.

There are problems with tidal power projects, however. Certain kinds of tidal generation plants are very large and expensive. Relatively few places in the world have enough tidal change to make tidal power worthwhile. Tidal dams that would enclose or block off bodies of water could upset the local ecology and interfere with local boat and ship traffic. Building solutions for these problems into a tidal power dam as part of its design adds to the expense of the project.

For these reasons, present-day tidal power projects are relatively small. Even so, several existing tidal dams can produce around 240 megawatts (MW) of power—enough to supply about 116,000 average American homes. Some planned projects could produce over four times this amount. There are several places in the world that could generate many gigawatts (GW) of electricity. While there are few tidal power plants in the United States, many locations are being considered (see Figure 14-9).

14A REVIEW QUESTIONS

1. Give two reasons why the Great Pacific Garbage Patch is a serious problem.
2. What are the three principal ways that the ocean moves?
3. What is the main cause of twice-daily tides around the world?
4. What causes the tidal bulge on the side of the earth farthest from the moon?
5. Why do most coastal locations experience two high tides separated by two low tides each day?
6. What are higher-than-usual tides that form when the sun's and moon's gravitational pulls work together?
7. Discuss one advantage and one disadvantage of using tides to generate electricity.
8. (True or False) All coastlines experience a high and a low tide a little more than twelve hours apart.

LIFE CONNECTION: OCEAN MIGRATIONS

When you first hear the word *migration*, what comes to mind? You probably think about flocks of birds flying south in formation toward warmer temperatures and more abundant food for the winter. But did you know that ocean animals migrate too?

Marine animals migrate for the same reasons that land animals do: to avoid winter weather, to find food, to give birth, and to raise their young in safe places. Many marine animals eat plankton, tiny organisms that drift along in ocean currents. Animals that eat plankton migrate along with the ocean currents to take advantage of an easy food supply. Clear waters provide safe environments for the young because mothers can see predators coming from far away.

Whales and sharks make some incredible migrations. One of the most amazing whale migrations is undertaken by the humpback whale. It bears its young in the warm, clear waters near the Caribbean. But clear waters mean no nutrients, no plankton, and so no food for the mothers. For months, they live off their blubber and go without eating as they take their babies on a nearly 3200-km journey to the cool waters off northern New England. Some go as far north as Iceland! They arrive just in time to feast on swarms of krill hatching in the nutrient-rich Arctic waters that support plankton growth. How do the whales know how to get there? Scientists speculate that at least some whales rely on ocean currents to find their way.

The blue shark mates in the northwestern Atlantic Ocean. Sometime after mating, the female blue shark travels eastward to give birth. Studies show that her journey takes her along the Gulf Stream and the North Atlantic Current, which form the North Atlantic Gyre ocean current, before she finally reaches the waters around Spain and Portugal to give birth.

Smaller ocean animals migrate too. Pacific leatherback sea turtles migrate over 25,000 km, one of the longest sea migrations known! They have been spotted as far north as Alaska and as far south as Chile. They follow a narrow migratory route, crossing several ocean currents as they go.

European eels hatch in the Sargasso Sea of the North Atlantic. This is an unusual current eddy located in the western side of the North Atlantic Gyre. The tiny larvae drift for about 300 days with the Gulf Stream and North Atlantic Drift currents to Europe and North Africa. Here they are called *glass eels* because they appear transparent. These juvenile eels swim into brackish estuaries and freshwater rivers to mature. When it's time to breed, they make the amazing journey back to the Sargasso Sea, and the cycle begins again.

Ocean currents also play a role in the migration of birds. Colder, nutrient-rich ocean currents that feed ocean animals also feed seabirds and their chicks. Some seabirds, like albatrosses and terns, can circumnavigate the globe following the great ocean currents and riding the global wind patterns.

Consider for a moment how important these inanimate natural features of the earth—ocean currents—are in the lives of many kinds of living things. How did these creatures come to rely on ocean migrations to survive? This is just another example of how God provides for His creatures in their design. How much more then can He also provide for us, whom He has made to bear His image!

The migratory route of a Pacific leatherback turtle

© 2011 Transnavicom,Ltd/Data SIO, NOAA, U.S. Navy, NGA, GEBCO/ Image IBCAO/ Image © 2011 DigitalGlobe/Image © 2011 TerraMetrics

14B Objectives

After completing this section, you will be able to

» contrast currents with other ocean motions.

» analyze the forces that create and affect currents.

» evaluate the effect of currents on weather and life.

14B

CURRENTS

What causes and affects currents?

14.7 SURFACE CURRENTS

You are planning a road trip from your house to your grandparent's house. How do you get there? Do you just start out in a straight line connecting your home with your destination? Of course not! You have to travel the roads and highways between. Most of your trip probably will be on the interstate because that is where you can travel the fastest.

Now imagine that you are a merchant ship captain planning a voyage between England and South America. Would you plan a straight-line track connecting your ship's port with your destination? If you did, your ship's owners would not be very happy with you. Attempting to sail a straight course would take longer and use more fuel than following the ocean's "highways."

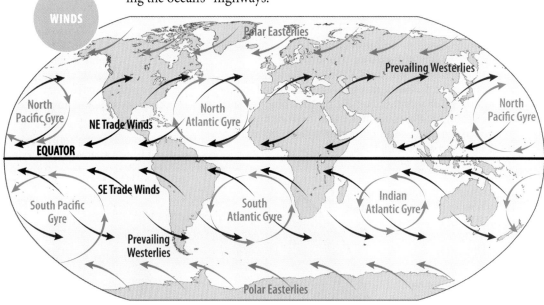

14-11

The winds depicted push the surface waters along, creating great circular gyres (blue) in each ocean basin north and south of the Equator. The winds and gyres turn due to the Coriolis effect (see pages 339–40).

The great currents are like ocean highways that move over the face of the earth. A *current* is a flow of seawater through an area of relatively stationary water around it. The most well-known currents flow along the surface of the ocean, but there are also currents that flow at middle depths and along the sea bottom.

Surface currents are set in motion by winds. They tend to follow steady global winds called *prevailing winds*. The two sets of prevailing winds that most affect the surface currents are the prevailing westerlies and the *trade winds* (see Figure 14-11). The trade winds blow at an angle from the east toward the Equator and force surface waters west. Westerlies blow from the west and toward the poles, moving surface waters east. Surface currents are usually long-lasting ocean features that can transport water long distances.

Ocean surface currents are affected by the rotation of the earth through the **Coriolis effect** (see next page). If the earth's surface were one big, uninterrupted ocean, currents would follow the same pattern as winds. But because continents separate the world oceans, currents must follow the shape of the ocean basins. Thus, currents tend to form great circling flows called **gyres**. Ocean gyres flow clockwise in the Northern Hemisphere between the Equator and northern continents. In the Southern Hemisphere, the gyres flow counterclockwise bounded by the Equator, the southern continents, and Antarctica (see Figure 14-12).

gyre (JIRE): (Gk. *giros*—ring, circle)

Ocean gyres in both hemispheres flow westward near the Equator. These currents are called the North and South Equatorial Currents. Eventually, they run into land and the ocean water piles up slightly. Most of the Equatorial Current flow turns north or south. But in each ocean basin there is a narrow current that flows eastward along the Equator between the two main Equatorial Currents. Called Equatorial countercurrents, they return surface waters to the east sides of the ocean basins.

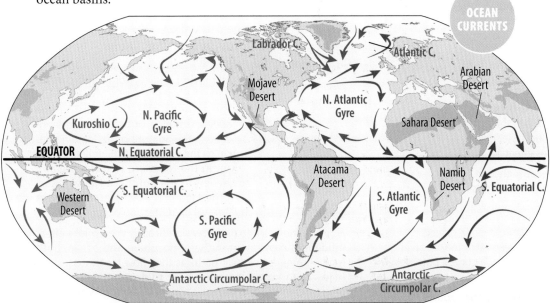

14-12

Major warm (red) and cold (blue) surface currents of the world's oceans. Desert regions affected by ocean currents are shown in tan shading.

The stretch of ocean just north of Antarctica is unusual because it is the only continuous band of ocean that encircles the earth in an east-west direction. Wind patterns create strong currents that circle Antarctica. Together they form the Antarctic Circumpolar Current. This current zone includes the southern portions of the Pacific, Atlantic, and Indian Ocean gyres. But it also transfers surface and subsurface waters between these main basins as it flows past Africa, South America, and Australia (see Figure 14-12).

There are many side branches off the main ocean gyres. These circulate water through the bays and seas surrounding the ocean basins, including the Arctic Ocean, which is separated from the other oceans by very narrow passages. They often form swirling patterns called *eddies* that are visible from low earth orbit (see Chapter opener photo).

eddy: (Poss. from Old Norse *id*—backward) + -dy; a whirling flow of water or wind

OCEAN MOTIONS 339

It is the five great ocean gyres of the world that trap drifting garbage from all kinds of sources—recreational boaters, wind-blown trash from coastal city dumps, merchant ships that lose cargo overboard during storms, ships dumping their trash, and even cities that purposely dispose of their garbage at sea. As these currents move, the Coriolis effect deflects the currents toward the middle of the gyres, where they carry their wastes. The floating garbage dumps present an extremely difficult challenge to clean up. Maybe you could be the scientist or government leader who begins to solve this problem.

THE CORIOLIS EFFECT

When you throw a baseball from left field to home plate, its path is straight to the catcher, right? Not really.

In 1835, the French engineer and mathematician Gustave-Gaspard Coriolis wrote a paper about this curiosity. He wrote that water spills through a water wheel in a line straight down. But if one imagined observing the water flow from a point on the water wheel (think of a bug holding on for dear life!), the water's path would appear to curve. Later, the military noticed that this principle applied to cannon balls fired over long distances. This is because the earth is rotating while the cannonball is in the air. The deflection of a moving object is known today as the Coriolis effect.

The Coriolis effect occurs because flying objects, or moving fluids such as winds and ocean currents, tend to follow a straight path. On a globe, a straight path forms a great circle (see Chapter 3). However, a great circle placed anywhere on the globe but on the Equator makes an angle to latitude lines. And that angle is why a curved path develops. Let's see how this happens.

The earth rotates from west to east, so the ground you are standing on is moving east at a certain speed. (You can't feel the speed because you, the air, and everything around you are moving at the same pace.) Your speed depends on your latitude. At the Equator, your speed is about 1670 km/h. At a latitude of 60°N, your speed is half that. And at the North Pole (90°N), your speed is zero. Now, imagine that a cannon ball is fired from your original location in any direction. The ball is moving over the earth at the speed and direction it was fired. But it also has the speed that is based on the latitude from where it was fired.

In the Northern Hemisphere, if the cannon were pointing anywhere south of its original latitude, the ball's eastward speed would be slower than the speed of the ground it moved over. So, to an observer on the ground, the cannon ball would drift to the right of a straight path. If the cannon were pointing north of its original latitude, the ball's eastward speed would be faster than the ground it moved over, and it still would seem to drift to the right.

All moving objects not in firm contact with the earth's surface experience the Coriolis effect. Increasing the speed, distance, or time of motion increases the amount of deflection relative to the earth. Objects deflect to the right in the Northern Hemisphere and to the left in the Southern Hemisphere. Only if the motion is exactly east or west at the Equator will no deflection occur.

You can imagine how significantly the Coriolis effect influences movement in the atmosphere and ocean. Both follow immense circular paths as they flow for thousands of miles. Even human activities have to deal with the Coriolis effect. Scientists and the military who use rockets and other kinds of projectiles must correct for this effect. Targeting computers include a Coriolis correction for the latitudes of the launcher and the target, as well as the direction and speed that the projectile will be moving.

So even though you might throw the baseball straight to the catcher, the ball will curve ever so slightly to the right of the direction you threw it (for a Northern Hemisphere player). But you probably shouldn't try to correct too much for the Coriolis effect when throwing that baseball home!

The difference in speeds due to Earth's rotation at different latitudes causes the Coriolis effect on a moving object not held to the earth's surface.

14.8 THE EKMAN SPIRAL

Early oceanographers noticed that cakes of Arctic ice moved at an angle to the right of the wind direction. Why didn't they move in the same direction as the wind? The famous Swedish physicist V. Walfrid Ekman suggested that this was due to the Coriolis effect, just like the motion of surface currents. These waters deflect at an angle of about 45° to the right of wind direction in the Northern Hemisphere (45° to the left in the Southern Hemisphere).

As a surface current moves, it drags deeper water along with it due to fluid friction. But because of the Coriolis effect, the direction of the current changes with depth. The deeper currents twist to the right or left of the surface current direction, depending on the hemisphere (see Figure 14-13). The deeper the current, the more it is deflected compared with the surface current. This change of current direction with depth is called the **Ekman spiral**. Current speed also slows as you go deeper because of friction in the water.

14.9 CURRENTS AND GLOBAL CLIMATE

Ocean currents affect the weather by carrying cold or warm water far from its source. Scientists who study global climate believe that ocean currents carry away about one-third of the solar energy deposited in the oceans near the Equator. Global winds pick up the rest of the energy from the surface waters and move it through the atmosphere. In the Northern Hemisphere, the *Gulf Stream* and the *Kuroshio* (or *Japan*) *Current* are warm currents that flow north from warm areas.

The Gulf Stream flows from the Gulf of Mexico toward Europe. Winds blowing from the west across this warm current (the prevailing westerlies) bring warm weather to northern Europe. As the current cools, it becomes the south-flowing Canary Current. Because of the Gulf Stream, England is warmer than New York City even though England is farther north. In a similar way, the Kuroshio Current flows along the east coast of Asia toward North America. As westerly winds carry energy from the current to warm western Canada, it becomes the cold North Pacific and Alaskan Currents. Western Canada is warmer than eastern Siberia as a result of the Kuroshio Current.

Along the eastern sides of ocean basins, cold currents flow from the polar regions toward the Equator. Lands near cold currents tend to be dry because cold air evaporates less moisture. For example, the cold California Current off the coast of Baja California keeps that Mexican state's climate dry. Both the **Atacama** Desert in South America, where rain falls only a few times each century, and the **Namib** Desert on the southwestern coast of Africa are regions that are downwind from cold currents (the Peru and the Benguela Currents, respectively).

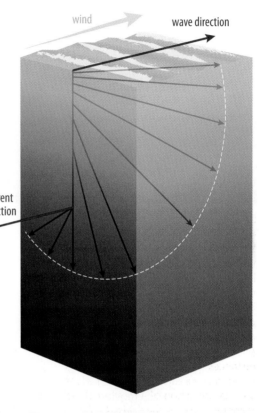

Surface current flow can extend down to about 100 m, depending on the surface current speed and available water depth.

14-13
An Ekman spiral. If the surface current is fast enough and the water deep enough, the deep flow can actually move in the opposite direction of the surface flow!

Kuroshio (koo ROW shee oh): kuro- (Jp. *kuro*—black) + -shio (Jp. *shio*—tide); the main ocean surface current that flows northeasterly along Japan's eastern coast

Atacama (AT uh KAM uh) **Desert**: 24°S 69°W
Namib (NAH mib): 24°S 15°E

14.10 UPWELLING AND DOWNWELLING

Seawater can also move between the surface and deep regions of ocean basins. The movement of dense, cold, salty water from the abyssal regions to the surface is called **upwelling**. These kinds of currents usually are rich in nutrients. They produce excellent conditions for phytoplankton growth, which is necessary to support all kinds of marine life. The world's greatest fisheries exist in places where upwelling occurs.

But what makes deep water rise to the surface like this? Upwelling exists in places where the sea level is lower than average. This happens along the Equator, for instance. The Equatorial currents twist away from the Equator because of the Coriolis effect and Ekman spiral flows. These currents drag water away from the Equator and pile it up in the middle of the major ocean gyres. The weight of the water in the middle of the gyre pushes downward and squeezes bottom water upward at the Equator, causing upwelling (see Figure 14-14).

Coastal upwelling can occur near coastlines when prevailing winds drive surface waters away from shore. Gravity forces the deeper waters toward the shoreline to replace the missing water. They rise toward the surface as the bottom becomes shallower.

Upwelling can occur in the middle of ocean basins too. As deepwater currents flow along the bottom, seamounts and underwater ridges can deflect these currents toward the surface, creating upwelling.

Downwelling is the opposite of upwelling. It occurs where sea level is higher than average as surface currents cause the ocean to pile up. The extra weight of the water can cause the deeper water to spread outward along the basin floor. The surface water then falls into the depths. This kind of downwelling is common in the middle of the ocean gyres as well as in smaller eddies.

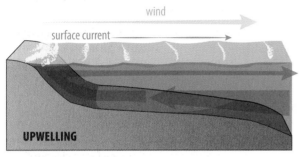

14-14 Upwelling occurs when surface waters carried away by winds are replaced by deeper waters rising to the surface.

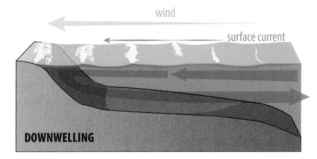

14-15 Downwelling occurs where the ocean surface is higher than average. As the water sinks because of its weight, it carries oxygen into the deep depths. But this water lacks nutrients.

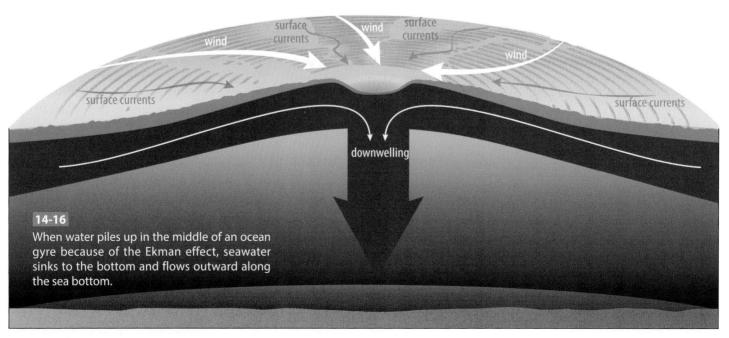

14-16 When water piles up in the middle of an ocean gyre because of the Ekman effect, seawater sinks to the bottom and flows outward along the sea bottom.

WORLDVIEW SLEUTHING: GREAT PACIFIC GARBAGE PATCH

We have all seen litter along the side of the highway. When you see one piece here or there you don't realize how much total litter there is. If you were to gather it all together, its volume would amaze you.

INTRODUCTION

You are an oceanographer working in Northern California. A local corporation is willing to fund worthwhile projects aimed at improving conditions in the Pacific Ocean. They have requested that proposals be sent to them for projects that they could fund. You propose to clean up the Great Pacific Garbage Patch.

TASK

Write a proposal to solve the problem of the Great Pacific Garbage Patch. The proposal must include background information on the issue, the issue's effect on the Pacific Ocean, and your proposed solution to the problem.

PROCEDURE

1. Research the Great Pacific Garbage Patch. Include what it is, how it formed, and how it was discovered. Do a keyword search on "Great Pacific Garbage Patch."

2. Research the impact of the Great Pacific Garbage Patch on people, animals, and the ocean itself. Do a keyword search on "Great Pacific Garbage Patch impact." You could also add "on animals," "on humans," or "on environment" to your search.

3. Research possible solutions to the Great Pacific Garbage Patch. Do a keyword search on "Great Pacific Garbage Patch solutions" or "Great Pacific Garbage Patch cleanup."

4. Write your proposal to include the three areas of research.

CONCLUSION

God has commanded us to use and care for the earth He provided to us. We should strive not to damage it by our actions; any damage we unintentionally cause should be repaired.

Coastal downwelling occurs where prevailing winds or currents pile up water on a shoreline or in a bay or gulf. If the water is trapped and cannot form a surface current, the only direction the water can go is down (see Figure 14-15).

Downwelling water usually has low amounts of nutrients because marine life in the upper zones of the ocean has already used them up. However, downwelling water from the surface is rich in oxygen (which deep-ocean life needs to live) from contact with the atmosphere and from oxygen-producing photosynthetic organisms. Most downwelling occurs in the polar regions. This water is very cold, so more oxygen can dissolve in the seawater.

14.11 DEEP CURRENTS

There are currents deep in the oceans that are much larger than surface currents. These **subsurface currents** are harder to detect, so we don't know as much about them. Winds can't drive these currents, which means that there must be other forces at work. Oceanographers have discovered that deep currents are caused by the Ekman spiral effect, downwelling (see Figure 14-16), or density differences. Most deep currents are caused by a combination of these factors.

14-17

Generalized view of the deep-water currents of the world

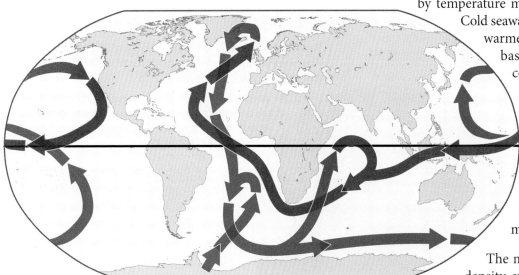

thermohaline (ther mo HEY leen): thermo- (Gk. *therme*—heat) + -hal- (Gk. *hals*—salt) + -ine; a property that depends on both temperature and salinity

Gravity drives currents that result from density differences. These are called **density currents**. You learned about the properties of seawater in Chapter 13. What can make seawater denser and heavier? When seawater is saltier, it is also denser. But seawater density is affected by temperature much more than by salinity. Cold seawater will sink and flow under warmer water. Because the ocean basins have sloped bottoms, cold water will flow downhill into the basins, forming deep currents that follow the bottom topography. This type of density flow is another example of a fluid convection current, just like the thermal plumes moving through the earth's mantle (see Chapter 5).

The most significant of the deep density currents are the deep circulation or **thermohaline currents**. These are slow currents that flow far beneath most major surface currents. They consist of cold, salty ocean water that has sunk to the depths by downwelling in the polar regions. Gravity drives these huge masses of water across the world's ocean basins, eventually mixing them with a deep current of very cold water that encircles the Antarctic continent. Thermohaline currents are much slower than surface currents. They move the same distance in a year that surface currents move in an hour. Oceanographers estimate that water in these currents may circulate for a thousand years or more before returning to the surface. They are still trying to understand the thermohaline currents and the effect they have on world climate.

14.12 OTHER DENSITY CURRENTS

In some places, a difference in salinity alone creates a density current. The saltier water sinks and flows under the less salty water. This is the main way that the Mediterranean Sea exchanges water with the Atlantic Ocean. The basin of the Mediterranean Sea is separated from the Atlantic basin by an underwater ledge where the two meet near the **Strait of Gibraltar**. The Mediterranean surface water, especially at the eastern end, is very salty because of rapid evaporation by the hot sun (the sea is close to the Equator). This extra-saline, dense water sinks to the bottom and fills the lowest parts of the basin until it spills over the ledge into the Atlantic. The less-saline surface water from the Atlantic flows into the Mediterranean Sea to replace it. Other seas separated from the main oceans by a restriction (as at Gibraltar) can have similar circulation patterns.

Fi Strait of Gibraltar: 35.9°N 5.7°W

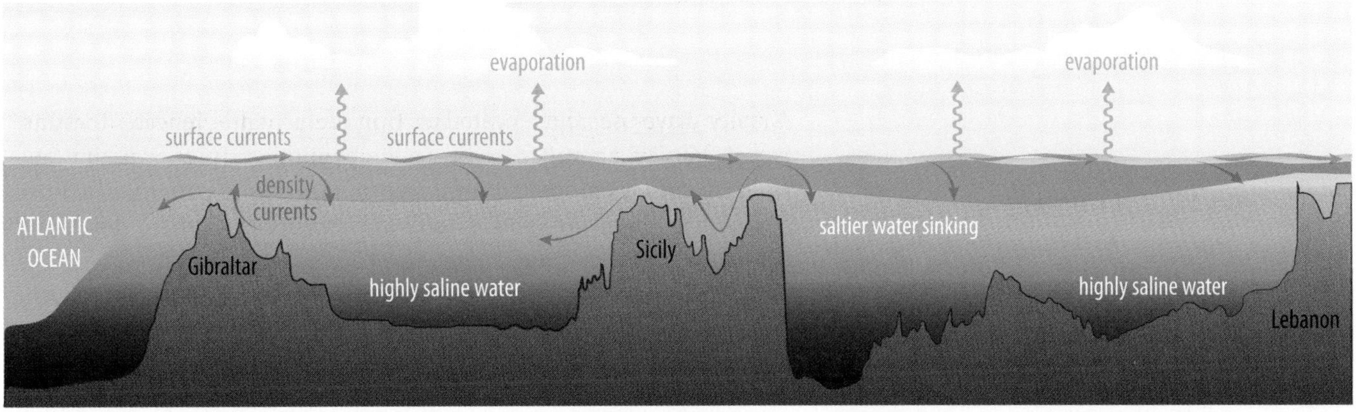

14-18
A density current exchanges water between the Atlantic Ocean and the Mediterranean Sea (vertical relief is exaggerated).

turbidity (tur BID ih tee): turbid- (L. *turbidus*—confusion, muddy) + -ity; describing a cloudy or sediment-laden condition

Another kind of density current, which is similar to an underwater mudslide or lahar, is a **turbidity current**. This sediment-laden mixture has a higher density than the surrounding water. It flows quickly down the slope along the ocean floor, eroding the sediment as it goes. In Chapter 13, you learned that submarine canyons might have been cut by turbidity currents.

14B REVIEW QUESTIONS

1. What two main factors control the speed and direction of the major ocean surface currents?
2. What do oceanographers call the great circling ocean currents?
3. Why does the Ekman spiral happen? Which way does it twist in the North Atlantic?
4. How do ocean currents affect global weather patterns?
5. What is upwelling, and why is it important?
6. Why are downwelling currents important to marine life in the abyssal zone of ocean basins?
7. Describe the differences between a surface current and a subsurface current.
8. What do oceanographers call global deep-water circulation?
9. Name three conditions that allow gravity to produce subsurface currents.
10. What kind of density current is caused by a muddy, rapidly flowing mixture of sediment and water? What events on land are similar to this?
11. (True or False) Water enters the Mediterranean Sea as a surface current and leaves that basin as a bottom density current.

14C

WAVES

Where do waves come from?

14.13 OCEAN WAVE PROPERTIES

When you think of the ocean, you probably picture ocean breakers pounding a sandy beach. But where do water waves come from?

14C Objectives
After completing this section, you will be able to
» define *wave terminology*.
» analyze the forces that create and affect waves.
» predict what kind of landforms will be created under specific sets of wave and land conditions.

A water **wave** uses the repeated motion of the matter in water to transfer mechanical energy from one place to another. In a wave, tiny suspended particles move through a repeating circular pattern. The wave energy lifts them up and forward, and gravity pulls them down and backward. The repeating pattern is called an *oscillation*. In waves passing over deep water, the water particles oscillate in one place as the wave moves along. You can see this in a lake or at the ocean. A floating object bobs and moves back and forth, but stays in one place (see Figure 14-19).

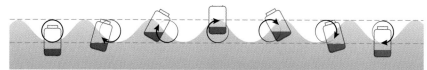

14-19
Water particles or objects in a wave follow a circular path.

You learned a little about waves in Chapter 2, but let's look at the parts of a water wave (see image below). Oceanographers describe a wave by its height and length. **Wave height** is the vertical distance between the top or crest of the wave and its bottom, or trough. The higher a wave is, the more energy it has. Wave height can be anywhere from zero in flat, calm water to many meters high in a storm or tsunami wave. High waves at sea are especially dangerous—large waves can damage and sink ships even today.

| The highest measured ocean wave ever was a 524 m tsunami wave that came ashore at Lituya, Alaska, in 1958! The highest deep-water wave actually measured at sea by a ship was a 34 m monster that formed in 1934. Tall waves like these usually appear briefly by themselves and are called *rogue waves*.

The length of a wave, or **wavelength**, is the distance between the same point on two side-by-side wave forms, usually crests or troughs. Wavelengths can vary from a few centimeters (ripples) to many kilometers (tsunamis).

| A wave's *amplitude* is the distance to either the wave's crest or its trough from a point midway between them. Wave height is a more useful quantity for studying water waves.

Marine scientists classify an ocean wave by comparing the depth of the water below the wave to the depth of the wave base. The **wave base** is the deepest depth below the ocean's surface that is affected by the wave's passage. Just as with surface currents, wave motion grows less with depth below the surface (see Figure 14-21 on the next page). At the wave base, the circular motion of particles in the water stops. When the ocean bottom is deeper than the wave base, the wave is considered a deep-water wave. The wave base of a deep-water wave is at a depth about half its wavelength.

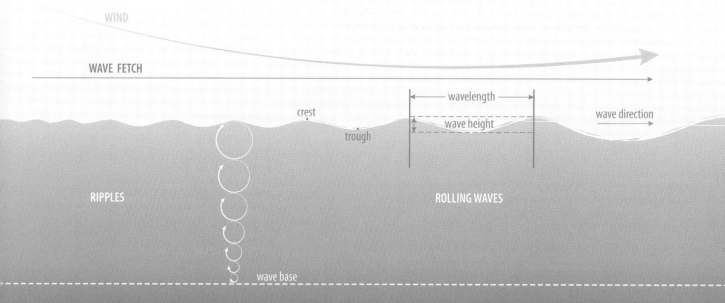

Wave period is the time between one wave crest and the next as they pass a stationary point. If you were sitting on a wooden dock timing a wave's period, you would start your stopwatch when one crest passes a post supporting the dock, and stop it when the next crest reaches the same post. Periods of typical waves are measured in seconds. The period of a wave is related to the **wave speed** (how fast a wave crest moves) and wavelength. Wave period is easier to measure than either wave speed or wavelength. In general, the longer the wave, the faster the wave moves.

A typical wave in the open sea may have a wave height of 1 m, a wavelength of 50 m, and a wave base of 25 m.

14.14 WAVE GENERATION

Winds produce most waves. As the wind blows along the surface of the ocean, it drags the surface water along in ripples. These disturbances add up to form wavelets, which combine to form waves. The distance that wind blows over the water's surface in one direction is the **fetch** of the wave. The longer the fetch, the more energy the wind can transfer to the water, and the higher the waves will be. Wind-driven waves develop over broad areas of the ocean and form a long series of waves that can extend for hundreds of miles.

Waves are the most direct result of wind energy. As with any two substances rubbing against each other, wind exerts friction on the water, dragging it along. Thus, currents will form along with waves.

But fetch is just one of the factors that affect the energy of a wave. The length of time the wind blows and its speed also affect wave height. The longer a wind blows in one direction and the faster it blows, the higher the waves will be and the more energy they will contain. Fetch also determines the wave's length, and thus its speed. Longer fetches mean longer and faster waves.

14-20

An internal wave develops between two liquids with different densities.

There are many other kinds of ocean waves. High splash waves form when something falls into the ocean, such as ice breaking off a glacier to form an iceberg, a meteorite impact, or a huge coastal rockslide. Internal waves move through the depths of the ocean at the surface between layers of water with different densities (see Figure 14-20). Tsunami waves, often wrongly called tidal waves, are huge waves caused by large earthquakes. They can travel at up to 800 km/h! When these waves come ashore, they can be very destructive. True tidal waves are the high and low tidal bulges that move around the earth as it rotates. Their wavelengths are about half the circumference of the earth!

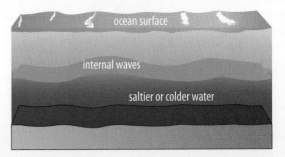

14-21

Wave fetch is the distance over which a steady wind blows as it creates waves. The longer the fetch, the longer and higher the waves.

14.15 SHALLOW-WATER WAVES

As deep-water waves approach the shore, they enter shallower water. When the water depth equals the wave base, the waves "touch bottom" and begin to change. The wave crests move closer together (the wavelength shortens), and the waves slow down. The circular motions of wave particles flatten, becoming more oval. Flattening the oscillations transfers some of the waves' energy upward to the surface of the waves. This lifts the waves higher, and wave height increases. Rather than smooth, rolling waves, they become more pointed.

14-22
Waves are transitional waves when they move into water that is shallower than the wave base. Waves become shallow-water waves when water depth is less than 1/20 of their wavelength.

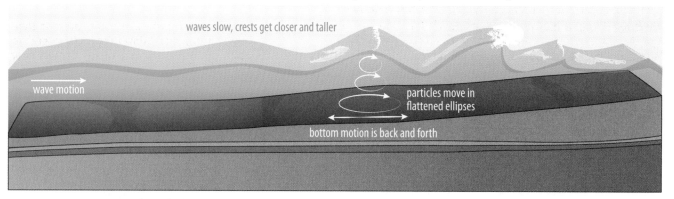

Closer to shore, the wave becomes a shallow-water wave. The water oscillations are highly flattened, and water particles slosh and move back and forth rather than in circles. Shallow-water wave speed depends completely on water depth, rather than on wavelength. The shallower the water, the slower the wave moves.

The remaining energy in shallow-water waves lifts them up into high, sharp crests. When a wave's height grows until it is more than one-seventh (14%) of its wavelength, it becomes unstable. The crest peaks sharply and then falls over on the forward side of the wave. Such a wave is breaking, and is called a **breaker**. The energy in a breaking wave forms new lower-energy waves in front of it that continue toward shore. Breakers may form several times before the water uses up its kinetic energy flowing up onto the beach.

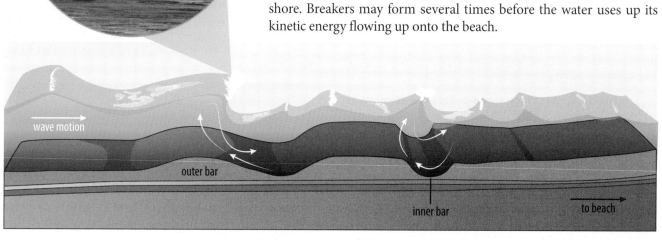

14-23
Breakers form as waves pass over shallower areas, such as near the shore. Friction with the bottom causes the wave to slow down and go higher. An unstable wave that falls over is called a breaker.

Long wave crests usually approach the shore at an angle. The end of the wave nearest to shore slows down faster due to friction with the shallow bottom. The other end of the wave in deeper water moves faster until it too slows down as it reaches the shallower water. This makes the wave bend, or refract. Refraction causes the wave to approach the coast nearly parallel to the shoreline.

Wave crests are rarely exactly parallel to the shoreline. Instead, even with refraction, they strike the shore at an angle and flow up onto the beach. When the water stops flowing, gravity pulls the water straight down the beach into the ocean. The repeated sawtooth motion of numerous waves causes the water and any sand and other debris it carries to move along the shore. This movement produces a current that flows parallel to the shore, called a **longshore current**. A longshore current is temporary and depends not only on the direction of waves but on wind direction and speed as well. These currents can have speeds up to several kilometers per hour. Though they are not particularly hazardous, beachgoers need to be aware that they could be fairly quickly carried along the beach away from where they entered the water to swim.

Wave diffraction also occurs where the shore forms a point into the ocean. Waves approaching the point will seem to wrap around it on all sides, focusing their energy on a fairly small section of the coastline. Points of land erode more quickly for this reason.

Water that flows onto a beach will quickly return down the slope of the beach to the ocean. Under some conditions, water flowing outward from the shore gathers into narrow, fast **rip currents** that flow offshore some distance. These surface currents usually flow through a gap in the breakers and the longshore bar under them. Rip currents are dangerous. See the box below.

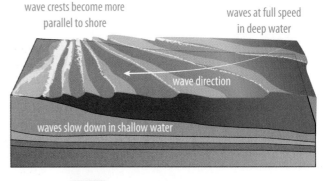

14-24
The leading end of a wave slows down in shallow water, which allows the trailing end to catch up. Waves become nearly parallel near the shore.

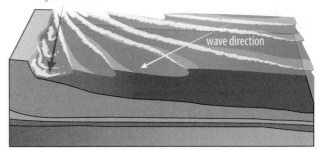

14-25
Longshore currents develop when waves approach the shoreline at an angle. These currents can cause beach erosion and deposition.

RIP CURRENTS: SWIMMERS BEWARE

Many beachside deaths are caused by occurrences of rip currents. These swift currents can knock a swimmer off his feet and carry him rapidly out to sea. Panicked swimmers often exhaust themselves by fighting against the current and eventually drown.

Rip currents occur when high waves dump large amounts of water on the beach. As the water builds up, it rushes back out into the ocean along narrow paths. These currents usually form in low points in the breakers or hollows in the shoreline, where water naturally flows out between breakers. They can flow at 1 m/s for several minutes, racing up to 760 m into the ocean before dying out.

You can easily locate a rip current. Look for strips of water where sand is being carried out by the current, or look for slight changes in the wave patterns as they meet a current. The bottom is usually deeper under a rip current, so a breaker crest will often disappear over a rip current channel.

If you are ever caught in a rip current, remember that rip currents are narrow. You simply need to swim parallel to the shore until you are free of the current. Then you can head back to the beach.

Rip currents generally flow perpendicularly away from the beach.

What problems associated with beach erosion do you see in this photo?

14-26
Groins protecting a harbor entry from sediment

14.16 WAVE EROSION AND DEPOSITION

In Chapter 13, you learned about the structure of a typical beach. In most cases, beaches consist of eroded sediments. The sediments can vary in size from very fine sand to gravel, large cobbles, or even boulders. Beach sediments can come from local rocks and soils, can be transported from other locations along a coastline, or can even come from deeper water offshore.

The slope or angle of a beach depends on the size of the sediment clasts that make up the beach. Fine-sand beaches can have slopes of less than 1°, while coarse cobble beaches are much steeper, up to about 24°. The coarser the clasts, the steeper the beach slope.

Waves shape beaches through erosion and deposition. The size of waves breaking on a beach can sort the beach sediments by particle size and density. So the slope and appearance of a beach can change, depending on recent wave activity. In some places, the size of waves varies with the season, so a beach may look different depending on the time of year.

Large, storm-driven waves can quickly erode the loose beach sediments and either deposit them offshore in shallow bars or transport the sediments along a beach and deposit them somewhere else. The rate of erosion is quite variable, and significant erosion can happen in short bursts of wave action. In some places, erosion is occurring at a rate of 0.5–1 m per year. Elsewhere, such as the outer banks of Virginia, beaches are eroding as much as 8 m per year. Beach erosion is expensive and dangerous for people who live close to the shoreline or work in buildings there. After just a few hours of battering by intense hurricane storm waves, buildings located 30–50 m from the shoreline can collapse into the sea. Land conservation efforts include building barriers called *groins* that jut out into the ocean from beaches that tend to erode. Groins slow and disrupt longshore currents, and cause them to drop their sediment loads on the beach rather than carrying them away.

Beaches are not the only type of coastal feature that waves create through erosion and deposition. Many coasts are rock cliffs. As waves erode the cliffs, they create beaches in front of the cliffs. In some places, the beach bottom focuses the wave energy at one spot in the cliff face, digging out a hole in the rock. With time, a *sea cave* forms. If a cave erodes through a narrow section of rock jutting out into the water, a *sea arch* results, similar to rock arches far inland. As wave action dissects coastal rock formations, short columns of rock separated by water are left behind. These are called *stacks*. A stack often results from the collapse of a sea arch. These coastal features form relatively rapidly over just a few decades or centuries, and they can be eroded away just as rapidly.

As you know, when waves erode rocks and sediments, the eroded materials are eventually deposited somewhere else. Just as with stream sediment transport, when longshore currents and eddies laden with sand encounter an obstacle or slow down, the sediment settles out. This often happens when a coastline suddenly changes direction, as at the ends of a bay. A sandy strip of land that extends from the shore across part of a bay is called a *spit*. The spit may eventually grow across the bay's mouth, closing it to the sea. This is a *bay barrier*. A spit with a sharp bend formed by current eddies around the end of the spit is called a *hook*. Waves in shallow coastal areas can deposit sand as an island parallel to the shore. The Carolina coastline has many of these barrier islands. A *tombolo* (TOM buh LOH) may form between a barrier island or stack and the shore when waves flow around the obstruction and deposit sediments there, creating a sandbar that connects it to the mainland.

14C REVIEW QUESTIONS

1. How does mechanical energy move along the surface of the ocean?
2. The energy of a wave is related to what wave property?
3. What is a wave's base? What is the depth of the wave base of a 160 m long deep-water wave?
4. How does a deep-water wave's speed relate to its wavelength?
5. What causes most surface waves? What other kinds of things could cause a large surface wave?
6. What two kinds of currents can occur as ocean waves come ashore?
7. What do you call a mass of rock cut off from the mainland by wave erosion? What sedimentary formation may connect such a rock to the shoreline?
8. (True or False) According to the "one-seventh wave-steepness rule," a wave with a wavelength of 25 m and a wave height of 3 m is stable.

Key Terms

tide	331
lunar tide	332
spring tide	334
neap tide	334
surface current	338
Coriolis effect	339
gyre	339
Ekman spiral	341
upwelling	342
downwelling	342
subsurface current	343
density current	344
thermohaline current	344
turbidity current	345
wave	346
wave height	346
wavelength	346
wave base	346
wave period	347
wave speed	347
fetch	347
breaker	348
longshore current	349
rip current	349

14-27
The Rock of Gibraltar photographed from a ship transiting the straits

CHAPTER 14 REVIEW

CHAPTER SUMMARY

» Studying the three kinds of motions of the earth's oceans—tides, currents, and waves—helps us to better use the oceans and to manage problems that people have created in them.

» The gravitational pulsl of both the moon and the sun on the earth's oceans cause tidal bulges in the oceans. Most locations experience two high and two low tides a day.

» Certain arrangements of the earth, moon, and sun, as well as the action of large storms, can cause unusual tides.

» Many countries are experimenting with tidal energy because it is non-polluting and renewable.

» The global surface currents are formed by a combination of the prevailing winds and the Coriolis effect.

» As one goes deeper below a surface current, the Coriolis effect deflects the subsurface flow to the right (in the Northern Hemisphere). The change of current direction with depth is called the Ekman spiral.

» Oceanic currents circulate warm and cold water between the Equator and polar regions. They significantly affect global and local climates.

» Upwelling and downwelling are important subsurface ocean currents that circulate nutrients and oxygen between the surface and deep waters.

» Subsurface currents form from the action of the Ekman spiral, from the direct action of gravity on piled-up water, or from differences in density due to temperature and salinity.

» Winds are the main cause of surface waves. Waves travel great distances, but the particles in waves oscillate in relatively stationary circular patterns.

» Waves are described by their wavelength, height, and period. The wave base is the deepest limit of a passing wave's influence on water particles. A wave's speed depends mainly on its wavelength until it reaches very shallow water.

» The energy or height of an ocean wave comes from wind blowing over a stretch of open water called its fetch. The longer the fetch and the stronger the wind are, the larger the waves.

» Certain unpredictable geologic events can cause large surface waves.

» Many locations in the ocean where water is stratified by temperature or salinity experience significant internal waves.

» Shallow-water waves approaching shore can create breakers, longshore currents, and rip currents.

» Waves naturally erode coastlines by abrasion and impact on bedrocks and sediments. Erosional features include sea caves, arches, and stacks.

» Waves can deposit sediments, forming various kinds of sandbars, spits, and tombolos.

14-28
This picture was taken along the Mediterranean coast in Greece. Notice the effect that erosion has had along the coastline.

REVIEW QUESTIONS

1. What is an ocean tide? How do we measure tides?
2. The moon's gravity causes the tidal bulge on the side of the earth facing the moon. What causes the tidal bulge on the opposite side of the earth?
3. How are the moon, earth, and sun positioned for a neap tide?
4. Why do major surface current gyres in the Northern Hemisphere flow clockwise?
5. The Equatorial currents flow westward until they run into land. List three things that can happen to the waters in these surface currents.
6. If a wind blows east across a stretch of the North Atlantic Ocean, in what direction will the resulting surface current move?
7. Explain how a cold surface current can affect nearby continental climate.
8. Why are zones of upwelling water found in the ocean near the Equator?
9. Explain how gravity can form density currents.
10. What are the most important density currents in the oceans?
11. Why does the Mediterranean Sea become so salty that it forms its own density current into the Atlantic Ocean?
12. How does a particle of water in a deep-water wave move in comparison to the wave itself?
13. Draw a profile of a series of several ocean wave forms and label a wave crest and trough. Indicate the wave height, wavelength, and wave base.
14. You are sitting on a beach watching waves move toward shore under a wooden dock. You note that the time between the first and sixth wave crests (five wavelengths) to pass a dock post is one minute. What is the period of one of these waves in seconds?
15. What condition within the ocean allows the formation of internal waves?
16. Why do wave crests bend to become more nearly parallel to the shore as the wave approaches land?

OCEAN MOTIONS

17. Which kind of beach would likely be wider—a fine-sand beach or a gravelly beach? Why? (*Hint*: Sketch the slope of each kind of beach and add the high- and low-water marks for the same heights of high and low tides.)

True or False

18. Ocean tidal bulges are actually waves with extremely long wavelengths.

19. The sun's gravitational pull can actually reduce the height of a high tide.

20. The United States is a world leader in tidal power generation.

21. The Great Pacific Garbage Patch has formed because surface currents in the gyres and eddies of the North Pacific Ocean collect the trash in one place.

22. One of Walfrid Ekman's discoveries was that surface currents do not flow in the same direction as the wind that causes them.

23. England is nearly as mild as New York because the prevailing westerlies carry warm air from nearer the Equator over the British Islands.

24. Deep-sea currents are usually in the opposite direction from surface currents.

25. The amount of energy in an ocean wave is directly related to the length and speed of the wave.

26. A wave approaching a shore has a height of 2 m and a length of 13 m. This wave is a breaker.

27. The Rock of Gibraltar was originally a rocky island in the Mediterranean that is now connected to the coast of southern Spain by a broad stretch of marine sediments. These sediments could be considered a tombolo.

Map Exercises

As you learned in this chapter, trash from all kinds of human activities has accumulated in huge mats of floating garbage in the Pacific Ocean. Where are these floating garbage dumps?

For this exercise, you will need a 3D Earth-viewing program like Google Earth with an active connection to the Internet. Your teacher will provide the three files you will need for this exercise.

28. Open the file Pacific_Ocean_Currents.kmz file in Google Earth. Describe the map overlay on the 3D globe.

29. Where in the Pacific Ocean would you expect to find the Great Pacific Garbage Patch?

30. Now open the file named Pacific_Garbage_Patch.kmz in Google Earth. Does this map overlay confirm your hypothesis in Question 29?

31. Open the file North_Atlantic_Currents.kmz in Google Earth. What do you notice about the Great North Atlantic Garbage Patch that is similar to the one in the Pacific?

CHAPTER 15
OCEAN EXPLORATION

- **15A** THE HISTORY OF OCEAN EXPLORATION — 356
- **15B** OCEANOGRAPHY IN ACTION — 360
- **15C** ENTERING AN ALIEN WORLD — 368

THE LAST FRONTIER

What is the last frontier? The West? No, we've explored that. The Antarctic? Scientists are stationed there year-round. Space? Robotic probes have explored most of the planets. And we have a continuously manned space station. So what, then? What area of the accessible universe has successfully resisted man's ability to explore it?

The oceans. We know a lot more about Mars's surface than we do about the bottom of the ocean. People have walked on the moon; none have walked on the ocean's deepest floors. Looking for alien life forms? Check the oceans, not space. The oceans contain 97% of all life on Earth, and we have not yet discovered all species that live there. Underwater microphones have picked up sounds from the deep ocean that scientists think come from an animal larger than the blue whale, the largest known animal. The oceans, their resources, and their importance to proper dominion are almost too great to imagine.

15A Objectives

After completing this section, you will be able to

» summarize the history of key advances in our knowledge of the world's oceans.

» identify the motivations behind these key advances.

15A THE HISTORY OF OCEAN EXPLORATION

How have we studied the oceans in the past?

15.1 STUDYING THE WATER PLANET

If the oceans are so mysterious and so essential for biblical dominion, why don't we know more about what covers over 70% of our planet? We do know a lot about the ocean's size, shape, surface currents, tides, and waves. We also know much about its composition and the life it holds. But even under normal circumstances, **oceanographers** often have to deal with difficult and uncomfortable working conditions on the ocean's surface, rocking in tiny ships that ride on huge waves driven by storms, often covered in ice or baking under the blazing tropical sun. When we study the ocean's depths, we run into even greater difficulties. We have to get instruments or observers deep into the ocean to collect data. Once there, we have to deal with frigid cold, crushing pressures, and pitch-black darkness. The technological challenges are enormous.

Collecting data at one place in one moment of time isn't useful all by itself. The ocean is huge, and conditions change all the time. Data from *millions* of locations in all three dimensions are necessary to give us a good idea of the ocean and its topography. We definitely have our work cut out for us!

15-1 Oceanographers have to work in all kinds of weather and waves at sea.

15.2 THE EARLIEST OCEANOGRAPHERS

You might wonder when people began studying the sea. The Bible doesn't mention the ocean between its creation (Gen. 1) and when the "fountains of the great deep" broke up (Gen. 7). We can only imagine that during the 1656 years between these two events, people probably learned to build ships to explore the world and use the ocean. Noah possibly used some of that knowledge to build the Ark.

After the Flood, Noah's descendants were content to settle inland in the Mesopotamian valley in present-day Iraq. It wasn't until people scattered following the Tower of Babel judgment several centuries after the Flood that exploration began in earnest. Eventually, people found their way to seacoasts. To reach distant lands, sailing and exploring the oceans became necessary.

The remains of the oldest-known boat have been found in a desert in Kuwait. Though scientists have dated the boat at about 7000 years old, it is probably 5000–5200 years old because it would have to be no earlier than the date of the Flood. (No known human artifact has ever been found that definitely came from before the Flood.) Egyptian murals show small ships from around that time. They indicate that people may have begun to travel on the Mediterranean Sea quite early in post-Flood history.

15-2
This tracing from an Egyptian tomb shows an example of one of the earliest-known seagoing vessels.

CASE STUDY: THE $6,000,000 CLOCK

How much would you pay for a clock? In 1714, the British Parliament offered £20,000 ($6 million in 2016) for a timekeeping device. Why would the British government spend that much money for a clock? Seven years earlier, the British Navy lost four ships and over 1500 men in the Scilly naval disaster. These lives were lost because the navigators didn't know where they were. In 1707, sailors could determine latitude using the sun's highest position during the day or constellations at night. However, determining longitude remained a challenge—they often approached land without knowing it.

To determine longitude, sailors must know the difference between local time and the reference time at their home port. They could determine local time from the skies, but they needed a clock to keep the time from their home port. While accurate clocks had been in existence for hundreds of years, they didn't work well on ships. The British navy needed a timepiece that would remain accurate over long voyages and resist changes in temperature, pressure, and humidity. It needed to be resistant to the salt air and remain unaffected by the constant motion of the ship on the sea.

John Harrison, a British clock maker, was very interested in building such a timepiece. He designed his first sea clock in 1730 and spent five years building it. He then spent the next thirty-six years improving its design. His final design, a sea watch, met all the requirements of the Longitude Act by Parliament. His watch remained accurate to allow sailors to determine their position to within a mile after travelling across the Atlantic Ocean. Harrison dedicated his life to benefit the sailors of Britain.

This is John Harrison's first sea clock, named H-1. While it was successful on a short voyage, much work was needed to perfect a useable timepiece for a trip across the ocean.

Questions to Consider

1. Why was it so difficult to determine longitude at sea?
2. Why was Britain willing to pay £20,000 for a clock?
3. Was this a good use of Harrison's time? Support your answer from a biblical worldview.

OCEAN EXPLORATION

Knowledge of the oceans increased as mariners sailed their ships farther from land. Simple navigation by following coasts and the North Star advanced to navigating by compass and using ocean currents. Cartographers created more complete maps of the earth's oceans and continents, but only after sea captains returned with the information. There were still big gaps in their knowledge. Cartographers often covered up unknown areas on charts with images of fantastic sea monsters to hide this fact! As people invented very accurate seagoing clocks and learned to navigate by the moon and stars, position fixes were more accurate, and charts also became more accurate. Traveling on the ocean's surface was far less risky when you knew where you were!

From the earliest voyages, sailors considered the ocean's surface a featureless expanse of water dotted by islands. Furthermore, our knowledge of the watery world below the surface was limited to bottom depths measured from ships or to what a swimmer could observe. Stories of divers' exploits are legendary. People have learned about shallow-water sea life and underwater formations from them. But surface dives were limited. It is unlikely that divers could go deeper than about 30 m. The ocean depths would remain a mystery for many centuries.

15.3 EARLY VOYAGES OF DISCOVERY

Stories of the early voyages of discovery fill many books, though the earliest records from voyages for the first few thousand years after the Flood are lost to history. Early on, daring people must have set out into unknown seas. Our historical records of explorers arriving in places like the British Isles, Australia, and the Pacific islands show that these places were already occupied with people and domesticated animals. They could have been reached only by sea.

Nearly every ship that sailed far from a coast during the 1400s to 1800s added to the knowledge of the seas, their shapes and arrangements, and currents and wind patterns. This time period is called the *Age of Exploration*. Men and their crews risked everything to open new paths to foreign lands. They suffered from fear of the unknown, lack of food and water, disease, shipwreck, storms, and attacks by native peoples in foreign lands. And perhaps even worse, they were out of the sight of land for months at a time.

What drove them to take such risks? They definitely enjoyed the thrill of discovery and the vision of getting rich. Merchants and governments wanted to open new routes of trade so that they could get things like gold, spices, and silks from faraway lands. Their accomplishments were truly extraordinary for their time.

15-3

The *Victoria* was one of the ships on Ferdinand Magellan's expedition, which was the first to circumnavigate the globe. These early ships were tiny, and their crews faced unimaginable dangers.

15.4 VOYAGE OF THE HMS CHALLENGER

During the nineteenth century, more and more people wondered whether the deep ocean bottom really was flat and lifeless. In 1870, Charles Wyville Thompson of Edinburgh University in Great Britain finally convinced the government to fund the first true oceanographic expedition. He was given a steam and sail warship named the HMS *Challenger*. Most of its guns and some rigging were removed to make room for oceanographic equipment. Thompson equipped the ship with two laboratories and a staff of scientists. He also installed steam winches, cables, and other equipment to drag instruments across the ocean bottom to take samples and make measurements.

In 1872, the **Challenger expedition** began a 127,580 km, four-year journey across the Atlantic, Indian, and Pacific Oceans. The crew charted the basins of these oceans. They analyzed water samples and bottom sediments. They cataloged 4717 new species of marine animals and plants! After taking thousands of depth measurements, they revealed how deep the ocean basins really are.

The expedition returned to England in 1876. So much new data was collected that it took several decades to publish all its scientific findings. The *Challenger* expedition was a turning point in our understanding of the oceans, opening up a new chapter in world exploration. It became the model for nearly all future scientific ocean expeditions.

The HMS *Challenger*

15A REVIEW QUESTIONS

1. Why are the oceans considered the last unexplored frontier?
2. What is the source of our earliest knowledge of the oceans?
3. What were three main reasons for the early voyages of discovery? What did these explorers discover about the oceans?
4. Why was the HMS *Challenger* expedition different from every other voyage of exploration up to that time?
5. What kinds of new equipment and methods did the *Challenger* expedition bring to ocean exploration that had never been tried before?
6. (True or False) Early explorers relied on accurate maps of the oceans (nautical charts) to help them discover new lands.

15B Objectives

After completing this section, you will be able to

» compare the methods we use to study the oceans.

» evaluate how technology improves ocean exploration.

15B
OCEANOGRAPHY IN ACTION

How do we explore the ocean today?

15.5 STUDIES IN OCEANOGRAPHY

Scientists take two approaches to oceanography. Some studies require information about the ocean at a certain spot. Others consider the ocean as a complex system, so information from many locations at the same time is needed. The ocean is also continuously changing—daily, during storms, with the seasons, and even over decades and centuries.

SERVING GOD AS AN OCEANOGRAPHER

JOB DESCRIPTION

Want to sail the seven seas? Search for strange new animals and unlimited mineral wealth? Scuba dive with schools of colorful fish or swim with dolphins? Then you want to be an oceanographer!

Oceanographers are scientists who study the oceans. There are lots of different fields in oceanography. You can do almost any science in the ocean that you can do on land. There are marine biologists, marine geologists, marine engineers, marine archeologists, chemical oceanographers, and physical oceanographers, to name a few.

EDUCATION

As you can imagine, the competition to get a job in oceanography is pretty stiff. Since oceanography involves so many different sciences, you have to study at least one science to prepare. A strong math background is important too. Most researchers in this field have master's or doctoral degrees, though ship crews and research assistants can have less education. You need to be able to use technology and write clearly to be a good oceanographer. Knowing a foreign language, staying physically fit, getting scuba certified, and being able to fix broken equipment can come in handy. And don't forget some shipboard experience. Got to have sea legs!

POSSIBLE WORKPLACES

Though most oceanographers are employed by government agencies or universities, many work for private companies exploring the ocean's resources. Developing sea-based energy production, harvesting mineral resources, and starting sea farms are tasks that need the skillful help of oceanographers. Many oceanographers also teach in coastal universities that have important oceanographic research programs.

But not all oceanographers spend a lot of their time at sea. People with similar abilities work in land-based labs to investigate what scientists out on the ocean bring back. They can also simulate conditions in the ocean with experiments that help us better understand different areas of marine science.

DOMINION OPPORTUNITIES

Oceanographers can provide useful answers to important questions. They help fishermen understand the life cycles of fish and how to sustain fish harvests. They monitor how the temperature and salinity of the oceans change. They create models of ocean currents. They study how beaches form and erode so that we can conserve beaches and wetlands. They study ways to harness the energy of waves, tides, and the heat in the ocean. They assist in understanding the role that oceans have in stabilizing global climate change. And they are learning to use marine animal products for medicines.

As you can see, the opportunities for lifelong work in oceanography are almost limitless. Considering the vastness of the ocean, its importance to life on Earth, and our responsibility to exercise biblical dominion, perhaps you should prayerfully seek a career as an oceanographer.

SMALL-AREA SAMPLING

When studying a small area of the ocean, scientists will sample and measure at multiple depths at one or more locations close together. But each sample point requires the ship to stop and lower instruments or sampling devices over the side. Some instruments can provide data right on the spot, such as the depth of the water. But scientists can't record most sample data or run the needed tests as they sample. They must first take the water, bottom, or biological samples to a lab on the ship or on shore where they can be analyzed.

LARGE-AREA SAMPLING

When studying large regions of the ocean, oceanographers try to sample a broad area in a short period of time. Plotting data on maps helps oceanographers find relationships among data and explain what they observe. GIS technology has made this task even easier with more powerful results. Wide-area surveys help oceanographers understand things like current flows and their effects on life, seawater chemistry, and the movement of thermal energy through the oceans. This work requires research ships to travel thousands of miles gathering samples. Crews can spend weeks or even months at sea. In recent years, modern technology has helped in the collection of some forms of data over wide areas, requiring less time at sea gathering such data.

Oceanographic ships, such as the Atlantis, *carry out most of the research on the oceans.*

15.6 BASIC OCEAN OBSERVATIONS

Oceanographers have to adapt to the special and often difficult conditions on ships as they collect data. Everything rusts. Everything gets wet. Rocking with storm waves, sweltering in the sun, freezing in rain or snow—the quest for knowledge must go on! They need some fairly rugged instruments (and people).

Studying the ocean means collecting a lot of different kinds of information. This includes sea-floor depth, seawater temperature and composition, salinity, currents, water clarity, and biological organisms at various depths. Oceanographers also want to know about the kinds of bottom surfaces, including their sediments, bedrock, topography, and forms of life. To get such information, oceanographers have developed rugged, relatively simple instruments that have changed little over time.

15-4

In the early days of oceanography, depth soundings were obtained using a lead line.

DEPTH SOUNDINGS

For centuries, navigators used a lead line (LED line) to take a **sounding** for determining water depth. To avoid running aground when approaching shore, a ship let out a rope or wire with a lead weight attached. If the lead line did not touch bottom after several fathoms (1 fathom = 1.83 m) were let out, the ship was safe from hitting bottom.

But a lead line was really hard to use for measuring deep depths. Just imagine trying to use this method to measure water 10,000 m deep. The *Challenger* expedition found that in deep places, the wire or rope often broke under its own weight! Data for deep soundings was often inaccurate. Deep currents grabbed the lead line or surface winds moved the ship, causing the line to curve, and adding many meters to its sounding depth. The invention of *sonar* (see page 366), technology that uses echolocation, fixed most of these problems.

WATER SAMPLING

Water samples at regular depth spacing provide important information about the properties of seawater. Fridtjof Nansen invented a water-sampling instrument in 1894 that is still in use today. The *Nansen bottle* is a rugged, simple steel cylinder, open at both ends, with spring-loaded stoppers. Figure 15-5 shows how it works. Usually, oceanographers clamp many bottles to the cable, one at a

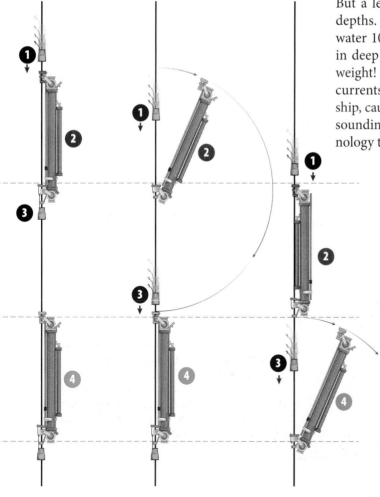

15-5

A messenger weight ❶ trips a mechanism on the Nansen bottle ❷ that shuts valves at both ends of the cylinder to capture the water sample. After tripping, a second weight ❸ is released, tripping the next bottle ❹ below.

time as the cable is lowered, to collect samples at many depths. After all the bottles are filled and sealed, the cable is hauled up and the bottles are removed one at a time. Nansen bottles include a specially designed mercury thermometer that retains the temperature reading at the time the bottle takes the water sample.

A more modern version of the Nansen bottle is the *Niskin bottle*, invented by an American ocean engineer in 1966. This sampling device is made of plastic to avoid contaminating the water sample with metal ions from the bottle. And rather than flipping, the bottle has a simpler mechanism that plugs both ends, trapping the water sample. This design allows up to thirty-six bottles to be mounted in a single frame hung from the end of a cable. Depth sensors shut the bottles at preset depths as the assembly is lowered. Instead of using thermometers, the apparatus measures water temperature using an electronic temperature sensor mounted on the sample frame. This modern system makes water sampling much easier and faster.

15-6
Niskin bottles mounted in racks called *carousels* permit quick sampling of seawater at many depths in the deep ocean.

MEASURING WATER CLARITY

Water clarity is essential for life because it determines how far light penetrates into the water. This is especially important for algae and other photosynthetic life forms in the photic zone. Sometimes the water gets cloudy with microscopic organisms or sediments. This can greatly reduce the amount of light reaching deeper waters.

For many years, oceanographers measured water clarity by lowering a special disk on a long pole or weighted cable. They painted the disk in alternating black and white quarters to improve visibility. This device was invented by Pietro Secchi, so it is called a *Secchi disk*. Oceanographers made their clarity observations by measuring how deep the disk could be lowered until they couldn't see it anymore. This depth was greater in clearer water. The main disadvantage of the disk was that one had to observe it from above the surface during bright daylight. Modern water clarity instruments use electronic **photocells** to measure the light received from a light source of known brightness after it passes through the water. These tools can be used at any depth, day or night.

15-7
The Secchi disk is used to determine surface water clarity.

A **photocell** is similar to a camera's light meter. It measures the brightness of the received light, and a small computer compares that value to the brightness of the known light to calculate water clarity.

OCEAN EXPLORATION   **363**

One of the many kinds of plankton nets

BIOLOGIC SAMPLES

Marine biologists want to know what kinds of organisms live in the ocean's depths. Many of these organisms are *plankton*. They are important because they are food for larger organisms. Marine biologists sample them with a *plankton net*. This device is a cone-shaped net made of a strong, fine-mesh material stiffened by a metal ring at its large end. A collection jar is clamped to the narrow end of the net. It is pulled through the water as the ship moves, and the water flowing through the net sweeps small, floating (*pelagic*) organisms into the jar. Weights clamped to the towline keep the net at a selected depth. When the net is hauled in, the bottle's contents are poured out into sample jars for examination.

Marine biologists are also interested in bottom-dwelling (*benthic*) animals. They collect them using a *grab sampler*. This is basically a metal box equipped with a spring-loaded trapdoor or clamshell shutters and attached to the end of a cable. After the box is lowered to the bottom, a messenger weight trips the doors shut, and it scoops up mud, plants, and small animals trapped inside. Then the sampler is hauled back to the surface. The marine biologist empties the sampler into a tray so that the organisms can be identified. All organisms are preserved for later study.

SEDIMENT CORES

Marine geologists frequently need to sample bottom sediments at various places. The sequence of sediments indicates something of the kinds and rates of sediment deposition at that location. To obtain a sample, a *sediment corer* is used. This instrument looks like a long steel pipe with heavy weights at one end (see Figure 15-9). Ocean-

15-8
A bottom sampler called an *Ekman grab sampler*

ographers lower it over the side of the ship, suspended by the weighted end on a cable. When released, gravity pulls it into the mud like a lawn dart. After it is pulled up, the corer holds a long plug of sediment that shows the different layers. The longest soft sediment cores extracted are nearly 59 m long! Biologists also examine sediment core samples for marine organism shells and even pollen grains.

15.7 TECHNOLOGY AND OCEANOGRAPHY

Most of the devices we just discussed are simple. They are lowered over the ship's side on a cable. They directly touch whatever they measure. They are tried and true, and we've been using ones like them since the earliest days of oceanography. But they are limited in what they can do. Advances in technology in the last half century have opened up a whole new world of data collection.

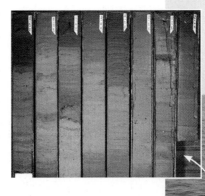

15-9
Oceanographers lowering a gravity sediment corer into the water (above). A set of core samples are shown on the left.

LIFE CONNECTION: A FARM ON YOUR ARMS

Imagine that you are really busy working and don't want to walk to the kitchen to make a sandwich. Instead, you decide to graze on a crop of nutritious food growing right on your own arms! Some zoologists suspect that a crab discovered on a 2005 deep-sea dive does just that.

Since the discovery of deep-sea hydrothermal vents in the 1970s, scientists have been intrigued by how life can survive in such an environment. An international team, using the submersible *Alvin* (DSV-2), explored portions of the Pacific-Antarctic Ridge in March 2005. They spent much of their time at a depth of 2200 m, south of Easter Island. Among their discoveries was a 15 cm long-crab whose walking legs and pincers were covered with dense mats of hairlike bristles. This species was nicknamed the "Yeti crab" after the legendary abominable snowman of the Himalayas.

Although scientists aboard the Alvin did observe the crabs feeding on mollusks damaged by the submersible, they believe that its hairy limbs are another source of food. Laboratory scientists examining the single specimen brought to the surface made an interesting discovery. The mats of bristles were covered with even finer bacterial filaments. These particular bacteria are able to produce food in the absence of sunlight. They do so by capturing sulfur atoms dissolved in the hot water from hydrothermal vents by a process similar to photosynthesis. In this case, however, no light is needed.

While only one crab was brought up for examination, scientists observed several live Yeti crabs in their natural habitat. Some were scrambling along pillow lavas near the vents. Others were located right beside the vents, gently waving their hairy limbs in the hot, cloudy water spewing from within the earth. This sulfur-rich water was the source of energy for the bacteria covering the crabs. This behavior and the unusual growth on their limbs led scientists to believe that the crabs can snack on these nutrient-rich bacteria. About 70 species of deep-sea crabs have been discovered, and a few others carry on this process. None, however, have such a thick growth of bristles.

Yeti crabs are just one of many amazing discoveries from a watery world less familiar to us than the surface of the moon. Our advances in technology have opened just a small window into this alien environment. Even here we see that God has designed, created, and continues to sustain His creatures (Ps. 104:25–30).

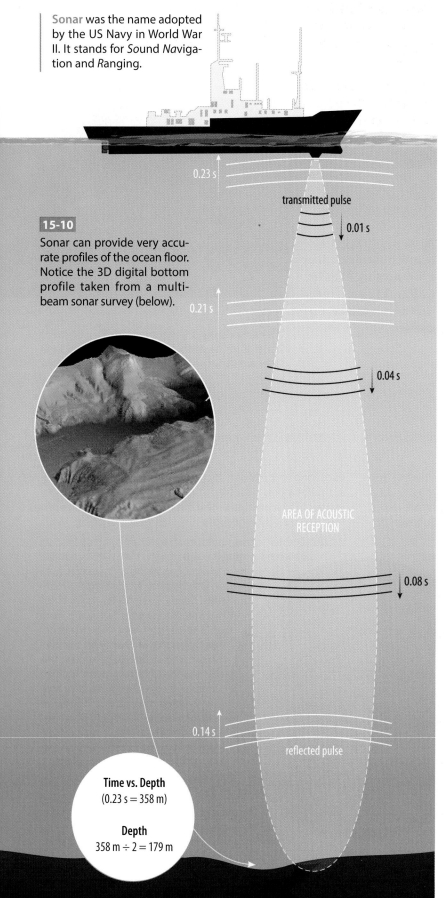

Sonar was the name adopted by the US Navy in World War II. It stands for *So*und *Na*vigation and *R*anging.

15-10 Sonar can provide very accurate profiles of the ocean floor. Notice the 3D digital bottom profile taken from a multi-beam sonar survey (below).

Time vs. Depth
(0.23 s = 358 m)

Depth
358 m ÷ 2 = 179 m

SONAR

The most important instrument invented to aid our understanding of the ocean basins is sonar. **Sonar** is a form of artificial echolocation. It was originally developed in World War I to hunt for enemy submarines. Just like a porpoise or a bat, the device sends out a pulse of sound that reflects off the bottom or off an object. A special underwater microphone detects the returning pulse. A receiver calculates and displays the object's distance by how long the round trip takes (see Figure 15-10). Special sonars called *bottom profilers* look straight down below the ship. They can quickly and continuously measure water depths along the ship's track. No need to use a lead line or even to stop the ship! Powerful bottom profilers can actually penetrate the upper sediment strata of the ocean bottom, making it easier for geologists to study these structures. The newest multi-beam sonars can produce detailed 3D digital maps of the ocean bottom.

WIDE-AREA DATA COLLECTION

Individual ships take time to move between ocean survey positions. Ocean conditions change during these delays. For studies that try to show how the ocean is at any one time, time delays produce errors. These studies need the same kinds of data from hundreds or thousands of sample points all over the ocean at the same time. No single ship or even a fleet of ships can provide such data coverage. Until just a few decades ago, accurate wide-area studies were not possible. We now have the devices and instruments to support this method of data collection.

Research voyages are expensive! Is there a way to collect ocean data without a ship and crew? When studies need frequent data collection over large areas, the most economical solution is to use automatic *buoys*. Buoys are large, sturdy, disc-shaped floats on which a variety of instruments is mounted. Usually they monitor weather, but they can also measure sea temperature, salinity, wave height, and even the currents that flow around them. They transmit measured data to central data processing stations by radio.

Most buoys are anchored to the bottom, so oceanographers know where the data is coming from. Other buoys drift with the currents and transmit their positions along with their data. All include radar reflectors and lights so that ships can avoid running into them.

Satellite technology has revolutionized our ability to monitor large areas of the ocean at the same time. Satellites analyze the ocean's surface in many ways. They use visual cameras, heat-sensitive infrared detectors, and even radar. Infrared surveys help oceanographers track warm and cold currents and measure the overall temperature of the sea surface. Radars work like sonar but use radio waves instead of sound. Shortwave radars can measure the ocean surface height to within a few centimeters. These measurements help us determine overall changes in sea level. But satellite radars can also detect bulges and dips in the ocean's surface that follow bottom features, like seamounts, canyons, and tectonic rifts. One satellite mission determines *sea surface salinity (SSS)* by measuring the brightness of radar reflections from the ocean's surface. Satellites are also a key part of the global positioning system (GPS). This system makes locating data far easier and more accurate than before the system was built.

All of these cool instruments sound impressive, don't they? But even with all these tools, we still know less about some parts of the ocean than we do about the far side of the moon.

15-11
A modern weather buoy operated by NOAA's National Data Buoy Center

15-12
Special satellites, such as NASA's Topex Poseidon, can map the ocean's surface height accurately to within a few centimeters.

15B REVIEW QUESTIONS

1. What are the two main approaches oceanographers take to studying the ocean?
2. List five things oceanographers may measure at a sample point.
3. What method have oceanographers used in the past to obtain a sounding? What method is used today?
4. Why would an oceanographer need to test for water clarity?
5. Why would a geologist want to use a sediment corer?
6. What is the most important advantage of sonar bottom profiling over the older method of lead-line soundings?
7. Which approach to oceanography (see Question 1) have modern data-collecting instruments, such as sonar, automatic buoys, and satellites, aided the most?
8. (True or False) Modern materials and digital electronics haven't changed the basic way that seawater samples are obtained from deep water.

15C Objectives

After completing this section, you will be able to

- » summarize the history of deep-sea diving.
- » explain the progress in ocean exploration vehicles.
- » evaluate the risks and benefits of ocean exploration.

15-13 Humans weren't designed to live underwater without special protections. Even this expert free diver will run out of air.

15-14 According to historical records, Alexander the Great was the first man to descend into the depths for the sole purpose of observation.

15C

ENTERING AN ALIEN WORLD

Why is it so difficult to study the ocean?

15.8 EARLY VISITS TO THE DEPTHS

Even with its challenges, the unobservable, mysterious ocean depths call to human curiosity. Through time, our ingenuity has solved one problem after another that prevented us from seeing into this alien underwater world. Today, though we can't freely live there, we can at least visit it, work in it, and use it. The biggest barrier to all these activities is providing air to breathe.

Of course, people knew that they had to take air with them to work underwater. This led to the invention of **diving bells**, which were large, bell-shaped enclosures. They were lowered to the sea floor from a ship and workers could exit through the open bottom to do work on the seabed. We do not know much about the first diving bell. Records indicate that in 332 BC Alexander the Great had a large barrel constructed out of glass from which he supposedly observed the bottom of the Mediterranean Sea. Many centuries later, in 1535, the Italian engineer Guglielmo de Lorena built the first practical, working diving bell. Made like a wooden barrel with a window, and weighted with bags of shot, it rested on the diver's shoulders. The only air available to the diver was trapped in the barrel as he descended to the bottom.

Over the following three centuries, scientists worked to increase both the safety and the length of working time underwater. In 1690, Sir Edmund Halley designed and built a heavy iron diving bell that could hold several divers. He invented a way to replenish the air in the diving bell by sending down barrels of fresh air from the water surface. This extended the time that a diver could work underwater. In 1789, John Smeaton improved this idea by using a pump to replenish the air continuously through a hose from the surface. Technology easily progressed from this to the diving helmet, which deep-sea divers used until recent times.

15-15 A century ago, divers wore canvas suits topped with a bronze helmet with view ports. They were supplied air by a hose connected to a tending vessel on the surface. This system was very difficult to use and prone to accidents. Many "hard-hat" divers died or were permanently paralyzed.

Even with such technology, helmet diving was extremely dangerous. Maintaining air pressure equal to sea pressure at the diver's depth was vital to breathing and moving. If there was too much pressure, a diver's suit became too stiff to move. If there was too little pressure, the diver couldn't expand his lungs against sea pressure to breathe and he would suffocate. If there was even less air pressure, the diver could be crushed by sea pressure. Hoses and safety ropes could get tangled or cut in sunken wreckage, trapping and drowning the diver.

In 1943, Frenchmen Émile Gagnan and Jacques-Yves Cousteau designed the first practical Self-Contained Underwater Breathing Apparatus (**scuba**). The scuba system included a tank of compressed air strapped to the back of the diver and an air regulator that allowed the diver to breathe. The diver was free as a fish! Even though scuba equipment greatly increases a swimmer's mobility and range of movement, the tank can hold only so much air. Recreational divers using a single tank can stay underwater less than an hour. Also, there are definite medical limits to how deep any diver can safely go. See the box on the next page to find out more about diving technology.

Today, divers for navies, oil companies, and other industries use gas mixtures containing helium, hydrogen, and oxygen for extremely deep dives. **Mixed-gas diving** is necessary because the usual gas mixture in the atmosphere is poisonous to divers under the extreme pressures at these depths. Divers have successfully worked in the ocean at depths greater than 530 m. Tests show that divers can work as deep as 700 m for short periods. But even with special gas mixtures, divers experience mental confusion, trembling, and other physical and mental problems at these extreme depths. Will people ever be able to live safely for longer periods of time in the ocean?

Jacques-Yves Cousteau (1910–97) was a world-renowned ocean explorer and engineer from France. He is best known for co-inventing the self-contained underwater breathing apparatus (scuba) and for his television specials that brought the world of oceanography into the homes of millions.

15-16
Scuba equipment was made practical by Jacques Cousteau (left) and Émile Gagnan.

15-17
A modern mixed-gas deep-water diver. One of the three cylinders holds pure oxygen; the others hold a mixture of helium, hydrogen, or even nitrogen. A microcomputer changes the amount of oxygen in the mixture as the diver changes depth.

DIVING TECHNOLOGY

Have you ever wished to join the fish and explore the ocean? It's likely that early humans had this desire. But just a few minutes in the water demonstrates that man wasn't created to live in the ocean. Unlike a fish's gills, human lungs can't extract oxygen from the water. And unlike aquatic mammals such as whales and dolphins, humans can't hold their breath for very long. Even human eyes don't work well in water. If you've ever opened your eyes in a swimming pool, you know this is true. God didn't create humans to live in the water, but He did give us the ability to solve problems. Over several thousand years, humans have learned how to visit the ocean.

The most important thing a human needs underwater is a good supply of air. If you've ever used a snorkel, you know that it lets you stay underwater for long periods of time. But snorkel tubes aren't very long for a reason. Go more than a few feet below the surface, and the water's pressure on your lungs is so great that your muscles can't expand your lungs, no matter how hard you try.

In the early 1800s, scientists tried pumping air down to a helmet that the diver wore over his head. It worked, but it was still easy to drown. It wasn't until 1837 that Augustus Siebe thought to enclose the diver in a watertight canvas suit attached to the helmet. Diving with Siebe's suit came to be called *hard-hat diving*. It was practical, but it required a lot of training and had many risks. The diver depended on an air hose to the surface and had to view the ocean through small glass ports in his helmet. Rather than swim, he had to walk over the bottom with heavy lead shoes.

But improved diving suits weren't the whole solution. As diving equipment grew more sophisticated, divers realized that there were medical problems to solve. Dive too deep and strange things begin to happen. The nitrogen in compressed air makes the diver become drunk (a condition called *nitrogen narcosis*). Oxygen becomes toxic and causes deadly convulsions. If the diver stays down too long and comes up too fast, compressed nitrogen dissolved in his blood turns into bubbles. The same thing happens to carbon dioxide in a soda after popping the top on the can. This painful and sometimes fatal condition is called *decompression sickness (DCS)*, or the *bends*.

In 1865, two French engineers created the first scuba system. It included a small air tank with an automatic breathing regulator. Called the *Aerophore* (see etching), it was the ancestor of the modern scuba rig. However, its air tank couldn't handle high pressures, so the diver had a very small amount of air.

In 1943, Émile Gagnan and Jacques Cousteau perfected the modern scuba system. Using a high-pressure air tank with an automatic *demand regulator*, the diver was completely untethered. A demand regulator provides air to the diver only when he breathes in. The regulator ensures that the air pressure is equal to sea pressure so that the diver can expand his lungs. Swimming around with fins on his feet and a plate-glass mask over his face, he could truly be a part of the ocean world. Over the next four decades, Cousteau went on to make underwater films. His films and television documentaries brought the wonders of the ocean to ordinary people.

But the increased mobility of the divers made serious medical accidents even more likely. Scientists and doctors studied the many medical problems and gradually developed solutions. Deep dives have to be kept short. Instead of ascending quickly, a diver comes up in stages so that his body can go through *decompression*. This allows the gases in his blood to be gradually exhaled instead of creating bubbles. Carefully calculated diving tables govern the stops. With these tools, a modern diver can avoid decompression sickness.

Divers avoid the other medical problems by breathing special air mixtures. On a deep dive, a diver might use a mixed gas called *trimix*, which contains oxygen, nitrogen, and helium. Because it contains less nitrogen and less oxygen than normal air, trimix helps the diver avoid nitrogen narcosis and oxygen poisoning.

But divers can still face problems. Smart divers never dive alone. The buddy system helps protect a diver from the many unexpected things that can happen underwater.

Thanks to many years of scientific research and the work of scientists and divers, diving today is both safe and affordable. Perhaps someday you too might want to become a certified diver.

15.9 UNDERWATER HABITATS

Many fictional stories have explored the idea of living underwater. To accomplish its purpose, an **underwater habitat** should

» protect the occupants from cold and water;
» provide a reliable supply of air, heat, power, and light;
» have space and facilities for people to carry out normal daily activities, such as working, sleeping, and taking care of personal hygiene;
» have space to store fresh water and food and allow for the preparation and eating of meals;
» provide access to and from the ocean.

This last feature can greatly affect habitat design and function. If access to the ocean is direct, then the air pressure inside the habitat has to be the same as the outside sea pressure at all times. Divers enter and exit the habitat through a "moon pool"—an opening in the floor where air meets water. This means that special gas mixtures need to be used if the habitat is at a deep depth, but divers can come and go as they please. Also, the habitat doesn't need to be built of strong, heavily framed metals, since pressures inside and out are the same.

If the interior of the habitat is at normal atmospheric pressure, then there are some significant differences. The habitat has to withstand the outside sea pressure. This requires thick metal walls and framing. Divers must enter and exit through a small separate room—called a *lock*—with pressure-proof doors. And divers must spend time in the lock decompressing when returning to avoid medical problems (see the box on the facing page). You now begin to understand that there are some difficult problems when it comes to living on the ocean bottom.

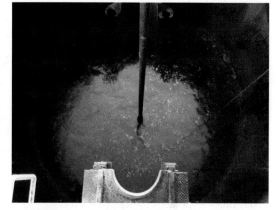

15-18
A moon pool allows direct access to the ocean from inside a habitat.

SEA HABITATS

In the 1960s, Jacques Cousteau and his team built a series of three habitats called *Conshelf*. Around the same time, the US Navy also built several experimental undersea laboratories called Sealab and Tektite to study deep-sea mixed-gas diving and living in isolation for long periods. NASA utilized the Tektite series of habitats for its space program.

More recently, smaller research facilities such as Aquarius and MarineLab have been used for a variety of undersea and space research. The most ambitious project is the Atlantica Expeditions, which its developer intends to be a permanent, growing, underwater community. Even so, many technological hurdles remain to be solved before people will live comfortably in the depths of the sea.

Marine scientists inside and outside the Aquarius research habitat

maximum human diving depth
700 m

15-19 Scuba divers can reach only a tiny fraction of the ocean's depths.

average ocean depth
3800 m

Bathysphere

15-20 William Beebe and Otis Barton with the *Bathysphere*

deepest trench
11,000 m

UNDERWATER VEHICLES

15.10 MANNED SUBMERSIBLES

In spite of all the advances in diving technology, human divers could still explore only a very tiny part of the ocean. Because of our physical limitations, 98% of the ocean volume remained out of reach for direct physical access, even with scuba or mixed-gas diving gear.

Overcoming this challenge required a special vehicle to protect the explorer from the alien environment of the ocean depths. The vehicle had to withstand sea pressure and provide a breathable atmosphere. To allow observation of the undersea world, it also had to include pressure-proof windows. Builders of early diving bells tried to solve some of these problems, but it wasn't until the twentieth century that technology allowed people to make the first significant trips into the ocean abyss.

THE BATHYSPHERE

For more than fifty years after the HMS *Challenger*, scientists were content to follow its model. Biologists sampled the life on deep-sea bottoms by dragging dredges through communities of organisms, then hauling them to the surface. After analyzing the lifeless and damaged organisms of some 1500 sampling nets, William Beebe thought of a better way. What if scientists could go down into the ocean and observe living things firsthand in their natural habitat? In the late 1920s, Beebe and engineer Otis Barton constructed a 1.5 m diameter steel ball, 3 cm thick, with three observation windows. They named it the ***Bathysphere***.

In 1930, the *Bathysphere*, with Beebe and Barton inside, was lowered on a steel cable from a crane barge into the dark waters near Nonesuch Island in the Bahamas. The two men completed a series of descents, the last being a 923 m dive. Beebe believed that each trip, where living things could be observed in their natural habitat, yielded more information than all his sampling nets combined. This success began a new era in oceanography and in our understanding of the oceans.

THE BATHYSCAPHES

The *Bathysphere* had two disadvantages—it couldn't go anywhere but straight down and straight up, and it couldn't collect and return with samples. In the late 1930s, Auguste Piccard began developing a mobile deep-diving vehicle called a **bathyscaphe**. It was a cross between a submarine and a small underwater blimp. The crew compartment was a steel sphere with walls 7 in. thick. The sphere had one pressure-proof window and an access hatch. It contained equipment to keep the air breathable, as well as other instruments. The sphere was mounted underneath a large buoyancy tank containing 106,000 L of gasoline to counteract the weight of the heavy sphere. This method worked because gasoline is less dense than seawater, similar to the principle observed in a hot air balloon. The buoyancy tank also contained batteries for electricity and barrels full of steel shot. With these the crew could release small amounts of weight to adjust the vehicle's buoyancy. Small electric motors with propellers moved the bathyscaphe through the water. It had no connection to the surface.

After proving the success of the basic bathyscaphe concept, Piccard built another, much-improved model, called the **Trieste**. Launched in 1953, the US Navy eventually purchased the vessel. *Trieste* conducted a series of dives, the deepest of which was 10,916 m into the Challenger Deep in the Mariana Trench. As far as we know, this is the deepest point in any ocean. During a 1960 dive, the craft encountered the greatest pressure known to exist in the ocean—about 1156 atm. *Trieste* had several kinds of sampling devices, including robotic arms with hand-like grippers and water samplers. Oceanographers first accomplished deep-sea photography with pressure-proof cameras and lights on the *Trieste*.

After its record-breaking dive in 1960, the US Navy redesigned the float chamber and made many improvements to the *Trieste*. In 1964, it was replaced by the *Trieste II* (DSV-1). This vessel was more seaworthy and made many important discoveries.

Bathyscaphes revolutionized deep-sea research, but they still had some limitations. They were slow and clumsy to operate. Crew size was limited to two or three, and the crew compartment was cramped. But the most limiting feature was the electrical batteries, which allowed only short visits to the bottom. The descent to the bottom for *Trieste's* visit to Challenger Deep took nearly 5 hours, but the crew had only twenty minutes to explore before they had to return to the surface. In 2012, Australian film director James Cameron descended to Challenger Deep in the deep submergence vehicle *Deepsea Challenger*. He was the first person to make a solo descent and was able to spend three hours at Challenger Deep.

MODERN DEEP SUBMERGENCE VEHICLES

Deep submergence vehicle (DSV) design and capabilities evolved as engineers attempted to build submersibles that could go deeper for longer periods and with larger crews. Numerous research DSVs have operated since the *Trieste*. Only seven nations have built DSVs. China and Australia are the most recent members of this exclusive club.

Manned Deep-Sea Exploration

Scientists and engineers acknowledge that the technical aspects of keeping a human safe deep in the ocean are more difficult than keeping him safe in space. The most significant problem is keeping the crushing pressure of the ocean at bay. In view of potential hazards to humans, should Christians be excited about manned deep-sea exploration?

15-21
This was the final form of *Trieste II* before its retirement in 1980. The vehicle is on display at the Naval Undersea Museum at Keyport, WA.

In 1964 the US Navy also used *Trieste II* to investigate the wreckage of the USS *Thresher*, the first American nuclear submarine to sink (with loss of the entire crew).

The depth of Challenger Deep recorded by the *Trieste* crew in 1960 was based on the installed instrumentation. More accurate measurements made in 2009 indicate that the depth is actually 10,911 m.

15-22
Deepsea Challenger carried James Cameron to Challenger Deep in 2012.

15.11 UNMANNED SUBMERSIBLES

DSVs give people lots of freedom to explore the ocean bottom. Scientists can get the most information this way. But placing humans in alien environments is always dangerous and expensive. An occupied DSV must devote space, weight, and power to keep its crew safe.

This is why researchers started using unmanned robotic submersibles all over the world. These vehicles are much smaller and more compact since they don't need a pressure-proof crew compartment and human life support systems. This makes them much less expensive to build and operate.

The earliest of these platforms were basically frames of metal tubes shaped like sleds, with pressure-proof instruments, cameras, and lights mounted on them. They were towed behind a ship. Their tether cables fed live video images to the mother ship, as well as real-time ocean data from the bottom. But these units could only follow the ship or be lowered straight down.

ROVs

Modern tethered submersibles are **remotely operated vehicles (ROV)**. An ROV is a nimble undersea robot. A team of pilots and scientists control its movements through data and power cables from a mother ship on the surface. ROVs are similar to the robotic rovers NASA sent to Mars. They gather information and samples using instruments under the direct control of operators. Oil companies routinely use ROVs to survey deep-water seabeds and build and repair deep oil well structures.

ROVs are limited by the need to have people control them and by the length of the control cable. While these limitations are acceptable for complicated ocean engineering and exploratory work, they actually make routine sampling and large-area surveys of the ocean more difficult. The tether prevents long distance movement of the ROV. And using a crew of pilots for this kind of work is expensive. Why couldn't we just send a submersible off on its own to collect the data?

15-23

The remotely operated vehicle (ROV) RovBuilder

AUVs

That is exactly what **autonomous underwater vehicles (AUV)** do. Guided by sophisticated artificial intelligence computer programs, these vehicles are similar to military torpedoes in shape and movement. Efficient battery-powered propellers move them, and their computer programs guide them along a specific track. As they "fly" up and down through the water, they collect temperature, salinity, and other kinds of data at planned points on the journey. Others may run straight or zigzag courses as they map the bottom topography with sonar. They can send data back to the researchers via satellite, or they may store it until researchers recover the vehicle.

> **ROVs and the 2010 Oil Spill**
>
> In 2010, during the massive oil spill in the Gulf of Mexico from the Deepwater Horizon well, ROVs were vital for monitoring and containing the leak until it was finally plugged. Most countries doing deep-sea research today are relying more on ROVs because they are safer, more economical, and just as capable as manned submersibles.

AUVs are just one type of the many kinds of **unmanned underwater vehicles (UUV)** in use today. Some drift with the currents, taking measurements of temperature, salinity, current speed and direction, and so on. They can report their position and data by sound pulses, or they rise to the surface to transmit their data by radio. After transmitting data, they descend back to a preprogrammed depth to continue their survey. Others glide through the depths by gravity on "wings," taking samples and surveying using sonar. And yet others may rest on the bottom, photographing the activities of bottom-dwelling organisms until summoned to the surface.

Considering the vastness of the oceans and their importance to life on Earth, oceanography is vital to understanding the earth better. Studying and managing its resources are key to proper dominion.

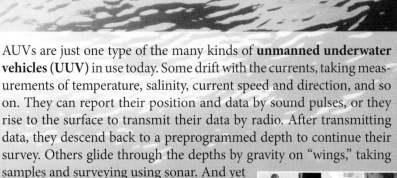

15-24
The *Jason* was one of the first practical ROVs (above). A US Navy sailor prepares to deploy the Bluefin autonomous underwater vehicle (AUV) (below).

15C REVIEW QUESTIONS

1. Describe the different ways that people tried to get air to underwater workers before the invention of scuba gear.

Key Terms

oceanographer	356
Challenger expedition	359
sounding	362
sonar	366
diving bell	368
scuba	369
mixed-gas diving	369
underwater habitat	371
Bathysphere	372
bathyscaphe	372
Trieste	373
deep submergence vehicle (DSV)	373
remotely operated vehicle (ROV)	374
autonomous underwater vehicle (AUV)	374
unmanned underwater vehicle (UUV)	375

15-25
La Chalupa is an undersea laboratory that protects occupants from the hazards of living and working underwater.

CHAPTER 15 REVIEW

CHAPTER SUMMARY

» The ocean covers the majority of the earth's surface and contains most of the life on Earth. Thus, oceanography is vital to pursuing proper dominion over the earth and its resources.

» Exploration of the oceans probably began within a few centuries after the Flood. Explorers discovered the shapes of unknown lands and the oceans separating them.

» During the Age of Discovery, explorers intended mainly to create trade routes and return wealth to their nations, rather than to scientifically study the oceans.

» The *Challenger* expedition (1872–76) was the first global deep-sea scientific voyage that focused largely on describing the ocean basins and marine life within the ocean depths.

» Oceanographers have two basic approaches to studying the physical ocean. Either they sample data at a single point from surface to bottom, or they try to look at wide areas of the ocean all at once. Sometimes their studies are combinations of these approaches.

» Basic ocean observations include water depth, seawater chemistry and salinity, temperature, clarity, and samples of marine life.

» More advanced ocean observations include sonar bottom profiling, ocean surface observations from satellites, and daily data reports from moored and drifting buoys.

» People have worked for thousands of years in the seas to obtain food, build structures, and salvage ships. But the working depths have been limited by the lack of air supply and by sea pressure.

» Advances in technology allowed people to overcome their natural limitations in the depths of the sea. First were diving bells, then helmet divers, and finally scuba and mixed-gas diving systems. The effects of sea pressure still limit how deep an unprotected diver can go.

» People have experimented with underwater habitats for research and recreation. Many of the same limitations with divers apply to living underwater, especially at great depths.

» Manned submersibles make it possible to visit the deepest places in the ocean, though only for short times.

» Manned deep submergence vehicles (DSVs) are giving way to unmanned underwater vehicles (UUVs) because they are cheaper to build and operate and human crews are not endangered.

REVIEW QUESTIONS

1. What evidence do we have that people probably began exploring the ocean very soon after the Flood?

2. Do the historical records, such as accounts of great feats of ocean exploration, tell us when some discovery was actually made for the first time? Explain.

3. What was the difference in the purposes for ocean exploration before the 1800s compared with later years?

4. According to historical records, which explorers made the following discoveries?
   a. Iceland
   b. Pacific Ocean
   c. ocean route to the East Indies
   d. ocean route around South America

5. List some qualities that all scientific instruments should have for data collection on the high seas.

6. What was one problem that early oceanographic ships like the HMS *Challenger* had when trying to measure the great depths of the ocean abyss?

7. Examine Figures 15-5 and 15-6. Explain why retrieving the cable using Niskin water samplers is much faster than using Nansen bottles.

8. After evaluating the method of taking a grab sample of ocean bottom organisms, identify the greatest drawback with using a grab sampler.

9. What are two important differences between an oceanographic buoy and a ship that make the buoy a more economical choice for taking daily sea and weather readings?

10. What was the main danger to workers using a diving bell?

11. How does scuba gear allow a diver to expand his lungs to breathe against the crushing pressure of the seawater?

12. What advantage would a deep-water habitat have for working on the sea bottom compared with divers swimming from a ship to the bottom and back?

13. What was the main reason that William Beebe wanted to build and use the *Bathysphere* in his work?

14. What was the most famous accomplishment of the DSV *Trieste*?

15. What would be the advantage of a nuclear-powered DSV, like the US Navy's NR-1, over a smaller battery-powered vehicle?

16. State one advantage and one disadvantage of an ROV.

True or False

17. The Bible tells us that Noah drew upon a long history of shipbuilding experience to construct the Ark according to the design God gave him.

18. Christopher Columbus was the first European to discover North America.

19. One of the main objectives of the *Challenger* expedition was to determine the depth and topography of the ocean basins.

20. Even with sonar bottom profilers, bottom-depth data can be slightly inaccurate because sound speed changes between the surface and the bottom as a result of seawater salinity and temperature variations.

21. For a helmet diver, maintaining correct air pressure inside his suit was the most important task for survival.

22. An underwater habitat with moon pool access to the ocean must be built extremely strong to resist outside sea pressure.

23. The main advantage of a manned DSV is the ability for human pilots and scientists to make on-the-spot decisions and observations that an AUV or even an ROV might miss.

15-26
Lake Sørvágsvatn on the Faroe Islands is a large lake located close to the ocean.

Location: 62.04 °N 7.22 °W

CHAPTER 16
SURFACE WATERS

16A STREAMS 379
16B LAKES AND PONDS 386

THREE GORGES DAM

The largest hydroelectric dam in the world is the Three Gorges Dam in China. It is a huge engineering project, damming a reservoir that is 660 km long. Its thirty-two main generators are capable of producing up to 3% of China's energy needs. It should be able to pay for itself in ten years. But there are hidden costs.

The Three Gorges Dam project has displaced 1.3 million people. It has increased landslides, flooded 1300 archaeological sites, and made several animal species nearly extinct. The dam itself is on a geologic fault. Some people are calling the dam an environmental catastrophe.

At the same time, the project has reduced flooding of the Yangtze River and erosion of its banks downstream. It will also help reduce air pollution as the coal-fired power plants it replaces are shut down. The dam could vastly improve the lives of millions of people.

So, is it worth it? Is this dam an example of good and wise dominion?

16A

STREAMS

What are the characteristics of the different types of streams?

16.1 SURFACE WATERS AND DOMINION

God has placed flowing water on the earth's surface for us to use. Flowing water has kinetic and potential energy that we can harness to produce electricity. We get *hydroelectric energy* (or *hydropower*) from water flowing through power dams built across rivers. Right now, hydropower supplies only 19% of the world's energy needs. But it accounts for 63% of the world's renewable energy. Hydropower doesn't pollute, and dams help control flooding in flood-prone rivers. Power-dam reservoirs store water for drinking and recreation.

Hydropower can be a good thing. But our efforts to exercise dominion will always have both positive and negative results in a fallen world. Part of exercising good and wise dominion is predicting harmful consequences of dominion and minimizing their effects. Though technology may seem to have great benefits, human life, health, and the ability to do essential work have first priority. This is especially important as scientists and engineers harness and manage our planet's surface waters.

16.2 STREAMS—FLOWING SURFACE WATERS

Flowing water that follows a distinct course over the land is a **stream**. Streams go by many names, depending on locale and culture. Some common names for streams, from smallest to largest, are *rivulet*, *brook*, *creek*, *stream*, *kill*, *branch* or *tributary*, *river*, and *waterway*. They usually flow continually, but in dry places they may be seasonal. They can be tiny trickles or broad, powerful rivers.

The water in streams can come from lots of different sources. Virtually all streams begin as fresh water. Rainwater and meltwater from snow or glaciers flow downhill under the influence of gravity. As you learned in Chapter 12, small trickles converge into larger streams, eroding streambeds and growing into rivers. Streams may also start as *springs* (see Chapter 17) when water flows out of the ground. Streams can start by flowing out of lakes. These are called *outlet streams*. Though streams begin as fresh water, they can become slightly salty as they pick up a load of sediment from the land they drain.

16A Objectives

After completing this section, you will be able to

» compare the different kinds of streams.

» sketch a stream from source to mouth and label its parts.

» analyze ways to wisely use streams.

16-1
The Three Gorges Dam

16.3 STREAM ANATOMY

Hydrologists, scientists who study Earth's water, examine several properties of streams. They study a stream's *elevation profile*, its *cross-section*, and its *drainage basin*. These three characteristics determine a stream's energy, the volume of water it carries, and how it affects the lands through which it flows. Let's look at each of these.

ELEVATION PROFILE

The **elevation profile** of a stream is a description of how the stream changes elevation from beginning to end. It defines its steepness and the energy of its water flow at any point along its route. The origin of a stream is its *headwaters* or **source**. The sources of most large rivers are usually high in mountainous areas. Near its source, a stream may be a mere trickle. Trickles join and grow to form a distinct stream, which flows quickly down the steep slopes. The steepness of a stream's channel is called its slope or **stream gradient**. The headwaters of a stream usually have a high gradient.

At the other end, most streams flow into another body of water. Rarely, they simply soak or flow into the ground. A stream's **mouth** is where it flows into an ocean, lake, pond, or another stream. For example, the mouth of the Amazon River is in the Atlantic Ocean. The mouth of the Jordan River is in the Dead Sea. The mouth of the Missouri River is in the Mississippi River. Most large rivers flow almost horizontally near their mouths. This means that they have a gradient of almost

16-2
High-gradient streams have steep rapids and waterfalls. They are often found in V-shaped valleys. Shown below is Lower Falls along the Yellowstone River.
Location: 44.718°N 110.496°W

A gradient is one tool scientists use to determine the way some measurable quantity changes over distance (e.g., see *geothermal gradient* in Chapter 8). A stream's gradient is the change in elevation of the streambed in meters for every meter the stream flows horizontally. The steeper the stream is, the larger its gradient.

16-3
Profile of a typical stream

16-4
Little River in Tennessee is a good example of a medium-gradient stream.

380 CHAPTER 16

zero. Since gravity drives flow, a stream can't flow any lower than its mouth. Thus, a stream's mouth is at its **base level**. The ultimate base level for most streams is sea level, though there are a few exceptions.

Old-earth geologists believe that a stream's base level controls its life span. In their model, all streams will erode their streambeds down to their base level, given enough time. Erosion is faster where their gradient is higher and slower where their gradient is lower. In an old-earth view, rivers can be hundreds of thousands or even millions of years old. For instance, geologists estimate that the Colorado River flowing through the Grand Canyon is around 5.4 million years old. But in a young-earth view of geology, no stream can be more than about 5500 years old.

A stream's gradient is controlled mainly by the soil and rock through which the stream flows. Soft soil or rock erodes quickly, while hard rock erodes slowly. A stream's slope can change from a low to a high gradient anywhere along its course, depending on the underlying rock and the rate of erosion. The best example of this is Niagara Falls. The Niagara River flows at a low gradient until it reaches the edge of the hard dolomite stratum over which it falls. It plunges to the lower riverbed made of softer sandstone. The falls erode the sandstone and shale lying under the dolomite riverbed. Gradually, the unsupported weight of the dolomite breaks the rock. As this happens, the upper riverbed erodes upstream.

STREAM CROSS-SECTION

If you could slice through a stream crosswise, you would be looking at a stream's **cross-section**. Streambed cross-sections can vary quite a bit. Narrow, V-shaped sections are typical of high-gradient, fast-flowing streams such as highland brooks, creeks, and mountain streams. They erode their streambeds downward faster than they erode their banks.

Broader, U-shaped channels are typical of relatively low-gradient streams, such as small and medium-sized rivers and their tributaries. These streams are generally slower and less turbulent, but they carry much more water. They deposit sediments in places where they slow down, such as on the inside of a bend. They erode their channel where currents are faster and more turbulent. Generally, erosion and deposition are about even.

For example, the Jordan River in Israel has a base level more than 415 m below sea level.

Maximum Age of Streams
Why can the age of modern streams be no older than about 5500 years?

Niagara Falls: 43.078°N 79.076°W

16-5
The Mississippi River is a low-gradient stream.
Location: 29.95°N 90.06°W

As a stream approaches its base level, downward erosion almost stops. To handle the volume of water from upstream, the river channel broadens out. The stream's cross-section becomes wide and relatively shallow. Flow and turbulence are very low. This makes the stream drop its sediments. If the base level is a large body of water, like a lake or the ocean, sediments can form a delta at the stream's mouth. You've probably heard of the Nile and the Mississippi River deltas. These areas are home to lots of wildlife and are very fertile.

As you can see, stream gradient and stream cross-section are closely related. These aspects of streams also control how they drain the land.

STREAM SYSTEMS AND DRAINAGE BASINS

Do you know what happens to rain after it falls on land? Some of it is absorbed by plants or the ground. Some evaporates back into the air. The rest enters streams, which transport excess water from the land's surface by flowing downhill. Smaller streams called **tributaries** (TRIB you TEHR eez) feed into larger streams. A stream and all of its tributaries make up a **stream system**. From above, a stream system looks like the veins in a leaf or the branches of a tree (see Figure 16-6).

A stream system can be classified by its complexity. Geologists identify the complexity of a stream system by assigning it an *order number*. Simple stream systems with no tributaries are *first-order streams*. As tributaries join the main stream, the order of the stream increases. A stream with at least one first-order tributary becomes a second-order stream. A stream with at least one second-order tributary is a third-order stream, and so on. The Amazon River is a twelfth-order stream.

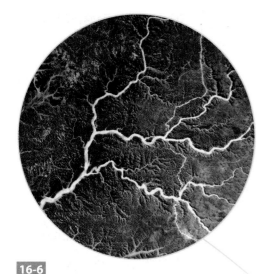

16-6
Many tributaries make up the drainage basin in the Tibesti Mountains in Chad.

Location: 20°N 17.5°E

Geologists call the tree-like pattern of a stream system a *dendritic pattern*. The word *dendritic* is taken from the Greek word *dendron*, which means "tree."

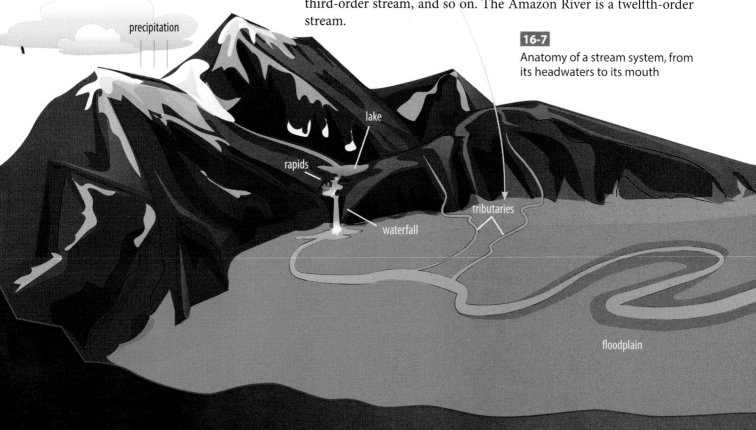

16-7
Anatomy of a stream system, from its headwaters to its mouth

The land that a stream drains is its **drainage basin**. The size of the drainage basin varies for different streams. For short rivers or tributaries, it may be only a few hundred square kilometers. But the Amazon River's drainage basin is more than 7.1 million km²—a third of South America!

A ridge of mountains, hills, or even just higher land separates one stream's drainage basin from another's. These form a basin **divide**. The mountain system that includes the Rocky Mountains forms the highest basin divide in North America, and is called the *Continental* or *Great Divide*. Land east of the Rockies drains into the Mississippi, Ohio, and other river systems. These streams flow into the Gulf of Mexico. Land west of the Rockies drains into the Pacific Ocean and the Gulf of California. The Appalachian and Smoky Mountains form an *Eastern Continental Divide*. Most of the land along the East Coast drains into the Atlantic Ocean. The Great Lakes and St. Lawrence River system drain the upper northeastern portion of the United States north of the *St. Lawrence Divide* (see Figure 16-9).

Triple Divide Peak: 48.573°N 113.517°W

16-8
Triple Divide Peak in Montana marks the meeting place of three drainage basins. The various slopes of the peak drain into the Pacific Ocean, the Atlantic Ocean, and Hudson Bay.

16-9
North American drainage basins and basin divides

16.4 STREAM FEATURES

Streams have features and structures important to those who use and manage them. Some features are simply interesting or attractive to people who love nature. But most of them reduce the usefulness of the stream or even cause problems. Hydrologists need to study these features to maximize their usefulness and prevent damage or loss of property.

High-gradient streams often contain *rapids*, or stretches where large rocks create turbulence. Rapids can be high up in the mountains near a stream's source or wherever the streambed quickly changes elevation. Rapids block boat and ship traffic.

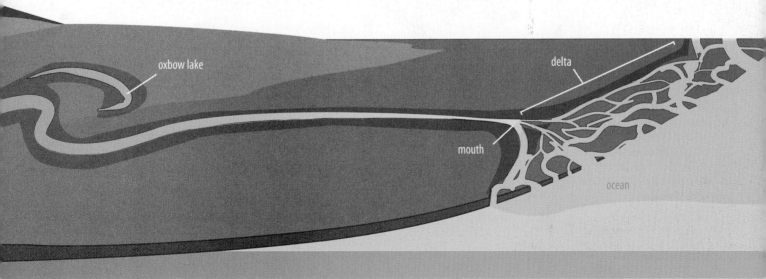

SURFACE WATERS 383

16-10
Looking down Angel Falls in Venezuela can be a dizzying experience. Angel Falls is the highest uninterrupted waterfall at 979 m tall.
Location: 5.97°N 62.536°W

meander (mee AN der): (L. maeander—from Maeandros, a winding river in Turkey)

16-11
An oxbow lake results when a meander is cut off by the stream. This lake is on the Amazon River in Brazil.
Location: 7.52°S 74.95°W

16-12
Some rivers routinely flood, creating fertile, level plain along the banks of the river. You can see the effect of a floodplain in the arid environment along the Nile River in Egypt.
Location: 26.1°N 32.75°E

A *waterfall* is an extreme example of rapids. Falls form where there is a very sudden change in streambed elevation. High falls can make a river impassable. The water flows over the edge of a streambed and drops vertically to the lower channel. This drop may be at the edge of a fault block or other tectonic elevation, or where the stream exits a glacial hanging valley. Falls can also form when streambed rocks erode unevenly; Niagara Falls is an example of this process. Rapids may result after a waterfall has eroded the lip of the streambed upstream. The tallest waterfall in the world is Angel Falls in Venezuela.

Streams can also produce *sand* or *mud bars*. Just like longshore bars at an ocean beach, deposition happens in places where water turbulence lowers enough for suspended sediments to settle out. Bars often form on the inside of bends in a stream, or where a stream broadens out as the bed becomes shallower.

When a stream's gradient approaches zero, the stream flows nearly horizontally. The stream no longer erodes downward. It uses most of its energy to slowly erode its bed sideways, much like a snake moves. This develops wide, looping bends in the stream called **meanders**. As erosion occurs on the outside of a meander, sediment deposition occurs on the inside of the bend. Over time, the meander moves sideways over the land surface. Meandering streams can cut broad river valleys between mountain ridges.

Occasionally, the bends at the ends of a meander loop will get so close together that the stream cuts through the narrow neck of land separating them. At first, the cutoff meander is just a side loop from the main stream surrounding an island. River sediment eventually fills in the ends of the cutoff meander, isolating it, and an *oxbow lake* results. Oxbow lakes and their sediment-filled remnants are common in flat river bottoms, such as the southern parts of the Mississippi River.

Rivers flood when more water enters the river than it can hold. Rivers that regularly flood create broad, fertile topsoil deposits called **floodplains**. These are excellent places to farm. The Nile River has perhaps the most famous floodplain (see Figure 16-12). Egyptian farmers have depended on its spring flooding to replenish the soil for thousands of years. However, flooding also makes it difficult for the farmers to build houses and barns in the floodplain.

All of these stream features create both opportunities for and barriers to using streams for our benefit, which is what biblical dominion is all about. Rapids and falls are permanent obstacles to boat and ship traffic along a stream. Building a system of locks and canals bypasses them. The Three Gorges Dam project includes a ship lock system that allows ships to reach the upper section of the Yangtze River (see Figure 16-13).

16-13
The Three Gorges Dam in China. Note the ship locks and canals to the north of the dam (orange box).

Location: 30.82°N 111.00°E

River bars are shipping obstacles that are always shifting with seasons or storms. Special navigators called *river pilots* help ship captains safely travel through these parts of rivers.

Meanders unnecessarily lengthen ship routes and can make certain stretches of rivers impassable for long barges or ships. Engineers will often "help" the process of meander cutoff by purposely digging an artificial canal to connect the two sides of a meander. In large rivers, this is safer and can greatly shorten the distance that a cargo ship or barge must travel to its destination.

Shipping isn't the only way to use streams. Local power companies often build small hydroelectric dams across the lower ends of rapids because the large-stream gradient helps create a deep, short reservoir behind the dam. The location for the Three Gorges Dam in China was selected because of the Yangtze River's profile at this point.

Where frequent flooding is a problem, engineers build high embankments or walls on both sides of a river channel. These *levees* keep the river from flooding the land. They are designed to keep it in its channel even when its waters are high. Then farmers can live and grow crops in the rich floodplains without having to be concerned about flooding. Again we see how solving one problem sometimes creates another. Without regular flooding, the soils in a floodplain aren't replenished, so farmers eventually have to use other methods to fertilize their land.

16.5 STREAMS AND SEASONS

A stream that flows year-round is a continuous or **perennial stream**. These streams get enough water from their sources to flow all the time. But not all streams are perennial. Many flow only after heavy rainstorms, during snowmelt, or during wet seasons. These are called **intermittent streams**. Topographic and geographic maps show intermittent streambeds as dotted blue lines. When dried up, these streams are simply empty channels that sometimes have puddles or moist, muddy spots in their beds.

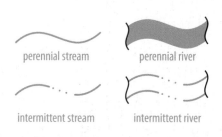

16-14
Topographic map symbols for small streams and rivers

Names for intermittent streambeds vary. In American deserts, the channel of a seasonally dry streambed is a *wash*. A similar feature in Arabian Africa may be called a *wadi*. In Spanish-speaking countries, an intermittently flowing stream is sometimes called an *arroyo*.

Intermittent streams in dry places such as the Sahara Desert often lie on top of streams that flow underground. They form well-watered green oases and sometimes even a water source for a herd of thirsty elephants or dusty desert travelers. However, many desert streams flow only after rare heavy rain storms. Their beds can be the channels for *flash floods*. People can die when they are suddenly caught in a flash flood while crossing a wash or wadi.

16A REVIEW QUESTIONS

1. Where is the largest power dam in the world? What is its source of water?
2. Identify three sources of water that can become a stream.
3. Sketch a typical stream from its source to its mouth. Label its source, mouth, and base level, and describe the stream gradient at three points along its length.
4. Which of the three stream cross-sections shown below is associated with a high-gradient stream?

5. What geographical feature describes the area drained by a stream system?
6. Which stream features make a stream impassable by large boats and ships? How do we sometimes solve this kind of a problem?
7. What is the main drawback to building levees along rivers that flood regularly?
8. Why are some streams intermittent?
9. (True or False) The Columbia River between Washington and Oregon shares the same drainage basin as the Ohio River.

16B

LAKES AND PONDS

How do lakes change over time?

16B Objectives

After completing this section, you will be able to

» relate a lake's chemical properties and anatomy to its geologic setting and elevation.
» categorize lakes by their properties.
» analyze different views of the origins of lakes.
» summarize the typical life phases of a lake.

Black Sea: 44°N 34°E

Caspian Sea: 42°N 51°E

Lake Superior: 48°N 88°W

16.6 WHAT IS A LAKE?

A **lake** is any isolated body of water that does not freely share water with the ocean. Just as there is no clear distinction between mountains and hills, there is no clear rule to classify seas, lakes, and ponds. Ponds are just smallish lakes. Landlocked seas are very large lakes. Their names are what people call them! Naming a landlocked body of water a lake or sea often depends on local historical tradition. In a few cases, bodies of salt water that meet the definition of a lake are named seas, like the **Black Sea** in Eastern Europe. Lakes range in size from small ponds, which can be less than 4000 m² in area, to the **Caspian Sea**, the world's largest lake with a surface area of 371,000 km². **Lake Superior** in North America is the world's largest

CASE STUDY: THE INFLUENCE OF RIVERS ON US HISTORY

In 1803, President Thomas Jefferson purchased 828,000 square miles from France. He quickly commissioned Captain Merriweather Lewis and Lieutenant William Clark to explore the new territory. Over the next two and a half years, they mapped routes through this rugged terrain.

So, how important were rivers in US history? Glance at a modern map of capitals and major cities and you will notice that many of them are located along the coasts and near rivers.

The first British settlement in the Americas was located at the mouth of the James River. The settlers chose that location for a number of reasons. It provided access to resupply from the ocean. The river provided a source of fresh water. The river also allowed easy access to the interior for exploration and trade.

If you follow the rivers, you'll observe the development of the American colonies. Rivers provided natural routes to the interior. The Northwest Passage was one of the most sought-after geographic features to connect the Atlantic and Pacific Oceans. Settlers often built towns along rivers for reasons similar to those of the Jamestown settlers. These locations also provided access to power for grain and lumber mills.

The Mississippi-Missouri River system, the fourth longest in the world, was key to early American history. Much like our highways today, the Mississippi was a primary route for travel and trade in the United States.

Questions to Consider

1. On the basis of what you have learned in this chapter, would you expect the Mississippi to have a high or a low gradient? Explain.
2. Would you expect to find rapids on the Mississippi? Explain.
3. What feature would you expect to find where the Mississippi enters the Gulf of Mexico?
4. Many of the early colonists were Christians seeking religious freedom. How would rivers factor into their efforts to fulfill the Great Commission?

---

freshwater lake in area (82,100 km²). Just like the oceans, lakes have depth too. Their volumes are important, because they store so much of the world's surface fresh water. **Lake Baikal** in southern Russia holds about 23,000 km³, nearly 20% of all the liquid surface fresh water in the world!

Most lakes have at least one *inlet stream* flowing into them, and at least one *outlet stream* flowing out. The water flow through a lake can hardly be noticed because the flow rates of its inlet and outlet streams are so small compared with the size of the lake. Precipitation and underground springs also feed lakes. The main source of water for a given lake may be one or more of these, depending on geology and climate. Water from lakes located above sea level may go all the way to the ocean by way of their outlet streams. Lakes below sea level can't drain to the ocean. Their outlet stream, if they have one, is trapped by the basin divide.

Lake Baikal: 54°N 109°E

Though lakes don't *visibly* flow, they actually *do* flow at the same rate as their inlet and outlet streams.

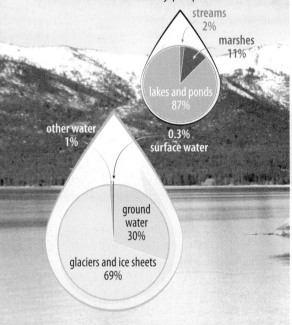

16-15 Surface water includes only a small portion of the earth's fresh water, but this water is the most accessible for use by people.

Bi Origin of Lakes

Respond to the following statement:

"After the Flood, most lakes re-formed in their pre-Flood locations."

If lakes are higher than sea level, they are usually fresh. That means that they have very little salt or dissolved minerals. Nearly all lakes are freshwater, though there are a few notable salty lakes. Lakes are an important focus of dominion in nature because they retain about 87% of the *liquid* fresh water on the earth's surface. Streams carry only about 2%, and wetlands make up the remaining 11%.

Lakes greatly affect their local environment. They provide a ready supply of water for plants and animals. Large lakes can modify the weather, producing cooler summers and warmer winters. They can supply water vapor for precipitation as long as they remain unfrozen. Historically, people have used large lakes as economical routes for transportation. Even today, large ships move cargo across the Great Lakes, connecting to the Atlantic Ocean through a series of locks and canals and the St. Lawrence Seaway. For these and many other reasons, we need to study lakes to know how to best use them.

16.7 ORIGIN OF LAKES

Where do lakes come from? Well, that depends on your worldview. According to an old-earth view of the planet's geology, lakes have always been a feature of the earth's surface. For at least the past 4 billion years, as tectonic plates formed supercontinents, there has always been some land surface above sea level that could have held lakes. Old-earth geologists suggest that most of the present-day continents have not been underwater for at least 300 million years.

Geologists think that most lakes are relatively young, within their view of Earth's history. The basins of a few large lakes around the world seem to be directly linked to great tectonic events. For instance, Lake Baikal fills a huge rift valley (see Chapter 7). So does Lake Tanganyika in Africa. Secular geologists believe that Lake Baikal is the oldest lake in the world, forming about 25 million years ago. Other tectonic lakes include flooded volcanic calderas. Crater Lake in Oregon is one of the largest of these.

A young-earth geologist will make the claim that no tectonic lake is older than about 5500 years because of the Flood. Recall that this event reshaped the surface of the whole earth through widespread tectonic action. It is highly unlikely that any lake in existence today still occupies its pre-Flood basin.

Glaciated regions contain the majority of other lakes in the world. These lakes and ponds number in the tens of thousands. Glacial lakes include tarns and kettle lakes (see Figure 16-16), and many ponds in small depressions formed by permafrost. Old-earth geologists believe that the last ice age ended around 10,000 years ago. They suggest that continental ice sheets shrank to their present sizes 8000 years ago. This means that most lakes in glacial features would be no older than this time span. Most are younger. Young-earth geologists believe that a single ice age ended around 800 years after the Flood, so these lakes could not be older than about 4500–5000 years. No matter whose story of the earth you believe, *most lakes formed in the recent geologic past.*

Where do lakes form? Well, a lake or pond can form anywhere there is a low spot in the terrain. Natural lakes formed in basins between mountain ranges, in low spots in broad, flat land, such as glacial outwash plains, and even in *sinkholes* caused by collapsed underground limestone cave systems (see Chapter 17). Many lakes formed in unusual places or under unique circumstances. People make lakes and ponds too. Power dam reservoirs are among the largest man-made lakes. Farmers often scoop out land or dam streams on their property to make ponds for watering their animals or for recreation.

16-16
Kettle lakes form in large depressions left behind when a glacier recedes.

16-17

The very unusual Dead Sea in Israel is located 430 m below sea level, making its shores the lowest land elevation on the earth. Its water is also almost ten times saltier than the oceans. Notice salt crystals forming on the rocks in this picture.

Location: 31.5°N 35.5°E

Fi Great Salt Lake: 41.2°N 112.7°W
Don Juan Pond: 77.56°S 161.28°E
Badwater Basin: 36.2°N 116.8°W

I **Isa Lake**

Isa Lake in Yellowstone National Park is very unusual. It straddles the Continental Divide. It has an outlet on its east end that drains its water to the Pacific Ocean—to the west! It also has a western outlet which drains water to the Gulf of Mexico, east and south of the lake!

16.8 UNUSUAL LAKES

There are some fascinating lakes around the world. Some are beautiful; some are colorful; some are just plain weird! Many unique plant and animal species call these lakes home. Let's check out a few of them.

Along the US eastern coastal plain and in the Midwest, mysterious small to medium-sized oval-shaped depressions contain water or marshy areas. They have also been found in many other places around the world. Nearly all of these receive their water only from rain or underground seepage. In the United States, they are called *Carolina bays*, *Maryland basins*, or *Grady ponds*. Lake Waccamaw in North Carolina is the largest of these water-filled depressions. It is nearly 8.4 km long, but averages only 2.3 m deep. There are more than half a million Carolina bays in the United States alone! See the box on page 392 for more information about these strange wet depressions.

Some unusual lakes are very salty. These lakes have inlets but no outlets. Dissolved minerals wash into the lakes and get concentrated as the water evaporates. These *salt lakes* form at the ultimate base level for an isolated drainage basin. Can you think of examples of lakes like these? Perhaps the Great Salt Lake in Utah comes to mind, nearly eight times as salty as seawater. Another one is the Dead Sea between Israel and Jordan. It occupies the lowest elevation of land on any continent. The saltiest lake in the world is Don Juan Pond in Antarctica. It is eighteen times saltier than the oceans! This is why it can remain liquid at a chilly −50 °C. The Don Juan Pond seems to be disappearing, perhaps because of changes in climate or ice cover. It is less than a foot deep.

Sometimes salty lakes are seasonal, just like some rivers. These lakes fill with spring meltwater or summer rains. For the rest of the year there is not enough water to keep them full. And they leave behind a bizarre landscape that looks like the moon's surface. One good example is Badwater Basin in Death Valley, California. When salty lakes disappear, they leave behind a *salt flat*. The salt crust can crack into hexagonal plates. Though an occasional heavy rain can resurrect the lake, it quickly evaporates in just a few weeks.

One of the weirdest landscapes in the world is the Salar de Uyuni in Bolivia, the world's largest salt flat. This landform covers an area of about 10,600 km² at an elevation approximately two miles above

RACE CARS AND SALT FLATS

The surface of salt flats like those at Bonneville are so level that high-speed vehicle testing can be done there to avoid crashes that can result from even the small bumps in most other road surfaces. The latest world's land speed record was set on the Black Rock Salt Flats in Nevada. There, in 1997, a British team was the first to break the sound barrier in a car equipped with two fighter jet engines! Geologists named the former lake that created the Black Rock Salt Flats *Lake Lahontan*.

Andy Green and the Thrust SSC Team vehicle that holds the land speed record of 763 mph (Mach 1.02)

sea level. When it gets a very rare rainstorm, the whole surface turns into one gigantic mirror. Nearby are several beautiful yet fantastical lakes—one red, one green, and one white. These lakes are permanent features. A combination of sediments and algae that feed on minerals gives these lakes their odd colors. Flamingos also live here, feeding on the algae. Interestingly, because the area is so high in elevation, the water freezes around the flamingos' legs every night!

Other examples of salt flats include the Bonneville Salt Flats in Utah. This feature formed after ancient Lake Bonneville, a huge North American lake that existed at the end of the Ice Age, suddenly drained into the Pacific Ocean. Miners often work in salt flats to remove concentrated salts and rare elements from the evaporites left behind.

Sometimes lakes form in the cones of inactive volcanoes. When you think of volcanoes, you probably envision fountains of lava, smoke, and dust. But they also emit lots of water vapor. And when a volcano stops erupting, the hydrothermal water from underlying magma can work its way up to its crater through joints in the fractured igneous rock. This is especially true if the volcanic cone collapses, forming a caldera. This water, along with precipitation, can fill the crater or caldera, forming a *volcanic* or *crater lake*. Water in these lakes may be acidic due to the magmatic gases present there. Crater lakes in inactive volcanoes stay full mainly from rain and melted snow. Their waters are often crystal clear because they carry little sediment. Crater Lake in Oregon filled the caldera of the old volcano Mount Mazama. It is the deepest lake in the United States. Lake Toba in Indonesia is the world's largest volcanic crater lake. Scientists suspect that the volcano that formed Lake Toba was huge, similar to the one under Yellowstone National Park. It was a supervolcano that likely spewed ash high in the atmosphere with an estimated VEI of 8.

16-18

Laguna Colorada, located in Salar de Uyuni, Bolivia, is named for its distinctive red color. It is a favorite feeding place for flamingos, as seen in this photo.

Location: 22.2°S 67.8°W

16-19

Crater Lake in the Cascade Mountains of Oregon is a beautiful lake in the caldera of a dormant volcano. At its western end is Wizard Island, a volcanic cone that formed within the caldera.

Location: 42.94°N 122.1°W

Fi
Black Rock Salt Flats: 40.86°N 119.14°W
Bonneville Salt Flats: 40.8°N 113.8°W
Lake Toba: 2.76°N 98.63°E
Lake Manicouagan: 51.4°N 68.7°W
Lonar Lake: 19.976°N 76.508°E

Where else do craters come from? Meteorites! And some of these craters are full of water. The largest meteoritic crater lake is the ring-shaped Lake Manicouagan in Quebec, Canada. The outer diameter of the lake is about 64 km and it contains 140 km³ of water! This lake even supplies an important power dam. Lonar Lake in India is a crater lake that is also salty.

LIFE CONNECTION: BOGS, BAYS, AND BLOODTHIRSTY PLANTS

In the 1930s, pilots flew over the southeastern United States, taking pictures of the ground. They compiled these pictures into one of the first-ever aerial surveys of the earth's surface. What do you think they found? One of their most surprising discoveries was tens of thousands of clear, oval depressions—*Carolina bays*. They ranged from 1 acre to 15 mi² in area. Many had prominent sand ridges along their eastern or southern edges. Even stranger was their alignment. They seemed to point like compass needles from southeast to northwest. Some of the depressions overlapped others. No one has counted them all, but some have estimated that there are more than 500,000 of them.

Of course, these low, boggy areas had been there all along. Some are ponds or lakes, like Lake Waccamaw in North Carolina. Landowners knew about them. In fact, farmers had drained many of them to grow crops in their damp soils. Engineers preferred building country roads around their edges instead of through them because their ground was too soggy. But we didn't know there were so many!

So if they don't always hold water, where does their name come from? Carolina bays get their name from trees that grow in them. The red bay, loblolly bay, and sweet bay trees frequently grow within their basins. Many other types of trees, evergreen shrubs, and grasses grow in these bogs but not in the surrounding landscape. Their distinct plants and animals make them very interesting to biologists.

In the decades since that first aerial survey, geologists, astronomers, and other specialists have tried to explain the origin of these landforms. Over the years, about nineteen competing theories have been proposed! While there's no one convincing theory, some are more believable than others. Some people think that wind blew them out. Some think that they formed when artesian springs bubbled up through soft sand. Others suggest that fish in spawning pools formed them when the coast was still under water. Some even suggest that they are craters made from fragments of an exploded comet.

The ecology of the Carolina bays is every bit as interesting as their geology. Because they are oases in drier areas, they attract many types of wildlife. Ecologists call the zone where one habitat merges into another an *ecotone*. Ecotones have unusually large numbers of different kinds of plants and animals. Most Carolina bays contain several of these ecotones along with the ecotone on their margins.

Most of these depressions contain what is known as a *shrub bog*. Bogs are poorly drained areas that usually contain soil composed mainly of peat. Peat is the compressed organic matter that builds up over many years of plant death and decay. While peat holds moisture well, it is acidic and low in nutrients. This means that only certain kinds of plants can survive here.

Fascinating plants that are insectivorous (insect-eating) are often associated with these boggy areas. They can survive in low-nutrient soils because they are able to capture and digest animal prey. Insects provide nitrogen and other key nutrients that other plants must get from the soil. As a result, insectivorous plants have an advantage over others in peaty, damp soils.

Sundew plants (top) are low-growing plants whose lollipop-shaped leaves are covered with hundreds of red, glistening hairs. When a small insect crawls or lands on them for a quick meal of nectar, it is quickly ensnared in the sticky tentacles. The leaf slowly curls around the helpless prey, which only gets more tangled as it struggles to escape.

The Venus flytrap (left) is perhaps the most familiar insectivorous plant. It has leaves shaped like clamshells that snap shut instantly when triggered by an insect walking across the surface. In a manner similar to that of sundews, it digests and absorbs its prey. Venus flytraps grow naturally only in the coastal region near the border of North and South Carolina. Carolina bays are a prime place to find them. In fact, the bizarre nature of this plant and its presence in Carolina bays led to one theory that the bays were created by aliens who also introduced the bloodthirsty plants!

16.9 THE LIFE PHASES OF A LAKE

Limnologists are scientists who study surface waters. They have observed that lakes and ponds go through a series of phases. You could say that they are born, get older, and die. Most lakes, especially small ones, will fill in and disappear, given enough time.

Some lakes are born when people or animals dam streams, or when people bulldoze depressions that water can fill. Farmers often make ponds on their property for watering livestock or for swimming. Other lakes form after rockslides or volcanic lahars dam streams. Limnologists first recognized the life phases of surface water bodies by observing beaver ponds.

Once a lake or pond forms, many things change it. Streams flowing into the lake carry sediments. These sediments naturally settle out because water slows down after it enters the lake. This factor alone would fill in lakes over time.

Living things affect lakes and ponds too. Plants in the photic zone of lakes grow and die. This adds organic matter to the bottom sediments. Just like land plants, aquatic plants anchor the bottom sediment to keep it from washing away by wave action near shore. Land plants and trees invade the shoreline, trapping soil that extends dry land into the body of water. Their growth causes the lake or pond to shrink. This effect is most noticeable in small ponds. Plant and algae growth can be very fast when lots of nutrients flow into the lake. Limnologists call this process **eutrophication**. Over time, eutrophication changes the kinds of organisms that live in a lake by increasing water turbidity, forming chemical toxins, and reducing oxygen levels.

People also extend the shoreline into the lake basin. Docks, piers, beaches, and marinas interfere with the natural water flow through lakes. This causes more sediment to collect. Human sewage and artificial fertilizer runoff from nearby farms or streams are the most important causes for lake and pond eutrophication. As you just learned, higher levels of artificial nutrients cause aquatic plants and algae to grow rapidly, trapping sediments.

All of this causes lakes and ponds to inevitably shrink. For very large lakes, sedimentation doesn't noticeably affect the overall volume of the lake. But for small ponds and lakes, the changes can be fairly fast and dramatic. Eventually, the lake becomes a swamp, then a wet, boggy area. Plants and trees trap wind-blown dust and dirt, creating topsoil. As more dead organic material collects, the soil surface rises above the level of the groundwater and dries out. All that will remain is a stream flowing through a fertile area of ground that was once a body of water.

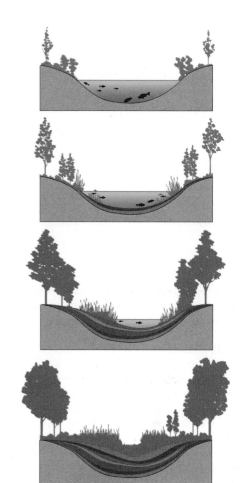

16-20
The aging of a pond results from sediment filling in the basin and the shoreline shrinking. Eventually, a dry lake results.

eutrophication (you troph i KAY shun): eutroph- (Gk. *eutrophia*—well nourished) + -ication—the process of making so

16-21
Lake eutrophication from artificial fertilizers can cause algal blooms (green scum), which increase water turbidity, reduce dissolved oxygen, and result in fish kills.

limnology: limn- (Gk. *limne*—lake) + -ology; science of lakes, rivers, and groundwater

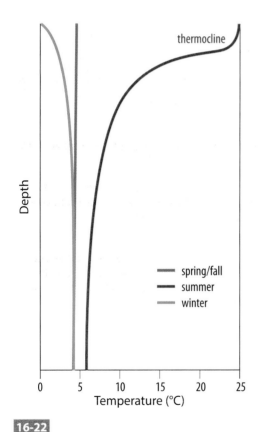

16-22

Summer, fall, and winter temperature profiles in a lake, showing how a seasonal turnover occurs

The transition from a young lake to dry land can take hundreds or even thousands of years. Climate change can slow the process if things get wetter. It can also speed it up if the climate gets drier, removing the source of water. Many factors affect the life phases of a lake.

16.10 LAKE SCIENCE—LIMNOLOGY

The study of lakes was first called **limnology**. Today, limnology includes the science of all forms of surface and underground water, including lakes and rivers. It began as a formal science in the late 1800s. François-Alphonse Forel, who studied Lake Geneva in Switzerland, is considered the father of limnology.

The methods and instruments of limnology are identical to those used in oceanography (see Chapter 15). Limnologists study the geology of lake basins, the physical and chemical properties of water, the movements of inland waters, and the biology of the organisms that live in and near bodies of water. However, limnologists focus mainly on how inland waters and communities of living things surrounding them affect each other. Oceanographers are concerned mostly about the oceans themselves.

Several physical features of lakes, especially large freshwater ones, are much more pronounced than in the oceans because these bodies of water are far smaller. One important seasonal process is called **thermal turnover**. It occurs twice a year in lakes located in middle and upper latitudes in both hemispheres. In the summer, these lakes develop a distinct thermocline, or thermal layer, just a few meters below the water's surface. The sun and warm summer air can raise the surface water temperature to around 25 °C or even more. Wave action keeps the surface waters well mixed. But below the wave base, the temperature quickly drops to 5–10° of freezing, even in the summertime.

In the fall, the sun angle is lower in the sky, and as the air cools, surface waters also cool. The surface waters become cooler than the deeper waters with the approach of winter. Cooler water is denser, so it sinks, and the deeper water rises to the surface. This sets up a vertical *convection current* called the *fall turnover*. When the water cools to 4 °C from top to bottom in the lake, convection stops. Liquid water is the densest it can be at this temperature.

When freezing temperatures arrive in winter, the surface water cools further, and when the surface temperature reaches 0 °C, the water freezes. Ice is actually less dense than its liquid form because of the unusual properties of water. This is why ice floats and lakes freeze over in the winter. Animals survive in the deep, cold, oxygen-rich waters, moving slowly. Some animals like frogs hibernate in the bottom mud. The lake continues to cool due to contact with the ice, and the ice grows thicker. The coldest water is at the *top* of the lake.

In the spring, the opposite process occurs. The ice melts and the sun warms surface water up to 4 °C. All the water in the lake is the same temperature, allowing mixing between all layers of the lake. This vertical convection current is the *spring turnover* that continues until the summer thermal layer forms. As in upwelling and downwelling in the ocean, nutrients and oxygen are exchanged between the surface and the bottom. This process benefits life at all levels of the lake.

Weather systems affect lakes too. A lake, especially a large one, can show signs that its water is sloshing back and forth in its basin. Observers see water levels repeatedly rising on one side of the lake while dropping on the other in a short time. This water movement is related to changes in atmospheric pressure over different parts of the lake. With changing air pressure, the downward force acting on different parts of a lake's surface changes too. Under the right conditions, this can set up a very long wave that sloshes back and forth for many hours or even days. This movement is called a **seiche**. Most seiches aren't very noticeable, but in some places, they can change lake levels several feet.

Limnologists are particularly interested in the way currents carry sediments and nutrients through a lake basin. While large lakes develop some deep currents similar to those of the oceans, most currents in a lake are wind-driven surface currents. This makes them temporary and changeable.

Studying lakes, streams, ponds, and rivers helps us to understand and use them better. We have a responsibility to care for plants, animals, and people that rely on the resources in surface waters.

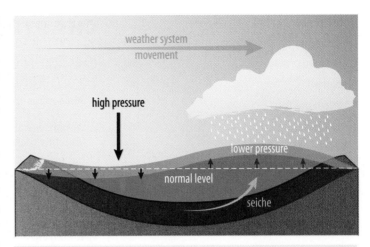
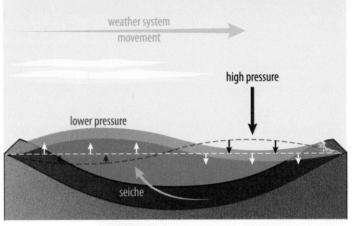

16-23
A seiche forms because of differences in air pressure above a body of water.

seiche (seesh): (poss. Ger. *seiche*—sinking); relating to changing of lake or sea levels due to atmospheric conditions

In Lake Erie, seiches have been observed up to 5 m between high and low water. The US Coast Guard has to warn boaters so that they don't run aground.

16B REVIEW QUESTIONS

1. What is a lake? What kind of water can it hold?
2. Why are lakes important to exercising biblical dominion?
3. How old are lakes in comparison to the age of the earth? How old can the oldest lake be according to a young-earth view of geologic history?
4. What differentiates unusual lakes from most lakes around the world?
5. Describe the life phases of a typical small pond after it forms.
6. What conditions can shorten the lifespan of a pond?
7. How is limnology similar to oceanography? How is it different?
8. What property of water drives the seasonal thermal turnover in lakes?
9. (True or False) A lake whose outlet stream flows faster than its inlet streams will eventually dry up.

Key Terms

Term	Page
stream	379
elevation profile	380
source	380
stream gradient	380
mouth	380
base level	381
cross-section	381
tributary	382
stream system	382
drainage basin	383
divide	383
meander	384
floodplain	384
perennial stream	385
intermittent stream	385
lake	386
eutrophication	393
limnology	394
thermal turnover	394
seiche	395

16-24 Located on the US-Canada border, Horseshoe Falls is one of three waterfalls that make up Niagara Falls.
Location: 43.078°N, 79.076°W

CHAPTER 16 REVIEW

CHAPTER SUMMARY

» Good dominion of surface waters involves maximizing their usefulness while minimizing the harmful effects of such actions.

» Streams are any flow of surface water confined to a channel. They can vary from tiny trickles to mighty rivers.

» Most streams have a high gradient with fast flow near their sources and then become less steep. Near their base level, streambeds are nearly flat.

» Old-earth geologists often use a stream's gradient to estimate its age.

» Stream cross-sections are related to their gradients and erosional power.

» Streams and their tributaries drain water from a drainage basin. This is called a stream system. A divide separates a stream system from other stream system drainage basins.

» Streams can have many common features, including rapids, waterfalls, bars, meanders, oxbow lakes, and deltas.

» Stream features create both opportunities for and obstacles to dominion. Well-designed engineering projects can often reduce the dangers and maximize the usefulness of streams.

» Most streams are perennial. Other streams are intermittent, occasionally appearing in otherwise dry streambeds.

» Lakes are bodies of water surrounded by land and cut off from the ocean.

» Lakes are important because they hold most of the liquid fresh water on the earth's surface.

» Limnology is the scientific study of all surface and underground water.

» Lakes are geologically young, no matter which story of Earth's history one considers. Young-earth geologists believe that no lake could have formed earlier than when floodwaters receded off the land.

» Many unusual lakes have formed across the world. Some are salty, while others occupy volcanic or meteorite craters. Some are intermittent lakes that form broad salt flats. And there are many whose origins are unknown.

» Most lakes, especially small lakes and ponds, go through a series of phases. They usually form quickly and then gradually fill in with sediment and organic matter. They shrink to become marshes, eventually leaving only a stream running through fertile land.

» Limnology is similar to oceanography in its methods, techniques, and equipment.

» Limnologists focus on the overall ecological community in lake research because lakes and ponds greatly influence living things around them.

» Seasonal turnovers, seiches, and currents are processes in lakes or ponds.

REVIEW QUESTIONS

1. After examining the pros and cons of the Three Gorges Dam project in China, what can you conclude about the results of trying to exercise dominion in a fallen world?

2. Through how many drainage basins could surface waters flow to the ocean from a given source water location?
3. What is the name of the highest ridge that separates the major drainage basins of North America?
4. What is the main difference between rapids and falls in a stream?
5. Why do meanders form?
6. Some streams flow only when filled by runoff from occasional distant thunderstorms. Why are these streams dangerous?
7. Which American river has a tributary that is longer than the river itself?
8. What are the possible sources of water for a lake?
9. Why is it unlikely that present-day lakes existed as they are now before the Flood?
10. What is a common reason for the existence of a highly salty lake?
11. What kind of mineral landform can be found in dry locations where a lake is either intermittent or has dried up over a long span of time?
12. What are the two main sources of water for crater lakes?
13. What is eutrophication of a lake? When can this process become harmful to a lake?
14. Why do smaller lakes have shorter "lifespans" than larger ones?
15. What action of lakes produces the summertime thermocline?

True or False

16. Only good things result from exercising biblical dominion.
17. All streams start at a higher elevation and flow downhill.
18. The Colorado River, which flows through the Grand Canyon, must be younger than about 5500 years.
19. A first-order stream has no notable tributaries.
20. Oxbow lakes are features of high-gradient streams.
21. The largest river basin in the world is drained by the Amazon River.
22. A body of water named a sea can actually be a lake.
23. Lake Baikal is 25 million years old.
24. The Dead Sea drains into the Arabian Gulf.
25. Eutrophication happens naturally in most lakes.
26. A seiche can occur in almost any landlocked body of water.

Map Exercises

To complete these exercises, you will need Google Earth or a similar 3D Earth-viewer program as well as Internet access.

Where Does the Water Go?

To which ocean will surface water flow if it starts from the following points?

27. 39.1°N 94.6°W
28. 67°N 101.7°W
29. 41.5°N 112.1°W
30. 47.7°N 117.5°W

16-25 These ponds are found in karst topography, which you will learn more about in Chapter 17.

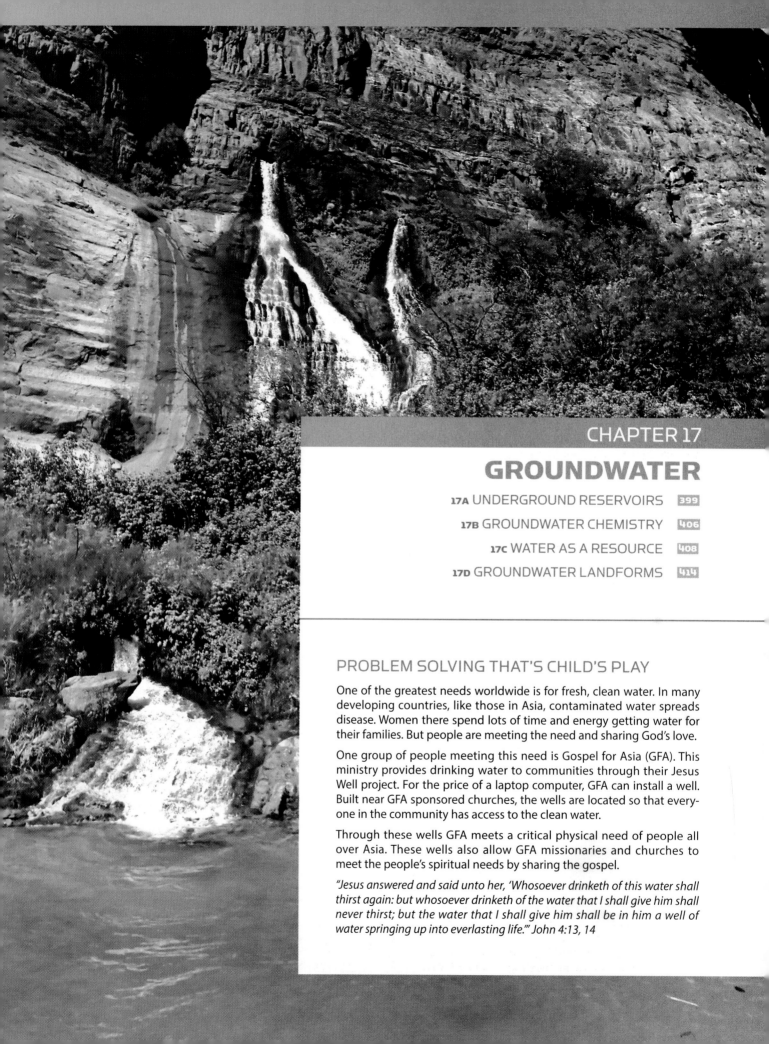

CHAPTER 17

GROUNDWATER

- **17A** UNDERGROUND RESERVOIRS — 399
- **17B** GROUNDWATER CHEMISTRY — 406
- **17C** WATER AS A RESOURCE — 408
- **17D** GROUNDWATER LANDFORMS — 414

PROBLEM SOLVING THAT'S CHILD'S PLAY

One of the greatest needs worldwide is for fresh, clean water. In many developing countries, like those in Asia, contaminated water spreads disease. Women there spend lots of time and energy getting water for their families. But people are meeting the need and sharing God's love.

One group of people meeting this need is Gospel for Asia (GFA). This ministry provides drinking water to communities through their Jesus Well project. For the price of a laptop computer, GFA can install a well. Built near GFA sponsored churches, the wells are located so that everyone in the community has access to the clean water.

Through these wells GFA meets a critical physical need of people all over Asia. These wells also allow GFA missionaries and churches to meet the people's spiritual needs by sharing the gospel.

"Jesus answered and said unto her, 'Whosoever drinketh of this water shall thirst again: but whosoever drinketh of the water that I shall give him shall never thirst; but the water that I shall give him shall be in him a well of water springing up into everlasting life.'" John 4:13, 14

17A

UNDERGROUND RESERVOIRS

How is water stored in the ground?

17.1 ESSENTIAL WATER

Almost three-fourths of the earth's surface is covered with liquid water. But over 97% of this water is in the oceans and salt lakes, water too salty for us to drink. Less than 2% of the earth's water is frozen in the polar ice caps and glaciers. Even though glacial ice holds the largest amount of fresh water on Earth, it's not easy to get that water to your glass! The second largest reserve of fresh water is **groundwater**, water found underground. Groundwater is the source of about 96% of the world's fresh water not locked up in glacial ice. Yet it accounts for less than 0.8% of all the water on our planet. Getting groundwater to your glass takes some effort too.

Clean drinking water is essential for life. God made the earth with an abundance of water, but it is surprising that only a very small part of it is drinkable. Why would a good God put His creatures in such a situation?

God gives us an opportunity to glorify Him through obeying the Creation Mandate. By giving us the challenge of meeting our physical needs, God encourages us to creatively use His world. As we exercise our God-given creativity and strength to carry out dominion over His creation, we glorify Him and have the opportunity to show love for others.

17A Objectives

After completing this section, you will be able to

» create a chart or graph that compares the major segments of the earth's water inventory.

» describe the water cycle.

» express the relationships between the terms *porous*, *nonporous*, *permeable*, and *impermeable* when used to describe rocks.

» discuss the geologic features, storage, and movement of groundwater applied to its availability as drinking water.

17-1

Only a small portion of the earth's water is groundwater, but this water is vital to the survival of life.

17-2 Children in Ghana obtain safe drinking water from a newly installed pump. Organizations such as Gospel for Asia seek to provide drinking water to regions that have none. By meeting this need, GFA missionaries also have the opportunity to share the gospel.

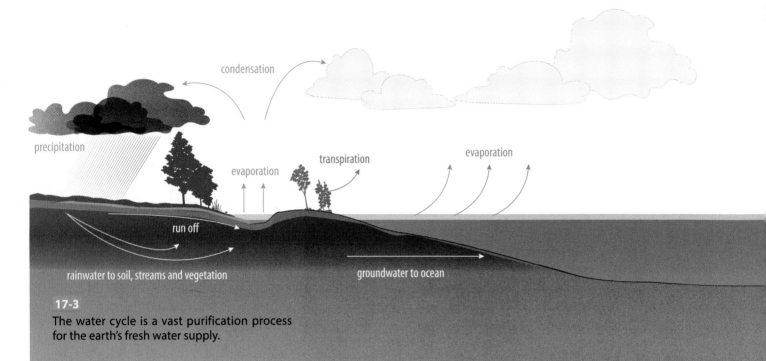

17-3
The water cycle is a vast purification process for the earth's fresh water supply.

> Earth scientists also call the water cycle the *hydrologic cycle*. The word "hydrologic" refers to any aspect of water in its natural setting.

percolate (PER ke layt): (L. *percolare*—to sift, strain, or filter)

17.2 THE WATER CYCLE

So where does all this groundwater come from? Groundwater is an important part of the **water cycle**. Water continually moves from the continents to the oceans, evaporates into the atmosphere, and moves back to the continents as wind-driven water vapor and clouds. The cycle is complete when the clouds drop rain and snow onto the continents. In one form or another, 20 *trillion* liters of moisture fall on the United States alone *each day*! The water cycle is required to purify and recycle the water we need.

Let's focus on the part of the water cycle when water falls on the continents. Where does the rainwater go? Some of it runs into streams (Chapter 16). Some forms puddles on the ground and evaporates back into the air. Plants absorb water from the soil, and they release some of this water back into the air through leaves and stems by *transpiration*.

The rest of the water *percolates* into the ground. It flows downward through soil until it reaches a depth at which water fills all the spaces between soil and rock particles. This is groundwater. You might think that this is a water graveyard, but it's not. Groundwater is an active part of the water cycle. Most groundwater can eventually return to the earth's surface.

17.3 THE WATER TABLE

The upper surface of the groundwater reservoir is called the **water table**. If you could peer into the ground with X-ray vision, the water table would look like a broad, wavy surface of water.

aeration (air AY shun): (Gk. *aer*—air) + -ation; mixing with air

Hydrologists have divided groundwater storage into several zones. The **zone of aeration** is the porous rock and soil above the water table. Here, all available spaces and pores between rock particles that do not

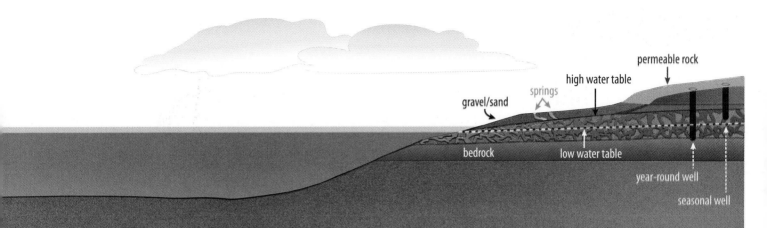

trap water are filled with air instead of water. The **zone of saturation** lies below the water table. Every available space is filled with water. Immediately above the zone of saturation is a narrow zone in which water tries to move upward from the water table by *capillary* action. This is similar to how water soaks upward into the raw edge of a paper towel or the surface of a dry sponge. The space in the ground where capillary action takes place is called the *capillary fringe*.

The water table's surface isn't actually flat. Beneath level ground, the water table's depth can vary with the rock and the availability of surface water. The water table often follows the contours of the land. It generally rises under hills and drops under valleys. But the water table's surface changes less than the earth's surface does. In some places, the water table is only a few meters below the surface. In other places, it is hundreds of meters below. At certain places called *springs*, the water table meets the ground's surface. Here, water flows from the ground, often continuously.

The height of the water table can change with the seasons. Generally, it is higher in the springtime when rain and melting snow are plentiful. This inflow of water to supply a groundwater reservoir is called *recharging*. It's similar to recharging your smartphone battery with electricity. In the dry season, the water table can drop several meters below the wet-season level.

Many farmers, homeowners, and industrial businesses draw water from wells. They need to have a reliable, year-round supply of water. Well drillers, home builders, and engineers need to know where the water table is before they drill a well. A well pipe will fill with water only if it extends below the surface of the water table. If the well is to produce water all year, it needs to be drilled below its dry-season level.

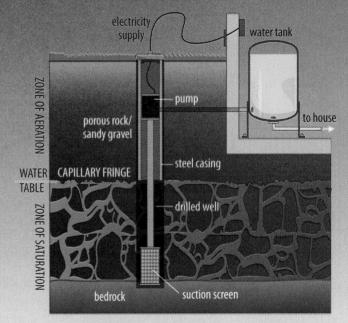

17-4

The zone of aeration holds some air in its spaces. In the zone of saturation, all spaces are filled with water. The water table is the upper boundary of the zone of saturation. Soil that contains water absorbed upward from the water table by capillary action is called the capillary fringe.

capillary: capillar- (L. *capillar*—hairlike) + -y; having fine, hairlike spaces

17.4 GROUNDWATER STORAGE

Most groundwater resides within tiny spaces and cracks in rocks. *Porous* rock has lots of these. Pumice is a good example of a porous rock. It is an igneous rock that contains numerous tiny gas bubbles, and looks like solid foam. Sandstone and even granite can be porous. However, chemical sedimentary rocks and igneous and metamorphic rocks that have not been fractured by tectonic forces are nonporous. This is because they formed that way, or their pores were squeezed shut or filled in by metamorphic processes. The *porosity* of a rock indicates the quantity and size of the pores in the rock. A higher porosity means that the rock is more porous. A rock must be porous to hold groundwater. But it's not enough just to be porous.

A rock must also have the property of **permeability** (PUR mee uh BIL ih tee). Porous rock that is *permeable* has pores or cracks that are connected and are large enough so that water can flow through them. Materials with high permeability include sand, gravel, and most sandstones. These materials have relatively large, connected spaces between them. Some types of limestone are also permeable because they have many small joints between grains or fossils.

A material that keeps water from flowing through it is *impermeable*. And even a porous rock can be impermeable! Pumice is an example. Though there are many pores in pumice, they don't go anywhere. That means that water can't flow through pumice. As you might expect, undisturbed clay, shale, most metamorphic rocks like slate and quartzite, and most igneous rocks are nearly impermeable to water. Some of these rocks (like shale, for example) can be pretty porous, but their holes, cracks, and pores are too small to let water through because of water's cohesion.

Impermeable rock can be a real barrier to groundwater. In some places, a thin layer of impermeable rock lies above a groundwater reservoir. Since water from the ground surface cannot seep down to the reservoir, it collects above the impermeable rock. This body of groundwater is called a *perched water table* (think of a bird perched above your head in a tree). Many hillside springs come from perched water tables.

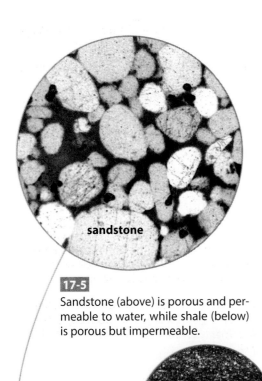

17-5
Sandstone (above) is porous and permeable to water, while shale (below) is porous but impermeable.

17-6
This cross-section of a hill shows how there can be two or more water tables in the same area.

BIBLICAL ORIGINS: THE ANALOGOUS DAYS THEORY

One purpose of the Creation story is to provide a work/rest pattern that man is to imitate. However, some Christians consider the six days of Creation to be God's work days, not six literal days.

According to this view, as analogous or symbolic days, God's work days can be any number of "earth days." Therefore, vast periods of time could actually have passed on Earth during one "work day" as God completed His creative acts. These periods of time are not stated anywhere in Scripture but can be inferred by studying nature.

The difference between the Analogous Days view and the Day-Age theory lies in the meaning of the word "day" in Genesis 1. The Day-Age theory considers the word itself to be an indefinite period of time for each real creation event. The Analogous Days theory considers the entire Creation narrative as a symbolic view of God's real creative acts in the world.

Questions to Consider

1. What do the recurring "evening and morning" statements in Genesis 1 indicate about the days of Creation?
2. Read Exodus 20:8–11. What does this passage suggest about the days of the Creation week?
3. If we accept the Analogous view of Creation, what effect may it have on our reading of other passages?

In some places, erosion can expose a tilted unit of sandstone sandwiched between layers of shale on a hillside. As surface water flows downhill, it can enter the porous sandstone. The sandstone becomes an **aquifer** as water flows by gravity underground into the stratum. If we drill a well at a lower elevation into the aquifer, the water may have enough pressure to spurt up out of the well. This creates an *artesian well*. Artesian wells were important sources of water in the early days of American history. But usage has depleted many aquifers, so water in most places must be pumped today.

> **aquifer** (AH kwi fer): aqui- (L. *aqua*—water) + -fer (L. *ferre*—to carry); a rock stratum that directs water from one place to another
>
> **Artesian** (ar TEE zhun) wells are named after the French province of Artois (formerly called Artesium), where Roman soldiers first observed such wells.

17.5 MOVEMENT OF GROUNDWATER

Groundwater tends to move downhill under the influence of gravity just as surface water does. But it moves more slowly—from 100 m per day to as little as 0.5 m in ten years! Usually groundwater moves in the same direction as surface water, though it is mostly controlled by the rock strata underground. Groundwater moving downhill may eventually come out of the ground as a spring, enter a stream or lake, or join an actual underground stream flowing through a cave (see Section 17D). Some caves have several underground streams at different levels. Eventually these streams feed into surface streams, lakes, or the ocean. Some may even trickle downward into deep rock formations and become heated as part of the hydrothermal cycle (see Chapter 8).

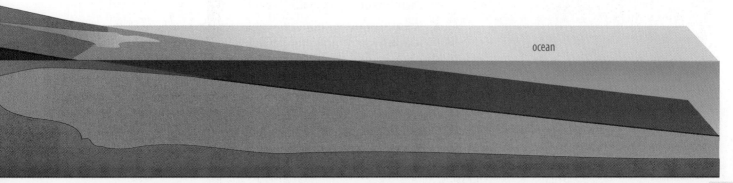

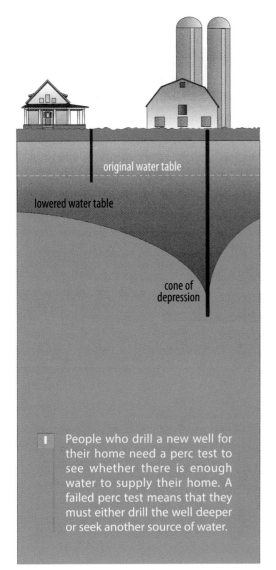

17-7

People have drawn water from the ground for thousands of years, often using open wells. With mechanical and electric pumps it is very easy to overpump a well.

People can control the height of the water table. Regularly pumping water from a well lowers the height of the water table immediately around it. This forms a **cone of depression**, a cone-shaped dip in the water table that is deepest at the well pipe. The shape and size of the cone of depression depends on how fast people pump the water and how quickly water percolates through the soil to resupply it. When people overpump a well, they create a large cone of depression. If the pumping rate is high and the water percolation rate through the rock is too low to keep up, the well could go dry.

Overpumping a well can affect the overall shape of the water table and the flow of groundwater around the well. Pumping too much water can lower the water table and pull water from neighbors' wells. If the direction of groundwater flow changes, nearby sources of contaminants like septic systems may actually flow toward a well instead of away from it. Wells near a seacoast can draw up salty water if the water table drops too much. But with a properly designed and located well, the problem of overpumping for normal water usage may happen only during long dry spells.

Groundwater is normally replenished as rain or meltwater percolates down to the water table. Water from streams or lakes can do this too. A **recharge zone** is an area on the land's surface that resupplies groundwater. These zones are often essential to a community's water supply many miles away. Recharge zones and aquifers also contribute to water purity by providing a place for beneficial bacteria to break down wastes and toxins (see Section 17C). In some places where natural recharge cannot keep pace with the demand, state and municipal governments use artificial ways to recharge groundwater.

17A REVIEW QUESTIONS

1. Using the data in Figure 17-1, create a chart or graph that compares the amounts of major categories of the earth's water inventory by percentage. Include the oceans, liquid fresh water, frozen fresh water, and groundwater. Be sure to label your diagram with the percentages.
2. Explain why the earth doesn't run out of purified water. Describe the major steps in this process.
3. Describe why a hillside spring exists.
4. How can rock store groundwater?
5. Draw a Venn diagram that shows the relationship between impermeable, permeable, porous, and nonporous rocks. Draw one circle for each and overlap them as appropriate. Write at least one rock example in each area of your diagram.
6. What happens to the water table around a water well that is used regularly?
7. How can groundwater be recharged?
8. (True or False) Groundwater is not an important source of fresh water that is useful for drinking.

DEPLETED AQUIFERS

Shlump! It swallowed a tree. Rhomp! Shlussh! Now it just swallowed two cars! No, it is not the monster that devoured Miami. It's a sinkhole opening at the surface. Sinkholes form when caves dissolved by groundwater collapse. They're a symptom of a lowering water table.

Sinkholes like this one in Florida create serious problems.

The level of the water table across the United States varies, depending on how much it rains or snows. We can also affect it by removing water. In times of drought, we can lower the water table too much. And the consequences can be disastrous.

One such consequence is sinkholes. These are more common when the water table is low because the water that normally supports the ground isn't there anymore.

A lot more sinkholes are developing in Florida lately. But this isn't the only problem that a low water table causes. In coastal regions, salt water can infiltrate from the ocean where the water table is low, making well water undrinkable. In desert areas like Arizona, huge cracks appear in the ground where groundwater is depleted. This happens because the ground doesn't sink uniformly as the water table drops. This sinking is called *subsidence*. Some cracks are hundreds of feet long and can extend hundreds of feet down to the water table.

Since we can't control droughts, we also can't completely avoid these problems. But we can manage how we use groundwater. One area where we have made great progress is in the agricultural usage of water for irrigation. In the United States, farm irrigation uses three to four times as much groundwater as public water systems in all the cities put together! But our farms are far more productive than most in the world. Ditch irrigation is simple, but a lot of water is lost to evaporation and to ground seepage. A better way to control water delivery to plants is by center pivot irrigation (see page 409). Water is sprayed onto the plants using large rotating irrigation rigs 400 m long or more. Automatic controls schedule watering for the cool of the day to minimize evaporation. Spray flow is controlled to avoid applying water to ground that isn't under crops.

These kinds of efforts to reduce water usage in industry and communities have reduced the pumping of some aquifers. In recent years, many communities have begun pumping treated wastewater back into aquifers to recharge them. With these kinds of efforts, we can conserve water and avoid the problems of depleted aquifers.

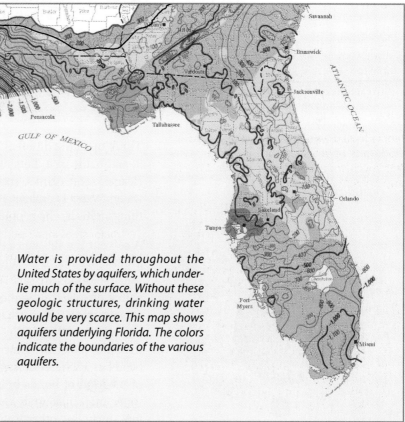

Water is provided throughout the United States by aquifers, which underlie much of the surface. Without these geologic structures, drinking water would be very scarce. This map shows aquifers underlying Florida. The colors indicate the boundaries of the various aquifers.

GROUNDWATER

17B Objectives

After completing this section, you will be able to

- relate the dissolving power of water to its physical and chemical properties.
- explain how the amounts and kinds of dissolved minerals in drinking water affect its hardness and usefulness.
- describe different methods for softening hard water.

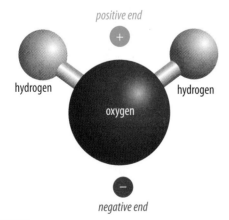

17-8
The amazing water molecule

17B
GROUNDWATER CHEMISTRY

Why is tap water not pure water?

17.6 WATER: THE UNIVERSAL SOLVENT

Think of all the different ways you use water. You drink it, cook with it, and bathe in it. Lots of products we use, such as latex paints, glass cleaners, soft drinks, soups, skin lotions, and insect killers, contain water. Water is amazing because it can mix with so many different things. This is why people call water the "universal solvent."

A **solvent** is a substance that dissolves other substances. Chemists call a substance that is dissolved in a solvent a *solute*. Common solutes in water solutions are salt, sugar, and minerals. The mixture of a solvent with a dissolved substance forms a *solution*. When a solvent dissolves something else, its particles surround the particles of what it dissolves and separates them. Why is water so good at this? It all comes down to the water molecule. Water molecules are shaped like boomerangs. The bend has a different electrical charge compared with the ends because of the kind of bonds holding the atoms together. This feature helps water molecules grab and hold lots of other atoms, molecules, and ions, including other water molecules, by electrical attraction. Water's structure accounts for all its special properties that you learned about in earlier chapters.

17.7 DISSOLVED MINERALS

The properties of the water molecule, H_2O, are truly unique among similar compounds. Christians can appreciate God's wonderful providence and design of this chemical so essential for life on Earth. Nowhere else in the known universe is there so much of this special compound. It is no accident that Earth is also the only place known to have life.

Absolutely pure water probably doesn't exist in nature. Although it may look pure, groundwater always contains dissolved minerals. Even rainwater contains dissolved gases, smoke, and dust particles. Remember from Chapter 12 that dissolved atmospheric carbon dioxide turns groundwater into a dilute acid, which eats away at rocks and minerals. The amount of dissolved minerals in groundwater can be measured and is reported by hydrologists as the **hardness** of the water. These dissolved minerals are usually harmless and may be beneficial. In fact, many resort areas called *mineral spas* advertise that their mineral waters are especially healthful.

Hydrologists call potable water—water that we can drink and cook with and that has lots of dissolved minerals—*hard water*. It contains dissolved compounds of calcium, magnesium, and other metals, such as iron. Lathering soap in such water is difficult because a chemical reaction with these minerals binds up the soap's molecules, blocking their dissolving action. So you need more soap to get clean. Also, a sticky scum forms instead of lather. This scum is what causes bathtub rings. It can also collect on clothing washed with soap or detergent. Iron can stain fabric and sinks or toilets with a rusty, yellow-brown color. Hard water also tastes metallic.

17-9

Hard water can leave deposits that are damaging as well as unsightly. The deposits on the showerhead below could slow the water flow or even stop it completely.

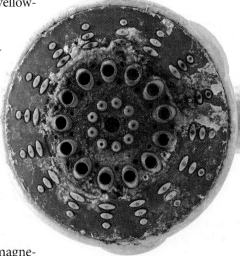

Some groundwater contains minerals dissolved from limestone or dolomite. When this water is heated or boiled, as it is in your hot water system, these minerals will precipitate to form insoluble carbonates. Carbonate deposits may form in teakettles and in hot water pipes. If these minerals build up long enough, the rocklike *scale* can clog pipes and damage plumbing parts. Calcium carbonate, $CaCO_3$, forms the mineral calcite and the rock limestone. Magnesium carbonate, $MgCO_3$, forms the mineral magnesite. When both ions are present in solution, they can form the mineral dolomite, $MgCa(CO_3)_2$.

Soft water, on the other hand, is naturally low in calcium and magnesium compounds. Soaps and detergents easily dissolve in soft water and produce lots of suds—so well, in fact, that you have to rinse your hair for a long time to get all the shampoo out. But when you do, your hair is squeaky clean! Soft water does not leave mineral scale in plumbing or teakettles. It doesn't taste metallic, either, though some people like the taste of mineral water.

17-10 Some people want to soften all the water in their house and will use a larger demineralizer. Others are interested in filtering their drinking water only, which can be done by passing the water through a filter.

Dissolved minerals in hard potable water can be removed by a process known as *water softening*. As we just mentioned, heating water containing certain compounds can unintentionally soften water, but with unwanted side effects. Hard water with compounds such as chlorides and sulfates does not soften with heat. However, water treatment systems containing tablets of borax, washing soda, or certain phosphate compounds can mix with this kind of hard water to precipitate the minerals as a soft sludge. Many industries and even some homes use this method to soften water before use or as it is used.

Hard water can be treated with special water filters called *demineralizers* or *ion exchangers*. These filters remove both suspended sediments and dissolved minerals. They work by absorbing the unwanted mineral ions and releasing replacement ions that don't cause problems. Water flows through a cylinder containing millions of tiny resin beads that attract the problem ions in water. Demineralizers improve the taste of drinking water and remove minerals that stain clothing, sinks, and toilet bowls. You can even buy water pitchers that have demineralizers built right in.

17B REVIEW QUESTIONS

1. What makes water so good at dissolving other things?
2. Why isn't groundwater pure water?
3. What kinds of problems does hard water cause?
4. List three ways to soften hard water.
5. (True or False) Water is a good solvent due to the shape of its molecule.

17C

WATER AS A RESOURCE

How can we wisely use water?

17.8 USING WATER

According to usage statistics, in 2013, Americans used 4300 L of water per person per day. In a year, the average family might use nearly eighty railroad tank cars full! In most places, water supplies can keep up. But in dry places or large cities, there's barely enough. And the situation is even more desperate in undeveloped countries. Swelling world population will put even greater pressure on available water supplies. Water is necessary for our way of life. How do people use water outside the home?

17C Objectives

After completing this section, you will be able to

» explain how we can use and conserve drinking water.
» identify the ways that drinking water can become polluted.
» relate the importance of drinking water and sewage treatment to modern, healthy living.

Gulp, Gulp, Gulp

One study estimates that average water usage per person in the poorest countries is less than 32 L per day, while Americans use more than a hundred times that amount. How should we feel about such statistics?

AGRICULTURE

We need water to produce food. Livestock need a lot of water. Cattle, sheep, and dairy owners will often drill wells to provide water to their animals. Some farmers on large cattle ranches still depend on windmills to pump groundwater into water tanks in isolated places. Large vegetable and fruit farms in dry but fertile places depend on some form of **irrigation** to water and grow their crops.

INDUSTRY AND TRANSPORTATION

We also need water for industry and manufacturing. You might not think of this, but steel manufacturing and fertilizer and other chemical production rely on lots of water. Water cools machinery and supplies steam in electrical generating plants. Nuclear power plants use water to transfer heat within the reactor plant and to generate electricity. A large fraction of industrial cooling water evaporates into the atmosphere and reenters the water cycle. Power dams require a reliable supply of water to operate.

Surface waters give us the ability to move lots of cargo economically. This is why most major non-coastal American cities are on rivers and lakes. During times of drought, lowered river and lake levels can interfere with cargo barge traffic.

17-11

Center pivot irrigation can be seen all over the western portion of the United States as patches of green circles in brown terrain. A long pipe supported by wheels and connected to a central water pipe hub slowly moves around the hub, spraying water directly onto the plants.

Location: 36.15°N 102.71°W

RECREATION

We also use clean, fresh water for swimming, boating, and fishing. Lots of people take a vacation to lakes and rivers. You're probably one of them! Though recreational uses of water resources don't seem to be essential, if water weren't available for water skiing, it wouldn't be available for drinking or producing energy either.

17.9 CONSERVING WATER

We need to monitor our ability to supply the water that people need as our population grows. Remember that having more people is a good thing. God commanded us to fill the earth with His image bearers (Gen. 1:28, 9:1). But this command also includes the responsibility to meet people's needs.

REDUCING WATER USAGE

As water consumers, we need to look closely at what we *need* to use compared with what we *want* to use. We may want to stand in a hot shower for half an hour, but do we really need to do that, especially if there is a water shortage due to drought or other water usage restrictions? Most people in the world use much less water than Americans do. While much of that water is for manufacturing and agriculture—two areas where the United States far surpasses other nations—we still use a lot more water than we really need.

Industrial processes have improved over the years so that we need much less water to do the same jobs. For example, paper manufacturing has halved the amount of water it uses. Instead of ditch irrigation (which wastes a lot of water), some farmers now use drip irrigation, as well as the center-pivot irrigation method (see previous page). These techniques bring water right to the plants in just the right amounts.

So how can you conserve water? Inexpensive plumbing fixtures can reduce the water you need for showers and toilets. Some fixtures can heat water only when and where you need it. This way, you don't have to run several gallons of water from the spigot to get hot water from a central water heater. But there are also some pretty basic things you can do. Do you take long showers? Do you keep the water running when you brush your teeth? Do you run your dishwasher when it has only a few dishes? How full is your washing machine when you do a

17-12 Drip irrigation delivers water right to the roots of crop plants.

load of clothes? The idea here is to *use only what you need*. In doing so, we are more likely to have enough water for everyone. Your family will have a lower water bill too!

REUSING WATER

Another way to conserve water is to use it for more than one purpose. For example, some homes can send water from the laundry, dishwasher, or sink to a *dry well* in the yard. This well distributes wastewater to a garden irrigation system or to water the lawn. This is commonly done in places that have extended dry seasons, like the American Southwest. Washing the family car on the lawn in dry localities directly reuses the water and helps keep the lawn green! Some industries can reuse cooling water by circulating it through cooling towers, instead of drawing water continuously from a river or lake. This is especially important if the water must be chemically treated to reduce corrosion of the cooling system components.

17.10 WATER POLLUTION

We pollute water as we use it. **Pollution** happens as we add something to any resource that makes the resource no longer usable for a particular purpose and can even make it poisonous. In order to protect water supplies and to ensure the health of our citizens, local, state, and national governments have established *water quality standards*. These rules define the limits for the pollutants in the water we use. What kinds of water pollutants exist?

17-13 Chemical pollution can contaminate drinking water, killing plants and animals needed for human food consumption. Wildlife is also affected.

CHEMICAL POLLUTION

The most common kinds of water pollutants are *chemical pollutants*. Many of the chemicals we use each day in the home end up going down the drain and into the sewer system. Most are designed to dissolve in water, but they still make it undrinkable. You should never dump motor oil, gasoline, oil paints, or pesticides down the drain. Most medicines should be made unusable in their original containers and thrown away in the trash rather than flushed down a drain.

Chemical pollutants also come from industries and farms, especially fertilizers and animal wastes. These enter streams and the groundwater system. These pollutants can cause rapid eutrophication of lakes and ponds.

17-14 Nuclear power plants avoid thermally polluting nearby streams by using cooling towers.

THERMAL POLLUTION

Water used for cooling in power plants and other industries can create *thermal pollution*. Warmer stream and lake temperatures reduce dissolved oxygen in the water. This can kill fish and aquatic plants. Warmer fresh water also helps undesirable organisms to grow. These can displace *native species* in a body of water. (A native species is any animal or plant that naturally lives in an area.) Many power plants now use cooling towers and evaporators that lower the temperature of coolant by using air, thus avoiding water thermal pollution.

MINERAL AND SEDIMENT POLLUTION

Dissolved minerals that get into groundwater can come from natural or man-made sources. Both must be removed to meet drinking water quality standards. People who draw water from wells may notice a cloudiness in their tap water after a period of heavy rain as the groundwater is recharged. Percolating water can often carry fine sediments along through pores in the rock. Sediments can fill up and block plumbing and septic systems and can also coat heat exchanger surfaces, reducing their efficiency.

BIOLOGICAL POLLUTION

Biological pollutants can be a serious threat. Bacteria naturally live in groundwater, and most are harmless or even beneficial, since many can break down contaminants into safe materials. But certain bacteria such as *Escherichia coli* (or *E. coli*) live in the intestines of people and most mammals. The presence of these forms of bacteria in water can be dangerous. When water quality managers find *coliform bacteria* in a drinking water supply, other disease-causing bacteria and viruses are probably present also. Coliform is a term that biologists and medical specialists use to describe contamination by most kinds of intestinal bacteria, not just *E. coli*. This can happen when sewage systems or surface water containing animal manure overflows into a stream, lake, or ocean. Recreational waters polluted with coliform bacteria are usually posted with "No Swimming" signs. Drinking water standards allow almost no coliform bacteria.

17-15 High coliform counts from sewage can affect the use of streams, lakes, and oceans. One form of coliform bacteria that is particularly hazardous is *E. coli*, pictured below.

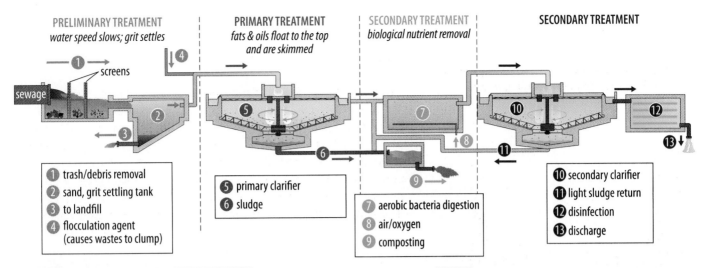

17-16
Raw water passing through a potable water purification plant will be treated in a multi-stage process.

17.11 WATER TREATMENT

You can see, then, that we really do pollute most of the water we use. After we use it for one thing, we usually can't use it for drinking, cooking, or other ordinary uses. But that water is not lost forever. Water can be recycled over and over again after removing solid wastes and undesirable chemicals. Even natural rural areas far from people can have water unsuitable for drinking. Organisms such as the protozoan *Giardia lamblia* live in clear brooks flowing through woods and they can cause an intestinal disease called "beaver fever." So don't drink the water!

Water purification naturally happens in the water cycle. Evaporating water leaves behind pollutants. Water vapor enters the atmosphere in a nearly pure state, though atmospheric pollutants can affect water droplets in clouds. You learned earlier that many kinds of natural soil bacteria break down chemical pollutants in groundwater, turning them into harmless substances. This is the idea behind household septic systems with *leach fields*. But for most communities, a more efficient method for providing clean drinking water is to recycle wastewater. Even if there are other natural water supplies, it's good and wise dominion to remove pollutants from wastewater before returning it to the environment or reusing it. Municipal sewage treatment plants use a process similar to that shown in Figure 17-16 to process wastewater. Sewage treatment plant standards for releasing water to a river are usually not as high as those for treating drinking water supplies. After you're done studying that diagram, go get a drink of water!

17C REVIEW QUESTIONS

1. List three ways that people where you live use water, other than for drinking, cooking, or cleaning.
2. What is the main idea behind water conservation?
3. What kinds of pollutants can affect water quality?
4. Use Figure 17-16 to describe the three basic ways that water treatment plants remove pollutants from water.
5. (True or False) Once water is polluted, it is lost forever as a source of drinking water.

17D Objectives

After completing this section, you will be able to

» evaluate old- and young-earth models for the origin of solution caves.

» explain where cave features come from.

» distinguish between a spelunker, a caver, and a speleologist.

» describe some features of karst topography.

GROUNDWATER LANDFORMS

What conditions are required to form caves?

17.12 WHAT ARE CAVES?

Have you even been in a cave before? There's something mysterious about caves. They make you want to look for strange creatures, hidden pools, or ancient drawings! They bring out the explorer in each of us.

A **cave** is a naturally occurring underground space. One cave may be easier to get to than another. But the big question is, how did caves get there? Some, especially those in desert areas, formed when

great slabs of rock were torn apart by weathering and gravity. This left spaces between the tilted slabs large enough for people to get in. Other caves were carved out by rivers or ocean wave action. Lava tubes, channels in rock through which lava flows, can become caves if they open at the lava surface. Glacier caves, or ice caves, form when meltwater hollows out ice.

The largest and most spectacular caves seem to have formed when great underground streams flowed through soluble rocks, such as limestone, dolomite, halite, and rock gypsum. These amazing caves are called **solution caves**. Because of their size and beauty, we'll focus on these kinds of caves in this section.

17.13 ORIGIN OF SOLUTION CAVES

Geologists are uncertain about exactly how solution caves formed. As the name implies, we understand that they form when water carries away minerals. While water dissolves these minerals well, acids do the job much more efficiently.

CURRENT THEORIES

One theory proposes that rain absorbed carbon dioxide (CO_2) from the atmosphere and the ground to form weak carbonic acid. The acid seeped into the ground through joints in the limestone. The carbonic acid dissolved the limestone, neutralizing the acid. As the limestone neutralized the acid, more carbonic acid from rainfall continued the process. Over time, this slow process eventually formed a cave under the force of gravity. One difficulty with this theory is that the neutralized water needs an outlet.

Is there a faster way to form a cave? Another theory states that hot acid from deep within the earth dissolved the limestone. Instead of moving downward from the surface, this acidic water moved upward. The acid in this process is sulfuric acid created by hydrogen sulfide (H_2S) reacting with hydrothermal water. The acid moved through joints in the limestone, aggressively dissolving the minerals and carving out caves. This process also needs an outlet but it is much easier for the water to exit at the surface. This stronger acid would produce caves in a much shorter amount of time.

CAVES AND WORLDVIEW

Do old- and young-earth scientists agree about cave formation? In some regards, yes. All **speleologists** (cave scientists) know that rain does become acidic when it absorbs carbon dioxide. They also know that carbonic acid dissolves the minerals in limestone through a very slow process. We all agree that solution caves form when acidic water dissolves the minerals in limestone. We also agree about the need for outlets to remove dissolved minerals so that the process continues. However, our view of history affects which process we believe likely happened in the past and how long it took.

Old-earth geologists believe that this process has always occurred and continues to occur at its same slow rate. Most hold to the carbonic acid theory and claim that this process formed caves over hundreds of thousands of years.

Young-earth geologists look at cave formation in the context of the Flood. They know that geologic forces during the Flood could provide the conditions for sulfuric acid formation. The Flood also could have provided the force to move acidic water upward through the limestone layers. A young-earth model also includes high levels of carbon dioxide and ample water to explain additional cave formation by carbonic acid erosion.

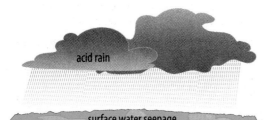

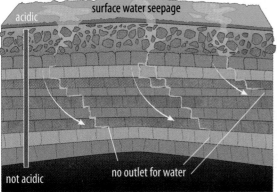

17-17
One theory of cave formation states that rain water absorbed carbon dioxide, which percolated downward through the limestone layers. Old-earth geologists assume that this process formed solution caves over hundreds of thousands of years.

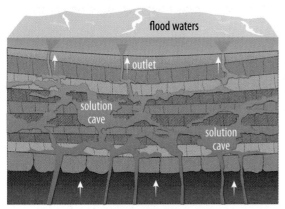

17-18
Some believe that sulfuric acid in hydrothermal water from deep in the earth may have eroded most of the solution caves that exist today during or shortly after the Flood. Creation scientists tend to accept this theory.

LIFE CONNECTION: CURIOUS CAVE CRITTERS

If you've ever toured a cave, you know there's not a lot of life there! Nearly all plants and animals depend on the sun to survive. The sun fuels photosynthesis to produce sugars, the base of most food chains. But there's no sunlight in a cave, except for a small zone near the entrance.

This is where you may find cave visitors, such as bats, bears, and snakes. They use the cave for temporary shelter but spend much of their lives elsewhere. Deeper in the cave, in the twilight zone, insects, spiders, centipedes, and salamanders may be more abundant. These can survive outside, but they can also spend their whole lives in the safety of the cave.

Most of a cave's volume is in the dark zone. Very few organisms can survive here in the cave's inky blackness and high humidity. Things that live here are permanent cave dwellers and probably couldn't survive in the outside world.

Animals that live in this pitch-black world get creative in their food choices. Because food is limited, most animals are quite small. Some scavenge food washed in by streams or floods. Guano (bat droppings) is a valuable food source for bacteria, fungi, and many small creatures. There are even mats of bacteria that can survive on iron or sulfur! Organisms like this are called *extremophiles*.

In this alien world, *speleologists* find some bizarre creatures. Many caves have ghostly pale flatworms and fish with little or no pigment in their bodies. No need to be flashy and colorful here! No need to even see, for that matter.

Many kinds of small, blind cavefish have eyes that are tiny or covered over by scales. In a world with limited food, it doesn't make sense to use energy for organs that aren't needed. Many of these special animals need less energy than similar species in the outside world.

Some cave insects and crayfish have very long antennae. Apparently, these help them to find their way around in the eternal night. Some crayfish have no eyes at all. For these creatures, an antenna is like a blind person's cane. Certain spiders and cave crickets have oversized legs that must, in some way, aid their movement and search for food.

The contrast between these permanent cave dwellers and similar creatures in the outside world raises a question: Did God create them like this? Or did they slowly develop these changes over many generations?

Since we believe that most caves are the result of the Flood and its aftermath, animals in caves didn't start there. God engineered living things with an amazing flexibility to survive in different environments. Populations of many of these cave species probably grew in caves with their limited food resources because there were fewer predators. Or perhaps they simply became isolated from the larger, sighted population. But no new kinds of life have evolved in caves. It is evident that most cave dwellers have lost features that their ancestors once had, not added new ones.

17.14 CAVE FEATURES

Caves and Judgment

Some of the most spectacular caves in our planet may have been formed through God's judgment in the Flood. What does this possibility suggest about God? How does studying caves relate to keeping the Creation Mandate?

speleothem (SPEE lee oh THEM): speleo- (cave) + -them (Gk. *thema*—deposit); secondary mineral deposits found in caves

After a solution cave formed, the water it contained often drained away as the water table lowered. Dissolved minerals—mainly calcium carbonate—became concentrated in pools left behind. Dripping or flowing groundwater carried dissolved minerals into these caves. This process continues today. When the water drips and then evaporates, the minerals come out of solution. Underwater or on cave walls and ceilings, the dissolved substances precipitate, gradually coating surfaces with a thin layer of calcite or other minerals. The strange and wonderful deposits formed this way are called **speleothems**. Because these formed *after* the original limestone strata in which the cave formed, speleologists call these formations *secondary limestone* deposits.

Before a cave dries out, it contains slow-moving streams and stagnant pools of water rich in minerals. As evaporation occurs, flat calcite or travertine deposits build up at the water's surface, forming *shelves*. Later, after the water level lowers, the shelves remain jutting out from cave walls. Also, strange, bulging, cloud-shaped deposits build up underwater. Unsurprisingly, these formations are called *clouds* by cavers.

Most speleothems form in air, long after the cave streams have drained away. Rock formations that build up from drips and trickles from the cave ceilings and walls are called *dripstone* or *flowstone*, depending on how the water moved. The most recognizable speleothems are **stalactites**—icicle-like projections—that hang down from ceilings and **stalagmites**—projections growing out of cave floors that form beneath the stalactites.

17-19 This image shows a variety of speleothems. While they all form by a similar process, they have vastly different appearances.

Most speleothems are made of calcium carbonate. Pure calcium carbonate is white, but impurities give it different colors. For example, iron oxide produces the reddish-brown color that many cave formations have. Yellows may come from dissolved organic matter from the land surface above the cave. Green colors come from copper compounds.

RAPID SPELEOTHEMS

Young-earth geologists have demonstrated that speleothems can form in very little time. In caves that have dried up, the growth rate of these structures is slow. However, stalactites and stalagmites grow rapidly in moist, active caves. For example, in Sequoyah Caverns, Alabama, geologists measured the growth of stalactites over a ten-year period at several centimeters a year. Undecayed carcasses of bats are sometimes found encased in stalagmite calcite at Carlsbad Caverns, New Mexico. If the stalagmite's rate of formation had not been fast, the bats would have decayed before fossilization. In one dramatic example, a curtain of stalactites formed from the ceiling of the foundation underneath the Lincoln Memorial in Washington DC. A photograph taken in 1968 showed that the stalactites had grown 1.5 m in the forty-six years since the monument was built.

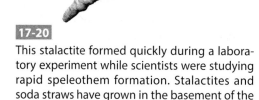

17-20
This stalactite formed quickly during a laboratory experiment while scientists were studying rapid speleothem formation. Stalactites and soda straws have grown in the basement of the Lincoln Memorial in Washington DC (below).

Using simulated cave conditions, scientists have studied the growth of stalactites in the laboratory. In one experiment, they installed a 50 cm stalactite having a mass of about 1.8 kg in an apparatus that supplied a solution of calcium carbonate and water that flowed over it. The stalactite added 35 g of calcite in eighty days. This deposition corresponds to an increase of 0.45 kg every three years. At this rate, the entire original stalactite could have formed in only twelve years. At the same time, a 17.5 g stalagmite grew beneath the stalactite.

17.15 KARST TOPOGRAPHY

In many regions of the world, thick strata of chemical sedimentary rocks lie under the earth's surface. These rocks have eroded to produce a distinct landscape called **karst topography**. The name comes from the striking landscape of the Kras Plateau (also known as the Karst Plateau) in Slovenia. Karst topography is heavily eroded and hilly. Florida, Indiana, Kentucky, Tennessee, and Virginia have areas of karst topography. Nearly every continent has regions containing *karstic* features.

One of the most distinctive features of karst topography is a *sinkhole*. A sinkhole develops when a section of a solution-cave ceiling becomes so thin that it can no longer support its own weight, and it collapses. On the land surface above the collapse,

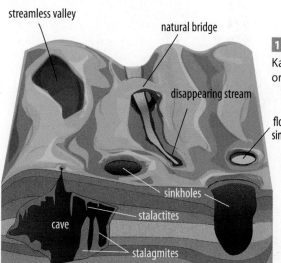

17-21
Karst topography includes some or all of these features.

The adjective "*karstic*" relates to karst topography.

A sinkhole pond

a bowl-shaped depression appears. This may be small or quite large. Some sinkholes have destroyed vehicles and buildings. In many places, sinkholes fill with water to form *sinkhole ponds*. Interestingly, the Arecibo Radio Telescope antenna in Puerto Rico was built inside a huge karst sinkhole (see page 622 for a picture of this telescope).

Karst topography generally includes a network of interconnected caverns, sinkholes, and cave openings. This forms a very irregular terrain. After collapse of the caves and extensive erosion, tall, tower-like hills litter the countryside. These are all that remain of the original limestone strata. Such terrain is called *tower karst topography* (see Figure 17-23 below). China has a large region of impressive tower karst hills.

A special kind of *natural bridge* occurs in karst topography. The bridge is a short section of a solution cave's ceiling, left over after the rest of the cave collapsed. One famous karstic natural bridge in eastern United States is located in the Natural Bridge State Park in the Southern Shenandoah Valley of Virginia. This bridge is 66 m tall and spans a 27 m gap.

17-22
Natural Bridge near Lexington, VA
Location: 37.6278 °N 79.5448 °W

17-23
Tower karst topography in the Guilin province of China
Location: 24.75 °N 110.54 °E

Karst topography often contains a *disappearing stream*. Water from a surface stream has seeped down to erode a joint in the underlying rock. If the downward trickle opens into a cave ceiling, the water flow quickly erodes a wider passage. Eventually, the stream follows the hole down into the cave and disappears from the surface. The abandoned streambed is a *streamless valley*. But it can also work the other way. Cave streams can erode through a cave wall. Or surface erosion can cut downward into a cave with a stream. In either case, an *appearing stream* forms.

17.16 CAVE EXPLORATION

Caves are begging for exploration. But you can't just wander into a cave and start poking around! Professionals call people like this **spelunkers**. Spelunkers don't know how to safely and responsibly explore caves. And they don't have the gear for it either. These people often deface caves with graffiti or damage speleothems. Their actions deprive other explorers the enjoyment of seeing these unusual features.

People who responsibly explore caves to learn about them, or just for recreation, call themselves *cavers*. Caving calls for planning and the use of specialized gear. Speleologists have to be good cavers. Exploring some caves requires an elaborate system of ropes, scuba-diving gear, bottled oxygen, or other special equipment. Though it can sometimes be dangerous, there are some incredible adventures in store for cave explorers!

spelunker (spee LUNG kur): spelunk- (L. *spelunca*—cave or grotto) + -er; one who visits a cave

17-24

Go have your own cave adventure!

17D REVIEW QUESTIONS

1. How are lava-tube caves, ice caves, and solution caves different?
2. Old-earth geologists rely on what two things to explain the formation of solution caves?
3. What four conditions that probably existed right after the Flood could have helped form solution caves?
4. What are two categories of speleothems that result from the slow buildup of mineral deposits on cave ceilings, walls, and floors as groundwater evaporates in air?
5. What icicle-like speleothems hang from cave ceilings? What kind often form directly beneath them?
6. Does geologic evidence suggest that speleothems require tens of thousands of years or more to form? Explain.
7. If you were hiking in an area of karst topography, what could you see?
8. Contrast the terms *spelunker*, *caver*, and *speleologist*.
9. (True or False) Speleothems can form only when the cave air is dry enough to allow groundwater to evaporate, leaving behind minerals.

SERVING GOD AS A SPELEOLOGIST

DESCRIPTION

Caves are places for adventures. People have discovered hidden rooms, massive fragile crystals, ancient artifacts, hidden glaciers, and unusual creatures in caves. No wonder people want to explore them! Professional cave scientists are called *speleologists*.

Speleologists spend lots of time in caves, where it's often wet, cold, claustrophobic, disorienting, and physically exhausting. They often have to rappel down sheer cliffs, scuba dive in underwater caves, crawl through muck, monitor poisonous gases, and wriggle through tight places, all in the name of science. But the thrill of discovery is worth it!

EDUCATION

What does it take to be a speleologist? Well, speleology is another one of those sciences that involves knowledge in lots of other areas, like biology, hydrology, chemistry, geology, climatology, mapping, and even archaeology! So if you want to be a speleologist, you could study any of these things. Most schools don't have degree programs in speleology.

But you don't have to wait until you're in your twenties to get into a cave! If you're interested, read everything you can get your hands on. Join the National Speleological Society. Stay fit and practice skills that you'll need for the sport of cave exploration. Take a tour of a cave or join a local caving club to plan your next cave adventure.

POSSIBLE WORKPLACES

Speleologists usually work for the government or for private science companies like National Geographic. But the result is the same—speleologists get paid to spend lots of time in caves all over the world.

DOMINION OPPORTUNITIES

Speleologists can responsibly and carefully collect data in fragile cave environments to help us understand and conserve caves. This helps us protect the unique resources we find there. Some caves are home to unusual animals, and the fantastic cave formations (speleothems) that can't be found anywhere else.

Speleologists help us understand karst topography better so that people can know how and where to build homes and businesses above ground. They've also discovered extremophile organisms in caves that produce substances that have the potential to fight off cancer and malaria. Speleologists can make our lives better while revealing the beauty of God's world deep underground.

CHAPTER 17 REVIEW

CHAPTER SUMMARY

» Almost all the earth's fresh water not trapped in glacial ice is groundwater.

» We need to find ways to supply fresh water for the earth's growing population as a way to love God's image bearers in obedience to Him.

» The water cycle is the natural, ongoing process by which water moves between the ocean, atmosphere, and land through evaporation, precipitation, and the flow of streams and groundwater.

» Groundwater percolates down through soil and permeable rock until it fills up all available spaces between particles. The upper surface of this zone of saturation forms the water table.

» The relationship of aquifers to the land's surface can produce springs, perched water tables, and artesian wells, among other groundwater features.

» Groundwater flows mainly under the influence of gravity, just as surface water does. The permeability and arrangement of rock strata control the direction and rate of groundwater flow.

» Drawing water from a well always forms a cone of depression in the water table at the well. If water is pumped at too high a rate, a deep cone of depression can form or the well can go dry.

» Overpumping a well can change the direction of groundwater flow, deplete nearby wells, and draw in pollutants.

» Water naturally returns to the groundwater reservoir by recharging. Groundwater can be artificially replenished by recharge wells.

» Groundwater has the ability to dissolve minerals. Water with large amounts of dissolved minerals is considered *hard*. Water lacking these minerals is *soft*. The measure of dissolved mineral content is hardness.

» Hard water can clog pipes with mineral deposits and reduce water quality. There are several methods for softening or demineralizing water.

» Fresh water is a vital resource. The value of water increases as it becomes more scarce.

» Water conservation is based on the idea that we use only what we really need and no more. Proper dominion demands that we search for ways to reduce the amount of water that we waste.

» Reusing water and recycling through water treatment conserves water.

» We normally pollute water as we use it. Water pollutants include chemicals, heat, bacteria and viruses, and sediments and minerals. All of these lower water quality and can make it unsuitable for various purposes.

» Water treatment plants remove pollutants so that they can safely return water to the environment or so that we can drink it. The output of treatment plants must meet government water quality standards.

» Caves may form in a variety of ways, but the largest and most spectacular caves were formed by flowing groundwater.

» Old- and young-earth theories for the origin of solution caves differ in their age, in how long they took to form, and with regard to the source of acidic groundwater that eroded the caves.

Key Terms

groundwater	399
water cycle	400
water table	400
zone of aeration	400
zone of saturation	401
permeability	402
aquifer	403
cone of depression	404
recharge zone	404
solvent	406
hardness	407
irrigation	409
pollution	411
cave	414
solution cave	415
speleologist	416
speleothem	418
stalactite	418
stalagmite	418
karst topography	419

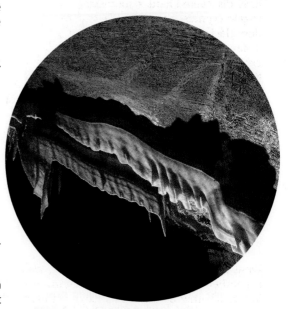

17-25

Cave bacon is a form of flowstone similar to curtains (or drapery) and has the distinctive color stripes seen above.

» Speleothems develop from the precipitation of dissolved minerals as groundwater evaporates. Some speleothems can form in or under pools of standing water, but most form in air as dripstone or flowstone.

» Old-earth geologists believe that most solution caves formed a long time ago, so they assume that speleothems have been growing for many thousands of years. Experiments and observation of recent speleothem deposition show that they can grow quite quickly.

» Karst topography includes sinkholes, natural bridges, streamless valleys, disappearing streams, and underlying mazes of solution caves, as well as the soluble rock in which these features occur.

REVIEW QUESTIONS

1. What are the benefits of projects such as Gospel for Asia's Jesus Wells?
2. Why is groundwater an important resource?
3. Sketch the main steps of the water cycle. Include in your sketch the land, the ocean, and the atmosphere.
4. If you were digging a hole into the ground, how could you tell when you reached the surface of the water table?
5. Why is it a good idea to drill a well during a dry season instead of during a wet season?
6. Differentiate between porosity and permeability.
7. To avoid creating a large cone of depression, what is the ideal relationship between the recharge rate and the rate at which a user draws water from a well?
8. Why do laundry and linen services prefer soft water rather than hard water?
9. How do farmers on large produce farms ensure that they can grow and harvest their crops even with a lack of rain?
10. Evaluate the statement, "There are too many people for the amount of water available. We need to reduce the birth rate and get rid of extra, unnecessary people."
11. Why is coliform bacteria in a water supply a problem?
12. What is the first step in treating water drawn from a reservoir on its way to becoming drinking water?
13. What is a key problem with the old-earth model for solution-cave formation?
14. Why are stalagmites often found directly under stalactites?
15. Give two examples showing that speleothems (stalactites and stalagmites) can form quickly.
16. How do geologists believe that tower karst topography developed?
17. How would life on Earth be different if bedrock were not at all permeable?
18. Why would caves have made good shelter for the growing human population shortly after the Flood?

True or False

19. Most of the world's liquid fresh water is in lakes, streams, and rivers.
20. Fresh water that does not flow to the sea or immediately evaporate back into the atmosphere is lost to the water cycle.

17-26
Shelfstone forms as a slow moving body of water evaporates, leaving mineral deposits in a level shelf.

21. The water table usually follows the changes of elevation of the overlying terrain.
22. Even impermeable rock can be highly porous.
23. Groundwater flows downhill by gravity just like surface water does.
24. Only pure groundwater can dissolve minerals and other earth materials.
25. The natural softening of hard water within a hot water plumbing system can actually be a problem for homeowners.
26. A lack of water in a region can be a problem even if we still have enough to drink.
27. We normally pollute water as we use it.
28. People can get sick from the *E. coli* bacteria in water.
29. The presence of any bacteria in groundwater is a sign that the water is polluted.
30. According to one young-earth model of cave formation, solution caves could have formed by groundwater flowing *against* the force of gravity.
31. A speleothem is a secondary limestone deposit because it formed after the sedimentary rock (in which the cave formed) was deposited.
32. Using carbon-14 dating methods, geologists have proved that speleothems require thousands of years to form.
33. Karst topography is a special type of terrain found only in the Kras limestone plateau in Slovenia.

CASE STUDY: WHITE NOSE SYNDROME

In 2006, bats from a particular cloud—the term for a group of bats—starting dying at an alarming rate. They had a fungal infection called *White Nose Syndrome (WNS)*. The disease spread quickly through the northeastern United States and along the Appalachian Mountains. It has been devastating for the bat population, killing over 90% of affected clouds within the first five years.

How does this connect with cavers? Scientists discovered that people were spreading the disease. People who entered caves with infected bats carried fungus spores out on their clothing, boots, and equipment. Cavers unknowingly carried these spores to the next cave, spreading the disease. In ten years, WNS had spread from New York to the Pacific Northwest.

Questions to Consider

1. Why do you think that bats live in caves?
2. Should we explore and study caves?
3. What is our obligation to the bats and other organisms that live in and around caves?
4. Why is "spelunker" considered a negative term?

This little brown bat has White Nose Syndrome. You can see the white fungal growth on its muzzle and wings.

5 | THE ATMOSPHERE

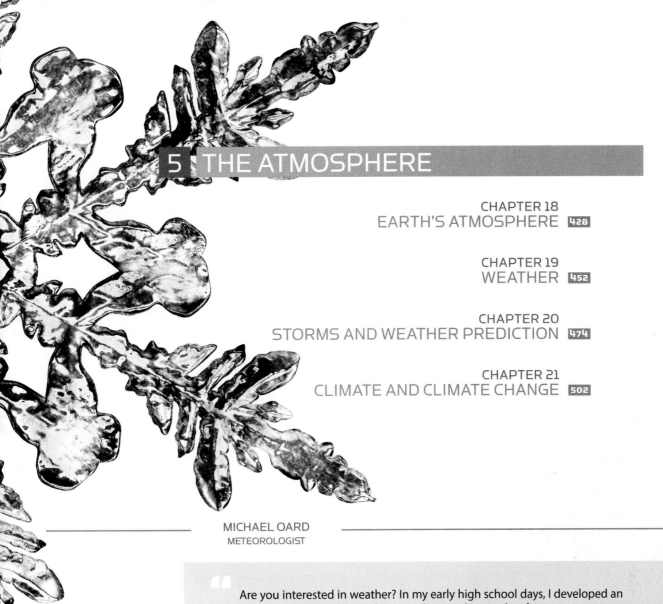

CHAPTER 18
EARTH'S ATMOSPHERE `428`

CHAPTER 19
WEATHER `452`

CHAPTER 20
STORMS AND WEATHER PREDICTION `474`

CHAPTER 21
CLIMATE AND CLIMATE CHANGE `502`

MICHAEL OARD
METEOROLOGIST

> "Are you interested in weather? In my early high school days, I developed an interest in meteorology, often wondering why weather forecasts are sometimes wrong.
>
> Since that time, I have pursued a career in meteorology and served as a weather forecaster with the National Weather Service for thirty years. As a meteorologist, I learned why, as a weatherman, I am sometimes wrong. Meteorology is a very complicated field with many unknowns. You may be surprised to learn that we still do not know exactly what causes tornadoes or lightning, for example.
>
> Weather forecasts depend on observations. These observations are taken at weather stations each hour, and more often in stormy weather. Surface and atmospheric observations are fed into one of the largest computers in the world. This computer uses equations to determine the current weather and to model what the weather might be in the future based on current conditions.
>
> So why are weather forecasts sometimes wrong? The two main reasons are that we do not have enough observations and our knowledge about what causes the weather is limited."

CHAPTER 18

EARTH'S ATMOSPHERE

18A WHAT IS THE ATMOSPHERE? `429`
18B SPECIAL ZONES IN THE ATMOSPHERE `439`
18C ENERGY IN THE ATMOSPHERE `445`

KILLER AIR

As Earth's population increases, our cities are getting bigger. God intends to see our planet filled with His image bearers. But our cities are having some growing pains. Air pollution is skyrocketing in the biggest cities around the world, cities like New York, Beijing, Mexico City, and New Delhi.

Air pollution forms when gases or particles that enter the air interfere with breathing or are hazardous in other ways. Sometimes these come from natural sources, like volcanoes, forest fires, radioactive radon, or even molds and pollen. But in the city, automobiles and fossil fuel power plants are the two main culprits. They are the source of the haze called smog that settles over so many of the world's cities.

Air pollution isn't just an inconvenience. It also kills people. And it doesn't just stay outside; sometimes indoor air pollution is even worse! Air pollution kills about 5.5 million people a year. Take a breath—we have work to do!

18A

WHAT IS THE ATMOSPHERE?

How does the atmosphere change with elevation?

18A Objectives

After completing this section, you will be able to

» describe how people can affect the atmosphere.
» identify evidence of design in the atmosphere.
» sketch the atmosphere's composition, temperature, and structure.
» trace the flow of carbon and nitrogen in the atmosphere.

18.1 AIR CARE

Our **atmosphere**, the envelope of gases that surrounds our planet, is crucial to life on Earth. Consider some of the other planetary bodies in our solar system. Jupiter's atmosphere is mostly hydrogen and helium, with clouds of ammonia ice crystals. Have fun trying to get a breath of fresh air there. You're not going to do much better on Mars or Venus, which have atmospheres made mostly of carbon dioxide. Venus's so-called air contains sulfuric acid! Closer to home, the moon is even worse. It hardly has a measurable atmosphere at all. Earth's atmosphere provides the gases that creatures here need to live. The Bible says that God created our earth to be inhabited (Isa. 45:18). But if our air gets too polluted, living things can get sick and die.

We need to take care of our air. Scientists are looking for ways to reduce the toxic particles and harmful gases that people release into the air. Cutting back on *emissions* from manufacturing, from cars and trucks, and from coal-fired power plants is part of this. Developing vehicles and power plants that don't use fossil fuels at all is another way, though some pollutants are unavoidable. But we should try to work on ways to filter and clean the air we breathe, both indoors and outside. First, though, let's learn about the atmosphere so that we can better understand the challenges we face.

Developing technology like this electric car can help reduce air pollution in cities around the world.

18.2 ORIGIN OF THE ATMOSPHERE

Some people credit the marvelous uniqueness and design of Earth's atmosphere to pure chance. To them, our amazing atmosphere is just one big coincidence. Do you remember their view of Earth's beginnings from Chapter 5?

THE OLD-EARTH STORY

According to old-earth scientists, the earth began forming more than 4.5 billion years ago. The early earth was a smaller, molten ball covered by a thin crust. Within several hundred million years of the planet's birth, a moon-forming collision occurred and melted the entire earth again. As the earth cooled from this collision, gaseous materials gathered into an atmosphere that was hot and contained mainly the light gases hydrogen and helium, which on their own can't support life. No free oxygen was present in the early atmosphere. The emissions of numerous volcanoes changed its composition. Water vapor together with carbon and nitrogen compounds began to collect. Ice from comet bombardments added water and other compounds to the oceans and atmosphere.

Sometime during the first billion years of Earth's existence, evolutionists believe that life appeared from nonlife. These life forms were *anaerobic bacteria* that could live in the absence of oxygen. Biochemists believe that oxygen would have destroyed the complex molecules needed to build even the simplest cell. Even so, ultraviolet light from the sun broke down some water vapor into oxygen and hydrogen. This oxygen quickly reacted with minerals in the rocks, so the simple bacteria were not harmed. Oxygen levels were probably less than 1% of today's levels. But along the way, oxygen-producing organisms called *cyanobacteria* evolved.

Around 2.4 billion years ago, the surface rocks had reacted with as much oxygen as they could, so the extra oxygen collected in the atmosphere, and the oxygen levels in the air jumped. This *oxygen catastrophe* killed off most of the anaerobic organisms, and paved the way for more advanced forms of life to evolve. Secular scientists believe that photosynthetic organisms are responsible for the modern oxygen-rich atmosphere we have today.

THE YOUNG-EARTH STORY

A biblical young-earth view of the atmosphere's origin is much simpler. To be completely honest, we don't *know* when God created the atmosphere because it is not mentioned in the Creation story in the Bible. We can assume that when God formed the firmament (or great expanse) between the waters above and the waters below on Day 2 (Gen. 1:6–8), He also created the atmosphere in preparation for the creation of plants on Day 3. Plants needed atmospheric carbon dioxide and water vapor to live, and all the other creatures and humans definitely needed to breathe during the remaining three days of the Creation week.

God made the atmosphere "very good," along with the rest of His creation. Even the Fall probably didn't noticeably change the atmosphere. But the Flood did. Volcanic activity and changes in the sizes and shapes of the oceans deeply affected the earth's atmosphere. We can still see the effects of these changes today in the atmospheric chemistry, weather patterns, and climate zones all over the world. But though God judged our world, it still bears signs of His design and loving care. The atmosphere is one grand exhibit of this.

Bi **Experiment in Evolution**

Have you heard of the Urey-Miller experiment? Two evolutionists did an experiment in which they simulated how life could have evolved in Earth's early atmosphere. They took a mixture of hydrogen, carbon dioxide, methane, water vapor, and ammonia and sparked electricity in it. They produced what they called "the precursors of life." Evolutionists called it a victory for their theory. But there are some major problems with this experiment.

18.3 ATMOSPHERIC REGIONS BY COMPOSITION

You are a pioneering meteorologist, taking a trip to explore Earth's atmosphere from the ground to space in person! You are one of the first passengers in NASA's new *space elevator*. But you need some information about the atmosphere before your trip. Will you need supplemental oxygen? How should you dress? Sounds like you will learn much about the layers of the atmosphere!

We can study the earth's atmosphere in several different ways. For example, we may be interested in its *composition*, that is, the kinds of gases it contains. In Chapter 2, you learned how everything is made of atoms. Different atoms make up different chemical elements. Nitrogen makes up more than three-fourths of the atmosphere near the earth's surface. It doesn't easily react with other chemicals, so it exists as a pure element in air and isn't poisonous. One of its main purposes is to dilute oxygen to the concentration that is best for life while avoiding a fire hazard! This is not an accident of Earth's geologic history; it is evidence of God's wonderful design.

18-2
First proposed in 1895, a space elevator would allow movement from Earth's surface to space without the use of rockets. This is an artist's rendering of an elevator car traveling to space.

CASE STUDY: SKYDIVING FROM SPACE

Looking down toward Earth from your space elevator, you might be tempted to jump if you had a parachute. Joseph Kittinger actually jumped from the upper atmosphere in 1960! Riding a balloon to 31,300 m, he skydived from this height. He set the record for the highest and fastest parachute jump in history as well as the longest free fall. It was fifty-two years before anyone even attempted to break these records.

On October 14, 2012, Felix Baumgartner began his day by donning a space suit. He rose to 39,040 m in a space capsule held aloft by a balloon. He jumped from this height to set the record for the highest skydive. Plummeting 36,529 m in 4 minutes and 20 seconds, he also set a new record for the longest free fall. He set another record for being the fastest skydiver at 1345 kph. Baumgartner broke the sound barrier during his free fall, exactly sixty-five years after Chuck Yeager accomplished the same feat in the Bell X-1 aircraft. However, Baumgartner failed to break the record for the longest time of descent. Joseph Kittinger participated in Baumgartner's record-breaking event as a member of mission control.

Both of these jumps show how far (or high) man will go to study the world God gave us. Data from each jump provided information about both the atmosphere and how people handle stresses in this environment. Engineers can use this information to design systems to protect people from these harsh surroundings.

Felix Baumgartner leaves the capsule for his record-breaking skydive in 2012. Notice the protective equipment that was needed even though they were still in Earth's atmosphere.

Questions to Consider

1. At 39,040 m, from which layer of the atmosphere did Felix start his jump?
2. Where within the layers of the atmosphere would these men have traveled through the coldest air? At approximately what altitude would this have occurred?
3. What is the composition of the air through which they traveled?
4. Why did they need space suits for their jumps if they were within the atmosphere the entire time?

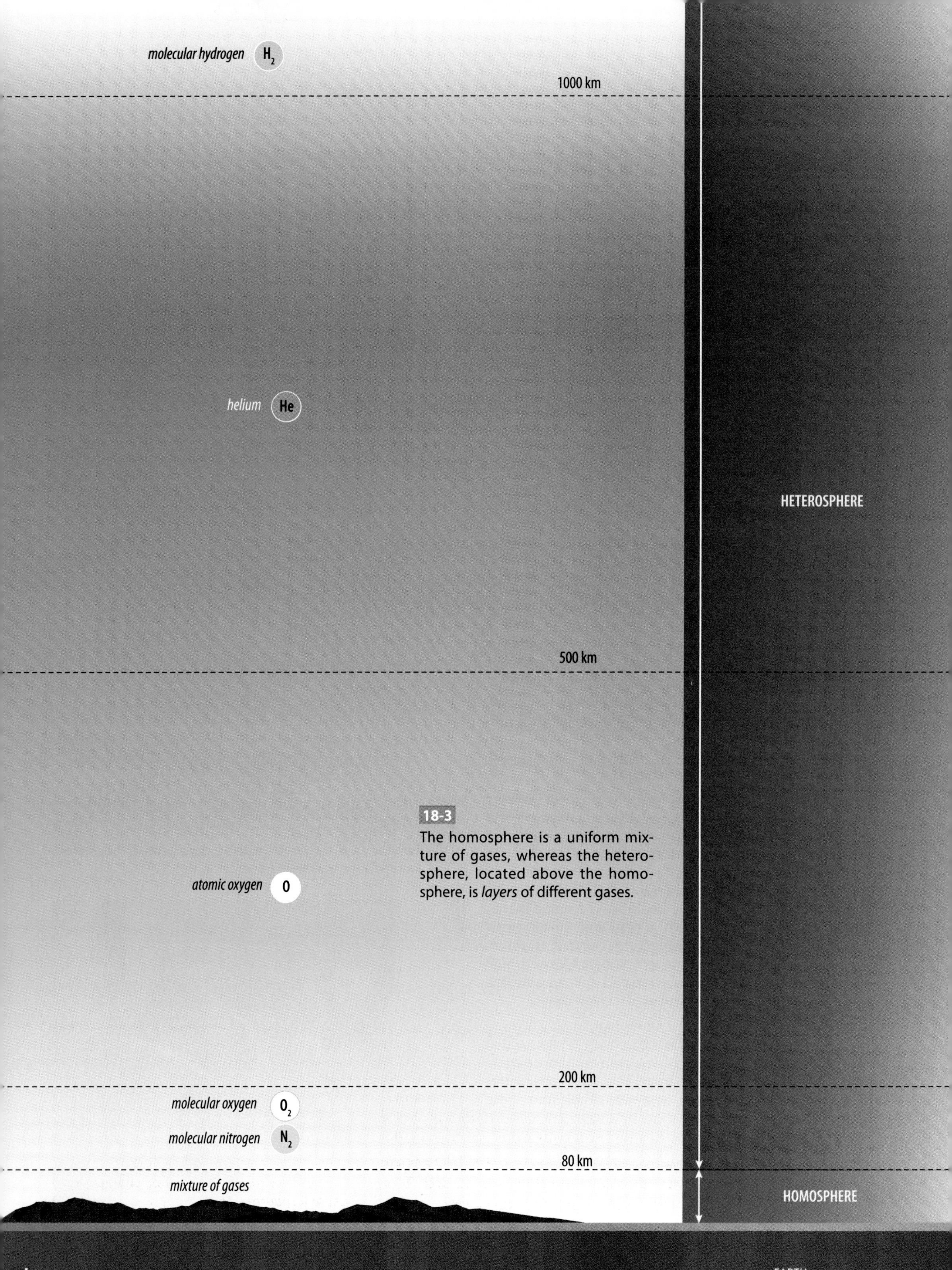

molecular hydrogen H_2

1000 km

helium He

HETEROSPHERE

500 km

18-3
The homosphere is a uniform mixture of gases, whereas the heterosphere, located above the homosphere, is *layers* of different gases.

atomic oxygen O

200 km

molecular oxygen O_2
molecular nitrogen N_2

80 km

mixture of gases

HOMOSPHERE

EARTH

We know that the atmosphere near the ground is a mixture of gases. Its volume is about 21% oxygen, 78% nitrogen, and 1% other gases such as argon, carbon dioxide, helium, and hydrogen. The portions of all these gases near the earth's surface are almost constant except for carbon dioxide and water vapor. Carbon dioxide may vary between nearly 0% to a little more than 2%. Water vapor may vary from almost 0% to 5%.

As we travel upward in our space elevator we notice that the composition of the atmosphere changes. All the gases that make up air are thoroughly mixed together in the lowest part of the atmosphere by winds and weather. They don't separate, settle out, or form layers. We call this uniform mix of gases the **homosphere**. It stretches upward to an altitude of 80 km. However, starting at about 3 km, you would need to breathe pressurized oxygen. Above 3 km, there is too little pressure for air to supply all the oxygen you need. Your brain starts acting fuzzy and your reactions slow down. That's why you are wearing a space suit for your trip!

As we travel above the homosphere, air composition changes. In this region, called the **heterosphere**, the atmosphere has several layers of different gases. Since there is little wind at these altitudes, gases settle out by their atomic weights—with the heaviest gases nearer the bottom. From lowest to highest altitude, they are molecular nitrogen (N_2), molecular oxygen (O_2), atomic oxygen (O), helium (He), and molecular hydrogen (H_2). You can see the arrangement of these layers in Figure 18-3 on the facing page. This order makes sense, since the nitrogen molecule is fourteen times as heavy as the hydrogen molecule.

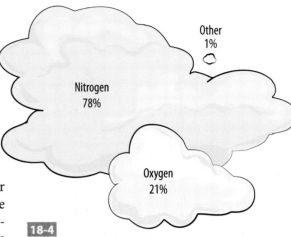

18-4
Air is mostly nitrogen, which serves to dilute the oxygen.

The prefix *homo-* means "like," "similar," or "uniform." The **homosphere** layer of the atmosphere is a nearly uniform mixture of gases.

The prefix *hetero-* means "separate" or "different." The **heterosphere** layers of the atmosphere contains separate layers of distinct gases.

18.4 TEMPERATURE LAYERS OF THE ATMOSPHERE

In addition to changes in composition, you also notice that the temperature probe on the exterior of the space elevator is indicating significant change as you travel upward through the atmosphere. The atmosphere has five distinct temperature layers. Air temperature changes within a layer, warming or cooling with altitude at different rates.

The lowest layer of the atmosphere is the **troposphere** (TROE puh SFEER). This zone begins at the earth's surface and extends upward to about 9 km over the North and South Poles, to about 11–12 km over the continental United States, and to 16–17 km over the Equator. This is where most of Earth's weather happens. As your elevator climbs, you will pass through layers of clouds. You can expect to run into brief rain showers. Gusts of wind will swing your space elevator back and forth on its carbon tether. In the troposphere, temperature drops with increasing altitude. The temperature lowers about 6.5 °C for every kilometer increase in altitude. Atmospheric scientists call this kind of steady change in temperature a **lapse rate**. At approximately 11 km up, the temperature stops changing. This upper boundary of the troposphere is the *tropopause*. Here, the temperature is a chilly −55 °C. Brrr! Your elevator's electrical heating and thermal insulation will pay off.

tropopause: tropo- (to turn or change) + -pause (L. *pausa*—stop)

EARTH'S ATMOSPHERE

The cold air at the top of the troposphere is dense and tends to sink. The warm air at the bottom of the troposphere is less dense, rises, and is replaced by the colder air settling from above. This process constantly mixes air in the troposphere. Rising warm air carries moisture that forms clouds and rain. Sinking cold air brings clear, cooler weather. This kind of thermal mixing happens only in the troposphere.

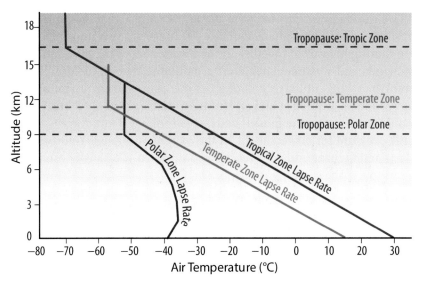

18-5
The steady drop in temperature as altitude increases is called the *lapse rate*. For regions of the atmosphere where the temperature *rises*, the lapse rate is *negative*.

stratosphere: (STRAT uh SFEER): strato- (L. *stratus*—a spreading out or layer) + -sphere

Birds in the Jet Stream?
Some birds catch a ride on winds called *thermals* up to the jet stream. This helps them save energy during migration. Learn more about winds and migrations in the Life Connection on page 463.

mesosphere (MES uh SFEER): meso- (Gk. *mesos*—middle) + -sphere

But your elevator is still climbing! You notice that the **stratosphere**, the second temperature layer, extends from the top of the troposphere to about 50 km. The stratosphere is free of clouds and dust. In this layer, the temperature *increases* from about –55 °C at the bottom, to about 0 °C at the top. You'll be able to see for hundreds of kilometers from this altitude. Aircraft flying long distances share the stratosphere with your vehicle to avoid turbulent storms in the troposphere. It can be windy here too. Some winds in the stratosphere blow as fast as 400 kph in paths called **jet streams**. Aircraft sometimes fly with these jet streams to increase their speed and to save fuel. Jet streams sometimes temporarily dip into the troposphere. The upper boundary of the stratosphere is the *stratopause*.

As the elevator travels into the **mesosphere**, you notice that air temperature again decreases with altitude. The coldest temperature of the atmosphere, –90 °C, occurs at the top of the mesosphere. Like the troposphere and the stratosphere, the mesosphere is a uniform mix of gases. Also like the lower layers, the mesosphere has winds. These winds often have high speeds and come from different directions. The mesosphere ends at the *mesopause* about 80 km above the earth. The troposphere, stratosphere, and mesosphere all lie within the homosphere composition layer (see Figure 18-3).

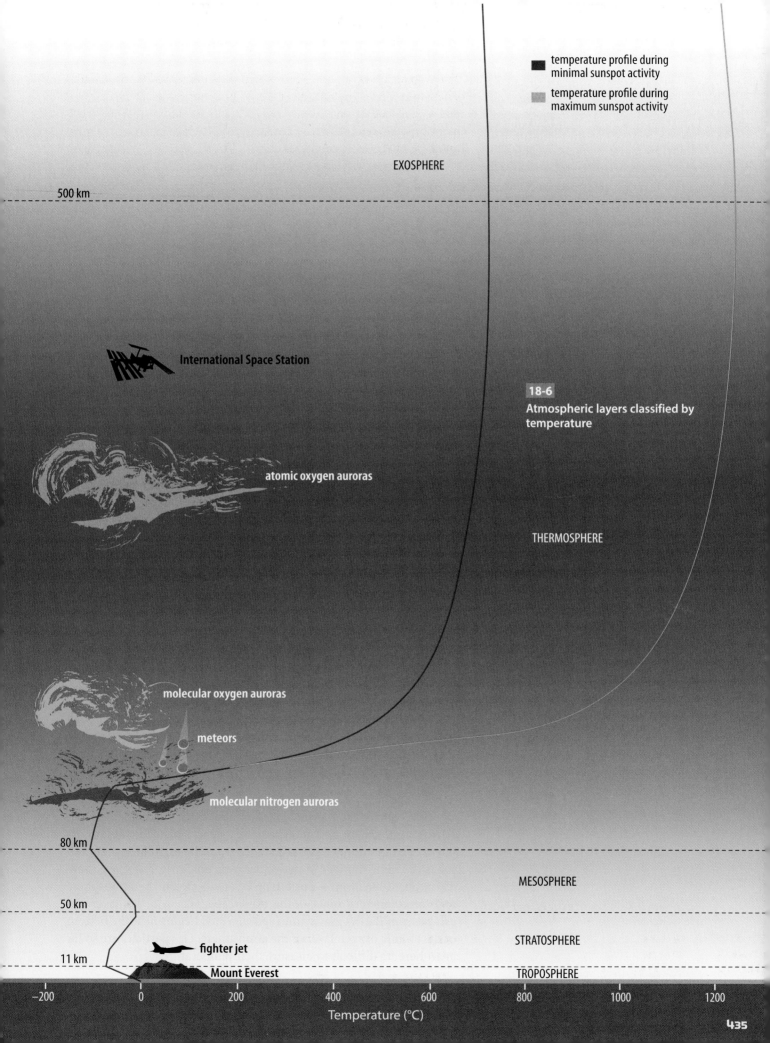

18-6 Atmospheric layers classified by temperature

18-7 Noctilucent clouds, strange bluish-white clouds, appear in the mesopause. Meteorologists aren't sure where they come from, but they're much different from clouds in the troposphere.

thermosphere (THURM uh SFEER): thermo- (Gk. *thermos*—warm) + -sphere

Spacecraft Cooling

Spacecraft traveling through the thermosphere during the day get hot from absorbing solar radiation. Because objects at this altitude must move at high speeds to stay in orbit, friction with the thin air makes them heat up even more.

To avoid absorbing excess heat, spacecraft are painted white or covered in mirrored skins. Both methods reflect heat away from the vehicle. Some spacecraft have fins that radiate heat into space on the shadowed side of the vehicle. Many satellites rotate to spread heating over their entire surfaces.

exosphere (EX uh SFEER): ex- (Gk. *exo*—outside) + -sphere

Passing through 80 km, your elevator enters the last true layer of the atmosphere—the **thermosphere**, which extends to the *thermopause* at about 500 km. The atmosphere is very thin here. Just 20 km above the mesopause, you enter true space at about 100 km. Higher than this altitude, an aircraft would have to move faster than orbital speed around the earth to get usable lift from the air. Scattered atmospheric atoms and molecules absorb the sun's full radiation. Daytime temperatures range from 650 °C to 1250 °C at the highest part of the thermosphere! In spite of the high temperatures, you would not feel warm in the thermosphere as long as you stayed in the shadow of your elevator vehicle. The molecules are so widely separated that too few would contact your suit to transfer heat that you could feel. So warmth from the surrounding daytime atmosphere is negligible. However, you would cook in direct sunlight without an insulated space suit!

Figure 18-6 shows daytime temperatures for the thermosphere. At night, the temperatures can be hundreds of degrees lower. Though the gas particles making up the thermosphere are widely scattered, they block some harmful radiation. They also protect the earth from small space rocks called *meteoroids* by burning them up as "shooting stars."

So what's above the thermosphere? There is no clear line between Earth's atmosphere and interplanetary space. Instead, the atmosphere becomes thinner as the distance from the earth increases until it has the same number of molecules per cubic meter as space itself. The transitional zone—not a true layer—between the atmosphere and interplanetary space is the **exosphere**. Gases in this layer can escape from Earth's gravity into space, which begins near the bottom of the atomic oxygen layer of the atmosphere. Gas particles are so far apart here it's difficult to measure their temperature.

18.5 CYCLES AND THE ATMOSPHERE

As you have learned in earlier chapters, key chemicals needed by living things are recycled in the natural environment. For example, organisms require carbon, nitrogen, and water, among other things. If these resources were not replenished by recycling, supporting the present amount of life on Earth would not be possible. In Chapter 13, we discussed how these nutrients flow from place to place in the marine environment. But that's just part of the big picture. What role does the atmosphere play in global nutrient cycles?

THE CARBON CYCLE

The flow of carbon from living things to the carbon reservoirs in the environment and back again is what the *carbon cycle* is all about. Atmospheric carbon dioxide is a key carbon reservoir in the carbon cycle. Plants use the gas during photosynthesis to build their tissues and produce fruits made of complex carbon compounds. When animals and people eat the plants, they take in the carbon compounds, which become part of *their* tissues. Some of the food we eat is broken down for energy, and, as part of this process, carbon dioxide is exhaled into the atmosphere.

After living things die and decay, their carbon compounds enter the soil carbon reservoir. Remember that carbon dioxide and carbon compounds can dissolve in surface water and groundwater, and can ultimately enter lakes and even the ocean. The freshwater and marine carbon cycles both take part in the overall movement of carbon through the environment. So there is a flow of carbon between living things, the earth, water, and the atmosphere.

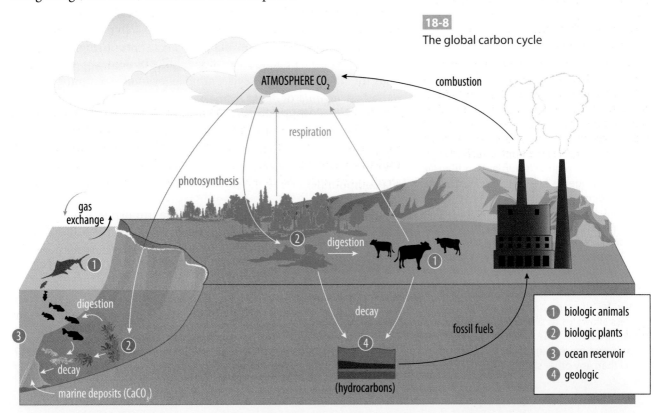

18-8
The global carbon cycle

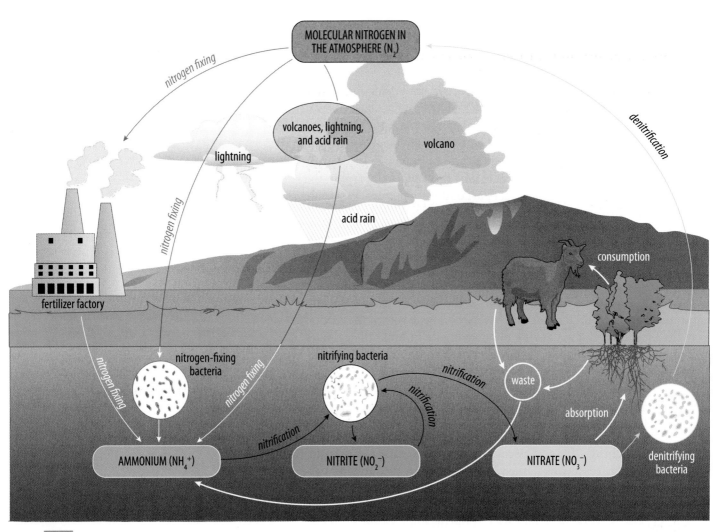

18-9

The global nitrogen cycle

THE NITROGEN CYCLE

The atmosphere is the largest nitrogen reservoir for the global *nitrogen cycle*. Air has lots of nitrogen, but most living things cannot directly use that element in life processes. Pure nitrogen does not easily take part in chemical reactions, like biological processes. As with the marine nitrogen cycle, atmospheric nitrogen must be converted to *organic* or *fixed nitrogen* compounds, such as ammonia (NH_3), by special *nitrogen-fixing* bacteria. These bacteria live in the soil, in special plants that provide a home for them, and in aquatic and marine environments. They take atmospheric nitrogen, and, through a series of chemical reactions, release ammonia or ammonia compounds.

Another source of fixed nitrogen in the atmosphere is lightning. Lighting breaks down nitrogen in the air and combines it with oxygen. The new compounds that form are another starting point for the nitrogen cycle.

Ammonia is the basic building block for many useful chemicals, such as those found in artificial fertilizers and cleansers. You probably have used glass cleaners containing ammonia because they easily clean greasy fingerprints and dirt from glass.

From Air to Asparagus

One of the main limitations on how much food people can grow is nutrients in the soil. Organic nitrogen is key, since it is plant food. In 1913 German chemists Fritz Haber and Carl Bosch introduced a method, called the *Haber-Bosch Process,* that turns nitrogen in the air into artificial fertilizer. This is the same thing that bacteria do! Suddenly, it was possible to grow much larger quantities of food than ever before.

Today, fertilizer made by the Haber-Bosch Process allows us to produce food four times more efficiently. This process probably has contributed more to growing the population today than any other single human discovery.

18A REVIEW QUESTIONS

1. What are some ways that we can reduce pollutants in the atmosphere?
2. Compare the two views of the atmosphere's origin.

3. Sketch and label the two different composition regions of the atmosphere, including the layers that make them up.
4. Explain how the troposphere is different from all the other temperature layers in the atmosphere.
5. How can we exercise wise dominion by using the jet streams in the stratosphere?
6. How is the mesosphere unique among the temperature layers of the atmosphere?
7. Which true layer of the atmosphere has the hottest daytime temperatures?
8. Why can't plants get nitrogen directly out of the air to use for building tissues?
9. (True or False) Atmospheric temperature decreases continuously and smoothly with altitude from the ground to the exosphere.

18B

SPECIAL ZONES IN THE ATMOSPHERE

How do special layers of the atmosphere protect life on Earth?

There are a few zones in the atmosphere that are pretty special. They do some very important things for the earth and help improve our ability to put dominion into practice. Their presence suggests that God designed them as features of a very good earth. Let's take a look at them.

18.6 OZONE LAYER

We started this chapter mentioning smog in large cities around the world. A lot of this smog contains a gas called *ozone*. It's actually a kind of oxygen, but you couldn't breathe it for long. Ozone in small amounts causes lung irritation and breathing problems. In high concentrations, it can kill. But did you know that high up in the stratosphere there is a layer of ozone? There it actually helps life!

18B Objectives

After completing this section, you will be able to

» relate special zones of the atmosphere to the other layers.

» explain how the special zones in the atmosphere are evidence of God's good design.

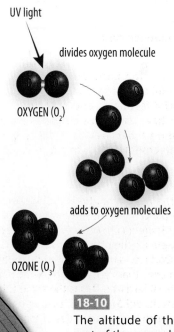

18-10 The altitude of the densest part of the ozone layer in the atmosphere. Ozone's unique properties shield the earth from the sun's ultraviolet light.

This region is called the **ozone layer**. It extends from about 20 km to 50 km above the earth's surface but it is densest in the middle of this band. An ozone molecule (O_3) has three atoms of oxygen bonded together. This is different from an ordinary oxygen molecule, which has just two atoms of oxygen (O_2). When an oxygen molecule in the stratosphere absorbs high-energy ultraviolet light from the sun, it breaks apart. Each of these lone atoms immediately joins another oxygen molecule to form ozone. Ozone is even more efficient than oxygen molecules at absorbing ultraviolet energy. Certain kinds of ultraviolet radiation are quite dangerous to life. They can kill single-celled organisms and cause cancers and other abnormalities in most animals and humans. The ozone layer is vital for living things because it shields them from much of the sun's harmful ultraviolet light. It converts this energy to heat. This is why the stratosphere is warmer than the troposphere below it (see Figure 18-6 on page 435).

When space probes began examining the earth's atmosphere in the early stages of the space program, they discovered a broad hole in the ozone layer over the South Pole. With further observation, scientists became concerned since it seemed to be growing wider. Even more research showed that the ozone layer over the rest of the world was thinning too. This could become a problem for all life on Earth. Read the box on the next page to find out more about the ozone hole and our efforts to correct this problem.

THE OZONE HOLE

Ozone is the poisonous gas that we can't live without. It forms a layer high in the stratosphere called the *ozone layer*. This layer actually protects life on Earth by absorbing harmful ultraviolet rays from the sun.

However, there is a problem. Every spring, the ozone layer develops a hole above Antarctica. It's a good thing there aren't too many people there! Scientists have linked this hole to a class of chemicals.

The main culprit is dichlorodifluoromethane (CCl_2F_2), otherwise known as Freon. It was developed in 1930 by Thomas Midgley Jr. and Charles Kettering of General Motors. People used it in refrigerators, air conditioners, dry cleaning solvents, aerosols, and some kinds of foam. Freon and other related chemicals called chlorofluorocarbons (CFCs) are odorless, tasteless, nonflammable, and noncorrosive. They are also not poisonous in small amounts.

In the 1950s, scientists studying the atmosphere discovered the ozone layer and its shielding effect. In the early 1970s, chemists found that ozone breaks down in the presence of CFCs. They predicted that atmospheric CFCs could dangerously deplete the ozone layer.

It takes several things for this to happen. First, you need clouds. In the spring, clouds of ice crystals form in the very high cold air of the stratosphere over Antarctica. These ice crystals speed up certain chemical reactions. To deplete the ozone, you also need CFCs. These chemicals contain chlorine, along with other elements. Sunlight is the third ingredient. If you remember, Antarctic winters see no sun. Not until springtime does the sun reach these areas of the world. When it does, energy in sunlight can break off a chlorine atom from a CFC molecule. This chlorine atom reacts with ozone molecules on the surface of cloud ice crystals to break them down into just plain old oxygen. One single chlorine atom can react over and over again. So even a little CFC can do a lot of damage.

These laboratory findings, seemingly supported by atmospheric observations, propelled the global community into action. The United States, along with several other nations, banned CFC production in 1978 in what was called the *Montreal Protocol*. As satellite evidence of ozone depletion continued to mount, other countries put bans on producing, selling, and buying CFCs in 1987 and 1993. The laws required that they be replaced with other chemicals, but these chemicals didn't work as well as Freon.

So what happened? Is the ozone hole getting smaller? Were scientists right about the link between Freon and the ozone hole? Has banning CFCs really made a difference?

Well, we're not quite sure about any of those questions. Scientists have been watching the ozone hole for several decades now. It has stayed about the same size, but that size is quite variable. In 2003, the hole seemed not to develop at all, perhaps because it was too warm. It may take a long time, maybe another twenty or thirty years or more for CFCs to disappear from the atmosphere. Only then will we be able to tell whether our efforts were actually effective in saving the ozone layer.

The ozone hole over Antarctica, as seen on October 3, 2015

It's possible that there are things about the ozone layer that we just don't understand yet. Time will tell whether this was good and wise dominion. Can man's activities really affect the earth and its systems in such a big way? Should we take immediate action to fix what we think are problems with the earth's systems? Do we understand Earth's systems well enough to make decisions about these things without understanding very long-term effects, such as the impact of solar cycles? These are questions that we must answer before we address what many people believe is the coming catastrophe of climate change.

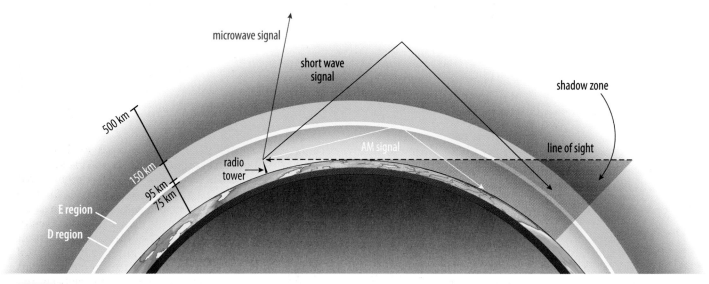

18-11
The structure and features of the ionosphere

18.7 IONOSPHERE

Above the ozone layer, there's another useful zone in the thermosphere, beginning around 75 km and extending to 500 km above the ground. Here, the sun's rays, as well as rays from space, break some of the atmospheric gas molecules into ions. This layer of ionized atoms forms the **ionosphere**. The ionosphere actually contains up to four layers that combine and change between day and night and with the seasons.

The ionosphere can reflect radio communication waves. Until the second half of the twentieth century, radio users depended almost entirely on this effect to communicate over long distances. Even today, some radio stations and amateur radio broadcasters send their messages around the curve of the earth by bouncing their radio signals off the ionosphere, especially at night. Some shortwave radio signals can return to Earth thousands of miles away from their transmitting antennae. For example, a Christian missionary radio station in Carriacou, an island just off the coast of Venezuela, has listeners in Finland and Russia without using communication satellites!

> The ionosphere doesn't reflect microwaves, the waves used in radar and satellite communications. They go right through. This is why NASA uses microwave signals to communicate with spacecraft.

18.8 MAGNETOSPHERE

Yet another amazing zone of the atmosphere is the **magnetosphere** (mag NEE toe SFEER). The magnetosphere extends thousands of kilometers above the earth's surface at distances equal to 10–25 earth radii or more (60,000–160,000 km). This region includes the exosphere of the earth's atmosphere.

James Van Allen discovered the magnetosphere in 1958 using information gathered by America's first artificial satellite, *Explorer 1*. More data revealed several rings or belts where most of the protons and electrons from the sun's solar wind end up trapped in the earth's magnetic field. These regions are fairly close to the earth compared to the outer boundaries of the magnetosphere. In honor of Van Allen's discovery, these regions of dangerous radiation are called the *Van Allen belts*. If you had continued your space elevator trip to the altitude of the Van Allen belts, you would have received a lethal dose of solar

LARRY VARDIMAN, ATMOSPHERE SCIENTIST

Larry Vardiman was at a turning point in his life. It was 1972, and he was a graduate student. He had joined the US Air Force in 1966 with a bachelor's degree in Physics. The Air Force decided to use him as a weatherman, sending him to do cloud physics research at Scott Air Force Base. Then he moved on to Colorado State University for a PhD in atmospheric science. The decision he had to make in his studies was this—would he remain faithful to the Bible?

The facts and theories that his professors were teaching him contradicted Scripture. After a spiritual struggle, Vardiman decided to hold to his belief in the Bible, even if it went completely against what he learned in the science classroom. Two weeks later, he met Henry Morris.

Henry Morris was one of the founders of the modern Creationist Movement. He and another key creation scientist, Duane Gish, encouraged Vardiman and confirmed the decision he had made. That encounter led him to a career in creation science.

After getting his doctorate degree, Vardiman became the director for the cloud seeding project in northern California. His goal was to bring rain to that drought-stricken area.

In time, Vardiman began to teach math and science at the Institute for Creation Research (ICR), originally near San Diego. ICR is a school that educates many creation scientists. There he began to research, write, and publish scientific articles in creationist journals. Eventually, he became the Chief Operating Officer for ICR. He's on the cutting edge of biblical origins science, especially in meteorology. Most recently, he's targeted the global warming controversy.

As an atmosphere researcher, Vardiman and one of his students tested the idea long held by many Christians that most, if not all, of the water for the Flood came from the collapse of a water vapor canopy. Their computer model of the earth's atmosphere suggested that under a canopy that could have supplied even a small fraction of the water needed for the Flood, living conditions would have been intolerably hot due to a runaway greenhouse effect. These findings show that even biblically based theories can be changed or discarded. Only the Bible is true and unchanging.

Our culture constantly attacks the Bible and its teachings. If you are ever in a situation that makes you question God's Word, do what Larry Vardiman did. The Bible is rock-solid, worthy of your faith and belief!

---

radiation as you passed through. Manned spacecraft that travel through these belts of intense radiation must either be heavily shielded or move quickly past them.

Do you remember Earth's specially designed feature of the magnetic field discussed in Chapter 4? Strong solar storms would be especially deadly if the charged particles could strike the earth unhindered. But the magnetosphere shields the earth from the brunt of these storms.

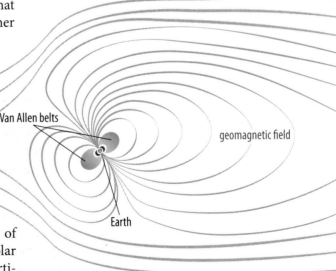

18-12

The magnetosphere and the Van Allen belts under normal solar conditions

18-13
An aurora display over the North Pole

aurora: (L. *aurora*—dawn)
borealis (bore ee AHL is): (Gk. *boreas*, northern)
australis (aw STRAHL is): (L. *auster*—southern)

When the solar wind is especially strong, as it is during a *solar flare* (see Chapter 22), some of the charged particles trapped in the magnetosphere escape near the earth's magnetic poles into the lower regions of the atmosphere. There they collide with gas molecules in the atmosphere, producing beautiful **aurora borealis** (the northern lights) and **aurora australis** (the southern lights) near the respective poles.

18B REVIEW QUESTIONS

1. Name three special layers or regions of the atmosphere.
2. In general, how do some of the special layers of the atmosphere help protect life on Earth?
3. Explain how ozone forms in the atmosphere.
4. Why are scientists concerned about the ozone layer?
5. Why wouldn't NASA want to use long-wave radio communications to stay in contact with astronauts in the International Space Station?
6. Where do most of the sun's solar wind particles end up trapped in the magnetosphere?
7. State three ways that space travelers can avoid too much dangerous radiation exposure from the Van Allen belts.
8. (True or False) The ionosphere is an important aid to modern microwave satellite communications.

18C

ENERGY IN THE ATMOSPHERE

How does energy from the sun affect the atmosphere?

18C Objectives

After completing this section, you will be able to

» sketch the flow of energy in the atmosphere.

» compare radiation, conduction, and convection.

18.9 THE RADIANT SUN

Almost all the energy at the surface of the earth comes directly or indirectly from the sun. The sun emits a wide variety of energy and particles. Most of the sun's radiated energy is in the form of *electromagnetic waves*—light energy. The sun also emits energetic protons and electrons that make up the *solar wind*. In Chapter 2, you learned that light, or radiant energy, comes in different forms (e.g., infrared and visible waves). These different kinds of waves vary in wavelength. Radio waves are the longest waves and gamma rays the shortest. Although the sun emits radiant energy at all wavelengths, most of the radiant energy received by the earth is in the visible and infrared wavelengths.

18.10 THE SOLAR CONSTANT

Although the earth receives only a tiny part of the sun's total energy output, it is enough to sustain all life on the planet. Astronomers call the rate of the flow of radiant energy from the sun that reaches the earth through space the **solar constant**. At the top of the atmosphere, 1.367 kilowatts (kW) of the sun's energy falls on each square meter of the atmosphere. But not all this energy makes it to the earth's surface.

Although the sun gives off a lot of visible light, the majority of our solar energy is from the invisible infrared band. As light passes through the atmosphere, most of the short wavelength rays and some portions of the infrared are absorbed. Clouds and airborne particles in the atmosphere *reflect* as much as 32% of the sun's energy. Clouds, dust, and water vapor *absorb* up to another 18%. This filtering process warms the atmosphere slightly. Under the least favorable conditions, up to half the incoming light does not even reach us! However, on a sunny day with the sun directly overhead, 60–75% of the incoming solar energy can reach the earth's surface.

The sun's energy that reaches the earth's surface on a typical sunny day with the sun directly overhead is about 0.95 kW/m² each second. Following the catastrophic eruption of Mount Tambora in 1815, volcanic dust reduced the surface solar illumination to 80% of that value. This very abnormal condition plunged the world into a "year without a summer."

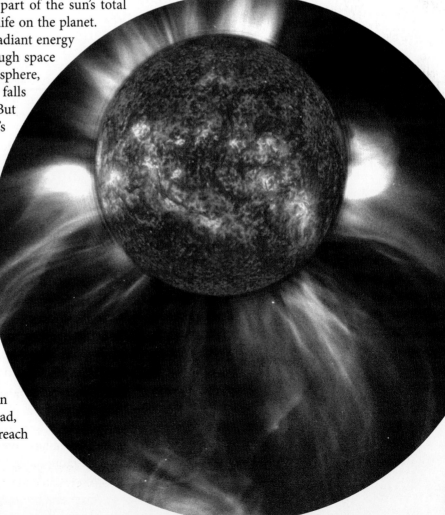

EARTH'S ATMOSPHERE

18-14 On a normal day, most of the sun's energy that gets to the earth is absorbed by land and water.

visible light

Air absorbs up to 18%

Clouds reflect up to 32%

Land and water absorb more than 50%

infrared radiation

When the sun's rays are nearly vertical to the earth's surface, the input of solar energy at the earth's surface is greatest. This can happen only within a narrow band of latitudes between 23½° north and south of the Equator, and then only twice a year at any given location. At other times of the year, and always at other latitudes, the sun's rays hit the ground at some angle, reducing the rate of energy absorbed by the ground, water, or ice. The farther a location is from the Equator, the more slanted the rays of the sun, even at its highest position in the sky. The more slanted the sun's rays, the less energy they deposit at the earth's surface.

18.11 WARMING THE ATMOSPHERE

As you just learned, the incoming solar energy doesn't heat the atmosphere very much where we need it—in the troposphere. So how does the atmosphere near the earth's surface absorb solar energy? There are three ways. Some of the light energy reflects off the earth's surface and heads back out into space. On its way through the atmosphere, the air absorbs some of this reflected energy. But the earth's surface absorbs most of the original incoming radiant energy and converts it to thermal energy. The warmed surface radiates infrared rays back into the atmosphere. Infrared rays are the same thing as thermal **radiation**. Gas molecules absorb heat from thermal radiation much more easily than they do from light rays.

A second method of heat transfer is direct contact between the surface of the earth and the atmosphere. Gas particles in the air are in constant motion and collide with water, rock, and soil particles, which the sun's radiation has heated. When these collisions occur, kinetic energy transfers to the gas particles. This process raises the temperature of the air in contact with these surfaces. Heat transfer by direct contact is **conduction**. It is the most efficient of the three methods.

The third method of heat transfer involves the vertical movement of warmed masses of air into cooler regions in the atmosphere. This process is **convection**. Convection is a common kind of density current similar to ones that exist in lakes, in the ocean, and within the mantle. Air warmed by radiation or conduction becomes less dense than the surrounding air. Cooler, denser air flows in under the warmer air, lifting it up into the atmosphere, forming a *convection current*. In addition to moving thermal energy to different levels of the atmosphere, convection currents are important in cloud and wind formation. You will learn about these in later chapters.

LIFE CONNECTION: UV LIGHT AND LIFE

Ellen was pumped! It was early June, 80° outside, and the sun was shining brightly—perfect weather for the first day at the beach. She and her friends met for lunch and some time to enjoy the sun and sand. A gentle sea-breeze kept them cool and the hours passed quickly. As she prepared to go home, Ellen noticed that her feet felt sore but thought it was just from the sand. By the time she got home, her feet were bright red, hot to the touch and very tender. She realized her mistake; she hadn't thought that she would need sunscreen this early in the season with the temperatures as low as they were.

Ellen was experiencing one of the negative effects of ultraviolet (UV) radiation. Only a small part of the radiant energy emitted from the sun is visible light—some of it is UV. Ultraviolet energy has wavelengths that are too short for the human eye to see. But we can definitely feel its effects—as sunburn!

Considering Ellen's experience, we often think that UV exposure is a bad thing. But that's not always true. For example, some UV exposure helps our skin to make vitamin D. You need about 10–15 minutes of sunlight a day to build all the vitamin D you need for that day. Vitamin D helps us absorb calcium. Therefore, UV indirectly affects your skeletal development, blood cell formation, immune system function, and many other processes. This is why your mom tells you to go outside and get some fresh air in the sun!

UV is helpful for more than just human nutrition. Studies have shown that it kills germs in the air. UV light also benefits plants. It helps them grow by increasing the amounts of certain hormones. It also helps them fight off pests and bacteria.

However, while small amounts of UV benefit plants, larger amounts will actually stunt their growth or even kill them. This affects not only plants in natural environments, but can also affect farmers' productivity. Because UV can impair plants' ability to perform photosynthesis, too much UV can lead to a smaller crop.

UV light also has bad effects on organisms other than plants. It impairs photosynthesis in some marine organisms. These organisms, called *phytoplankton*, are the basis of the marine food chain, as well as a major source of the oxygen we breathe. So UV can have a snowball effect on ocean organisms right up the food chain.

For humans, the potential problems of overexposure to UV radiation go beyond sunburn. It can prematurely age skin, damage eyes, and suppress the immune system. Probably the scariest effect is its role in causing some types of cancer. Too much UV causes changes (called *mutations*) in a person's DNA, possibly leading to various cancers. Skin cancer is the most common form of cancer in the United States—5.4 million cases every year! A fifth of Americans will develop skin cancer at some point in their lives. Thankfully, you can do a lot to keep from getting skin cancer. The best way is to limit how much time you spend in direct sunlight.

A little ultraviolet light is good for your health. But too much can cause some major health problems. So go catch a few (but not too many) rays!

18.12 THE GREENHOUSE EFFECT

Carbon Dioxide = Air Pollutant?

Many climate scientists and environmentalists think that carbon dioxide is a pollutant because it is a greenhouse gas. They directly link carbon dioxide with global warming and climate change. Not too many of them talk about water vapor this way though. You've probably heard of people talking about how to "reduce your carbon footprint." Living things need carbon from carbon dioxide. We'll discuss this more in Chapter 21 on climate and climate change.

As we have seen, land and water absorb most of the sun's energy that reaches the earth after it passes through the atmosphere. But the earth then indirectly warms the atmosphere through the **greenhouse effect**. The shorter wavelengths of light strike the ground after passing through the clear atmosphere, just like light passing through the glass of a greenhouse. The warmed earth changes the shorter-wave solar energy to longer-wave infrared energy. The infrared energy radiates back toward space, and the lower atmosphere traps and absorbs much of this energy. This is very similar to what happens in a greenhouse. Certain gases that make up the atmosphere, mainly water vapor and carbon dioxide, contribute most to absorbing heat. These kinds of gases are called **greenhouse gases**.

The atmosphere helps Earth's inhabitants in more ways than just providing oxygen. Assuming that we could even live without an atmosphere, we would be scorched in the daytime by the sun's full radiation. At night, the earth's temperature would drop far below freezing, just like it does on other planets. On the other hand, if the atmosphere absorbed all the sun's energy, none of it would get to us. Think of it—no sunny days! God's love of beauty and of His creatures has provided us with an atmosphere that is just right for life.

But we need to keep it that way. We need to create models that help us understand the atmosphere and protect its features. This is not easy. God made the earth, its atmosphere, and its processes complex. There's lots to do, but we have seen some success. Cities such as Los Angeles and Mexico City, that had been plagued with smog, now benefit from cleaner air because of atmospheric modeling.

18C REVIEW QUESTIONS

1. What kinds of radiant energy make up most of what the earth's surface receives?
2. What is the solar constant? Where is it measured?
3. Make a sketch that shows what happens to the sun's radiant energy as it passes through the earth's atmosphere to the earth's surface.
4. Compare the three different ways that the sun warms the atmosphere.
5. Describe the greenhouse effect.
6. Which two gases contribute most to the greenhouse effect?
7. (True or False) All other conditions remaining the same, temperatures on the ground under cloudy skies could be warmer than under clear skies because of the greenhouse effect.

The different layers of the atmosphere appear as different colors in this picture taken by astronauts on the International Space Station.

Key Terms

atmosphere	429
homosphere	433
heterosphere	433
troposphere	433
lapse rate	433
stratosphere	434
jet stream	434
mesosphere	434
thermosphere	436
exosphere	436
ozone layer	440
ionosphere	442
magnetosphere	442
solar constant	445
radiation	447
conduction	447
convection	447
greenhouse effect	448
greenhouse gas	448

18-15 When most people think about our atmosphere they think of the weather that occurs there. At an elevation of almost 6300 ft, the Mount Washington Observatory in New Hampshire is often above the clouds.

CHAPTER 18 REVIEW

CHAPTER SUMMARY

» One of the challenges of Earth's growing population is minimizing air pollution in large cities.

» We have a God-given responsibility to wisely care for and use our atmosphere.

» The atmosphere is the envelope of gases that surrounds the earth. Life on Earth would not be possible without it.

» Old-earth scientists believe that Earth's original toxic atmosphere appeared very shortly after the planet formed. After 2 billion years or so, oxygen produced by organisms gradually made the atmosphere able to support more complex forms of life, and biological evolution took off.

» Biblical creationists believe that the earth's atmosphere was probably created no later than Day 2 of Creation week. The atmosphere's ability to protect and sustain life is a compelling case for design by our Creator.

» The two key elements in our atmosphere are oxygen and nitrogen.

» The two regions of the atmosphere differentiated by their composition are the homosphere and heterosphere.

» The regions of the atmosphere that have distinct temperature profiles are the troposphere, stratosphere, mesosphere, thermosphere, and exosphere.

» Nutrients in the earth such as carbon and nitrogen are recycled among land, water, air, and living things.

» The ozone layer and the magnetosphere are special regions lying in the earth's extended atmosphere that absorb or trap harmful solar radiation and particles.

» The ionosphere is a special layer that is useful for certain kinds of global radio communications.

» Not all the sun's energy that reaches the earth makes it to the surface. Nearly half of this energy can be absorbed or reflected by clouds, dust, and atmospheric gases before it reaches the earth's surface.

» The three ways that the atmosphere can absorb solar energy are radiation, conduction, and convection.

» Carbon dioxide and water vapor are two greenhouse gases that contribute most to warming the earth's atmosphere.

» The earth's atmosphere shows good design in the many ways that it supports life on Earth. We have a dominion responsibility to protect and preserve the atmosphere.

REVIEW QUESTIONS

1. What is air pollution? Why is it a problem?
2. What makes Earth's atmosphere different from those surrounding other bodies in our solar system? Why is this significant?
3. In one word for each, state the underlying basis for the two theories of the atmosphere's origin.
4. From where do old-earth scientists think the oxygen in our atmosphere first came?

5. Create a bar graph that compares the percentage by volume of the gases in the atmosphere.
6. Why are lighter gases higher up in the atmosphere?
7. Use Figure 18-6 to find what temperature layers lie in the heterosphere.
8. What is a lapse rate?
9. What happens to atmospheric temperature above the tropopause in the stratosphere?
10. Defend the following statement: "When we study the earth's atmosphere, we are led to praise God for His care for His creatures."
11. What is the key carbon reservoir in the carbon cycle?
12. Where does organic or fixed nitrogen come from?
13. Using Figure 18-6, identify which temperature layers overlap the ozone layer and the ionosphere.
14. Do any of the special layers of the atmosphere completely block, absorb, or reflect visual light wavelengths? How do we know?
15. Why is the earth's atmosphere not completely transparent to incoming solar radiant energy?
16. In general, how does the solar constant change from the top of the atmosphere to the surface?
17. Make three columns titled *Radiation*, *Conduction*, and *Convection*. Write two everyday examples in each column that are different from those mentioned in the text.
18. How does carbon dioxide contribute to the "greenhouse effect"?
19. What features in the atmosphere does God use to protect His creatures on Earth from extremes of heat and cold?

True or False

20. Christians should be concerned that the world is getting overpopulated with people.
21. Christians should be interested in technology that reduces air pollution.
22. The old-earth model for the origin of the atmosphere is closely related to the nebular hypothesis.
23. All the atmospheric temperature layers with the lowest temperatures have winds and a uniform mixture of gases.
24. The thermosphere is where most of the weather that we experience occurs.
25. The continual mixing of warm and cold air occurs only in the troposphere.
26. The exosphere is mostly hydrogen.
27. The carbon and nitrogen cycles are basically the same but involve different elements.
28. The amount of solar energy reaching the earth's surface on any given day is the same at the Equator as at the Arctic Circle.
29. The sun's energy reaches the earth mainly through conduction.
30. Carbon dioxide is an air pollutant because it is a greenhouse gas.

18-16
Winters can be particularly brutal on Mount Washington, New Hampshire. Rime ice covers the observatory.

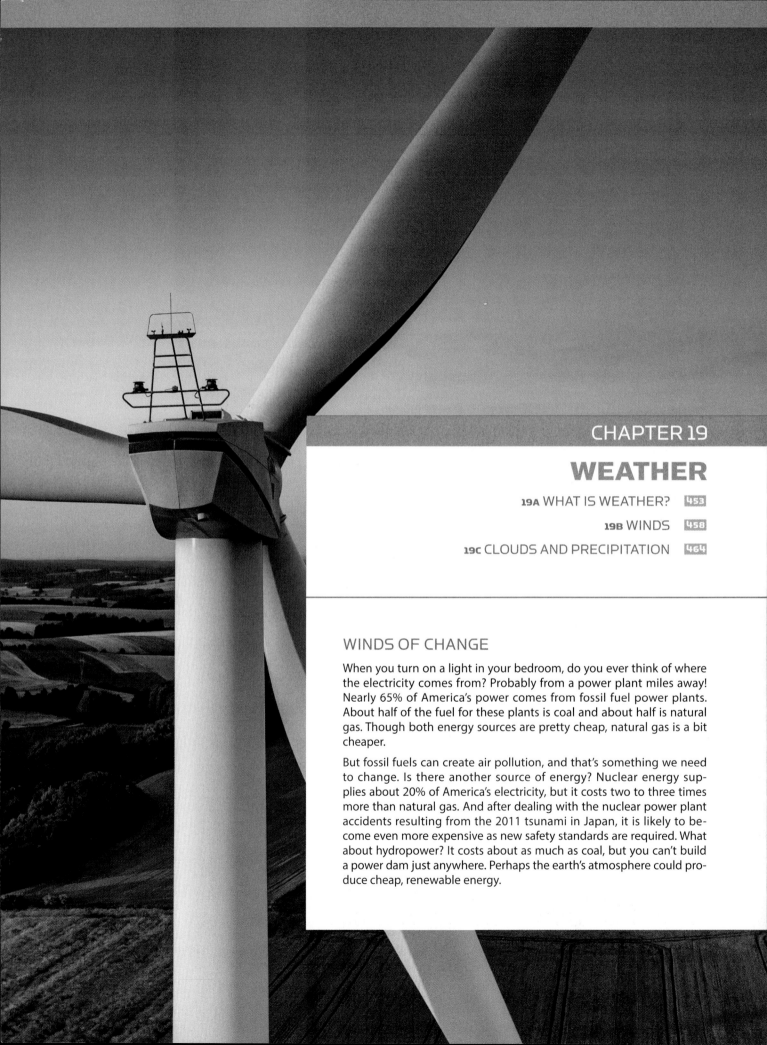

CHAPTER 19

WEATHER

19A WHAT IS WEATHER? 453
19B WINDS 458
19C CLOUDS AND PRECIPITATION 464

WINDS OF CHANGE

When you turn on a light in your bedroom, do you ever think of where the electricity comes from? Probably from a power plant miles away! Nearly 65% of America's power comes from fossil fuel power plants. About half of the fuel for these plants is coal and about half is natural gas. Though both energy sources are pretty cheap, natural gas is a bit cheaper.

But fossil fuels can create air pollution, and that's something we need to change. Is there another source of energy? Nuclear energy supplies about 20% of America's electricity, but it costs two to three times more than natural gas. And after dealing with the nuclear power plant accidents resulting from the 2011 tsunami in Japan, it is likely to become even more expensive as new safety standards are required. What about hydropower? It costs about as much as coal, but you can't build a power dam just anywhere. Perhaps the earth's atmosphere could produce cheap, renewable energy.

19A

WHAT IS WEATHER?

How do scientists collect weather data?

19.1 WIND AND WEATHER

Just like coal, fresh water, land, and metals, air is a natural resource. Maybe you're not used to thinking of air this way. But as we've seen, we need to take care of our atmosphere. And we are always looking for new energy resources to support the world's growing human population. Is there some way to harness energy from the atmosphere to create electricity?

One way to both keep air clean and harness its energy at the same time is by using the growing technology of *wind power*. Wind-generated electricity doesn't produce hazardous wastes or chemical air pollution, and it's available in many places around the world. Engineers use computers to design and build efficient wind turbines, the largest of which are twice the height of the Statue of Liberty! The turbines are elevated to expose their blades to smoother winds far above ground. Hundreds of wind turbines connected together in facilities called *wind farms* can supply power to lots of homes. The Department of Energy estimates that the United States could supply up to 20% of its energy needs with wind power by the year 2030. This would provide electricity at a fraction of the cost of coal, gas, and nuclear energy, without the hazards. But we need to take on many challenges before we can reach this goal. And it all starts by understanding our atmosphere's wind and weather.

19A Objectives

After completing this section, you will be able to

» evaluate the risks and benefits of wind power.

» describe the weather data that meteorologists collect.

» compare the different aspects of weather with one another.

Decisions, Decisions

In spite of wind power's benefits, many people don't like wind farms. They don't like the noise, visual clutter, and effects on bird populations. And mechanical things break, so there are reliability and safety concerns too. What should we consider when weighing the benefits and drawbacks of wind power?

19-1 The world's largest wind turbines can generate 8.0 MW of electricity, enough to power almost 6500 average American homes per turbine.

meteorology (MEE tee or AHL uh jee): meteor- (Gk. *meteoron*—having to do with the sky) + -ology (Gk. *logos*—word, discussion, subject)

19-2
Meteorologists have many tools for collecting data. This weather balloon is one such tool.

19.2 WEATHER DATA

Weather is the condition of Earth's atmosphere at any given time and place—and it changes a lot. The scientific study of the atmosphere is called **meteorology**. Meteorologists use instruments to collect data so that they can describe the weather. This description includes information about temperature, atmospheric pressure, wind speed and direction, precipitation, and humidity. Meteorologists also attempt to model the atmosphere to predict future weather from existing weather conditions.

TEMPERATURE

Probably the first thing you notice about the weather on a clear day is the temperature. Do you remember from Chapter 2 what temperature is? Temperature is an indication of the amount of thermal energy in the air. Temperature is probably one of the main reasons why you check weather forecasts. The unprotected human body can tolerate only a small range of temperatures. Outside temperature determines what clothes to wear. This makes temperature an important part of understanding weather. Meteorologists measure temperatures using *thermometers* and *thermographs*. They usually indicate temperatures in degrees Fahrenheit (°F) or degrees Celsius (°C).

How do we know what the temperature 0 °C is? Why isn't this the same temperature as 0 °F? Scientists anchor all temperature scales to special physical conditions that can be trusted to remain constant anywhere. The differences in the scales resulted from the different ways that the scales were originally developed.

In 1714, German scientist Daniel Gabriel Fahrenheit developed a temperature scale with its zero point at the coldest temperature that he could create using a freezing saltwater solution. Following a suggestion by Isaac Newton, he chose 96° to represent his own body temperature. He found that pure water boiled near 212° and froze near 32° on his first thermometer. Noticing that there were 180 degrees between these temperatures, he recalibrated his thermometer so that the freezing point of pure water was exactly 32 °F and its boiling point was exactly 212 °F.

BOILING POINT OF WATER
*212 °F   *100 °C
Fahrenheit   Celsius
*32 °F   *0 °C
FREEZING POINT OF WATER
*reference temperatures

Several decades later in 1742, Swedish astronomer Anders Celsius developed a metric temperature scale—the Celsius scale—that was based on the same two properties of water, but with 100 degrees between them. Interestingly, he set the boiling point of pure water at 0 °C and the freezing point at 100 °C. (Carolus Linnaeus reversed this illogical order in 1744.) Comparing the number of degrees between the freezing and boiling points of water on both scales, you can see that a degree Fahrenheit is smaller than a degree Celsius. But even though the temperature of air may have different number values on different scales, it still has the same amount of thermal energy.

19-3 Atmospheric pressure decreases as altitude increases because the height of the column of air above the ground is shorter.

PRESSURE

Like other forms of matter, the gases in the atmosphere have weight because of the earth's gravity. The weight of these gases bearing on a surface is called **atmospheric pressure**—the force exerted by air on a standard area, like a square meter or a square inch. If the atmosphere surrounding the earth never changed and was the same temperature and thickness everywhere, the atmospheric pressure would be constant. But when air warms, it expands and becomes less dense. This lessens the weight that the air exerts on the ground surface, lowering air pressure. When air cools, it contracts and becomes denser. This increases the weight that the air exerts on the ground—and higher pressure results. So air pressure can change with changes in air temperature.

Air pressure also changes with elevation. This happens because as you go farther up in elevation, there is less air above you, so it can exert less weight. That means that there is lower atmospheric pressure at the top of a mountain than at sea level.

Air pressure is measured in units of force per area, like pounds per square inch (lb/in^2, or psi) or newtons per square meter (N/m^2). This metric unit is used so often in the physical sciences that it was given a special unit name, the *pascal (Pa)*. Normal sea level air pressure is 14.7 psi. In the metric system, it is 101,325 Pa. Meteorologists found this value to be pretty clumsy to use, so they defined a unit called the *bar* that is equal to exactly 100,000 Pa. So standard atmospheric pressure (1 atm) is about 1.013 bar, indicated more commonly as 1013 millibars (mb). The highest air pressure ever measured was 1086 mb in frigid Mongolia in 2001. The lowest *sustained* pressure recorded was 870 mb in the eye of a typhoon in 1979. The lowest *sea-level* air pressure ever recorded was 850 mb in the center of a tornado! *That* would make your ears pop!

The instrument we use to measure air pressure is a <u>barometer</u>. The first true barometer was a glass tube sealed on one end and filled with the liquid metal mercury. The pressure of the air at sea level supports a column of mercury about 760 mm high. Modern barometers use sealed flexible metal chambers hooked to a meter mechanism (Figure 19-4).

Mountain Cookin'

Recipes often have special high-altitude directions because of the relationship between altitude and atmospheric pressure. At higher altitudes (thus lower pressure), water boils at lower temperatures. Food takes longer to cook because the water isn't as hot as at lower altitudes. Breads and cakes also rise more because air pressure is lower.

barometer: baro- (Gk. *baros*—weight, press down) + -meter (a measuring instrument); any instrument that measures air pressure

19-4 An aneroid barometer contains a sealed thin-walled chamber that flexes with changes in air pressure. The motion is connected to the indicator needle.

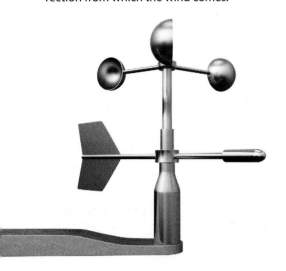

19-5
The anemometer (upper part of image) measures the speed of the wind. Wind vanes (lower part of image) work by pointing toward the direction from which the wind comes.

anemometer (AN eh MOM eh tur): anemo- (Gk. *anemos*—wind) + -meter (Gk. *metron*—measure); a wind speed measuring instrument

Atmospheric pressure constantly changes as huge volumes of warm and cold air move across the earth's surface. These bodies of air are called *air masses*. Warm, cloudy, and damp weather often occurs with a low-pressure air mass. Cool, clear, and dry weather often comes with a high-pressure air mass. You'll see how air mass temperature and pressure affect weather in Chapter 20, where you will also learn about storms.

WIND

Wind is moving air. It's an important aspect of weather. Meteorologists monitor wind speed and direction because of its ability to change weather and do damage. You will study its damaging effects in Chapter 20. We want to know two aspects of wind—wind direction and speed. *Wind direction* is the direction *from which the wind is coming*. For example, a wind traveling from west to east is a west or westerly wind. People long ago defined wind direction this way because of how weather (or wind) vanes work. The vanes turn to point into the wind, not toward the direction the wind is going. Wind direction is reported by the points on a compass (north, northeast, east, etc.) or in compass degrees (000°, 045°, 090°, etc.).

Wind speed is simply how fast the wind is moving. Do you remember how speeds are reported? Meteorologists may use kilometers per hour (kph) or miles per hour (mph), but most often they report wind speeds in nautical miles per hour, or *knots* (1 kt = 1.15 mph). They use an instrument called an *anemometer* to measure wind speed.

Wind is associated with changing weather because a moving air mass *is* wind. Wind brings changes in temperature and moisture. Wind also helps air evaporate moisture. This is why you feel cooler standing in front of a fan. Any amount of wind can make air seem cooler. In hot weather, a breeze is a welcome relief! It helps our body's cooling mechanism work better. In cold weather, a wind is less welcome. In a cool breeze, your skin is losing heat as fast as if you were standing in still air at a lower temperature. This is called the **wind-chill factor**. A strong wind can lower the temperature that you feel on your skin by more than 20 °C!

PRECIPITATION

We've discussed much about water—ocean water, surface water, and groundwater. But all of this water either is the source of precipitation or comes from it. **Precipitation** is any form of water that falls to the earth's surface from the sky. Rain is the most common form of precipitation in most places, though it can also be snow, hail, sleet, and even fog drip. In the United States, precipitation is usually measured in millimeters or inches of liquid water using one of the many different kinds of *rain gauges*. You'll learn more about these forms of precipitation in Section 19C.

HUMIDITY

Another important weather condition is **humidity**. This is a measure of the amount of water vapor in the air. *Absolute humidity* is the total grams of water per cubic meter (g/m^3) of air. The amount of water vapor in the air depends on the rates of evaporation and condensation. At higher temperatures, evaporation happens faster than condensation. This means that we expect more moisture in the air at higher temperatures.

Relative humidity is the ratio of the water vapor in the air compared with the amount needed to saturate the air. Saturated air can't hold any more water vapor. We typically use percent to express the ratio of water vapor actually in the air to saturated air. For example, if the humidity is 67%, then the air contains 67% of the water needed to saturate the air. If the humidity is 100%, the air is saturated and the rates of condensation and evaporation are the same. This means that no more water will evaporate. If the temperature drops when the humidity is 100%, condensation will occur faster than evaporation. Water will change from a gas to liquid form. Humidity is measured with a hygrometer.

19-6

A rain gauge collects falling rain and measures it in a cylinder that is similar to a graduated cylinder. Some rain gauges use a tipping bucket that weighs rain as it falls.

Hot, Humid, and Miserable

People can perspire to get rid of excess heat. In hot weather, we cool off when moisture evaporates from our skin. If the relative humidity is high, condensation occurs at almost the same rate as evaporation and so the air removes very little perspiration from our bodies. This makes us feel hot and sticky. But if the relative humidity is low, the rate of evaporation is much higher than the rate of condensation and the air removes the perspiration quickly, making us feel cooler.

19A REVIEW QUESTIONS

1. Suggest some of the challenges that lie ahead for using wind power on a large scale.
2. What is the study of the atmosphere called?
3. List five conditions used to describe weather.
4. The weight and density of atmospheric gases have the greatest effect on which of the measurable weather conditions?
5. What two things happen to air in the atmosphere when it cools?
6. What is the average sea level pressure (1 atm) measured in millibars? in pounds per square inch?
7. What is the most common form of precipitation? List three other forms.
8. Compare absolute humidity with relative humidity.
9. Dry air can be the same temperature as more humid air, but it feels cooler. Why?
10. (True or False) When a weather forecaster says that there is a low-pressure air mass moving through your area, you can expect the air temperature to drop when it arrives.

19-7

This is a *hygrometer*, an instrument used to measure relative humidity. *Psychrometers* are also used. Humidity is the most difficult weather data to measure accurately.

> **19B Objectives**
>
> After completing this section, you will be able to
>
> » explain what factors affect winds.
> » locate and name the major global wind belts.
> » identify sources of local winds.

19B

WINDS

What determines the speed and direction of the wind?

19.3 WIND CREATION

Masses of air develop over separate regions of the world having different pressures, temperatures, and humidity. These weather differences can also create winds. Let's look at the three factors that determine wind speed and direction.

PRESSURE EFFECTS

The atmosphere regularly develops higher pressure over one region of the earth compared with a lower pressure over another. When this happens, air moves from higher to lower pressure, creating a wind. Meteorologists call this push the *pressure gradient force*. Remember the definition of *stream gradient* from Chapter 16? A gradient is how fast some measurement or value changes as you change distance. The **pressure gradient** is the rate that the air pressure changes with distance. Just as a larger stream gradient makes water flow faster, a larger pressure gradient produces a stronger wind. And just as water flows downhill, wind blows down the pressure gradient—from higher to lower pressure.

If this were the only thing affecting winds, they would always flow in straight lines, and mainly from the poles toward the Equator. But wind patterns are much more complicated than that. Several other factors affect the speeds and directions of the important wind patterns.

CORIOLIS EFFECT

In Chapter 14, you learned how the Coriolis effect influences ocean currents and just about any motion over the earth's surface. Winds are affected too. The Coriolis effect deflects a wind perpendicular to its path over the earth's surface. Winds in the Northern Hemisphere deflect to the right, and in the Southern Hemisphere to the left. The effect is stronger the farther away from the Equator you go. The amount of deflection also depends on wind speed. If the air is calm, there is no deflection. If the wind is blowing a gale, the effect is significant. In fact, all winds would be gales if it weren't for another factor that affects global winds—friction.

19-8
The pressure gradient is the slope of the change in pressure with distance over the earth's surface. The steeper the slope, the greater the pressure gradient force, and the stronger the winds that form. Air always moves from areas of high pressure to areas of low pressure.

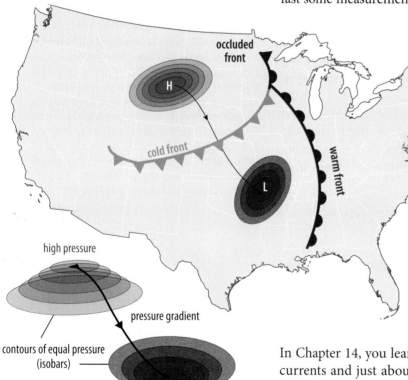

19-9

Trees, buildings, and hills exert friction on wind, especially near the earth's surface.

FRICTION EFFECTS

Forces are pushes or pulls. Anything on Earth that moves always has a *friction* force acting on it. You're probably all too familiar with friction. It makes the lid of a jar hard to remove, but it also keeps your feet from slipping out from underneath you when you stand! Friction acts on moving air too. Trees, mountains, hills, buildings—all these are obstacles to winds and slow them down. Even turbulence within the air slows down wind. Fluid friction, or *drag*, pushes opposite to the direction wind blows and is related to wind speed. The greater the wind speed is, the greater the drag. Friction has the greatest effect on wind closest to the ground, and that is where the greatest turbulence occurs. This is why engineers mount wind turbines on high towers to maximize wind speed and minimize turbulence. Since the ocean surface is basically flat compared to land, wind farms don't have to be built as tall in offshore locations.

CYCLONES AND ANTICYCLONES

Now let's put all these things together to see how a regional wind develops. Let's assume that a mass of cold air several kilometers thick is motionless in the Northern Hemisphere. The cold air mass has a relatively high pressure. A large region of low-pressure air is many miles to the south of the high-pressure air mass. The pressure gradient force begins to push the higher-pressure air southward and it gains speed. But as the air moves, the Coriolis effect begins to deflect this wind to the right, or in a clockwise direction. Friction also increases, especially for surface winds. Eventually, the pressure gradient force balances the Coriolis effect, and its speed is limited by friction with the ground.

But as the winds slow, the pressure gradient force exerts itself, especially nearer the centers of strong low-pressure systems. The force deflects the wind to the *left* toward the low-pressure area and it speeds up. Both friction and the Coriolis effect increase and nearly, but not quite, balance the pressure gradient force. The winds end up circling the low-pressure zone counterclockwise. Meteorologists call such a low-pressure area a **cyclone** or a *cyclonic wind pattern*. Therefore, winds flowing in a clockwise direction outward from a high-pressure area are called an *anticyclonic wind pattern*, and the area itself is an **anticyclone**.

cyclone: (Gk. *kukloma* or *kykloma*—coil of a snake; possibly from Gk. *kuklos*—to whirl about)

In the real world, centers of high and low air pressure are always moving in relation to each other, so winds are always shifting between them. In addition, there are many local factors that create surface winds and determine their speed and direction. We will see this in Subsection 19.5.

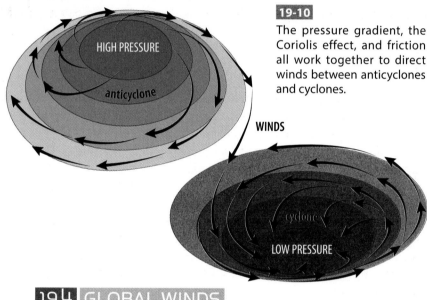

19-10
The pressure gradient, the Coriolis effect, and friction all work together to direct winds between anticyclones and cyclones.

19.4 GLOBAL WINDS

On a grand scale, the sun's uneven heating of the earth causes global pressure differences. Since the Equator gets the most sun with near-vertical rays, the air at the surface warms the most and has the lowest density. The air is forced upward by inflowing cooler air from regions north and south of the Equator. The rising, low-density air makes the height of the atmosphere greater at the Equator than elsewhere. Equatorial air high in the troposphere flows "downhill" toward the poles. A combination of factors causes the moving air to descend near 30° north and south latitude. The falling air forms high-pressure bands at these latitudes.

THE TRADE WINDS

The descending air at 30° latitude divides, part flowing back toward the Equator and the rest flowing toward the pole. The horizontal flow toward the Equator replaces the rising equatorial air. The Coriolis effect deflects the winds toward the west in both hemispheres. These

19-11
Cross-section of the troposphere showing the general circulation of air between the poles

reliable breezes form the **trade winds** that blow from the northeast and southeast in their respective hemispheres. As you may recall, the trade winds drive the equatorial ocean currents toward the west in each ocean basin.

THE PREVAILING WESTERLIES

The Coriolis effect deflects the winds flowing poleward from 30° latitude toward the east. These winds become the **prevailing westerlies**. They are important for directing air masses across the United States. Although winds sometimes come from other directions due to temporary weather conditions, most of the winds in the United States are from the west and southwest. They bring warm, humid air from the Gulf of California and the Gulf of Mexico to the middle and eastern states. They also bring hot, dry air from southern California and the Mexican desert to the southwest and midwestern states. In the oceans, the westerlies also drive the east-flowing currents in the great ocean gyres.

THE POLAR EASTERLIES

The westerly winds of the mid-latitudes create a general flow of air toward the polar regions, where they meet cold surface air from the poles. The dense, cold, descending air above the poles tries to flow toward the Equator. The Coriolis effect deflects these winds toward the west, forming the **polar easterlies**. The cold polar winds bring cold, dry air over the high-latitude regions of the continents in both hemispheres. There they meet the warm, humid prevailing westerlies. This collision of different air masses at around 60° latitude causes storms. The air here rises up into the troposphere. Part flows back toward the Equator to complete the mid-latitude circulation path and the rest flows back to the poles to reenter the polar circulation paths.

PERMANENT HIGHS AND LOWS

Because of rising or falling air, there are places in the world with almost permanent high and low pressure. These zones often have light and variable surface winds, or no winds at all. At the Equator are the *doldrums*, a low-pressure zone caused by rising warm air. At 30° latitude (either north or south) are the *horse latitudes*, zones of high pressure caused by descending air. Both of these places are dangerous for sailing vessels because there's little or no horizontal wind. The only times that strong winds blow in these zones are when storm systems such as hurricanes pass through.

At 60° latitude, the warm prevailing westerlies meet and rise over the dense polar easterlies, and the underlying low-pressure area is known as the *subpolar low*. North and south *polar highs* form over the poles from falling cold air.

Colonial Weatherman

Have you ever noticed that weather in the United States tends to move eastward? Benjamin Franklin was the first to report that storms moved in this direction across the American colonies.

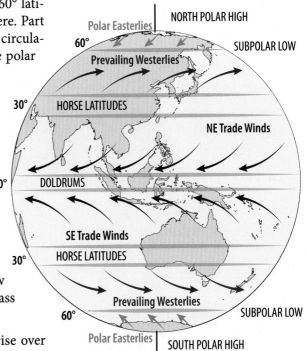

19-12

Major global pressure belts and winds

19.5 LOCAL WINDS

Global wind patterns control the movement of air circulating through the atmosphere between the Equator and the poles. But the winds you feel from day to day are most likely driven by more local factors. *Local winds* blow as a result of changes in temperature or pressure resulting from heating or cooling of fields or forests, the presence of a city or lake, or air flowing over a mountain range or near an ocean coastline. Storm systems also strongly affect local winds.

On a sunny day at the shore, for example, land heats up faster than the water does. This warms the air over the land, making it less dense than the air over the sea. The cooler sea air flows toward the shore to displace the rising warm air, creating a pleasant *sea breeze*. When night comes, the land cools more rapidly than the water does. When the land is cooler than the water, the air over the land also becomes cooler and denser than the air over the sea. The air from the land then moves from the shore to the sea to displace the less dense air over water. This kind of wind is called a *land breeze*.

In a similar way, rocky mountain slopes tend to heat rapidly in warm seasons compared with relatively sheltered, wooded valleys. The less-dense mountain air rises and is replaced by the cooler, denser valley air, forming *valley breezes* that flow uphill. At night, or in cooler seasons, the opposite occurs. Bare rocks rapidly radiate their heat into the sky. The air touching the rocks quickly cools by conduction. The cooler, denser air flows down the slopes into the valleys that are generally warmer because trees and fields tend to hold heat better. The cool winds flowing downhill are *mountain breezes*.

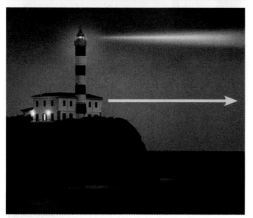

19-13
During the day, local winds called sea breezes blow toward land from the ocean (top). At night, the winds reverse their flow and land breezes occur (bottom).

19-14
Valley breezes (left) and mountain breezes (right) are winds caused by differences in air density acting on mountain slopes.

Under some conditions, high pressure over mountains can drive cold, dry air down the slopes into lowlands lying under low-pressure areas. The relative humidity of the air becomes even less as air pressure increases. These very dry winds, called **chinooks** in the American West and **foehns** in the Alps, can rapidly sublimate or melt snow cover.

Chinook (shih NOOK) winds are named after the native Chinook people in the Pacific Northwest who first identified them.

foehn (FANE)

Other local winds form where air masses come together. These formations are called *weather fronts*. Strong local breezes can build up when the sun simply heats the ground. Some local winds can be dangerous. We'll look at these kinds of winds brought by thunderstorms, tornadoes, and hurricanes in Chapter 20 on Storms and Weather Prediction.

A Waterfall of Wind

Air on high plateaus or other flat highlands quickly cools at night. It can flow under gravity's influence like a waterfall down onto the lower elevations. This is called a *fall wind*. If channeled into narrow valleys, its high speeds can be destructive.

LIFE CONNECTION: WINDS AND MIGRATION

Mount Everest is just the tallest of many mountains that make up the Himalayas, a range that hinders thousands of birds migrating to the Tibetan plateau for the summer. But there is one bird that uses its understanding of winds to navigate around, and possibly over, that peak. The bar-headed goose is one of the highest-flying birds in the world. While scientists have tracked them flying at 7000 m, there have been reported sightings of bar-headed geese flying over Mount Everest at 8848 m. These birds conduct their migration to take advantage of changes in air density and wind conditions.

Scientists don't understand everything about why birds migrate or how they know where to go. Evidence indicates that some birds use the sun and stars as guides. Some even seem to use the earth's magnetic field. Many birds like the bar-headed goose use the winds during their migration.

The American bald eagle, for example, has a very complex migration. It soars in the air until it finds a long column of rising air called a *thermal*. It then hitches a ride on this thermal up to the jet stream. It doesn't even need to flap its wings! The eagle will eventually glide back down toward the ground, where it will catch another thermal. The eagle can repeat this process as it travels to its new home. Gliding like this conserves precious energy for the bird during long, hard migrations.

Birds are not the only animals that ride the wind. Many insects also use winds in migration. The monarch butterfly is an amazing example of this. Some of these insects begin their journey in southern Canada and fly up to 4800 km to Mexico! During this journey, the butterflies must cross several mountain ranges, including the Appalachians. Sometimes they use winds that travel up and over the mountain ridges to gain altitude over the mountains. This is not simply a matter of coincidence. Many insects catch a ride only on favorable winds that take them to the right place.

The abilities of many kinds of animals to migrate long distances and around seemingly impassable barriers are truly amazing. We don't know whether animals had to migrate when they were first created, but if not, God certainly designed in them the ability to develop this essential skill. These creatures have a sense of direction that we humans can only wish for!

The monarch butterfly uses winds to migrate 4800 km to its winter grounds in Mexico.

19B REVIEW QUESTIONS

1. What causes air to move?
2. What three factors work together to determine regional wind speed and direction?
3. What do we call the rate at which air pressure changes with distance between high- and low-pressure areas?
4. What causes the friction that opposes winds?
5. Which clock direction do anticyclonic winds move in the Southern Hemisphere?
6. Why does the atmosphere near the Equator form a belt of low pressure? What is this area called?
7. Explain why the global surface winds from around 30° latitude don't blow directly from the north or south toward the Equator. What are the names of these generally steady winds?
8. What winds are deflected toward the east at latitudes between 30° and 60°?
9. What kinds of regional surface winds normally exist in zones where air moves vertically, either rising or falling?
10. What type of local breeze often occurs near the ocean coastline at night?
11. (True or False) The Coriolis effect has little influence on the vertically descending air in the horse latitudes.

19C Objectives

After completing this section, you will be able to

- explain how clouds form.
- relate clouds, air temperature, and humidity to precipitation.
- compare the different forms of precipitation.
- classify clouds by altitude, shape, and potential for precipitation.

19C
CLOUDS AND PRECIPITATION

How do clouds play a part in weather?

19.6 CLOUD FORMATION

Water can be a solid, liquid, or gas in the atmosphere. When it is a gas, or water vapor, it is invisible to the eye, but we feel its presence as humidity. When it is liquid water or solid ice in the atmosphere, it often forms *clouds*.

THE HUMIDITY FACTOR

Before you learn how clouds form, let's look at how relative humidity affects the state of water in air. Air temperature directly affects how much water vapor is in the air. Warmer air tends to have more moisture than cooler air does. This is because the warmer air allows for a higher rate of evaporation compared with the rate of condensation.

As air with a certain mass of water vapor cools, the rate of condensation approaches the rate of evaporation. In other words, its relative humidity increases. When the air temperature drops to the point where the rate of condensation equals the rate of evaporation, the amount of water is at its maximum value and the relative humidity is 100%.

What happens when the air temperature cools further? The rate of condensation is higher than the rate of evaporation. Water molecules begin sticking together, forming tiny water droplets. This is why the temperature at a relative humidity of 100% is called the **dew point**. Water droplet condensation continues as long as relative humidity is at 100%. The resulting liquid or solid water droplets form clouds or mist.

CLOUDS

A **cloud** is a mass of extremely tiny liquid water droplets or ice crystals in the air. Where does this water come from? Air warms as it touches the earth's surface heated by the sun. At the same time, absolute humidity increases as water vapor enters the air from plant transpiration, the breath and activities of animals and people, evaporation from surface water, or subliming snow or ice. For a cloud to form, the air has to cool to the dew point for the amount of water vapor that the air is holding. A cloud of any kind can exist only if the temperature of the air in the cloud is at or below the dew point of the air.

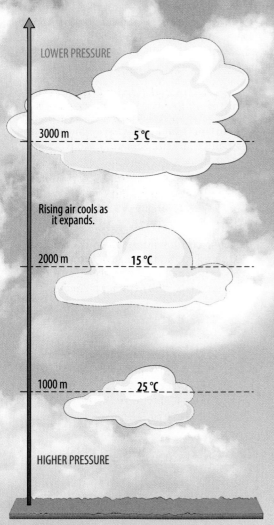

19-15
The energy needed to cause air to expand as it rises into lower pressure comes from its thermal energy, causing the rising air to cool.

So if cooling is required to reach the dew point, how does a parcel of air cool to form a cloud? You learned in Chapter 18 that heat moves by conduction, radiation, and convection. All of these methods can cool air to form a cloud. But air can cool in another way that you may not expect—by simply rising.

If warm, humid air is forced upward, it cools. (We'll study ways that air can be forced upward in the next chapter.) You know that atmospheric pressure lessens with altitude. As it rises, the humid air mass pushes outward on the atmosphere and it expands because the air pressure inside the parcel is slightly higher than outside. The energy needed to expand comes from the kinetic energy of the parcel's air molecules. Thus, as air rises, the average kinetic energy of its molecules decreases. This means that its temperature drops as it gains altitude. See Figure 19-15 for an illustration of this process. This is the most common method of cooling in cloud formation.

When pure, still, humid air reaches its dew point, its water vapor can condense, but only with great difficulty. Condensation can be helped if there is something present for the vapor to condense on. **Condensation nuclei** are microscopic particles suspended in the air that easily attract water molecules. These particles can be common materials such as dust, pollen, soil, salt crystals, or smoke. Water droplets quickly grow around these particles. If the dew point is below freezing, water vapor deposits as ice crystals on these particles, which are then called **freezing nuclei**.

SUPERCOOLED CLOUDS

Sometimes clouds of liquid water droplets can exist in the troposphere at temperatures far below the freezing point of water. Some cloud droplets remain liquid even down to temperatures of –40 °C. Water in this condition is called *supercooled water*. When the atmosphere is quiet, supercooling can occur. Clouds in this condition are very unstable. Just a slight agitation or turbulence is needed to rapidly convert the water to ice.

A Jet-Sized Freezing Nucleus

Supercooled clouds can be a serious problem for airplanes. Water droplets easily freeze on airplane wings as a plane flies through them. This is called *icing* and results from the turbulence of the plane's passage. Icing on wings changes their shape, reducing lift, and adds weight to the plane. Both conditions make it harder for the plane to fly. Many early aircraft crashed due to catastrophic icing.

Today, aircraft are treated with chemicals before taking off in icing conditions. Some planes are equipped with deicing systems that break up ice buildup during flight.

19-16
Deicing chemicals are used shortly before takeoff as a safety precaution to keep supercooled water in the lower atmosphere from freezing to airplane surfaces.

19.7 CLOUD TYPES

Meteorologists name clouds by their shapes ⓞ, their altitudes ⓞ, and their potential to rain on your picnic.

HIGH CLOUDS: 6–12 KM ① ciro-

① cirrus—high, wispy curls

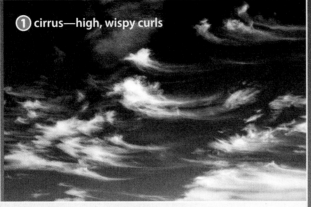

Cirrus (Ci) clouds are high, thin, wispy clouds made of ice crystals. They are often called *mare's tails*.

Cirrocumulus (Cc) clouds appear as small white puffs. They sometimes form in rows of clouds called a *mackerel sky*.

MIDDLE CLOUDS: 2–6 KM ② alto-

② cumulus—piles or billows

Cumulus (Cu) clouds are white and puffy with some vertical development. You can find them at all height levels. They can be associated with fair weather or light rain or snow.

Altocumulus (Ac) clouds appear as patches or layers of puffy, mid-level clouds. They may form repeating rows (also called a mackerel sky). Wisps of rain and snow can fall from altocumulus clouds.

LOW CLOUDS: 0.8–2 KM ③ strato-

③ stratus—thin layers or sheets

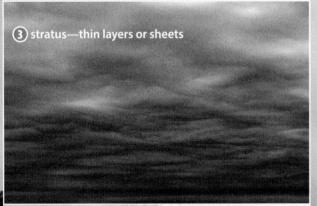

Stratus (St) clouds are low layers of gray, sheet-like clouds. Light to heavy rain can fall from stratus clouds. Stratus clouds form at ground level, causing **fog** to develop.

Stratocumulus (Sc) clouds are dark and flattened on the bottom but billowy on the top. Light to heavy rain can fall from stratus clouds.

Cirrostratus (Cs) clouds are high, thin sheets made of ice crystals. They can scatter light to form large circles around the sun or moon.

Altostratus (As) clouds are flat layers of gray mid-level clouds. The sun will not cast a shadow if there are altostratus clouds. Light rain and snow can fall from altostratus clouds.

Nimbostratus (Ns) clouds are low, dark layers of clouds that bring rain. Steady rain or snow falls from nimbostratus clouds.

Meteorologists not only name clouds by their shapes and altitudes but also by their likelihood to produce precipitation. A form of the word nimbus is included in the cloud name, when appropriate. Nimbus clouds are dense, moisture-laden clouds.

Cumulonimbus (Cb) clouds extend through all three levels. These billowy clouds often bring thunderstorms and produce rain and hail.

VERTICALLY DEVELOPED CLOUDS: 0.3–10 KM

④ cumulo-

467

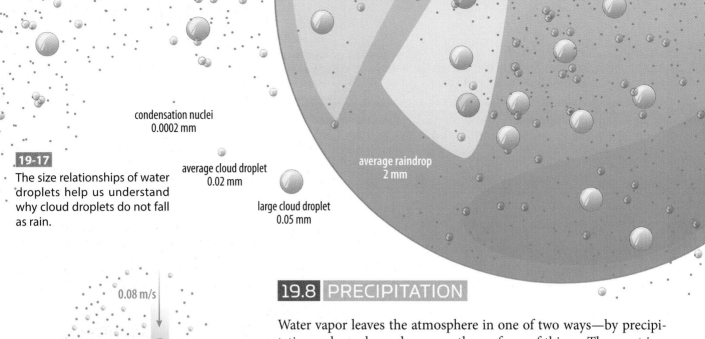

19-17 The size relationships of water droplets help us understand why cloud droplets do not fall as rain.

19.8 PRECIPITATION

Water vapor leaves the atmosphere in one of two ways—by precipitation or by a phase change on the surfaces of things. The most important of these processes is precipitation. Water droplets in a cloud, though still a liquid, are really small. They're only about 0.02 mm in diameter! So what keeps these droplets up in the sky where clouds are located? They are supported by the constant motion of air molecules and air turbulence.

But if the relative humidity is high enough, larger and more numerous cloud droplets form. Eventually, the air molecule collisions can't hold them up and the droplets begin to fall slowly. As they fall, they join with other droplets, and their sizes grow rapidly. This process is an example of **coalescence**.

When cloud droplets grow large enough, they fall as some form of *rain*. Droplets of a fine *drizzle* are about 0.5 mm in diameter. Raindrops in a heavy shower may be as large as 5 mm in diameter. An average raindrop is hundreds of times larger than a cloud droplet (see Figure 19-17).

As temperatures in the clouds drop to below freezing, several things may happen. As we mentioned earlier, supercooled cloud droplets may form and chill to many degrees below freezing. As coalescence occurs, they begin to fall. If the drops remain liquid all the way to the ground, they may suddenly freeze if the ground or surface is below freezing itself. This results in *freezing rain*.

If the supercooled raindrops freeze solid as they fall, they form ice pellets called *sleet* if they make it to the ground. As is often the case, ice pellets can form in vertically developed clouds like cumulonimbus. These clouds have strong internal *updrafts* that carry freezing raindrops back up to very high, frigid altitudes. As they go up, the ice pellets coalesce with falling supercooled raindrops. Then they freeze, fall in downdrafts, and collide with more raindrops, layering ice like an onion. This cycle of ris-

19-18 Coalescence is how raindrops build by joining many smaller drops.

coalescence (кон uh LEH sence): co- (L. *com*—together) + -alescence (L. *alescere*—to grow)

ing and falling repeats, forming *hail*. When the hailstones are too heavy to be carried by the updrafts, they fall to the earth. Hailstones are much larger than raindrops. The largest can grow to several inches in diameter. Hail this size can cause serious damage to buildings and cars.

In cold, humid clouds, water vapor can deposit directly as a solid on freezing nuclei, forming tiny ice crystals. These crystals grow rapidly as water molecules arrange themselves in unique and fantastic patterns. The results are snowflakes. *Snow* is the most common type of solid precipitation. Because snow often falls as delicate six-sided flakes, the density of fallen snow is far lower than rain. On the average, it takes 25 cm of snow to produce the same amount of water as 2.5 cm of rain. Most of the earth's fresh water is locked up in glaciers and ice caps that first began as fallen snow.

Snow may begin to melt as it falls, especially if the air near the ground is warmer than higher up where the snow formed. The mixture of snow and rain can also form sleet. In the United States, weather broadcasters describe such precipitation as a *wintry mix*.

19.9 DEW AND FROST

The same processes that cause water vapor to form rain or snow high in the air can occur right at ground level. But instead of forming precipitation, the liquid or solid water accumulates on solid surfaces. Condensation of water vapor occurs at the dew point. When a cool surface causes the temperature of a nearby film of air to drop below this point, droplets of *dew* form on the surface. Dew forms on grass, cars, bicycles, spider webs, or any other object that is cooler than the surrounding air. The dust particles that exist on most surfaces, as well as the objects themselves, provide condensation nuclei for dewdrops. Heavy dew forms when the air is calm. Moving air delays or prevents the formation of dew.

19-19
Notice the layers—rings of clear and white ice—in this hailstone. Typical hailstone diameters are ¼ to ¾ in., but this one is larger than a baseball. A record setting 8 in. diameter hailstone landed in South Dakota in July 2010.

19-20
Wilson Bentley, a Vermonter who became the world's expert on snowflakes and snowflake photography in the early 1900s, called them "a masterpiece of design." He thought it likely that no two snowflakes are alike.

Dew forms when water vapor in the air condenses on a surface with a temperature below the dew point.

19-21
Frost is not a form of precipitation. It is water that has changed directly from vapor to solid. Frost can form spectacular patterns on cold surfaces.

In cold, subfreezing weather, the maximum possible absolute humidity is far less than at warmer temperatures, which often forces the dew point below freezing. Less water evaporation takes place when it is cold as well, so relative humidity is generally low in winter weather. Various conditions, like an unfrozen body of water, can raise relative humidity in cold air to the dew point. When this happens, water vapor deposition occurs, forming *frost*. Frost often occurs when the soil is still above freezing, but the air above the soil is below freezing. Cold grass, leaves, and metal surfaces draw water vapor out of the air as ice crystals. In poorly insulated windows, the cold windowpanes can create intricate, delicate frost patterns out of the humid air within the house.

19C REVIEW QUESTIONS

1. What are the physical states in which water can exist in the atmosphere? Give an example of how we can sense each state.
2. What term refers to the temperature at which the relative humidity is 100%?
3. What can happen to water vapor in saturated air if the temperature drops to the dew point of −7 °C and there are dust particles present?
4. What is liquid water colder than 0 °C called?
5. Name and briefly describe the three categories of cloud shapes.
6. Give the prefixes or names for the four cloud height categories.
7. Name the kind of cloud that forms a broad flat layer around 4 km above the ground.
8. What changes of state can happen to water in the atmosphere to form precipitation?
9. Explain how raindrops form.
10. What forms when water vapor condenses on an object? What forms when water vapor deposits on a colder object?
11. (True or False) Dew will form when nighttime temperature drops to 5 °C, at which temperature the relative humidity is 84%.

CHAPTER 19 REVIEW

CHAPTER SUMMARY

» Weather is the current condition of the atmosphere.

» Meteorology is the study of the atmosphere.

» The most important measurable properties of weather are temperature, atmospheric pressure, humidity, wind speed and direction, and precipitation.

» Different kinds of instruments are needed to measure each kind of weather data.

» Atmospheric pressure differences drive winds on Earth.

» Pressure gradients, the Coriolis effect, and friction affect wind speed and direction.

» Global winds result mainly from uneven heating of the earth's surface by the sun and by the earth's rotation (the Coriolis effect).

» The Northern and Southern Hemispheres both have three main belts of winds—the trade winds, prevailing westerlies, and polar easterlies.

» Winds can exist at high altitude or at the surface of the earth. They can also blow vertically as well as horizontally.

» Relatively permanent belts of low or high pressure exist where the flows of air are mainly vertical.

» Winds can affect small areas where land meets sea and where mountains meet valleys. Temporary weather conditions and storms can generate local winds.

» Water in the atmosphere can be a gas, liquid, or solid, and can change from one state to another.

» Humidity and temperature control the development of clouds.

» Liquid water or ice crystals can exist in the atmosphere as clouds. Clouds can be layered (stratus), heaped (cumulus), or wispy (cirrus).

» Clouds may be classified by altitude as high (cirro-), middle (alto-), low (strato-), or vertically developed (cumulo-).

» Clouds are named by their shape, the altitude where they are formed, and sometimes by their potential for precipitation (nimbus or nimbo-).

» Water leaves the atmosphere mainly by precipitation. Water can also condense as dew or deposit as frost on cool surfaces at ground level.

» Raindrops grow by coalescence. As they fall, they may grow and break up many times before hitting the ground.

» Water in the solid state falls as sleet, hail, or snow.

Key Terms

weather	454
meteorology	454
atmospheric pressure	455
wind	456
wind-chill factor	456
precipitation	457
humidity	457
relative humidity	457
pressure gradient	458
cyclone	460
anticyclone	460
trade wind	461
prevailing westerly (wind)	461
polar easterly (wind)	461
dew point	464
cloud	464
condensation nucleus	465
freezing nucleus	465
fog	466
coalescence	468

19-22

Meteorologists need data from all over the earth, including what can be obtained over the ocean. Weather buoys like the one pictured here are used to remotely collect weather data from far out at sea.

19-23 Wall clouds—distinctive cloud formations associated with severe weather—form ahead of cumulonimbus clouds.

REVIEW QUESTIONS

1. What is the difference between weather and meteorology?
2. What are the two reference points of both the Celsius and Fahrenheit temperature scales?
3. Toward which direction is an east wind blowing?
4. Name the direction of the wind if a weather vane is pointing at 225° by compass.
5. If the absolute humidity of the air on a summer's day is 15 g/m³ and 25 g/m³ would saturate the air, what is the relative humidity of the air?
6. What is the most important factor in the development of a regional wind?
7. What causes winds to deflect as they blow over long distances?
8. What factor doesn't affect high-altitude winds as much as surface winds?
9. What direction would regional winds near the surface at 20°S tend to blow?
10. What wind system influences most of the weather in your country?
11. What part of the day would you expect winds to be calm at the seashore?
12. What condition must exist in a mass of air for a cloud to form?
13. Why do air masses cool as they are forced upward?
14. How can supercooled water be a problem?
15. How would you classify fluffy, high-level clouds?
16. Why is the cloud name *altocirrus* incorrect?
17. Compare the size of a cloud droplet with an average raindrop.
18. Compare freezing rain and sleet.
19. Describe how hailstones grow.
20. What are the minimum differences in weather conditions needed to form frost instead of dew?
21. Do you think humans will one day be able to control the weather with any degree of success? Why or why not?

True or False

22. A temperature of 212 °F is 112° above 100 °C.
23. Atmospheric pressure in warm masses of air is higher because they occupy more volume.
24. Moving air feels cooler than still air.
25. Wind speed and direction depend on three forces acting on air masses—the pressure gradient force, the Coriolis force, and the force of friction.
26. The existence of temperature differences in the earth's atmosphere is the main cause of global wind patterns.
27. Water can change state directly from ice to water vapor.
28. Air masses can cool without a loss of heat.
29. What we see as a cloud is its water vapor.
30. *Cirrocumulus* is a valid name for high-altitude, puffy clouds.
31. Freezing rain, hail, and sleet occur at different temperatures of the same precipitation process.
32. Snow forms when liquid water freezes.

Map Exercise

33. Using the link provided by your teacher, suggest four places in the world that would be good places for a wind farm. Be sure to explain your choices.

MOUNT WASHINGTON: HOME OF THE WORLD'S WORST WEATHER?

At a height of 6288 ft, Mount Washington is located in the White Mountains of New Hampshire. On its peak there is an observatory, established in 1934, that is operated by a private educational and research organization. The purpose of the observatory is to collect weather data for forecasting, for educating the public, and for learning more about weather. The data that has been collected there over the years is "extreme."

Even when we look at the average conditions on the mountain, we see that they are unusual. Normal temperatures range from 5 °F in January to 49 °F in July, while the lows range from −4 °F in January to 44 °F in July. It snows throughout the year at the summit with trace amounts each July and an average of 1.1 in. every June. An average 281 in. of snow falls each year. The wind blows with an average speed of 35 mph.

If the normal weather seems extreme, how do the extremes seem? The record high temperature on the mountain is 72 °F in August 1975; the record low of −47 °F was reached in January 1934. In February 1969, a storm buried the mountain with 173 in. of snow, which contributed to a record 566 in. for the year. That works outs to over 14 ft of snow in February and 47 ft for the year! The wind has gusted to over 100 mph in every month of the year and once reached 231 mph in April 1934. Mount Washington may have the worst weather in the world.

Questions to Consider

1. Why would an observatory be built in such a challenging location?
2. Why do you think it is so cold at the observatory?
3. The winds on Mount Washington typically blow from west to east. In what direction would a wind vane on Mount Washington typically point?
4. Does weather research fulfill the Dominion Mandate?

19-24
Ice covers the observatory on Mount Washington. Members of the team have to go out and remove the snow and ice from the sensors.
Location: 44.27 °N 71.304 °W

WEATHER

CHAPTER 20

STORMS AND WEATHER PREDICTION

20A AIR MASSES AND FRONTS `475`

20B SEVERE WEATHER `481`

20C WEATHER FORECASTS `495`

AND TODAY THE WEATHER WILL BE...

So what are you planning to do this weekend? Go on a hike? Take in a soccer game? Are you going shopping? Or are you planning to stay cozy inside and read a book? Well, the weather is probably a big factor in your plans. And with a tap on a screen, you can know what the weather will be like today, tomorrow, or even next week! Have you ever thought about how much science goes into that?

Around the world and around the clock, meteorologists are gathering data and making sense of it. Ocean buoys, weather stations, and satellites measure and relay information about air temperature and pressure, wind speed, precipitation, humidity, and cloud cover. And all of this helps you know whether you should pack that picnic lunch, load up your canoe and tent, or just stay home and out of the weather!

20A

AIR MASSES AND FRONTS

What creates weather?

20.1 WHY PREDICT THE WEATHER?

Weather prediction is something that's become a vital part of our modern society. Keep in mind everything that happens when your school is canceled due to stormy weather. The National Weather Service issues an advisory or warning that is based on approaching weather. Your local stations communicate this to your community. Your principal monitors the reports to decide whether to cancel school. You get notified by phone, radio, TV, or the Internet. Meteorologists can even predict what time the storm will hit!

Think about the science and planning that goes into evacuating an area that will be hit by a strong hurricane. Do you evacuate or not? Officials must consider many serious factors before making this decision. Where will the storm actually hit? Which way should people flee? How do you locate and transport all the elderly or people with special needs? When do you turn the utilities off, and what will happen when you do that? How do you maintain security? You don't want to evacuate an area unnecessarily. Lives and the area's economy depend on accurate predictions and effective communications, no matter how severe the weather. So let's take a look at what creates weather and how weather forecasters make sense of it.

20.2 AIR MASSES

Our weather changes because air is always moving. Weather forecasters watch air masses and their movement to predict and report the weather. To a meteorologist, an **air mass** is a huge body of air in the troposphere with similar temperature and humidity throughout. Air masses can cover hundreds or even thousands of square kilometers, and they can extend vertically to heights of several kilometers. Air masses are key to understanding weather. Within an air mass the weather is generally similar, but when air masses meet, they don't usually mix, at least not right away. Instead, precipitation and storms can form where they collide.

SOURCES OF AIR MASSES

An air mass gains its properties from the ground or sea beneath it as the air slowly moves over the earth's surface. The air mass takes on the humidity and temperature of areas that we call *source regions*. Source regions may be land or ocean. They are not normally windy places, because the air mass must remain in contact with the earth's surface long enough to take on its temperature and humidity.

20A Objectives

After completing this section, you will be able to

» explain how air masses move with weather.
» identify air masses by their source regions.
» connect weather to the interaction of two or more air masses.
» describe processes that produce precipitation.

20-1

Scientists use a two-letter designation for air masses. The first letter is a lowercase *c* (**continental**—dry) or *m* (**maritime**—moist) to indicate the humidity. The second letter is an uppercase *A* (**arctic**—very cold), *P* (**polar**—cold) or *T* (**tropical**—warm) to show the temperature. This image shows the air-mass source regions that affect North America.

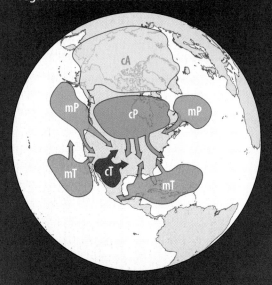

Meteorologists classify air masses by the humidity and temperature that they have obtained from their source regions. Each category has a letter that identifies the characteristic.

Air mass categories identify five different source regions usually recognized by meteorologists: *continental arctic* (cA), *continental polar* (cP), *continental tropical* (cT), *maritime polar* (mP), and *maritime tropical* (mT). Interestingly, maritime arctic (mA) air masses rarely, if ever, form, although with the Arctic ice cap melting, we may see more of these in the future.

AIR-MASS WEATHER

As air masses move, they carry their source region's weather conditions. For example, a continental tropical air mass from Mexico brings dry, warm weather when it moves into the United States. Continental polar air masses from Canada bring dry, cold conditions. During the summer, continental polar air masses bring clear, cool conditions. In winter, though, they often bring gripping cold spells that last for weeks. The maritime polar air masses from the oceans often bring in humid, cool air. Maritime tropical air masses bring humid hot spells to the central and southeastern United States in the summer. Frequent thundershowers and occasional tornadoes accompany these air masses. However, during winter months they may bring mild and cloudy weather.

Meteorologists label an air mass by comparing its temperature to the temperature of the ground over which it moves. *Warm air masses* are warmer than the ground and *cold air masses* are cooler. A polar or arctic air mass is a cold air mass, and a tropical air mass is a warm air mass.

Continental polar air masses bring cool, clear weather.

20.3 FRONTS

When two different air masses meet, they don't easily mix unless they have similar temperature and humidity. Instead, they form a boundary called a **front**. A front forms because of the differences in density between the two air masses. Colder, denser air masses move under warmer, less-dense air masses like a wedge. Fronts are action-packed places—they are where most storms form.

If the colliding air masses stop moving, or move parallel to the front between them, the front itself doesn't move over the ground. Such a condition is a *stationary front*. But air masses are usually on the go. One air mass moves into an area carried by prevailing wind patterns. It pushes the current air mass out, so the front between them moves. A moving front is named by the kind of air mass replacing the existing one over a region.

If a warm air mass overtakes a cold air mass, the front is a *warm front*. Because of its forward motion and lower density, the warm air mass gently rises above a thin wedge of cooler air. As the air rises, it can cool to its dew point. Wide systems of stratus clouds develop across the frontal area (see Figure 20-2).

This dangerous-looking storm cloud formed at the front of a fast-moving cold air mass.

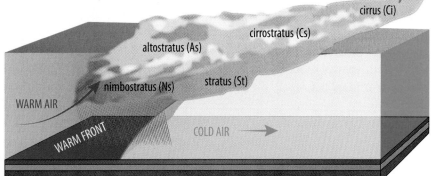

20-2 A fast-moving warm air mass moving up over a cold air mass produces a warm front.

There are many indications to an observer on the ground when a weather front passes through. For a warm front note the following:

» Temperatures rise.
» Pressure slowly decreases and then remains steady.
» Winds often shift from the southeasterly to the southwesterly direction in North America.
» Dew point slowly rises and then is steady.

If a cold air mass replaces a warm air mass, a *cold front* is moving in. Cold fronts often advance more quickly because cooler, denser air is more effective at pushing the warm air from a region. The angle of the face of the frontal area is also steeper because friction with the ground slows the ground air compared with the air higher up. Fast-moving cold air masses (more than 32 kph can force warm air up quickly, forming storm clouds (see Figure 20-3). Indications that a cold front is passing through include the following:

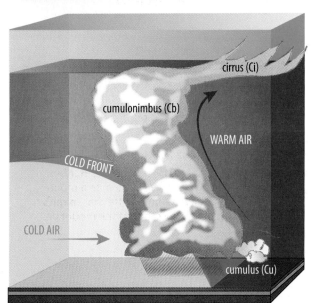

» Temperatures suddenly drop.
» Pressure rises quickly and then more slowly.
» Winds begin southwest to southeast in North America, but quickly swing to the northwest as the front passes, often with gusts.
» Dew point suddenly falls and then drops slowly.

Sometimes a cold air mass catches up to a warm front—a warm mass that is displacing a cold air mass. When this happens the cold air masses trap the warm air mass between them. This causes the warm air mass to rise over the other air masses and lose contact with the ground. This is an *occluded front*, which usually causes precipitation as the warm air water vapor cools and condenses.

20-3 When a fast-moving cold air mass moves under a slow-moving warm air mass, it produces a cold front.

On weather maps, fronts have special symbols (see Figure 20-5). The rule for their use is that the edge symbols point in the direction in which the air masses are moving. In general, there's precipitation along fronts unless the air is dry. Let's explore how this happens.

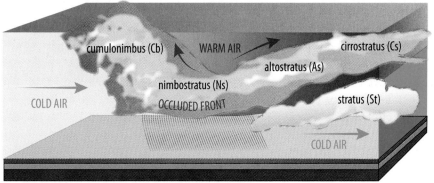

20-4 An occluded front forms when warm air is trapped between two cold air masses and loses contact with the ground.

20-5 Weather map symbols for the four types of fronts

20.4 AIR MASSES AND PRECIPITATION

When you think of weather prediction, you probably think most about planning your activities around rain or snow. So why do clouds form and bring precipitation? It's mainly about lifting air.

CONVECTION

Sometimes air is lifted by *convection*. Atmospheric convection works just like convection in the mantle of the earth or in density currents in the ocean. On a sunny day, the sun heats the ground, and the ground warms the air just above it by conduction. Cooler, denser ground air displaces this warmer, less-dense air by flowing in under it. The warmer air rises rapidly and cools to its dew point. This process produces clouds, and eventually enough water vapor may condense to form precipitation. Clear-air cumulus and cumulonimbus clouds form by convection. There are many places in the world that have a brief shower or thunderstorm almost every afternoon caused by convection.

MOUNTAINS

Air is lifted when warm, moist air crosses over a mountain. The mountain forces the air to rise and cool. At the dew point, clouds form and precipitation begins before the air mass reaches the peak and spills over the other side. Weather that forms in this way comes from **orographic** lifting. As the air flows down the other side of the mountain, it warms and its relative humidity falls. Clouds vanish and the air becomes very dry. The environmental conditions on the windward side of a mountain (facing the wind) are often far different from the lee side (sheltered from the wind), which is in a *rain shadow* (see Figure 20-7). This effect creates many of the world's large deserts, such as the **Atacama Desert** in Chile and the **Tibetan Plateau**.

20-7
Orographic precipitation on the eastern slopes of the Andes Mountains produces lush vegetation and vineyards. The rain shadow to the west of the mountains produces desert-like conditions.

20-6
Convection clouds can often be seen over flat regions on hot afternoons.

orographic: oro- (Gk. *oros*—mountain) + -graphic (Gk. *graphos*—shows or describes). Orographic clouds often wrap over a mountain in a thin mantle that mimics the shape of the mountain.

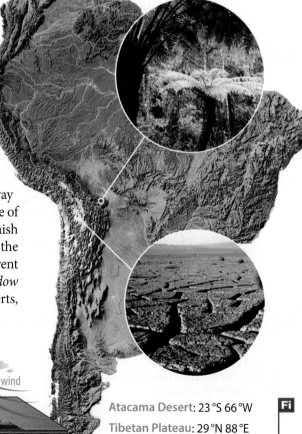

Atacama Desert: 23°S 66°W
Tibetan Plateau: 29°N 88°E

FRONTS

As we mentioned earlier, air is lifted when air masses meet. This is most dramatic at a cold front. The colder air, acting like a wedge, forces the warmer air to rise and cool. This effect is called **frontal wedging**. If the warmer air cools to its dew point, it forms clouds and can release its moisture as precipitation. The appearance of clouds and precipitation is probably the most noticeable evidence for a passing front.

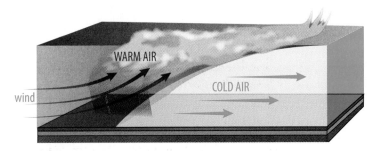

20-8 Precipitation is caused by warm air rising due to convection currents, orographic lifting, and frontal wedging (above).

COLLIDING WINDS

The fourth process that lifts air results from the collision of horizontal air currents. It is closely related to convection, and it often occurs over strips of flat land between bodies of water, as it does in Florida. This process is called **convergence**. Masses of moving air, as with opposing sea breezes, flow inward toward a heated section of land from opposite directions. The only way for the colliding air currents to go is up, thrusting large volumes of heated humid air high into the atmosphere. Again, the rising air cools to its dew point and vertically developed clouds form. Short-lived, strong thunderstorms often result.

20-9 Convergence of wind currents also causes precipitation.

20A REVIEW QUESTIONS

1. Why is weather prediction important?
2. Describe the properties of an air mass.
3. How does a source region affect a mass of air moving slowly over it?
4. What do meteorologists consider when they label an air mass that moves into a region of the country?
5. Name, give the symbol for, describe, and identify a source region for the five usual kinds of air masses.
6. What happens when two air masses meet?
7. What changes occur when a warm air mass overtakes a cold air mass?
8. How does the slope of a cold front differ from the slope of a warm front? Why?
9. What time of day would you expect convective thunderstorms to form?
10. What effects on the local environment can occur as prevailing winds blow moist air over a high ridge of mountains?
11. (True or False) Rapidly rising barometric pressure and cooling temperatures indicate that a cold front has passed through your area.

DOPPLER RADAR

Have you ever noticed that when the sound of an ambulance approaches and then passes you, its siren seems to change? At first, the pitch of the siren is higher, and then it becomes lower. This is known as the Doppler effect. The same thing would be heard if the ambulance were still and you were moving. Why does this happen?

When a source of sound waves or your ear is moving relative to the other, the speed at which the sound waves enter your ear is affected. This sounds like a change in pitch. Meteorologists use the same effect to help monitor weather with technology called *Doppler radar*.

This device sends out pulses of microwave radio signals that bounce back from clouds and precipitation. In between pulses, a receiver at the station listens for echoes. Just as with sound, the "pitch" of the reflected radar waves is shifted by moving raindrops and clouds. The receiver processor detects these shifts and displays them.

Using computers, meteorologists analyze these radar echoes to detect thunderstorms, tornadoes, hurricanes, and heavy precipitation within a radius of 460 km. The weather service uses 159 special Doppler radars called NEXRAD radars. The acronym *NEXRAD* stands for NEXt generation RADar. The weather service first started using them in 1988. They are a big improvement over the 1950s-era radars they replaced.

A Doppler radar image of the characteristic hook (arrow), indicating the presence of a tornado

These radars can distinguish between snow and rain. They can also separate clouds of water droplets from clouds of ice crystals. The computers can determine wind speed and direction within storms and display this data right on the radar screen. They can show the structure of massive wind storms such as hurricanes. These instruments give us advance warning of severe weather and dangerous conditions, helping us to save people's lives.

Modern Doppler weather radar transmitters and receivers are housed in domes to protect them from the elements.

20B

SEVERE WEATHER

How do severe storms form?

Air masses can cause some agitation where they meet. They often generate **storms**—severe disturbances in the atmosphere. Storms can involve changes in all the weather factors you learned about in the last chapter. They may bring precipitation, strong winds, and changes in air pressure and temperature. In this section, you will study four types of severe weather: *winter storms*, *thunderstorms*, *tornadoes*, and *hurricanes*.

20B Objectives

After completing this section, you will be able to

» classify storms and explain how they form.

» describe the major hazards of each kind of storm.

» identify the key actions to take to remain safe in each kind of storm.

Snow can clog roads, reduce wheel traction, and limit visibility for drivers. This can often cause accidents.

WINTER STORMS

20.5 UNDERSTANDING WINTER STORMS

At the beginning of the chapter, we discussed some of the things that happen if you have a weather-related school closure. Snow days are great—no school! But they're great only if you are warm and dry inside. Sometimes **winter storms** that bring severe winter weather create big problems, even killing people who are without proper shelter.

Winter storms can include any form of freezing or frozen precipitation, usually when the general air temperature is below freezing. A *wintry mix* consists of periods of rain, ice, and snow. If all three fall at the same time, we call it *sleet*. Snow is the most common form of solid precipitation. For snow to fall, water vapor must form crystals by deposition within cold strata in the lower atmosphere. For snow to accumulate on the ground, the entire air mass must be below freezing so that it doesn't melt as it falls. Fluffy snow piles up when the ground and other surfaces are also below freezing.

20-10
A nor'easter churns off the New England coast, bringing large snowfall and high winds.

The name **Nor'easter** comes from the strong northeast winds blowing in from the North Atlantic, carrying moist air. These storms often occur on the north side of a cyclonic low-pressure center located near the mid-Atlantic states.

Sometimes an area can get a lot of snow, piles and piles of it! This happens especially in the northeastern section of the United States when a cyclonic winter storm called a *Nor'easter* sweeps moisture in from the Atlantic Ocean and drops it as snow. The counterclockwise rotation of the storm system produces strong winds that can create deep snow drifts as the snow piles up.

An area on the leeward side of a large body of open water can also get a lot of snow. An air mass moving across the water picks up moisture, and then drops it as snow on the leeward side of the body of water as the air rises and cools. In the northern United States east of the Great Lakes, this kind of precipitation is called *lake effect snow*, and it can be quite heavy.

Deadly Ice Storm

In 1998, an ice storm that lasted for eighty hours interrupted electricity to more than 4 million people in Ontario, Quebec, and northern New England. Power was off for as long as a month in some places. Nearly thirty people froze to death and 100,000 had to evacuate their homes after loss of power. This single storm destroyed miles of power transmission wires and many thousands of trees and had a severe economic impact on the entire region.

20.6 HAZARDS OF WINTER STORMS

Winter storms can be dangerous. Ice reduces the friction between tires and roads, making it harder to control vehicles. When snow begins to choke roads and highways, accidents can happen. Snow can interfere with the movement of emergency vehicles, and it can close airports. If snow is heavy enough, tree limbs can break, making it dangerous to be near them, and they in turn can take down power lines or damage houses. In snowstorms, you should dress warmly and avoid overexertion. On long winter trips in snow, you should take water, high-energy snacks, and extra blankets—just in case your car gets stranded.

Things can get especially hazardous if freezing rain develops. It can glaze road surfaces with ice and weigh down power lines. Ice storms actually damage more trees than snow storms, making them more dangerous than snow, because they can cause widespread power outages that last for days. People are stuck indoors with limited ability to stay warm.

During and following a heavy ice storm, you should avoid walking under trees or power lines that are heavily coated in ice. Broken branches and electrical wires can fall suddenly, endangering people underneath.

20-11
Winter storms that bring freezing rain can be dangerous because they can cause long power outages during the frigid winter.

20-12
Ice builds up fast during a winter ice storm.

When a large thunderstorm is approaching, the rain is often preceded by a blast of noticeably cooler air called a *gust front*.

One, Two, Three…Kaboom!

Did you know that you can figure the distance to lightning in your head? Light reaches you almost instantaneously, but sound travels at only about 1/3 km/s. If you count the seconds between a lightning flash and the sound of thunder, you can calculate how far the sound has traveled. Divide the number of seconds by 3 or 5 to see how many kilometers or miles away the storm is. For example, if 15 seconds pass between the lightning and the thunder, the sound has traveled 5 km. You usually can't hear thunder more than 16 km away.

THUNDERSTORMS

20.7 UNDERSTANDING THUNDERSTORMS

There are more thunderstorms happening every day than any other kind of storm. **Thunderstorms** bring thunder, lightning, and usually heavy precipitation. Most are small, less than a few kilometers in diameter. Though they usually form in the summer, they can occur in the winter along the edges of cold fronts. A string of thunderstorms along a cold frontal system is called a *squall line*.

All thunderstorms form in moist, unstable air and progress through three stages.

❶ Developing Stage

Moist, unstable air rises through convection, orographic lifting, frontal wedging, and convergence. As the air is lifted, it cools and condensation begins. This releases energy that continues the lifting process. At this point, the vertical motion within the storm is primarily an updraft.

❷ Mature Stage

Once there is enough condensation, water droplets coalesce and begin to fall as rain. Updrafts continue to carry moisture and energy upward, but the falling rain produces downdrafts. Updrafts sometimes lift some of the raindrops back to the top of the storm, creating hail. Some hailstones repeatedly move up and down within the storm and grow to very large sizes. The air moving upward and downward causes a buildup of electrical charge. When enough charge collects, lightning is produced.

❸ Dissipating Stage

As the storm loses energy, downdrafts dominate as much of the moisture and energy drain out of the storm in the form of rain and possibly hail.

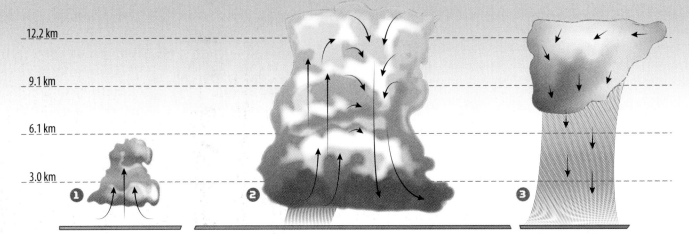

484   CHAPTER 20

20.8 HAZARDS OF THUNDERSTORMS

Though thunderstorms can bring wind gusts and even hail, their greatest hazard is lightning. **Lightning** is an electrical discharge either between clouds or between a cloud and the ground. Where do these charges come from?

Water and air particles in the cloud swirl around and collide. Electrons accumulate on the water droplets, creating charged particles. A positive charge builds up at the top of the cloud, and a negative charge at the bottom. This produces a positive charge on the surface of the earth. As the charge difference increases, a gigantic spark that we call lightning occurs. For lightning to jump only 2.5 cm, it takes about 25,000 volts! That's five times the shock you get from static electricity when touching a metal doorknob. Yet lightning bolts are often thousands of meters long. They can have a charge difference of hundreds of millions of volts.

Lightning is not only powerful; it's hot—really hot. The temperature within the lightning bolt can reach 28,000 °C! This tremendous heat rapidly expands air along the lightning's path with an explosive force, creating a partial vacuum. An instant later, the air collapses to fill the vacuum. Both of these processes produce the sound we call *thunder*.

You can see that the energy in lightning can be hazardous to buildings and living things. Lightning is electricity, and like the electricity in your home, it is attracted to things that it can flow through easily. These are called *conductors*. It is also attracted more to tall, thin objects like trees, flag poles, and standing people than to low, flat ones, like the surface of a field. Lightning causes damage mainly by heating whatever material it passes through with huge amounts of electrical energy. In trees, water vapor in the trunk and branches flashes to steam and explosively blows the wood apart. In buildings and trees, the heat can start fires too. When animals or people are struck by lightning, the electrical current can damage nerves, tissues, and major organs. Direct strikes by lightning are almost always fatal.

Thunderstorms are occasionally accompanied by hail. You learned in Chapter 19 that hailstones can grow to enormous sizes. Imagine being hit by a falling chunk of ice the size of a softball moving at more than 30 m/s! Even smaller hailstones can do serious damage to buildings, cars, trees, and people.

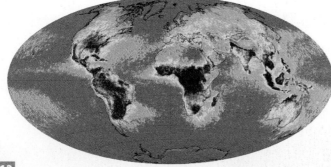

20-13

Look at the places in the world that get the most lightning strikes (red shading). Make some generalizations about what you observe.

TORNADOES

20.9 UNDERSTANDING TORNADOES

Sometimes really violent thunderstorms spawn tornadoes. A **tornado** is a narrow, rapidly spinning column of air extending downward from a cumulonimbus cloud. Condensing water vapor and debris lifted into the air by the low-pressure center usually creates a visible funnel. Hurricanes, tropical storms, and convective storms created by strong cold fronts can produce tornadoes. These devastating rotating storms occur over all continents except Antarctica, but they are most common and destructive in the southern and central sections of the United States—an area called *Tornado Alley*.

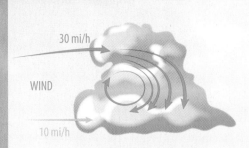

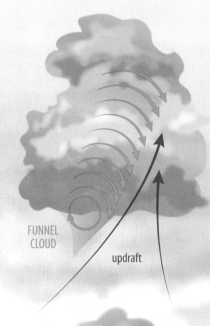

20-14

Tornado Alley in the United States

Tornadoes are associated with massive rotating storms called *supercells* and form only under unusual circumstances. These supercell storms create a condition called *wind shear*, which is a large change in wind speeds at different altitudes. Wind shear creates a large horizontal rotating mass of air. The updraft within the thunderstorm tilts the mass of air vertically. The rotating air grows downward toward the ground, creating a funnel cloud. When the funnel extends downward and finally touches the ground, it is considered a tornado. Because tornadoes are relatively infrequent and destructive, meteorologists consider them separate windstorms from the storms that create them.

At a distance, a tornado often looks like a gigantic elephant's trunk or a thin rope dangling from a cloud. A very large one looks like a wide, black cylinder connecting ground and cloud. An average funnel is about 400 m in diameter at its base, the part in contact with the ground, but can range from 1 m to 5 km wide. This base is important because it is the most destructive part of a tornado. Sometimes several funnels extend down from the same cloud, causing great destruction. Multiple tornadoes from the same storm system in different places are not unusual.

Tornadoes can happen in any month, but in the southern United States most tornadoes occur in March through May. In more northern states, they occur most frequently in the summer. Most tornadoes touch down between 3:00 p.m. and 9:00 p.m., though they can appear at all hours of the day and night.

A tornado may stay in one place, but it usually moves and can travel as fast as 100 kph! This is why you should never try to outrun a tornado in a car. Its direction of movement is usually from the southwest to the northeast, but it can move in other directions as well. At the same time, a tornado's motions can frequently complicate the efforts of people trying to dodge it. It may sway from side to side and can lift up and touch down many times along its path. Sometimes the funnel is invisible until it actually touches down, but its funnel is usually white or gray. Its color often turns black as it touches the earth, picking up dust and debris and anything in its path. Lightning flashing twenty or thirty times per minute may accompany the tornado, and lightning may flash continuously inside the funnel.

Tornado Watch

The National Weather Service (NWS) has developed radar systems that can give people about ten minutes of warning. The NWS can often give a thirty-minute warning that tornado formation is likely.

Tornado Chasers

Tornado chasers are meteorologists and professionals who track and pursue tornadoes so that we can learn more about them. They take pictures, make measurements, and put instruments in the paths of tornadoes. They are experienced and know the risks. "Kids, don't try this at home!"

20.10 HAZARDS OF TORNADOES

Killer Twisters

In 1974, 148 tornadoes touched down in thirteen different states in a sixteen-hour period, killing 319 people and injuring 5484 others along a 4000 km storm path.

Enhanced Fujita-Pearson Scale

Rating	Sustained wind (kph)
EF0	105–137
EF1	138–177
EF2	178–217
EF3	218–266
EF4	267–322
EF5	>322

Tornadoes are often rated on just the Enhanced Fujita-Pearson scale (for example, EF3), since meaningful measurements of path length and width during the storm may not be possible.

The aftermath of a tornado near Tuscaloosa, Alabama, demonstrates the destructiveness of a tornado.

Each year about 1200 tornadoes strike the United States, causing about 80 deaths and 1500 injuries. The high-speed, rotating wind and the updraft are the two great destructive features of a tornado. The force of a tornado wind can take steel beams and spear trees with them. Many times, after a tornado has passed, people find that pieces of straw have been driven like nails into trees or telephone poles. The updraft within the funnel can lift people, animals, automobiles, railroad cars, and even houses into the air. Rarely will a tornado simply move an object some distance and set it down unharmed. It usually destroys whatever it touches.

A tornado's path of destruction is usually narrow, but its whirling wind speeds may be as much as 515 kph—higher than in any other storm. Tornado strength or intensity is rated on the six point *Enhanced Fujita-Pearson (EF) scale*. See Appendix I for a detailed description of this scale. The EF rating is an estimate of the wind speed that is based on the damage caused by the tornado, taking into account construction materials and methods. Dr. Pearson further improved the original Fujita scale by adding indicators that describe the width of the destruction path and the distance that the tornado was in contact with the ground.

Low pressure within the funnel of a tornado is not the destructive force it once was thought to be. In 2003, a researcher successfully placed a pressure-recording instrument in the path of an F4 tornado that measured a pressure drop of only 0.1 atm. Pressure inside even a powerful tornado is just a little lower than at sea level. So tornadoes don't cause buildings to explode in low pressure. Other research and field observations have shown that a tornado's high winds simply push the building's windward walls inward. The tornado's updraft lifts the roof off the building. The other walls fall outward due to the winds.

Hurricane Isabel as seen from the International Space Station. You can see the eye at the center of the storm.

HURRICANES

20.11 UNDERSTANDING HURRICANES

Hurricanes, the largest storms of all, are giant cyclonic windstorms that form over tropical or subtropical oceans. Hurricanes go by other names in other countries. In the Western Pacific, they are called *typhoons*. In the Indian Ocean, they are known as *cyclones*. Unlike tornadoes, these storms can cover nearly 1,000,000 km²! By definition, they have winds of at least 119 kph, but sometimes winds exceed 320 kph! And they can last for up to three weeks! Hurricanes bring not only high winds but also high tides and heavy rains, causing flooding and heavy loss of life and property. This is exactly why we need to understand them.

A HURRICANE IS BORN

Tropical hurricanes need two things to form: warm water (at least 26.5 °C) and the Coriolis effect. The perfect place for this is over oceans within 10–15° latitude of the Equator. They don't start closer because the Coriolis effect is not strong enough to rotate the winds in a cyclonic pattern. They generally don't start farther north or south because the waters are usually not warm enough. (Those that do are called *extratropical hurricanes*.) Hurricanes don't form over land because friction resists the cyclonic circulation of the air. In the Northern Hemisphere, they can develop from June through October. This is when weather conditions are just right. There are only six regions in the Atlantic, the Pacific, and the Indian Oceans where tropical hurricanes are usually born. These are called *tropical cyclone basins*.

20-15

Cross-section of a hurricane showing the eye and the internal wind patterns. Cool, dry air enters from above the storm, creating the calm eye of the storm. Warm, moist air spirals in toward the eye near the bottom, upward as it nears the eye, and back out in spiraling fashion as it rises.

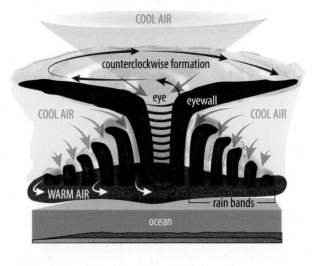

STORMS AND WEATHER PREDICTION

NAMING HURRICANES

If a cyclonic storm's winds reach 63 kph or more, meteorologists categorize it as a **tropical storm** and assign it a name. This is especially true in the Atlantic Ocean and eastern Pacific Ocean near the United States. In general, the name of the storm reflects the cultural region where it occurs. This policy became standard after studies indicated that Asian people didn't heed warnings of storms with unfamiliar Western names.

A committee of the World Meteorological Organization creates lists of tropical cyclone names that are culturally appropriate for the different regions of the world. Most storm lists are in alphabetical order, alternating male and female names. The lists are active for a year and are recycled every four to six years depending on how many lists exist for each region. If one particular storm is especially devastating, its name is removed from the list and another takes its place.

THE LIFE OF A HURRICANE

In the tropical oceans, stagnant, overheated, moisture-laden air begins to rise and cool in a low-pressure area. Clouds form and rain begins. Cooler surrounding air rushes toward the low-pressure area formed at the surface by the rising warm air. In the Northern Hemisphere, this wind motion sets up a counterclockwise cyclonic spiral flow. The air-system rotation funnels more air into the low-pressure center, which is sustained by the rising, warm, moist ocean air. As the low pressure in the center drops, the pressure gradient force (see Chapter 19) grows, increasing the speed of the inrushing air. The cycle progresses as more winds rush in and the spiral whirls more vigorously. Eventually, when winds attain a speed of 119 kph, the tropical storm becomes a hurricane. Hurricanes are rated on the five-category **Saffir-Simpson scale**. (See a more detailed description of the scale in Appendix I.)

Photographs from earth orbit show loose spiral arms at the edges of the storms. These spiral arms become closer together and winds increase in speed as they approach the center. Near the center, the winds are strong. Air rapidly rises and cools to produce extremely heavy rainfall. But within a circular wall of towering cumulonimbus clouds lies the relatively calm center of the hurricane, called the **eye**. In the eye, clouds thin and the rain often stops completely. Winds are calm or light and shifting. You might even be able to see the sun during the day or stars at night from inside the eye of a hurricane.

Many hurricanes cross the coastline and head inland. Due to friction with land, a hurricane's winds usually slow down at this point. Land also lacks warm, moist air that fuels hurricanes. This loss of heat robs the hurricane of the energy that it needs to maintain the low-pressure center. When the wind speed drops below 119 kph, the hurricane becomes a tropical storm again. Cooler land air drawn into the storm helps condense the remaining moisture, and the cyclonic storm usually dies as heavy showers over a large area.

The **Saffir-Simpson scale** for hurricane damage was developed in the early 1970s by structural engineer Herbert Saffir and by Robert Simpson, who was director of the National Hurricane Center during that time.

Saffir-Simpson Hurricane Categories

Category	Sustained wind (kph)
1	119–153
2	154–177
3	178–208
4	209–251
5	≥252

HURRICANE HUNTERS

How do we know so much about hurricanes? Much of the credit must be given to "Hurricane Hunters." These are teams of brave pilots and meteorologists who fly specially equipped aircraft through these violent storms to directly obtain data that could not be obtained any other way. This major scientific contribution is the joint effort of three organizations: the National Hurricane Center (NHC), the National Oceanographic and Atmospheric Administration (NOAA), and the United States Air Force (USAF).

Hurricane hunter aircraft fly in and around hurricanes. The disk beneath the aircraft is a high-tech radar for studying hurricanes.

HURRICANE MOVEMENT

Tropical storms and young tropical hurricanes generally move west, blown by trade winds. As the speed of the storm center builds, the Coriolis effect curves it to the right in the Northern Hemisphere. Without other influences, the storm will eventually curve back into the mid-Atlantic (or Western Pacific for typhoons). But hurricanes don't exist in isolation from other weather factors. The jet stream and other nearby high- and low-pressure systems can influence the motion of a hurricane, often in unpredictable ways.

Figuring out the future path of a hurricane can be tricky. Despite studying hundreds of hurricanes and creating complex storm models with supercomputers, meteorologists have a hard time predicting hurricane movements. Hurricanes have stopped, backed up, curved to the left, or even made loops and crossed the same location twice! Nothing is certain when dealing with a hurricane.

The colored areas on this map (left) indicate the computed likelihood that areas will experience tropical storm-force winds.

You may have seen maps on TV weather programs that show the most probable track of a storm. They also include maps showing where the storm has higher and lower probabilities of making landfall. These maps are created by complex computer models. It is the science behind these models that saves people's lives.

This color-enhanced infrared satellite photo (above, right) shows Hurricane Katrina over the Gulf of Mexico in August 2005. Katrina caused more damage than any other hurricane in American history.

Weather satellites have vastly improved our ability to track hurricanes. Meteorologists can remotely measure ocean and cloud temperature, wind speed and direction, cloud density, and many other factors that help predict the growth and movement of these powerful storms. They can update their data almost in real time to make their models even more accurate. Satellite photographs of a hurricane allow meteorologists to precisely track a storm's center.

LIFE CONNECTION: WHAT HAPPENS TO ANIMALS DURING HURRICANES?

We often hear about the people hurricanes kill and the damage they cause. But have you ever wondered how hurricanes affect animals? Can animals somehow detect that a major storm is coming? If so, how? And how do they cope with the aftereffects of hurricanes?

Many animals seem to sense an approaching storm. They seek shelter before it even hits. For example, sharks often disappear from the coastline just before a hurricane strikes. Scientists believe this incredible occurrence has to do with pressure changes. As a hurricane builds, air and water pressure drop. Some sharks can apparently detect this change in pressure and head for open water where they can weather the storm.

Dolphins have similar abilities. Scientists hypothesize that dolphins may react to changes in water salinity. Hurricanes dump huge amounts of fresh water into the oceans. All this fresh water makes the ocean in that area less salty. Dolphins, accustomed to seawater, may somehow detect and respond to this change by leaving the area before the storm hits.

Ocean-dwelling animals are not the only ones affected. Birds often encounter strong storms during their migrations. Sometimes they will seek alternate routes to avoid the storm. If they cannot avoid it, they seek shelter in caves or other safe areas.

The aftermath of hurricanes is often more devastating to animals than the hurricanes themselves. These aftereffects come in many different forms. Lowered salinity, sedimentation, debris, and low oxygen levels are just a few of the many negative effects that hurricanes have on sea creatures.

Land animals can be even more affected. In coastal areas that suffer extensive destruction, many pets and even farm livestock are left homeless and without shelter. Often they are separated from their owners. They wander the wreckage, sometimes injured, looking for food. Wild animals in coastal lands are in the same predicament. The storm surge can quickly drown most ground-dwelling animals. After a hurricane has destroyed coastal habitats, the remaining animals must move to other locations or perish. On the other hand, scavengers and carrion eaters have a bountiful supply of food for many days after a storm.

Most people are distressed at the suffering of wild and domestic animals, who don't understand what has happened to them. So there are many people and organizations that focus on rescuing animals and reuniting pets with their owners after a storm or other disaster. This might seem to be of little importance compared to relieving the suffering of people and rebuilding after a disaster. But it is just another aspect of being good stewards of the living things that God commanded us to care for.

20-18
Staying behind during an evacuation for a hurricane just isn't worth it!

20.12 HAZARDS OF HURRICANES

A hurricane is coming. A thin haze of cirrus clouds covers the sky. The air is calm, hot, and humid, and air pressure is typically near normal. Then, atmospheric pressure begins to fall sharply, the high cirrus clouds overhead are followed by cirrostratus and cirrocumulus clouds, and the wind increases. Immediately before the fury begins, blue-black nimbus clouds roll and tumble overhead. Rain may begin hours or even several days before the high winds appear, depending on the size and shape of the storm.

Far out at sea, the high winds produce huge rolling waves called the *storm swell*. These waves move at speeds up to 50 kph and can break on the shore several hundred kilometers ahead of a hurricane. These waves are quite destructive and can quickly erode beaches and damage shore structures like groins and piers.

A hurricane's full destructive power is in its high winds, heavy precipitation, and wave action. A hurricane can spawn tornadoes, hail, and torrential rains. Wind may come from almost any direction. It can break off the tops of trees like matchsticks, twist heavy steel billboard frames like pretzels, and peel the roofs off buildings like a can opener. A hurricane can have hundreds of tornado-like eddies embedded within it that add to its destruction.

STORMS AND WEATHER PREDICTION

Shoreline areas suffer most from a hurricane. In addition to the high storm swells that come crashing ashore, the winds on the right side of the storm's direction of motion (the *dangerous semicircle*) tend to push water ahead of the storm. When combined with the lower air pressure within a hurricane, which causes local sea level to rise, the ocean piles up against the shore, raising the average sea level several meters. This higher-than-normal water level is called a **storm surge**. Add to the storm surge a daily high tide (especially a spring tide), plus the action of storm swells, and entire coastal communities can be completely swamped and devastated.

If your area is directly threatened by a strong hurricane, the best thing your family can do is evacuate far inland or to someplace far to the left of the storm's direction of motion. Many people stay to "ride out the storm," mistakenly thinking that the storm effects won't be as bad as predicted, or that they need to be present to protect their possessions after the storm has passed. No property is worth risking your life (or the lives of your rescuers)!

We have a duty, both to God and to our neighbors, to understand severe weather so that we can prevent loss of life and minimize damage where possible. How can we do this? Making observations, developing workable models, and setting up a network to communicate warnings can help save lives. This is the work of God-glorifying science.

The storm swell from Hurricane Dennis pounded the shore of Panama City Beach, Florida, in 2005.

20B REVIEW QUESTIONS

1. List the four types of storms that you have studied in this section. Describe two things that they have in common.
2. What precautions should your family take when driving long distances in snow?
3. Which of the storms discussed in this chapter is the most common?
4. What happens in a cumulonimbus cloud that is directly responsible for the formation of lightning?
5. How far away is a lightning flash if it takes twelve seconds to hear the thunder?
6. Where do most tornadoes occur in the world?
7. State the two main ways that tornadoes cause damage and describe how the damage happens.
8. Considering the name used, tropical storm Barry would likely be located in what part of the world? Where would tropical storm Choi-wan be happening?
9. What is the center of the hurricane called? Describe this region of the storm.
10. What three things cause a hurricane's winds and energy to diminish over landmasses?
11. Describe the typical path of a Northern Hemisphere Atlantic hurricane. How can actual hurricane paths differ from the typical?
12. Why do we need to understand how severe weather forms?
13. (True or False) The Enhanced Fujita-Pearson and Saffir-Simpson scales classify tornadoes and hurricanes by the amount of damage they do, not by their sizes.

WORLDVIEW SLEUTHING: SEVERE WEATHER RESPONSE

INTRODUCTION

You have volunteered for the county disaster preparedness organization. The county wants to provide pamphlets to the public to prepare for different types of hazardous weather. You have been assigned to create one of these pamphlets.

TASK

Select one of the four types of severe weather described in this chapter—winter storm, thunderstorm, tornado, or hurricane. Research the severe weather; be sure to include basic information about the storm, long-term preparation, and immediate response. You must create a pamphlet to educate the public on the particular type of severe weather and how to avoid hazards associated with it.

PROCEDURE

1. Research the basics of the storm that you have selected. For example, if you chose tornadoes, you could do a keyword search for "tornado" or "severe weather tornado."

2. Research the recommendations for preparing before these storms happen by doing a keyword search, for example, on "tornado preparedness."

3. Research the recommendations for responding when these storms occur by doing an Internet search using the keywords "what to do in a tornado."

4. Plan your pamphlet production and collect any photos that you need, being careful to give proper credit. Try to give equal space to each of the three topics.

5. Produce your pamphlet and show it to another person for feedback.

CONCLUSION

Forecasters can provide ample warning for some storms while others occur unexpectedly. As in any emergency, prior planning and understanding proper responses to severe weather saves lives. The science of weather prediction is something that Christians can value because it helps protect people and better use God's earth.

20C

WEATHER FORECASTS

Why is predicting the weather so difficult?

20.13 THE STATION MODEL

Meteorologists handle weather data pouring in from all over the world. This information could come from automated and manned weather stations, airports, ships, automated buoys, aircraft, and satellites. Meteorologists take this data and make sense of it. They plot weather data from each weather station in a symbolic form called a **station model**. These models use special notations and numbers to communicate information in a simple and standard way on weather maps.

Figure 20-19 shows an example of the station model. Detailed descriptions of the symbols and numbers used in a station model can be seen in Lab Manual Appendix D. The circle in the middle represents the location of the weather station on the map at the time of the report. Land and moored buoy stations are stationary. Weather data from ships, aircraft, and other mobile stations is placed on a map at the unit's location at the time of the report. The interior of the circle is used to show the amount of cloud cover at the station. Sections of the circle are blacked in, depending on the amount of sky obscured. If there is wind, a line (staff) points from the circle toward the direction *from which*

20C Objectives

After completing this section, you will be able to

» describe weather station models.

» explain how weather data is used to construct weather maps.

» evaluate the probable accuracy of a weather forecast.

20-19

To a meteorologist, reading a station model is a straightforward task. Each number, symbol, and location provides specific weather information. This information is used to make weather predictions.

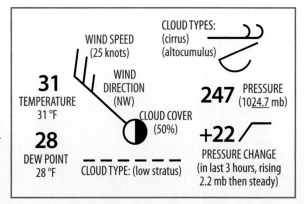

STORMS AND WEATHER PREDICTION

the wind comes (just like a wind vane). The number and lengths of angled lines (flags) attached to the staff indicate the wind speed. Arranged around the station model are symbols and numbers representing cloud types seen, temperature and recent changes, pressure and recent changes, visibility, and many other factors as reported in the data.

A single station model doesn't provide much help figuring out the bigger weather picture or making predictions. But when many station models are plotted on a weather map, weather forecasters can begin to see patterns emerge.

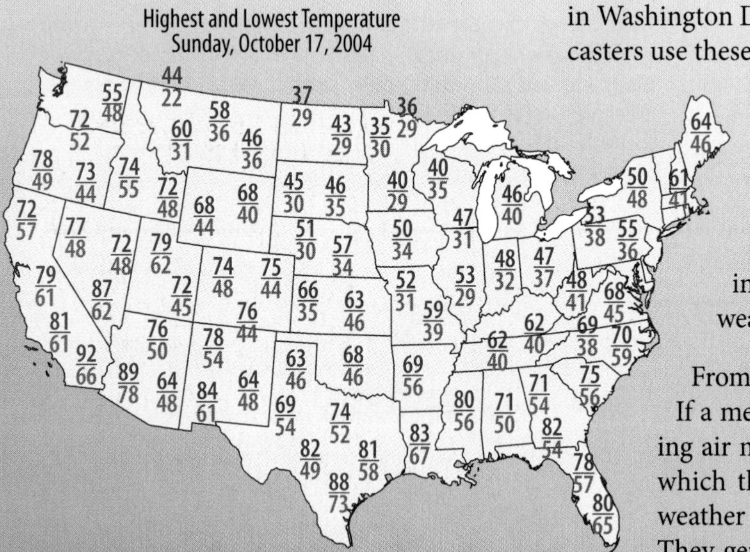

20.14 WEATHER MAPS

The National Weather Service has many centers that form a network to analyze weather station reports. The Weather Prediction Center (WPC), one of the National Centers for Environmental Prediction in Washington DC, creates many of the weather maps. Weather forecasters use these maps every day to tell you what the weather will be.

WPC thematic weather maps plot combinations of temperatures, pressures, winds, centers of high and low pressure, cloud cover, fronts, precipitation, or storm activity. The combinations of thousands of pieces of data on any given map can relate processes in the atmosphere that help make predicting future weather possible. These maps visually organize the data.

From this big picture, weathermen can identify air masses. If a meteorologist knows the direction and speed of a moving air mass, he can predict the weather for the area toward which that air mass is moving. WPC supercomputers use weather data twice a day to create the latest forecast picture. They generate mathematical models of the atmosphere that project what the weather will do over the next several days. These tools help meteorologists make weather predictions.

20-20
Highest and lowest temperature maps (see example below) and precipitation areas and amounts maps (see example on facing page) show recent weather that can be used for predicting weather over the next day or so.

Highest and Lowest Temperature
Sunday, October 17, 2004

Many weather maps point out areas that have similar weather conditions. Lines like those on a topographic map connect these points. Some weather maps show lines of equal temperatures called *isotherms*. For example, a winter weather map may have an isotherm connecting points with temperatures of 20 °C, and others for 10 °C and 0 °C. If they wrap around a region like contour lines around a hill, they can show the limits of a polar air mass.

isotherm: iso- (Gk. *isos*—equal) + -therm (Gk. *therme*—heat)

Other weather maps show lines of equal pressure—*isobars*. These maps help meteorologists locate centers of high and low pressure, which can help identify wind speed and direction, as well as the movement of weather systems. The spacing of isobars determines the pressure gradient force, which controls wind speed. Isobars closer together indicate the presence of stronger winds. Widely spaced isobars indicate weaker winds.

isobar: iso- + -bar (Gk. *bare*—weight, pressure)

A line connecting places having equal rainfall is called an *isohyet*.

isohyet: iso- + -hyet (Gk. *uetos*—rain)

There are four principal weather maps or charts prepared by the National Weather Service daily. Since these charts present a summary view (or synopsis) of the weather data for a given time frame, they are called **synoptic weather maps**. These maps are available to everyone online at the WPC website. The most familiar map is the Surface Weather Map, which shows weather station data and analysis at 7:00 a.m. eastern time for a particular day. The map uses the letters *L* and *H* to indicate the locations of the centers of well-defined low- and high-pressure areas, respectively. The standard weather front symbols given in Section 20A show the locations of fronts. To avoid cluttering the map, only a few of the total number of reporting stations are included. The other three synoptic maps are the Highest and Lowest Temperature Map, the Precipitation Area and Amounts Map, and the 500 mb Height Contours Map.

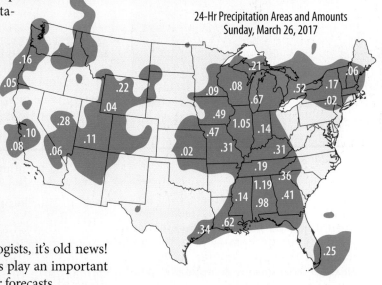

24-Hr Precipitation Areas and Amounts
Sunday, March 26, 2017

By the time all this information gets to meteorologists, it's old news! Can meteorologists still use it? Yes! Weather maps play an important part in understanding climate and making weather forecasts.

20-21

A meteorologist working to create the next weather forecast

GIS Weather Maps

Weather programs and news stations have been using GIS principles in presenting weather forecasts for many years. The digital maps they present often show local NEXRAD radar images, zones of precipitation, pressure centers, fronts, and cloud systems. These are all different data layers that together show the weather picture more effectively than any single layer could.

20.15 GIS AND WEATHER

Meteorologists have naturally adapted the GIS concepts you learned about in Chapter 3 to weather analysis and forecasting. Instead of looking at four separate maps covering the same region on a wall or a computer screen, they layer them so that relationships can be easily seen among pressure, temperature, precipitation, and cloud cover. Additional layers describing kinds of precipitation, locations of supercell storms, icing conditions, and so on, help weather forecasters understand the present and future weather picture. The WPC provides weather data in GIS files that can be displayed in 3D Earth viewers such as Google Earth. This is free of charge to the public.

GIS also helps with storm emergency response. Map layers (showing population centers, evacuation routes, utilities, emergency services such as fire stations, hospitals, and police) help government officials evaluate the need for and the effectiveness of evacuation orders. In the case of small intense local storms like tornadoes, these kinds of maps can identify locations for sending emergency responders immediately and spot which services may have been knocked out by the storm.

20.16 MAKING A FORECAST

In the United States, the various centers of the National Weather Service issue weather forecasts for the nation. Eight times per day, the WPC website publishes surface weather analyses. The Climate Prediction Center (CPC) of the NWS produces 6–10-day, 8–14-day, 30-day, and 60-day outlooks. Some of the short-range outlooks are daily. The monthly and seasonal outlooks are published monthly. Each new prediction takes into account the most up-to-date information. Your local meteorologist uses this information in his daily forecasts.

The 30-day and seasonal outlooks give helpful long-range weather information. They include things like the amount of precipitation expected during the period, the average high and low temperatures, weather hazard assessments, and drought outlooks. Because there are so many variables that affect weather, the further out the prediction is, the less accurate it will be. Long-range forecasts are much more general and less reliable than short-term predictions. Basically, weather predictions are a combination of studying weather maps, examining the output of atmospheric computer models, and applying a meteorologist's professional experience.

Men and women all over the world use weather data to better understand the activities of our restless atmosphere. They make it their life's work to serve others by studying weather data. They produce forecasts that help others plan their activities or even save their lives. Clearly, a Christian should consider meteorology a proper response to God's commands to pursue dominion over the earth as a way of showing love for others.

NATIONAL WEATHER SERVICE

The mission of the National Weather Service (NWS) is to "provide weather, water, and climate data, forecasts, and warnings for the protection of life and property and enhancement of the national economy." Its goal is to help people prepare for and respond to severe weather.

Within the NWS are centers that specialize in some of the storms discussed in this chapter. The National Hurricane Center (NHC) monitors tropical weather systems as they develop over the oceans. The NHC monitors hurricanes and predicts their paths. This enables the NHC to provide warning of these severe storms. Scientists at the Storm Prediction Center predict and track blizzards, severe thunderstorms, and tornadoes.

While the time frames differ for each type of storm, we can be thankful that scientists at the NWS are monitoring weather and providing as much warning of severe weather as they can.

Questions to Consider

1. What is the purpose of the NWS?
2. Why does the NHC track hurricanes far out in the ocean?
3. Which type of severe weather do you think is the most difficult to predict and warn people about?

20C REVIEW QUESTIONS

1. List three sources for weather data used by the National Weather Service to develop weather forecasts.
2. What types of information are indicated by the staff and flags that are part of the standard meteorological station model?
3. Why do meteorologists use weather maps instead of looking at large tables of geographic coordinates and numbers?
4. What are the lines on weather maps that connect locations having the same temperature? same pressure? same amount of rainfall?
5. How do maps showing isobars help meteorologists? What usually results when isobars are close together?
6. Give an example of a weather map prepared daily by the National Weather Service and describe it.
7. On a surface weather map, what could a weather station model located 300 km west of the coast of California in the Pacific Ocean represent?
8. What GIS principle of mapping is most effectively used when creating weather maps?
9. How does the reliability of weather forecasts change as the forecast period moves further into the future? Explain.
10. (True or False) Weather maps are not really useful to meteorologists for predictions because by the time they get the maps, the data is old news.

Key Terms

Term	Page
air mass	475
continental (air mass) (c)	476
maritime (air mass) (m)	476
arctic (air mass) (A)	476
polar (air mass) (P)	476
tropical (air mass) (T)	476
front	477
orographic lifting	479
frontal wedging	480
convergence	480
storm	481
winter storm	482
thunderstorm	484
lightning	485
tornado	486
hurricane	489
tropical storm	490
eye (hurricane)	490
storm surge	494
station model	495
synoptic weather map	497

Lenticular clouds often form as moist air is forced up and over mountains and other landforms.

CHAPTER 20 REVIEW

CHAPTER SUMMARY

» The movement of air masses is key to understanding weather.

» The source of an air mass determines its conditions. The five kinds of air-mass source regions are arctic, polar, tropical, continental, and maritime.

» Air masses may be cold or warm, and they take on the humidity of their source regions.

» Air masses interact to create fronts that may be warm, cold, stationary, or occluded.

» Clouds and precipitation form by convection, orographic lifting, frontal wedging, or convergence.

» Winter storms bring severe winter weather in the form of lots of snow or freezing rain. Large amounts of snow can be caused by Nor'easters or lake effects.

» Thunderstorms are the most common violent storms generated by air mass interactions. They usually produce lightning and rain and can also drop hail and spawn tornadoes.

» Tornadoes are rapidly rotating wind systems associated with violent storms. Tornadoes cause damage mainly through high wind speeds and strong updrafts. They are the most dangerous kind of storm.

» Hurricanes are giant cyclonic windstorms that form over the tropical and subtropical ocean. The heavy rainfall, high wind speeds, and storm surge they create can be very destructive.

» Weather data can be communicated using station models.

» As weather data is gathered, it is plotted and analyzed on weather maps.

» Meteorologists use a combination of digital atmospheric models, weather maps, and professional experience to make predictions that can be distributed to the public. Nearly all these maps and predictions are available for free on the Internet.

REVIEW QUESTIONS

1. What is one of the most serious decisions that can come from a weather forecast?

2. What would be the weather characteristics and direction of movement of a typical air mass moving slowly over the North Atlantic? What type of air mass would it be?

3. If a squall line of thunderstorms has passed overhead, followed by cool, clear, sunny skies, what probably has happened to produce these conditions?

4. Explain how a range of mountains can influence the formation of clouds and precipitation, and give the name for this effect.

5. What is likely the most dangerous kind of winter storm, a deep snowstorm or a heavy ice storm? Explain.

6. What weather conditions can produce a cumulonimbus thunderstorm? What weather effects often accompany a thunderstorm?

7. Why were meteorologists surprised to find that pressure inside the funnel of a tornado dropped only to 0.9 atm?
8. Discuss the differences between a tornado and a hurricane.
9. What issues are involved in the evacuation order of an area for a predicted disaster, such as a hurricane?
10. Why does God permit destructive storms, especially those that result in huge losses of life?
11. How does the National Weather Service obtain weather data from remote, unpopulated locations?
12. Compare the usefulness of a single weather station's data and the overall weather picture to a single puzzle piece.
13. If the same isobar passes through both Boston, Massachusetts, and New York City on a weather map, what can you conclude?
14. Why is the weather map that you see on TV or in a newspaper considered a synoptic map?
15. Why is a one-day forecast more reliable than a seven-day forecast?
16. Do you think that humans will one day be able to control the weather with any degree of success? Why or why not?

True or False

17. Air masses moving over the Hawaiian Islands are probably identified as *mT* on a weather map.
18. An occluded front forms when a cold air mass is trapped between two warm air masses and loses all contact with the rest of the atmosphere.
19. When air is lifted by any means, clouds and even precipitation can form.
20. High-pressure systems are the main reason that Nor'easters are such important sources of snow in the Northeast United States.
21. Thunderstorms can occur at any time of year.
22. You are completely safe from lightning inside a sturdy house.
23. The best tactic to keep a house from exploding from the great drop in pressure during a tornado is to open a window in the house.
24. An Atlantic hurricane is heading inland across the South Carolina coast. Observers at a seashore on the left side of the hurricane's track should expect to see lower-than-normal sea levels as the storm passes.
25. According to the station model in central Pennsylvania (on the map to the right), the wind is blowing from east to west.
26. Today's meteorologists depend mostly on their experience and intuition when creating local weather forecasts.

Map Exercise

27. Study the daily weather map on the right. On the basis of your understanding of the general movement of weather systems across the United States, what kind of change in temperature, pressure, and precipitation do you predict will occur for the Charleston, South Carolina, area over the next few days?

Mammatus clouds form in very unstable air and often accompany severe weather.

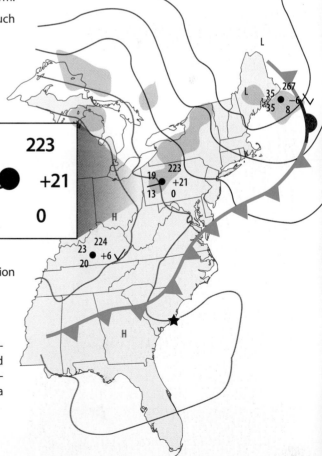

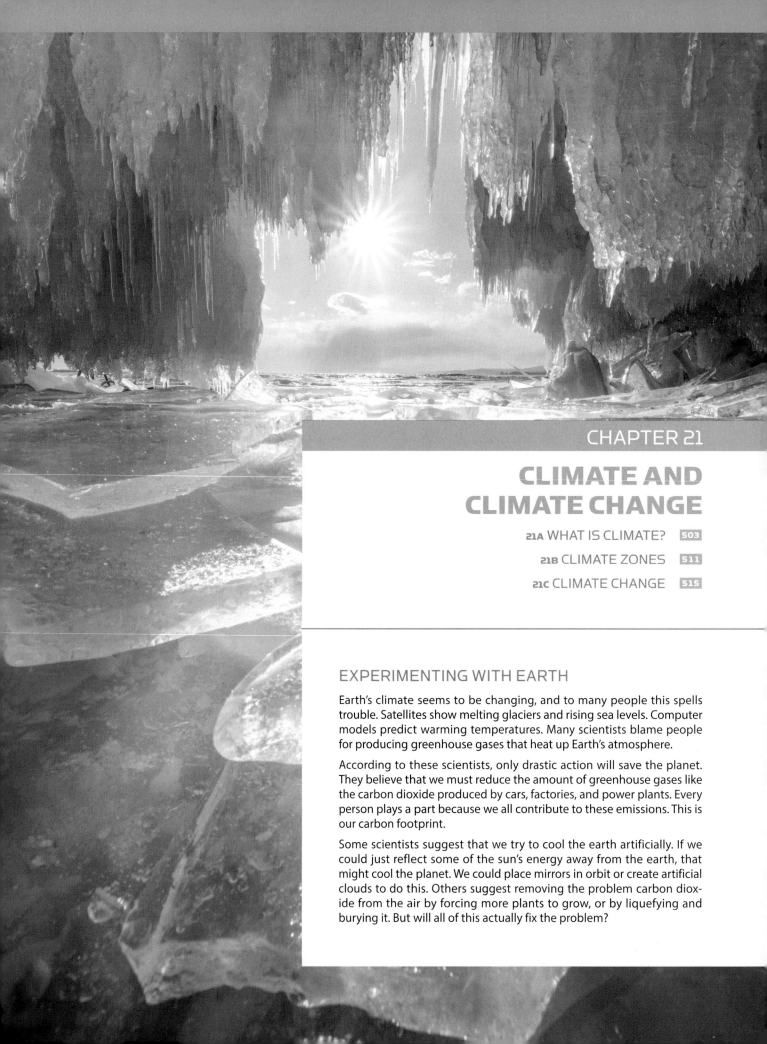

CHAPTER 21

CLIMATE AND CLIMATE CHANGE

21A WHAT IS CLIMATE? `503`

21B CLIMATE ZONES `511`

21C CLIMATE CHANGE `515`

EXPERIMENTING WITH EARTH

Earth's climate seems to be changing, and to many people this spells trouble. Satellites show melting glaciers and rising sea levels. Computer models predict warming temperatures. Many scientists blame people for producing greenhouse gases that heat up Earth's atmosphere.

According to these scientists, only drastic action will save the planet. They believe that we must reduce the amount of greenhouse gases like the carbon dioxide produced by cars, factories, and power plants. Every person plays a part because we all contribute to these emissions. This is our carbon footprint.

Some scientists suggest that we try to cool the earth artificially. If we could just reflect some of the sun's energy away from the earth, that might cool the planet. We could place mirrors in orbit or create artificial clouds to do this. Others suggest removing the problem carbon dioxide from the air by forcing more plants to grow, or by liquefying and burying it. But will all of this actually fix the problem?

21A

WHAT IS CLIMATE?

What is climate, and what factors affect it?

21.1 CONSERVING EARTH

21A Objectives

After completing this section, you will be able to

» contrast climate with weather.

» analyze how different factors may affect climate.

At the heart of the climate change debate is the question of how we should treat Earth. God made this planet our home, and He wants us to take care of it. That doesn't mean that we can do anything we want with Earth's resources. It also doesn't mean that we should protect those resources at all costs. So should we use Earth or protect it from use? We should do both! **Conservation** is the preservation and wise use of earth's resources with a mind for the future. The work of conservation involves protecting some resources that are limited, have big effects on people, or are key parts of Earth's environment. Conservation also includes using other resources in managed ways, like creating the national parks.

Christians should be the most passionate conservationists because wisely using God's world is one of the first commands God gave to mankind (Gen. 1:28). He cares about His world and wants us to use it in ways that allow the people He made in His image to flourish and grow. A growing human population is part of the Creation Mandate. Whether we recycle, adopt a highway, plant a garden, or do animal rescue, we're taking care of God's creation for His glory.

Thinking about God's world in light of His Word brings clarity to the climate change debate. We must make decisions that best use the earth's resources, save money, and, most importantly, help people. And to do all these things we need to understand Earth's climate and what affects it. So let's get started!

21.2 CLIMATE VERSUS WEATHER

We've looked at weather and how meteorologists make sense of it. How is climate different from weather? Scientists who study climate look at the same data as those that study weather—mainly temperature, humidity, and precipitation.

Weather is a description of what the earth's atmosphere is doing in the *present*, or in the near future. **Climate** is the average weather for an area over a much longer period of time. *Climatology*, the study of climate, looks at trends in weather over a span of many years. *Climatologists* are scientists who study climate. Meteorologists tell you whether you need to put on a winter coat for the day, but climatologists tell you what kind of wardrobe you will need!

> Hey Mom! In earth science class we learned that the earth is warming ... so I need a new wardrobe ... right!?

21.3 FACTORS THAT AFFECT CLIMATE

Do you ever notice the weather in places other than where you live? What do you observe? Though weather changes every day, Vermont has snowy, cold winters and Florida has mild, warmer winters. But why is that true? And why does the state of Washington get so much rainfall while parts of California don't? After all, they're both on the US Pacific coast. *Average temperatures and precipitation define climates.* But many conditions control these factors for a given region of the world. Let's understand what these conditions are before looking at the different kinds of climate zones.

TEMPERATURE

Latitude. Surface air temperature depends on the amount of thermal energy in the air. The sun's rays can warm the air directly, but most heat comes from contact with ground or water heated by the sun. The sun's rays heat the earth's surface most efficiently when the rays are vertical. Vertical rays travel the shortest path through the atmosphere, so fewer are absorbed. Also, vertical sunlight packs the most rays into a given area of the earth's surface, so more heating can occur in a given amount of time. The angle that the sun's rays make with the earth's surface depends on *latitude*. Latitude is one of the most important conditions affecting regional temperature and climate.

Calendar. But latitude alone doesn't control the sun's angle. The time of year, or *season*, does too. **Seasons** are repeated, natural changes in temperature, precipitation, and the amount of daylight. You remember from Chapter 4 that the earth's rotational axis is tilted 23½° from a line perpendicular to the plane of its orbit. This means that the sun's rays are vertical at 23½° latitude twice during Earth's orbit. When the sun is directly over 23½°N latitude, it is the summer solstice—first day of summer—in the Northern Hemisphere and the winter solstice in the Southern Hemisphere. Approximately six months later the sun's rays are vertical at 23½°S latitude, which marks the winter solstice—start of winter—for the Northern Hemi-

1 Microclimates

Sometimes a mountain, a city, or a body of water can change the climate in a small area. This is called a *microclimate*. We mentioned this in the chapter on mountains. But did you know that cities can have this effect too? Cities are usually several degrees warmer than the surrounding countryside. This is called the *urban heat island effect*.

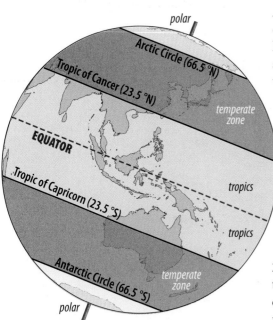

21-1 Latitude on the earth's surface affects climate.

21-2

When the sun's rays are at a higher angle to the ground, the sun's energy is more concentrated. Additionally, less energy is lost as the rays travel a shorter distance through the earth's atmosphere.

sphere and summer solstice in the Southern Hemisphere. Cartographers have given these parallels special names. The northern one is the *Tropic of Cancer*; the southern, the *Tropic of Capricorn*. The latitudes between these two parallels lie in the *tropics*, the warmest temperature belt on the earth's surface.

The earth's tilt defines two other parallels that climatologists use to identify temperature belts. At the summer solstice, the sun's rays shine across the North Pole and graze the earth at 66½°N on the *far* side of the world (see Figure 21-3). This parallel is 23½° from the pole and is called the *Arctic Circle*. The same thing happens in the Southern Hemisphere at its summer solstice, which defines the *Antarctic Circle*. The earth's surface and atmosphere between these parallels and their poles lie within the *polar zones*. These climates are really cold because the sun's rays are either at a large angle from the vertical, or totally absent for nearly half the year! Latitudes between the tropics and the polar zones lie within the *temperate zones*. The temperate belts are comparatively mild, neither too hot nor too cold on the average. Most of us live in a temperate zone.

The seasonal tilt of the earth also controls the length of day. Days are longer than nights in summer so more heating takes place. In winter, the days are shorter than nights, so the earth cools for longer periods.

The Tropics of Cancer and Capricorn were named about 2000 years ago. At that time, the sun's position fell within the constellation *Cancer* when it was at its *northernmost* angle in the sky. Its position fell within the constellation *Capricornus* when it was at its *southernmost* angle in the sky. You will learn about constellations in Chapter 24.

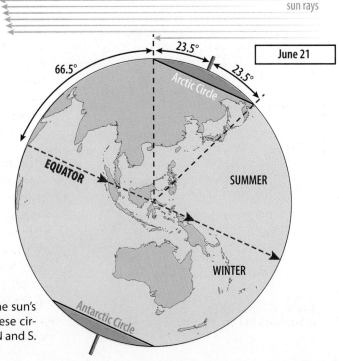

21-3

The Arctic and Antarctic circles are defined by how far the sun's rays extend beyond the pole at the summer solstice. These circles of latitude are 23½° from the poles, or latitude 66½°N and S.

Though Mount Kilimanjaro in Africa is in the tropics, it still has glaciers on top!

Elevation. Recall from Chapter 18 that air temperature in the troposphere usually drops with altitude (the lapse rate). This means that places near sea level at a given latitude are generally warmer, while places at a higher elevation are generally cooler. This is why you can hike in British Columbia's mountain snow or go swimming at its ocean beaches during the month of June!

Water. The presence of large bodies of water, like the ocean and large lakes, directly affects climate. You learned in Chapter 14 that the major ocean currents warm lands far from the tropics and cool lands far from polar regions. Great Britain has similar average temperatures as New England because of the influence of the Gulf Stream Current, even though it lies 10 degrees latitude farther north. Similarly, the West Coast of the United States is cooler than Japan because of the influence of the cold California Current.

THE CANOPY THEORY

Imagine the rain pounding on the Ark's roof for forty days and nights. At times it must have sounded like a rushing roar. Have you ever thought about how much water that involved? Where did it all come from?

For generations, many Christians have suggested that all this rain came from a canopy of water vapor above the atmosphere. Somehow the atmosphere held this layer in place. Christians have pointed to Genesis 1:7 as the proof text for this belief. The canopy supposedly shielded out harmful solar radiation and made the world's pre-Flood climate more uniform. And its weight increased atmospheric pressure, especially oxygen pressure close to the ground. They concluded that these conditions allowed animals and plants to grow much larger and people to live far longer. Another consequence was that the earth's climate was tropical before the Flood. But is this view physically possible or even biblical?

The concept of a water vapor canopy was actually first proposed by an old-earth philosopher named Isaac Vail in 1874. He proposed that millions of years ago, seven layers of different basic substances surrounded the molten, evolving earth, like the rings of Saturn. Drawing from ancient mythologies, he suggested that as the earth cooled, these layers collapsed one at a time. These collapses were millions of years apart.

Vail believed that each of these layers formed a distinct geologic layer. The final, lightest canopy was water. It collapsed during the Flood. These ideas were supposed to unite uniformitarian and creation science. Many Christians have no idea that this was the start of the canopy theory that they believe today. But during the early decades of the modern creation science movement, the canopy theory was often taught as a biblical certainty.

What does the Bible say? We read in Genesis 1:7, *"And God made the firmament, and divided the waters which were under the firmament from the waters which were above the firmament: and it was so."*

How should we understand this verse? The key word in this passage is the Hebrew word *raqiya,* translated *firmament.* It can also mean *expanse* or *the visible arch of the sky.* It is most likely that "the waters which were above the firmament" refers to the clouds. The word *raqiya* refers to the place where God put the sun, moon, and stars (Gen. 1:14), but also to the place where the birds fly (Gen. 1:20). Some have proposed that this must imply that the waters must be above our atmosphere and above space since both are referred to as *raqiya*.

It is important to note that Genesis is written from the author's perspective. In everyday experience people look up into the sky (raqiya) and see where the birds fly and the place where God put the celestial lights. With this perspective in mind and on the basis of the use of *raqiya* in the context, we should likely not conclude that the word could refer to some sort of ceiling, dome, or canopy as Vail proposed. "Sky" seems to fit the biblical sense of the word *raqiya* with clouds or atmospheric waters being the waters "above."

In 1990 and 1998, atmospheric scientist Larry Vardiman and other creationary scientists decided to put the canopy theory to the test. They used supercomputers to model the earth's climate, assuming that a water vapor canopy surrounded the earth. They initially assumed that the canopy contained enough water to create the Flood. The results were astounding!

The model showed a runaway greenhouse effect, similar to what exists on Venus today. Average surface temperatures reached several hundred degrees. Vardiman's team reduced the amount of water vapor in the model until the greenhouse effect would produce hot surface temperatures in which only extreme forms of life could barely survive. The collapse of this vapor canopy could produce only 2 m of water. To maintain a livable but still equatorial 63 °C surface temperature, the canopy could have rained only 10 cm when it collapsed!

On the basis of these computer model results, many creation scientists today don't believe that an atmospheric water vapor canopy is an accurate idea taken from Genesis 1:7. Considering the history of the canopy theory, they're probably right.

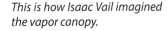

This is how Isaac Vail imagined the vapor canopy.

21-4

The northern leeward side of the Himalayan mountain range in Tibet is very dry compared to the lush windward side in India. This is an example of the rain shadow effect.

| monsoon (Arab. *mawsim*—season)

PRECIPITATION

Besides temperature, precipitation is the other major factor in contributing to the climate at a given location. In previous chapters, you learned that humid and dry air masses control weather far from where they form. Marine air masses are driven by *prevailing winds* over land, where precipitation occurs. This is why the Pacific Northwest in the United States and Canada has frequent rainy weather and why Great Britain is famously known for its rainy climate. Some *seasonal winds* control precipitation as wind patterns change directions through the year. A good example of these is **monsoon** winds in Asia. These winds blow for six months of the year from the Indian Ocean into southern and southeast Asia. The heavy rains they bring are followed by six months of very dry weather as the winds reverse and blow outward from the Gobi desert in central China.

Air masses also produce precipitation as they flow over mountain ranges. This *orographic precipitation* on a mountain's windward side can be quite heavy. But very dry conditions can result from a *rain shadow* on the leeward side of mountains as air descends and warms. These differences in precipitation can create dramatically different climates on opposite sides of the same mountain!

In summary, the climate in a particular place in the world results from annual averages of temperature and precipitation. The yearly amounts can vary from year to year, but over time, they are pretty much the same. In the next section, you will learn about the important types of climate zones and where they are located around the world.

21A REVIEW QUESTIONS

1. What good for the earth's climate might come from injecting large quantities of liquefied carbon dioxide into rock strata deep underground? What might happen if climatologists' models are incorrect about the influence of carbon dioxide on global warming?
2. How are climate and weather different, given that they rely on the same data?
3. How do the following affect the temperature of a given location?
   a. latitude
   b. calendar (season)
   c. elevation
   d. water
4. As you go from the Equator to the poles, how does general climate change?
5. In what two ways can large bodies of water affect climate?
6. How do mountains affect climate?
7. (True or False) The climate in two places in different parts of the world will always be the same if the locations are at the same latitude.

BIBLICAL ORIGINS: THE FRAMEWORK HYPOTHESIS

Some Christians see in Genesis 1 a descriptive pattern or framework of creative acts. They believe that the Genesis 1 narrative does not intend to describe accurately how God created, but only that He did create all things.

According to the Framework view, the Creation story contains two groups of three "days." The first three days represent the appearance of the three "kingdoms" of creation. During Day 1, God created the earth and heaven, as well as darkness and light. In Day 2, God created the expanse that forms the astronomical heavens and the sky. And on Day 3, God separated the dry land from the sea.

During Days 4 through 6, God created the "rulers" of the kingdoms. On Day 4, He created the sun, moon, and stars to rule the day and the night. On Day 5, God formed the flying creatures and the sea creatures to rule the sky and the seas. Finally, on Day 6, He created the land creatures, as well as humans, to rule the land. People who hold to the Framework view believe that God rested and is still resting on the seventh "day."

According to the Framework view, Genesis 1 is organized by literary parallelism, not as a literal step-by-step history. Only those creative acts that are necessary to support later events imply a time sequence. So, for example, some people believe that the creation of day and night (Day 1), and the sun, moon, and stars (Day 4) actually occurred at the same time.

Questions to Consider

1. The Framework view claims that the parallels between Days 1–3 and Days 4–6 are so remarkable that the days of Genesis 1 must not be chronological. However, there are several unparalleled elements between Days 1–3 and 4–6. How many can you find?
2. There are interesting parallels between Days 1–3 and Days 4–6. However, these parallels are subtle features regarding how Genesis 1 is structured. Identify the most obvious feature of how Genesis 1 is set up. What does this feature suggest about Genesis 1?
3. Do you think that the Framework view would have become popular if secular science had no problem with Genesis 1? What does your answer suggest about the Framework view?

SERVING GOD AS A CLIMATOLOGIST

Behind Lonnie Thompson spreads a mountain glacier from which he has collected ice core samples.

JOB DESCRIPTION

Lonnie Thompson may have spent more time than anyone else in the world above an elevation of 18,000 feet. So what is he, a mountain climber? No. A Sherpa? No, but he does spend months at altitudes at which mountain climbers spend only a few days.

Lonnie Thompson is a climatologist. He takes ice core samples from melting glaciers at both tropical and subtropical latitudes. He's trying to preserve history in the ice.

So what is a climatologist, anyway? A climatologist is a scientist who studies Earth's climate and the factors that seem to affect it. Climatology is a branch of atmospheric science. Scientists in this field study tree rings, take ice cores, study hurricanes, and create and analyze climate models. They focus not on weather, but on trends in weather.

EDUCATION

If you're interested in being a climatologist, you'll need to study the physical sciences. You'll probably need a degree in atmospheric science. But it doesn't stop there. Courses in biology and oceanography as well as mathematics and computer science can come in handy. Climatologists spend a lot of time working with models. And you'll need to get out there and interact with other experts. Perhaps you could even climb some of those mountains with a climatologist.

POSSIBLE WORKPLACES

There are three workplace choices for a climatologist: the field, the lab, or the classroom. Climatologists have jobs similar to research meteorologists, though they focus on using weather data in different ways.

DOMINION OPPORTUNITIES

A climatologist's work is impressive, can be difficult, but is vitally important. His data is valuable and is needed to understand the changes to the world around us. Climatologists conduct research that help people plan food production. They study changes in weather patterns that could signal increases in severe weather. Their research also is needed in the climate change discussion.

We know that weather impacts our lives every day. Changes in climate will impact people well into the future. You could be a part of that exciting research, if you choose to help others as a climatologist.

21B

CLIMATE ZONES

How do scientists classify climates?

21.4 CLASSIFYING CLIMATES

If you were a pioneering climatologist, how would you classify climates around the world? Would you look at the differences in vegetation or animals? What about a region's location near bodies of water?

In 1884, climatologist Wladimir Köppen designed a system that divided the earth's surface into **climate zones**. Köppen's system focused on the plants in different places of the world. He thought that this was the best way to see the long-term effects of climate in a region. He later realized that he needed to make his system more measurable. So he adjusted his categories to consider annual precipitation and average temperature. Another scientist, Rudolf Geiger, helped him produce the system that we use today. In Figure 21-5, you can see how climatologists look at this information for specific places in the world.

Many climatologists classify climates using six major categories. These categories are based mainly on precipitation and temperature information:

» temperate continental climate
» polar climate
» temperate marine climate
» tropical rainy climate
» dry climate
» highlands climate

Many of these zones include local conditions that create subzones of the climate zones. Notice that areas of the world that are far apart can have similar climates. Climate zones do not depend on country borders, but on similar climatic conditions.

You may get the impression from the climate zone map on the next two pages that climates change suddenly as you move from one zone into another. But they don't usually do that. For example, if you were walking across Europe, you wouldn't know exactly when you moved from a temperate marine zone into a highland zone. One zone gradually merges into another. Let's take a trip around the world, looking at climate.

21B Objectives

After completing this section, you will be able to

» identify six major kinds of climates.
» give examples of the different kinds of climates.

Climate Graphs

The graphs in Figure 21-5 show two kinds of data on the same graph. What are they? Can you figure out which graph shows a polar region? Which shows a tropical rainforest? How can you tell?

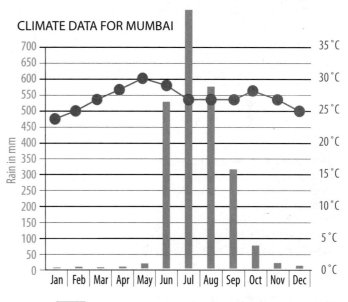

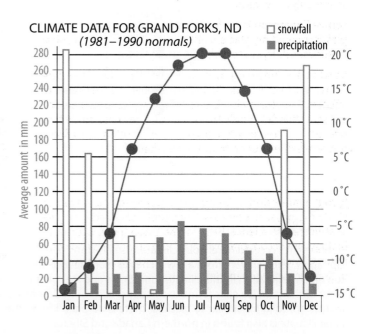

21-5

Two climates and two sets of precipitation and temperature data for one year

21.5 INTO THE ZONES

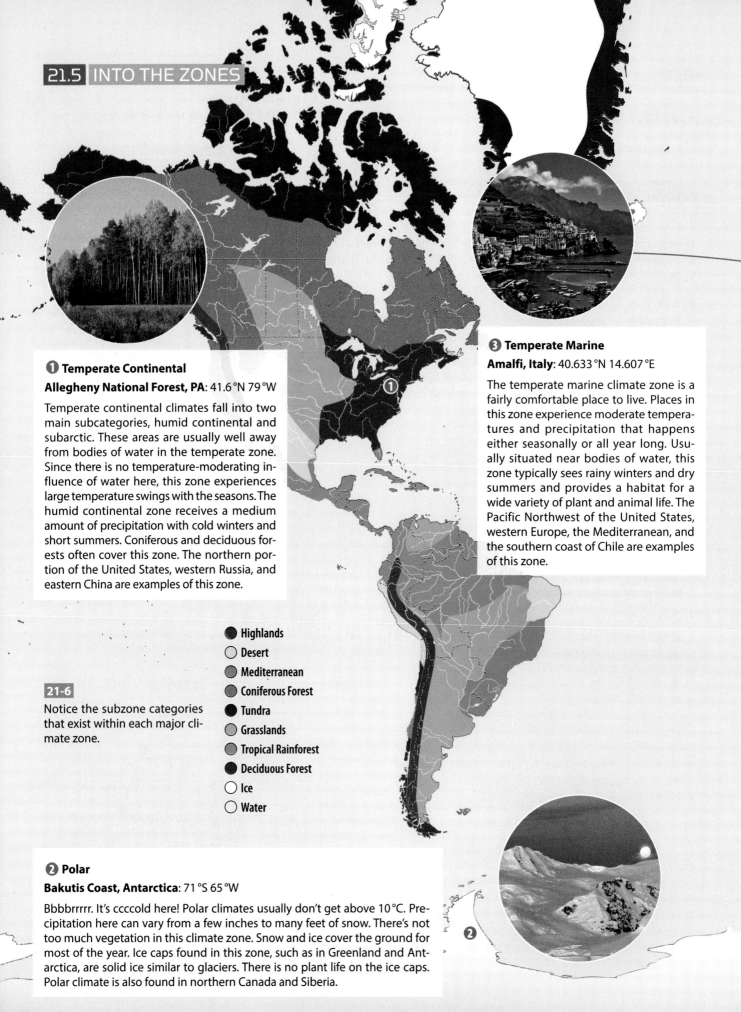

① Temperate Continental
Allegheny National Forest, PA: 41.6°N 79°W

Temperate continental climates fall into two main subcategories, humid continental and subarctic. These areas are usually well away from bodies of water in the temperate zone. Since there is no temperature-moderating influence of water here, this zone experiences large temperature swings with the seasons. The humid continental zone receives a medium amount of precipitation with cold winters and short summers. Coniferous and deciduous forests often cover this zone. The northern portion of the United States, western Russia, and eastern China are examples of this zone.

③ Temperate Marine
Amalfi, Italy: 40.633°N 14.607°E

The temperate marine climate zone is a fairly comfortable place to live. Places in this zone experience moderate temperatures and precipitation that happens either seasonally or all year long. Usually situated near bodies of water, this zone typically sees rainy winters and dry summers and provides a habitat for a wide variety of plant and animal life. The Pacific Northwest of the United States, western Europe, the Mediterranean, and the southern coast of Chile are examples of this zone.

- Highlands
- Desert
- Mediterranean
- Coniferous Forest
- Tundra
- Grasslands
- Tropical Rainforest
- Deciduous Forest
- Ice
- Water

21-6 Notice the subzone categories that exist within each major climate zone.

② Polar
Bakutis Coast, Antarctica: 71°S 65°W

Bbbbrrrrr. It's ccccold here! Polar climates usually don't get above 10°C. Precipitation here can vary from a few inches to many feet of snow. There's not too much vegetation in this climate zone. Snow and ice cover the ground for most of the year. Ice caps found in this zone, such as in Greenland and Antarctica, are solid ice similar to glaciers. There is no plant life on the ice caps. Polar climate is also found in northern Canada and Siberia.

❻ Highlands
Ghyaru, Nepal: 28.64°N 84.14°E

While the other five climate zones can occur somewhere near sea level, abruptly rising mountains can significantly alter the climate compared with the surrounding lowlands. Climatologists group these mountain microclimates together into the highland climate zone. Due to the high elevation, these areas are typically much colder than the surrounding lowlands. The yearly amount of precipitation can vary a lot between places. Highland climates follow mountain chains, such as the Rockies, Andes, Alps, and Himalayas.

❹ Tropical Rainy
Ninh Binh, Vietnam: 20.2°N 105.9°E

Tropical rainy climates are both very warm and very wet, which is why we find these places near the Equator. Many tropical zones have lots of rain all year long. Tropical rainforests such as those in Indonesia and in the Amazon have this kind of climate. We find many insects, lush vegetation, and exotic animals. In fact, half the world's species of plants and animals are here! Other tropical rainy locations have yearly cycles of wet and dry weather, such as India and Vietnam. They get heavy seasonal rains from monsoon winds. Savannas, open grassy areas with scattered trees and low shrubs, also fall into this category.

❺ Dry
Peterman, Australia (Outback): 25°S 131°E

Dry climates produce little precipitation or have greater evaporation than precipitation. Though they appear at different latitudes, dry places are usually far from large bodies of water or on the leeward sides of high mountain ranges. Dry climate zones fall into two subcategories: semiarid zones receive up to twice the precipitation as arid zones. Dry climate zones can also be either hot or cold. The arid zones include the world's deserts, such as the Sonora, Saharra, and the Outback. The Gobi desert and the Atacama deserts are in rain shadows of mountain ranges. The Mongolian steppes and the prairies of the United States are in a semiarid zone.

21B REVIEW QUESTIONS

1. What do climatologists consider when categorizing an area's climate?
2. Create a table with three columns. Write "Climate Zones" as the heading for the first column and list the climate zones. Write "Precipitation" and "Temperature" as the headings for the second and third columns. Fill in the table with the values "low," "medium," "high," or "variable" for each climate zone.
3. You are taking an epic bicycle trip across the Australian continent from the west to the east coast. Describe the climate zones you travel through by referring to pages 512–13.
4. How do you think climates affect where you can find certain kinds of animals?
5. Give an example of a highland climate.
6. Compare temperate marine and temperate continental climates.
7. (True or False) Many arid and semiarid zones exist in rain shadows of mountains, even in temperate latitudes.

21C

CLIMATE CHANGE

How does climate change relate to a Christian worldview?

21.6 CLIMATE CHANGE AND GLOBAL WARMING

Climate within individual zones can change over time in different places on Earth's surface. However, climatologists have found evidence that the earth's climate as a whole has been different in the past. More importantly, they think that it may be changing rather quickly today.

Regional climate zones are defined by their seasonal temperatures and precipitation. So a change of climate could involve rising or dropping temperatures or more or less precipitation. Usually both change in some way. These kinds of changes then affect the plants and animals that live in a given zone. Many scientists believe that part of the recent changes in climate zones seen across the world have resulted from an increase in the world's average temperature. This is called **global warming** and most people think that this is something we need to stop.

In any discussion of climate change and global warming, we need honest answers to four questions:

» Is the earth warming?
» If so, are people causing it to warm?
» If it's warming because of human activities, is that bad?
» Will solutions that politicians and scientists suggest actually fix the problem?

We cannot answer all these questions in this textbook. But you can learn some scientific evidence and how worldview affects its interpretation.

21C Objectives

After completing this section, you will be able to

» analyze potential causes for climate change.
» critique worldview assumptions behind global climate models.
» evaluate current fears of climate change.
» formulate a Christian perspective of climate change.

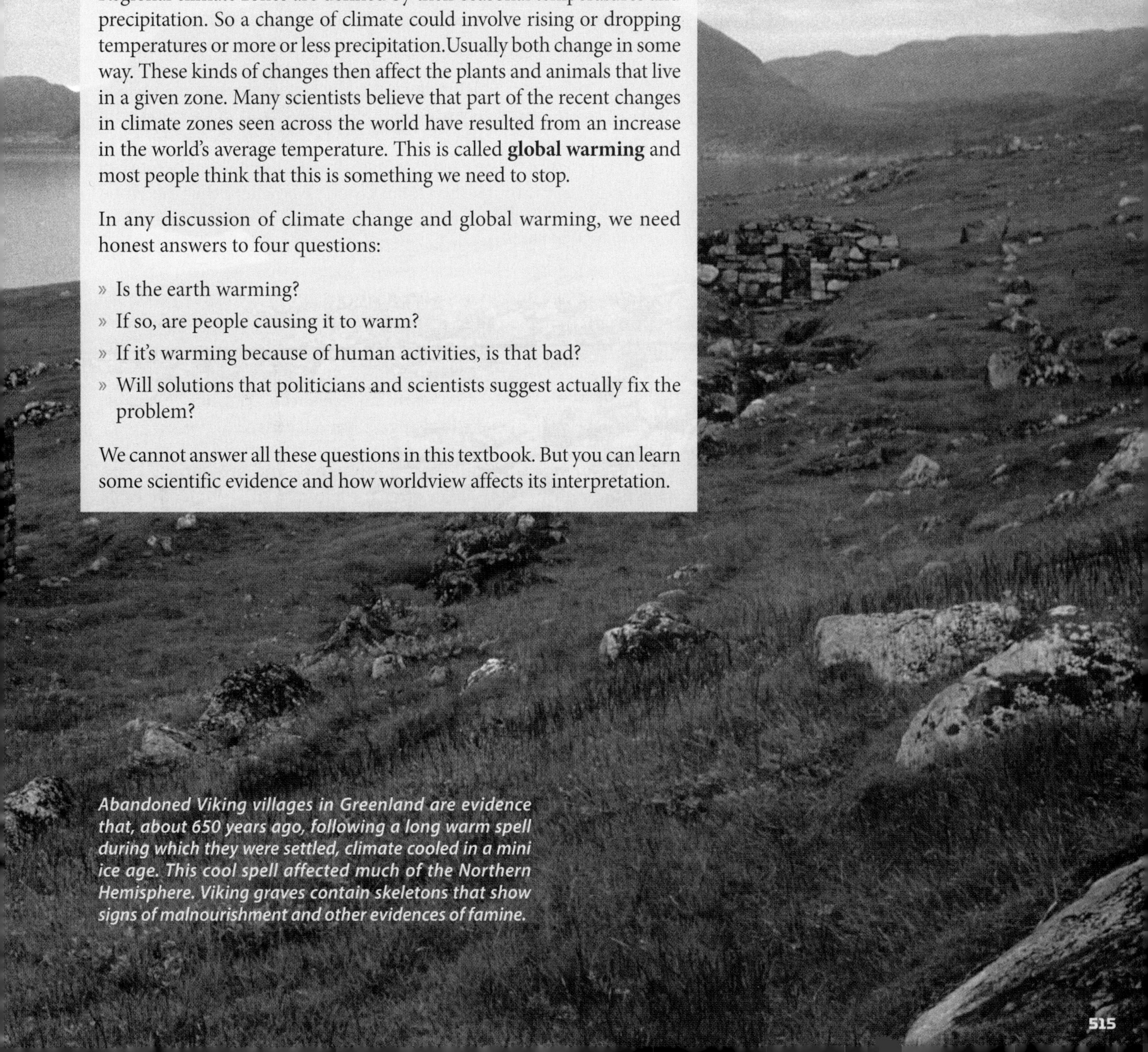

Abandoned Viking villages in Greenland are evidence that, about 650 years ago, following a long warm spell during which they were settled, climate cooled in a mini ice age. This cool spell affected much of the Northern Hemisphere. Viking graves contain skeletons that show signs of malnourishment and other evidences of famine.

21.7 A HISTORY OF CLIMATE CHANGE

The history of Earth's climate is a history of change. Both the old-earth naturalistic view and the young-earth biblical view of Earth history recognize that the climate has changed in the past. How climate changed and why it has changed have different explanations, depending on one's worldview.

In the old-earth view, global climate depended on the amount of continental surface area exposed above the ocean surface, the amount of heat received from the sun, volcanic activity, and even the kind of life existing on Earth during a given geologic time frame. Secular scientists believe that Earth has gone through at least five extended glacial periods during the past 4 billion years. They think that during several of these periods, thick ice sheets may have covered nearly all the earth's land surface. Between glacial periods were very warm times. According to evolutionists, animals and plants evolved and adapted to these changing conditions or became extinct.

A young-earth view of history involves global climate change too. God created a world completely suited for His purposes. It was a world with a climate ideal for life. But our sin brought God's curse on the earth and all its processes. The earth's climate drastically changed because of God's judgment through the global Flood. We believe that the Flood set up the conditions for a global cooling period that produced the Ice Age. Centuries later, the atmosphere cleared and the earth warmed, ending the Ice Age. Historical records suggest that the earth has continued to warm. But many cold spells, or "mini ice ages," produced famines that affected human populations and altered the course of human history. These periods were interrupted by warmer ones that allowed people to move into previously cold, uninhabitable regions of the world. With the gradual warming, ice sheets continue to shrink. Deserts have grown where lush vegetation once thrived.

Though we notice climate change in the past and present, we know from God's promise to Noah after the Flood that, "*While the earth remaineth, seedtime and harvest, and cold and heat, and summer and winter, and day and night shall not cease*" (Gen. 8:22).

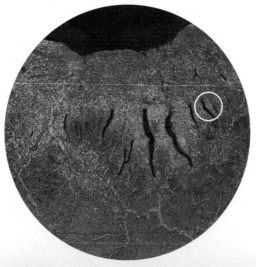

21-7
The Finger Lakes of New York were carved out as glacial ice migrated south. After the glaciers receded, the long, narrow depressions in the earth filled with water.

THE MELT ZONE

In 1914, Earnest Shackleton set out with a band of fearless explorers to Antarctica in an ice-resistant ship called the *Endurance*. His goal? To be the first to cross Antarctica. After arriving near the continent, things went quickly, uh, "south."

Sea ice waged war on his ship, eventually crushing it before the explorers could make it to land. In one of the greatest survival stories in history, Shackleton's whole band of men overcame incredible odds to make it safely back to England. Their story shows the challenges of exploration in polar regions.

Ice rules in the polar zones. Coastal glaciers move out to the sea and float when the water gets deep enough. This thick platform of ice is called an *ice shelf*. The largest ice shelf in the world is the Ross Ice Shelf (loc.: 80°S 180°E), fed by eight large glaciers and many smaller ones.

Ice can break off an ice shelf or any glacier entering the ocean to form large floating chunks called *icebergs*. This is the process of *calving*. Any glacier that reaches the sea may calve, but most icebergs come from huge continental glaciers. There are only two places in the world where these exist—Greenland and Antarctica.

If you had been on Shackleton's ship in 1915, you would have seen more and more of these icebergs as you approached Antarctica. Some are huge.

The largest iceberg in history was a monster with the unassuming name *B15A*. It was 11,000 km² in area, about the size of Long Island, New York! It was the largest fragment of an even larger piece of ice that would eventually break off the Ross Ice Shelf in 2000. It rose nearly 50 m above water and was 500 m thick. That means that most of the iceberg was hidden below the water. Even considering this, B15A would have looked like a sixteen-story building jutting up out of the water!

Ice floes located on the Weddell Sea, Antarctica

Seawater in the bays around Antarctica and the entire Arctic Ocean freezes over during the long winters. This sea ice can grow to several meters thick in places. Arctic polar bears use these ice sheets to get out farther into the water to hunt for seals.

In the spring and summer, wind and waves break up these thin, floating ice sheets into pieces called *ice floes*. Floes are more fragile than icebergs.

But things are changing in Greenland and Antarctica, the lands of ice. They are becoming melt zones. Glaciers are thinning and shrinking in both these places. Lakes of meltwater are forming and growing on the tops of these glaciers. Most scientists blame global warming.

What many people don't understand is that factors other than temperature can control the growth and shrinkage of glacial ice. Glaciers grow when more snow falls in winter than melts in the summer. So precipitation is as great a factor as temperature.

The effects of changes in polar ice may surprise you. Many people are concerned that melting glaciers may raise sea levels to dangerous heights, which would be devastating to coastal regions. Others can see benefits resulting from melting ice caps. The long-sought Northwest Passage from the Atlantic to the Pacific north of Canada may actually open up. Icebreakers and merchant ships may be able to steam through Arctic ice in the summer. Around Greenland, oceans that are now ice-free for more of the year than they were before are exposing long-hidden sources of oil. This is bringing money to Greenland, a land that has been economically tied to its mother country, Denmark. Now there is a growing political movement in Greenland for it to become its own country.

So will we be sending ships over the North Pole and growing crops on the coasts of an actually green Greenland? The polar regions may yet give us some surprises like they gave Shackleton.

Iceberg B15A in the Ross Sea

CLIMATE AND CLIMATE CHANGE

Seasonal Animal Adaptations

How do animals deal with seasonal changes in climate? Some animals, such as chipmunks, turtles, and snakes, *hibernate*. This minimizes the food they'll need during hard winter months. The Alaskan wood frog freezes solid during the winter, then thaws out and starts hopping again come spring! Some animals in habitats that become very hot and dry enter an inactive state where their bodies shut down. This is called *estivation*.

El Niño: (Sp.—the child), sometimes called *El Niño de Navidad*, which means *the Christmas child*. This system sometimes develops around Christmastime.

La Niña

While switching back to normal, the pressure changes that cause the ENSO can overshoot, creating higher pressure in the eastern Pacific atmosphere and lower than normal pressure in the western Pacific and Indian Ocean. A condition called *La Niña* ("the girl") can develop. This drives the warm water far into the western Pacific and brings cold waters farther west than normal. Australia and Indonesia can experience flooding and South America's western coast has droughts and very low temperatures.

21.8 SHORT-TERM CLIMATE CHANGE

Certain types of climate change affect global temperatures for short periods of time. Seasons, natural cycles, and important geologic events are responsible for these. Let's take a closer look.

SEASONS

As you learned in Section 21A, seasons are regular, short-term changes in climate. These shifts in weather occur because of the way that the earth's rotational axis is tilted. At different points in the earth's orbit, either the Northern Hemisphere or the Southern Hemisphere is exposed to the more nearly vertical rays of the sun. Even though the most efficient heating occurs at the solstices, the hemispheres continue to warm (and cool) until a few months later when heat gain and heat loss are equal.

OBSERVED CLIMATE CYCLES

Earth has longer cycles in its winds and the weather they cause. About every five years, there's a shift in the winds and currents in the southern Pacific Ocean, called the El Niño **Southern Oscillation (ENSO)**. For a span of nine months to two years, the ENSO temporarily changes climate in the Pacific Ocean basin and affects weather all over the world. How do things change?

In a normal year, strong trade winds blow warm water from South America westward toward Asia and Australia. This is the normal Southern Equatorial Current flow that brings precipitation to Indonesia and the eastern coast of Australia. The warm water along the Equator is replaced by deep cold water upwelling off the coast of South America and the cold current from the Southern Ocean near Antarctica.

When an ENSO develops, high pressure in the Indian Ocean and Western Pacific weakens or reverses the trade winds. Warm humid air blows east and brings flooding rains to equatorial South America, Mexico, and southwestern United States. Warm surface waters move into the equatorial coastal areas, stopping deep-water upwelling. The lack of the cold, nutrient-rich seawater drives fish away, putting fishermen out of work. Severe droughts occur in Australia and Indonesia. The ENSO can even affect jet streams in the stratosphere, the monsoons of Asia, and hurricane development in the Atlantic.

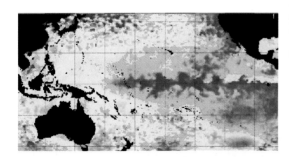

21-8

Sea surface temperature in a normal year (above) and during an El Niño year (right)

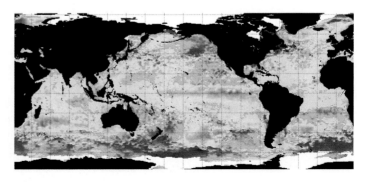

Longer cycles of pressure and temperature changes, ranging from ten to thirty years apart, also affect the Pacific and Atlantic Oceans. Climatologists actually see shorter climate cycles on top of longer cycles in these oscillations. The cycles can sometimes reinforce each other, and sometimes they weaken each other. Coupled with the ENSO, global climate cycles can produce variable weather from year to year. But the average temperatures over many centuries inferred from tree-ring studies and other indirect evidence indicate a relatively steady or slightly cooling global temperature until very recently.

VOLCANISM

In 1991, Mount Pinatubo in the Philippines blew its top. This volcanic explosion, one of the largest of the twentieth century, spewed millions of tons of ash into the air. How do you think that affected the earth's climate? Climates cooled around the world because the ash suspended high in the atmosphere blocked some of the sun's energy from reaching the earth's surface. Temperatures were lower than average for about two years! That's a big change resulting from such an isolated volcanic event. And this was not the first time this had happened. Remember that following the explosive eruption of Mount Tambora in 1815, millions of people starved from crop failures around the world, and some places in the temperate zone had snow every month of the year! Large regional dust storms and forest fires can have similar effects.

21-9
Climatologists believe that they can detect changes in precipitation and temperature by comparing the spacing of tree rings among hundreds of trees in an area.

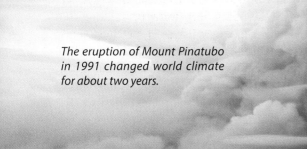

The eruption of Mount Pinatubo in 1991 changed world climate for about two years.

21.9 LONG-TERM CLIMATE CHANGE

We've seen how some things can cause temporary changes in Earth's climate. But these are not causing scientists concern. What they are worried about is one-way global warming that will last for decades or even centuries. Most scientists agree that during the last half of the twentieth century, the earth experienced a slight but quick increase in average world temperature, perhaps a little more than half a degree. But they're concerned that it could change more dramatically. Let's look at some of the possible causes of long-term climate change.

POSSIBLE NATURAL CAUSES

> **Solar Activity and Climate**
>
> Some scientists think that there may be a connection between sunspot activity and Earth's climate. They associate low numbers of sunspots with global cooling. One extended time of low sunspot activity from 1645 to 1715 is called the *Maunder Minimum*, which coincided with the Little Ice Age in North America and Europe. Some scientists also link solar rays with cloud formation.

Global warming seems to hinge on factors that control how much of the sun's energy is absorbed by the troposphere from the earth. In Chapter 18 we learned that the earth gets energy from the sun. Changes in the sun can generate changes on Earth. The sun goes through cycles of energy output that repeat every decade or so. These cycles are linked to the number of sunspots that are visible.

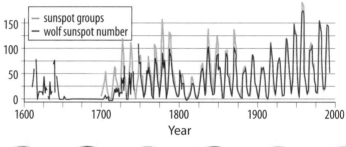

11 Aug 1980　14 Aug 1981　23 Aug 1982　11 Aug 1983　14 Aug 1984　10 Jul 1985　15 Aug 1986　24 Jul 1987　29 Jul 1988　18 Aug 1989

21-10

Sunspot activity changes over time. Changes in the amount of energy leaving the sun can affect the temperature in Earth's atmosphere.

There are also changes in the earth's tilt and in the shape of its orbit that change the timing and the latitudes where the peak amounts of solar energy are absorbed by the earth. But secular scientists believe that these cycles repeat with periods of tens of thousands to hundreds of thousands of years, so they may or may not be involved in current warming trends.

Another factor capable of causing long-term climate change is global volcanism. Old-earth geologists believe that there have been long periods of widespread volcanism throughout geologic time. These episodes are closely related to the ice ages in the secular view of climate history, which they think lasted for hundreds of thousands of years.

Clearly, most of these natural causes of climate change would affect the earth over *very* long periods of time. In any case, scientists have been gathering climate data relating to natural processes for less than 200 years, so we may not understand the relationship of these things to climate very well.

A RECOVERING EARTH

The earth has experienced drastic changes that have affected its climate. Young-earth scientists agree that when God flooded the earth to complete His judgment on man's sin, the upheaval triggered a series of climate changes. The warm oceans and cloudy skies that were likely remaining after the Flood plunged Earth into the Ice Age. The Bible is silent about this geologic period of time. But the ice sheets, glaciers, and icebergs remaining today, as well as the geologic evidence of glacial erosion, give testimony to this stage of Earth's history. And the warming trend we see may be a sign of a world still recovering from the Flood's effects.

POSSIBLE HUMAN CAUSES

We've seen that the earth's climate as a whole probably warmed by about half a degree during the second half of the twentieth century. This is a significant jump compared to the previous 1000 years. Indirect factors such as tree rings and ice core data indicate that temperatures were either relatively stable or even cooling slightly over that period.

Assuming that a 0.5–0.8 °C temperature rise has occurred, the big question to answer is whether humans are causing it. Many scientists say human activities that increase carbon dioxide in the atmosphere cause global warming because of the greenhouse effect. This is called the **greenhouse hypothesis**.

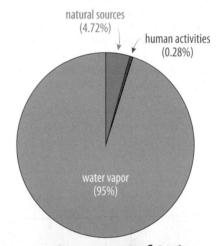

21-11
This pie chart shows the actual contribution humans make to greenhouse gases. By itself, this graph doesn't tell the whole story about global warming, but it is one factor to keep in mind.

Scientists think that people can increase carbon dioxide in the atmosphere by burning fossil fuels. We burn fossil fuels when we drive our cars, use electricity from a fossil-fuel power plant, or heat our homes with gas, oil, or wood. They think that we can also increase carbon dioxide by affecting ways that carbon dioxide is naturally removed from the air. Plants use carbon dioxide during photosynthesis to make plant tissues, wood, and food for us. So if plants, especially trees, are destroyed, they can no longer remove carbon dioxide from the air. If more carbon dioxide is released into the air than the remaining plants can take in, its level might increase. Large-scale logging of trees, called *deforestation*, is happening around the world. People cut down trees to use them for lumber or to make room to farm crops. People need lumber and farmland. But deforestation can become a problem when it isn't done wisely.

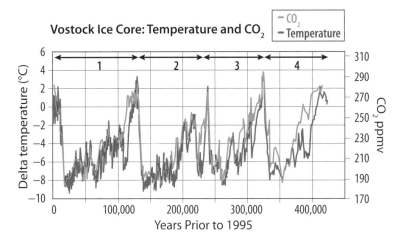

Carbon dioxide seems to be rising in the atmosphere. And people's activities *do* produce carbon dioxide. But it is hard to tell how these two facts are related to global warming. The problem is not a simple one of cause and effect. If natural cycles are causing global warming, there's really not much we can do. If people are causing it, then we need to find out whether proposed solutions will be effective and avoid creating more problems.

21-12
This graph shows the changes in both atmospheric CO_2 levels and temperature over time.

CASE STUDY: CO_2 IN THE ATMOSPHERE

Companies in the United States and Canada are developing CO_2 removal systems to combat global warming.

In April 1970, NASA launched the *Apollo 13* spacecraft in history's third moon landing mission. However, two-thirds of the way to the moon an exploding oxygen tank changed the mission radically. The landing would never occur and the mission was now a rescue mission. NASA and the crew faced many challenges, one of the greatest being excess CO_2 in the spacecraft. The crew had to build a filter system to remove CO_2 from the air.

There are climate scientists today who believe that the earth is facing a similar challenge. NASA's most recent atmospheric measurement, in January 2017, showed 406 particles of CO_2 per million particles in the air. While that doesn't sound like a large amount, the number had never exceeded 300 until the 1960s, and it has been steadily increasing since then. Some climate scientists believe that CO_2 is one of the primary driving forces behind global warming.

Scientific companies in the United States and Canada are developing technology to remove CO_2 from the air. They have proposed a number of possible uses for the amounts that they remove, such as to feed livestock, to fuel vehicles, or even to recharge oil fields.

Questions to Consider

1. Why was NASA concerned about the CO_2 levels in the *Apollo 13* spacecraft?
2. If CO_2 is a problem on the earth, what are some possible solutions beyond building CO_2 removal systems?
3. How should a Christian approach the issue of CO_2 in the atmosphere?

21.10 A PROBLEM OF MODELING

Back in the early 1990s in the state of Arizona, an organization built a $200 million experimental "Earth in a bottle." Called *Biosphere 2*, it contained a sealed atmosphere and miniature copies of five different Earth biomes. Its purpose was to create an isolated, self-sustaining system of plants, animals, and humans to verify that such a model would work for a space colony. So what was the result? Biosphere 2 did not complete most of its intended objectives. But it did show how complex the real earth environment is. It is almost impossible to model such a complicated system on such a small scale.

You've learned enough about our atmosphere to appreciate how difficult it is for climatologists to model Earth's climate. There are so many variables—water vapor, clouds, sea ice, snow cover, vegetation, living things, the oceans, sunlight, winds, precipitation. The list could go on. A few climate scientists estimate that there are more than 1000 variable climate factors. Just think about how much the weather forecasts change in just a few days. We don't completely know how all these things interact with each other. Even the scientists who work on climate models admit this.

So can we predict how current trends in climate are going to affect us in the future? There is no supercomputer powerful enough now or in the foreseeable future to make the necessary calculations. And that's even if we could identify all the factors, create the program, and obtain the data to feed into the program. The problem is just that huge. Scientists normally perform experiments to verify their models. But how are we supposed to experiment with the earth's climate?

Biosphere 2 showed just how complex the real biosphere, Earth's systems together with its life, actually is.

21.11 CLIMATE CHANGE AND WORLDVIEW

A Christian worldview requires us to approach the global warming question with care and wisdom. Christians should not deny that global warming might be occurring. We should not think that no climate scientist could report true data. We must fully support solid, well-reasoned scientific research. While we currently see an increase in global temperatures, we don't have enough evidence to understand the cause. We don't know whether people are causing these changes or whether Earth's systems have enough flexibility to manage the change to self-correct.

We need to do more research and collect more data to better understand Earth's climate. We should focus on wisely using Earth's resources to support a growing world population in ways that are economically smart. Let's explore alternative energies. Let's preserve life by managing wildlife and developing technology that saves and improves people's lives. Let's address resource problems like clean drinking water. This is an approach to conservation that pleases God. This is science at its best.

21-13
There are many ways that Christians can have a positive impact on the environment.

21C REVIEW QUESTIONS

1. How is global warming related to climate change?
2. How do both young- and old-earth scientists believe that the history of Earth's climate is a history of change?
3. How does Earth's climate change every year?
4. What climate changes happen during an ENSO?
5. Explain how volcanoes can temporarily change Earth's climate.
6. According to secular scientists, what are two general conditions that can produce long-term climate changes?
7. What are two kinds of human activities that some scientists believe contribute to climate change?
8. How accurate are models of climate change? What does that say about the certainty of their predictions?
9. What is the most radical solution proposed by environmentalists to the potential problem of human-caused global warming?
10. (True or False) Scientists agree that the earth seems to be warming.

LIFE CONNECTION: ARE POLAR BEARS ON THIN ICE?

The poster child for global warming has become the polar bear. Polar bears live in the Arctic. There's lots of floating ice off the coasts of Canada, Greenland, and Siberia. Polar bears are among the largest bears and land carnivores, along with the Kodiak bear. They're born on land but they spend most of their time in the sea. And though they can swim great distances, polar bears use floating ice, or ice floes, to rest and to get out farther from land to hunt for seals, which make up most of their diet. So if Earth's climate is warming, the polar bear is on "thin ice," right?

Not so fast! Let's look at how scientists get data about these animals. It's really hard to track polar bears. We've been following them by tracking radio transmitting collars since the mid-1980s, and it's difficult to monitor them over large areas. We don't have a very good record of past populations either.

To track just one bear, researchers must first find one using a helicopter. They shoot it with a tranquilizer dart, fit it with a radio transmitter, and then release it. That's a lot of work, just for one bear! And there are an estimated 20,000 polar bears! So do we really have a grasp of what's going on?

There are nineteen subspecies of polar bears that scientists know about. Right now, researchers don't have any population information on seven of these subspecies. They estimate that three have stable populations, eight have declining populations, and one actually has a growing population. Local Inuit people have observed more polar bears around Arctic towns. Though the Inuits think that this means there are more bears, scientists disagree. They say that this means that the bears can't get to their normal food sources. Official organizations like the United States Geological Survey have even predicted that two-thirds of polar bears could completely disappear by 2050 because of global warming.

Interestingly, most of these same scientists are evolutionists. They say that the history of the polar bear is one of adaptation to changing climates. Perhaps we should be less concerned about whether species will be able to adapt to climate change. Let's focus our resources on caring for and using the world instead.

Are polar bears in trouble? If they are, are we to blame?

CLIMATE AND CLIMATE CHANGE

Key Terms

conservation	503
climate	504
season	504
climate zone	511
temperate continental (climate)	512
polar (climate)	512
temperate marine (climate)	512
tropical rainy (climate)	513
dry (climate)	513
highlands (climate)	513
global warming	515
El Niño Southern Oscillation (ENSO)	518
greenhouse hypothesis	521

Icebergs break off from glaciers and ice shelves. Some scientists worry that global warming will result in more icebergs, which threaten shipping.

CHAPTER 21 REVIEW

CHAPTER SUMMARY

» Climate is the average seasonal weather in a location or region.

» The energy of the sun, latitude, season, bodies of water, and topography all affect an area's climate.

» The earth's surface can be divided into climate zones that are based on a region's vegetation, precipitation, and temperature.

» The six major climates according to the Köppen climate system are *temperate continental*, *polar*, *temperate marine*, *tropical rainy*, *dry*, and *highlands*.

» One kind of climate change is global warming, which involves increasing average temperatures all over the world.

» There is much evidence for major climate changes in the past. Explanations for how and why climates changed come from one's worldview.

» Earth's climate experiences short-term changes through the seasons, weather cycles, and volcanism.

» Most scientists agree that the earth's average temperature warmed in the second half of the twentieth century.

» According to secular scientists, Earth's climate could have experienced climate changes over many thousands of years through natural changes in the sun's energy, the earth's patterns of rotation and orbital motion, or prolonged volcanic episodes.

» When evaluating the causes and effects of global warming, Bible-believing Christians need to remember that the earth may still be recovering from the devastating effects of a global flood and the following Ice Age.

» Some climate scientists suggest that high levels of greenhouse gases, especially carbon dioxide from human activities, are causing global warming.

» Some scientists have observed a sharp increase in the warming trend and an increase in carbon dioxide levels due to advances in human industry and transportation. But there is no scientific proof connecting these two observations.

» Earth's climate is highly complex. That makes predicting the causes and long-term effects of climate change very difficult.

» Nothing in the Bible denies that global warming could occur. It seems to be happening as indicated by many observations.

» We should focus on predictable problems in conserving resources and carefully consider solutions that could have unintended consequences.

» We need to use the earth's resources wisely, without causing unnecessary pollution, to preserve life, and to deal with problems that affect human lives and welfare.

REVIEW QUESTIONS

1. Describe a Christian view of conservation.
2. How do meteorologists and climatologists handle data differently?

3. Explain how latitude and seasons work together to affect the amount of the sun's energy received by a place on Earth.
4. Why does land elevation often affect temperature?
5. Why is considering a region's vegetation useful when classifying climate?
6. Which climate zones appear mostly in the tropics?
7. Which climates appear mainly in the temperate zone?
8. In what climates would you be most likely to find animals that have adapted to cold, snowy weather?
9. In what climate zone do most of the plants and land animals reside?
10. Which of the four global warming questions listed on page 515 can be answered by operational science?
11. How do young-earth scientists believe Earth's climate has changed? What do they believe are the reasons for this change?
12. Why does a climate change seasonally?
13. Describe the effects of an ENSO on South America.
14. How would a series of widespread forest fires have a short-term effect on climate?
15. List three of the possible natural causes of long-term climate change according to old-earth scientists.
16. What two human activities do some scientists say could produce or increase global warming? To what are both of these related?
17. Why is it difficult to tell whether people are actually causing global warming?
18. List several variable factors that could affect the earth's climate.
19. What are some of the predicted effects of catastrophic global warming?
20. What are some ways that a warming Earth might be good?
21. What should Christians do about the potential dangers of climate change?
22. How should a Christian respond to evidence for global warming?

True or False

23. The main factors that help us classify a climate are the strength and amount of sunlight that an area receives.
24. The history of Earth's climate is a history of change.
25. Most scientists agree that there's been a sharp increase in average world temperature and atmospheric carbon dioxide in the past century.
26. Changes in the earth's orbit and rotation have been very influential on Earth's climate since the Flood.
27. Nearly all scientists agree that global warming is happening because of people's activities.

Map Exercise

28. Look at the map to the right. It shows the locations of worldwide weather stations that give data to determine average world temperature. The colors represent the number of years that they have been providing data. What do you notice? How do you think this data affects the accuracy of inferences about the change of average world temperature?

Wildfires occur around the world. Some scientists believe that global warming has increased the number of wildfires.

Length of Station Record (years)
0 10 20 30 40 50 70 90 110 130>

6 THE HEAVENS

CHAPTER 22
THE SUN, MOON, AND EARTH SYSTEM 530

CHAPTER 23
OUR SOLAR SYSTEM 559

CHAPTER 24
STARS, GALAXIES, AND THE UNIVERSE 587

CHAPTER 25
SPACE EXPLORATION 618

DR. RON SAMEC
ASTRONOMER

> "Have you ever thought about what stars are like? What makes a planet different from a star? Most students imagine stars as suns like our own, and they picture planets like our Earth. How are our sun and its family of planets any different from the billions of other heavenly bodies out there?
>
> For one thing, the mass of stars ranges from less than 8/100 of the sun's to a whopping 80–90 times its mass or more. Many stars have cool, red surfaces, while a few are very hot. The sun is a moderately hot star with a surface temperature that averages 5400 °C. Most stars differ from our sun in one more important way. God has made most stars in the heavens binary stars. Astronomers estimate that 60–90% of all stars revolve around companion stars.
>
> And how are the planets different? In this unit, you will learn that four of the planets seem to consist mostly of dense gases. At least three of these planets—Jupiter, Saturn, and Neptune—emit more energy than they absorb from the sun. This is a real problem for secular planetary scientists, but it is no problem for creationist scientists who believe that this extra energy is simply left over from God's creation of these young worlds. They are only thousands, not billions, of years old!
>
> God created the earth, the sun, and the heavens to sustain life and return glory to Him. Give all praise to His name!"

CHAPTER 22

THE SUN, MOON, AND EARTH SYSTEM

- **22A** THE SUN — 531
- **22B** THE MOON — 539
- **22C** THE SUN, MOON, AND EARTH AS A SYSTEM — 545

STONEHENGE

In Wiltshire, England, an ancient and mysterious circle of massive stones juts toward the sky. This is Stonehenge, the most significant stone monument in England and possibly the world. The more we learn about this prehistoric place, the more questions we have.

What do these stones mean? Archaeologists who have intensely studied this site for decades still don't know. We do know that the people who built Stonehenge were students of the sky. It was built so that it aligns with both sunrise on the summer solstice and sunset on the winter solstice. Because of this, some people wonder whether Stonehenge was a celestial observatory of sorts. Perhaps this 4000-year-old monument will guard its secrets for a long time to come.

22A

THE SUN

How does the sun change, affecting life on Earth?

22.1 A SPECIAL STAR

The sun is just an average star among billions in the whole universe. Or is it? It's our star, and that makes it special because we are special. God created it to be a source of light and heat for the ones who bear His image here on Earth. In Chapter 4, you learned that it has just the right properties and is at just the right place to do its job. It is one of the innumerable gifts that God has given us. Among its many purposes, it is to be a signpost for seasons, days, and years (Gen. 1:14). This is probably one of the reasons why the early Britons built Stonehenge. Their need to know the solstices and months of the year was vital to their survival. Crops had to be planted and gathered at the right times of year to maximize their winter food supply. Clearly, using such a timekeeping device would be a result of good and wise dominion. But Stonehenge was likely built as a place to worship the sun and moon too. The huge and complex arrangement of stones far exceeds the simple need for a calendar—it's more like a temple to serve the **celestial** deities. Even today, natural man tends to view himself as insignificant before the awesomeness of the heavens. But we aren't the servants of the sun. The sun serves us as God intended.

What do we know about the sun itself? For one thing, it's enormous! Its diameter is about 1.4 million km, or 109 times the diameter of the earth! Well, you may ask, why doesn't it look larger? The sun is about 150 million km away! The distance from the earth to the sun has a special name—the **astronomical unit (AU)**. The Italian astronomer Giovanni Cassini first determined the earth-sun distance scientifically in 1672. The astronomical unit is a convenient way to express vast distances within our solar system. So the sun is big but it's far away. Yet it's still the largest source of energy for Earth. Let's take a closer look at our brilliant ball in the sky.

22A Objectives
After completing this section, you will be able to
» describe the sun's structure, activity, and energy.
» summarize the sun's influence on Earth.

The term *celestial* indicates anything having to do with astronomical objects.

400   450   500
WAVELENGTH (NM)

22.2 THE SUN'S ENERGY

To life on Earth, the sun's light and heat are the most important things that it provides us. About 93% of the energy leaving the sun is in some form of light or electromagnetic (EM) waves, such as radio waves, visible light, and x-rays. High-energy protons and electrons carry away the other 7% of the energy leaving the sun.

THE SOLAR SPECTRUM

Physicists describe light energy as EM waves because of the way the electrical and magnetic energies act within a packet of light. Do you remember the parts of a wave from Chapter 2? The distance from the crest or top of one wave to the crest of the next is the wavelength. Solar EM waves have wavelengths from many kilometers for radio waves to **picometers** for gamma rays. EM wavelengths vary nearly continuously between these two limits. The entire span of EM waves emitted by the sun is the **solar spectrum**.

22-1
The visible light spectrum varies continuously from violet through indigo, blue, green, yellow, and orange to deep red. There are no distinct colors—they blend smoothly from one into another.

A **picometer (pm)** is one-trillionth of a meter, or one-billionth of a millimeter.

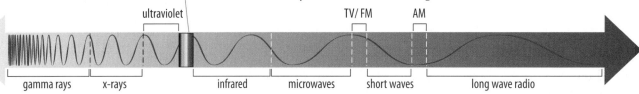

22-2
Visible light and other electromagnetic waves emitted by the sun differ in wavelength—the distance from one wave crest to the next.

A **nanometer** is 1000 times larger than a picometer—one-millionth of a millimeter. Thus, wavelengths of visible light are extremely tiny.

THE VISIBLE SPECTRUM

Most of the solar spectrum is invisible to us, though we can sense some of its waves as heat (infrared waves). However, tucked into the middle of the solar spectrum is a band of waves that our eyes and brain interpret as *visible light*. The cells in our eyes are tiny, so the wavelengths they are sensitive to are even smaller. Visible light wavelengths are measured in **nanometers** (nm). The colors we see depend on the wavelength of light. Visible light varies from about 390 nm (deep violet) to about 700 nm (deep red). The visible light from our sun includes most wavelengths of the visible spectrum. However, elements in the sun's atmosphere absorb some wavelengths.

UV, IR, AND OTHER BANDS

Other kinds of light lie just outside our ability to see them, but some animals are sensitive to these forms. Ultraviolet (UV) waves have wavelengths that range from 10 nm to 400 nm. They are too short for the human eye to detect. These are the rays that cause sunburn. Some birds and insects seem to be sensitive to UV light. Infrared (IR) waves have wavelengths ranging from the edge of visible red light at 700 nm to around 300,000 nm. They are far too long for the eye to detect, though we can feel them. Some animals like certain snakes can see infrared, which they use for hunting. The sun also emits

Harnessing Solar Energy
People around the world need more energy. *Solar energy* may be part of the solution. Though it is not concentrated and is not available 24-7, solar energy is inexhaustible and nonpolluting. It can be especially useful for heating water and for heating and cooling homes. Solar energy could also be used to break down water into hydrogen and oxygen for electric cars. Learning more about using the sun for the benefit of mankind is another way that we can obey the Creation Mandate.

radio waves, which have wavelengths much longer than infrared waves, and x-rays and gamma rays, which have wavelengths much shorter than ultraviolet waves.

THE SUN'S COLOR

Two factors determine the color of the sun we see: the intensity of the light waves emitted by the sun and the sensitivity of our eyes to those colors. The sunlight is most intense in the green-yellow-red region of the spectrum. Our eyes can detect only red, green, and blue light, but our brains combine these colors into many other colors. Our eyes are most sensitive to green and red light, which our brains combine to make yellow. So our visual system (eyes, nerves, and brain) is most receptive to yellow light. You can see how God made our eyes to work best under the sun He created for us.

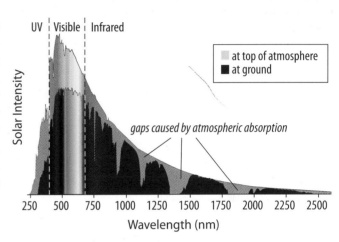

22-3
Most of the sun's visible spectrum lies in the yellow-green band.

22.3 THE SUN'S COMPOSITION

What is the sun made of? You might think that it is too far away to find out. But astronomers have a tool called a *spectroscope* that "fingerprints" light from stars and planets (see the box on page 600). When an element becomes hot enough, it gives off separate, distinct colors that only that element produces (see Figure 22-4). The sun's spectrum helps astronomers catalog which elements are present at its visible surface. With this information, they can determine its composition.

The sun contains at least 79 of the 92 naturally occurring elements found on Earth, but in quite different amounts. The most abundant element in the sun is hydrogen. The second most abundant is helium. These two elements make up about 98% of the sun's mass. Heavier elements make up the rest.

22-4
Elements in a light source emit specific wavelengths in a spectrum (thin, bright lines, top image). This image was obtained from a spectrograph. The sun emits a near-continuous spectrum. The sunlight that we see has some missing wavelengths due to elements in the sun's atmosphere, which absorbs specific wavelengths (bottom image).

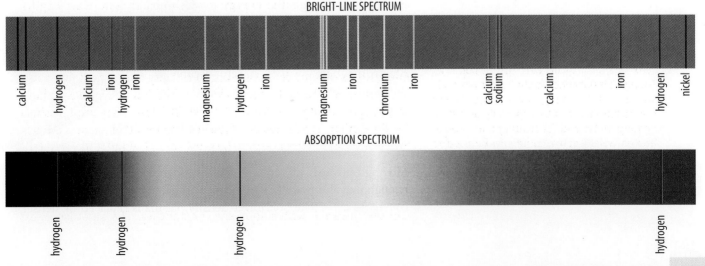

THE SUN, MOON, AND EARTH SYSTEM

22.4 THE SUN'S STRUCTURE

INTERIOR

Astronomers can't see the inside of the sun, only its surface. Just as with the earth, these scientists have to create a model of its interior to figure out what is going on inside. They use assumptions that are based on theories from nuclear physics. The most widely accepted model divides the sun's interior into three sections: the core, the radiative zone, and the convective zone.

The *core*, the innermost region, is where the nuclear reactions that generate the sun's energy take place. According to the current model, the diameter of the core is about one-fourth of the sun's total diameter. In this region, nuclei of hydrogen atoms combine under intense heat and pressure to form the nuclei of helium atoms in a process called **nuclear fusion**. Fusion releases tremendous amounts of energy, like one continuous hydrogen bomb. If our model of the sun's interior is correct, then the sun converts 655 million tons of hydrogen to 650 million tons of helium each second to release this energy! (The difference in mass becomes energy.) Calculations indicate that the core could be as hot as 15 million °C! These conditions are unimaginable, but atoms here form a *plasma* of dense and intensely energetic ionized particles.

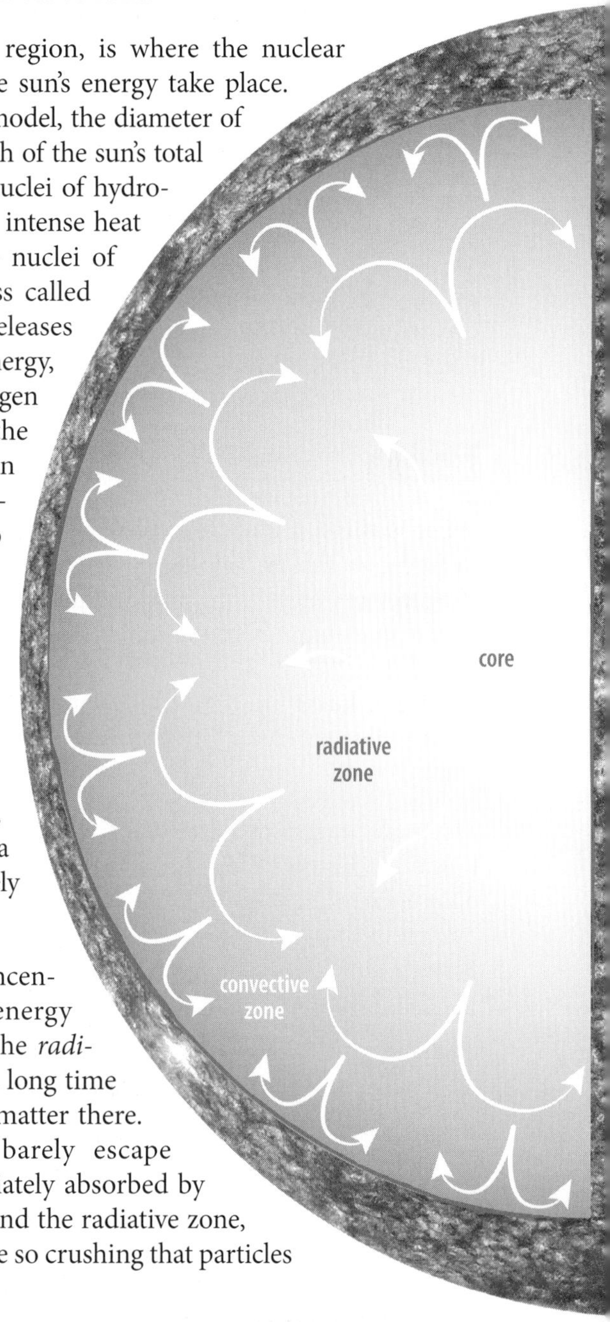

After leaving the core, concentrated electromagnetic energy moves outward through the *radiative zone*. This can take a long time because of the density of matter there. Electromagnetic waves barely escape one atom and are immediately absorbed by another. Within the core and the radiative zone, the pressure and density are so crushing that particles don't move very much.

Moving outward still, energy reaches a depth where pressure is low enough that atoms are free to move. This happens at the bottom of the *convective zone*. The hot, lower-density plasma is replaced by denser,

22-5
Model of the sun's interior. Energy leaves the core by conduction and travels through the radiative zone as electromagnetic waves (radiation). Once it reaches the convective zone, the plasma is free to move and rises to the surface in convection currents.

cooler plasma from the surface, creating innumerable convection currents. This process brings the sun's nuclear energy to its surface.

SURFACE

The visible surface of the sun is called the **photosphere**. And it's hot, though not as hot as the interior. Its temperature is between 4800 and 6000 °C. Even if you could tolerate the heat (no spacesuit could insulate you), you couldn't stand on the sun's surface. It's a fluid and you would quickly sink into its depths. And oh, by the way, you would weigh 28 times as much as on Earth!

photosphere: photo- (Gk. *phos, photo-*—light) + -sphere. This is the visible, light-producing surface of the sun.

The sun's surface is not plain and featureless. It looks like a seething pot of boiling water as mounds of hot plasma rise from the sun's interior. These mounds are *granules*. They make the sun's surface look bumpy in some images. You can also find *sunspots* and *solar flares* here. See the box on page 537 for photos and brief descriptions of these and other interesting solar features.

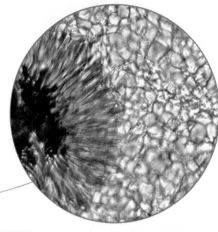

22-6
The uniform regions of this image are granules. An average granule is 1000 km or more in diameter. The dark section is a sunspot.

ATMOSPHERE

The sun's atmosphere extends out into space from the photosphere. It is made of plasma particles that are too energetic for the sun's gravity to hold. Its atmosphere has two parts. Just above the photosphere is the thin **chromosphere**. It is about 2000 km deep. Temperature changes from 6000 °C at the photosphere to around 20,000 °C at its upper surface. Pointed masses of hot plasma called *spicules* leap 10,000 km or more from the chromosphere's surface. The chromosphere has clouds just like Earth. But these are clouds of plasma called *plages* and *faculae*.

22-7
The sun's chromosphere

At 2000–3000 km above the photosphere, the chromosphere dissipates into the **corona**. The corona is the ever-changing outer atmosphere of the sun. The corona's temperature soars into the millions of degrees. It has no clear-cut outer edge. Astronomers observe **prominences** in the corona, a spectacular display of solar fireworks. Prominences are streams of material that appear to rise into the corona from the chromosphere and then gradually fall back.

corona (kuh ROH nuh): (L. *corona*—crown or wreath)

prominence (PRAHM ih nence): (L. *prominentia*—a projection). The largest known prominence occurred in 1946. It reached a height of 483,000 km and was 113,000 km wide!

THE SUN, MOON, AND EARTH SYSTEM

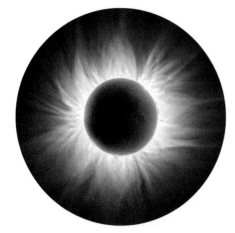

22-8
The corona as seen during a total solar eclipse

Though the sun's atmosphere is usually invisible against the glare of the sun, astronomers have found ways to learn about it. When the moon passes between the sun and the earth during a *total solar eclipse*, it just covers the sun's disk, blocking out its glare. Then the chromosphere and corona become clearly visible. These eclipses are rare, so astronomers use a tool called a *coronagraph* to block out the sun's disk at other times when observing it with a telescope or satellite.

The sun's corona becomes the **solar wind**, a flow of mostly protons, electrons, and small atomic nuclei from the sun. The solar wind extends far into space, as much as three times the radius of the solar system from the sun. Here at Earth, the solar wind blows by the earth's magnetic field, which protects us from the wind's dangerous radiation. The strength of the wind depends on the sun's activity. Particles in the solar wind can affect the earth's ionosphere.

22.5 THE SUN'S ACTIVITY

Of all the stars in the universe, the sun is the only one close enough for astronomers to see surface detail. We can use these observations to learn what other stars might be like.

SUNSPOTS

Sometimes small dark areas called **sunspots** appear on the sun's surface. These are cooler areas within the photosphere. A typical sunspot is about 4200 °C, still hotter than lava! Sunspots can last from a few hours to a few months. They typically have about the same diameter as the earth. Most exist for days or weeks. Sunspots often occur in pairs as local magnetic disturbances in the surface of the sun. One spot is the magnetic north pole, and the other, the magnetic south pole of the disturbance. The magnetic field slows down the movement of plasma from inside the sun. This could explain why sunspots are cooler than their surroundings.

22-9
Sunspot cycles over the last forty years. On average, going back to the late 1800s, the peaks occur about eleven years apart.

The number of sunspots reaches a peak (a solar maximum) about every eleven years (see Figure 22-9). This is called a *sunspot cycle*. At the peak of a cycle, the sun emits more energy, even though the sunspots are cooler and darker. The opposite is true at times of fewest sunspots (a *solar minimum*). These changes have a small but measurable effect on Earth's climate. The sunspot cycle affects our planet in other ways too. During active periods, the likelihood for damaging solar

22-10
A large group of sunspots

  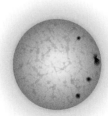

22-11
Sunspots near the sun's equator move faster than those near its poles. This sequence of illustrations demonstrates the sun's rotation.

flares increases, and most other visible features of the sun's surface are more numerous. The amount of ultraviolet (UV) energy emissions changes with the solar cycle. During solar minimums, the sun produces less UV radiation. This means less ozone forms in our atmosphere, which allows more UV radiation through to the earth's surface.

Astronomers can observe sunspots slowly moving across the sun's disk from the left *limb* (edge) to the right limb of the sun. A sunspot that disappears at the right limb reappears at the left limb in about two weeks after passing behind the sun. This helps astronomers understand how the sun rotates. Because the sun is a ball of plasma, its entire mass does not all rotate at the same speed. It takes about twenty-five days for the sun to rotate at the equator, and twenty-eight days at latitudes of 40° N and S. Near the poles, it takes thirty-six days! The sun rotates on an axis that is tilted 7° to the plane of the earth's orbit.

UP CLOSE WITH THE SUN

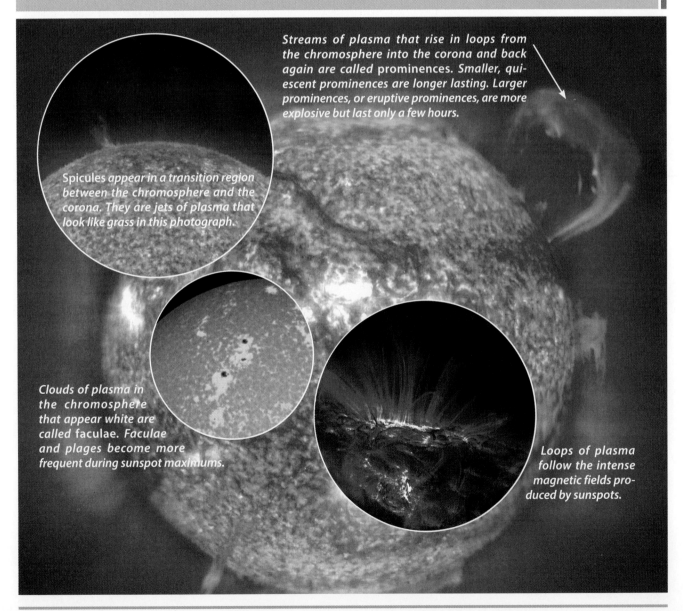

Spicules *appear in a transition region between the chromosphere and the corona. They are jets of plasma that look like grass in this photograph.*

Streams of plasma that rise in loops from the chromosphere into the corona and back again are called **prominences**. *Smaller, quiescent prominences are longer lasting. Larger prominences, or eruptive prominences, are more explosive but last only a few hours.*

Clouds of plasma in the chromosphere that appear white are called **faculae**. *Faculae and plages become more frequent during sunspot maximums.*

Loops of plasma follow the intense magnetic fields produced by sunspots.

THE SUN, MOON, AND EARTH SYSTEM

Aurora australis taken from the International Space Station. Auroras occur a few days after a solar flare and appear in the northern and southern skies near the North and South Poles.

Sometimes the sun has severe disturbances. A **solar flare** is a sudden energetic explosion of the sun's matter that emits bursts of rays and particles. They occur more often during the solar maximums. Solar flares seem to form when magnetic fields associated with sunspots come together and violently destroy each other. If they involve a large portion of the sun's surface, an extremely large mass of particles blows outward in what is called a *coronal mass ejection (CME)*. These kinds of solar flares are the most damaging. Radiation given off by any solar flare can disrupt radio communications on Earth and damage satellites and electrical power grids. Astronauts in the International Space Station (ISS) must take refuge in a heavily shielded portion of the station so that they don't get sick from the radiation. It would be like getting several thousand chest x-rays in a short time span if they didn't! These storms can also cause spectacular displays of polar auroras and can even affect the weather.

22A REVIEW QUESTIONS

1. Give two reasons why the sun is important to us here on Earth.
2. What is the solar spectrum?
3. Why is the visible light band of the solar spectrum visible while the other forms of electromagnetic energy are not?
4. State two reasons why the sun appears somewhat yellow to us.
5. What are the main chemical elements in the sun?
6. Create a sketch of the sun's interior and label it.
7. How does the temperature of solar plasma change as you proceed from the sun's surface into interplanetary space? Provide approximate temperature values to support your answer.
8. What are sunspots?
9. How is sunspot activity related to other kinds of solar activity?
10. (True or False) God intended to use the sun to remind us how insignificant we are in the universe.

22B

THE MOON

What is the structure and surface of the moon like?

22.6 THE MOON'S STRUCTURE

The moon is the closest astronomical body that we can see in the sky. It is Earth's only natural **satellite** and companion. As it orbits the earth approximately once each month, it accompanies Earth around the sun. Though the moon doesn't create light as a star does, we can still see it because it reflects sunlight. Scripture tells us that God created the moon to be "the lesser light to rule the night" (Gen. 1:16).

22B Objectives

After completing this section, you will be able to

» sketch the moon's structure.
» describe the moon's surface.

satellite (L. *satellitem*—guard, attendant); any natural or artificial object that orbits a planet. Natural satellites are called *moons*.

A planet's equatorial bulge is described with a value called *flattening* by astronomers. The larger the number, the greater the bulge. The moon's flattening is only 0.00125, while Earth's is 0.00335.

SIZE AND SHAPE

The moon is much smaller than the earth. Its average diameter is only 3474 km, less than the width of the United States. Its surface area is a little more than that of Africa. But it is still the largest moon in the solar system in relation to its planet. The moon is almost a perfect sphere. There is less than a 4 km difference between its equatorial and polar diameters. This probably is due to the moon's very slow rate of rotation; otherwise, it would bulge at its equator as the earth does.

INTERIOR

When the Apollo astronauts visited the moon between 1969 and 1972, they left seismometers there. And these seismometers measured moonquakes! Just as with the earth, the data generated helps us model the interior of the moon. The moon's interior seems to have a core, mantle, and crust, much like Earth (see Figure 22-14). The inner core is probably solid, with a liquid outer core. The core transitions into the mantle, which geologists believe is similar to Earth's but with a higher iron content. The thin crust is composed of low-density rocks. Chemical analyses of the moon's crust and mantle have been made from rock samples brought back by the Apollo moon missions.

MASS AND GRAVITY

Scientists first estimated the moon's mass around 200 years ago. With better instruments and direct measurements of gravity by astronauts and satellites, we now know that the moon's mass is only 1.2% that of Earth's. With this information, and knowing that the moon has a volume of only 2% of Earth's, geologists can estimate its average density to be about 3.3 g/mL, less than Earth's 5.5 g/mL.

22-12
The moon's volume is only 2% that of Earth's.

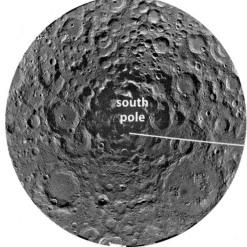

22-13
Lunar geologists believe that water and ice may hide in the deep shadows of these craters near the moon's south pole. If enough water exists here, it might be able to supply a lunar colony.

22-14
Model of the moon's interior

The moon's mass and size determine its surface gravity. Astronomers were able to estimate the surface gravity long before the moon missions and lunar satellites could measure it directly. The moon's gravity is about one-sixth of that at the earth's surface. On the moon, if you threw a baseball straight up, it would take about six times as long to hit the ground as it does on Earth. You could throw the ball a lot farther too. Satellites have shown that the moon's gravity varies quite a bit over its surface. Large regions of the moon's crust are made of denser minerals, which probably formed as molten material from the mantle flooded the surface after huge meteorite impacts.

ATMOSPHERE AND MAGNETOSPHERE

The moon has almost no atmosphere. Its weak gravity simply can't hold one. Some estimate that the whole mass of the moon's atmosphere is less than 10 tons! Gases from the interior of the moon probably replenish it. Matter at the moon's surface also releases gas particles as the solar wind slams against it. The composition of its atmosphere is mostly lighter metal atoms. Normal atmospheric gases on Earth (such as oxygen and nitrogen) are almost completely absent. Lunar satellites have detected water vapor in the moon's atmosphere, especially near the moon's poles. This may mean that there is ice in the deeply shadowed craters there.

The moon has a very weak magnetic field, less than one-hundredth that of Earth's. However, it doesn't have definite north and south magnetic poles as the earth's magnetic field does, so there is no earthlike magnetosphere.

22.7 THE MOON'S SURFACE

People have observed the moon's surface since the sixth day of Creation. However, it wasn't until the invention of the telescope that we were able to discern anything other than patches of light and dark on its surface. People believe that they can see images in these patterns, like the man in the moon. Depending on one's cultural background, these images could include a woman, a frog, a skull, or even a toad. Humans did not observe the far side of the moon until the twentieth century, when a Soviet space probe returned photographs to Earth.

CRATERS

The moon's surface is barren, a stark contrast to our life-filled Earth. Scattered over the moon's surface are millions of round, shallow depressions ranging from microscopic size to hundreds of kilometers in diameter. These are called *craters*. Steep, nearly circular walls that project above the surrounding terrain outline many craters.

After much telescopic and geologic study, astronomers have concluded that practically all lunar craters were formed by the impacts of flying space rocks—*meteoroids* and *asteroids*. These cosmic bodies varied in size from grains of sand to huge mountain-sized objects. You will learn about these in Chapter 23. Moving at speeds around 30 km/s or more, these objects created craters at impact when their kinetic energy instantly changed into thermal energy. The heat vaporized the object and some of the moon's surface, and the gases blasted out the crater. The materials that are blasted out of a crater during a meteorite impact are called *ejecta*.

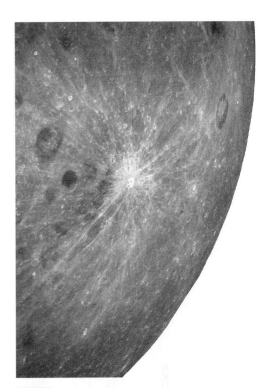

22-15
The crater Giordano Bruno with its system of rays

crater: (Gk. *krater*—bowl). Craters were given their identity in 1609 by the Italian scientist Galileo, the first person to observe the moon with a telescope.

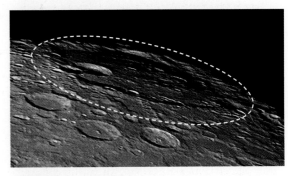

22-16
Crater Bailly is the largest walled crater visible on the near side of the moon.

You can view the moon in Google Earth just as you can view Earth! Go to View|Explore, then click on Moon.

Tycho: 43 °S 11 °W (Lunar)

THE SUN, MOON, AND EARTH SYSTEM

Fi **Elliptical crater**: 1.9 °S 47.6 °E (Lunar)
Davy Crater Chain: 11.1 °S 6.6 °W (Lunar)

22-17
A view of Mare Tranquillitatis (Sea of Tranquility)

mare (MAH ray) (L. *mare*—sea); p. maria (MAH ree ah)

22-18
Important surface features of the moon's near side

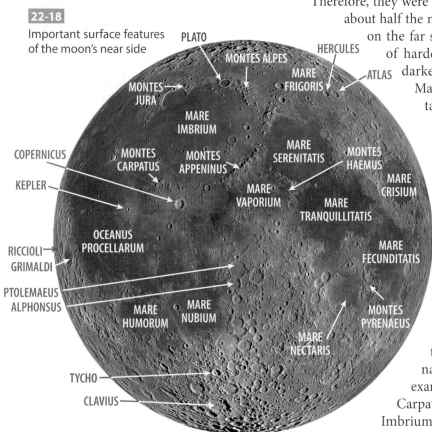

Fi **Mare Imbrium**: 33 °N 16 °W (Lunar)
Montes Alpes: 46.4 °N 0.8 °W (Lunar)

Among the largest distinct craters visible from Earth is Crater Bailly, which is 303 km in diameter. The most spectacular crater is Tycho, located near the moon's south pole. Tycho itself is only 82 km in diameter. But it is the bright streaks or **rays** radiating from Tycho that make it special. Some are over 1600 km long! Sometimes you can even see them with the unaided eye. The rays are made of material that is lighter in color than the underlying surface and that seems to have blasted out of the crater.

There are many odd craters on the moon. Some craters have central peaks while others have none. The crater walls of some have steps, while others are quite smooth. The rims of flooded or "ghost" craters can barely be seen rising above a smooth plain. Many craters are elliptically shaped, indicating that the objects creating them struck at a shallow angle. And in many places, a long crater chain formed, probably from a comet that broke up into many pieces before impacting.

LUNAR SEAS

Other depressions and low areas on the moon's surface do not have distinct walls or rims. Ancient observers thought that these generally darker areas on the face of the moon were large lakes or seas. Therefore, they were called *maria* (plural of **mare**). Maria cover about half the moon's near side, but there are almost none on the far side. They are broad lowland plains made of hardened basaltic lava that makes them look darker than the surrounding upland regions. Maria are roughly circular, and arcs of mountains create sections of a rim around some. Most lunar geologists suggest from these features that the maria basins are impact craters formed when huge asteroids or comets crashed into the moon. Following the impact, magma welled up from the mantle and flooded the giant craters. The largest such mare is **Mare Imbrium**, 1100 km in diameter.

MOUNTAINS AND VALLEYS

Between the maria and craters are highlands and mountain ranges similar to those on Earth. Several of these have been named after Earth's mountain ranges. For example, the **Alpes**, Apenninus, Caucasus, and Carpatus Mountains form part of the rim of Mare Imbrium. These mountains rise abruptly from the plains of the mare to heights as much as 8 km. Their steep slopes would be difficult to climb, especially because climbers would have to wear bulky spacesuits! Their tectonic history is even more difficult to describe than Earth's. Most seem to be related to the rims of the huge impact basins, while other mountains formed from disruptions in the crust.

The moon also has interesting mountains that lunar geologists believe are extinct volcanoes. These volcanoes could be classified as the shield type here on Earth. Few if any show a distinct crater, so some geologists believe that they may actually be **domes** of magma pushed up from below.

Another type of major feature of the moon's surface is a long, narrow valley called a **rille**. The shapes and depths of rilles are related to their geologic origins. Lunar geologists classify them as straight, sinuous or winding, and arcuate (arc-shaped). *Straight rilles* probably formed as grabens in rift valleys, similar to those on Earth. *Sinuous rilles* may be collapsed lava tubes, since most can be found leading from lunar volcanoes. **Hadley Rille,** explored by the *Apollo 15* mission, is of this type. *Arcuate rilles* probably formed when the flood basalts in mare basins cooled and shrank away from their margins, creating gaps.

22-19

Photo of the Montes Apenninus, a mountain range on the edge of Mare Imbrium, taken by the Lunar Reconnaissance Orbiter

22-20

Hadley Rille, photographed from lunar orbit (above), was explored by *Apollo 15*'s Lunar Rover (left).

Gruithuisen Dome: 32.9 °N 39.7 °W (Lunar)
Hadley Rille: 25.7 °N 3.1 °E (Lunar)

Lunar features seem to have remained unchanged for as long as we have been observing them. This is because there is no weathering by rain or wind. And there seems to be little in the way of plate tectonics as there is on Earth. Lunar seismometers show only occasional moonquake activity within the crust. The only way that the moon's surface visibly changes today is by meteorite impacts.

22B REVIEW QUESTIONS

1. Why is the moon nearly spherical?
2. How do we know about the moon's interior?
3. In what two ways is the moon's thin atmosphere replenished?
4. What do we call the nearly circular depressions on the moon's surface? Who gave them this name? How did they form?
5. How did rays around certain craters form?
6. According to lunar geologists, how did most of the maria form?
7. Do geologists believe that the moon's rilles all formed the same way? Explain your answer.
8. (True or False) The lunar seas are broad plains of frozen water.

HIDDEN FIGURES

Visualize in your mind a typical engineer or mathematician working at NASA in the 1950s. What do you see? Probably a white man. But don't tell that to Katherine Johnson. She and the other mathematicians working with her at NASA were black and female, and they helped America win the space race.

Through the 1940s and 1950s, black women were finally entering the white-collar American job force. But because of segregation laws at the time, they worked in separate offices, used a "colored girls" bathroom, and ate at segregated tables. NASA was in the throes of the space race when Katherine Johnson started working there. Her job was that of a human computer, a mathematician working to model the trajectory required to get Americans into orbit and beyond to the moon. She helped craft the equations that were used to calculate these trajectories. In a day when people were beginning to rely on electronic calculators, John Glenn said of Katherine Johnson, "Get the girl to do it. I want this human computer to check the output of the electronic computer, and if she says they're good, you know, I'm good to go."

Glenn thought of Katherine as an essential part of his mission's success, despite the culture in the still-segregated South. The racism that Katherine faced didn't keep her from doing what she was good at and doing what she loved in order to help people.

22C

THE SUN, MOON, AND EARTH AS A SYSTEM

How do the sun, moon, and earth interact?

22.8 MOTIONS OF THE SUN, THE EARTH, AND THE MOON

Imagine just for a moment that the sun, the earth, and the moon didn't move. Not even one inch. How would things be different? Some parts of Earth would never see the light of day. And the parts that did would get really hot! There would be no change in seasons—every place would have just one kind of weather. It would be really hard to tell time without some kind of artificial technology. Half of the places on Earth would never get a glimpse of the moon. And there would be no tides or other changes that we associate with the moon. In fact, it's quite possible that life as we know it would not be possible on Earth.

But God has made the sun, the earth, and the moon to move around each other in a regular, efficient, yet beautiful ballet. And it is some of the relationships of these three bodies that we're going to explore in this section.

22C Objectives

After completing this section, you will be able to

» describe how sun and earth interactions cause seasons.

» identify and explain the moon's phases.

» analyze how sun, moon, and earth interactions create eclipses.

» differentiate between ocean tides and earth tides.

The **apparent motions** of the sun, moon, planets, and stars through the heavens are the result of observing them from our location here on a moving, rotating earth. Viewed from elsewhere, their motions appear totally different.

ecliptic (ih KLIP tik): (Gk. *ekleiptikos*—directly or through). The ecliptic is the line marking the apparent path of the sun against the background stars throughout the year as the earth moves around the sun.

SUN MOTION

To observers on Earth, the sun seems to move through the sky on a daily basis. That is *apparent motion* due to the earth's daily rotation. The sun's position also changes among the stars throughout the year. If you observed one star carefully every night, you would notice that it rises above the eastern horizon about four minutes earlier each evening. If we could observe the stars in relation to the sun, we would see that the sun appears to follow a path through the stars. This is called the **ecliptic**. After a year, the sun appears to return to the same place. This is also an apparent type of motion because our point of view in the *celestial sphere* changes as the earth revolves around the sun.

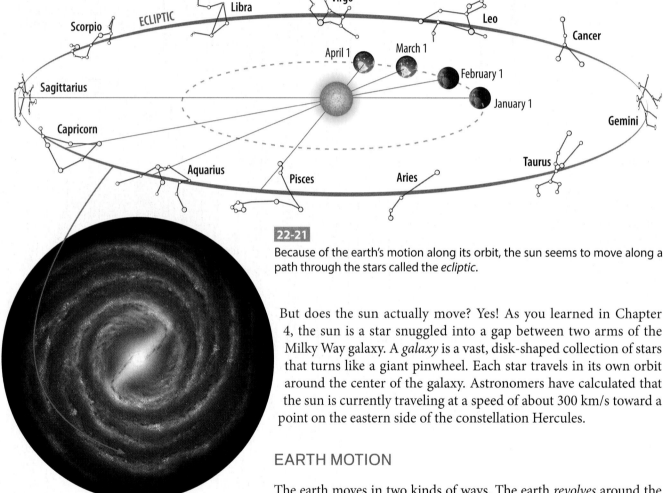

22-21
Because of the earth's motion along its orbit, the sun seems to move along a path through the stars called the *ecliptic*.

22-22
The sun is located between two spiral arms of the Milky Way galaxy, about three-fifths of the galaxy's radius from its center.

But does the sun actually move? Yes! As you learned in Chapter 4, the sun is a star snuggled into a gap between two arms of the Milky Way galaxy. A *galaxy* is a vast, disk-shaped collection of stars that turns like a giant pinwheel. Each star travels in its own orbit around the center of the galaxy. Astronomers have calculated that the sun is currently traveling at a speed of about 300 km/s toward a point on the eastern side of the constellation Hercules.

EARTH MOTION

The earth moves in two kinds of ways. The earth *revolves* around the sun once a year. This kind of motion is its **revolution**. The path that the earth follows is an elliptical orbit. The ecliptic against the background stars marks the plane of the earth's orbit. This surface is the *ecliptic plane*. Astronomers use the ecliptic plane as one way to orient the planets and other objects in the solar system. The sun and the earth are the only two bodies in the solar system that always lie exactly on the ecliptic plane. If you observe the earth's orbit from **above** the ecliptic, the earth revolves counterclockwise around the sun.

When we say that a point of view is from "**above**" an object's orbital plane, we mean as viewed from the same side of the plane as the sun's north pole. Since the orbital planes of all the planets in the solar system lie close to the earth's ecliptic plane, this is a good way to orient your viewpoint.

The kind of motion of the earth most obvious to us is its **rotation** around its *axis*, the imaginary line around which it spins. You learned earlier that this axis is tilted 23½° from a line perpendicular to the ecliptic plane. We see the sun, moon, planets, and stars rise in the east and set in the west. Observed from above the North Pole, the earth rotates counterclockwise, or from west to east. Observers in the Northern Hemisphere see stars rotate counterclockwise around a point above the geographic North Pole. Southern Hemisphere observers see the sky rotate clockwise around a point above the South Pole.

MOON MOTION

The moon has the same kind of motions as the earth: it revolves around the earth as it rotates on its own axis. The plane of the moon's orbit is tilted 5.4° to the ecliptic plane. The moon's orbit is an ellipse, just like the orbits of all other objects in our solar system. The moon orbits in a counterclockwise direction when viewed from above its orbital plane. The average distance from the center of the moon to the center of the earth is about 384,400 km. The moon's point of closest approach, called the *perigee*, is about 362,600 km, center-to-center. At the most distant point of its orbit, called the *apogee*, the moon is about 405,400 km away. If you look carefully, you may notice that the moon actually looks slightly larger at its perigee than at its apogee.

The moon rotates far more slowly than the earth. It rotates once on its axis in the same time it takes to completely orbit the earth. Therefore, it keeps essentially the same face toward us all the time. We will discuss how long that actually takes shortly. Interestingly, because of its elliptical orbit, we can see slightly beyond the edge of the moon's face at certain times due to a slight wobble that its orbit seems to cause.

22-23
From this time-lapse photo you can conclude that something is moving, but is it the earth or the sky?

perigee: peri- (Gk. *peri*—toward) + -gee (Gk. *gaios*—earth); the point closest to the earth

apogee: apo- (Gk. *apo*—away from) + -gee; the point farthest away from the earth

perigee 362,600 km | apogee 405,400 km

22-24
The moon's orbit is inclined more than five degrees to the plane of the earth's orbit.

22.9 PHASES OF THE MOON

Have you noticed how the moon changes shape? Obviously it's not the moon itself that is changing, but the amount of its sunlit surface that we can see. But why does this happen? This apparent change in shape is called a **lunar phase**. This is another kind of effect resulting from the positions of the sun, the earth, and the moon relative to each other.

22-25 The abrupt change from light to dark along the moon's terminator occurs because the moon has practically no atmosphere (left). By comparison, the earth's denser atmosphere makes the change appear more gradual (right).

Before we begin discussing the phases of the moon, we need to clarify some terms. As you just learned, the same surface of the moon always faces us. This is the *near side* of the moon. The opposite side that always faces away from us is the *far side*. The surface of the moon facing the sun is the *light side*, while the shadowed surface away from the sun is the *dark side*. The line between the light and dark sides of any astronomical body is called the **terminator**. As the light side covers more of the near side, the less of the dark side we can see. A moon's phase relates to how much of the near side is lit by the sun. As this lighted area grows larger, the moon is *waxing*. As the area grows smaller, it is *waning*.

waxing—gradually increasing in size, number, strength, or intensity

waning—gradually decreasing in the same

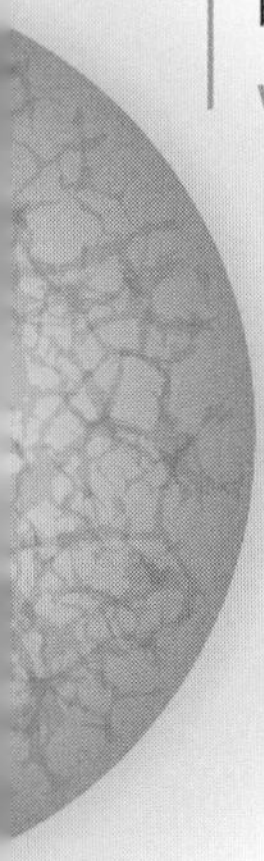

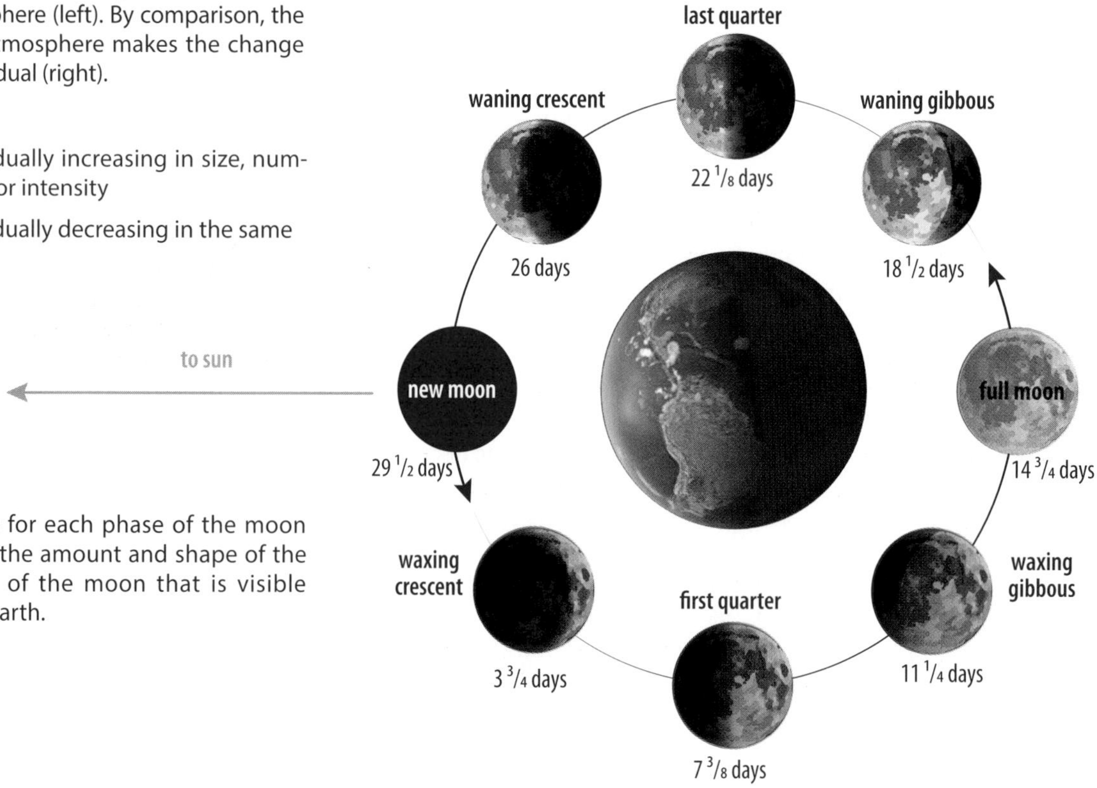

22-26 The name for each phase of the moon describes the amount and shape of the light side of the moon that is visible from the earth.

22.10 TIME AND SEASONS

The relationships between the sun, moon, and earth also help us keep time. Astronomical timekeeping is a very ancient practice and is one of the reasons for the existence of Stonehenge. Let's look at the year, month, and day.

THE YEAR

The earth takes a *year* to orbit the sun, right? But how do you determine when that happens from here on Earth? The simplest method is to measure the time it takes the sun to complete one full cycle along the ecliptic. Since this motion is apparent only to an observer on Earth, it takes one full orbit to complete. Astronomers call this period a *sidereal year*.

sidereal (sigh DER ee all): sidere- (L. *sidereus*—constellation, star) + -al (having to do with); referencing something to a star or constellation

There are several other ways to calculate an *astronomical year*, but they are somewhat more complicated to describe. Also, they don't differ from a sidereal year by more than a few minutes. We use the *Gregorian year*, which consists of 365.2425 days. A *calendar year* is 365 days long and is simply the time between the same calendar day from one year to the next. Because there is a mismatch between calendar and Gregorian years, a *leap year* occurs every fourth year by including an extra day to bring both calendars back together. Even these adjustments aren't perfect, so we have *century leap years* (where an extra adjustement is made) to keep the astronomical and Gregorian calendars in agreement.

THE MONTH

A *month* is the time it takes the moon to complete one orbit of the earth. In fact, the words "moon" and "month" come from the same root word because the month is traditionally associated with the moon. As with a year, there are several ways that astronomers can determine a month. The time between new moons can define one orbit. This period is about 29.53 days and is called the *lunar* or *synodic month*.

However, the moon completes a 360° orbit with reference to distant stars after only 27.3 days. This is the *sidereal month*. As a method of timekeeping, it is not very useful. The lunar month is more than two days longer than a sidereal month. This is because the earth is moving around the sun, so it takes longer for the moon to catch up to its new moon position with the sun each orbit.

The lengths of calendar months vary because it is impossible to make a lunar month fit evenly into a calendar, Gregorian, or sidereal year. So the Gregorian calendar contains a mixture of 30- and 31-day months, with 28-day February to fill the gap. February receives 29 days during leap years.

THE DAY

A *day* is one full rotation of the earth. But as with the year and the month, we need to select a reference point to define the day. The sun is always due south of a Northern Hemisphere observer at local noon, so that seems to be a good place to start. A *solar day* is the time from one local noon to the next, or about 24 hours. Because the earth's orbit is elliptical and the earth's speed varies throughout its orbit (see Chapter 23), the length of the solar day can vary by many seconds through the year. Until recently, astronomers used the *mean solar day* for time calculations. This was an average of the length of a solar day obtained by observing distant galaxies. A mean solar day is about 86,400 seconds long. There are about 365.24 solar days in a solar year.

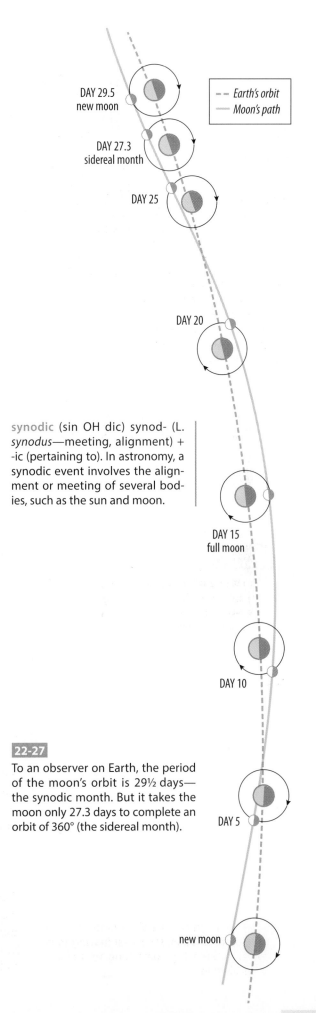

synodic (sin OH dic) synod- (L. *synodus*—meeting, alignment) + -ic (pertaining to). In astronomy, a synodic event involves the alignment or meeting of several bodies, such as the sun and moon.

22-27
To an observer on Earth, the period of the moon's orbit is 29½ days—the synodic month. But it takes the moon only 27.3 days to complete an orbit of 360° (the sidereal month).

THE SEASONS

The seasons are a direct result of the earth's motion around the sun. The summer and winter solstices, as well as the spring and fall equinoxes, are defined by four distinct points in the earth's orbit. As you learned in Chapter 21, the solstices occur when the earth's axis leans directly toward or directly away from the sun. The equinoxes occur when the sun's rays are perpendicular to the earth's Equator. Its axis is tilted neither toward nor away from the sun. The equinoxes happen three months after the previous solstice. These astronomical events are used to mark the beginnings of seasons. The summer solstice (June 21 in the Northern Hemisphere, December 21 in the Southern Hemisphere) and winter solstice (December 21 in the Northern, June 21 in the Southern) are the beginning of summer and winter, respectively. The spring equinox (March 20 in the Northern Hemisphere, September 22 in the Southern Hemisphere) and fall equinox (September 22 in the Northern, March 20 in the Southern) mark the beginning of their seasons.

22-28
In each hemisphere, the number of daylight hours is greatest when that hemisphere is tilted toward the sun.

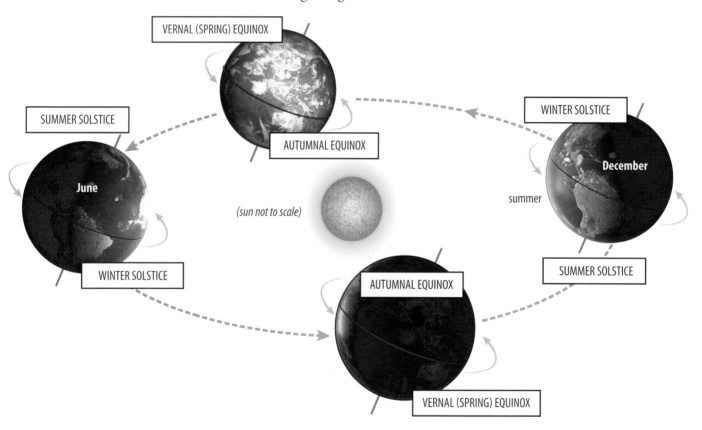

22.11 ECLIPSES

The sun, moon, and earth produce another kind of fascinating event—an eclipse. An *eclipse* happens when one celestial body cuts off the light from the sun to another body. The same thing happens when your brother or sister gets between you and the TV while you're watching it! Within the sun-moon-earth system, there are two types of eclipses—solar and lunar.

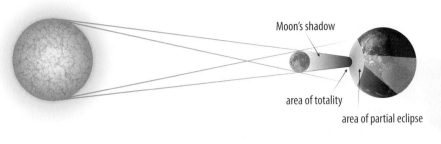

SOLAR ECLIPSES

A **solar eclipse** occurs when the moon blocks some of the sun's light from reaching the earth. This can happen only at a new moon when the moon is between the sun and the earth. But it doesn't happen at every new moon. Why not? The moon's orbit is tilted slightly to the ecliptic plane, so there are only two points in the earth's orbit where the moon's orbital plane and the ecliptic line up with the sun. To produce a solar eclipse, the moon also has to be at the right place in its orbit between the sun and the earth at these points. This happens so rarely that solar eclipses are important astronomical and cultural events.

22-29
A solar eclipse can occur only when there is a new moon (moon is between the sun and the earth) and the moon's orbit is in the same plane as the earth's orbit. In a total solar eclipse, the moon cuts off all the sunlight reaching the earth. Solar eclipses are difficult to observe because we can't look directly at the sun without damaging our eyes.

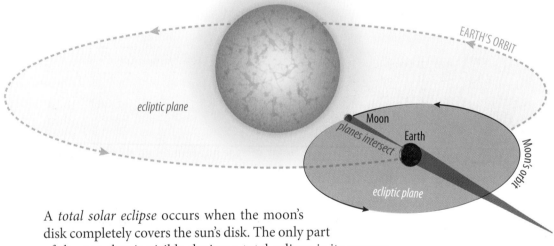

A *total solar eclipse* occurs when the moon's disk completely covers the sun's disk. The only part of the sun that is visible during a total eclipse is its corona. Only observers in the path of the moon's main shadow can see a total eclipse as the shadow moves across the earth's surface. The main shadow, or *umbra*, of the moon creates a disk of darkness only about 160 km wide on the earth. In any given location, a total eclipse lasts just a few minutes before the sun's face begins to appear. Observers that are not within the umbra of the moon see only a *partial solar eclipse*, or no eclipse at all. If the moon is near its apogee, it may be too distant to block the sun's disk completely. In this kind of eclipse, a ring-shaped portion of the sun's disk shows around the moon. This is called an *annular eclipse*.

annular (from *annulus*): (L. *anulus*—ring); the property of being shaped like a ring

If you have the opportunity to observe a solar eclipse, be sure to use proper eye protection. You should never look directly at the sun except through special dark sun-viewing glasses (standard sunglasses are *not* safe). A brief exposure to even a small part of the sun's direct rays during an eclipse can cause temporary blindness or even more serious injury.

So what happens during a total eclipse? For about an hour, the moon's disk moves steadily across the face of the sun. You probably wouldn't notice it change much until it was almost completely covered. Then the sky begins to darken. Plants and animals behave as though night were coming. Flowers close; birds roost in the trees. Farm animals may head for the barn. Just before totality, you may be able to see (with proper eye protection) some sunlight blazing between the mountain peaks and moon craters. When the moon completely covers the sun, the sky is as dark as a moonlit night. The temperature drops. A breeze may blow. Stars become visible. This can last from just an instant to 7½ minutes. Then, without warning, a shaft of light appears on the moon's limb. Quickly the sky brightens, and normal daytime conditions return. In about another hour the moon moves away and leaves the sun's disk completely uncovered.

LUNAR ECLIPSES

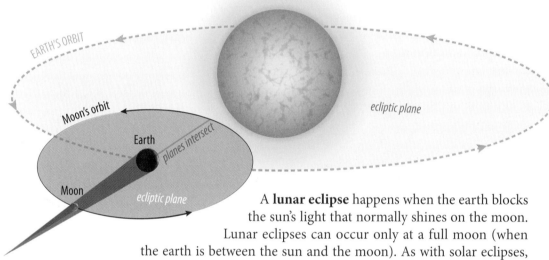

22-30

A lunar eclipse can occur only when there is a full moon (Earth is between the sun and the moon) and the moon is in the same plane as the earth's orbit. In a total lunar eclipse, the earth cuts off all the sunlight from reaching the moon.

A **lunar eclipse** happens when the earth blocks the sun's light that normally shines on the moon. Lunar eclipses can occur only at a full moon (when the earth is between the sun and the moon). As with solar eclipses, they don't happen at every full moon because of the tilt of the moon's orbit. Unlike a solar eclipse, anyone can see a lunar eclipse if the moon is visible from his location.

Eclipses of the moon are easy to observe. You don't need any special eye protection. The moon's west-to-east orbital motion carries it through the earth's shadow and out again. Even as it passes through the earth's umbra, the moon is still visible because sunlight refracts through our atmosphere and gives it a dull, coppery-red color. Total lunar eclipses can last nearly four hours from start to finish. The longest period that the moon can be completely in the earth's umbra is less than two hours. If the moon just grazes the earth's shadow, a partial lunar eclipse occurs.

22.12 TIDES

You're standing on the coast of New Brunswick, looking out at the Bay of Fundy. The tide is coming in. In about six hours, the water rises 15 m! Boats resting on hard ground rise and begin to float. Sea stacks that once rose from land become islands. Rapids and whirlpools develop as water floods through chokepoints of land. Ocean water flows up rivers that feed the bay, creating true tidal waves. And this is all because of the gravitational interactions of the sun, moon, and oceans.

Ocean tides are the most obvious effects of the gravitational pulls of the sun and the moon on Earth. But did you know that these same forces affect the earth's crust too? To a geophysicist, *earth tides* are far more important to the earth and moon's motions than ocean tides are.

Tidal forces affect the earth-moon system in important ways. They deform Earth's crust just as they do the oceans. Earth tides don't work quite the same way as ocean tides. The gravitational pulls of the moon and sun set up various kinds of oscillations in the crust that don't match their daily pulls as the earth rotates. These different oscillations can add together at times to create a bulge in the crust as large as 55 cm! Earth tides can cause earthquakes and even affect volcanoes. The variation of the earth's height can affect sensitive physics instruments, and even has to be factored into GPS altitude calibrations.

You learned in Chapter 2 that no force acts by itself. If the moon exerts a force on the earth, then the earth exerts an equal force on the moon. Tidal forces on the moon cause moonquakes. Earth's gravity pulls on the moon at its apogee, and that force makes it move slightly faster in its orbit as it falls back toward the earth. The higher speed causes it instead to slowly drift farther from Earth. At the same time, the earth's spin gradually slows. All these factors result from *tidal forces* between the earth and the moon.

22-31
One of the types of tidal motions in the crust of the earth. Red is up and blue is down.

We can predict and use ocean tides. We can use the movements of the sun, moon, and the earth to tell time. Both the predictability and usefulness of the motions of these heavenly bodies tell us something about God. He made the universe an orderly and predictable place that obeys the natural laws that He spoke into existence at Creation. He also made the universe with intricacy and complexity so that it isn't a simple matter for us to exercise dominion for His glory. But the effort is well worth it to discover His secrets and wisdom!

22C REVIEW QUESTIONS

1. What do we call the path of the sun's apparent motion through the heavens?
2. Do the sun, moon, and earth all experience rotations and revolutions? Explain.
3. When viewed from above the ecliptic plane, what seems to be the common direction of revolution and rotation for the sun, moon, and earth system?
4. If the same surface of the moon always faces us, how can we say that the moon is rotating?
5. Diagram and label the phases of the moon in order, beginning and ending with the new moon.
6. What are the two reference points that scientists use to determine the astronomical length of a year, a month, or a day?
7. What relationships create the seasons?
8. At what point(s) in the lunar cycle can a solar eclipse take place? At what point(s) for a lunar eclipse?
9. Give three ways that understanding the interactions of the sun-moon-earth system helps us exercise good and wise dominion.
10. (True or False) Tides apply only to large motions of the world ocean.

Key Terms

astronomical unit (AU)	531
solar spectrum	532
nuclear fusion	534
photosphere	535
chromosphere	535
corona	535
prominence	535
solar wind	536
sunspot	536
solar flare	538
ray	542
mare	542
rille	543
ecliptic	546
revolution	546
rotation	547
lunar phase	547
terminator	548
solar eclipse	551
lunar eclipse	552

22-32
This image of a solar eclipse, taken in Minnesota in 2011, shows sunspots on the surface of the sun.

CHAPTER 22 REVIEW

CHAPTER SUMMARY

» The sun is an average-sized but special star, about 109 times as big as the earth and 150 million km away.

» Most (93%) of the sun's energy is emitted in the form of electromagnetic waves, which include radio, infrared (IR), visible light, ultraviolet (UV), x-rays, and gamma rays.

» The human eye is designed to work best at the most intense wavelengths in the sun's visible light spectrum.

» The sun's composition is 98% hydrogen and helium. The other 2% is made up of most of the known natural elements.

» We believe that the sun's interior consists of three sections: the core, the radiative zone, and the convective zone, which releases solar energy at the photosphere. In the core, hydrogen is converted into helium by nuclear fusion.

» The sun's atmosphere consists of a thin chromosphere and a very extensive corona. The corona becomes the solar wind that extends far beyond the solar system.

» Sunspots are magnetic features on the sun's surface. Their numbers indicate the activity of the sun. During peak sunspot periods, solar flares can affect communications, electrical generation and transmission, and even weather on the earth.

» The moon is nearly a perfect sphere. Scientists infer from lunar seismic data that the moon has a two-phase core, a mantle, and a crust.

» The moon is much smaller than the earth, with only 1.2% of Earth's mass and a surface gravity about one-sixth that of Earth's.

» The moon's atmosphere is so thin that it is nearly undetectable. Its magnetosphere is weak and shows no distinct north or south poles.

» Lunar topographic features include craters (of variable sizes), maria, mountains, and rilles.

» The sun revolves around the center of the Milky Way galaxy and rotates on its axis.

» The earth revolves around the sun and rotates on its axis.

» The moon rotates on its axis and revolves around the earth. Its period of rotation matches its period of revolution—the same surface always faces the earth.

» The positions and motions of the sun, moon, and earth can determine important periods of time astronomically.

» Seasons on Earth occur because of the tilt of the earth's rotational axis and its orbit around the sun, which change the amount of heating of a given hemisphere through the year. The start of each season is marked by a solstice or an equinox, both of which are defined by the direction of the tilt of the earth's axis toward the sun.

» A solar eclipse occurs when the moon blocks the light from the sun to the earth at a new moon.

» A lunar eclipse occurs when the earth blocks the light from the sun to the moon at a full moon.

» The gravitational pulls of the sun and moon create both ocean tides and earth tides.

REVIEW QUESTIONS

1. How is the sun a special star?
2. Can we see most of the sun's electromagnetic energy? Explain.
3. How do astronomers know about the sun's composition?
4. Where does the sun's energy come from? How is it released into space?
5. Describe the appearance of the sun's photosphere.
6. What kinds of things happen in the sun's atmosphere?
7. How does the sun's activity affect the earth?
8. Make a sketch of the current model of the moon's interior. Label each section.
9. Why do craters, mountains, and even astronauts' footprints on the moon remain crisp and clear over time?
10. What causes the apparent motion of the sun along the ecliptic?
11. Describe the general shape of the moon's orbit around the earth and any special positions in its orbit.
12. Why don't lunar eclipses occur with every full moon?
13. How would life on Earth be different if there were no moon?
14. Why are earth tides much smaller than ocean tides?

At times planets are visible by the unaided eye. In this photo, we can see the moon, Mars, Jupiter, Mercury, and Venus. Their size and brightness as we observe them depends on their actual size, their distance from us, and the reflective capability of their surfaces.

True or False

15. Our eyes can detect yellow light waves.
16. The most common element in the sun is hydrogen.
17. The hottest part of the sun itself is its corona.
18. The sun rotates faster at its equator than it does at its poles.
19. Dramatic changes in the sun's magnetic field at its surface seem to produce sunspots, solar flares, and prominences.
20. The nearly perfect spherical shape of the moon is an indication that it is spinning very rapidly.
21. You would weigh half as much on the moon as you do on Earth if you were dressed the same way in both places.
22. The moon has no atmosphere.
23. The sun's motion through the heavens as viewed from Earth is an optical illusion.
24. The earth's orbit is inclined 23½° to the ecliptic.
25. A solar eclipse can occur only at a new moon.
26. The simplest method to compute a year is to count a certain number of new moons.
27. During a lunar eclipse, the moon blocks the sun's light.
28. You can still see the moon during a total lunar eclipse.
29. The moon is slowly drifting away from the earth.

Map Exercises

For these exercises, you will need access to a computer with the Google Earth program. Start the program and navigate to the Moon view (View|Explore|Moon).

30. Turn on the sunlight viewer (View|Sun) in Google Earth. A toolbar similar to the image below should be displayed in the upper left corner of the 3D viewer. Part of the moon should be in shadow.

    Slowly slide the pointer on the time line until the moon is a new moon (the near side is fully in shadow). Where is the sun in the 3D viewer? (*Note*: You may have to drag the moon view up or down to see the sun.) How does this relate to what you learned about new moons?

31. Explore the Apollo moon mission landing sites. In the sidebar, make sure that the Apollo Missions layer is checked; uncheck all the other layers. Zoom in on each of the six landing sites. Click on the balloons and photos near each to learn about those missions.

CHAPTER 23
OUR SOLAR SYSTEM

23A MODELING THE SOLAR SYSTEM — 560
23B THE PLANETS — 567
23C NON-PLANETARY OBJECTS — 577

TO PLUTO AND BEYOND

In 2006, NASA launched a space probe to the frontier of our solar system. This probe, called *New Horizons*, explored Pluto and beyond. It was the first probe ever to go to Pluto and its moon Charon. NASA launched the craft, about the size of a grand piano, toward Jupiter at 58,500 kph—the highest launch speed of any space probe. Jupiter gave it an extra boost of speed by gravity assist so that it traveled the rest of the way at 83,000 kph!

That great speed allowed *New Horizons* to arrive at Pluto on July 14, 2015, just 9½ years after launch. After collecting and sending 6.25 GB of data from Pluto, *New Horizons* continues to travel. It will next visit the Kuiper Belt—the solar system's suburbs—which contains thousands of icy rocks. Astronomers have only recently discovered these mysterious members of our solar system. Some are about the same size as Pluto. *New Horizons* will help us better understand our celestial neighborhood.

Artist's rendering of the view from within the Kuiper Belt. After visiting Pluto in July 2015, New Horizons continues on to the Kuiper Belt. NASA expects it to arrive there in 2019.

23A Objectives

After completing this section, you will be able to

» analyze models of the solar system.

» discuss the cultural significance of the adoption of the heliocentric system.

» describe the properties of planetary orbits.

The **Kuiper Belt** is named after Dutch astronomer Gerald Kuiper (KI per) (1905–73), who first proposed the existence of a reservoir of comets in the outer reaches of the solar system.

23A

MODELING THE SOLAR SYSTEM

How do we know that the sun is the center of our solar system?

23.1 WORLDVIEW AND EXPLORATION

Our **solar system** includes the sun and the objects that its gravity holds around it. What are its outer boundaries? It depends on whom you ask. Some might say it ends where the sun's gravity is no stronger than the gravity from other nearby stars in the Milky Way galaxy. Others might say that the edge of the solar system is where the inward *galactic wind* pressure equals the outward *solar wind* pressure.

So why did we hurl a $650-million space probe toward the outer reaches of our solar system? Out of curiosity. Scientists want answers to the question, "How did our solar system form?" Most astronomers view Pluto, its moons, and the objects of the **Kuiper Belt** as 4.6-billion-year-old solar system refuse—leftover matter from planet building after the big bang. Think back to the nebular hypothesis for the solar system's origin (Chapter 5). Some astronomers believe that they will be able to identify the composition of the earliest building blocks of the solar system by studying Pluto and the Kuiper Belt objects. They also think that the Kuiper Belt is the source of comets. One biological evolutionary theory suggests that some comets may have crashed to Earth long ago, bringing organic chemicals that triggered the origin of life. *New Horizons* intends to look for these compounds in the comet nursery.

So the *New Horizons* mission plans to gather evidence for a secular, deep-time model for the origin of the solar system and of all life on Earth. But does that make it worthless to a young-earth Christian? Certainly not! On a practical level, every advance in space technology gives us new tools to study creation or to improve our lives. But God also wants us to be curious and to discover the secrets that He has hidden in the universe. By learning more about the universe that He created, we can learn more about Him.

23.2 THE COPERNICAN REVOLUTION

It is awe-inspiring to look up at a night sky studded with stars. Have you ever noticed that when you look at the sky, you feel like you are at the center of a large dome? All the stars seem to be attached to the arched ceiling. Over time, it appears to turn. That perspective was the starting point for man's earliest understanding of the solar system.

THE GEOCENTRIC THEORY

What can be more stable than the earth? A very few ancient Greek philosophers suggested that the earth itself might move, but they were ridiculed into silence. Almost no thinking person in the centuries before Christ believed the earth could move. If anything moved in the night sky, ancient observers reasoned that it was the sky itself that was moving and not the earth. The star-filled dome must rotate around the earth like a hollow, black shell. The sun and moon, often viewed as gods, also moved around the earth in their own paths. But ancient observers noticed that a few bright objects didn't move with the stars. They seemed to wander in regular but baffling paths. These were called **planets**. If the starry heavens were on a distant, hollow sphere, then each of the other heavenly objects must be moving on a sphere of its own that turns about the earth. The spheres must be hollow, invisible shells so that one can see through them to the stars beyond. Because philosophers believed celestial things to be perfect, an important philosophical assumption of the time, the paths of planets had to be *circular*.

This ancient model of the universe is called a **geocentric theory**. The Greeks had a well-developed geocentric model of the universe by the fourth century BC. The philosophers Plato and Aristotle had the most influence in developing this system. They proposed that the celestial objects were arranged in the following order from nearest to most distant: the moon, the sun, Venus, Mercury, Mars, Jupiter, Saturn, and finally the fixed stars.

planet (Gk. *planetos*—wanderer)

geocentric (JEE oh SEN trik): geo- (Gk. *geo*—earth) + -centric (Gk. *kentron*—center); a model of the universe with the earth at its center

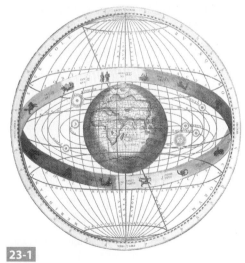

23-1

The geocentric system

retrograde: retro- (L. *retro*—back) + -grade (L. *cradus*—step)

While this model seemed reasonable to most of the ancient philosophers, there were problems. The geocentric theory couldn't explain two observations. Sometimes planets appeared to be bright and close, and at other times, smaller and farther away. It also couldn't explain why certain planets appeared to slow down, stop, and then back up as viewed against the background stars, an action known as *retrograde* motion. These things shouldn't happen in a perfect geocentric universe!

To understand retrograde motion, think about being in your car in traffic. If you are moving faster than another car, the other car seems to move backward. Assume that your car remains still while all other objects move around it. If other cars appear to move backward, the only explanation is that they are moving backward. Similarly, if Earth can't move, then planets that seem to move backward must actually be moving backward.

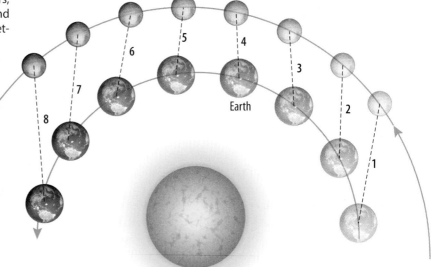

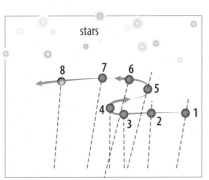

23-2
This image shows the relative positions of Earth and Mars over a period of time. The dotted line represents someone viewing Mars from Earth. Because Earth is moving faster, it passes Mars, making Mars appear to stop (position 4) and then move backward (to position 5). This is retrograde motion.

Ptolemy, a Greek philosopher in the second century AD, tried to solve these problems by shifting each planet's sphere, or *deferent*, so that the earth was no longer at its center. This adjustment would allow the planets to be closer and more distant at times. To explain the retrograde motion, he offered that the planets actually moved in smaller circles, called *epicycles*, that moved within the shell of their deferents. Ptolemy also changed the order of the objects so that Mercury and Venus stayed close to the sun. The Ptolemaic system was very difficult to use, but provided fairly good predictions of astronomical positions to astrologers, who were the main users of this information.

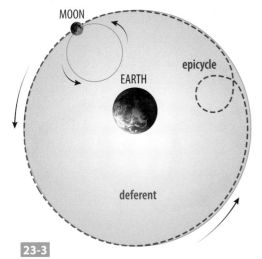

23-3
Ptolemy added epicycles, small loops in a planet's motion, in an attempt to correct the geocentric model.

Regrettably for the advancement of astronomical knowledge, several important Christian philosophers during the Middle Ages argued that the Ptolemaic model was supported by the Bible. They reasoned that it proved the central importance of man and the earth in God's creation. They also believed that it harmonized the observed motions of the heavenly bodies with certain Bible passages. In later centuries, Islamic astronomers developed more accurate geocentric systems that did not rely on deferents and epicycles.

THE HELIOCENTRIC THEORY

The fourteenth, fifteenth, and sixteenth centuries brought the great rebirth of learning called the *Renaissance*. A renewed interest in every area of culture swept across Europe. There was also a reawakening of interest in the Bible. With Gutenberg's movable-type printing press, the Bible and many other books became more available. The printed books stimulated new thinking and reopened the question, "Does the earth move?"

Polish astronomer Nicolaus Copernicus was born during this exciting period in history. A full-time student until age 33, Copernicus studied mathematics, astronomy, medicine, law, and theology at several prominent universities. Later, Copernicus seriously began to study the heavens. He observed the motions of the planets and pored over Ptolemy's and other philosophers' writings on this subject. He believed that Ptolemy's system was too complicated to represent the actual arrangement of the heavens. He began to seek a simpler model of the universe.

For years Copernicus collected data in a large work called *On the Revolutions of the Celestial Spheres*, or *The Revolutions* for short. Published in 1543, after he died, it made a stunning suggestion—perhaps the earth and planets moved around the sun, placed at the center of the universe. Only the moon, Copernicus thought, orbits the earth, and the earth rotates on its axis. Astronomers call this view a **heliocentric theory**.

Even today, there are religious people who hold to a geocentric theory. The change of viewpoint about the earth's position in the universe deeply affected people's thinking about themselves and their significance. For some, it greatly reduced the influence of the Bible on science and it opened the door to naturalism. This drastic change in thinking became known as the *Copernican Revolution*. The Copernican theory still gives us a much more workable model to explore the solar system. And as Christians have reflected on it, the position of the earth relative to the sun does not truly determine the importance of humanity in God's program.

23-4

This painting by Huens shows Copernicus working on the heliocentric model. His model in the background shows the sun in the center being orbited by Earth.

heliocentric (HEE lee oh SEN trik): helio- (Gk. *helios*—sun) + -centric (center)

23.3 DESCRIBING HOW PLANETS MOVE

The Copernican heliocentric theory was simpler than the Ptolemaic, but it still needed work. Many of the inaccuracies of the geocentric model remained because Copernicus assumed that the planets' orbits were circular. In the century following Copernicus's death, several scientists helped to greatly improve the accuracy of his model.

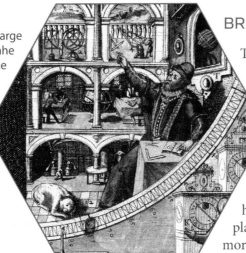

23-5
Tycho Brahe's mural of a large quadrant in Denmark. Brahe (1546–1601) was one of the most accurate astronomical observers in history.

BRAHE

Tycho Brahe (BRAH hee), a Danish astronomer, didn't completely accept the heliocentric theory. His view of the solar system was a combination of both the geocentric and heliocentric models. But he was the most accurate astronomer of his time. He built very precise instruments in palace-like observatories to support his observations. One of his assistants, Johannes Kepler, would use his detailed data to determine the true shape of planetary orbits. This made the heliocentric model more accurate and workable.

GALILEO

At the same time that Brahe was making observations, Galileo Galilei used his new telescope to study the four largest moons of Jupiter. As he observed them orbiting Jupiter, he concluded that not every object in the heavens orbits the earth. Galileo found this interesting, but it had little impact on the two models. The geocentric model didn't suggest or prohibit the idea of moons orbiting other planets. Galileo also observed Venus over a long period and saw that Venus had phases, just like the moon. The geocentric model predicted only crescent phases for Venus. As Galileo observed other phases for Venus, he became convinced that scientists should replace the geocentric model.

23-6
Italian mathematician and scientist Galileo (gal e LAY oh) Galilei (gal e LAY) (1564–1642) was known for his astronomical studies and for his discovery of the law of inertia. In this fresco by Giuseppe Bertini, Galileo shows the Doge of Venice how to use a telescope.

KEPLER

Brahe's assistant Johannes Kepler was a Christian and a brilliant mathematician. Kepler saw his work as fulfillment of the Creation Mandate. Shortly after joining Brahe's institute, he began to study Brahe's observations of Mars. Other members of Brahe's staff had already tried to make sense of them and had failed. After several years of work involving hundreds of pages of calculations, Kepler formulated three laws that successfully described planetary motion. In 1609, he published the first two laws in a book called *New Astronomy*. His third law came later, in 1618.

23-7
Johannes Kepler (1571–1630) was a German mathematician, astrologer, and astronomer.

Kepler's major contribution to the heliocentric theory was his mathematical proof that a planet's **orbit** isn't a circle but an **ellipse**. An ellipse is like a slightly flattened circle. While mathematicians construct a circle from its center, they construct an ellipse using two points inside the ellipse called *foci* (plural of *focus*). Kepler's three laws of planetary motion state the following:

1. Planets move in ellipses with the sun at one focus.

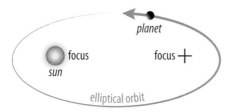

23-8
Kepler mathematically proved that planet orbits are elliptical, not circular. (The ellipse is exaggerated for clarity.)

2. A planet moves faster when closer to the sun and slower when farther away.

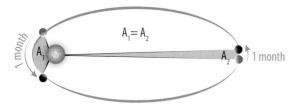

23-9
Kepler's second law. Because a planet's speed changes in its orbit, a line connecting a planet and the sun sweeps out equal areas in equal times.

3. The time planets take to orbit the sun (their *orbital period*) and their average distances from the sun always have the same mathematical relationship. A distant planet will have a longer period than a planet that is nearer the sun.

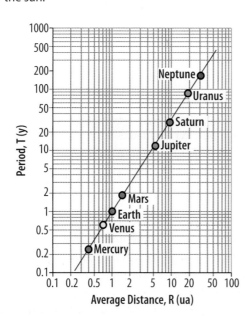

23-10
Kepler's third law. The time a planet takes to orbit the sun (its period) is mathematically related to its average distance from the sun.

These laws are important because they describe how everything in the solar system moves around the sun, including space probes. Astronomers and space engineers depend on these laws to plan space missions like *New Horizons*.

perihelion (per uh HEE lee yun): peri- (Gk. *peri*—near) + -helion (Gk. *helios*—sun)

aphelion (uh FEL yun): ap- (Gk. *apo*—away from) + -helion (sun)

The orbits of all the planets lie very close to the earth's ecliptic plane. Most planets have orbits that are nearly circular. But because their orbits are actually ellipses, their distance from the sun changes. When a planet is closest to the sun, it is at its *perihelion*. At its *aphelion* the planet is farthest from the sun.

NEWTON

So by the mid-1600s, astronomers believed that the earth moved, and they knew the general shape of the orbits that planets follow. But what keeps the planets circling the sun? Galileo had shown that moving objects tend to follow a straight path unless something forces them to move from that path. What force was acting on the planets to make them follow an elliptical course? Many strange theories were suggested, but it was the great English physicist Isaac Newton who finally provided the answer.

After carefully studying the motions of the moon around the earth, Newton developed a mathematical law that described the gravitational force. It is gravity that holds the planets and other members of the solar system in their orbits. The force of gravity causes a less massive object, such as the moon, to fall toward a more massive object, such as the earth. Its orbital speed causes it to continually miss the earth as it falls toward it. Newton demonstrated that a combination of gravity and orbital speed keeps the planets in orbit.

23-11
Isaac Newton (1642–1727), the English mathematician and scientist who developed the law of gravitation

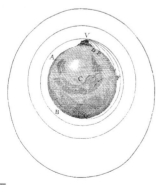

23-12
Newton suggested that a cannonball fired from a high mountain (V) could orbit the earth if its sideways speed were high enough to cause it to miss the planet as it fell toward it.

23A REVIEW QUESTIONS

1. According to most deep-time secular theories, how did the solar system begin?
2. Besides the stars, what seven heavenly bodies could the ancient astronomers observe with the unaided eye?
3. What is a geocentric theory?
4. What person in the second century AD is most noted for improving and promoting the geocentric theory?
5. State at least one major problem with the geocentric theory.
6. Describe the Copernican heliocentric theory.
7. Why have some historians viewed the Copernican Revolution as an important change in European thinking?
8. What observation did Galileo make that convinced him to replace the geocentric model?
9. What did Kepler discover about the shape of planetary orbits?
10. What object(s) did Newton observe in his development of the law of gravitation?
11. (True or False) The main reason that Copernicus questioned the geocentric theory was its complexity.

23B

THE PLANETS

How do the planets in our solar system compare with each other?

23.4 WHAT IS A PLANET?

In 2005, a group of astronomers at the Mount Palomar Observatory in California were studying some routine telescope images of a field of stars. They discovered a small dot that changed position very slowly over several photos. After further study, they came to an astounding conclusion that threw modern astronomy into upheaval.

They had discovered a "tenth planet"! Far beyond Pluto's orbit was an object they thought was bigger than Pluto. They went on to discover two more objects similar in size to Pluto. As these discoveries accumulated, astronomers began to question what a planet really was. In 2006, the International Astronomical Union (IAU) met to redefine and reclassify the objects in our solar system.

So what *is* a planet? At the 2006 IAU convention, those attending defined a planet as a body in the solar system that

» orbits the sun,
» has a spherical shape, and
» does not share the space of its orbit with any other object of significant size, other than moons.

According to this classification, there are only eight planets: Mercury, Venus, Earth, Mars, Jupiter, Saturn, Uranus, and Neptune.

So what happened to Pluto? Well, Pluto was reclassified by the IAU as a dwarf planet. **Dwarf planets** are objects that

» orbit the sun,
» are rigid and can be nearly spherical,
» *can* share their orbits with other bodies, and
» are *not* satellites (moons).

The main difference between a dwarf planet and a planet is the dwarf planet's neighborhood. Pluto and other dwarf planets do not have enough mass, and therefore enough gravity, to clear out other nearby orbiting objects.

Dwarf planets are also not satellites of planets. As you learned in Chapter 22, we commonly call a natural satellite a **moon**. A moon is a rocky object that orbits another object in the solar system. Moons, along with anything else in the solar system that isn't a planet or dwarf planet, are classified as **small solar system bodies (SSSB)**. We will look at these and dwarf planets in Section 23C.

23B Objectives

After completing this section, you will be able to

» categorize objects in the solar system.
» describe the position, appearance, size, composition, motion, and special features of the planets in our solar system.
» contrast other planets in the solar system with Earth.

Models and Classification

One activity of science is classifying what is observed. As scientists make breakthroughs, they have to adjust their models. Sometimes this means that they change the way they classify things. This shouldn't bother us. This is how scientists make their models more workable. So it's OK that Pluto isn't a planet anymore!

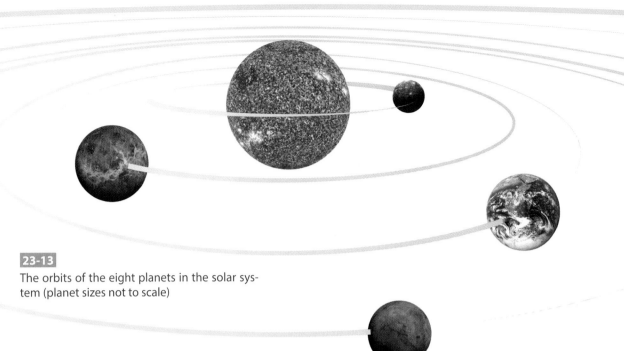

23-13
The orbits of the eight planets in the solar system (planet sizes not to scale)

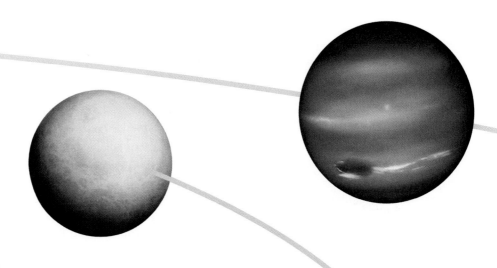

CLASSIFICATION BY PROPERTIES

The more we learn about the planets, the more amazing they seem. Each is a unique, beautiful, and impressive display of God's creativity. Though they are all different, astronomers can classify the planets into basic groups by their characteristics. Planets about the same size and density as Earth are called **terrestrial planets**. These are Mercury, Venus, Earth, and Mars. The planets larger than Earth have low densities and are made mostly of light elements, gases, and ices. These planets—Jupiter, Saturn, Uranus, and Neptune—are called the **Jovian planets**. Their dense, gaseous atmospheres are so deep that space probes can't reach their surfaces—if they have surfaces. Because they are mainly made of elements that are gases on Earth, these planets are sometimes called the *gas giants*. Their immense sizes and masses result in gravities greater than Earth's. It is worth noting here that none of the solar system planets except Earth has significant amounts of oxygen and liquid water—substances that are crucial for life.

terrestrial (tuh RES tree ul): (L. *terra*—earth); earthlike. The terrestrial planets are also called the *inner planets* because they are closer to the sun than to the asteroid belt.

Jovian (JOE vee un): Jov- (L. *Jovis*—early name for the Roman god Jupiter) + -ian (related to); like the planet Jupiter

23-14

Terrestrial planets are those that are about the same size and density as the earth. Jovian planets are much larger and less dense than the earth.

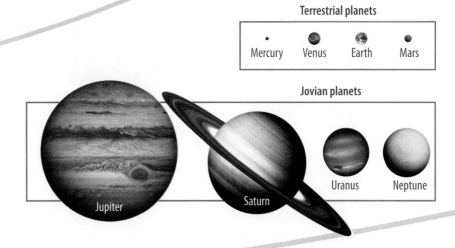

OUR SOLAR SYSTEM   569

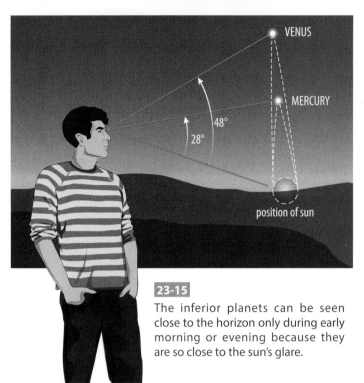

23-15 The inferior planets can be seen close to the horizon only during early morning or evening because they are so close to the sun's glare.

CLASSIFICATION BY POSITION

Planets can also be classified by their location in the solar system compared to Earth's orbit. Their positions affect where they appear in the sky as well as their apparent motions. Planets closer to the sun than the earth are called the *inferior planets*. As viewed from the earth, these planets are never far from the sun. For a good portion of their orbits, they are either in front of the sun and lost in its glare or behind the sun and out of sight. When they are far enough to one side or the other to be seen, they show phases like the moon. You can see the inferior planets just a few hours before sunrise or after sunset, when the glare of the sun is below the horizon. Their phases are visible through a small telescope. Astronomers can view the black silhouettes of these planets as they cross in front of the sun—events called *transits* (see Figure 23-16). Only inferior planets can transit the sun.

LIFE CONNECTION: JUST ADD WATER?

Many news releases about space exploration over the past few decades share one common ingredient—water. Whether the story is about the Mars rovers, comets, the moons of Jupiter or Saturn, or even planets orbiting distant stars, H_2O is the key in the search for life beyond Earth.

Water holds such an attraction for scientists that they will take it any way they can find it. Martian rocks with strata are seen as evidence that water deposited the mineral layers. Astrogeologists study the nuclei of comets, sometimes nicknamed "dirty snowballs." In 2005, NASA's *Deep Impact* mission blew a chunk out of Comet Tempel 1 just to see what chemicals would be ejected.

Geyser-like plumes of vapor from cracks in Saturn's moon Enceladus are imaginatively interpreted as evidence that a rich, warm, organic soup lies trapped beneath its frozen south pole. In 1995, scientists announced the first discovery of a planet in a distant solar system. The continuing discovery of hundreds of additional *exoplanets* since that time raises hopes that perhaps the latest one will show evidence of having water vapor in its atmosphere.

In case you haven't noticed, water is critically important for life as we know it. All organisms, from bacteria to baboons to blue whales, require water for their survival. Most modern scientists view water as the telltale sign that life could or did exist in a location. It is seen as a nursery for the origin of all life and a substance that ensures life's continuation.

Water vapor fountains erupting from the ice crust of Saturn's moon Enceladus

Many secular scientists are convinced that there was no God to supernaturally speak life into existence (Gen. 1:20, 24, 26). Their best guess is that life on Earth probably arose from nonliving materials in a warm ancient sea, or perhaps deep in the earth bathed by warm hydrothermal fluids. They are so convinced of this that they think it would be arrogant to assume that it could have happened only on our planet.

Another motivation to find water beyond Earth is the failure of scientists to demonstrate how life first arose right here at home. Despite their best efforts, scientists have been unable to create life from nonlife by simulating earthlike conditions. Consequently, some of them reason that since they *know* it occurred, it must have started somewhere else. Perhaps life began on another watery planet or on one of its moons and traveled here. Against all odds, it traveled to our Earth on a comet, meteorite, or even space dust.

So, should the confirmation that water exists on another planet or distant moon cause you to question your faith in the Bible's Creation account? While water is necessary for all life, it alone is not sufficient. You need hundreds of compounds structured in just the right way. This collection must obtain usable energy from its environment and have the ability to replicate itself. Even the simplest life yields undeniable evidence of a Creator.

Planets farther from the sun than the earth are called the *superior planets*. Usually, the superior planets slowly move from west to east across the star field. But the earth moves faster in its orbit than they do, and it passes them once per year. As we move ahead of them, they appear to move backward against the stars. This is known as *retrograde motion*. After a few weeks, they again begin to move to the east. This was the retrograde motion that early astronomers had so much trouble explaining with geocentric systems. Superior planets also show most of their disks as viewed from Earth because we spend most of our time in "front" of them as they are illuminated by the sun.

23.5 THE TERRESTRIAL PLANETS

23-16
A telescopic photograph of Venus transiting the face of the sun

MERCURY

Mercury is the planet closest to the sun (0.39 AU). It was named after the speedy messenger of the Roman gods. Photos of Mercury sent back by space probes show a planet covered with craters and maria, similar to our moon.

Mercury is also the smallest planet. Though Mercury has a density similar to Earth's, its gravity is too weak to hold much of an atmosphere. It has a short year but it also rotates slowly. A Mercury year has only 1.5 Mercury days! Its orbit is more elliptical than that of any other planet.

Mercury has no known moons. Its slow rotation and closeness to the sun cause surface temperatures to soar as high as 425 °C at its equator on the day side! On the night side and in shadowed craters at the poles, temperatures can plunge to as low as −190 °C. Mercury also has a magnetic field similar to Earth's, though only about one-hundredth as strong. Planetary geologists believe that this comes from within the planet's relatively large liquid metal core.

Magnetic field: yes
Diameter: 0.38 Earth
Mass: 0.055 Earth
Density: 5.4 g/cm^3
Gravity: 0.38 g
Tilt: 0°
Day: 58.6 d
Year: 88.0 d
Moons: 0

VENUS

Venus is the second planet from the sun (0.72 AU). It was named after the Roman goddess of beauty and love. As Earth's nearest neighbor, Venus regularly outshines all other celestial bodies except the sun and the moon. A view of its surface is hindered by an opaque atmosphere, but radar studies and space probes show its surface to be heavily marked with volcanic features. It has relatively few impact craters. This indicates to planetary geologists that its surface is geologically young or has been smoothed over by periods of widespread tectonic activity.

Planetary Data

Each planet description in this subsection includes the planet's important properties. *Gravity* data is for gravity at the visible surface of the planet in multiples of Earth's surface gravity (1 *g*). *Tilt* is the angle the planet's rotational axis is tilted from a perpendicular to its orbital plane. *Magnetic field* indicates that the planet has a geomagnetic field (originating from the planet's interior). *Day* and *Year* are in units of Earth time.

Magnetic field: no
Diameter: 0.95 Earth
Mass: 0.82 Earth
Density: 5.2 g/cm³
Gravity: 0.90 g
Tilt: 177°
Day: 243 d
Year: 224.7 d
Moons: 0

Of all the planets, Venus is most like the earth in size and mass. However, its axial tilt is 177°, assuming that it rotates counterclockwise like the other planets. Another way to think of this is that it appears to rotate backward. Astronomers think its composition is similar to Earth's but its interior is likely much hotter.

Like Earth, Venus has enough gravity to hold an atmosphere—a hot, dense, nasty atmosphere. The lower atmosphere is mostly carbon dioxide, while above are thick clouds of sulfur dioxide and sulfuric acid. Atmospheric pressure at the surface is nearly 100 times that of Earth's. Standing there would be like being under 1 km of water! A runaway greenhouse effect produces the highest surface temperatures of any planet—averaging 460 °C! This atmosphere reflects about 90% of the sun's energy, giving the planet a dazzling white appearance. Space probes have detected a weak magnetic field. This seems to originate in the upper atmosphere from its interaction with the solar wind, rather than from the planet's interior.

EARTH

Earth is the third planet from the sun (1.00 AU). Compare the information with that of the other planets.

Magnetic field: yes
Diameter: 1.00 Earth
Mass: 1.00 Earth
Density: 5.5 g/cm³
Gravity: 1.00 g
Tilt: 23.5°
Day: 1.0 d
Year: 365.2425 d
Moons: 1

MARS

Mars is the fourth planet from the sun (1.52 AU). It gets its name from the Roman god of war. It is the only terrestrial planet that is also a superior planet. It's easy to find in the sky because of its reddish color. Though not as bright as Venus, Mars can sometimes be brighter than Jupiter.

Magnetic field: no
Diameter: 0.53 Earth
Mass: 0.11 Earth
Density: 3.9 g/cm³
Gravity: 0.38 g
Tilt: 25.2°
Day: 1.03 d
Year: 687 d
Moons: 2

Mars is much less dense than Earth, so its internal structure and composition must be quite different. Mars has a thin, dusty atmosphere composed almost entirely of carbon dioxide, with a little nitrogen and trace amounts of other gases. Surface air pressure is less than 1% of Earth's. Average surface temperature is a chilly −63 °C. The planet has no geomagnetic field.

Mars has two tiny moons, **Phobos** and **Deimos**, that could be large rocks captured from the asteroid belt. Through orbiters and robotic rovers on its surface, we've learned a lot about Mars's numerous impact craters, volcanoes, polar ice caps, and dust storms. It has the largest volcano in the solar system (**Olympus Mons**). Many erosional features suggest to astrogeologists that Mars may once have had large quantities of liquid water flowing over its surface. Strong Martian winds power dust storms that can last for weeks.

Phobos (FOH bahs) and Deimos (DY mahs) are the names of the two sons of the Greek god of war, Ares.

Olympus Mons: 18.65 °N 133.81 °W (Martian)

CASE STUDY: JOURNEY TO MARS

Mark Watney is a crewmember on a manned mission to Mars. When he and the rest of the crew attempt to evacuate due to a massive storm, Mark is injured and assumed dead. However, Mark survived the storm and was stranded, struggling to remain alive until rescue could come over two years later. This gripping tale of adventure and survival sounds like the plot of a Hollywood movie, doesn't it? That's because it is just that.

How far-fetched is this story? Is there any serious thought of sending astronauts to Mars? NASA's Journey to Mars program is looking to accomplish that very thing. This program has three major phases. The first involves extensive experimenting on the International Space Station (ISS), the "Earth Reliant" stage. Next, NASA will enter the "Proving Grounds" phase with a series of deep-space tests outside of Earth's orbit. Finally the agency hopes to progress to the "Earth In-dependent" phase when astronauts will live and work on Mars with only occasional support from Earth.

The idea of a Mars colony certainly captures the human imagination, but it may surprise you to know of NASA's planning and progress toward that goal.

NASA astronauts are already testing equipment that may be used on a manned mission to Mars.

Questions to Consider:

1. What do you think may be some challenges to colonizing Mars?
2. Why do you think NASA is starting with missions on the ISS?
3. How long would a trip to Mars take?

I imagine that this erosion was caused by a huge amount of flowing water over the surface of Mars.

How come you can't imagine that happening on Earth?!

OUR SOLAR SYSTEM

Brightness

As you might suspect, a planet's brightness is related to how big it is and how far away it is. But another factor is a planet's ability to reflect the sun's light, since it doesn't make its own. This is called the planet's *albedo* (al BEE doh). Albedo is expressed as a decimal percentage. Venus has an albedo of 0.65, meaning that its dense atmosphere reflects 65% of the light it receives. The object with the highest albedo in our solar system is Saturn's moon, Enceladus. It is a shiny, ice-encrusted object with an albedo of 0.99!

23.6 THE JOVIAN PLANETS

We noted earlier that the Jovian planets are special because of their composition of mainly light elements and their position far beyond the asteroid belt. But they are unusual for another reason. Jupiter, Saturn, and Neptune emit up to 70% more energy than they receive from the sun. No one knows why this is true, and there are several theories to explain the data. Secular astronomers suggest that the radiated energy is left over from the planets' origin in the solar nebula. Interestingly, Uranus, which is often described as a twin of Neptune, emits almost no excess energy. Creationists view this problem as evidence for creation of the individual planets. You will see that the Jovian planets' differences are greater than their similarities.

JUPITER

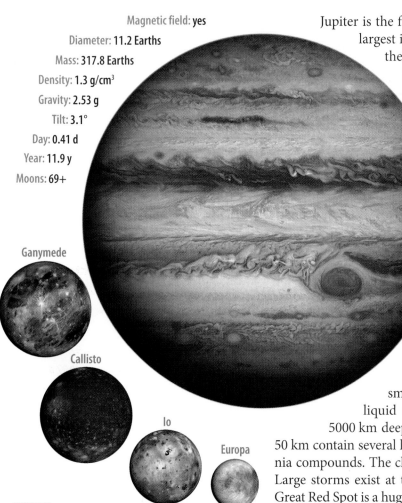

Magnetic field: yes
Diameter: 11.2 Earths
Mass: 317.8 Earths
Density: 1.3 g/cm³
Gravity: 2.53 g
Tilt: 3.1°
Day: 0.41 d
Year: 11.9 y
Moons: 69+

23-17
The Galilean moons of Jupiter arranged in order from largest to smallest. Ganymede and Callisto are both larger than the planet Mercury.

Jupiter is the fifth planet from the sun (5.20 AU) and is the largest in our solar system. The planet is named after the supreme Roman god. It is usually the third brightest nonluminous body in the sky after the moon and Venus.

Jupiter is truly gigantic in both size and mass. Because its rotational speed is so high, it has a noticeable equatorial bulge. Jupiter contains almost 2½ times the amount of matter of all the other planets in our solar system combined. It is composed mainly of the light elements hydrogen and helium, so it has a density about one-fourth that of Earth's. This is one reason why its surface gravity isn't huge. Even so, a 120-lb student would weigh 300 lb on Jupiter!

Geologists believe that Jupiter's interior structure is really unusual, consisting of a small rocky core surrounded by a deep layer of liquid metallic hydrogen. The atmosphere is about 5000 km deep and is mostly gaseous hydrogen. The upper 50 km contain several layers of clouds made of ammonia and ammonia compounds. The clouds form bands that race around the planet. Large storms exist at the margins between the bands. The famous Great Red Spot is a huge cyclonic storm several times the size of Earth.

Jupiter has many moons. The four largest were first discovered by Galileo with his telescope in 1610, and so are called the Galilean moons. They are Ganymede, Callisto, Io (EYE oh or EE oh), and Europa. Jupiter also has several faint dust rings. Its magnetic field is the strongest of any planet in the solar system and probably originates in the large mass of metallic hydrogen in the planet's interior.

SATURN

Saturn is the sixth planet from the sun (9.6 AU) and one of the four gas giants. It is named after the Roman agricultural god. Saturn is nearly as bright as the brightest stars, even though it is very distant from Earth.

Saturn is second in size to Jupiter. It has a prominent equatorial bulge: its equatorial diameter is 10% greater than the distance from pole to pole. Interestingly, Saturn is less dense than water. If you could find a large enough pond, Saturn would float in it! Its interior structure and composition are probably similar to Jupiter's, with an atmosphere about 1000 km thick. It has fast-moving bands of clouds similar to Jupiter's and has one seasonally occurring storm called the *Great White Spot*.

At least sixty-two moons orbit Saturn but probably many more, not including the chunks of ice that make up its rings. Its largest moon is Titan, the only moon in the solar system with a dense atmosphere. Titan is larger than Mercury, and liquid hydrocarbons exist on its surface. Saturn's ring system is probably its most amazing feature. There are about twelve main rings. The ring system averages only about 20 m thick and is made mostly of water ice particles. Saturn has a simple magnetic field much weaker than Jupiter's and not even as strong as Earth's.

Magnetic field: yes
Diameter: 9.4 Earths
Mass: 95.2 Earths
Density: 0.7 g/cm³
Gravity: 1.07 g
Tilt: 26.7°
Day: 0.44 d
Year: 29.5 y
Moons: 62+

URANUS

Uranus (YUR en us) is the seventh planet from the sun (19.2 AU) and the third gas giant. Uranus is named after the Greek sky god. Though it was visible, ancient astronomers couldn't recognize Uranus as a planet because it was very faint and moved too slowly. The English astronomer Sir William Herschel discovered it in 1781 with an optical telescope. Astronomers sometimes classify both Uranus and Neptune as *ice giants* because their atmospheres and interiors contain much larger amounts of icy substances than Jupiter and Saturn.

Uranus is the third-largest planet after Jupiter and Saturn but is fourth in mass. One of the most unusual aspects of Uranus is its axial tilt, which is 97.8°. It seems to be lying on its side.

Magnetic field: yes
Diameter: 4.0 Earths
Mass: 14.5 Earths
Density: 1.3 g/cm³
Gravity: 0.89 g
Tilt: 97.8°
Day: 0.72 d
Year: 84.0 y
Moons: 27

Astronomers believe that the structure and composition of Uranus is quite different from the larger gas giants. Its interior is unknown, but geologists think that there may be either no core or a small, solid, rocky one surrounded by a thick mantle of water, ammonia, and organic compound ices. A gaseous atmosphere at least 300 km thick contains hydrogen, helium, and methane. Uranus has the coldest atmosphere in the solar system. Several layers of ice clouds exist in this region. Astronomers believe that the planet's blue color may come from light scattered off the water or methane ice clouds. Besides its extreme axial tilt, Uranus has another surprise. The poles of its magnetic field are tilted nearly 60° from its spin axis. Furthermore, the center of its field doesn't even pass through the center of the planet! This strange situation has suggested to creationary geologists that the liquid-metal core model as the source of a planet's magnetic field can't possibly be correct.

In 1977, astronomers using ground telescopes discovered five faint rings around Uranus. Data obtained from the *Voyager 2* probe and the Hubble Space Telescope have increased the number to thirteen rings. Uranus has twenty-seven known moons. The five largest are Miranda, Ariel, Umbriel, Titania and Oberon. Eighteen of the moons orbit near the plane of the planet's equator and take part in shaping the extensive ring system. The remaining nine orbit at odd angles, and even in the opposite direction from the rest.

NEPTUNE

Magnetic field: **yes**
Diameter: **3.9 Earths**
Mass: **17.1 Earths**
Density: **1.6 g/cm³**
Gravity: **1.14 g**
Tilt: **28.3°**
Day: **0.67 d**
Year: **164.8 y**
Moons: **14**

Neptune is the eighth and most distant planet from the sun (30.1 AU). Neptune is named for the Roman god of the sea. It was discovered when astronomers noticed that something seemed to be exerting a gravitational pull on Uranus. Using Newton's law of gravitation, mathematicians calculated where the unknown planet should be. In 1846, an astronomer searched for and found the new planet less than 1° from where it was predicted to be. This was a major victory for showing the usefulness of Newton's law of gravitation.

Neptune is the third most massive planet. Except in the amount of energy it radiates, Neptune resembles Uranus. It is also nearly as cold as Uranus. Neptune's density is higher than either Saturn or Uranus, so it has the second-highest surface gravity after Jupiter.

Neptune has fourteen moons, but only two, Triton and Nereid, are observable from Earth. It has at least four faint rings. Recent observations show that at least some of the rings seem to be disappearing.

23B REVIEW QUESTIONS

1. What evidence forced astronomers to redefine the solar system objects in 2006?
2. How does Earth differ from all the other planets?
3. Which planets undergo solar transits as viewed from the earth?
4. What is a terrestrial planet?
5. Describe a Jovian planet.
6. What is the name given to the apparent backward motion of a superior planet as the earth passes it?
7. List the eight planets in order from the sun outward.
8. Which planet
   a. has the largest diameter?
   b. is the hottest?
   c. is the coldest?
   d. is the densest?
   e. is the smallest?
   f. is the brightest?
   g. has the most rings?
   h. has the largest moon?
9. What is unusual about Uranus?
10. (True or False) The planet that has the most mass also has the highest density.

23C

NON-PLANETARY OBJECTS

Besides the sun and planets, what other bodies exist in the solar system?

23.7 DWARF PLANETS

Let's review the IAU definition of a dwarf planet from Section 23B. Dwarf planets are objects that (1) orbit the sun, (2) are rigid and can be nearly spherical, (3) can share their orbits with other bodies, and (4) are not moons. Dwarf planets differ from planets in that third characteristic. A dwarf planet lacks sufficient gravity to clear other objects from near its orbit.

You know of one dwarf planet already—Pluto. Percival Lowell, a wealthy American businessman and astronomer, built his own observatory in order to search for another large planet (like Neptune) that people suspected existed farther out. But he died before finding it. In 1929, a young astronomer named Clyde Tombaugh joined the observatory staff. Very accidentally, he discovered a moving speck on photographic plates near where Lowell estimated that his "Planet X" should be. The celestial object was named Pluto.

Pluto was an oddball right from the start. The plane of its orbit is tilted more than 17° to the ecliptic, while all the others have orbital inclinations less than 7°. Its orbit is more elliptical than any other planet's.

23C Objectives

After completing this section, you will be able to

» classify non-planetary objects in the solar system.
» describe the small bodies in our solar system.
» explain where non-planetary objects may be found in the solar system.

23-18

New Horizons took this image of Pluto during its study of the dwarf planet in July 2015.

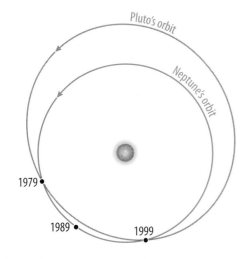

23-19
The orbit of Pluto in relation to Neptune. For about twenty years of a 248-year orbit, Pluto is closer to the sun than Neptune is.

And Pluto's perihelion is closer to the sun than Neptune's perihelion. So, for part of its orbit, it is closer to the sun than Neptune is. Pluto's density is much less than that of a terrestrial planet, but its composition seems to be more an even mixture of rock and ices. Pluto is also special because it forms a *binary* or *double planet* with its largest moon Charon, which is more than half the size of Pluto.

But then astronomers discovered several even more distant objects than Pluto. And when they discovered Eris, which is larger than Pluto at an average distance of twice that of Neptune from the sun, astronomers decided to revise their definitions of solar system objects. There are now five dwarf planets including Pluto. The first discovered was Ceres. Before the change in classification, it was the largest *asteroid*. The other three, Haumea, Makemake, and Eris, are members of the Kuiper Belt. This region is a ring of innumerable objects that extends from the orbit of Neptune to as far as 55 AU from the sun. Astronomers classify Kuiper Belt objects as *transneptunian objects* because they are found beyond the orbit of Neptune.

23.8 SMALL SOLAR SYSTEM BODIES

Besides the sun, the planets, planetary moons, and the dwarf planets, there are many small objects in the solar system. These other celestial bodies are so small that even if we could gather them together, they would not make one good-sized planet. Yet, in spite of their unimpressive size, these lightweights sometimes put on a spectacular display. If you have ever seen a comet or meteor, you have experienced one of these marvels of God's handiwork. These smaller members of the solar system generally fit into three groups: asteroids, comets, and meteors. These objects are called small solar system bodies (SSSBs).

ASTEROIDS

The largest SSSBs are the **asteroids**. Asteroids are not like planets at all. For one thing, they are much smaller. Sizes can range from the size of a house up to a third of the size of the moon. Their weak gravity can't hold an atmosphere. They can have very odd shapes, often appearing like lumpy potatoes. Their composition is irregular; many are not solid rock at all but a loose collection of boulders covered in a thick layer of gravel and dust. Some asteroids have one or more small, natural satellites.

Most asteroids have orbits between those of Mars and Jupiter. This region is known as the *asteroid belt*. It's more like an "asteroid donut." But many asteroids are not a part of the main asteroid belt. Asteroids are members of families when they have similar orbits and composition or move in groups, especially if they are not in the main belt. Two large families of asteroids are the **Trojans** and the **Greeks**. These families lie in the same orbit as Jupiter. The Greek family is ahead of Jupiter in its orbit and the Trojans are behind. These asteroids are locked into these positions because of the interaction between the gravitational pulls of Jupiter and the sun.

> The dwarf planet Ceres, formerly classified as the largest asteroid, has enough mass to form a spherical shape. In fact, its mass alone is equal to more than a third of all the asteroids combined.

> Asteroids in the **Trojan** and **Greek** families are named after characters of the opposing sides of the war described in the Greek epic poem *The Iliad*.

There are at least three asteroid families that exist in Earth's vicinity. These are *near-earth asteroids (NEAs)*. Many even cross Earth's orbit, though very few actually come close to Earth. Some people are worried that some of these NEAs could hit the earth, creating great devastation. You might be worried about that too, considering all these objects whizzing by! See the box on the next page for a discussion of how we should respond to this potential hazard. Collisions between objects in the solar system are extremely rare because of the vast distances between them.

23-20

The relative sizes and orbits of the nine largest asteroids

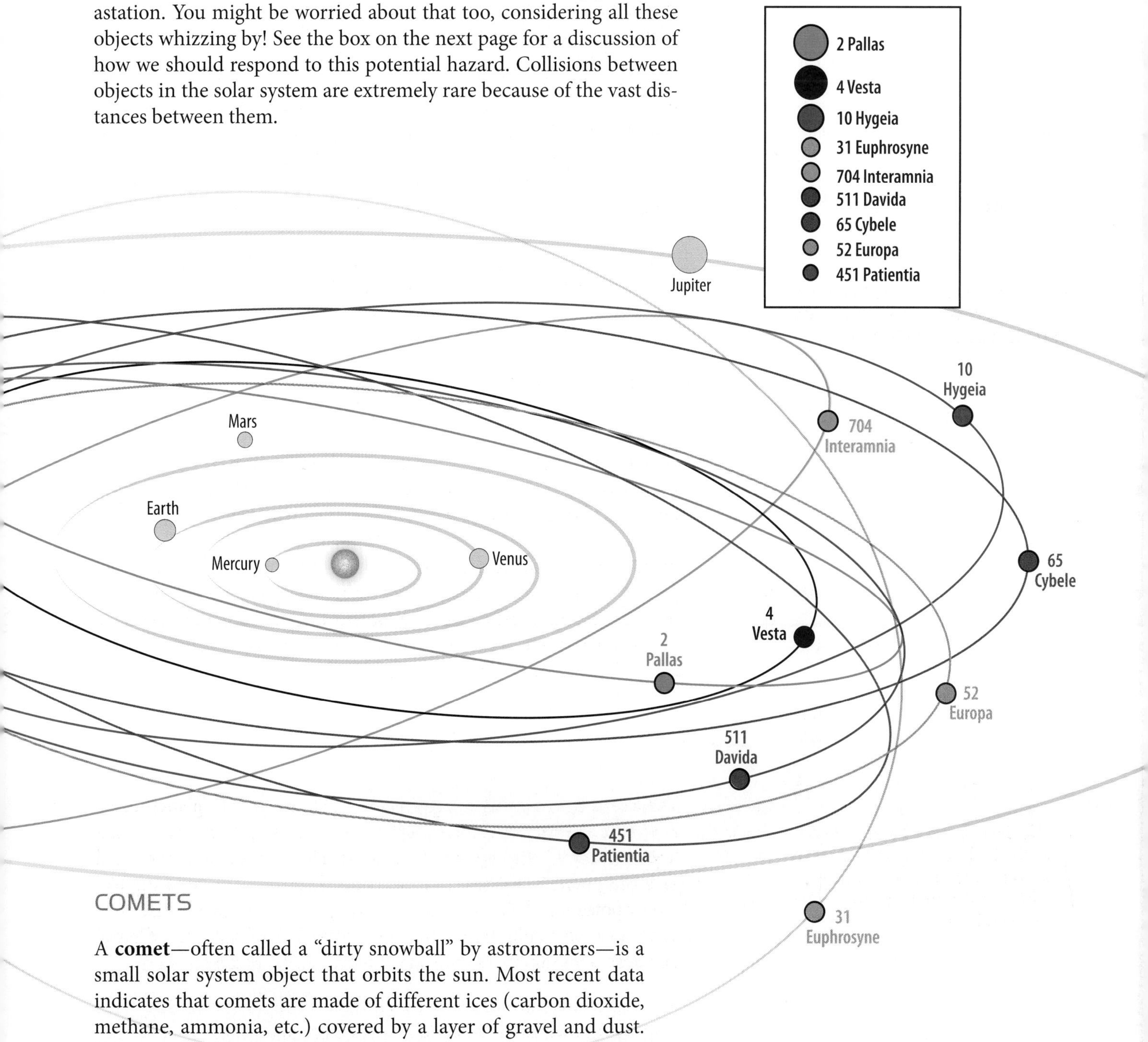

2 Pallas
4 Vesta
10 Hygeia
31 Euphrosyne
704 Interamnia
511 Davida
65 Cybele
52 Europa
451 Patientia
Jupiter
Mars
Earth
Mercury
Venus
10 Hygeia
704 Interamnia
65 Cybele
4 Vesta
2 Pallas
52 Europa
511 Davida
451 Patientia
31 Euphrosyne

COMETS

A **comet**—often called a "dirty snowball" by astronomers—is a small solar system object that orbits the sun. Most recent data indicates that comets are made of different ices (carbon dioxide, methane, ammonia, etc.) covered by a layer of gravel and dust. When a comet's orbit carries it toward the sun, the ices sublimate and then blow off as gas, carrying bits of rock and dust with it. The sunlight and solar wind push against the fine matter and carry it away, forming one or more tails that reflect the sunlight. Many comets get brighter than the planets, and some even outshine the moon for a short time. Later, when the comet moves away from the sun, the materials cool and solidify, and the comet fades from sight.

ASTEROID STRIKE!

At 7:17 a.m. on June 30, 1908, in a remote forest of Siberia, a massive explosion split the sky. It was the equivalent of a powerful nuclear bomb. "The sky was split in two, and high above the forest the whole northern part of the sky appeared to be covered by fire," said one frightened witness. "My shirt was almost burned on my body. At that moment, there was a bang in the sky and a mighty crash." The force threw him down, knocking him unconscious.

He had witnessed what is known as "the Tunguska Event." A mysterious fireball exploded in the atmosphere, knocking down trees over 2150 km² and incinerating 500 reindeer in their tracks. Horses fell over 650 km away. People heard the blast 1000 km away. A measurable change of air pressure passed around the globe twice. Tremors were felt around the world, and a reddish haze lit the skies for days.

So what was it? Years later, researchers found a circle of flattened trees. In the soil and burned resins of the trees, they found microscopic diamonds and carbon spheres that they believed formed from the intense heat of an impact. But they found no crater or meteorite. Many teams have since visited the site. They have offered some wild explanations of what hit the earth— an exploding spaceship, for example, or a piece of antimatter, or a black hole. Most meteorite specialists accept the conclusion that a million-ton comet or comet fragment, perhaps a football field in width, smashed into the atmosphere at a speed of 120,000 kph. The comet exploded in the air, leaving no crater. Modern soil studies support this hypothesis. They revealed the presence of tons of powdery material that could be the remains of the shattered nucleus of a comet. If so, this is the only historical case of a comet striking the earth.

Some of the devastation from the unexplained Tunguska event

Geologists know that large objects have struck the earth in the past because we have found at least 178 meteorite craters all over the world. Could a large impact happen again? Many scientists believe that a large comet impact near the end of the Pleistocene in North America wiped out most of the humans and large animals living there. This is known as the Younger Dryas impact, which preceded a 1000-year cooler climate period. These dates and timespans are according to the secular geologic time scale.

To avoid such a devastating event in the future, astronomers worldwide scan the skies, looking for errant asteroids and comets. They're detecting more asteroids than ever before. One such program is NASA's Near-Earth Asteroid Tracking (NEAT) program. Its purpose is to detect all asteroids more than 1 km in diameter that approach the earth. For many people, this effort isn't enough. They want to see a system of Earth- and space-based radars set up that look continually outward into space. Some call this proposed early warning system *Spaceguard*, named after a science fiction system that did the same job.

Are we in danger of being struck by an asteroid? No one knows for sure, but expert astronomers think that it is extremely unlikely. Most asteroids don't orbit in the same plane as the earth's orbit. This means that the only way an asteroid could hit the earth would be if its orbit intersected the earth's orbit. But it could happen. In recent years, several unknown asteroids have appeared out of the glare of the sun and passed the earth inside the moon's orbit.

How damaging would a direct hit from such an object be? It depends. Dozens of objects up to tens of meters in diameter pelt the earth's atmosphere each year. Astrophysicists believe that asteroids up to 100 m in diameter would burn up in the atmosphere before striking the ground. An object 100–500 m in diameter could cause significant local devastation if it hit a populated area or caused a tsunami. Objects larger than 500 m could be more destructive than the combined force of many hydrogen bombs detonating at the same time. Physicists believe that the Tunguska impact was as powerful as a 10–30 megaton bomb. Even a small body in solar orbit carries a great deal of kinetic energy because of its tremendous speed.

While astronomers believe that such impacts have a very low probability of occurring, all such events are controlled by the Lord. This is reassuring for the Christian. But we still need to learn all we can about small solar system bodies to exercise good and wise dominion over His Earth.

Comets have two main parts: the *head* and the *tail*. The head includes a *nucleus*, which is the solid part of the comet, and the *coma*, the gaseous envelope that surrounds the nucleus as the ices sublimate. The nucleus contains most of the comet's material. Its diameter is usually less than 80 km. Besides reflecting light, some gases in the coma and tail may glow or *fluoresce*, producing lovely blue or reddish tails. The tail forms only when the comet enters the inner solar system, and it points away from the sun due to light pressure.

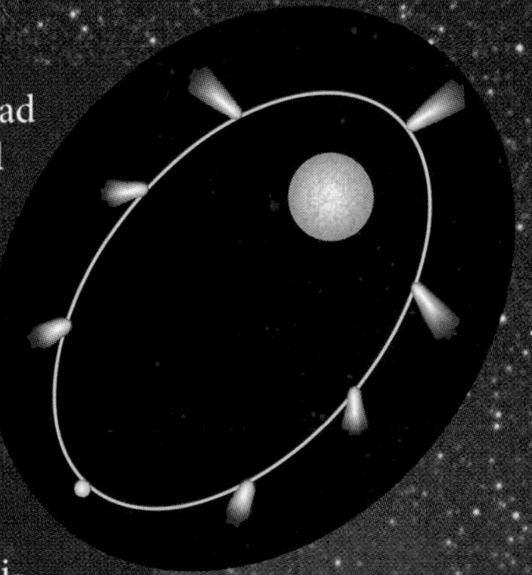

23-21
A comet's tail points away from the sun. The tail is longest when it is nearest the sun.

Ancient peoples thought that comets were supernatural visitors. One of the first men to treat comets as ordinary celestial objects was **Edmund Halley**. He noted a regular pattern in the appearance of one bright comet from historical records and successfully predicted its next appearance to be many years after he died. Halley's Comet was named in his honor. Do you want to see this comet? Its next appearance should be in 2061.

Edmund Halley (HAL ee) (1656–1742) was a prominent English astronomer who lived at the same time as Isaac Newton.

ion tail

dust tail

The parts of a comet showing the two types of tails it can develop

coma

Many comets have very long and narrow elliptical orbits, similar to Halley's Comet. Most bright comets appear at regular time intervals and are called *periodic comets*. Periods between appearances can range from many months to several centuries. Some comets seem to appear only once, but their orbital periods may be so long that we haven't recorded more than one appearance. Some deep-time astronomers believe that comets like these come from reservoirs of comet nuclei. Currently, many astronomers think that short-period comets come from two reservoirs, the Kuiper Belt (which *New Horizons* will visit) and the scattered disk. Some astronomers still think that they come from a reservoir farther out called the *Oort cloud*. They speculate that this cloud formed early as the solar nebula condensed, and its objects were never pulled into the solar system. They suggest that the gravitational field of the solar system changes as the larger planets orbit the sun. An extra-strong pull at the right time may nudge an object out of its distant orbit and toward the sun. There are many aspects of this theory that don't fit the secular model of the solar system's origin.

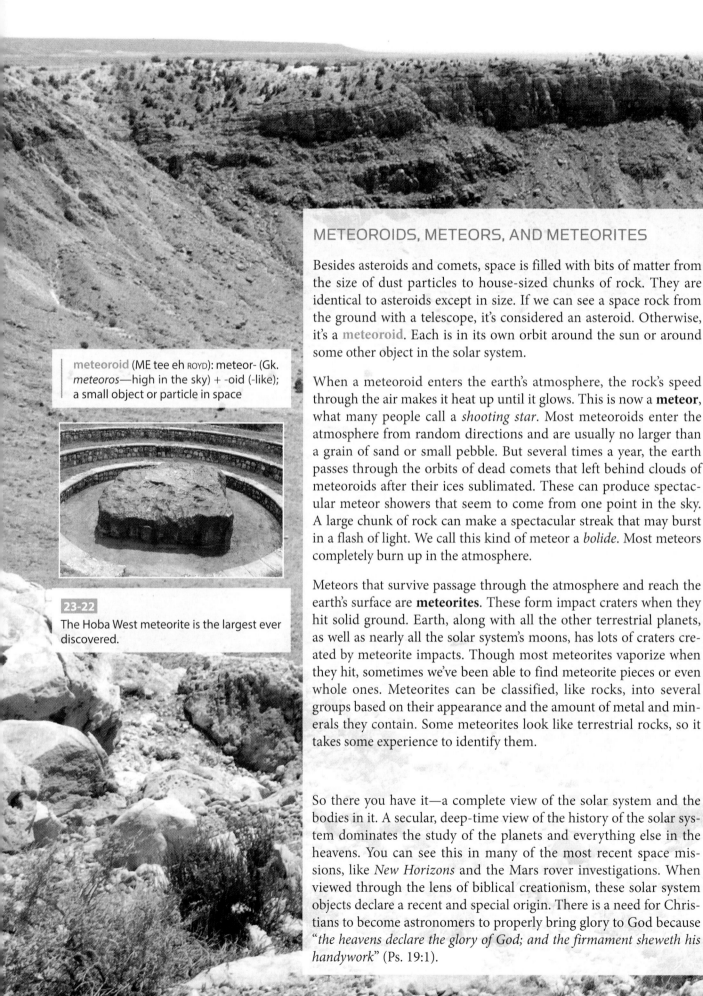

meteoroid (ME tee eh ROYD): meteor- (Gk. *meteoros*—high in the sky) + -oid (-like); a small object or particle in space

23-22
The Hoba West meteorite is the largest ever discovered.

METEOROIDS, METEORS, AND METEORITES

Besides asteroids and comets, space is filled with bits of matter from the size of dust particles to house-sized chunks of rock. They are identical to asteroids except in size. If we can see a space rock from the ground with a telescope, it's considered an asteroid. Otherwise, it's a **meteoroid**. Each is in its own orbit around the sun or around some other object in the solar system.

When a meteoroid enters the earth's atmosphere, the rock's speed through the air makes it heat up until it glows. This is now a **meteor**, what many people call a *shooting star*. Most meteoroids enter the atmosphere from random directions and are usually no larger than a grain of sand or small pebble. But several times a year, the earth passes through the orbits of dead comets that left behind clouds of meteoroids after their ices sublimated. These can produce spectacular meteor showers that seem to come from one point in the sky. A large chunk of rock can make a spectacular streak that may burst in a flash of light. We call this kind of meteor a *bolide*. Most meteors completely burn up in the atmosphere.

Meteors that survive passage through the atmosphere and reach the earth's surface are **meteorites**. These form impact craters when they hit solid ground. Earth, along with all the other terrestrial planets, as well as nearly all the solar system's moons, has lots of craters created by meteorite impacts. Though most meteorites vaporize when they hit, sometimes we've been able to find meteorite pieces or even whole ones. Meteorites can be classified, like rocks, into several groups based on their appearance and the amount of metal and minerals they contain. Some meteorites look like terrestrial rocks, so it takes some experience to identify them.

So there you have it—a complete view of the solar system and the bodies in it. A secular, deep-time view of the history of the solar system dominates the study of the planets and everything else in the heavens. You can see this in many of the most recent space missions, like *New Horizons* and the Mars rover investigations. When viewed through the lens of biblical creationism, these solar system objects declare a recent and special origin. There is a need for Christians to become astronomers to properly bring glory to God because *"the heavens declare the glory of God; and the firmament sheweth his handywork"* (Ps. 19:1).

Barringer Meteorite Crater near Winslow, AZ. Though it's not the largest crater on Earth, it's certainly visible!

Location: 35.03 °N 111.02 °W

23C REVIEW QUESTIONS

1. Why isn't Pluto classified as a planet anymore?
2. Name the five dwarf planets.
3. Between what two planets are most asteroids located? Name some other places where they may be found.
4. Identify the main parts of a comet.
5. What was Edmund Halley's great contribution to the study of comets?
6. Distinguish between meteoroids, meteors, and meteorites.
7. (True or False) Because there are so many objects whizzing around the solar system at different angles and orbits, collisions happen frequently.

SERVING GOD AS AN ASTROGEOLOGIST

Some scientists that work for NASA are practically Martians. But they are quite human. When working, they often wear two watches, one that keeps Earth time, and the other that keeps Mars time. Their job requires them to live on Martian time, which has a day 40 minutes longer than Earth. Why do they do this?

These scientists work at NASA's Jet Propulsion Laboratory (JPL). In 2004, NASA landed two robotic roving vehicles on the surface of Mars—*Spirit* and *Opportunity*. The geologists at JPL work with the rover control teams to analyze the data that the rovers collect about Mars's surface.

JOB DESCRIPTION

Scientists who do this kind of work are called *astrogeologists*. Astrogeology is a combination of astronomy and geology. Using remote sensors, astrogeologists study the rocks of planets, moons, and other objects in the solar system to learn more about their history, composition, and structure. They also examine impact craters and volcanism on rocky bodies in the solar system to learn about tectonic and meteoric processes. When available, astrogeologists can study actual rock and dust samples from meteoroids, comets, and the moon that have been obtained by manned missions and space probes.

EDUCATION

As with so many of the careers you've read about in this textbook, you can begin now to prepare for a career as a planetary geologist. Study advanced math and sciences like biology, chemistry, physics, geology and environmental science. A bachelor's degree is essential, and a master's degree is even better. Computer and communication skills help too. Being able to use GPS and GIS and having the skill to develop computer models help aspiring astrogeologists get ahead of the competition.

POSSIBLE WORKPLACES

Astrogeologists can be involved in teaching and research positions. Most work for universities, but some work for private research institutions or for government agencies like NASA or JPL. When private commercial space missions become common, probably within your lifetime, new fields of space exploration and consulting will open up.

DOMINION OPPORTUNITIES

Astrogeologists, especially those involved in exciting missions to Mars, Mercury, and Pluto, enjoy the thrill of exploration, and especially the questions that come up with each new discovery: "How did this form?" "What is the composition of that object?" "How does this finding fit into a model of solar system formation?"

Though Christians have answers from God's Word to some of these questions, we can still engage in the work of astrogeology to learn more about our neighborhood in the universe. We'll continue to learn of God's vast creativity and of why our home planet is unique (Isa. 45:18).

CHAPTER 23 REVIEW

CHAPTER SUMMARY

» Astronomers look to the *New Horizons* mission for data about how the planets evolved from the solar nebula. Christians can interpret and use this data from a Christian worldview.

» Geocentric models placed an unmoving Earth in the center of the universe with the sun, moon, planets, and stars revolving around it.

» The heliocentric (or Copernican) model placed the sun in the center of the solar system with the earth and other planets revolving around it.

» Tycho Brahe made extremely accurate and precise observations of planetary movement. His assistant Kepler used these observations to discover that the planets move around the sun in elliptical orbits.

» Galileo was the first person to use a telescope for astronomical observations. He became a strong advocate for the heliocentric model of the solar system.

» Newton developed the law of gravitation, which describes the force that holds the solar system together.

» A planet is a body in the solar system that orbits the sun, has a spherical shape, and does not share its orbital space with any other body of significant size other than moons.

» The eight planets are Mercury, Venus, Earth, Mars, Jupiter, Saturn, Uranus, and Neptune.

» Dwarf planets are objects that orbit the sun and are not moons, are spherical, and can share their orbital space with other bodies.

» Small solar system bodies include everything in the solar system that's not a planet or dwarf planet—moons, asteroids, comets, and meteoroids.

» A planet can be classified by its properties and by its position relative to Earth.

» Terrestrial planets have average densities about the same as Earth's, are made mainly of rocky materials, and have either relatively shallow, thin atmospheres or none at all.

» Jovian planets are large, massive planets like Jupiter, made of mostly light elements with deep, dense atmospheres.

» Inferior planets are closer to the sun than Earth, exhibit phases, and transit the sun.

» Superior planets are farther from the sun than Earth and exhibit retrograde motion.

» The five dwarf planets are Pluto, Ceres, Haumea, Makemake, and Eris.

» Asteroids are much smaller than planets having very weak gravitational fields. Thus, they have no atmosphere. Most orbit the sun at a distance between the orbits of Mars and Jupiter.

» Comets are small objects made mostly of frozen carbon dioxide, ammonia, methane, and dust particles. When near the sun, they develop one or more highly reflective and glowing tails.

» Meteoroids are celestial bodies orbiting the sun that are smaller than asteroids. Meteors glow white hot when they pass through Earth's atmosphere. Meteorites are meteors that strike the earth's surface, usually forming a crater.

Key Terms

solar system	560
planet	561
geocentric theory	561
heliocentric theory	563
orbit	565
ellipse	565
dwarf planet	567
moon	567
small solar system body (SSSB)	567
terrestrial planet	569
Jovian planet	569
asteroid	578
comet	579
meteoroid	582
meteor	582
meteorite	582

23-23
False color image of Saturn's rings

23-24
A Pioneer mission spacecraft heads out of the solar system.

REVIEW QUESTIONS

1. What is the ultimate purpose of the *New Horizons* mission?
2. Who wrote *The Revolutions*, and why was it important?
3. How did observation play a role in confirming the Copernican theory?
4. The heliocentric theory is generally accepted as fact today, but at first it was ridiculed. Many currently accepted facts were at first unpopular. Likewise, many readily accepted ideas were later proved false. How should a Christian react to new ideas?
5. How are the planets' orbits similar?
6. Classify each of the following as a planet, dwarf planet, or small solar system body (SSSB).
    a. Ceres
    b. Luna (Earth's moon)
    c. Mercury
    d. Halley's Comet
    e. Titan
    f. Apollo asteroids
    g. Pluto
    h. meteorite
7. Make a classification chart that relates planets, terrestrial planets, Jovian planets, dwarf planets, moons, small solar system bodies, asteroids, comets, and meteoroids.
8. How do terrestrial planets and gas giants (Jovian planets) differ?
9. Explain why Mercury does not have a significant atmosphere.
10. How have we come to know of or to discover the planets?
11. Backward-spinning planets, tipped planets, and some backward-revolving moons are problematic for the nebular theory. Explain why.
12. How are dwarf planets and asteroids different?
13. What are the differences between a comet and a meteoroid?
14. How old will you be when Halley's comet returns?
15. Why is it difficult to distinguish between asteroids and meteoroids?
16. Should Christians fear collisions of comets, asteroids, or meteors with Earth? Explain your answer.

True or False

17. According to the Ptolemaic geocentric theory, all stars are the same distance from the earth.
18. Copernicus's heliocentric theory of the solar system was simpler than the Ptolemaic system but still had some of the same problems.
19. Planets remain at the same distance from the sun as they orbit it.
20. Pluto is not a planet.
21. All the terrestrial planets exhibit solar transits.
22. All the inferior planets lack moons.
23. All the superior planets have moons.
24. Astronomers can find asteroids anywhere in the solar system.
25. Asteroids are so small that they don't exert any detectable gravity.
26. Comets are visible to the unaided eye only when they're in the inner solar system.
27. A meteor is a meteoroid that has entered Earth's atmosphere.

CHAPTER 24
STARS, GALAXIES, AND THE UNIVERSE

- 24A STARS — 588
- 24B GAS TO GALAXIES — 601
- 24C THE UNIVERSE AND ITS ORIGIN — 606

EYE IN THE SKY

High above the atmosphere's blur, a telescope soars over Earth, capturing razor-sharp snapshots of the depths of space. The Hubble Space Telescope (HST), launched in 1990, is a school-bus-sized astronomical wonder. It has produced images of objects in the universe in visible-light wavelengths and in some nonvisible ones too. Its precision and clarity make it priceless. The HST has opened our eyes so that we can better appreciate the vastness and variety of the universe. But the data it has provided has raised thousands of new questions about the nature and origin of everything.

NASA has been planning a new space telescope with even greater capabilities to replace the HST. In October 2018, NASA will launch the James Webb Space Telescope (JWST). The JWST will seek to shed light on many of the questions raised by the HST and generate new ones as we probe the depths of the universe.

24A Objectives

After completing this section, you will be able to

» find stars in the sky using constellations.

» describe ways that stars are named.

» identify and describe the common properties of stars.

» compare the sun with other stars.

» classify stars by their luminosity and color.

» describe the common remnants of stars after they die.

24-1
In 2018, NASA plans to launch the replacement to the Hubble Space Telescope, the James Webb Space Telescope. This is a model of the telescope.

24A
STARS

What have we learned about stars through observation?

24.1 LOOKING OUTWARD

The Hubble Space Telescope (HST) revealed a universe bigger than we can imagine and in far more detail than ever before: vast, shimmering nebulas, dazzling clusters of thousands of stars, exploding stars larger than a hundred suns, webs of unnumbered galaxies. Do you feel yourself shrinking?! What are we anyway? Is a person just a tiny speck of life in a vast, impersonal universe?

Though we're small, our importance in the grand scheme of things is central. God has crowned us with the honor of dominion over His creation. To help us meet this responsibility, He gave us a curious nature. We naturally want to study what He has made and find answers to questions about everything. As you learned in Chapter 22, all things in the heavens were made to serve us and to declare God's glory.

Tools like the HST do a very good job showing the immensity of the observable universe. But what can scientists do with such images? They can't test the stars or do controlled experiments with them. Astronomers build theories on the basis of what they can measure in the images, the laws and theories of physics, and their own presuppositions. But presuppositions come from our worldviews. Our interpretation of what the HST images are showing us depends on our view of the history of the universe. Keeping this in mind, let's see what we can know about the universe from what we can observe. Stars fill the sky, so we will start with them first.

24.2 CONSTELLATIONS AND NAMING STARS

God created the stars for signs and for seasons (Gen. 1:14). The heavens have drawn man's gaze from the earliest moments of history. Early observers of the skies saw pictures in the patterns of the stars, just like you might see objects in the changing shapes of clouds. In general, any recognizable pattern of stars is called an **asterism** and usually has a name. Certain notable asterisms are **constellations**.

The names of most constellations are quite ancient. Both in Job's and Abraham's times, people recognized patterns of stars as constellations. Several are mentioned by name in the Old Testament (Job 9:9; 26:13; 38:31–32; Amos 5:8). The Greeks, Arabs, East Indians, and Chinese all knew many of the modern constellations, although they gave them different names. Classifying groups of stars this way helped ancient people to track time, keep records, and navigate on land and sea.

asterism: (Gk. *asterismos*—marking with stars)

constellation: con- (L. *com*—together) + -stella- (L. *stella*—star) + -tion (about); a distinct combination of stars

24-2

The ancient constellations depicted people, animals, and objects. Often their names were related to mythological tales. Shown at right is the constellation Orion.

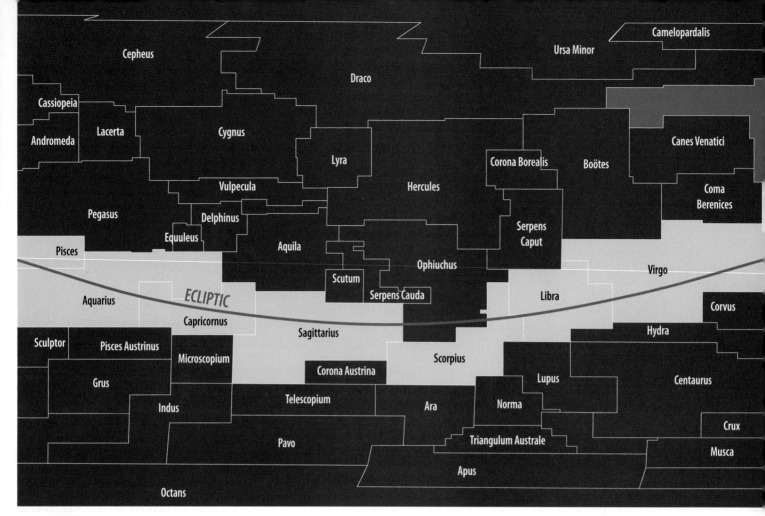

24-3 The modern constellations. The constellations of the zodiac lie along the ecliptic.

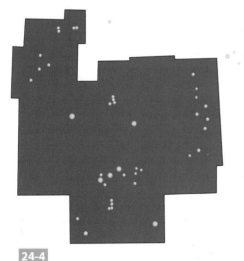

24-4 Modern constellations represent regions of space and all the stars within them. Astronomers have maintained the ancient names for the same regions of space. Note this artist's conception of the modern constellation Orion.

Early astrologers, who relied the most on constellations, illustrated them with fanciful images of humans, beasts, and other things. At the time of the Romans, there were forty-eight named constellations. Early in the twentieth century, astronomers redefined constellations as regions of the sky separated by north-south and east-west line segments for a total of eighty-eight constellations (see Figure 24-3).

Why are constellations useful? Aren't they just interesting patterns of stars in the sky? Well, think of how someone could find your house knowing only your address. They could look on a map to find your state and city, then zoom in on your street, and finally use the house number to locate the house you live in. This is why astronomers use constellations—to locate stars and other distant celestial objects.

In 1603, Johann Bayer developed a way to find stars in the sky using something like an address. Bayer assigned each star a Greek letter along with a Latin form of its constellation name. Bayer began by assigning Greek letters to the stars in a constellation in order of brightness. In his system, alpha (α) was usually the brightest star in the constellation, beta (β) the next brightest, and so on down the Greek alphabet. For example, the brightest star in the constellation Centaurus is Alpha Centauri. Some constellations had many more visible stars than could be named with the twenty-four Greek letters, so Bayer also used first the capital letters of the Roman alphabet and then the lowercase letters as needed.

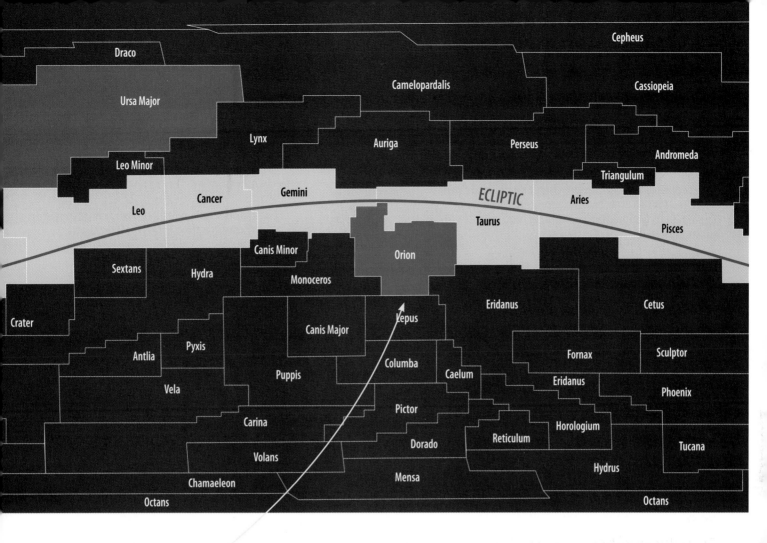

Sometimes Bayer relied on position to assign star names. For example, the star at the lip of the Big Dipper is not the brightest in the constellation—Ursa Major—but it is a logical star to start with. That makes this star Alpha Ursae Majoris. Stars can also be located using celestial coordinates, similar to latitude and longitude. See the box on the next page to learn about this system. Professional astronomers can use a variety of naming systems that are based on different catalogs like the *New General Catalogue* (see Subsection 24.6). Newer catalogs rename the stars in older catalogs with the newer system. Today, stars are named either as sequential numbers within a catalog system or by their celestial coordinates. This is necessary because the tens of thousands of stars being discovered are too many to name using the older systems.

Most stars that were visible to the ancients also have an individual or *proper name*. For example, the star at the end of the Big Dipper's handle is named Alkaid. But it is the seventh named star in Bayer's system, so it is also called Eta Ursae Majoris, because *eta* is the seventh Greek letter (η). Similarly, the star at the end of the handle in the Little Dipper (Ursa Minor, or Little Bear) has the proper name Polaris, or the North Star. But this star also has the Bayer name Alpha Ursae Minoris because it is the brightest star in that constellation.

Ursa (UR suh) Major: Ursa (L. *ursa*—bear) + Major (L. *majoris*—larger); the Greater Bear. The main asterism in the constellation is called the *Big Dipper*.

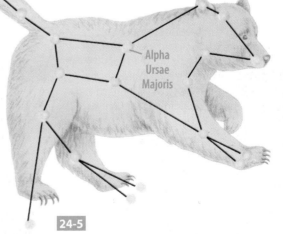

24-5

Ursa Major is a large constellation. Many people will recognize part of Ursa Major—the Big Dipper.

MAPPING THE SKIES

Get out a globe and find Paris. How did you find it? Well, you probably found Europe, then France, and then searched for the little dot that represents Paris. But if you had its geographic coordinates, you could go straight to it.

Astronomers use a similar system to find their way around the heavens. The sky appears to us like the inside surface of a sphere. The celestial coordinate grid is projected on the *inside* of the celestial sphere rather than on the outside like a globe of the world.

Let's get oriented. The celestial equator is a projection of the earth's Equator on the celestial sphere (see Chapter 3 to review projections). If you stood on the Equator, the celestial equator would be directly overhead. The celestial poles are directly above the geographic North and South Poles. To an observer on the ground, the celestial sphere appears to rotate around these points as the earth rotates.

Celestial latitudes, or parallels, are simply projections of the geographic latitude lines onto the celestial sphere. So far so good. Celestial coordinates are fixed in the heavens, just as the geographic coordinates are fixed on the earth's surface.

Celestial latitude is called *declination (DEC)* and is measured in degrees. Astronomers use a plus sign for objects north of the celestial equator and a minus sign for stars south of it.

But the earth turns under the sky, so you can't use geographic longitude for celestial longitude. Instead, astronomers chose a fixed longitude line called the *prime hour circle* for 0° celestial longitude. It is a semicircle that connects the

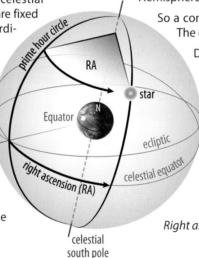

Declination (DEC)

north and south celestial poles and passes through a fixed point on the ecliptic. That point is called the *vernal equinox*, which is another name for the spring equinox. In the celestial coordinate system, the vernal equinox is in the direction that the sun lies at the spring equinox. Currently, the vernal equinox is in the southern section of the constellation Pisces. Celestial longitude marks coordinate angles to the east of the Prime Hour Circle.

Celestial longitude is called *right ascension (RA)*. It is a term left over from the very early days of astronomy and navigation. RA is measured in hours, minutes, and seconds. RA ranges from 00h to 24h, and the position of the RA coordinate is an object's *hour angle*. Hour angles increase to the east of the Prime Hour Circle (to the left as viewed from the Northern Hemisphere).

So a complete celestial coordinate is DEC and RA. The celestial coordinates for the star Sirius are

DEC −16° 42m 58.0s RA 6h 45m 8.9s

or

DEC −16.7161° RA 6.7525h (decimal).

Astronomers use celestial coordinates to find stars, galaxies, quasars, nebulas, and clusters that remain relatively fixed in the celestial sphere. This method is far more precise than just saying an object is in a certain constellation.

Right ascension (RA)

24.3 PROPERTIES OF STARS

By looking at the night sky, a person can see that stars are different. Some are bright. Some are dim. Some are reddish or bluish, and others look pure white. Classification is an important aspect of science, and classifying stars is no exception. As the Bible says, "*There is one glory of the sun, and another glory of the moon, and another glory of the stars: for one star differeth from another star in glory*" (1 Cor. 15:41). So stars are different. Many properties help us classify stars.

BRIGHTNESS

Probably the first difference you notice among stars is their brightness. A star's brightness compared with other stars as viewed from Earth is its **apparent magnitude**. Hipparchus, an early Greek astronomer, was the first to create a system for assigning *stellar* magnitude. He divided the stars into six ranks of brightness. The brightest stars were of the first rank or first magnitude, slightly dimmer stars were second magnitude, and so on. The sixth magnitude included the faintest stars that he could see. So the brighter stars have smaller magnitude numbers.

Stellar comes from the Latin word for "star" and refers to anything having to do with stars.

Today, astronomers still use Hipparchus's approach, but with changes to make it more precise. They use a mathematical system to calculate magnitude and have divided the scale into smaller steps. A mathematical system that is based on brightness measured by instruments does not depend on one person's judgment or eyesight. They also extended both ends of the scale to describe objects brighter and dimmer than those Hipparchus had considered.

Each stellar magnitude is 2.51 times brighter than the next larger (more positive) number on the scale (see Figure 24-6). An increase of two magnitudes is 2.51 × 2.51 = 6.3 times brighter. As astronomers refined Hipparchus's scale, they realized that some stars were too bright to be included with other first-magnitude stars. So former first-magnitude stars and very bright objects like the sun and the moon now have magnitudes near zero or even negative values. At the other end of the scale, objects far fainter than those Hipparchus could see with just his eyes have magnitudes from +7 to more than +30.

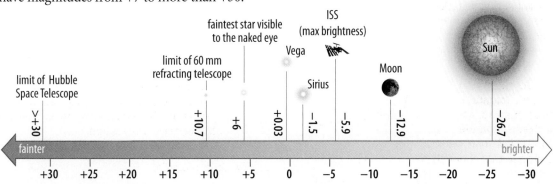

24-6

The modern stellar magnitude scale

The apparent brightness of a star depends on two main factors—distance to the star and its *absolute magnitude*. Distance reduces the light intensity. A distant car headlight appears dim, but up close it can be too bright to look into. The same thing happens to stars. A very bright but very distant star can have a lower apparent magnitude than a closer but less bright star. **Absolute magnitude** is the apparent magnitude that a star would have if we placed it at a standard distance. Absolute magnitude is a measure of the actual brightness of a star. This way, the star is not so bright that its glare alters its appearance. Other factors can reduce the apparent brightness of a star, such as the presence of dust or gas in between that scatters the starlight. Knowing absolute magnitude allows scientists to calculate the distance to the star.

The apparent magnitude of the sun is −26.7. Its absolute magnitude is +4.83.

Some unusual stars change in brightness regularly because they expand and contract. The first *variable star*s to be observed were the type called **Cepheid variables**. As they grow larger, they emit more light and thus become brighter. When they shrink, they emit less light and become dimmer. The time these variable stars take to go from one luminous maximum to the next is related to their absolute magnitude. Knowing their actual brightness, astronomers can compute how far away the stars are using basic rules of optics. Thus, variable stars are useful as "celestial yardsticks." More than anything else, we owe our knowledge of the size of the universe to Cepheid variables.

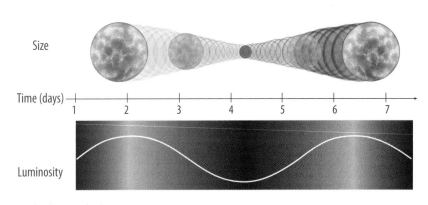

24-7
Cepheid variable stars were first discovered in the constellation Cepheus. They are brighter when they expand and dimmer when they contract.

DISTANCE

It's difficult to appreciate how far away stars really are. The nearest star, Proxima Centauri, is 40.2 trillion km away! When you're talking about distances like these in units of kilometers or miles, it leaves your mind numb. One unit for distances between stars is the **light-year (ly)**. A light-year is the distance that light travels in a year, about 9.5 trillion km. Astronomers typically use an even larger unit called the parsec. A **parsec (pc)** is equivalent to 3.26 ly, or 31 trillion km. A distance of 10 parsecs is the standard distance at which apparent magnitude of a star is measured. So Proxima Centauri is 1.30 pc (4.24 ly) away. You can see that 1.30 pc is easier to work with than 40,200,000,000,000 km!

"Nearby" means longer distances to an astronomer than to most people. If the distance to a star or other object is only a few light-years, then the object is "close." Some objects appear to be billions of light-years away. These are "distant"—billions of trillions of miles away!

24-8
This is the Sombrero galaxy, which is about 15,300 parsecs across. On this image, 1 mm represents 110 pc.

So how do astronomers calculate the distance to a star? For relatively "close" stars, they use a method similar to the way you judge distances with your own eyes—parallax. **Parallax** is the apparent change in position of a distant object as one's viewing position changes. Figure 24-9 illustrates how parallax works as you view a scene from one eye and then the other. The distance between your eyes causes the difference in views.

24-9
When viewing a pencil against the "star" on the board with one eye and then the other, the pencil seems to shift position with respect to the distant "star."

In astronomical parallax, astronomers can see a change in the apparent position of a star as observed through a telescope from different positions in the earth's orbit. Using geometry, they can compute the distance to the star. This method works only for objects less than about 100 parsecs away. Objects farther than that show no noticeable parallax shift.

24-10
Stellar parallax. The circles show telescopic views of the same star from opposite sides of the earth's orbit.

Astronomers use variable stars for measuring distances to relatively nearby galaxies. For more distant stars and for galaxies far away from our own, astronomers use a mathematical rule. This rule relates how the light spectrum we see from them shifts toward the red end of the visible spectrum with increasing distance from Earth. This property of distant starlight is called *red shift*. The box on page 600 discusses how this rule is used.

MOTION

You learned in Chapter 22 that the earth's rotation makes all the stars seem as if they are moving in circular paths around the poles. But this is just apparent motion caused by our change in viewpoint on a rotating Earth. Parallax shift is another kind of apparent motion. But do stars actually move? Yes. Just as the sun orbits the center of the galaxy, so do the other stars in our galaxy as they follow their own orbits. We can often observe this motion because nearby stars orbit faster or slower than the sun's motion.

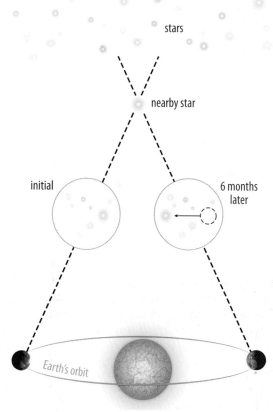

When astronomers look out into the sky, they say that they are looking along an imaginary *line of sight*. When stars move toward or away from Earth along the line of sight, they display *radial motion*. Radial motion is determined by observing the shift of light toward the red end or the blue end of the spectrum. The Doppler effect (see the box on page 481)—the apparent change in frequency and wavelength due to a moving source or detector—causes this shift. *Red shift* indicates the rate of motion away from us and *blue shift* tells us the rate of motion toward us. The star's motion across the line of sight, or *proper motion*, is more challenging to measure because it requires many observations over long periods. By combining the star's radial motion and proper motion, scientists determine the star's *real motion*.

24-11 A star's real motion in relation to the earth is a combination of radial motion and proper motion. Barnard's star (below) shows proper motion from 1950 to 2010. This star is only about 6 ly away, and it has the largest proper motion of any known star.

24.4 CLASSIFYING STARS

Astronomers classify stars on the basis of important properties of the stars themselves—a combination of *luminosity, temperature, color, size,* and *composition.* Remember from Chapter 22 that most stars, including our sun, are made of about three-quarters hydrogen, a quarter helium, and a small percentage of heavier elements. We know this from spectroscopic analysis of a star's photosphere (see the box on page 600). The proportions of these elements affect the temperature and color of the star's surface. But the amount of matter or mass in a star is the most important factor in determining its classification. Large hot stars that are mostly hydrogen are blue or blue-white. Medium mass stars or those with a larger percentage of helium are yellow to orange. Cool stars are either very small or have a large percentage of helium and other heavier elements. These stars tend to be orange to red. In practice, astronomers divide the span between two letters into tenths from 0 to 9. Lower numbered stars in a class are hotter. For example, our sun is a class G2 star. Betelgeuse (BEET el jooz), a huge red star, is classified as an M2. Sirius is a hot white A1 star.

Luminosity is the main property used in classification. It depends on the temperature and size of the star. It is a physical property closely related to absolute magnitude. Luminosity depends on both the size and temperature of a star. Large, massive stars are much more luminous than smaller stars at the same temperature. On the other hand, small, high-temperature stars are less luminous than large, cooler stars because the larger stars have more surface area. Astronomers may indicate a star's luminosity with a Roman numeral on the scale I, II, III, IV, or V, with V being the most luminous. Our sun is a G2V type star.

Luminosity and color may help classify stars by size, though there is no reliable relationship. Brighter, hotter stars often are very large, but bright, cool stars can be large as well. Betelgeuse is a red, cool star 500 times larger than the sun. Very large stars like these are called **supergiants**, with diameters greater than 100 suns. Stars can come in all sizes and luminosities, including *giants*, *subgiants*, *subdwarfs*, and *colored dwarfs*. Our sun is a largish, yellow dwarf star. Though it appears quite average, it serves a special purpose. It is exactly what is needed for life on Earth.

The possible combinations of stellar temperature, color, size, and luminosity seem to be endless. But when astronomers plot all this data on a graph, they can see interesting relationships. The **Hertzsprung-Russell (H-R) diagram** (Figure 24-13) shows one way of organizing this information. Certain kinds of stars seem to clump together on the diagram. The long diagonal band through the center contains the majority of stars. For this reason, astronomers call it the **main sequence**, and stars with these characteristics are *main-sequence stars*. It starts up in the top left corner with hot, bright, blue, massive stars, and ends in the lower right corner with cool, dim, red, low-mass stars. But what about those other groups of stars?

24.5 STELLAR AGING AND DEATH

Astronomers know that stars give off light and other forms of energy from nuclear fusion in their cores, just like our sun. Main-sequence stars use hydrogen to form helium and a few other light elements. But when a star uses up its hydrogen fuel, physicists believe that some important changes take place in a relatively short period of time. The kinds of changes depend on the mass of the star.

24-12
Relative sizes of stars

24-13
The Hertzsprung-Russell diagram illustrates the relationships of star properties. The diagram is named after the two astronomers who developed it about 1910.

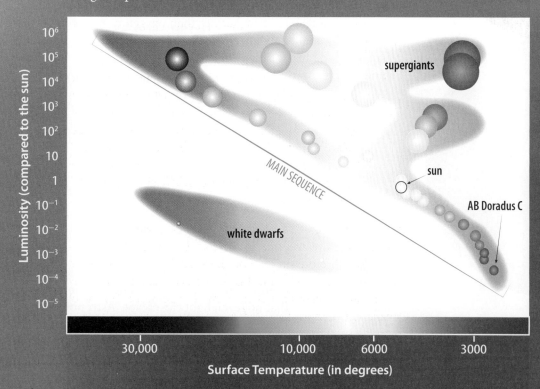

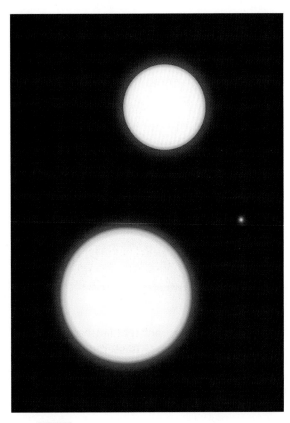

24-14

Three stars are shown above: at the bottom is IK Pegasi A, at the top is our sun, and IK Pegasi B—a white dwarf—is the small dot in the center.

nova (L. *nova*—new). Most novas first appear as a new star in the sky.

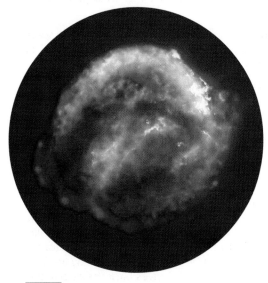

24-15

This is a NASA image of the Kepler Supernova Remnant taken in the x-ray, optical, and infrared wavelengths.

The mass of a star is usually measured in relation to our sun's mass. So a low-mass star has at least one-tenth the mass of our sun, and can be up to four-tenths of a solar mass ($m_\odot$)—the mass of our sun. This unit is used to estimate the masses of stars, nebulas, and galaxies. The sun's mass is 332,950 times the mass of the earth. Astronomers believe that old, very-low-mass red dwarfs at the lower-right end of the main sequence just slowly burn out after a long time.

As a star about the size of the sun begins to run out of hydrogen, it starts using helium in the fusion process. The extra heat produced causes the star to expand into a **red giant**. These stars are above and to the right of the main sequence in the H-R diagram. As the helium is used up, the star shrinks slightly and becomes more luminous. When the helium in a red giant of less than 8 solar masses is gone, the outer layers of plasma blow off as an expanding shell of glowing gas and dust called a *planetary nebula*. The dead core of the star collapses into a *white dwarf*, which we find in the lower left region of the H-R diagram.

A **white dwarf** can be half to four times Earth's diameter but contains nearly the same mass as our sun. It has no source of energy, so it slowly cools and dims to a *black dwarf* over time. But if a white dwarf is a member of a binary star system, and it can pull matter from its companion star, eventually it may gain enough hydrogen to ignite fusion on its surface. This causes a violent nuclear explosion, producing a temporary bright flash called a **nova**.

Stars larger than about 8 solar masses have a quick and violent death. After using up their helium, physicists believe that stars like these expand to form bright blue, white, or yellow supergiants with lots of heavier elements in their interiors. These are stars located at the top of the H-R diagram, off the main sequence. The heat from the fusion of this matter can cause some stars to expand and shed their outer layers. As fusion between elements heavier than oxygen or nitrogen occurs, several things could happen, depending on the mass of the star. They usually explode in a brilliant flash, forming a **supernova**, and leave behind a white dwarf. Very large stars are usually destroyed by the blast.

If a white dwarf's mass is more than about 1½ solar masses, it will collapse into a city-sized **neutron star**. Neutron stars quickly stop shining, but matter falling into the dark stars, pulled by their extreme gravity, can be heated until the matter glows. Neutron stars spin rapidly, and some emit narrow beams of radio waves. If the beam sweeps past Earth, the object seems to pulse in radio waves. Because of this, these objects are called *pulsars*.

Physicists believe that if the supernova leaves a remnant with more than 2–3 solar mass, its collapse won't stop at a neutron star. It will continue falling inward, forming a really strange object called a **black hole**. A black hole is an unimaginably dense object. The crushing gravity is so intense that not even light can escape. Astronomers can't see black holes, but they can observe their effects. As the loose matter in space around one falls into it, the matter can emit radiation in all bands of the electromagnetic spectrum.

24A SECTION REVIEW

1. Scientifically, what is one of the biggest limitations of studying stars?
2. How do astronomers identify the general regions of the sky where stars may be found and named?
3. How did Bayer name the brightest or main star in a constellation?
4. How does the apparent brightness of stars change as their stellar magnitude numbers increase?
5. What is the difference between the apparent magnitude and the absolute magnitude of a star?
6. What unit of distance is more convenient for measuring the distance to stars than miles or kilometers?
7. What three methods do astronomers use for determining the distance to stars and other objects in interstellar space?
8. What are the two main properties that astronomers use to classify stars?
9. What do the groups of stars *not* on the H-R diagram main sequence represent?
10. Using Google Sky™, another sky-viewing program, or a star chart, find the constellation Orion. Give the historical name for the star Alpha (α) Orionis.
11. (True or False) The light from Proxima Centauri that we see today started toward Earth 4.2 light-years ago.

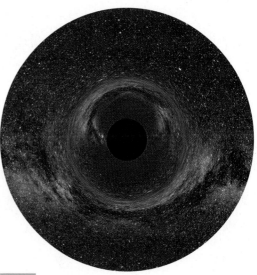

24-16

Though astronomers can't see black holes, they can see their effects, such as *gravitational lensing*, shown in this artist's illustration. Gravity is so intense that it bends light like an optical lens.

WORLDVIEW SLEUTHING: EXTRATERRESTRIAL INTELLIGENCE

In May 2015, a group of Russian scientists received a radio transmission. That in itself wasn't a shocking event, except that it came from an unexpected place. The scientists believe that the signal came from a solar system 29 pc away. That means that there must be intelligent life out there, right?

INTRODUCTION

You are a young astronomer working for NASA and your supervisor has assigned you to research this historical event and prepare a presentation.

TASK

Produce a presentation for NASA on the existence of extraterrestrial life. The presentation should be three minutes long and cover extraterrestrial life as well as the search for extraterrestrial intelligence and radio transmissions received from space, including the May 2015 transmission received by the Russians.

PROCEDURE

1. Read the Astrobiology box on page 605.
2. Do background research on extraterrestrial life using "extraterrestrial life" as your search term.
3. Do background research on current work being done in the search for extraterrestrial life by doing a keyword search for "search for extraterrestrial intelligence" or for "SETI."
4. Do research on the recent radio transmission from space by doing a keyword search for "A SETI Signal 2015" or "space messages 2015."
5. Plan your presentation and collect any photos that you need, being careful to give proper credit. Try to give equal space to each of the three topics.
6. Produce your presentation and show it to another person for feedback.

CONCLUSION

Many people have wondered about the possible existence of extraterrestrial life. Though it is possible that there is life on other planets, it is probably limited to nonintelligent life if it exists at all. Many people who reject the Bible's story of Creation wonder whether life was planted here by life from other planets. But that simply moves back the question—where did *that* life come from?

FINGERPRINTING STARS

Stars are really, really far away. This makes them difficult to analyze because all we can do is look at them. But astronomers have a tool as useful as a detective's fingerprint kit. It's called a *spectroscope*.

The light that stars produce comes from the way that the hot elements glow in their photospheres. Each kind of element glows with a particular color and usually in several colors. Spectroscopes are instruments that can break down the light from stars into distinct colors. From this information, called *spectroscopy*, astronomers can tell what elements are present in the atmosphere of the star. So how do spectroscopes work?

Spectroscopes separate wavelengths of light by using a thin, transparent plate called a *diffraction grating*. This plate has thousands of fine grooves etched on its transparent surface. The grooves cause the light waves to bend, and the amount they bend depends on their wavelengths. Short wavelengths of light toward the blue end of the spectrum are bent the least; the longer wavelengths toward the red end are bent the most. The grating spreads the colors out into a spectrum by wavelength.

Knowing this information, what would we expect a star's spectrum to look like? Stars are made mostly of hydrogen, which, when heated, emits only specific wavelengths of light. We may expect that the spectroscope would show only those specific wavelengths. Then again, stars are made not only of hydrogen, so we could also expect to see fainter lines for helium, sodium, iron, nitrogen, carbon, and oxygen, which are all present in stars. However, due to the density and pressure with stars, we don't see these specific wavelengths.

A star's photosphere produces light at nearly all visible wavelengths. But as this light streams outward through its chromosphere, atoms of elements there absorb the wavelengths that they would normally produce in a bright-line spectrum.

Hubble Space Telescope took this image of Sirius A and B. The small star to the lower left is the white dwarf Sirius B.

This is a basic property of all incandescent matter. The *emission spectrum* of a hot element is the same as its *absorption spectrum*. So Earth-based observers using solar spectroscopes actually see dark lines against a colorful *continuous spectrum*. This absorption spectrum is also called a *dark-line spectrum*.

Dark-line spectra are therefore very useful for analyzing a star's composition. But they can also be used to tell the distance to a star and how fast it is moving toward or away from us! The wavelength of each line in the spectrum for an element is known very precisely. But astronomers routinely find all the lines of a given element to be shifted to longer (redder) or shorter (bluer) wavelengths in proportion to the wavelength of each line.

There are two ways to explain the *red shift*. The star could be moving away from us, so the star's motion makes the wavelengths we see coming from the star seem longer (and redder). This is caused by the *Doppler effect* (see the box on page 481). The greater the speed of the star away from Earth, the redder it appears (or bluer if the star is moving toward us). The other reason for the red shift could be that the light waves were actually stretched out because of changes in the fabric of space. This is called *cosmological red shift*. Secular astronomers believe that this occurred as the universe expanded after the big bang. Creationist astronomers have a similar explanation but from a different cause. The astronomer Edwin Hubble was able to develop a rule that equated the red shift of the stellar spectra to the distance of the stars from Earth. The greater the red shift, the greater the distance.

Spectroscopy in astronomy is an important research tool. With the development of inexpensive digital cameras, even amateur astronomers can participate in this fascinating field.

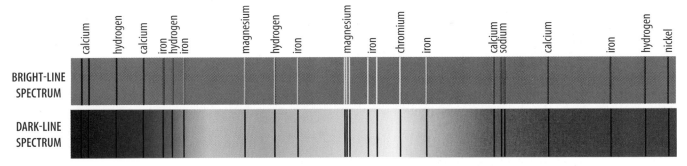

The top image shows a bright-line (emission) spectrum, which is produced as elements in a light source emit specific wavelengths (thin bright lines). The bottom image is an absorption (dark-line) spectrum, which is formed as white light passes through gases of different elements. Notice that in both images, the bright and dark lines align because particular elements will absorb the same wavelengths that they emit.

24B

GAS TO GALAXIES

Besides stars, what else is in the universe?

24.6 NEBULAS

24B Objectives

After completing this section, you will be able to

» describe and classify various objects in the universe.
» differentiate between a double star and a binary star.
» identify and classify various kinds of star clusters.
» summarize the history leading to the recognition of what a galaxy is.
» classify galaxies by their shapes and sizes.

If you look at the night sky with your unaided eyes, the stars appear to be individual points of light. But are they? And if you look at the belt of the constellation Orion, you see a smudge of light hanging like a sword in its sheath. What is that? It looks like a stationary comet, but comets move. Charles Messier was concerned that objects that clearly weren't stars might be confused with new comets that also looked like glowing splotches before they developed tails. In 1781 he drew up a catalog of 103 of these objects and assigned each one a number beginning with an "M" to identify it. Nebulas were one common kind of object in his catalog. A **nebula** is a cloud of dust or gas that can be seen against a background of stars. Messier also cataloged clusters of stars and other non-stellar objects that looked like faint smears of light.

In 1887, another astronomer, Johan Dreyer, published his first *New General Catalogue (NGC)*, listing over 13,000 nebulas and other non-stellar objects. Dreyer assigned his own number to each of Messier's objects, and many professional astronomers often prefer to use Dreyer's NGC numbers.

There are many kinds of nebulas, and astronomers group them by their appearance or their origins. Most seem to be simply clouds of dust or gases that fill space among the stars. Some like the Orion Nebula (M42) reflect or glow from nearby starlight. Others appear as black smudges of smoke against a bright field of background stars. Many are planetary nebulas or supernova remnants.

24-17
The Orion Nebula (in the lower left corner) that forms the "sword" hanging from Orion's belt above is named *M42* in Messier's catalog.

STARS, GALAXIES, AND THE UNIVERSE

24.7 DOUBLE AND BINARY STARS

Take a close look at the middle star of the Big Dipper's handle. On a clear, dark night, you should be able to see that there are actually two stars close together. The brighter is Mizar and the dimmer is Alcor. Such close pairings of stars are called *visual double stars*. But are they actually double stars? Does gravity hold them together?

Astronomers must carefully observe double stars, usually for many years, to answer this question. If the stars in a visual double actually do revolve around each other, then they are called a **binary star**. Astronomers believe that most of the stars in our galaxy are true binaries. The brightest star in the sky, Sirius, is a binary star with a white dwarf.

The first true binary star to be discovered was Algol, a bright star in the constellation Perseus. A larger orange *companion star* orbits the smaller blue-white *primary star*. It regularly comes between the primary and Earth, like an eclipse. The regular dimming of Algol is a property of all *eclipsing binary stars*. Most binary star systems are not arranged so that they form eclipsing binaries when viewed from Earth. More recent observations have shown Algol is actually a three-star system.

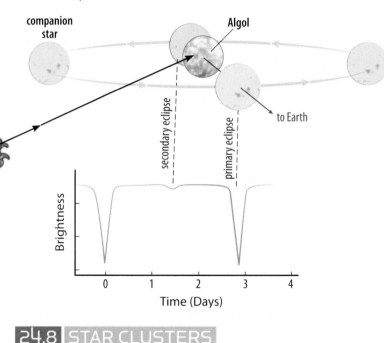

24-18
Algol is one of the stars that mark Medusa's head in the constellation Perseus.

24.8 STAR CLUSTERS

Some fuzzy objects Messier cataloged were not nebulas, but actually groups of many stars. Since his time, astronomers have discovered thousands of star clusters. A **star cluster** is a group of stars that are close enough to be held together by gravity and that have the same proper motion. A cluster is *not* considered an asterism since the stars in asterisms are generally not affected by each other's gravity. The stars in an *open cluster* are relatively far apart and distinct. An open cluster has no definite shape. The Pleiades (M45) in the constellation Taurus is a bright open cluster visible to the unaided eye.

A *globular cluster* is a spherical region of space filled with thousands of closely spaced stars. The stars in a globular cluster are so close together that they appear like a glowing blur in low-power telescopes. High-power telescopes can see individual stars. Star clusters are common in our galaxy and have been found in other nearby galaxies.

24.9 GALAXIES

Early observers of the heavens saw a blotchy band of light that stretched across the sky. To many it seemed to be a glowing path among the stars. It has many names in the languages of the world, but most European cultures call it the *Milky Way*, which originally came down in history from the Greek name for this feature. Apart from mythology, Greek philosophers believed that it was the glowing breaths from stars where the earth touched the celestial spheres.

Galileo proved with his telescope that the Milky Way actually was made of innumerable faint stars. In 1750, the English astronomer Thomas Wright first suggested that the Milky Way that people called the **galaxy** might be a vast, flattened disk of stars. If so, he thought, then perhaps the large, fuzzy nebula in the constellation Andromeda and the Magellanic Clouds (visible only in the Southern Hemisphere) might also be distant masses of stars. Some astronomers began calling these disk-shaped nebulas *island universes*, but the majority believed that they were located within the Milky Way Galaxy. Until well into the twentieth century, most people thought that the Milky Way *was* the universe. After much debate, Edwin Hubble finally observed variable stars in these nebulas that were too distant to be within the Milky Way. Improved telescopes showed that there were numerous galaxies and that the universe was much larger than anyone had imagined.

Modern astronomers estimate that there may be more than 2 *trillion* galaxies in the visible universe. And these may be separated by an average of many millions of light-years of nearly empty space. These numbers are truly mind-blowing!

24.10 CLASSIFYING GALAXIES

Galaxies are loose masses of many billions of stars, dust, gases, and all the other objects we just mentioned. They seem to be held together by gravity, and they rotate around a center of mass. Many astronomers believe that the central source of gravity in galaxies and other large groups of stars may be immense black holes. Or it could be something that we have never observed. Some astronomers believe that galaxies may be held together by something called *dark matter*. The center of the Milky Way Galaxy is shrouded in great clouds of dust, so we can't directly see what lies there. Infrared images of the galactic center show a monstrous, hot, glowing object there. Most astronomers believe that it is a black hole equal to 2 million solar masses! Our galaxy is about 100,000 ly in diameter and 2000 ly thick at its center. There may be 100–400 billion stars in the galaxy.

24-19

Globular cluster M80 is located 10,000 pc from Earth and could contain 500,000 stars.

The name *Milky Way* and the word *galaxy* are closely associated in Greek mythology.

galaxy (Gk. *galaxias*—milky). The Greeks called the Milky Way *kyklos galaktikos*—the milky circle.

24-20

This is an image of our Milky Way galaxy using infrared wavelengths. Astronomers now think that the Milky Way is a barred spiral galaxy. The central bar seen in this image is 27,000 ly long.

STARS, GALAXIES, AND THE UNIVERSE 603

The Hubble Space Telescope confirmed that galaxies are numerous and fill the visible universe.

Astronomers classify galaxies by their shapes. Galaxies like the Milky Way and Andromeda are flattened disk shapes with bulges in the middle. Dusty arms loaded with stars twist away from the center like a pinwheel. These are called *spiral galaxies*. Spiral galaxies can have two or more arms. Many spirals have a bar-like thickened region connecting opposite arms. These are called *barred spiral galaxies*. Most astronomers agree that the Milky Way is of this kind. Galaxies that look like smooth, glowing disks or flattened spheres of stars are *elliptical galaxies*. The largest galaxies known are ellipticals. *Irregular galaxies* don't have any regular pattern or shape. They contain lots of dust and gases in between stars. The two Magellanic Clouds in the sky are irregular dwarf galaxies that orbit the Milky Way. Astronomers have also discovered galaxies that don't easily fit into any other category (e.g., ring galaxies, colliding galaxies, etc.).

24-21
Spiral Galaxy M51 (above) is 12 million ly away from Earth. NGC 1316 (below left) is an elliptical galaxy and I Zwicky 18 (below right) is a dwarf irregular galaxy.

LIFE CONNECTION: ASTROBIOLOGY

Life in space—is it a question for scientists or for science fiction writers? Sounds more like science fiction! However, NASA has been studying astrobiology since 1959 and the last time we checked, NASA has more scientists than science fiction writers.

NASA has funded several exobiology projects including the Search for Extraterrestrial Intelligence (SETI) in 1971 and the *Viking 2* mission in 1976. Even today, many tasks of the *Curiosity* rover on Mars relate to exploring for possible life on Mars. The European Space Agency (ESA) and the Russian Federal Space Agency have planned missions in 2018, while NASA has a mission planned for 2020.

Why do scientists think that life could exist beyond the earth and its atmosphere? Probabilities provide the basis of one line of reasoning. As we already know, the universe, even just the observable universe, is huge. A scientist by the name of Frank Drake developed a formula to predict the number of extraterrestrial worlds that we could find. By looking at habitability zones and other conditions, some scientists believe that there could be 10,000 intelligent forms of life in the universe.

Evidence from Earth itself has helped convince scientists that life could exist in the harsh environment of space. Remember the box on page 180 on the Pompeii worm? The worms live in the extreme conditions of the black smokers at the bottom of the oceans. Scientists have found microbes living deep in the ice of the Arctic. Scientists reason that if life can survive in the extremes of Earth, then why couldn't it exist in space?

So what evidence have we found? Most of the evidence relates to the presence of water. Since water is necessary for life as we know it, most astrobiological searches have been for water. The thinking is that if there is water, there could be life. We have found evidence that Mars and some asteroids may have water. While we have found water, we have yet to find evidence of life, present or past.

NASA continues the study of astrobiology as one focus of its space program. Various rovers like the one pictured below look for signs of life on Mars.

24-22
The quasar HE 1013-2136 (inset), estimated to be 10 billion ly away

This is an artist's rendering of the accretion disk of quasar ULAS J1120+0641. It is assumed that a black hole, with a mass of 2 billion solar masses, is at the center.

24C Objectives

After completing this section, you will be able to

» differentiate between the two competing cosmogonies.

» discuss evidence used to support the Big Bang theory of cosmogony.

» summarize various attempts to solve the starlight/distance problem in a creationist theory of cosmogony.

» validate the significance of Earth and humans in a vast universe.

QUASARS

Astronomers can't always explain everything they observe! *Quasi-stellar objects (QSO)*, or **quasars** (KWAY zars), are just one example of this. In the 1960s, radio astronomers discovered strong radio emissions from an area with no bright stars. They isolated the emissions to visually faint, starlike objects. Studying their light, scientists decided that the objects must be moving away from Earth at astonishing speeds and be very, very far away. But more importantly, they were amazingly bright. Their absolute magnitudes were greater than an entire galaxy's, but they appeared like stars! Further research has indicated that they may be very energetic centers of distant galaxies surrounding a monster black hole. Sky surveys have discovered more than 200,000 quasars. They remain poorly explained residents of our universe.

24B REVIEW QUESTIONS

1. What was the original reason for identifying and naming nebulas?
2. Are all double stars binary stars? Explain.
3. How can astronomers tell whether groups of stars actually form a star cluster?
4. Until the twentieth century, how far did most people believe the observable universe extended?
5. What is a galaxy? What kind of galaxy is the Milky Way?
6. Are quasars related more to galaxies or to stars? Explain.
7. (True or False) Galaxies are held together by centrifugal force.

24C

THE UNIVERSE AND ITS ORIGIN

How did the universe form?

24.11 EVIDENCE IN THE UNIVERSE

Before we begin, let's consider two closely related terms—**cosmology** and **cosmogony**. We studied cosmology earlier in this unit—the study of the structure of the universe, including the structures of stars, planetary systems, and galaxies. Now we will consider cosmogony, which refers to the study of the universe and its origin.

All scientists collect evidence, create hypotheses, and work to develop theories, hoping to advance knowledge in their field. The evidence that they gather is the same regardless of worldview. However, their worldview will affect their hypotheses, theories and interpretation of the data. So let's look at the evidence related to the universe and its origin.

RED SHIFT AND AN EXPANDING UNIVERSE

People have studied the stars from the earliest days of history. Scientists have learned about stars just by observing them with their eyes. As we developed new tools—telescopes and satellites—we have learned about galaxies, nebulas, pulsars, and quasars. While these tools have answered many questions, they have raised many more. We learned in Section 24A that we can identify the elements in stars by the light that we observe.

As scientists looked farther into space with their advanced instruments, they noticed something interesting. Vesto Slipher, an astronomer at the Lowell Observatory, noted that the light from galaxies appeared to have longer wavelength than expected. He observed that light shifted toward the red end of the spectrum. This shift is similar to the *Doppler effect*, the apparent change in frequency due to a moving source or detector, which we have discussed before. Unlike the Doppler effect, **cosmological red shift** is the apparent lengthening of the wavelengths due to the expanding universe, not due to the motion of the stars themselves.

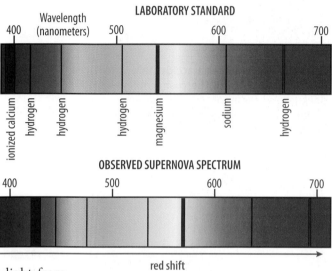

24-23
The spectrum of a nearby star (top) compared with the red-shifted spectrum of the same star if it were at a great distance (bottom)

This discovery had a huge effect on astronomy. If the light from almost all objects in space is red shifted, then they must all be moving away from the earth. This means that *the universe must be expanding*.

Scientists understood that the universe was expanding, but there was more to come. Edwin Hubble studied the red shift of galaxies, wanting to know whether the shift amount related to the distance from the galaxy. He created a model, now called *Hubble's law*, which showed greater red shift for objects that were farther away. These calculations demonstrated that the farther the object was from Earth, the faster it was moving away.

Hubble's law allowed astronomers to find distances beyond those that stellar parallax allowed. As astronomers built better telescopes, they looked farther and farther out into space. Hubble's law allowed them to measure the immense distance to those galaxies. Today, astronomers have calculated the observable universe to extend 4.22 billion parsecs (13.77 billion light-years) from Earth.

COSMIC MICROWAVE BACKGROUND RADIATION (CMBR)

Scientists continue to build better telescopes and other instruments so that they can look deeper into space. Scientists wondered whether the light from farthest out into space would be red shifted the most. Ralph Alpher and Robert Herman hypothesized that this light would no longer be visible light, but would be shifted to the microwave or even radio wave regions.

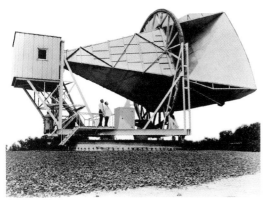

24-24 This microwave antenna accidentally discovered the existence of the cosmic microwave background radiation while engineers were trying to identify radio noise in telephone systems.

On May 20, 1964, scientists at the Bell Telephone Laboratories were conducting experiments in radio astronomy and satellite communication. They received an unexpected microwave signal. They discovered **cosmic microwave background radiation (CMBR)**—the predicted long-wavelength light from the edge of the observable universe. Astronomers continue to study this radiation and observe CMBR in all directions. Initially the radiation appeared evenly spread throughout the universe.

STELLAR AGING

In the universe overall, we see a tremendous variety of stars. We see huge supergiants and tiny dwarf stars. We see yellow, blue, and red stars. We see what appear to be young and old stars. Through the process of fusion, stars consume one element and form others. Knowing how much of each element is present in the stars, astronomers can estimate their age. In other words, *stars age*. But the way that the stars, galaxies, and universe have changed over time and how long these changes have taken depend on one's interpretation of the evidence. And interpretation is based on worldview.

24.12 THE BIG BANG THEORY OF COSMOGONY

The discoveries of red shift and expansion of the universe led scientists down a familiar road. Just as geologists followed slow-moving tectonic plates back from our current continents to Pangaea, astronomers followed rapidly moving light back from the galaxies today to the big bang. Georges Lemaître first introduced the concept of the **Big Bang theory** in 1927.

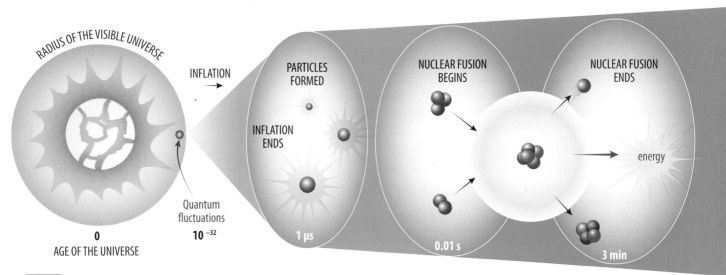

24-25 This is a timeline of the Big Bang theory. To the left is the beginning of the big bang when the universe was super hot and dense. Notice that at 380,000 years, the light that caused the CMBR is emitted.

This theory states that the universe began 13.77 billion years ago as a singularity—all the universe condensed in a single, infinitely small, dense, super-hot state. Initially everything in the universe was energy and then the universe began to expand. As it expanded and cooled, subatomic particles formed, followed by nuclei of light elements (hydrogen and helium). The universe expanded farther, allowing neutral atoms to form. At about 380,000 years, the universe spread far enough that light (photons) could remain for extended periods. According to the nebula hypothesis, the universe continued to cool, allowing stars to form. Over billions of years, the universe changed to its present form.

TESTING THE BIG BANG THEORY

Red Shift

As we have discussed throughout this book, the power of a model is demonstrated by its workability. Does a theory predict things that we can observe? Obviously, the Big Bang theory implies an expanding universe and red-shifted light, both of which fit nicely with observations. But how does CMBR factor into the old-earth story?

CMBR

CMBR is a prediction of the Big Bang theory that scientists have also observed. According to the theory, as the early universe expanded, it was initially very dense. The singularity was so dense that all matter immediately absorbed any emitted *photons* (light). However, once the universe expanded sufficiently, its density would be low enough that light (photons) could exist without being absorbed. Scientists believed that they should be able to detect this radiation from this very early stage of the universe. They expected the radiation to be evenly distributed. They also expected it to be from the farthest points of the universe and therefore be red shifted into the microwave or radio wave spectrums. The discovery of the CMBR was a validation of the Big Bang theory for the secular scientists.

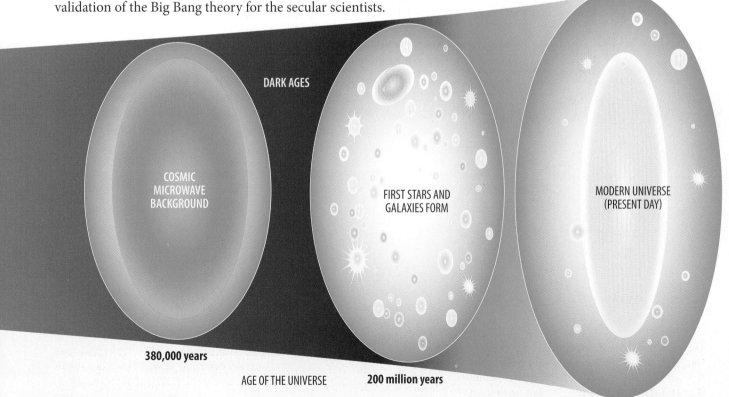

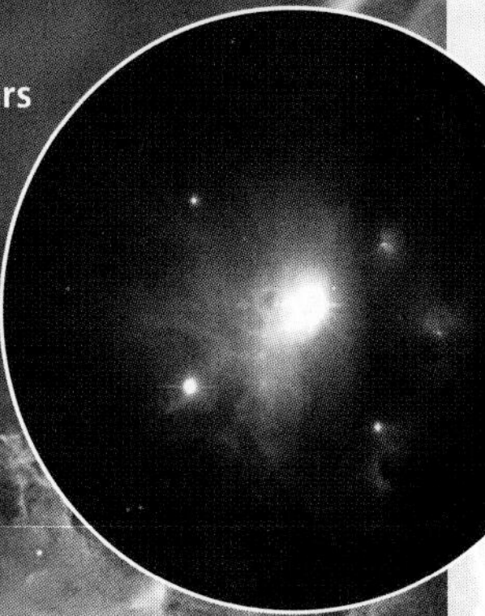

24-26
The three reddish stars to the right in this image are protostars, while the bright white stars are thought to be young stars.

Stellar Aging

Stellar aging closely connects with the Big Bang theory. According to the theory, everything in the universe formed out of initial singularity. As the universe expanded and cooled, some of the matter condensed, according to the nebular hypothesis. This matter formed a *protostar*—a hot, dense collection of matter that became a star once nuclear fusion began. These stars formed with different initial masses, placing them in different regions of the HR diagram. As stars used their fuel, they transitioned along the HR diagram until their stellar death. This process of stellar birth, aging, and death is referred to as **stellar evolution**.

EXTENDING THE BIG BANG THEORY

As new data comes in, scientists must adjust the models, hypotheses, and theories that they have developed over time. *Dark energy* and *dark matter* presented the need for two modifications to the old-earth story of the universe. They were not, as many think, attempts to fix problems in the Big Bang theory.

As scientists study an expanding universe, they must answer some questions. Does the universe expand at a constant rate? Is the expansion slowing or is it accelerating? Some scientists expected the expansion to slow due to gravity. However, what does the data say? The data indicates that the rate of expansion is accelerating. In order for matter to accelerate outward, there must be something pushing outward that is stronger than the gravity pulling inward. Scientists hypothesize that *dark energy* provides the outward push.

Dark matter, on the other hand, more closely connects to the structure of the universe. Scientists know that gravity is the force that holds everything together. Scientists observed that galaxies had more mass than they could see with electromagnetic radiation. Therefore, they hypothesized that there must be matter—**dark matter**—that we cannot observe by standard methods. Current observations agree with their concept of dark matter. Today scientists at the Large Hadron Collider (see Chapter 2) continue to try to observe the particles that make up dark matter.

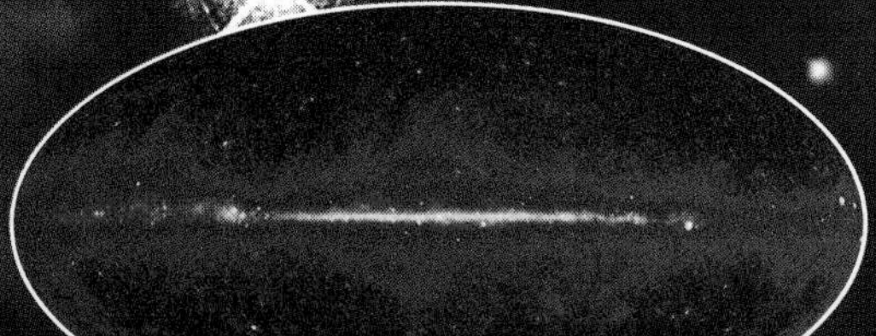

24-27
The photo above shows the Milky Way galaxy with the locations of proposed concentrations of dark matter in purple.

ANALYZING THE BIG BANG THEORY

Secular cosmologists must answer a major question: Where did stars and galaxies come from? Scientists understand that gravity holds stars and galaxies together. Additionally, they hypothesize that gravity pulled all matter together according to the nebular hypothesis. However, if everything was moving apart from the big bang, what triggered the initial collapse of matter to form stars and galaxies?

24-28
This is an all-sky picture showing the distribution of the CMBR. NASA developed this image from nine years of data from the Wilkinson Microwave Anisotropy Probe (WMAP). Notice that the CMBR is not uniform.

Another related question has to do with the CMBR. When scientists first observed the CMBR, it appeared evenly spread throughout the universe, as predicted. However, as instruments have improved, they have begun to see variations in the radiation. Think of an astronaut approaching the earth from space. Initially the earth appears as a smooth sphere, but as he gets closer, he begins to see mountains and valleys. The closer scientists look at the CMBR, the less uniform it appears. They now hypothesize that the variations in the CMBR explain how regions of space may have condensed into stars and galaxies. They believe that the denser regions in the CMBR match up with where these structures formed. However, the scientists don't know what caused these variations themselves.

The third question still facing secular cosmologists is the matter of the universe's end. The Big Bang theory's explanation of the universe's beginning raises this question. The fate of the universe depends on the gravity and the expansion force—dark energy. If gravity dominates, then the universe will stop expanding and collapse back in on itself. If the two forces balance, then the universe will just keep expanding forever. If the dark energy dominates, then the universe will expand faster and faster. Scientists at NASA have a space probe—the Wilkerson Microwave Anisotropy Probe—collecting data to try to answer this question.

DANNY FAULKNER, ASTRONOMER

Just like a telescope, the Bible brings astronomy into focus for Dr. Danny Faulkner. He takes his Bible seriously, seriously enough to swim against the current of secular worldviews in his field.

Dr. Faulkner taught at the University of South Carolina-Lancaster for twenty-six years. He now conducts research and writes as a member of Answers in Genesis. He has published several papers and books on creationism, physics, and astronomy. He spends many nights a year studying the sky from a local observatory. His specialty is stars, especially binary stars. He has studied the skies for as long as he's studied the Bible.

When he was six years old, Faulkner turned to Christ for salvation. His father was the minister of a small local church. As he grew both in the Lord and in his knowledge of astronomy, he continued to apply the Bible to his field of interest. He received a bachelor's degree in mathematics, and then went on to complete a master's degree in astronomy and physics. His education culminated in a doctorate in astronomy.

Faulkner sees his work in astronomy as God's calling on his life. He enjoys the opportunity to use the gifts and abilities that God gave him to advance our understanding of the universe. While many astronomers view the scientific evidence from a naturalistic worldview, Faulkner sees God's glory in the heavens. Moreover, he takes every opportunity to influence people to see the universe from a biblical worldview. He views the universe as a beautiful place that is the work of God, a Supreme Artist. At the heart of this universe is Earth, incredible and unique in its design to support life and to serve God's image bearers.

24.13 CREATION THEORIES OF COSMOGONY

Creation week was a week of miracles. God's Word says, *In the beginning God created the heaven and the earth* (Gen. 1:1). God created the universe on Day 1. On Day four, God created the sun, moon, and stars. *And God said, Let there be lights in the firmament of the heaven to divide the day from the night; and let them be for signs, and for seasons, and for days, and years* (Gen. 1:14). We know that God created the heaven and the earth in six days. Using biblical genealogies, young-earth scientists understand that Creation occurred about 7000 years ago. The challenge is to propose and test explanations within a biblical worldview that fit the scientific evidence.

TESTING, EXTENDING, AND ANALYZING CREATION THEORIES

Cosmogony is a relatively new field of study for young-earth scientists. Because of this, creationists are testing and refining several theories to see which one best explains the evidence. *The biggest challenge for creationist astronomers to solve is the starlight travel problem.*

Here's the problem. Scientists know that light travels at the speed of light through space. Astronomers believe that the farthest star that we can currently observe is 13.77 billion light years away. If light travels at the speed of light, then it should take 13.77 billion years for that light to arrive from that distance. How can light travel for 13.77 billion years in a universe that is only 7000 years old? To solve this problem, creationist scientists have proposed a number of possible solutions.

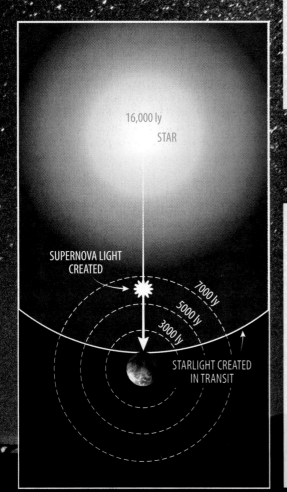

Light Created in Transit

This explanation assumes that God created all heavenly bodies on Day 4 around 7000 years ago. The universe then was as big as it appears to be today. Our original parents saw the stars immediately because God created the light rays already arriving at Earth, no matter the star's distance from Earth.

One of the problems with this idea is that it suggests that we really can't trust our senses to provide us reliable information since the starlight we see didn't come from real events. That's a problem for explaining stellar aging, since the universe hasn't existed long enough for the emissions of aging stars to reach us moving at the speed of light. It is possible that the stars were created in every stage of life: from what seem to be hot, "young" stars (according to a secular view of astronomy) to cool, very "old" stars. But this problem really comes to *light* when you consider a nova. In order for us to observe a nova from a star 16,000 ly away, the light we observe couldn't have left the star when it exploded.

Time Slowed by Gravity

This might sound odd to you, but experiments have shown that the strength of gravity can affect how fast time passes. A model called the *White Hole cosmology* suggests that during the Creation week God created the universe as a huge ball of matter centered on the earth. The intense gravity at Earth's location slowed time. Then God separated matter from Earth when He created the firmament. He created the stars and galaxies from the waters above the firmament (Gen. 1:7) and then spread them out. Billions of years passed out in the universe during Day 2 through Day 4 on Earth because distant gravity was far less than at Earth. This is one of the most popular creation theories of cosmogony because it allows light from stars to communicate real data and events. This theory explains red shift (image right), CMBR, and stellar aging.

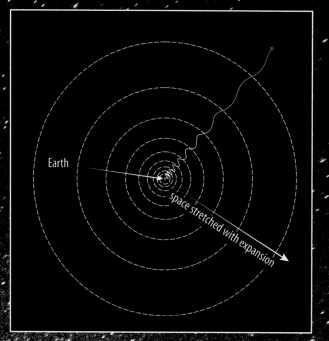

Dasha Solution

Danny Faulkner (see box on page 611) proposed the third solution—the Dasha solution. This response proposes that during Creation week, God used accelerated processes in His creative work. For example, plants could have been created as seedlings that grew. But they grew fast enough to provide Adam with food on Day 6. Similarly, the sun, moon, and stars may have been created in a way that allowed the actual light to travel by rapid supernatural processes so that they could fulfill their role "for seasons, and for days, and years" (Gen. 1:14). This theory also explains the evidence in the universe.

24-29 Supernova 1994D in galaxy NGC 4536 is 58.76 million ly away. The Dasha solution suggests that God may have sped up the light from this supernova during Creation week so that we could observe it in 1994.

It is important to remember that both the secular and creationist astronomer have the same evidence to work with as they try to explain the origin of the universe. Both have struggled to understand the evidence that we observe. We all have to decide to put our faith in something or someone else. Our worldview determines what we judge to be a worthy object of our faith. A Bible-believing Christian concludes that God's Word is the only authoritative source of truth that is worthy of belief.

24.14 OUR PLACE IN THE UNIVERSE

Let's take a trip to the edge of the universe. You are traveling in an advanced-design spaceship that can travel at nearly the speed of light—300,000 km/s. You whiz past the moon in about 1 second. About eight minutes later, you are at the sun. Four more hours of travel takes you past Neptune's orbit. To get to the closest star, Proxima Centauri, would take over four years. After about 25,000 years of travel, you could reach the center of the Milky Way Galaxy. If you wanted to travel outside our galaxy to the Andromeda Galaxy, it would take 250 million years to get there! From there, you could view the entire Milky Way. Traveling for 1 billion years would give you a panoramic view of the universe around the Milky Way. At this distance, you are still less than 10% of the distance to the edge of the observable universe. Our home galaxy would be part of a cluster of more than thirty galaxies called the Local Galactic Group. In addition, this group is a very small part of a network of trillions of galaxies that fill the universe.

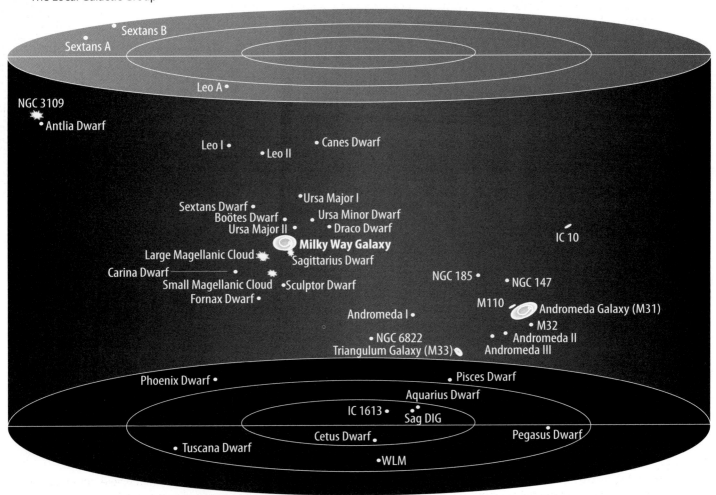

24-30 The Local Galactic Group

Surrounded by all this is the fragile earth, home to small, feeble man and all the life known in the universe. How important are we in such a huge place? The universe isn't infinitely large to show how insignificant we are—it is huge to show how great our Creator God is. God designed Earth uniquely for life and then made man in His image (Gen. 1:26). Through the work of the Holy Spirit, He made regenerate people His dwelling place (John 14:16–18). By God's perfect plan we are the center of all creation (Ps. 8:3–6).

24C REVIEW QUESTIONS

1. Assuming that the universe was created, what is one likely reason for its immense size?
2. What do we call the study of the origin of the universe? What is the starting point for people who work in this area of knowledge?
3. What are the two kinds of evidence used by secular cosmologists to justify a very large universe?
4. Why does a large universe suggest a very old universe?
5. Both cosmological red shift and the CMBR are the result of what prediction of the Big Bang theory?
6. Why does the phrase "stellar evolution" mean something different from "biological evolution?"
7. What do Christians know for certain about the creation of the universe?
8. Which creationist cosmogony seems to do the best job of explaining the evidence for a large, old universe in view of the young creation described in the Bible?
9. What is the key difference between secular and Creation-based cosmogony?
10. (True or False) We can assume that the universe operated according to today's natural laws throughout Creation week.

"Stand still, and consider the wondrous works of God." (Job 37:14)

BIBLICAL ORIGINS: THEISTIC EVOLUTION

Theistic evolution is another attempt to bring the Bible into agreement with scientific theories. It is the most drastic compromise between belief in Creation and naturalism. Those who believe this theory propose that God used naturalistic processes to create things. Supposedly, God initiated biological evolution by creating matter and/or life. He then allowed it to follow its "natural course."

This view twists the entire Biblical history of God's Creation (not just the Genesis account) in order to make God's words fit a theory. Many defenders of this position believe that the literal biblical narrative of Creation is a myth. Some say it is a fictional story that is based on Babylonian fables. Others claim that it is a form of symbolic poetry.

Theistic evolutionists place themselves in a logical dilemma. We should not be so quick to question a natural reading of the Bible. Theistic evolution's main flaw is the same as that of all the other views of origins besides the six-day literal view: death before sin. The theory of evolution is a theory that progresses through death. In the Bible's story, the presence of death and suffering trace back to one historical event—Adam's fall. Jesus, the second Adam, has come to rescue us from sin and the death that it brings.

Questions to Consider

1. Does theistic evolutionary theory agree with what the Bible actually says?
2. If theistic evolution were true, then where did sin come from, and why do we need a redeemer?
3. Do non-Christians accept theistic evolution as a valid theory?

Key Terms

Term	Page
constellation	589
apparent magnitude	593
absolute magnitude	593
Cepheid variable	594
light-year (ly)	594
parsec (pc)	594
parallax	595
supergiant	597
Hertzsprung-Russell diagram	597
main sequence	597
red giant	598
white dwarf	598
nova	598
supernova	598
neutron star	598
black hole	598
nebula	601
binary star	602
star cluster	602
galaxy	603
quasar	606
cosmology	606
cosmogony	606
cosmological red shift	607
cosmic microwave background radiation (CMBR)	608
Big Bang theory	608
stellar evolution	610
dark matter	610

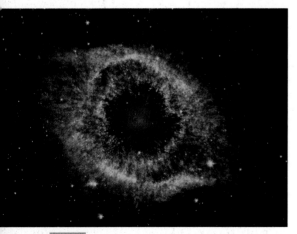

24-31 The Helix, NGC 7293, is a planetary nebula located in the constellation Aquarius. It is one of the planetary nebulas that are closest to the earth.

CHAPTER 24 REVIEW

CHAPTER SUMMARY

» A constellation is a specific area of the sky named for a pattern of stars, or asterism. Astronomers have subdivided the sky into eighty-eight constellations.

» Stars can be named by a variety of methods. Most visible stars have very ancient names.

» Important star properties include apparent and absolute magnitude, luminosity, distance, motion, color (temperature), and size or mass.

» Stars are classified by their color and their luminosity. These properties place them somewhere on the Hertzsprung-Russell (H-R) diagram.

» Most stars fit into the main sequence on the H-R diagram. Other regions of the diagram contain stars in advanced stages of stellar life.

» Stars change as their hydrogen is used up in fusion. When all its hydrogen is consumed, the star "dies" in a way that depends mainly on how much mass it has.

» Remnants of dead stars include planetary or other kinds of nebulas, white dwarfs, neutron stars, and black holes.

» Nebulas are immense clouds of gases and dust. They have different appearances depending on their composition and the presence of stars to illuminate them.

» Binary stars are pairs of stars that are gravitationally linked. Usually one orbits around the other. A majority of stars in the galaxy are members of binary pairs.

» Star clusters are groups of stars that are held together by gravity. They orbit or lie within galaxies.

» Galaxies are loose masses of stars, dust, and gases. Astronomers classify them as spiral, barred spiral, elliptical, or irregular on the basis of their shape. Other kinds also exist.

» Quasars are extremely luminous objects that are brighter than an entire galaxy but are much smaller in size. Most seem to be very distant.

» The universe is huge, with lots of empty space. Yet man has purpose and meaning because he is central to God's plans for the universe.

» People use cosmogony to try to describe how the universe came into existence and how it changes over time.

» The universe seems very old and very large. The Big Bang theory is widely accepted as the way the universe came into existence. As interpreted by secular cosmologists, much observational evidence seems to support this theory.

» The Big Bang theory does not agree with the biblical account of the origin of the heavens. Therefore, Christians must look for a different explanation that is based on the Bible.

» Christians do not have to develop a detailed cosmogony that meets the standards of secular science since it is clear that the universe was miraculously made under conditions that cannot be studied using scientific methods.

REVIEW QUESTIONS

1. Why is the HST such a good tool for exercising biblical dominion?
2. What is the main reason for using constellations in modern astronomy?
3. Which is brighter, a star with a magnitude of –0.3 or one with a magnitude of +0.3?
4. Why are parallax shifts for estimating the distance to stars of limited usefulness?
5. What are the two kinds of motion that describe stars' real motion?
6. For stars on the main sequence of the H-R diagram, what two properties have the most influence on luminosity and color class?
7. How does the sun compare to other stars in mass, color, and luminosity?
8. What condition in a star leads to stellar death?
9. According to our models for stellar aging, what is the main difference between the final stages of a star of 1 solar mass and one of 10 solar masses?
10. What is the difference between a black hole left over from a star and a galactic black hole?
11. What gave astronomers the first clue that true binary stars existed?
12. What are the differences between asterisms, constellations, and clusters?
13. What much larger objects did many of Messier's nebulas turn out to be?
14. Is the Milky Way the largest galaxy in the universe? Justify your answer.
15. How far away from Earth is the center of the Milky Way?
16. How do astronomers use red shift to determine distance to a star or galaxy?
17. For what reason did cosmologists propose the existence of dark matter?
18. Why is a Christian cosmogony that assumes that light was created in transit a problem for a large, young universe?

True or False

19. The brightest star in a constellation is always given the Bayer designation *alpha*.
20. The apparent magnitude of the sun is greater than its absolute magnitude.
21. Most stars have a composition similar to that of the sun.
22. Astronomers agree that all known black holes are supernova remnants.
23. Double stars are pairs of stars that appear close together in the sky.
24. The Milky Way is an elliptical galaxy.
25. The idea that the universe has always existed is not supported by many important scientific laws.
26. A Christian cosmogony has no problem with the formation of protostars from a nebula.
27. What secular astronomers call stellar evolution is just changes in stars over time.
28. Placing photos of stellar objects in a sequence from protostars to dusty disks, to stars, to supergiants, to white dwarfs is the same thing that paleontologists do when they line up a series of fossils to demonstrate biological evolution.

The horse's head in the horsehead nebula is a dust cloud blocking the bright red light from an emission nebula.

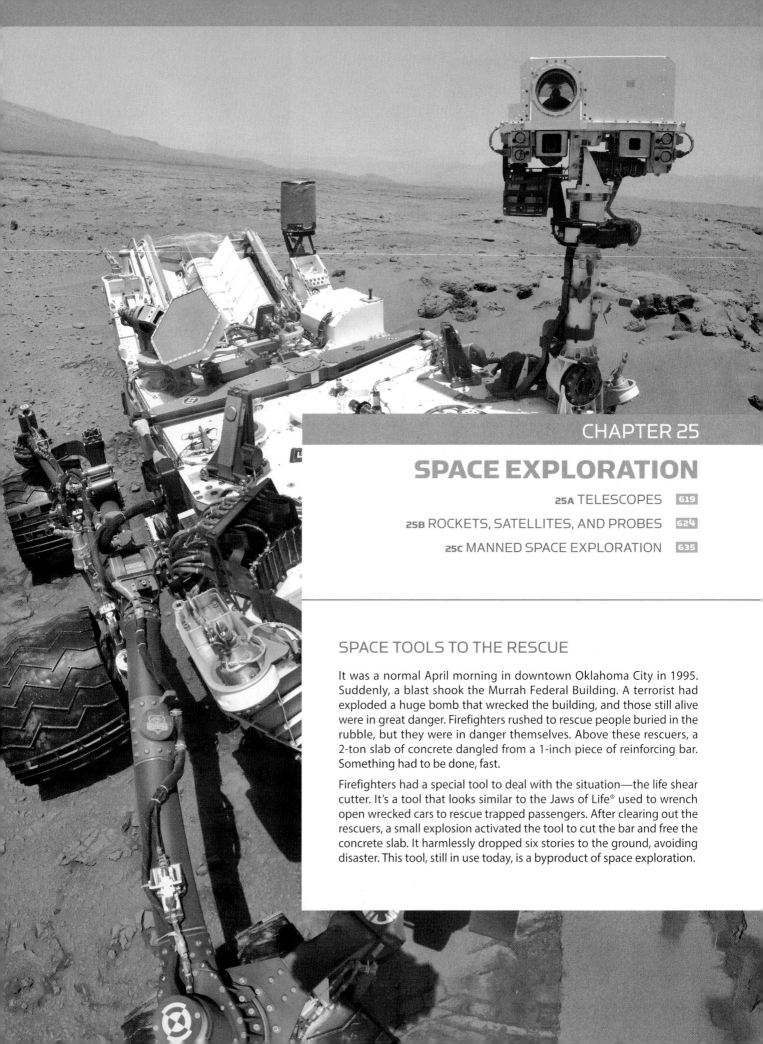

CHAPTER 25
SPACE EXPLORATION

25A TELESCOPES 619
25B ROCKETS, SATELLITES, AND PROBES 624
25C MANNED SPACE EXPLORATION 635

SPACE TOOLS TO THE RESCUE

It was a normal April morning in downtown Oklahoma City in 1995. Suddenly, a blast shook the Murrah Federal Building. A terrorist had exploded a huge bomb that wrecked the building, and those still alive were in great danger. Firefighters rushed to rescue people buried in the rubble, but they were in danger themselves. Above these rescuers, a 2-ton slab of concrete dangled from a 1-inch piece of reinforcing bar. Something had to be done, fast.

Firefighters had a special tool to deal with the situation—the life shear cutter. It's a tool that looks similar to the Jaws of Life® used to wrench open wrecked cars to rescue trapped passengers. After clearing out the rescuers, a small explosion activated the tool to cut the bar and free the concrete slab. It harmlessly dropped six stories to the ground, avoiding disaster. This tool, still in use today, is a byproduct of space exploration.

25A

TELESCOPES

How do telescopes work?

25.1 TECHNOLOGY IN THE SPACE AGE

If you ever watched a recording of a shuttle launch, you might have noticed small flashes of light at the base of the boosters during the lift-off from the launch platform. The cutting tool used in the Oklahoma City bombing rescue is based on these explosive hold-down bolts that were used to release the shuttle from Earth (see Figure 25-1). But the life shear cutter isn't the only technology brought to us through space exploration.

The household smoke detector was originally made for the first American space station. Padding in the helmets of football players was originally developed by NASA for aircraft passenger seats. Materials for artificial limbs are the same as the foam insulation used by the space shuttle's external fuel tank. Cordless power tools were invented for Apollo astronauts. In many ways, the same technology used for space exploration allows us to better exercise biblical dominion here on Earth by helping people and saving lives.

The heavens have always attracted the attention of people. God created the heavens to serve us (Gen. 1:14–18) and to declare His glory and majesty. As early civilizations developed observational tools, space exploration in the form of astronomy began. In 1609, Galileo became the first to turn an optical instrument to the heavens. *Telescopes* are instruments that form enlarged images of distant objects, making them appear closer. The telescope revolutionized the study of the heavens. Let's take a look at the different types and the ways that astronomers use them.

25A Objectives

After completing this section, you will be able to

» evaluate the importance of space exploration.

» classify telescopes by their structure.

» explain the function and limitations of various kinds of telescopes.

25-1
Technology like the explosive hold-down bolts (inset) for the space shuttle led to life-saving devices like life shear cutters (below) used by rescuers.

SPACE EXPLORATION 619

25-2

A refracting telescope uses an objective lens to gather light and an eyepiece to magnify the image produced by the objective lens (see diagram below). The world's largest refractor, with a 40 in. objective, is located at Yerkes Observatory in Williams Bay, WI.

25.2 OPTICAL TELESCOPES

Optical telescopes form images by gathering visible light using lenses or mirrors. There are two main parts of an optical telescope: the eyepiece and a main lens or mirror that work together to gather, magnify, and focus light. Optical telescopes are divided into two main types according to how light is gathered before it passes through the eyepiece.

REFRACTING TELESCOPES

Galileo's telescope was a refracting telescope. The lenses in a **refracting telescope** refract or bend light to produce an image. A simple refractor contains two lenses, the *objective lens* and an *eyepiece*. They are mounted in a light-proof tube to keep them aligned. The objective lens gathers the light and directs it to the eyepiece. The larger the objective lens, the more light it can gather and the more detail you can see. The ability to separate fine details in an image is called *resolution*. Resolution refers to how small an angle the telescope can see. The diameter of the telescope determines the resolution. Telescopes with a larger-diameter objective lens have a better resolution. The job of the eyepiece is to magnify the image produced by the objective lens. Telescopes with better resolution have larger objectives, not eyepieces producing higher magnification.

Refracting telescopes have two limitations—size and color distortion. Glass objective lenses larger than about 1 m in diameter are so heavy that gravity bends the lens, distorting the images. Color distortion happens in all refracting telescopes. This distortion happens because the lenses refract different wavelengths of light at slightly different angles. This causes each color to focus at a slightly different location. This problem can be partially corrected by combining two or more special lenses, but it can't be eliminated. These problems led to the development of a different kind of optical telescope.

REFLECTING TELESCOPES

English physicist Isaac Newton built a telescope that avoided color distortion by not using lenses to gather light! He developed the **reflecting telescope**, which uses a dish-shaped mirror as the main light-gathering element. The light enters the front of the tube, reflects off the curved *primary mirror* to a flat *secondary mirror*, which directs the light out the side of the tube into the eyepiece.

Reflecting telescope mirrors can have much larger diameters than refracting telescopes. Since the light doesn't pass through the mirror as it does a lens, the mirror mount can support the mirror so that it doesn't sag. Thus, their light-gathering power is far greater. They also avoid the color fringe problems created by large refractor objective lenses. However, Newtonian telescopes have open tubes, so their mirrors collect dust and dew. They have more optical components than refractors, which can disperse or absorb light at each surface, and the metal mirrored surfaces can become discolored from corrosion. Very large Newtonian reflectors are also awkward to use because of the position of the eyepiece at the upper end of the elevated tube. These problems led to some other reflecting telescope designs.

The Cassegrainian reflector has a main and a secondary mirror, but the secondary reflects light back through a hole in the center of the main mirror to the eyepiece. When using a Cassegrainian reflector, the observer stands below and behind the telescope. A related design is the Schmidt-Cassegrainian. This is a *composite telescope*, which uses mirrors and lenses in a sealed tube. Besides providing a wider view of the sky, the sealed interior of the telescope keeps dew and dust off the main mirror.

But even reflectors have a size limit. The large 5 m Hale telescope in California has a single-piece glass mirror that weighs nearly 15 tons! Better designs and materials that reduced the mirror weight have resulted in single-piece main mirrors as large as 8.4 m in diameter.

Since light-gathering power is the most important feature of telescopes, astronomers and engineers have found ways to increase the effective diameter of reflecting telescopes. Truly huge reflectors have main mirrors made of small segments mounted on a large, sturdy lattice framework. Each mirror is individually adjustable by computer. Segmented mirror telescopes adapt to the changing atmospheric conditions above them. Such systems are called *adaptive optics*, and they eliminate much of the atmospheric blurring. The largest segmented mirror telescope is the Gran Telescopio Canarias located in the Canary Islands of Spain. Its mirror has thirty-six hexagonal segments and is 10.4 m in diameter!

Another method of increasing the light-gathering power of reflector telescopes combines the images from two or more telescopes at different locations. The result is as if the image was produced by a much larger telescope. The two 10 m Keck telescopes (see Figure 25-4) when combined like this have an effective diameter of an 85 m telescope! These advances and other related technologies have given us a better view of the heavens from the ground than ever before.

25-3

A Newtonian reflecting telescope uses a curved mirror to gather light. The light is reflected to a flat mirror that directs it into the eyepiece (see diagram below). The Subaru telescope, the world's largest single mirror reflector, with an 8.2 m primary mirror, is located on Mauna Kea, HI.

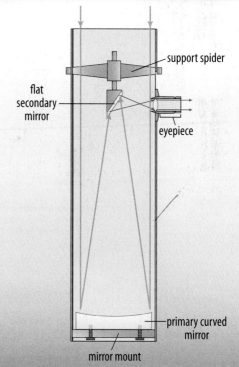

25-4

Numerous telescopes are located on the dormant volcano of Mauna Kea. The two Keck telescopes are on the right in this image.

I Radio telescope antennas must be large for two reasons. Many astronomical radio sources are weak and a large antenna can collect more energy. Larger antennas can also collect longer waves, so they can search for a larger range of signals.

25-5
The 305 m fixed-disk radio telescope antenna near Arecibo, Puerto Rico, has its receiver on a suspended support that can be positioned to steer the beam anywhere within 20° of the point directly above.

25.3 RADIO TELESCOPES

So far we have discussed only optical telescopes. But celestial objects can emit electromagnetic energy as radio waves, microwaves, infrared light, ultraviolet light, x-rays, and gamma rays. Our atmosphere blocks most of these waves, but radio waves can pass through. **Radio telescopes** are telescopes that "see" radio wavelengths.

Radio wavelengths measure from 1 mm to many kilometers in length. Most radio telescopes are huge dish-shaped antennas, made that way so that they can receive a wide variety of radio wavelengths and extremely weak signals. The dish reflects the radio waves into a small receiver horn that carries the radio signals to analysis equipment. As with an optical telescope, the larger the diameter of the dish, the better the resolution of the radio signal. A radiotelescope "image" is built by plotting the positions of radio signal strength data as colored pixels in the same way that a digital photograph is made. The largest single radio telescope is the Arecibo radio telescope, with a 305 m diameter dish! (See Figure 25-5.) Multiple dishes can work together to greatly increase the effective size of the antenna. The *Very Large Array* in New Mexico has twenty-seven antennas 25 m in diameter. Together, they can be equivalent to a radio telescope 36 km in diameter!

25.4 SPACE TELESCOPES

Except for visible light and some radio waves, the atmosphere blocks most of the rest of the electromagnetic spectrum. To observe interesting things in infrared, ultraviolet, or high-energy objects emitting x-rays and gammas, you have to get above the atmosphere. Since the early 1970s, astronomers have been sending special observatory satellites into orbit around the earth or as instruments orbiting the sun. Free of the haze and distortion of the restless, moving atmosphere, these telescopes also have free access to the entire electromagnetic spectrum of the universe. Astronomers can learn more about stars, galaxies, and nebulas if they are able to analyze all the energy that these objects emit.

25-6
The Hubble Space Telescope

You probably have seen the amazing images captured by the Hubble Space Telescope (HST). But this is just one of many space observatories currently operating in orbit. The Italian Space Agency has an x-ray and gamma ray telescope in orbit called **AGILE**. NASA and the European Space Agency (ESA) launched the infrared Herschel Space Observatory in 2009, the largest space telescope up to this time. NASA's **GALEX** mission is investigating the structure of galaxies in the ultraviolet (UV) band. In October 2018, NASA will add the James Webb space telescope (JWST) to this collection of space exploration platforms. All these telescopes are revealing exciting things in the distant reaches of space.

AGILE—*Astro-rivelatore Gamma a Immagini LEggero*

GALEX—GALaxy Evolution EXplorer

25A REVIEW QUESTIONS

1. How do people directly benefit from space exploration?
2. What does a telescope need to form a bright, detailed image of a distant object? What part of the telescope is used to obtain this?
3. Compare refracting and reflecting telescopes.
4. What are some limitations of refracting telescopes?
5. What are some limitations of Newtonian reflecting telescopes?
6. Name two methods for increasing the light-gathering power of reflectors.
7. State two reasons why radio telescope antennas tend to be very large.
8. Give two advantages for having a telescope in space.
9. (True or False) Space exploration has improved the life of nearly everyone in the United States.

25B Objectives

After completing this section, you will be able to

- explain how a rocket works.
- identify the challenges of exploring the solar system.
- contrast satellites, probes, and landers.
- explain how satellites, probes, and landers are used.

25B

ROCKETS, SATELLITES, AND PROBES

How can we explore space without leaving Earth?

25.5 ROCKET SCIENCE

Have you ever fired off a model rocket or bottle rockets during a Fourth of July celebration? If so, you've experienced rocket science! Rockets are the key to getting space exploration off the ground, so to speak. A **rocket** is a device powered by a *reaction engine*, which is similar to a jet engine. However, a rocket engine does not need a supply of outside air, so it can work in a vacuum. What is a reaction engine and how does it work? You actually learned about this principle in Chapter 2. Isaac Newton explained that for every force acting on an object, the object exerts an equal force or *reaction* in the opposite direction on whatever caused the original force. A reaction engine uses this principle by forcefully ejecting a gas in one direction (the action), and the matter in the gas exerts a force on the engine in the opposite direction (the reaction).

Rockets eject hot gases by burning fuel inside a *combustion chamber*. The expanding gases escape through a narrow *nozzle* that makes the gases move even faster in one direction. The gases *must* exert a force on the rocket because of Newton's law, and the rocket moves in the opposite direction from its *exhaust*. Rockets don't move because they push against the atmosphere; they move because of the rocket's reaction force. The rocket principle works anywhere—in air, underwater, and even in the near vacuum of space.

Historians are certain that it was the Chinese who invented rockets, probably at least 800 years ago. These were propelled by black powder chemical explosives and were used for both fireworks and warfare. The technology slowly spread west to India, the Middle East, and Europe through trade. Early rockets were unguided, using only a stick to aim them. In the 1700s, engineers attempted to improve their accuracy by making them spin during flight. It wasn't until the twentieth century, though, that rocket development really "took off"!

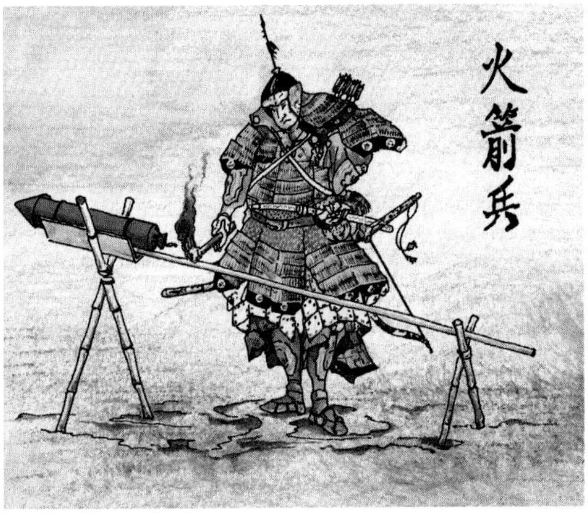

25-7
The Chinese are believed to be the inventors of the first rockets.

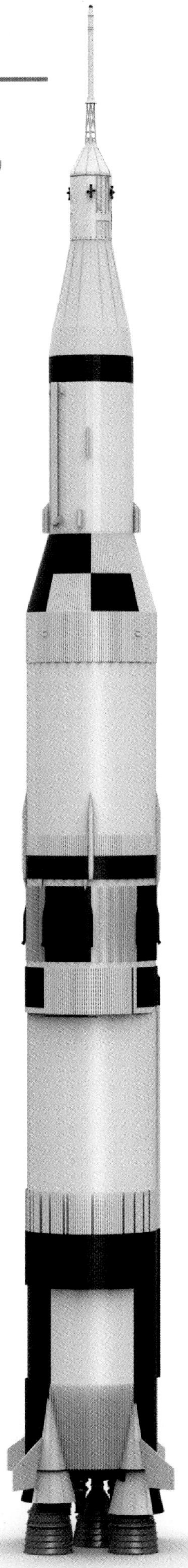

In the early 1900s, three men independently suggested that rockets could carry people into space. Russian schoolteacher Konstantin Tsiolkovsky published a paper about rockets but did not experiment with them. Hermann Oberth, a physicist, started the German Society for Space Travel to research rockets. American physicist Robert H. Goddard performed many experiments with rockets. Goddard invented a rocket that burned gasoline and liquid oxygen as fuel. This method was a significant change from the solid-chemical rocket fuels used up to that time. His rockets also had electrically powered *gyroscopic* guidance systems. By 1935, Goddard had rockets that could fly higher and more accurately than any solid-fuel rockets. His contributions to rocketry were so great that many call him the Father of Modern Rocketry.

As part of the treaty ending World War I, the defeated Germans had to agree not to use long-range artillery. Instead, they established a rocket research program under the leadership of rocket scientist Wernher von Braun. During World War II, his team designed the V-2 ballistic missile, with which Germany was able to attack England from up to 200 miles away.

25-8
NASA is working on the next generation of launch vehicles. Shown above is the Space Launch System rocket.

When Germany was defeated in 1945, the communist Soviet Union and the rest of the Allies raced to capture the German missiles and scientists. The Soviets gained control of many of the former rocket technicians and V-2 rockets. But von Braun and most of his scientists surrendered to the United States forces and were able to bring some V-2 rockets with them. These events led to the beginning of America's space program and the international race to explore space. The modern National Aeronautics and Space Administration (NASA) is von Braun's legacy.

25.6 ARTIFICIAL SATELLITES

In astronomy, a **satellite** is a smaller body that orbits a larger one. Moons are natural satellites. Man-made objects that orbit a planet or moon are *artificial satellites*. On October 4, 1957, the Soviet Union (the Union of Soviet Socialist Republics—USSR) launched the world's first artificial satellite, *Sputnik 1*. A stunned United States successfully orbited its first satellite, *Explorer 1*, a few months later on January 31, 1958. The space race was on!

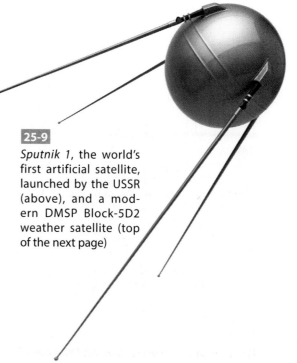

25-9
Sputnik 1, the world's first artificial satellite, launched by the USSR (above), and a modern DMSP Block-5D2 weather satellite (top of the next page)

SPACE EXPLORATION 625

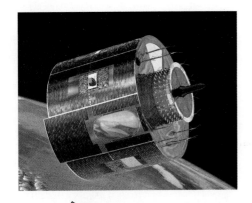

There's a lot of science that goes into getting a satellite where you want it. You need a powerful rocket and a complex guidance system. A satellite also needs to have just the right speed—too much, and it will fly off into space; too little, and it will fall out of the sky and burn up in the atmosphere. Proper orbital speed depends on how far above the earth you want the satellite to orbit. In lower orbits, the satellite needs to go faster because gravity is stronger. Satellites farther out orbit more slowly because gravity is weaker.

SATELLITE ORBITS

The purpose of a satellite determines where and how it should orbit. There are two main features of an orbit: inclination to the Equator and altitude. Orbits that are roughly parallel with the Equator are *equatorial orbits*. Satellites can communicate with only a narrow band of latitudes around the Equator unless they are very high. Satellites that pass over both poles are in *polar orbits*. These satellite sensors sweep out north-south paths as the earth rotates under them. Orbits that are tilted at steep angles to the Equator serve special purposes for users located at high latitudes on the ground.

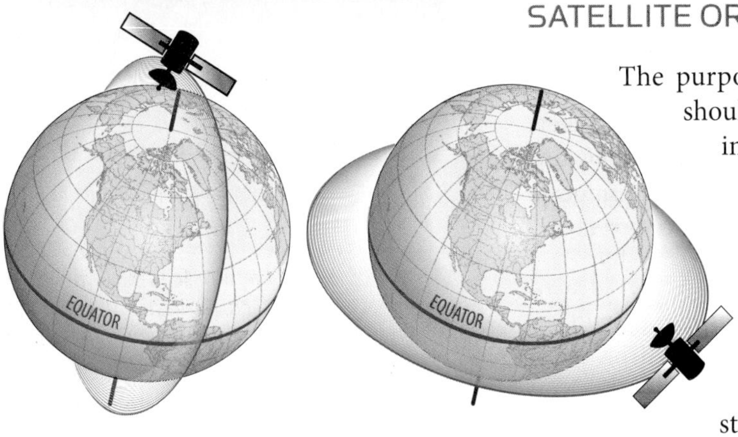

25-10
Orbits may pass over the poles (left) or be roughly parallel to the Equator (right).

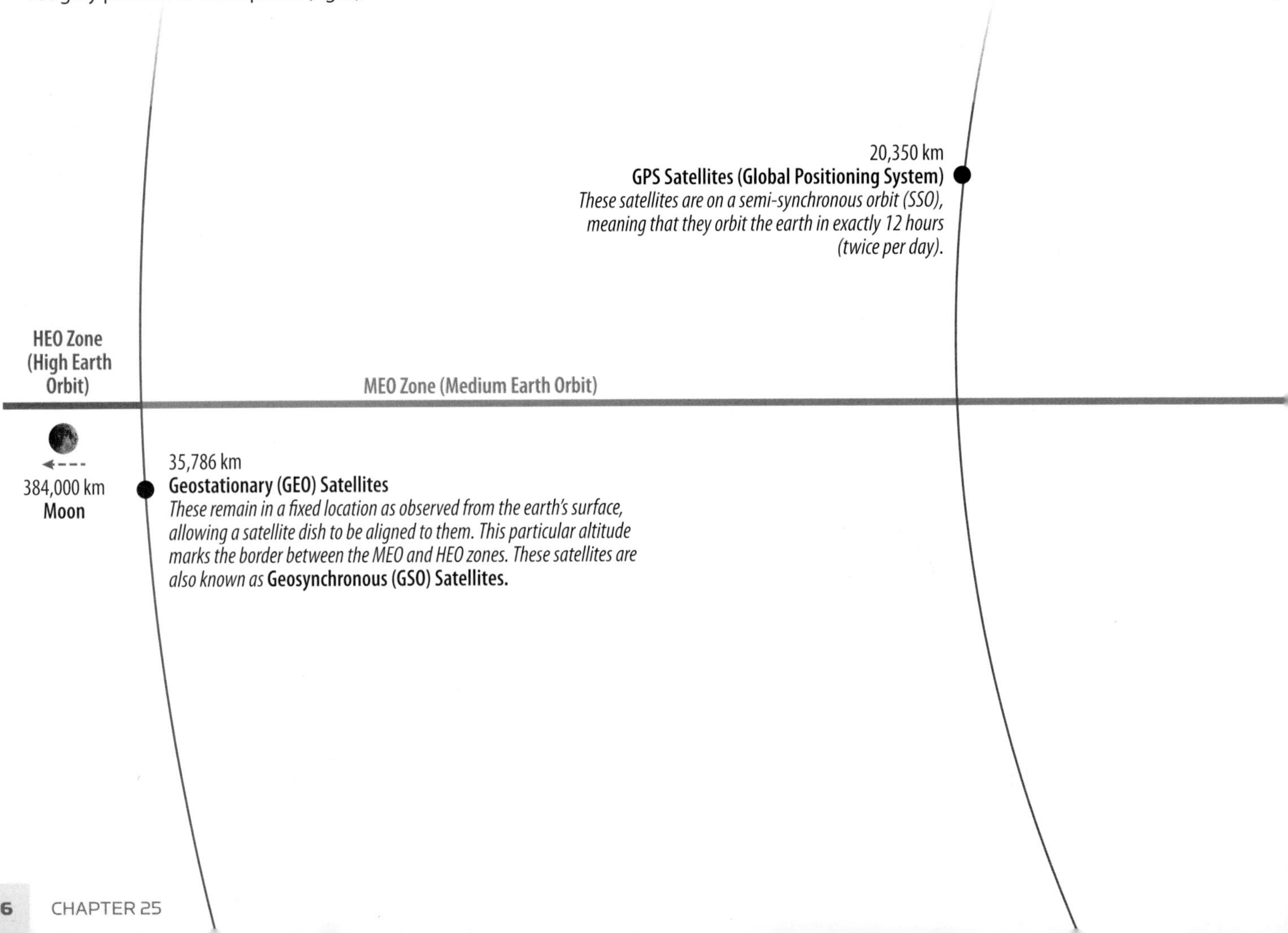

20,350 km
GPS Satellites (Global Positioning System)
These satellites are on a semi-synchronous orbit (SSO), meaning that they orbit the earth in exactly 12 hours (twice per day).

HEO Zone (High Earth Orbit)

MEO Zone (Medium Earth Orbit)

384,000 km
Moon

35,786 km
Geostationary (GEO) Satellites
These remain in a fixed location as observed from the earth's surface, allowing a satellite dish to be aligned to them. This particular altitude marks the border between the MEO and HEO zones. These satellites are also known as **Geosynchronous (GSO) Satellites.**

Orbit altitude determines how fast the satellite must move and how many orbits a day it makes. A *low earth orbit (LEO)* is just far enough above the atmosphere that drag does not affect a satellite (above 160 km). These orbits are the least expensive to use because less powerful rockets can reach them, and radio signals don't need to be strong. However, more satellites are needed if a service must be continuous because only a small area of the ground can "see" an LEO satellite at any given time. The time for a single orbit in an LEO is also very short (around 90 min), so a lot of the earth's surface can be covered in a day.

If a satellite needs to remain in a single place in the sky as viewed from a ground station, then a *geostationary*, or *geosynchronous*, *orbit* is needed. Think satellite TV! A ground station sends the program to the satellite, and then the satellite beams the signal to an entire continent. Each homeowner's dish antenna must point to one of these satellites to receive a signal. The same goes for a weather satellite that continuously observes the weather over one section of the earth. These equatorial orbits are 35,786 km above the earth. At this distance, the satellite takes exactly 24 hours to orbit the earth, so it remains stationary relative to a point directly below it on the earth's surface. Geostationary orbits are much more expensive to reach and require larger rockets. The longer distance requires stronger radio signals as well.

The altitude of a satellite will determine whether it moves relative to the earth. It also determines how much of the earth it can observe at any given time.

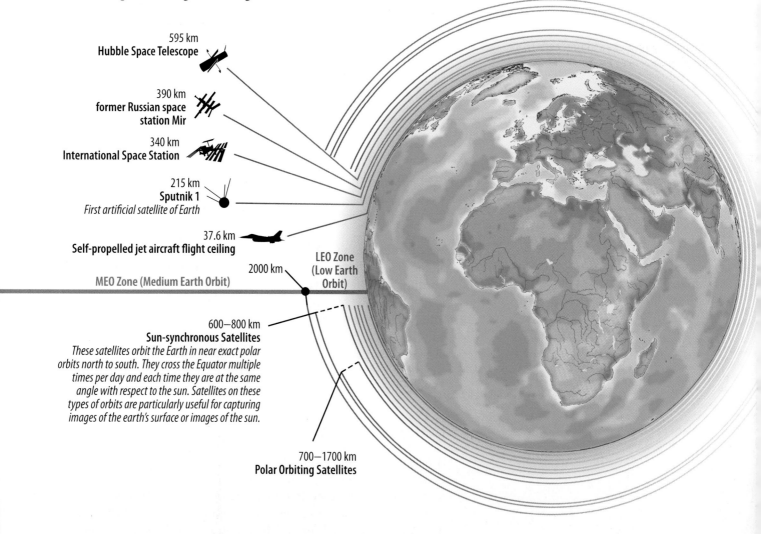

SATELLITE USES

We have launched artificial satellites for all kinds of purposes. They fall into four broad categories, but some satellites serve more than one purpose. Satellites are mainly used for science, business, public service, and the military.

Scientific research includes monitoring the earth's ground vegetation, climates, ozone, sea surface salinity, ice cover, and many other factors of interest to scientists. Businesses use satellites mainly for paid communication services (TV, satellite, and phone systems). Many companies also obtain and sell high-resolution photos of the earth's surface for GIS applications and Earth-viewing programs such as Google Earth. Public service satellites include weather and navigation systems. Weather satellites also receive and rebroadcast emergency locater beacons from ships and aircraft in distress. The military is an important user of satellites, mainly for communications and for keeping tabs on potential enemies. The US Air Force also runs the GPS. Overall, you would have a hard time living a normal life without satellites! Radio, television, the Internet, GPS, cell phones—the list could go on. And the possibilities of future satellite technology are almost endless.

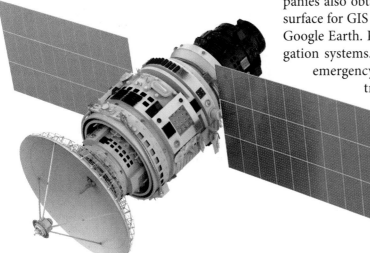

25-11 Satellites come in all shapes and sizes, as varied as their purposes.

25.7 PROBES

Most space vehicles orbit Earth as artificial satellites. But others orbit the sun or other planets. Sometimes their path of travel doesn't orbit a body at all. These vehicles usually follow a one-way path from Earth. They are called **probes** because they probe the frontiers of space.

Space Junk

You might think that putting all this stuff in space might clutter things up a bit. And you're right! Ground stations are keeping track of tens of thousands of pieces of orbiting junk larger than a bolt. Each year, the International Space Station has to be moved in its orbit to keep from coming too close to debris that could damage it.

SOUNDING ROCKETS

When studying the atmosphere, scientists run into a problem. High-altitude research balloons can rise to only 37 km, while the lowest LEO satellites must stay above 160 km. So one type of scientific probe used to investigate the region between these altitudes is the *sounding rocket*. These suborbital rockets carry instruments into the upper atmosphere and then parachute back to Earth to be used again. They send back data on cosmic radiation, temperature, air pressure, wind, atmospheric chemical composition and contaminants, and the geomagnetic fields. They sometimes do experiments on the go.

LUNAR PROBES

After learning how to put satellites in orbit, the space race reached even further as the Soviet Union and the United States began developing vehicles designed to investigate more distant places. The next logical step was the moon. A lunar mission is far more difficult than just getting into earth orbit. The moon is more than a quarter of a million miles away. Compared to the rest of space at that distance, it is a *very* small target! And even if you have a rocket large enough to get there, the vehicle needs to be able to control its speed and course precisely so that it goes into orbit around the moon. Poor timing, or the wrong speed or direction, and the probe misses the moon completely or crashes into it. In the early days, the rockets were powerful enough, but controlling them was still a challenge. The computers at that time were heavy, simple, slow, and unreliable. You have far more computer speed and power in your cell phone today than rocket computers did in those days!

Amid many mission failures, the Soviet Union placed several unmanned landers on the moon starting in 1959. The first completely successful American unmanned lunar probe wasn't until 1964. The Soviets even landed a probe on the surface of the moon in 1966 while the Americans were struggling just to get to the moon with successful probes.

Since those early years, the United States, Soviet Union, Japan, China, and India have all sent lunar orbiters to map and study the moon. Several orbiters dropped probes that impacted the lunar surface. Future plans include new robotic rovers and even a mission to return rock and soil samples. Several countries have plans for establishing human bases on the moon. These missions are providing detailed information of the moon's surface.

25-12
A mockup of *Luna 9*, the first probe to *soft land* on the moon in 1966

PLANETARY PROBES

Even as Americans were trying to successfully reach and study the moon with probes, the Soviets as well as the Americans were looking beyond the moon to the solar system. The technical challenges were immense. As before, powerful rocket propulsion and accurate guidance systems were needed. But traveling the hundreds of millions of miles to another planet meant using massive rockets carrying lots of fuel or finding some other solution.

Exploring the superior planets means sending a vehicle "uphill" in the sun's gravity field. If we were to launch the rocket straight away from the sun, it would require many tons of fuel. One serious limitation of rocket propulsion is that the engine has to boost its fuel as well as the payload. As the rocket uses fuel, there is less mass to push, so the rocket gains speed. So one solution that engineers developed for this problem long ago was the use of *multistage rockets* and *booster rockets*. When the fuel is used up in the first stage, the engines and the weight of the rocket body containing the empty tanks can be dropped off. This makes the next stages more efficient. Booster rockets are usually solid-fuel rockets attached to the side of the main rocket. When the fuel is gone, the booster is dropped off.

25-13
The moon's water vapor distribution (blue) measured during the Indian *Chandrayaan-1* mission in 2009

25-14
This image shows the interstage separation between the S-1C and the S-2 stages during the *Apollo 6* mission.

But even with large multi-stage rockets, getting enough of a push to directly head for the outer solar system would require vehicles too heavy to get off the ground. So physicists looked at the problem and realized that you don't need to go straight "uphill." Just like hiking up a mountain, it is often easier to hike at an angle up the hill than straight up. So by launching the probe in the same direction as the earth orbits, you can use the earth's speed to boost the probe on its way. The rocket needs only to add a little speed to cross over to the next planet.

Missions to more distant planets require other strategies. These use a technique called *gravity assist*. You can gain a lot of speed by shooting a probe inward around the sun, using the sun's gravity to speed it up. The extra speed whips the probe outward toward the distant planets on the other side. Mission planners can also send a probe around a massive planet like Jupiter, trading a tiny amount of the planet's speed to boost the probe on its way, like a cosmic crack-the-whip. This is the method that the *New Horizons* mission used to reach Pluto and beyond.

Reaching Venus and Mercury presents different problems. These inferior planets are "downhill" from us in the sun's gravity field. A poorly planned path would send a probe flying by these planets because it

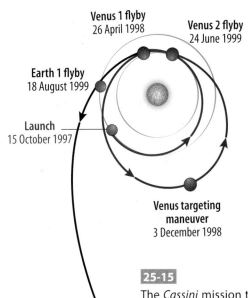

25-15
The *Cassini* mission to Saturn used multiple gravity assists to reach the outer solar system.

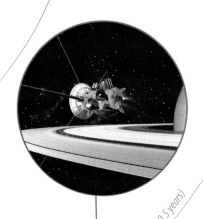

would be moving so fast. Only small rockets are needed to launch these probes away from Earth. The sun's gravity does the rest of the work. Even so, it requires fancy footwork to get a probe into orbit around an inner planet. The *Messenger* probe that orbited Mercury had to make seven loops around the sun, passing Earth once, Venus twice, and Mercury three times before entering orbit around Mercury. Each pass was needed to slow the probe's speed using planetary gravity so that it could finally use its engine at the end of the trip.

Once a probe reaches its destination, it can meet its mission goals in several ways. It can fly by its target, as have several comet probes and *New Horizons*. It can crash into the target. This method was used to investigate comet Tempel 1 as part of the *Deep Impact* mission (2005). Probes can go into orbit around the planet as the *Messenger* mission to Mercury (2011), the *Mars Reconnaissance Orbiter* (2006), and the *Cassini-Huygens* (2004) and *Galileo* (1995) missions to Saturn and Jupiter, respectively. Some probes are designed to return samples to Earth. The *Stardust* mission to comet Wild 2 sampled the comet's coma in 2004 and then returned the dust samples to Earth in 2006. One other category is the open-ended mission, where a probe just keeps on going out into interstellar space. There are currently two active probes on such missions—*Voyager 1* and *Voyager 2*. After touring the outer planets in the 1970s, mission controllers sent them on different paths out of the solar system to find the limits of the solar wind. *New Horizons* is heading out on an interstellar path after visiting Pluto in 2015 on a flyby mission.

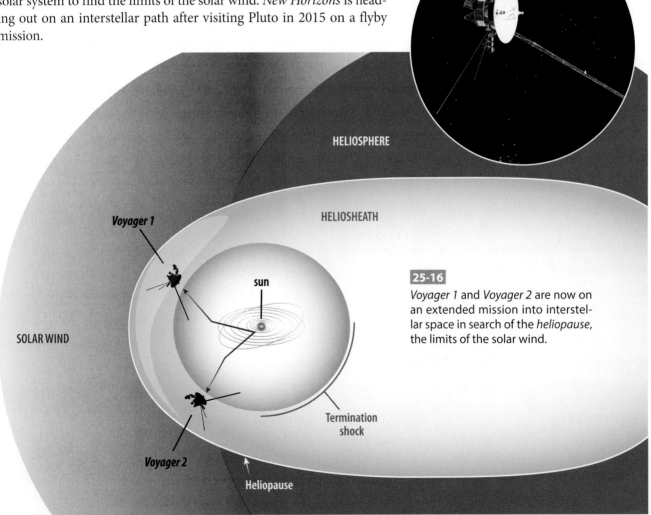

25-16
Voyager 1 and *Voyager 2* are now on an extended mission into interstellar space in search of the *heliopause*, the limits of the solar wind.

SPACE EXPLORATION

25.8 LANDERS AND ROVERS

LANDERS

Planetary scientists really like getting up close and personal with planets, so one of their favorite kinds of probes is a lander. A **lander** is a probe that is designed to land softly on a planet and then transmit photographs and environmental data back to Earth. The first successful planetary lander was a Soviet probe to Venus in 1970. It lasted for 23 minutes before the heat and pressure of the atmosphere destroyed it. Since then, we have successfully placed more landers on Mars than on any other celestial object. The two American Viking landers returned the first photos of the Mars surface in 1976.

Several lander missions were unique. The probe *NEAR Shoemaker* was originally intended to approach and orbit the asteroid 433 Eros in 1999. After completing its mission, its engineers landed the probe on the asteroid—something it wasn't designed to do! It transmitted gamma ray information for several days afterward. The *Cassini-Huygens* mission to Saturn included the *Huygens* lander. It *was* designed to land on Saturn's largest moon, Titan, the only moon in the solar system with a real atmosphere. It sent back to Earth many fascinating photos of strange organic liquid streams and evidence of liquid erosion as it dropped through Titan's cloud-covered atmosphere on parachutes in 2005.

| NEAR—Near Earth Asteroid Reconnaissance

25-17

A view of the Mars polar lander sitting on the surface of Mars

ROVERS

Robotic roving probes provide scientists with the most flexibility for exploring a planet's surface short of actually going there. **Rovers** are mobile landers that are remotely controlled by operators on Earth. They take pictures, sample a wide variety of environmental data, and do chemical and physical studies of the rocks and soil over which they travel. Two of the most successful were NASA's Mars rovers *Spirit* and *Opportunity*. These rovers took on personalities of their own as the world eagerly viewed their amazing images and followed their daily activities for years. *Spirit* became stuck in loose soil in 2009. After repeated attempts to free the rover, it went silent in 2010. NASA officially ended its mission in 2011. Currently, *Opportunity* is investigating Endeavor Crater.

A new generation of rovers is already roaming across Mars's surface. The *Curiosity* rover arrived on Mars in August 2012. Its nuclear power source will be more reliable than the solar power of previous rovers. Since landing, *Curiosity* has kept busy collecting samples and conducting testing of surface materials and the atmosphere.

The *Spirit* and *Opportunity* Mars exploration missions were originally planned to last ninety days. *Spirit* continued to operate for over five years. *Opportunity* has been exploring since 2004.

Spirit location: 14.6°S 175.525°E (Martian coordinates)

Opportunity location: 2.329°S 5.344°W (Martian coordinates)

Curiosity location: 4.714°S 137.356°E (Martian coordinates)

25-18

Curiosity is investigating the surface and atmosphere of Mars. Apparently even Mars rovers take selfies.

25B REVIEW QUESTIONS

1. What is a rocket engine and how does it produce motion?
2. What was the fuel of the earliest rockets? How were they aimed?
3. What major improvements did Robert Goddard make to rocket design?
4. How did world history in the twentieth century influence the development of space exploration?
5. What is the relationship between satellite altitude and orbital speed?
6. What is the major advantage of an LEO? Identify a major disadvantage.
7. What is the difference between a satellite and a probe?
8. Discuss different ways that space probes accomplish their missions.
9. (True or False) Mars is the only celestial body where we have successfully used landers.

SERVING GOD AS AN AEROSPACE ENGINEER

"Houston, we've had a problem." It was April of 1970, and three Americans were in grave danger. Jim Lovell, Jack Swigert, and Fred Haise of *Apollo 13* were on their way to the moon for the third moon landing when an oxygen tank exploded in the service module. The loss of oxygen—and the loss of electrical power that the fuel cells were supposed to produce using it—meant aborting the mission. The new mission was simply to get home alive!

Such an emergency required on-the-spot creativity. *Aerospace engineers* at NASA came to the rescue. They found ways for the crew to improvise, using materials like storage boxes and duct tape to replace equipment destroyed by the explosion. Along with a little seat-of-the-pants flying, the crew made it home safely in the face of extreme odds.

JOB DESCRIPTION

Aerospace engineers design and build spacecraft as well as any other flying vehicle. They can be separated into two groups—aeronautical engineers and astronautical engineers. Aeronautical engineers design vehicles used in Earth's atmosphere. Astronautical engineers focus on spacecraft used outside Earth's atmosphere. Because there's no room for error, aerospace engineers work in teams to develop and test planes, rockets, and spacecraft.

EDUCATION

Many universities have degree programs in aerospace engineering. The tasks of this science overlap with other fields of engineering, such as chemical, electrical, and mechanical engineering, so study in these areas is important too. To work for NASA, you'd probably need an advanced degree in engineering or mathematics.

POSSIBLE WORKPLACES

Not all aerospace engineers work for NASA though. Private aircraft manufacturers and new commercial space transportation companies hire aerospace engineers too. Many engineers teach at universities as well. In spite of the glamour of the field, it is very hard to start a career in aerospace engineering because it is very specialized.

Aerospace engineers can find themselves in a quiet lab or office, on a noisy airstrip, or at a launch pad on land or at sea! Some can work in factories or assembly plants. Some spend all their time in the classroom teaching the next generation of engineers.

DOMINION OPPORTUNITIES

Exploring space has benefits, but it has risks too. The aerospace engineer's job is to minimize the human and financial risks and maximize the benefit and success of the vehicle's mission. This is the work of increasing the usefulness of creation.

25C

MANNED SPACE EXPLORATION

Are the benefits of manned space exploration worth the risks?

25.9 RACE TO THE MOON

For the Soviet and American leaders around 1960, the ultimate goal was putting a man in space. The early successes of the Soviet Union in space and with planetary probes left the Americans doubting their ability to keep up. Then, in April 1961, the Soviets put the first man in orbit! US government leaders believed that losing the space race would mean political disaster and loss of influence around the world in the face of the growth of communism under the Soviets.

But just a month later, in May 1961, President John F. Kennedy gave a speech to Congress that gripped the nation: "I believe that this nation should commit itself to achieving the goal, before this decade is out, of landing a man on the moon and returning him safely to the earth." No one had ever given such a technological challenge in the history of the world! Suddenly, the economic resources and engineering know-how of the United States were committed to achieving a goal that captured the imagination of the American people.

25.10 CHALLENGES OF MANNED SPACE FLIGHT

HARDWARE

The problems with entering and living in space are similar to the ones that explorers faced entering the depths of the oceans (Chapter 15). People need air and protection from a hostile environment. Space vehicles must be airtight, strong enough to be launched from and return to Earth, and able to carry all that human passengers need to complete their mission. These include air, food, water, electricity, tools and instruments, and many other things.

REACHING ORBIT

Getting into space with all this needed equipment meant building a rocket powerful enough to lift it at least into orbit. That wasn't too hard. But the rocket couldn't be so powerful that it would flatten the human occupants like bugs inside their spacecraft during the launch. So new engine and rocket designs were needed. Astronauts had to be trained to remain conscious so that they could at least monitor instruments under the extreme forces of a rocket ride into orbit.

25C Objectives

After completing this section, you will be able to

» summarize the challenges of sending humans into space.

» summarize the history of manned space exploration.

» evaluate the risks and benefits of manned space exploration.

25-19

On September 12, 1962, President Kennedy gave his "We choose to go to the moon" speech at Rice University.

The period of history between the end of World War II and the breakup of the Soviet Union around 1990 was called the *Cold War*. The space race was just one area of competition between the Soviet Union and the United States for political influence across the globe.

25-20
Astronauts need to train for living and working in microgravity—what many people call *weightlessness*.

MICROGRAVITY

Living in space is also a real challenge. In a *microgravity environment*, there is no "down," and nothing stays put by itself. You have to get used to feeling like you are continually falling. Liquids don't flow or pour—they float as spherical blobs if they escape a container. Moving around a large compartment takes practice so you don't crash into things. And what happens to your health when you live for a long time in microgravity? For humans to be successful in space, they had to learn about living there first.

REENTRY

Beyond the problems of getting into orbit and working there, how does an astronaut get back to Earth? You can't just parachute! One of the greatest challenges in manned space exploration has been returning to Earth safely. In an LEO, a spacecraft is moving at about 7800 m/s. If it attempts to dive into the atmosphere at this speed, it will burn up! This happens for two reasons: friction and conservation of energy. In Chapter 2, you learned about two different types of mechanical energy—kinetic and potential. Orbiting spacecraft have lots of both because of their high speeds and altitudes. And to return to the ground in one piece, they have to get rid of all of it.

25-21
Glowing plasma created during reentry demonstrates conservation of energy.

The atmosphere is the only matter available that can take the energy away from the spacecraft. As this happens, kinetic and potential energy are converted into thermal energy. The air gets so hot that it turns into plasma, like a blowtorch. This is heat that could burn up the spacecraft if it isn't designed properly. Returning spacecraft are protected by heat shields made of special materials to insulate the vehicle. They are also designed to keep the shield facing in the right direction during reentry. The insulating materials on early spacecraft slowly burned off the heat shield, transferring the heat back to the atmosphere. The space shuttle used insulating ceramic tiles that simply kept the heat and plasma out of the vehicle.

Landing spacecraft involved another design decision. Early American vehicles were designed to land at sea because we had the ships to recover them, and the landings were "softer" and thus safer. Soviets chose to recover their vehicles on land because of their large land area, and it was a simpler, though riskier, method. They could also keep tighter control of the landing area for political and security purposes. Later, during the US space shuttle program, the shuttle glided to a landing at one of NASA's several airfields.

25.11 STEPPING STONES TO THE MOON

NASA decided to work up to the moon mission in steps, starting with relatively simple vehicles that could carry one man. They learned what they could and applied that knowledge to the next type of spacecraft, which was larger and carried two men. Finally, the actual spacecraft that was to go to the moon with a crew of three was put through its paces in LEO before it was declared safe for the historic voyage. Let's take a brief look at each of these programs.

25-22
The challenge of space exploration didn't end with getting to outer space. Even after making it back through the atmosphere, astronauts had to land and be recovered.

PROJECT MERCURY

America's first manned spacecraft was the Mercury capsule. It carried one astronaut in very cramped quarters. Alan Shepard became the first American in space on May 5, 1961. Shepard did not orbit the earth; he simply traveled up above the atmosphere and back down in a suborbital flight that tested the launch and reentry systems. John Glenn was the first American to orbit the earth in another Mercury capsule. There were a total of two suborbital and four orbital manned flights in Project Mercury.

A suborbital flight is any rocket launch that does not gain enough speed to enter orbit. A vehicle in suborbital flight immediately returns to the earth's surface.

25-23
John Glenn (1921–2016) entering the *Friendship 7* capsule, in which he would become the first American to orbit the earth on the Mercury 6 mission

25-24
John Glenn relaxing after safely returning from his orbital flight

CASE STUDY: NO SAFE RETURN

"This nation should commit itself to achieving the goal, before this decade is out, of landing a man on the moon and returning him safely to the earth." With these words, President Kennedy catapulted the United States into a race to beat the Soviet Union to the moon. People were willing to sacrifice to become astronauts. Some even sacrificed their lives when they failed to return safely home.

Though NASA takes extreme precautions to protect astronauts, people have died in training, in aircraft accidents, in rocket testing, and in actual space flights. The first space-related fatalities were three *Apollo 1* astronauts who died in a fire inside the capsule during pre-flight testing in 1967. NASA's most public loss was the space shuttle *Challenger* disaster in 1986. The nation watched on live TV as the shuttle exploded 73 seconds into its mission. All seven crewmembers, including an elementary schoolteacher, died in the accident. NASA lost a second space shuttle in 2003, when the shuttle *Columbia* broke apart during reentry. Damage to ceramic tiles, designed to protect the shuttle from the heat of reentry, caused this accident.

In these three accidents, sixteen Americans lost their lives while advancing our understanding of space. Attracted by the adventure, the astronauts yet understood the dangers. They decided that the benefits of understanding space outweighed their own personal risk.

Questions to Consider

1. Why is space exploration so challenging and risky?
2. Why should we study space?
3. Is sending people into space worth the risk?

On January 28, 1986, the fuel tank of the shuttle Challenger *exploded 73 seconds into its flight because of a faulty booster rocket. All seven crew members, including schoolteacher Christa McAuliffe, died in the accident.*

PROJECT GEMINI

The next step toward putting a man on the moon was developing a spacecraft that could carry several crewmen. A moon mission was more than one person could manage by himself. The Soviet Union again led the way with *Voskhod 1*, a three-man capsule launched in 1964. A cosmonaut (Soviet astronaut) was also the first person to work outside a spacecraft in orbit. These excursions are called *extravehicular activities*, or *EVAs*.

The United States launched its own multi-crewman spacecraft under Project Gemini during the years 1965–66. The two-man Gemini crews tested procedures that would be needed to get to the moon. The most important was meeting and docking (or joining) two spacecraft in space. The first American EVAs were completed during the Gemini program. The ten Gemini missions showed that America was nearly ready to head for the moon.

PROJECT APOLLO

Project Apollo was the actual moon landing program. The moon mission launch vehicle was the three-stage *Saturn V*, the most powerful rocket ever built by the United States. After reaching earth orbit and discarding the first two stages, the third stage would send the lunar payload on its way. The moon mission spacecraft consisted of three parts: the three-man *command module*, the attached (unmanned) *service module*, which formed the fourth stage of the rocket, and the *lunar excursion module (LEM)*, which carried two astronauts from lunar orbit to the moon's surface, while the third remained in orbit in the command module.

Preparation for the first moon landing went forward, but the first Apollo mission ended in disaster. Three astronauts died during a catastrophic fire inside the *Apollo 1* capsule while conducting a test on the launch pad. After correcting the problems, NASA completed the remaining test flights without serious problems. On July 20, 1969, years of feverish activity and anticipation paid off as Neil Armstrong and Edwin Aldrin walked on the moon while Michael Collins orbited overhead in the *Apollo 11* command module. They returned safely to Earth on July 24, 1969. In ten Apollo lunar missions, a total of twelve American men walked on the moon—still the only humans to have done so.

25-25
Neil Armstrong and the United States flag on the surface of the moon

25-26
Skylab was America's first space station. An overhead view of Skylab in orbit shows that a solar panel is missing on the left side. The solar panel and a shield were lost during its launch. The blanket on the lower section is a substitute shield erected by Skylab's second crew.

25.12 SPACE STATIONS

After man had made it to the moon, the next goal was to set up a long-term presence in space. The way to do this is to have a manned satellite that stays in orbit for a long time. Spacecraft need to be able to dock with such a satellite so it can be resupplied. This special satellite is called a **space station**.

THE SALYUTS

After the United States won the race to the moon and was busy with the conclusion of the Apollo program during the 1970s, the USSR placed in orbit the first space station, Salyut 1. Six other stations followed. All the stations in this series eventually fell from orbit and burned up in the atmosphere.

SKYLAB

In 1973, the United States launched its first space station, Skylab, which was more like an orbital workshop. In all, three crews of three men each occupied Skylab, the last staying for eighty-four days. While in the station, the astronauts did experiments and observed their bodies' reactions to long-term microgravity. In July 1979, almost six years after the last crew left it, Skylab fell harmlessly from orbit into the Indian Ocean and uninhabited portions of Australia.

MIR

In 1986, the USSR launched a new space station, Mir (Russian for "peace"). Although the first module was no larger than the Salyuts, Mir was more advanced. It could be used as a part of an expandable space colony or as living quarters for a trip to Mars. Several modules were added, providing increased living space and research facilities.

After the Soviet Union collapsed, Mir continued to be the only space station for another decade. Many nations helped the Russians to keep it operating. Eventually, the facility began to fail due to age and lack of funds to maintain it properly. In 2001, Russian technicians skillfully maneuvered the 135-ton hulk to a spectacular but safe splashdown in the Pacific Ocean. It was time for a new space station.

The International Space Station (ISS)

INTERNATIONAL SPACE STATION

In order to reduce the total costs to any one country, the United States and Russia decided to jointly build a space station that could be used by both countries. Delays and soaring costs led to other countries joining the project. Currently, twenty-six nations are participating, including, Japan, Canada, and those of the European Space Agency. Because of this multinational involvement, the new space station was named the **International Space Station (ISS)**. Begun in 1998, most of the modules and major pieces of the ISS were delivered by the American space shuttles.

The ISS is the largest artificial satellite of Earth. Its crews have made many new and exciting discoveries in biological and materials sciences, as well as in space medicine. Its microgravity environment provides unique opportunities for investigating many scientific and manufacturing processes. This research is producing products that are changing our lives. The ISS is scheduled to continue operating until at least 2028. With the end of the space shuttle program in 2011, the ISS crews are transported by Russian Soyuz spacecraft.

25.13 THE SPACE TRANSPORTATION SYSTEM

After the Skylab project was completed, the American space program focused on developing an economical, reusable space vehicle—the *Space Transportation System (STS)*. Most people know these vehicles as **space shuttles**. NASA needed a vehicle to carry large satellites and probes into orbit and do maintenance on them. NASA also wanted a vehicle that was large enough to build and maintain space stations and transport their crews and supplies. Making the vehicle reusable was supposed to keep program costs low.

The shuttle *Enterprise* was flight tested as early as 1977 but never flew into space. The first orbital mission was the space shuttle *Columbia* in 1981. Three other orbiters were built to support the STS program—*Challenger*, *Discovery*, and *Atlantis*. The *Endeavor* was built later as a replacement vehicle.

The space shuttle on the launch pad had four major parts: the orbiter itself, a large, orange, external liquid-fuel tank, and two solid-fuel booster rockets. Both the orbiter and the booster rockets were reusable. The main advantage of the shuttle was its large cargo bay with clamshell doors that opened in orbit. Cargo, satellites, and experiments could be carried there. The shuttle also had living quarters for up to seven astronauts.

The STS program never lived up to early promises to be an economical orbital transportation system. The orbiters were extremely complex to maintain. The space shuttles were all retired in 2011 after thirty years of service.

25.14 THE FUTURE

What does the future hold for manned space exploration? Many people wonder whether humans should be going into space at all. Does the risk to life and limb justify exploring such a hostile environment? These concerns are similar to those of deep-sea exploration.

God created us to have a natural curiosity about the universe. Space exploration appeals to our sense of adventure and our curiosity. As we've seen, some of the best results of space exploration have been new technologies that help improve people's health, make us safer, and help us work more efficiently. But both the way that we explore space and the reasons that we do it are affected by our worldview.

Space exploration needs to be done in a way that is most economical and least risky. The US government cannot afford a new, more capable STS. It seems as though the future replacement for the space shuttle may come from private companies, such as SpaceX and Virgin Galactic. Private companies seem to be able to minimize expenses in ways that a government organization like NASA can't. They probably will avoid excessively complex designs, and they will make their rockets safe enough to be profitable.

Though unmanned space exploration is safest, there are many advantages to human observers and workers on the scene. Humans can make important, instantaneous decisions beyond the abilities of a robotic or autonomous vehicle. A Christian worldview can justify manned space exploration as long as lives are not needlessly placed at risk. But atheistic worldviews like communism often minimize the value of human lives. Over the years, the Soviets/Russians have taken risks that other nations have found unacceptable in their space programs.

We need to explore space in ways that maximize the benefits and minimize the risks. The value of long missions to Mars or of setting up bases on the moon at our current level of technology is questionable. These plans as they stand are very risky. There just don't seem to be good scientific or economic reasons to justify endangering human lives and exposing them to the conditions they will have to endure in space and while on Mars or the moon.

25-27 Colonizing Mars at this point in history is probably *not* worth the risk to human life.

Space exploration needs to be done for the right reasons. Soviet cosmonaut Yuri Gagarin has been quoted as saying, "No, I didn't see God, I looked and looked but I didn't see God," after he orbited the earth. However, when the crew of *Apollo 8* orbited the moon on Christmas Eve, 1968, they read to their global television audience Genesis 1:1–10. Both Yuri Gagarin and the Americans were intelligent and well educated. Both were looking at the same evidence—the vast expanse of outer space and the good earth. Yet they came to very different conclusions. Why?

Secular scientists look for natural explanations for everything. But a Christian can truly appreciate the special nature of our planet, the solar system, and the universe. Through the knowledge we gain, we can better exercise proper dominion and obey the two great commandments—to love the Lord our God with all our hearts, souls, and minds, and to love our neighbors as ourselves (Matt. 22:37, 39).

New Opportunities

The shuttle astronauts included women, who had been excluded from earlier US manned space programs. In 1983, Sally Ride became the first American female astronaut in space, on *Challenger*. Astronaut Eileen Collins became the first female shuttle pilot and eventually commanded the shuttle mission that placed the heavy *Chandra X-ray Observatory* in orbit. Scientists who were not professional astronauts, both Americans and those from other countries, have also flown on space shuttle missions. Congressmen and senators have flown in the space shuttle, including the former astronaut Senator John Glenn (who was 77 years old when he flew on *Discovery* in 1998), and there have been plans to put teachers and journalists into space.

25C REVIEW QUESTIONS

1. What factors fueled the race between the United States and the Soviet Union to put humans into space?
2. What was perhaps the greatest challenge of human space flight?
3. Other than their size and complexity, what was the main difference between the Mercury, Gemini, and Apollo space capsules?
4. Who was the first American in space and in what year did he accomplish this?
5. What were the goals of Project Gemini?
6. Why did the Apollo spacecraft need to be able to hold more than one person?
7. How many manned modules were involved in an Apollo moon mission? What were their functions?
8. What did the Soviets do after the United States won the race to put a man on the moon?
9. What factors finally brought an end to the space station Mir?
10. How did the International Space Station get its name?
11. Make a summary statement that describes how we should pursue manned space exploration.
12. (True or False) NASA's Space Transportation System proved to be a relatively safe and economical way to carry cargo and passengers back and forth between Earth and orbit.

LIFE CONNECTION: CRITTERNAUTS

When the United States decided to send humans into space, questions sprung up immediately. How would space flight affect people? Could a man withstand the forces of liftoff? Could he swallow in microgravity? How would he react with no direction being "down"? Could he even breathe? Living animals had to go into space before we dared to risk the lives of people.

The first flights were pretty simple. The Americans started with fruit flies in a series of successful suborbital V-2 flights in 1947. Next, they tried monkeys and mice in 1948–49. The recovery systems didn't work well, and the monkeys died in the crashes, but the mice survived. During these flights, scientists closely observed animals with TV cameras and sensors to see how they fared. The rhesus monkey Yorick gained some fame as the first monkey to survive a suborbital flight in 1951.

The Soviet Union began its flight testing using dogs in suborbital flights. The first successful orbital flight by an animal was *Sputnik 2*, carrying the dog Laika in 1957. Sadly, the Soviets hadn't developed an orbital reentry system yet, so Laika died after a few hours in orbit. The Soviets launched many dog flights up until their first human orbital flight in 1961. The dogs Belka and Strelka were the first animals to orbit the earth and survive recovery in 1960. Dogs were selected because they were considered less nervous than monkeys and were more easily trained.

The first space chimpanzee, named Ham, made a suborbital flight for the United States in 1961 just before Alan Shepard's historic flight. Ham splashed down in the Atlantic Ocean to test the ocean recovery system. The chimp Enos was the first to orbit in 1961, just before John Glenn's flight.

In 1968, the Soviets launched the first animals on a mission around the moon in their Zond-series moon vehicles. One mission successfully returned, one was a one-way flyby, and a third failed during reentry. The Americans took some pocket mice and nematodes (roundworms) around the moon on the *Apollo 16* and *Apollo 17* missions in the early 1970s.

After the Americans won the race to the moon in 1969, animals were launched into space mainly for pure science. How does the microgravity environment affect living things? What are the short- and long-term effects of the high radiation there?

In 1966, the Soviet dogs Veterok and Ugolyok set the first orbital endurance record of 22 days. This record was not surpassed by humans until the *Soyuz 11* mission in 1971.

Several other nations besides the Soviet Union/Russia and the United States have sent animals on space voyages. China, Argentina, France, Japan, and Iran are members of this exclusive club. In the decades to come, we can expect other advanced nations to join.

The variety of animals that have gone into space is amazing. Besides the primates, dogs, and other mammals already mentioned, guinea pigs and cats have also flown. Some other vertebrates have included swordtail fish, carp, frogs, and tortoises. Invertebrates, being smaller and easier to maintain, have also logged many hours in space. Various spiders and insects ranging from cockroaches to wasps have broadened the frontiers of our knowledge about space travel.

Many of these animals bore the brunt of early space flight failures. Monkeys and mice experienced horrendous forces on launch and reentry that no humans could have lived through. Many vehicles carrying animals crashed or burned up in the atmosphere or simply never returned to Earth.

It is regrettable that animal lives were lost during these early space flights. One Soviet scientist was devastated after the loss of two dogs during a failed reentry. Several of the American space monkeys were permanently memorialized in various ways. These animals were used to prepare the way for human spaceflight. Because humans bear the image of God and have dominion over the animals, it is proper that we use animals for such purposes rather than risk human lives. This kind of work is completely consistent with a Christian worldview.

Key Terms

refracting telescope	620
reflecting telescope	620
radio telescope	622
rocket	624
satellite	625
probe	628
lander	632
rover	633
space station	640
International Space Station (ISS)	641
space shuttle	642

CHAPTER 25 REVIEW

CHAPTER SUMMARY

» Space exploration is a way to glorify God through the discoveries we make and to help people by adapting space technologies to medicine, materials, and other uses.

» Refractors, the first telescopes invented, use lenses to gather light and to magnify and resolve image details.

» Reflecting telescopes use a large mirror to gather light instead of the objective lens used in refracting telescopes.

» Composite optical telescopes combine features of both refractors and reflectors to produce a more compact telescope.

» Radio telescopes create images of objects in the universe from the radio waves they emit.

» Space telescopes are located above the atmosphere where they can detect wavelengths that are blocked by Earth's atmosphere, avoid the blurring effects of moving air, and operate day and night.

» Rockets are vehicles propelled by reaction engines that work according to Newton's principle of action-reaction.

» Rockets were invented by the Chinese long ago but reached advanced stages of development only in the twentieth century.

» After World War II, both the United States and the Soviet Union used German rocket technology to advance their space exploration programs.

» Artificial satellites are used for communications, navigation, science, and military purposes. The kind of orbit they are placed in depends on their purpose.

25-28
People living anywhere require supplies. When you live in the ISS, getting supplies can be challenging. The crew of the ISS uses the robotic arm to capture the Japanese cargo vessel *HTV-6*.

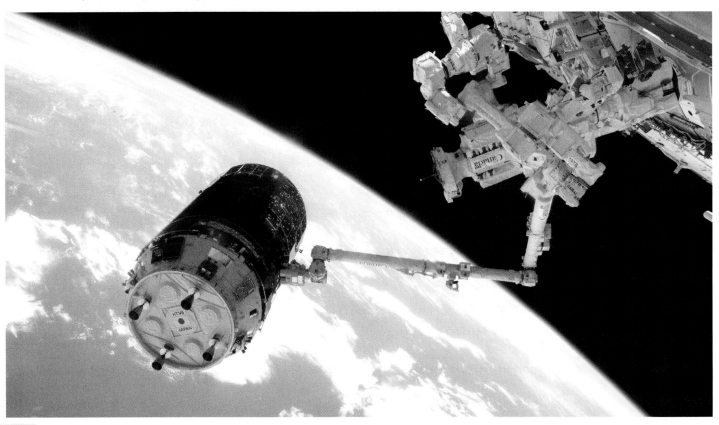

- » Probes are vehicles that leave the earth to investigate the upper atmosphere and interplanetary or interstellar space. They can accomplish their missions by orbiting their targets, by crashing into them, or by flying past them and ultimately flying out of the solar system.

- » Landers and rovers are probes that allow observation of a planetary body at its surface. Rovers are mobile vehicles that can be controlled by operators on Earth.

- » The race to land a man on the moon was a struggle between the United States and the Soviet Union for world leadership and was one aspect of the Cold War between democracy and communism.

- » The American manned space program began with a series of increasingly complex projects leading to the Apollo moon missions.

- » The Soviets led in many aspects of manned space exploration until the Americans landed on the moon first. Losing that race, the Soviets focused on orbital space stations.

- » The increasing costs of building an orbital outpost resulted in a group of countries contributing to the construction of the International Space Station.

- » The American Space Transportation System (shuttle program) was essential to the construction of the ISS, but was retired in 2011 after thirty years of service.

- » Future manned space exploration projects need to produce real scientific and economic value to offset the risks to human lives and tremendous cost that space travel involves.

25-29
The space shuttles were instrumental in the construction of the ISS. Here the space shuttle *Endeavor* is approaching the ISS with supplies.

REVIEW QUESTIONS

1. What are the advantages of telescopes over other forms of space exploration?
2. What do telescopes use to gather light and magnify and focus images?
3. What kinds of energy can telescopes use to form images?
4. Why are reflecting telescopes typically larger than refractors?
5. Why are radio telescope antennas so large?
6. What happened at the end of World War II that shaped the space race between the Soviet Union and the United States?
7. To stay in orbit at an altitude of 320 km, a certain satellite must travel at a speed of 27,720 kph. To orbit at an altitude of 600 km, would the same satellite have to orbit at a faster or slower rate?
8. What kind of orbit would be necessary for a satellite that continuously monitors the weather over the Indian Ocean?
9. Explain the difference between a space probe and an Earth satellite.
10. How do sounding rockets differ from other probes?
11. What is the advantage of using a multistage rocket?
12. What is the greatest challenge in sending probes to the superior planets? to the inferior planets?
13. Which celestial object has been the most extensively explored by landers?

14. What are the two main considerations when designing the launch vehicle to take a human being into orbit?
15. What was the purpose of Project Mercury?
16. Describe the main components of the Apollo mission spacecraft (not the launch vehicle) and the purpose of each.
17. What was the next goal of space exploration after landing a man on the moon?
18. What was the purpose of the Space Transportation System?
19. What new kinds of approaches to space transportation are expected to drastically lower the cost of manned space flight?

True or False

20. Magnification is the most important property of a telescope.
21. Using two 8 m telescopes together can produce a better image than either can by itself.
22. No single space telescope can observe all the wavelengths of energy that a star emits.
23. Rockets move because they push against the atmosphere.
24. Astronauts in the ISS experience more gravity in LEO than a weather satellite does in a geostationary orbit.
25. Several countries continue to study the moon to prepare for building a lunar base.
26. Every planet has been visited by at least one space probe.
27. All Apollo lunar missions made successful moon landings.
28. Every Soviet/Russian space station has burned up in the atmosphere.
29. The US space shuttle has been essential to the completion of the International Space Station.

Essay

30. As a Christian considers the risks to human space travelers, how should he view manned space exploration?

APPENDIX A

UNDERSTANDING SCIENTIFIC TERMS

You may find science a little intimidating because of all the long, unfamiliar words that scientists use. However, you can figure these terms out by breaking them down into simple parts that have meaning to you. When you see a difficult scientific word, look at the entries in this appendix to help you understand that term. These word parts may come at the beginning, end, or in the middle of the term, depending on their meaning.

For example, if you ran into the words "pyroclastic flow," you could separate "pyroclastic" into three parts, *pyro-*, *-clast-*, and *-ic*. *Pyro-* means "fire," *-clast-* means "sharp" or "broken rock," and *-ic* means "of" or "related to." So pyroclastic flow is of or related to sharp or broken rock fire—the burning hot rock material that flows from some volcanoes.

Some roots can be used only as prefixes, others only as suffixes. Still others can be used as either. For example, a *helio*centric model is a model in which the sun is at the center of the solar system, while peri*helion* refers to the position in a planet's orbit at which it is nearest the sun.

Pyroclastic flow

A

a, an (Gk.)—not, without
ab (L.)—away from
ac, ad, ag (L.)—to, toward
acous (Gk.)—hear
aer, aero (Gk.)—air
alter (L.)—change
amal (Gk.)—soft
amphi, ampho (Gk.)—on both sides
ante (L.)—before
ant, anti (Gk.)—opposite, against
aqua (L.)—water
aster, astr, astro (Gk.)—star
audio (L.)—hear
aut, auto (Gk.)—self

B

bar (Gk.)—weight; pressure
batho, bathy (Gk.)—deep, depth
bi (L.)—two, twice, double
bio, bios, biot (Gk.)—life

C

calc, calci (L.)—calcium
centi (L.)—a hundred
centr, centri, centro (Gk.)—center
chem, chemi, chemo (Gk.)—chemical
chrom (Gk.)—color
chron, chrono (Gk.)—time
clast (Gk.)—sharp or broken
cline (Gk.)—sloping
co, com, con (L.)—with, together
cosm, cosmo (Gk.)—universe, world
cycl, cyclo (Gk.)—circle, wheel

D

de (L.)—loss; removal
deci (L.)—tenth
div (L.)—apart
duce, duct (L.)—to lead
dyna (Gk.)—power

E

eco (Gk.)—house, abode
electr (L.)—electric
en, end, endo (Gk.)—within, inner
epi (Gk.)—upon, over, beside
equ, equa, equi (L.)—equal
ex, exo (Gk.)—out, outside, without
extra (L.)—outside, more, beyond, besides

F

fissi (L.)—split, divide
flam (L.)—fire
folia (L.)—thin strata
fund (L.)—basis
fusi (L.)—to join together

G

gene, genea, geneo, genesis, geno (Gk.)—birth, origin, family, beginning
genic, genous (Gk.)—producing
geo (Gk.)—earth
glob, globo, globus (L.)—ball, globe
gnos (Gk.)—knowledge
grad (L.)—step, walk, slope
graph, grapho, graphy (Gk.)—to write
grav (L.)—heavy
gyro (Gk.)—spinning

H

halo (Gk.)—salt
helio, helion (Gk.)—sun
hemi (Gk.)—half
hetero (Gk.)—other, different
homo, homeo, homio (Gk.)—like, same, resemble
hydr, hydra, hydro (Gk.)—water
hyper (Gk.)—over, beyond
hypo (Gk.)—under, beneath

I

ic (Gk.)—of or relating to
inter (L.)—between
ism (Gk.)—belief; process of
is, iso (Gk.)—equal

K

kine, kinema, kinemato, kines, kinesi, kinet, kineto (Gk.)—move, moving, movement

L

litho (Gk.)—stone
log, logo, logus, logy (Gk.)—word, study of
lun, luna (L.)—moon

M

macr, macro (Gk.)—large
magneto (Gk.)—magnetic
mal (L.)—bad
mar (L.)—sea
medi, media, medio (L.)—middle
mes, meso (Gk.)—middle
met, meta (Gk.)—between, with, after, change
meteor (Gk.)—in or of the atmosphere
meter, metry (Gk.)—measure
micro (Gk.)—small
mill, mille, milli, millo (L.)—one thousand
mit (L.)—to send
mono (Gk.)—one
morph, morpha, morpho (Gk.)—form, shape
mult, multi (L.)—many

N

nan, nani, nano, nanus (Gk.)—small, dwarf
nomy (Gk.)—the science of
nuc, nucle, nucleo (L.)—central part

O

ocul, oculi, oculo, oculus (L.)—eye
opt, opti, opto (Gk.)—eye, vision
organ (Gk.)—living
oro (Gk.)—mountain
orth, ortho (Gk.)—upright; perpendicular
osis (Gk.)—condition; disease
ox, oxy (Gk.)—oxygen

P

paleo (Gk.)—ancient
pan (Gk.)—all
par, para (Gk.)—beside
pause (Gk.)—to stop
pend (L.)—hanging
peri (Gk.)—around; near
petro (Gk.)—rock
phon, phono (Gk.)—sound
phos, phot, photo (Gk.)—light
phyt, phyto, phytum (Gk.)—plant
poly (Gk.)—many
post (L.)—after
pre (L.)—before, in front of
pro (L.)—before, in front of
prot, prote, proto (Gk.)—first, original
pyro (Gk.)—fire

R

radi, radia, radio (L.)—spoke, ray
retro (L.)—backward

S

sal (L.)—salt
scient (L.)—knowledge
scope, scopy (Gk.)—to see, watch
sect (L.)—to cut
seism (Gk.)—earthquake
semi (L.)—half
sol (L.)—sun
son (L.)—sound
spec (L.)—see; look at
speleo (L.)—cave
sphere (Gk.)—ball; globe
stasis (Gk.)—stand still
stella (L.)—star
strat (L.)—layer
sub (L.)—below; under
super (L.)—above; over
syn (Gk.)—together

T

tel (Gk.)—distant
terra (L.)—earth
tetr, tetra (Gk.)—four
therm, thermos (Gk.)—heat
top, topo (Gk.)—place
tran, trans (L.)—across, through
trop, tropae, trope, tropo (Gk.)—turn, change
trud (L.)—to thrust

U

uni (L.)—single; one

V

vacu (L.)—empty
vari, vario (L.)—difference
vect (L.)—to carry
vitre (L.)—glass; glass-like
volu (L.)—bulk; amount

APPENDIX B

MATH PRINCIPLES AND GRAPHING

Math

Scientists frequently use very large and very small numbers. They prefer to write these numbers as concisely as possible, especially if they are round numbers. So they developed a system of unit prefixes that tell the size of the factor of ten that should multiply the main number in the measurement.

Example: 1 kilometer = 1 km = 1 × 1000 m = 1000 m

Metric Prefixes

Prefix	Symbol	Factor	Power of 10	Example
giga-	G-	× 1,000,000,000	10^9	gigabyte (GB), used for measuring digital data
mega-	M-	× 1,000,000	10^6	megajoule (MJ), work done by a bulldozer
kilo-	k-	× 1000	10^3	kilometer (km), distance on Earth's surface
hect-	h-	× 100	10^2	hectare (ha), measure of land area
deka-	da-	× 10	10^1	dekapoise (daP), a unit of viscosity or thickness of a fluid
(base)	—	× 1	10^0	gram (g), a standard unit of mass
deci-	d-	× 1/10	10^{-1}	decibel (dB), sound loudness
centi-	c-	× 1/100	10^{-2}	centimeter (cm), distances in a laboratory
milli-	m-	× 1/1000	10^{-3}	millivolt (mV), heart pacemaker signal
micro-	μ-	× 1/1,000,000	10^{-6}	micropascal (μPa), sound wave pressure
nano-	n-	× 1/1,000,000,000	10^{-9}	nanometer (nm), the size of atoms

Calculations

Don't tell your math teacher this, but math is for science! Here are some math concepts that you'll need in your earth science class.

Ratios

It's really helpful to be able to compare the sizes or numbers of things in science. One way to do this is to find the *ratio* of one quantity to another.

Let's do an example: Find the ratio of girls to the total number of people in a group, maybe your class or family. To do this, take the number of girls and divide that by the number of everyone in the group. If there are 6 girls and 9 people total, then divide 6 by 9. You get 6/9, or 2/3 after reducing. So the ratio of girls to everyone in the group is 2 to 3, or 2:3. In other words, there are two girls for every three people in the group.

Averaging

You and your friends discovered one day that the number of candies in each bag of M&Ms® isn't the same. You want to know how many candies you should expect to be in a bag. You need to find the average number of candies per bag. An average value falls somewhere in the middle of the group of values. It's a value that is typical of the group.

Averaging is straightforward. Just add together all the values in the group—the number of candies in each bag of M&Ms, for instance. Then divide the sum by the total number of values. Suppose you and your four friends counted 38, 42, 40, 37, and 40 candies in the bags in your group. To find the average of these, add the numbers together and divide by 5, the number of bags.

$$38 + 42 + 40 + 37 + 40 = 197$$

197 candies (total) ÷ 5 bags = 39.4 candies per bag (average)

But there aren't fractions of a candy in a bag! What do we do now?

Rounding

The solution to your averaging dilemma is called *rounding*. Since there are only whole candies in a bag, it makes sense to round the number to the ones place. Look at the first decimal place to the right, the tenths place. If that number is 5 or larger, then add 1 to the ones place and drop all decimal numbers. If the tenths place value is 0 to 4, just drop the decimal numbers and leave the ones place unchanged. In this case we round 39.4 candies to 39 candies.

So you should expect about 39 candies per bag of M&Ms. The larger the number of bags you check, the more likely it is that your average will represent a typical value of candies.

GRAPHING

Scientists often like to compare two or more groups of numbers to see whether they are related in some way. When scientists plot data to compare different quantities, they make a *graph*. The simplest graphs compare two changing quantities, called *variables*. Usually, one of the variables changes in a regular way, or the scientist can't control it (e.g., time). This is called an *independent variable*. It doesn't depend on anything in the data. The other variable is expected to change in some way related to the independent variable. Its value *depends* on the first variable, so it is called the *dependent variable*. That makes sense! The values of the independent and dependent variables are called the *coordinates* of the data. To plot the data, you use an *ordered pair* of coordinates, where the first number in the pair is the independent data coordinate and the second is the dependent data coordinate. You may have graphed ordered pairs in the form (x, y) in a math class.

Scientists usually plot the independent variable on the horizontal axis, with increasing values to the right. The dependent variables are typically plotted on the vertical axis, increasing upward. These are not hard and fast rules, and many graphs are arranged differently to improve the clarity of the plot.

Useful graphs include a title describing the graph's purpose and labels identifying the quantities and units used on each axis. Numbered scales and a grid are included to help you estimate values of the variables plotted on the graph.

Graphs come in different forms. A simple plot of points on a graph is called a *scatterplot*. This is the starting point for many graphs. Scientists like to detect trends in the data and to create an equation that describes the trend. They draw a *best-fit curve* through the pattern of dots. The kind of curve depends on how the data changes. We still call it a best-fit curve even if it doesn't actually curve. Curves that are straight lines are called *linear graphs*. These are fairly rare in nature. Most trends in nature range from slightly curved to really wavy! These graphs, logically enough, are called *nonlinear graphs*.

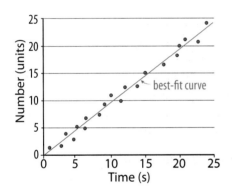

A scatterplot with a linear best-fit curve   A scatterplot with a nonlinear best-fit curve

One of the most important things to learn from a graph is the rate that the dependent variable changes in comparison to the independent variable. This rate is the *slope* of the graph. A steep slope shows that things are changing quickly. A horizontal slope shows that the data is not changing at all. A rising curve from left to right has a *positive slope*; a dropping curve has a *negative slope*.

You can sometimes use scatterplots and trend lines to obtain information not measured in the data set. If you follow the trend line between two data points, these values are estimated—not measured. Obtaining unmeasured data this way is called *interpolation*. Scientists often try to predict the values of the dependent variable beyond the range of the measured independent variable data points. This method of analysis is called *extrapolation*. Extrapolating data depends heavily on assuming that the trend will continue as observed within the measured data.

Slope

We use slope often in earth science for things like streams, terrain, and pressure gradients. To calculate the slope of a line, select two points on the line and then divide the change of its vertical measure (y-coordinate) by the corresponding change in its horizontal measure (x-coordinate). In equation form:

$$\text{slope} = \frac{\text{change in vertical measure}}{\text{change in horizontal measure}}.$$

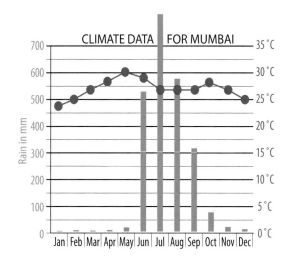

In a *bar graph*, dependent data is plotted as vertical bars at each independent data value. *Area graphs* fill in the area of a graph. When several dependent variables are plotted on a single graph, the areas can be compared to show their relationships visually. *Pie charts* are another type of area graph. They are especially useful for showing percentages of a whole. You will get lots of practice working with graphs in your Lab Manual!

A bar graph (left), pie chart (below, left), and an area graph (below, right)

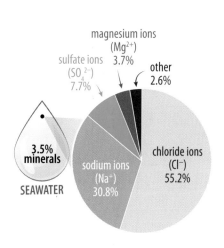
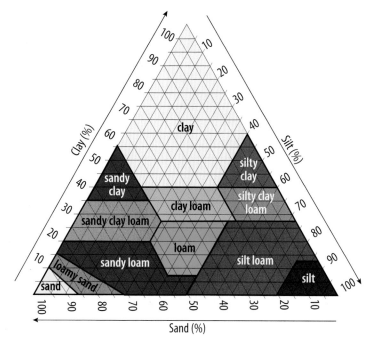

APPENDIX C

PERIODIC TABLE OF THE ELEMENTS

Group	1A	2A	3B	4B	5B	6B	7B	8B	8B	8B	1B	2B	3A	4A	5A	6A	7A	8A
	1	2	3	4	5	6	7	8	9	10	11	12	13	14	15	16	17	18
1	1 Hydrogen H 1.008																	2 Helium He 4.003
2	3 Lithium Li 6.94	4 Beryllium Be 9.012											5 Boron B 10.81	6 Carbon C 12.01	7 Nitrogen N 14.01	8 Oxygen O 16.00	9 Fluorine F 19.00	10 Neon Ne 20.18
3	11 Sodium Na 22.99	12 Magnesium Mg 24.31											13 Aluminum Al 26.98	14 Silicon Si 28.09	15 Phosphorus P 30.97	16 Sulfur S 32.06	17 Chlorine Cl 35.45	18 Argon Ar 39.95
4	19 Potassium K 39.10	20 Calcium Ca 40.08	21 Scandium Sc 44.96	22 Titanium Ti 47.87	23 Vanadium V 50.94	24 Chromium Cr 52.00	25 Manganese Mn 54.94	26 Iron Fe 55.85	27 Cobalt Co 58.93	28 Nickel Ni 58.69	29 Copper Cu 63.55	30 Zinc Zn 65.38	31 Gallium Ga 69.72	32 Germanium Ge 72.63	33 Arsenic As 74.92	34 Selenium Se 78.97	35 Bromine Br 79.90	36 Krypton Kr 83.80
5	37 Rubidium Rb 85.47	38 Strontium Sr 87.62	39 Yttrium Y 88.91	40 Zirconium Zr 91.22	41 Niobium Nb 92.91	42 Molybdenum Mo 95.95	43 Technetium Tc (98) *	44 Ruthenium Ru 101.1	45 Rhodium Rh 102.9	46 Palladium Pd 106.4	47 Silver Ag 107.9	48 Cadmium Cd 112.4	49 Indium In 114.8	50 Tin Sn 118.7	51 Antimony Sb 121.8	52 Tellurium Te 127.6	53 Iodine I 126.9	54 Xenon Xe 131.3
6	55 Cesium Cs 132.9	56 Barium Ba 137.3	57 Lanthanum La 138.9	72 Hafnium Hf 178.5	73 Tantalum Ta 180.9	74 Tungsten W 183.8	75 Rhenium Re 186.2	76 Osmium Os 190.2	77 Iridium Ir 192.2	78 Platinum Pt 195.1	79 Gold Au 197.0	80 Mercury Hg 200.6	81 Thallium Tl 204.4	82 Lead Pb 207.2	83 Bismuth Bi 209.0	84 Polonium Po (209) *	85 Astatine At (210) *	86 Radon Rn (222) *
7	87 Francium Fr (223) *	88 Radium Ra (226) *	89 Actinium Ac (227) *	104 Rutherfordium Rf (267) *	105 Dubnium Db (268) *	106 Seaborgium Sg (269) *	107 Bohrium Bh (270) *	108 Hassium Hs (277) *	109 Meitnerium Mt (278) *	110 Darmstadtium Ds (281) *	111 Roentgenium Rg (282) *	112 Copernicium Cn (285) *	113 Nihonium Nh (286) *	114 Flerovium Fl (289) *	115 Moscovium Mc (290) *	116 Livermorium Lv (293) *	117 Tennessine Ts (294) *	118 Oganesson Og (294) *

Atomic number: 92
Symbol: U
Name: Uranium
Atomic mass: 238.0
* Radioactive
- rounded to four significant digits
- mass number of isotope with longest known half-life indicated by ( )

Lanthanide series

| 58 Cerium Ce 140.1 | 59 Praseodymium Pr 140.9 | 60 Neodymium Nd 144.2 | 61 Promethium Pm (145) * | 62 Samarium Sm 150.4 | 63 Europium Eu 152.0 | 64 Gadolinium Gd 157.3 | 65 Terbium Tb 158.9 | 66 Dysprosium Dy 162.5 | 67 Holmium Ho 164.9 | 68 Erbium Er 167.3 | 69 Thulium Tm 168.9 | 70 Ytterbium Yb 173.0 | 71 Lutetium Lu 175.0 |

Actinide series

| 90 Thorium Th 232.0 | 91 Protactinium Pa 231.0 | 92 Uranium U 238.0 | 93 Neptunium Np (237) * | 94 Plutonium Pu (244) * | 95 Americium Am (243) * | 96 Curium Cm (247) * | 97 Berkelium Bk (247) * | 98 Californium Cf (251) * | 99 Einsteinium Es (252) * | 100 Fermium Fm (257) * | 101 Mendelevium Md (258) * | 102 Nobelium No (259) * | 103 Lawrencium Lr (266) * |

Legend:
- Alkali metals
- Alkaline-earth metals
- Transition metals
- Post-transition metals
- Metalloids
- Nonmetals
- Halogens (also nonmetals)
- Noble gases
- * Radioactive isotopes

APPENDIX D

TOPOGRAPHIC MAP SYMBOLS

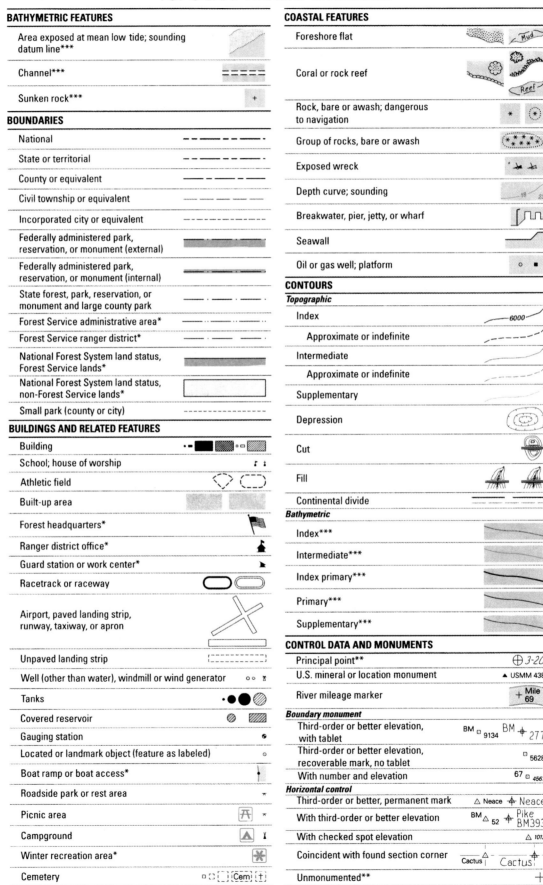

BATHYMETRIC FEATURES

Area exposed at mean low tide; sounding datum line***	
Channel***	
Sunken rock***	

BOUNDARIES

National	
State or territorial	
County or equivalent	
Civil township or equivalent	
Incorporated city or equivalent	
Federally administered park, reservation, or monument (external)	
Federally administered park, reservation, or monument (internal)	
State forest, park, reservation, or monument and large county park	
Forest Service administrative area*	
Forest Service ranger district*	
National Forest System land status, Forest Service lands*	
National Forest System land status, non-Forest Service lands*	
Small park (county or city)	

BUILDINGS AND RELATED FEATURES

Building	
School; house of worship	
Athletic field	
Built-up area	
Forest headquarters*	
Ranger district office*	
Guard station or work center*	
Racetrack or raceway	
Airport, paved landing strip, runway, taxiway, or apron	
Unpaved landing strip	
Well (other than water), windmill or wind generator	
Tanks	
Covered reservoir	
Gauging station	
Located or landmark object (feature as labeled)	
Boat ramp or boat access*	
Roadside park or rest area	
Picnic area	
Campground	
Winter recreation area*	
Cemetery	

COASTAL FEATURES

Foreshore flat	
Coral or rock reef	
Rock, bare or awash; dangerous to navigation	
Group of rocks, bare or awash	
Exposed wreck	
Depth curve; sounding	
Breakwater, pier, jetty, or wharf	
Seawall	
Oil or gas well; platform	

CONTOURS

Topographic

Index	
Approximate or indefinite	
Intermediate	
Approximate or indefinite	
Supplementary	
Depression	
Cut	
Fill	
Continental divide	

Bathymetric

Index***	
Intermediate***	
Index primary***	
Primary***	
Supplementary***	

CONTROL DATA AND MONUMENTS

Principal point**	3-20
U.S. mineral or location monument	USMM 438
River mileage marker	Mile 69

Boundary monument

Third-order or better elevation, with tablet	BM 9134 BM 277
Third-order or better elevation, recoverable mark, no tablet	5628
With number and elevation	67 4567

Horizontal control

Third-order or better, permanent mark	Neace Neace
With third-order or better elevation	BM 52 Pike BM393
With checked spot elevation	1012
Coincident with found section corner	Cactus Cactus
Unmonumented**	+

CONTROL DATA AND MONUMENTS – continued
Vertical control
Third-order or better elevation, with tablet	BM × 5280
Third-order or better elevation, recoverable mark, no tablet	× 528
Bench mark coincident with found section corner	BM + 5280
Spot elevation	× 7523

GLACIERS AND PERMANENT SNOWFIELDS
- Contours and limits
- Formlines
- Glacial advance
- Glacial retreat

LAND SURVEYS
Public land survey system
Range or Township line	
Location approximate	
Location doubtful	
Protracted	
Protracted (AK 1:63,360-scale)	
Range or Township labels	R1E T2N R3W T4S
Section line	
Location approximate	
Location doubtful	
Protracted	
Protracted (AK 1:63,360-scale)	
Section numbers	1 - 36 1 - 36
Found section corner	
Found closing corner	
Witness corner	WC
Meander corner	MC
Weak corner*	

Other land surveys
- Range or Township line
- Section line
- Land grant, mining claim, donation land claim, or tract
- Land grant, homestead, mineral, or other special survey monument
- Fence or field lines

MARINE SHORELINES
- Shoreline
- Apparent (edge of vegetation)***
- Indefinite or unsurveyed

MINES AND CAVES
- Quarry or open pit mine
- Gravel, sand, clay, or borrow pit
- Mine tunnel or cave entrance
- Mine shaft
- Prospect
- Tailings
- Mine dump
- Former disposal site or mine

PROJECTION AND GRIDS
Neatline	39°15' 90°37'30"
Graticule tick	55'
Graticule intersection	
Datum shift tick	

State plane coordinate systems
Primary zone tick	640 000 FEET
Secondary zone tick	247 500 METERS
Tertiary zone tick	260 000 FEET
Quaternary zone tick	98 500 METERS
Quintary zone tick	320 000 FEET

Universal transverse mercator grid
UTM grid (full grid)	273
UTM grid ticks*	269

RAILROADS AND RELATED FEATURES
- Standard gauge railroad, single track
- Standard gauge railroad, multiple track
- Narrow gauge railroad, single track
- Narrow gauge railroad, multiple track
- Railroad siding
- Railroad in highway
- Railroad in road
- Railroad in light duty road*
- Railroad underpass; overpass
- Railroad bridge; drawbridge
- Railroad tunnel
- Railroad yard
- Railroad turntable; roundhouse

RIVERS, LAKES, AND CANALS
- Perennial stream
- Perennial river
- Intermittent stream
- Intermittent river
- Disappearing stream
- Falls, small
- Falls, large
- Rapids, small
- Rapids, large
- Masonry dam
- Dam with lock
- Dam carrying road

TOPOGRAPHIC MAP SYMBOLS

RIVERS, LAKES, AND CANALS – continued

Feature	
Perennial lake/pond	
Intermittent lake/pond	
Dry lake/pond	
Narrow wash	
Wide wash	
Canal, flume, or aqueduct with lock	
Elevated aqueduct, flume, or conduit	
Aqueduct tunnel	
Water well, geyser, fumarole, or mud pot	
Spring or seep	

ROADS AND RELATED FEATURES

Please note: Roads on Provisional-edition maps are not classified as primary, secondary, or light duty. These roads are all classified as improved roads and are symbolized the same as light duty roads.

Feature	
Primary highway	
Secondary highway	
Light duty road	
Light duty road, paved*	
Light duty road, gravel*	
Light duty road, dirt*	
Light duty road, unspecified*	
Unimproved road	
Unimproved road*	
4WD road	
4WD road*	
Trail	
Highway or road with median strip	
Highway or road under construction	
Highway or road underpass; overpass	
Highway or road bridge; drawbridge	
Highway or road tunnel	
Road block, berm, or barrier*	
Gate on road*	
Trailhead*	

* USGS-USDA Forest Service Single-Edition Quadrangle maps only.

In August 1993, the U.S. Geological Survey and the U.S. Department of Agriculture's Forest Service signed an Interagency Agreement to begin a single-edition joint mapping program. This agreement established the coordination for producing and maintaining single-edition primary series topographic maps for quadrangles containing National Forest System lands. The joint mapping program eliminates duplication of effort by the agencies and results in a more frequent revision cycle for quadrangles containing National Forests. Maps are revised on the basis of jointly developed standards and contain normal features mapped by the USGS, as well as additional features required for efficient management of National Forest System lands. Single-edition maps look slightly different but meet the content, accuracy, and quality criteria of other USGS products.

SUBMERGED AREAS AND BOGS

Feature	
Marsh or swamp	
Submerged marsh or swamp	
Wooded marsh or swamp	
Submerged wooded marsh or swamp	
Land subject to inundation	

SURFACE FEATURES

Feature	
Levee	
Sand or mud	
Disturbed surface	
Gravel beach or glacial moraine	
Tailings pond	

TRANSMISSION LINES AND PIPELINES

Feature	
Power transmission line; pole; tower	
Telephone line	
Aboveground pipeline	
Underground pipeline	

VEGETATION

Feature	
Woodland	
Shrubland	
Orchard	
Vineyard	
Mangrove	

** Provisional-Edition maps only.
Provisional-edition maps were established to expedite completion of the remaining large-scale topographic quadrangles of the conterminous United States. They contain essentially the same level of information as the standard series maps. This series can be easily recognized by the title "Provisional Edition" in the lower right-hand corner.

*** Topographic Bathymetric maps only.

APPENDIX E

GEOLOGIC TIME SCALE

GEOLOGIC TIME SCALE 659

APPENDIX F

MODIFIED MERCALLI INTENSITY SCALE

Mercalli Magnitude	Effects Observed
I	Not felt except by a very few under especially favorable conditions.
II	Felt only by a few persons at rest, especially on upper floors of buildings. Delicately suspended objects may swing.
III	Felt quite noticeably by persons indoors, especially on the upper floors of buildings. Many people do not recognize it as an earthquake. Standing vehicles may rock slightly. Vibration similar to the passing of a truck. Duration estimated.
IV	Felt indoors by many, outdoors by few during the day. At night, some awakened. Dishes, windows, doors disturbed; walls make cracking sound. Sensation like heavy truck striking building. Standing vehicles rocked noticeably.
V	Felt by nearly everyone; many awakened. Some dishes and windows broken. Unstable objects overturned. Pendulum clocks may stop.
VI	Felt by all; many frightened. Windows, dishes, glassware broken, books fall from shelves, some heavy furniture moved; a few instances of fallen plaster. Damage slight.
VII	Difficult to stand upright. Furniture broken. Damage negligible in buildings of good design and construction; slight to moderate in well-built ordinary structures; considerable damage in poorly built or badly designed structures; some chimneys broken. Noticed by persons driving vehicles.
VIII	Damage slight in specially designed structures; considerable in ordinary substantial buildings, with partial collapse. Damage great in poorly built structures. Fall of chimneys, factory stacks, columns, monuments, walls. Heavy furniture overturned.
IX	General panic. Damage considerable in specially designed structures, well-designed frame structures thrown out of plumb. Damage great even in substantial buildings, with partial collapse. Buildings shifted off foundations.
X	Some well-built wooden structures destroyed; most masonry and frame structures destroyed along with foundations. Track rails bent.
XI	Few, if any, (masonry) structures remain standing. Bridges destroyed. Track rails greatly bent.
XII	Damage total. Lines of sight and level distorted. Objects thrown into the air.

Credit: US Geologic Survey, Department of the Interior/USGS

APPENDIX G

LANDFORM REGIONS OF THE CONTINENTAL UNITED STATES

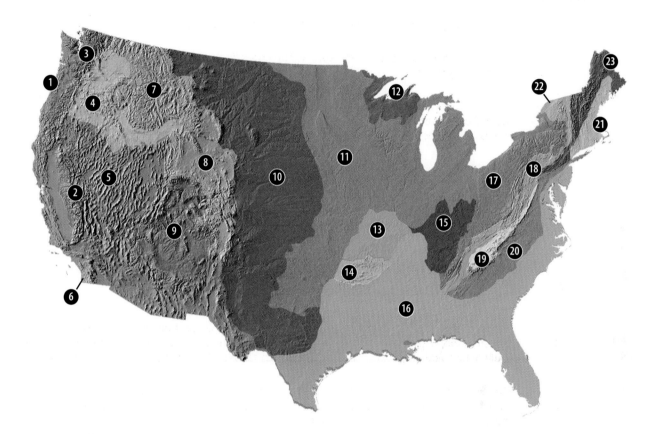

1. Coastal Ranges
2. Sierra Nevada Range
3. Cascade Range
4. Columbia Plateau
5. Great Basin and Range Province
6. Lower California
7. Rocky Mountains
8. Wyoming Basin
9. Colorado Plateau
10. Great Plains
11. Central Lowland
12. Superior Upland
13. Ozark Plateau
14. Quachita Mountains
15. Interior Low Plateaus
16. Coastal Plain
17. Applachian Plateau
18. Valley and Ridge Province
19. Blue Ridge
20. Piedmont Plateau
21. Southern New England
22. Adirondack Province
23. Northern New England

APPENDIX H

RELATIVE HUMIDITY

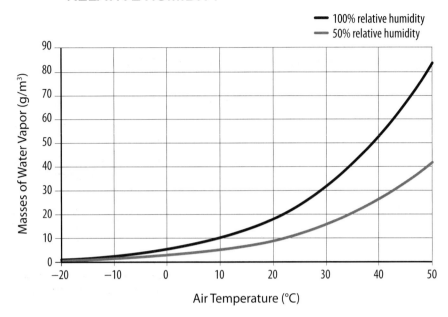

Relative Humidity (%)

		Differences between wet-bulb and dry-bulb temperatures (°C)													
		1.0	2.0	3.0	4.0	5.0	6.0	7.0	8.0	9.0	10.0	11.0	12.0	13.0	14.0	15.0
Air temperature (dry bulb) (°C)	0°	81	64	46	29	13										
	1°	83	66	49	33	17										
	2°	84	68	52	37	22	7									
	3°	84	70	55	40	26	12									
	4°	85	71	57	43	29	16									
	5°	86	72	58	45	33	20	7								
	6°	86	73	60	48	35	24	11								
	7°	87	74	62	50	38	26	15								
	8°	87	75	63	51	40	29	19	8							
	9°	88	76	64	53	42	32	22	12							
	10°	88	77	66	55	44	34	24	15	6						
	11°	89	78	67	56	46	36	27	18	9						
	12°	89	78	68	58	48	39	29	21	12						
	13°	89	79	69	59	50	41	32	23	15	7					
	14°	90	79	70	60	51	42	34	26	18	10					
	15°	90	80	71	61	53	44	36	27	20	13	6				
	16°	90	81	71	63	54	46	38	30	23	15	8				
	17°	90	81	72	64	55	47	40	32	25	18	11				
	18°	91	82	73	65	57	49	41	34	27	20	14	7			
	19°	91	82	74	65	58	50	43	36	29	22	16	10			
	20°	91	83	74	66	59	51	44	37	31	24	18	12	6		
	21°	91	83	75	67	60	53	46	39	32	26	20	14	9		
	22°	92	83	76	68	61	54	47	40	34	28	22	17	11	6	
	23°	92	84	76	69	62	55	48	42	36	30	24	19	13	8	
	24°	92	84	77	69	62	56	49	43	37	31	26	20	15	10	5
	25°	92	84	77	70	63	57	50	44	39	33	28	22	17	12	8
	26°	92	85	78	71	64	58	51	46	40	34	29	24	19	14	10
	27°	92	85	78	71	65	58	52	47	41	36	31	26	21	16	12
	28°	93	85	78	72	65	59	53	48	42	37	32	27	22	18	13
	29°	93	86	79	72	66	60	54	49	43	38	33	28	24	19	15
	30°	93	86	79	73	67	61	55	50	44	39	35	30	25	21	17

APPENDIX I

TORNADO/HURRICANE CATEGORY SCALES

Enhanced Fujita-Pearson Tornado Categories

RATING	3 SECOND GUST		DAMAGE (from original Fujita scale)
	KPH	MPH	
EF0	105–137	65–85	Damage is light, such as damaged trees and tree limbs, damaged signs and chimneys.
EF1	138–177	86–110	Damage is moderate. Winds are strong enough to push over unanchored manufactured homes and destroy attached garages. Roof surfaces may be peeled off and some tree trunks may be snapped.
EF2	178–217	111–135	Damage is considerable, such as manufactured homes destroyed and residential roof structures damaged. Large trees toppled and debris can become airborne.
EF3	218–266	136–165	Damage is severe. Winds are strong enough that roofs and walls may be torn from buildings. Small structures and non-reinforced brick buildings destroyed. Most trees uprooted.
EF4	267–322	166–200	Damage is devastating, well-built houses destroyed, structures moved off their foundations. Cars blown some distance and large debris airborne.
EF5	> 322	> 200	Incredible damage. Reinforced concrete buildings damaged. Cars become airborne. Trees debarked and strong residential structures leveled off their foundations.

Credit: National Oceanic and Atmospheric Administration, Storm Prediction Center, US Department of Commerce

Saffir-Simpson Hurricane Categories

CATEGORY	SUSTAINED WIND SPEEDS		DAMAGE
	KPH	MPH	
1	119–153	74–95	Very dangerous winds will produce some damage. Residential roof surface and siding damage. Some tree branches snapped and shallow-rooted trees toppled. Power line damage.
2	154–177	96–110	Extremely dangerous winds will produce extensive damage. Well-constructed homes experience roof damage. Many shallow-rooted trees toppled. Week-long power outages possible.
3	178–208	111–129	Devastating damage. Well-built homes experience severe damage, including roof structural damage. Many trees uprooted or snapped. Week-long power and water outages possible.
4	209–251	130–156	Catastrophic damage will occur. Well-built homes experience severe damage, including loss of roof and some walls. Most trees uprooted or snapped. Month-long power and water outages possible. Many locations uninhabitable for weeks or months.
5	> 251	> 156	Catastrophic damage will occur. Many well-built homes destroyed. Most trees uprooted or snapped. Month-long power and water outages possible. Most of the area uninhabitable for weeks or months.

Credit: National Oceanic and Atmospheric Administration, National HurricaneCenter, US Department of Commerce

APPENDIX J

SOLAR SYSTEM DATA

Physical and Orbital Properties of the Sun, Planets, and Moon

Property		Sun	Mercury	Venus	Earth
Average orbital radius (AU)		NA	0.39	0.72	1.00
Diameter	km	1,391,980	4880	12,104	12,742
	mi	865,121	3025	7522	7919
Volume, compared with Earth's		1,300,000	0.056	0.866	1.00
Mass, compared to Earth's		333,000	0.055	0.815	1.000
Density (g/cm^3)		1.408	5.427	5.243	5.514
Average equatorial temperature (°C)		5500	67 (−170 to 430)	462	15 (−89 to 58)
Orbital period		NA	87.97 d	225 d	365.2425 d
Average orbital speed (km/s)		NA	47.3	35.0	29.8
Rotational period		25 d (equator) 34 d (poles)	58.6 d	−243 d (retrograde)	23.9 h
Number of satellites		Unknown 8 planets	0	0	1 moon
Atmospheric content		73.5% hydrogen 24.9% helium 0.8% oxygen trace of many others	42% oxygen 29% sodium 22% hydrogen 6% helium 0.5% potassium	96.5% carbon dioxide 3.5% nitrogen	78% nitrogen 21% oxygen 1% other
Gravity, compared to Earth's (g)		27.94	0.38	0.90	1.00
Inclination of axis from perpendicular to orbit (°)		7.25 (to ecliptic)	0.0	177	23.5
Albedo		NA	0.14	0.69	0.37

	Moon	Mars	Jupiter	Saturn	Uranus	Neptune
	0.00257	1.52	5.20	9.55	19.22	30.10
	3474	6792	142,984	120,536	51,118	49,540
	2159	4221	88,650	74,732	31,770	30,790
	0.020	0.151	1321	764	63.1	57.7
	0.0123	0.107	318	95.2	14.5	17.1
	3.344	3.934	1.326	0.687	1.27	1.638
	−53 (−173 to 117)	−63 (−143 to 35)	−108	−139	−197	−201
	27.32 d	1.88 y	11.9 y	29.5 y	84.0 y	164.8 y
	1.022	24.08	13.1	9.69	6.80	5.43
	27.32 d	24.62 h	9.93 h	10.6 h	−17.2 h (retrograde)	16.1 h
	NA	2 moons	69 moons 4 rings	62 moons many rings	27 moons 18 rings	14 moons 5 rings
	trace gases	96% carbon dioxide 1.9% argon 1.9% nitrogen	90% hydrogen 10% helium traces of methane, ammonia	96% hydrogen 3% helium trace of methane	83% hydrogen 15% helium methane ammonia	80% hydrogen 19% helium 1% methane
	0.17	0.38	2.53	1.07	0.89	1.14
	6.69	25.2	3.13	26.7	97.8	28.3
	0.14	0.17	0.52	0.50	0.51	0.41

GLOSSARY

A

abrasion The process by which wind-driven sand erodes exposed rock. This process is similar to artificial sand blasting.

absolute magnitude A star's brightness as it would appear at a standard distance from that star.

abyssal plain The relatively flat, deep floor of an ocean basin.

acid test A test used to identify certain minerals by observing their reaction with dilute hydrochloric acid to produce gas bubbles.

active volcano A volcano that is currently erupting or has erupted during recorded history.

actual height The height of a mountain above its base. A mountain's base is the bottom of the mountain relative to the surrounding terrain. The base may be at, above, or below sea level.

air mass A huge body of air in the troposphere with approximately uniform temperature and humidity.

anthracite The densest and hardest form of coal; very dark in color and more than 90% carbon. Anthracite burns with the hottest and least smoky flame of any form of coal.

anticyclone The high-pressure weather system from which winds spiral outward in a clockwise direction in the Northern Hemisphere or counterclockwise in the Southern Hemisphere.

aphanitic (rock) A fine-grained igneous rock with crystals too small to see even with a microscope; often found in extrusive igneous rocks.

aphelion The point in an orbit around the sun at which an astronomical body is farthest from the sun.

apogee The point in an orbit around the earth at which the orbiting body is farthest from the earth.

apparent magnitude A star's brightness as viewed from Earth.

aquifer A stratum of permeable rock that stores and transports groundwater sandwiched between impermeable strata.

arctic air mass (A) A frigid air mass originating over the poles.

asteroid A rocky object in orbit around the sun that meets the International Astronomical Union (IAU) definition of a small solar system body (SSSB). Unlike meteoroids, they can be large enough to be observed from Earth.

astronomical unit (AU) A unit of astronomical distance equal to the earth's average distance from the sun (150 million km); a useful unit for measuring distances within and near the solar system.

atmosphere The envelope of gases that surrounds the earth or any other astronomical body held in place by that body's gravity. Also used as a unit of measure for atmospheric pressure, abbreviated *atm*: 1 atm is defined as the normal barometric pressure at sea level.

atmospheric pressure The weight of atmospheric gases.

atoll A ring of low coral islands and reefs surrounding a central lagoon.

atom The smallest neutral particle of an element that retains the characteristics of that element. Atoms are composed of a positive nucleus containing protons and usually neutrons, surrounded by negatively charged electrons.

atomic number A number that defines each element. It is the number of protons in an atom of that element.

autonomous underwater vehicles (AUV) An unmanned, untethered, submersible vehicle that uses artificial intelligence to control its operations. AUVs operate without any direct human control and are an economical way to collect data from deep in the oceans over a long period of time.

autumnal equinox The moment at which the sun is directly over the Equator, marking the start of autumn. *Equinox* means "equal night."

B

base level The lowest elevation to which a stream can flow.

bathyscaphe The earliest untethered deep-sea ocean exploration vehicle that provided a limited ability to explore the sea bottom.

Bathysphere The first deep-water manned submersible. It consisted of a hollow steel ball that was equipped with windows and was lowered and raised on a steel cable.

beach The portion of the coast that is shaped by sea level and wave action. It includes the berm, the shore, and any longshore bars produced by breaking waves.

benthic organisms An animal or plant that lives on or is anchored to the ocean bottom.

Big Bang theory A naturalistic model for the origin of the universe. It claims that the universe began about 13.5 billion years ago from a single dense point of matter that violently expanded into its present size through completely natural processes.

binary star A pair of gravitationally connected stars that revolve around the same center of gravity. Astronomers believe that most stars in our galaxy are binary stars.

biological weathering The process by which the actions of plants, animals, and other organisms break down rocks.

bituminous coal Moderately dense, hard coal; about 70% carbon and dark brown to black in color

black hole An astronomical object so massive and dense that its intense gravity prevents even light from escaping. Black holes can form from the collapse of a supernova remnant. Extremely massive black holes may be at the center of most galaxies.

blue shift The apparent shift in the spectrum of stars to shorter wavelengths at the blue end of the spectrum. This effect can be explained by the star's movement toward the earth.

boiling The change of state from a liquid to a gas at the boiling point. Boiling is a form of vaporization that occurs when the pressure in the liquid exceeds the atmospheric pressure.

boiling point The temperature at which a liquid changes to a gas.

breaker An unstable wave whose crest falls or breaks into foam. Waves typically break when a wave's height becomes more than one-seventh of its wavelength.

buoyant force The upward force that a fluid exerts on an object in the fluid. The magnitude of the buoyant force is equal to the weight of the displaced fluid.

C

carbonate A mineral family composed of minerals that contain one or more metal atoms and carbonate. The carbonate ion (CO_3^{2+}) is an important rock-forming substance that also takes part in the carbon cycle and produces carbon dioxide when treated with an acid during the acid test of identification.

carbon cycle The flow of the element carbon and its compounds between Earth's carbon reservoirs. These reservoirs include the crust, atmosphere, ocean, plants, and animals. The carbon cycle replenishes the carbon compounds required by living organisms.

cardinal direction One of the four key geographic directions: north, south, east, or west.

cartography The art and science of making maps.

catastrophic plate tectonics A model of the plate tectonics theory that proposes that the continents we see today formed during the one-year period of the Flood. The rapid motion during the Flood accounts for all the tectonic features that we see today.

cave A naturally occurring underground hollow in earth, rock, or ice.

Cepheid variable An unusual star that regularly changes brightness as it expands and contracts.

***Challenger* expedition** The first true oceanographic expedition beginning in 1872 and lasting four years. The expedition utilized a British warship, the HMS *Challenger*, modified for the voyage. The expedition set the standard for all future oceanographic voyages until well into the twentieth century.

change of state The process by which a material changes from one state of matter to another by adding or removing energy.

chemical change The change from one pure substance to another by the rearrangement of atoms in a chemical reaction. We typically recognize chemical changes by a change in color, change in energy, formation of a gas, and/or formation of a solid.

chemical weathering Chemical changes that dissolve rocks or break them down into smaller pieces.

chromosphere The lowest layer of the sun's atmosphere starting at the photosphere and extending out to 2000 km from the surface.

cinder cone volcano A small, steep-sided volcano made mainly of cinder-like pyroclastic materials. Cinder cones generally form during brief, explosive emissions of cinders and ash.

clast A piece of broken or eroded rock. Clasts can range in size from microscopic particles to boulder-sized fragments.

clastic sedimentary rock A rock made from eroded fragments or clasts of other rocks.

cleavage A property of some mineral crystals that easily split along distinct planes. Cleavage varies from no cleavage to perfect cleavage. Mineralogists use cleavage to help identify minerals.

climate The average weather for an area over the span of many years.

climate zone An area of the world that has a particular kind of climate that is based mainly on precipitation and temperature.

climatology The study of long-term weather conditions.

cloud A mass of extremely tiny liquid water droplets or ice crystals in the atmosphere.

coal An organic sedimentary rock formed from fossilized terrestrial plant matter. The three main types of coal—classified on the basis of the extent of metamorphism—are lignite, bituminous coal, and anthracite.

coalescence The process by which droplets join together and grow to form raindrops or hail.

cold front A boundary where a cold air mass displaces a warm air mass.

comet An astronomical body in orbit around the sun that meets the International Astronomical Union (IAU) definition of a small solar system body (SSSB) and is made of rock and/or ice. When a comet is close enough to the sun, it forms a coma and tail that are often visible from Earth to the unaided eye.

compound A pure substance formed by chemically combining two or more elements.

compound mineral A mineral made of a chemical compound. Mineralogists group chemically similar compound minerals into mineral families.

compression A contact force that acts to squeeze or crush an object or substance.

condensation The change in state from a gas to a liquid caused by removing energy. Condensation occurs at the dew point and is the opposite of vaporization.

condensation nucleus A microscopic particle that is suspended in the air and on which water easily condenses to form droplets.

conduction The movement of thermal energy through a medium. The transfer of thermal energy by direct contact between two substances. An example is the heating of the atmosphere when it contacts the ground.

cone of depression A cone-shaped dip in a water table at a well pipe; produced when water is pumped out at a higher rate than it is replenished.

conservation The preservation and wise use of natural resources, usually with consideration for the current and future needs of humans.

constellation One of eighty-eight groups or patterns of stars used to subdivide the heavens to locate and name individual stars. The names of constellations are based on elements from historical culture, such as Greek mythology.

contact metamorphism Changes in the composition, texture, and structure of rocks (metamorphism) caused mainly by contact with high-temperature magma or lava. This kind of metamorphism still occurs today in areas of active volcanism.

continental air mass (c) A low-humidity air mass originating over land.

continental drift The slow movement of continents as tectonic plates shift over millions of years according to the continental drift model.

continental drift theory A model that explains the shape and arrangement of present continents and their features as the product of the slow motion of tectonic plates over millions of years.

continental rise A thick deposit of loose sediments that creates a transition from the continental slope to the deep, relatively flat ocean floor.

continental shelf The shallow, submerged edge of continents extending from the beach.

continental slope The steeper incline from the lip of the continental shelf into the deep ocean basin.

contour line Any line on a relief map that connects points of equal elevation above or below sea level.

convection The transfer of thermal energy by the flow of a fluid caused by differences in density. Warmer, less dense material moves upward as it is displaced by a cooler, more dense fluid.

convergence The lifting of air into the atmosphere when horizontal air currents from opposite directions collide.

convergent boundary The margin between two tectonic plates that are moving toward each other. Converging boundaries are often locations of subduction and mountain building.

coral reef A massive underwater geologic feature that forms in relatively shallow tropical waters from the secretions of coral polyp colonies.

core In Earth, the dense, extremely hot central region of the earth's interior consisting of two parts: a liquid outer core and a solid inner core. The core makes up about 15% of the earth's volume. In general, a core is the center of any celestial object. In a star, for example, the core is where nuclear fusion powers the star.

Coriolis effect The deflection of the path of a moving object over the earth's surface due to the earth's rotation. Objects in the Northern Hemisphere are deflected to the right; those in the Southern Hemisphere are deflected to the left. The Coriolis effect is an important factor in ocean currents and atmospheric winds.

corona The hot outer atmosphere of the sun that extends far out into space to become solar wind. Temperatures rise to millions of degrees. It is the location of solar prominences.

Cosmic Microwave Background Radiation (CMBR) Electromagnetic energy in the microwave band that appears to come from all directions in space; predicted by the Big Bang theory as stretched light from the original big bang.

cosmogony The scientific study of the origin of the universe.

cosmology The study of the universe, specifically how the universe began and how it has changed over time. Cosmology is deeply affected by a scientist's worldview.

crater A bowl-shaped depression on an astronomical body's surface. Craters can be caused by volcanoes, geysers, or meteorite impact.

Creation Mandate God's command in Genesis 1:28 to manage the earth wisely and to use it for His glory and man's benefit.

cross-section A view of an object in which it is sliced in half. This allows the viewer to see the internal structure of the object and the arrangement of its parts.

crust The solid, relatively low-density outermost layer of rocky planets and natural satellites. It lies under loose surface materials and the oceans.

cyclone The low-pressure weather system toward which surrounding winds spiral counterclockwise in the Northern Hemisphere and clockwise in the Southern Hemisphere. It is sometimes used informally or locally for various kinds of cyclonic storms, such as hurricanes in the southwestern Pacific Ocean.

D

dark energy A hypothetical form of energy that permeates all space and causes the universe to expand by exerting a gravitational repulsion that counteracts ordinary gravitational attraction.

dark matter Theoretical matter that makes up approximately 27% of the universe. This matter is not observable via electromagnetic radiation and therefore cannot be seen by normal scientific means. Hypothesized to account for the observed gravitational effects within galaxies.

data Any information scientists collect by observing nature.

deep submergence vehicle (DSV) A manned submersible designed to carry a small crew to extreme ocean depths for scientific research. DSVs allow on-the-spot observation and sample collection by human crews providing some mobility, though they are limited by a battery power supply.

deep-time view The idea that the earth is ancient—billions of years old. The term represents the secular approach to the history of geology, the earth, and the universe according to the principle of uniformitarianism.

deflation The action of wind over a large area to remove small clastic particles, leaving behind cobbles and larger rocks that form desert pavement.

density current A current resulting from a difference in density between two masses of water in contact with each other. Higher density water will flow under the less dense water, displacing it. Density differences can result from differences in temperature, salinity, or the amount of suspended sediment.

deposition (change of state) The change in state from a gas directly to a solid caused by removing energy. Deposition is the opposite of sublimation.

deposition (geology) The process in which solid particles drop from a moving fluid to the bottom of the fluid. Deposition usually occurs when the flow rate decreases.

depositional mountain A mountain landform produced by the deposition of volcanic materials or sediments by ice, water, or wind. Examples include volcanoes, drumlins, eskers, kames, and sand dunes.

desalination The process of removing salt and other chemicals from seawater to make it drinkable and/or to exploit chemical resources in seawater.

descriptive data Data describing an observation using words. Descriptive data depends more on the observer's judgment, so it may not be as reliable or repeatable as measured data. Also known as *qualitative data*.

desert pavement Surface material consisting of cobbles and larger rocks left behind after wind has removed small clastic particles through deflation.

dew point The temperature at which the relative humidity of an air mass is 100%. If the air mass's temperature drops further, condensation will begin.

diluvial geologic time scale A creation model of the earth's geologic history that classifies and arranges rocks in the sequence of their formation relative to the biblical flood. The diluvial geologic time scale divides the earth's history into the Creation event, the time before the Flood, the Flood event, the time immediately after the Flood (including the Ice Age), and recent time.

dip The angle of slope of a fault face or stratum, measured from the horizontal plane; also called *dip angle*. Along with the strike, dip helps indicate the orientation of the fault surface or stratum at a given location.

dip-slip fault A fault whose main motion is parallel to its dip direction.

divergent boundary The margin between two tectonic plates that are moving away from each other. Diverging boundaries are the locations of sea-floor spreading and rift valleys.

divide A ridge of mountains, hills, or elevated land that separates one drainage basin from another.

diving bell A large, bell-shaped submersible lowered from a ship in order to explore or work on the sea bottom. Replenishing air in the diving bell was one of the most difficult problems with this device.

Doppler effect The apparent change in the wavelength (and frequency) of a wave due to the relative motion between the observer and the source. Waves will appear longer if the source and observer move apart; they appear shorter if the observer and source move closer together.

dormant volcano A volcano that has not erupted within recorded history. Seismic activity beneath it may indicate that it could erupt in the future. The difference between a dormant and an active volcano is artificial and often misleading, since most of the world's geologic history is incomplete or only recently documented.

downwelling The downward movement of surface water in places where seawater has piled up due to wind or current action.

drainage basin The land area drained by a stream and bounded by the basin divide.

dry climate A climate zone that gets little precipitation, or where the amount of evaporation is greater than the precipitation received; includes arid and semiarid regions like deserts and steppes.

dune A depositional landform that results from wind-deposited sand and soil. Blowing sand continually reshapes dunes, resulting in their drift over time.

dwarf planet An astronomical body in orbit around the sun that has enough mass to be rigid, is nearly spherical, but has too little gravity to have cleared its orbit of other objects of significant size according to the International Astronomical Union (IAU). A dwarf planet can share its orbit with other bodies.

dynamic metamorphism Changes in the texture and structure of rocks (metamorphism) that occur when conditions rapidly change in rocks. This process occurred more in Earth's history than in the present, though it may still occur in rocks near very active faults.

earthquake A measurable movement of the earth's crust. Earthquakes release energy stored by the accumulation of stress along faults from tectonic activity.

earth science The knowledge and practices of observing the earth and its processes. Earth science involves using appropriate tools to create models that describe and explain how the earth works.

echolocation The process used by certain animals to navigate using the echoes of sound waves that they emit.

ecliptic The sun's apparent path among the stars as observed from Earth; equivalent to the plane of the earth's orbit.

Ekman spiral The varying deflection of current with increasing depth beneath the water's surface current due to fluid friction and the Coriolis effect.

electromagnetic energy The combined action of electrical and magnetic energies in the form of waves; also called *radiant energy*. Comprises a spectrum that includes radio, microwave, infrared, visible, and ultraviolet light as well as x-rays and gamma rays.

element A pure substance made of only one kind of atom.

elevation The height of a point on the earth's surface.

elevation profile A description of how a stream changes elevation from beginning to end.

ellipse A geometric figure that appears as a flattened circle. It is constructed from two points called *foci*. The orbit of a planet around the sun is an ellipse with the sun at one focus.

El Niño Southern Oscillation (ENSO) A shift in pressure systems, winds, and currents in the equatorial and southern Pacific Ocean regions that repeats approximately every five years. An ENSO may last from about nine months to two years and results in significant temporary changes in climate around the world.

energy The ability to do work. Energy exists in many forms and can transfer between objects or change form.

epicenter The epicenter is the location on the earth's surface directly above an earthquake's focus.

Equator The great circle of latitude perpendicular to the earth's rotational axis. The Equator is 0° latitude and divides the earth into the Northern and Southern Hemispheres.

erosion The process of wearing away rock and transporting sediments.

erosional mountain A mountain landform carved out by extensive erosion, usually from a plateau; also called a *residual mountain*. Examples include buttes and mesas.

eutrophication A natural or artificial increase of nutrients in a lake or stream. Eutrophication causes blooms of algae and bacteria, reduces oxygen levels, increases turbidity, and can change the kinds of organisms living in the body of water.

evaporation The change of state from a liquid to a gas below the boiling point but above the freezing point. Evaporation is a form of vaporization that occurs when particles near the surface of the liquid gain sufficient energy to escape the liquid.

exfoliation A mechanical weathering process in which changes in temperature or the removal of the weight of overlying rock breaks down thin layers or slabs of rock.

exoplanet Any planet outside our solar system.

exosphere A temperature layer in the earth's atmosphere above the thermosphere (above 500 km) where the atmosphere thins into interplanetary space.

extinction The process by which a species, family, or larger group of organisms dies out.

extinct volcano A volcano that has never erupted within historical times, shows no seismic activity, and is often heavily eroded.

eye (hurricane) The low-pressure center of a hurricane containing relatively clear, calm, rising air. The eye usually has light winds and low precipitation and pressure.

F

fault A joint or crack in rock where the rock on both sides of the crack has moved relative to each other. Faults may range in size from microscopic to many kilometers long.

fault-block mountain A mountain landform produced by tectonic forces moving large blocks of crust relative to each other.

fetch The distance that wind blows over the water's surface in one direction. Fetch affects the height, length, and speed of the waves that are produced by wind.

first law of thermodynamics Scientific law that states that matter and energy cannot be created or destroyed, but can only transfer between objects or transform to a different form. Also called the *law of conservation of matter and energy*.

flood basalt A thick layer of igneous rock that covers a large area of the earth's surface; also called a *basalt trap*.

floodplain A broad, level area surrounding a low-gradient river created by regular flooding of the river. Floodplains usually have very fertile topsoil that has been deposited by floods.

fluid A state of matter in which the particles can flow. Liquids, gases, and plasmas are all fluids.

focus The point of an earthquake's origin deep within the earth. One of two points that define an ellipse, such as a planet's orbit.

fog Stratus clouds at ground level.

fold mountain A mountain landform created by folded rock strata.

foliated metamorphic rock Metamorphic rock containing flattened or flaky mineral crystals, giving it a banded or layered appearance.

force A push or a pull on an object.

fossil Any remains or trace of a formerly living organism preserved by natural processes.

fossil fuel A natural energy source that is believed to have formed from fossilized organisms. Fossil fuels include coal, petroleum (oil), and natural gas.

fracture The characteristic manner in which some minerals break; a property of minerals used for identification.

freezing The change in state from a liquid to a solid caused by removing energy. Freezing is the opposite of melting.

freezing nucleus A microscopic particle suspended in the air on which water easily deposits to form ice crystals.

front The boundary where at least two different air masses meet.

frontal wedging The lifting of a warm air mass above a cold air mass acting as a wedge.

frost wedging A mechanical weathering process in which water trapped in cracks and pores of rock freezes and expands, creating fragments of rocks.

G

galaxy A huge mass of billions of stars, dust, and other astronomical objects held together by gravity.

gas A fluid state of matter in which the particles move so fast and are so far apart as to have little interaction with each other. Gases have both a changeable shape and volume.

geocentric theory An ancient model of the solar system that placed the earth at the center with the sun and planets orbiting the earth in perfectly circular paths.

geographic coordinate The latitude and longitude used to locate a point on the earth's surface.

geographic information system (GIS) A system that uses digital map data, programs, hardware, and data storage systems to create and display information on maps. GIS also involves trained users who gather, plot, and analyze geographic information. GIS is particularly useful in that users can add or remove layers of data to show geographic relationships.

geologic column The theoretical collection of rock layers in the order in which they formed. According to the principle of superposition, the lowest layers are oldest while the uppermost layers are youngest, having formed later. Although scientists have never found a complete column in any one location, geologists believe that by matching portions of partial columns from different locations they know the complete column.

geologic time scale A model of the earth's geologic history that classifies and arranges rocks in the periods of their supposed formation. The secular geologic time scale arranges rocks from 4.5 billion years ago to the present. Geologists assign ages to rocks in the column using radiometric dating and fossils according to the assumptions of their worldview.

geology The study of Earth's structure, materials, and processes.

geothermal energy The energy generated and stored in the earth that can be harnessed to generate electrical power and heat.

geothermal gradient The rate at which temperature rises with increasing depth into the earth's interior.

geyser A hot spring that forcefully ejects hot water and steam from the ground at regular intervals.

glacial drift Sediments deposited by a glacier.

glacial till Unsorted glacial drift consisting of clay, sand, pebbles, cobbles, and boulders.

glacier A large mass of dense, compacted snow and ice that flows downhill under the influence of gravity. Glaciers were and are important means of land surface erosion and deposition.

Global Positioning System (GPS) A system of satellites, computers, and receivers that is able to determine the geographic coordinates of a receiver on Earth.

global warming A form of climate change involving a general increase of average global temperatures. Most climate scientists believe that the world has recently entered a period of global warming; many also believe that human activity causes this warming.

grain Pieces of mineral crystals, rock fragments, and even fossils in a rock that determine its texture.

gravity The attractive force between any two objects based on their masses. The strength of gravity decreases rapidly as the two objects move apart.

greenhouse effect A process by which a planet's atmosphere keeps energy from being lost to space. The greenhouse effect protects life on Earth by retaining sufficient energy to maintain liquid water.

greenhouse gas Natural atmospheric gases such as water vapor, carbon dioxide, and methane that warm the atmosphere by the greenhouse effect.

greenhouse hypothesis The hypothesis that global warming is produced by human activities, such as burning fossil fuels. Also known as *anthropogenic global warming*.

groundwater Natural water found underground that accounts for 96% of the world's unfrozen fresh water.

gyre A large, circling ocean current created by prevailing winds and deflected by the Coriolis effect and continental coastlines.

H

habitability zone The region around the sun or another star in which conditions are right for life to exist like it does on Earth.

halide A mineral family of salt compounds. Halides include the mineral halite, which is chemically the same as table salt (NaCl).

hardness In mineralogy, a mineral's ability to resist scratching that relates to the strength of its crystal structure. Mineralogists use the Mohs Hardness Scale to assign values of hardness to identify minerals. In groundwater chemistry, it is a description of the amount of dissolved minerals in groundwater. Harder water contains more dissolved minerals.

heliocentric theory A model of the solar system that places the sun at its center with the earth and the other planets orbiting around it.

Hertzsprung-Russell (H-R) diagram A graph used to classify stars on the basis of their color/temperature class and their brightness, or luminosity. Since luminosity is related to size and temperature, luminous stars tend to be hotter or larger, and dimmer stars tend to be smaller or cooler.

heterosphere A region of the earth's atmosphere in which there are distinct layers of different gases.

highlands climate A microclimate representing one of the lowland climate zones closer to the poles at some higher elevation further from the poles. The microclimates found on mountains are often significantly different from the surrounding lowland climate zones.

hill A natural elevation of the earth's surface rising to a summit. Hills are lower than mountains as defined by local and historical traditions. There is no scientific distinction between hills and mountains.

historical science The investigation of events that happened in the unobservable past by observing evidence in the present. Historical science depends heavily on a scientist's worldview.

homosphere A layer of the earth's atmosphere in which there is a uniform mixture of gases.

horizon (soil) A relatively uniform layer of soil. Fully developed soil typically has three horizons.

humidity A measure of the amount of water vapor in the air. Absolute humidity is measured in grams of water per cubic meter of air. Relative humidity is the ratio of the amount of water vapor in the air compared to the maximum that could be vaporized at the current temperature.

humus Decayed organic matter in the soil's A-horizon that holds water and gives plants nutrients.

hurricane An immense cyclonic windstorm that forms over tropical or subtropical oceans; hazardous to ships at sea and extremely damaging if it moves over land; also called a *typhoon* or *cyclone*.

hypothesis An initial explanation for a scientific problem. It is the starting point for testing the validity of an explanation.

I

ice age A period of Earth's history in which glaciers covered large portions of the earth. Also known as a *glacial period*. Secular models include numerous ice ages, while creation scientists believe that *the* Ice Age occurred after the Flood.

igneous Having to do with molten or formerly molten earth materials.

igneous rock A rock formed from solidified magma or lava.

impermeability The property of a substance that prevents other material from passing through it. In geology, the property of some rock to prevent water from passing through it.

index fossil A fossil that regularly appears in certain layers of the geologic column. Paleontologists use index fossils to assign relative ages to rock layers.

inferior planet Planets whose orbits lie between the earth and the sun; includes Mercury and Venus.

intensity A measure of the concentration of energy or the effects of the energy released. In earthquakes, it is a measure of the damage that an earthquake causes using the Modified Mercalli Intensity (MMI) scale. See Appendix F.

intermittent stream A stream that does not flow year-round; flows only after heavy rainstorms, from snowmelt, or during wet seasons.

International Date Line (IDL) An imaginary line through the Pacific Ocean which marks the starting point for each new day. The IDL follows 180° longitude except where doing so would divide portions of the same country.

International Space Station (ISS) Currently the only permanently manned space station. Five space agencies run this joint program, which provides a facility for scientific research in the microgravity environment of space.

intrusive volcanism Any process or feature that involved the forcible entry of magma into a preexisting rock formation.

ion A charged atom or group of atoms caused by a charge imbalance, which is produced by an unequal number of protons and electrons.

ionosphere A region of variable height in the earth's atmosphere where solar radiation breaks gas molecules into ions. The ionosphere reflects long-wave radio signals but is transparent to high-frequency radio signals, a useful property for communication around the globe.

irrigation Any artificial method for supplying water to crops; essential for producing crops from dry but fertile land.

island arc A long, curved string of volcanic islands formed by submarine volcanoes along the margins of converging tectonic plates.

isostasy The balance between the downward weights of rock, water, and ice, and the upward, buoyant force exerted by the mantle. Isostasy explains why less dense continental crust is thicker under mountains, and why dense, thin ocean crust is underwater.

J

jet stream A belt of high-speed winds in the stratosphere that flow across the upper and middle latitudes.

joint A crack in a rock that shows no indication of motion on either side of the crack. Joints in rock usually occur in groups and are related to the direction of the stress that formed them.

Jovian planet A planet similar in size and composition to Jupiter; includes the outer planets (Jupiter, Saturn, Uranus, and Neptune).

K

karst topography Distinct topography produced by the erosion and collapse of solution caves in thick strata of chemical sedimentary rocks; includes tower karst topography, sinkholes, natural bridges, disappearing streams, streamless valleys, and solution caves.

L

lake Any isolated body of water that doesn't freely empty into the ocean.

lander A space probe designed to land softly on an astronomical body. Landers transmit photographs and environmental data from the body's surface back to Earth.

landform A shape or structure on the earth's surface such as a mountain, valley, beach, plain, desert, plateau, or lake.

landslide The rapid downward movement of earth material.

lapse rate The steady drop of temperature with altitude in the troposphere. The average lapse rate is 6.4 °C/km.

latitude, lines of Also known as *parallels*. Any line on a map or globe parallel to the Equator, including the Equator itself; they are all circles that are perpendicular to the axis of rotation of the earth. Latitudes range from 0° at the Equator to 90° north (N) at the North Pole, and 90° south (S) at the South Pole.

lava Molten rock that has escaped from the earth's interior. Magma is molten rock in the earth's interior.

law A description of the relationship between two or more variables that is based on repeated observations. A law describes something that happens without any explanation as to why it happens. Laws are often mathematical.

leap year A year in the Gregorian calendar having 366 days. In a leap year, an extra day is added to the end of February to align the calendar to the astronomical year.

lightning An atmospheric electrical discharge that occurs either between clouds or between a cloud and the ground.

light-year (ly) The distance that light travels in a year; about 9.5 trillion km.

lignite Dark brown, soft, crumbly coal that is about 30% carbon and includes liquid organic substances. Lignite can contain fibrous fossilized plant matter.

limnology The science of surface and underground fresh water, including lakes and rivers.

liquid A fluid state of matter in which the particles are free to move but are held close to each other. A liquid substance has an unchanging volume but a changeable shape.

lithosphere The upper mantle and the crust that move together with tectonic plates.

loam An especially fertile topsoil containing about 40% sand, 40% silt, and 20% clay; often contains humus.

loess A thick deposit of fine, windblown dust and silt that provides an excellent base for rich soil.

longitude, lines of Also known as *meridians*. A line of longitude is any line on a map, perpendicular to parallels, that indicates the direction of true north. Scientists also define it as a semicircle on a globe that connects the North and South Poles. Longitude ranges from 0° at the Prime Meridian to 180° west (W), and 180° east (E).

longshore current An ocean current that flows parallel to the shore and that is caused by waves or wind that approach the shore at an angle.

lunar eclipse An astronomical event that occurs when the sun, earth, and moon are aligned so that the earth blocks the sunlight falling on the moon; can occur only at a full moon.

lunar phase The regular, repeatedly changing appearance of the shape of the lighted face of the moon visible from Earth; caused by the relative positions of the earth, moon, and sun.

lunar tide A tide produced by the gravitational interaction between the earth, moon, and oceans.

luster The amount and quality of light that a mineral reflects; can be used to help identify minerals.

M

magma Molten rock in the earth's interior. When magma moves to the earth's surface it is classified as lava.

magnetosphere The geomagnetic field that surrounds the earth originating in the earth's interior and extending to 160,000 km. The magnetosphere protects living things from the damaging effects of radiation from the sun.

magnitude The size or intensity of some measurable quantity. In seismology, it is a measure of an earthquake's energy. In astronomy, it is a measure of a star's brightness.

main sequence A region in the Hertzsprung-Russell (H-R) diagram containing most stars in the universe. The main sequence is the long, narrow band that stretches diagonally across the H-R diagram.

mantle Hot, plastic rock that fills the interior of many rocky planets and some natural satellites. Located between the crust and the core. In the earth, the mantle makes up 84% of the planet's volume.

map A model of an area of the earth's surface.

map projection The transfer of geographic information from a 3D globe onto the 2D surface of a map. Map projections always distort geographic information.

map scale The ratio of a distance represented on a map to the same distance on the earth's surface. Two numbers separated by a colon indicate map scales. A large scale fraction indicates that the scale of the map is large, meaning that the area of the earth's surface that it represents is small.

mare A broad, lowland plain on the moon made of hardened lava; darker and more level than surrounding highland areas on the moon's surface.

maritime air mass (m) A high humidity air mass originating over the oceans.

mass The measure of the amount of matter in an object.

mass wasting The process by which gravity transports rocks and soil downhill. Creeps, slumps, landslides, and rockslides are examples of mass wasting.

matter Anything that has mass and takes up space; a physical substance.

meander A wide, looping bend in a low-gradient stream.

mean sea level (MSL) The calculated average height of high and low tides in a particular location. It is the reference height used by cartographers, mariners, and aircraft pilots to measure elevation, ocean depth, and altitude.

measured data Data that usually consists of numbers with units. Scientists produce measured data using instruments. Measured data is usually more objective than descriptive data. Also known as *quantitative data*.

mechanical weathering The physical process that breaks down rock into smaller pieces without changing its chemical composition.

melting The change of state from a solid to a liquid at the melting point that is caused by adding energy. Melting is the opposite of freezing.

melting point The temperature at which a solid changes to a liquid.

mesosphere A temperature layer of the earth's atmosphere above the stratosphere and below the thermosphere, extending from about 50 km to about 80 km, the top of the homosphere. It contains the coldest temperatures in the atmosphere.

metamorphic rock Rock formed when igneous, sedimentary, or other metamorphic rocks have been altered by extreme temperature, pressure, or chemical action.

metamorphism The process that alters the composition, texture, and structure of rocks by extreme heat, pressure, or chemicals.

meteor A rocky object from space that falls through the earth's atmosphere and heats up until it glows. It is commonly called a *shooting star*.

meteorite A meteoroid that is large enough to have survived the fall through Earth's atmosphere to hit the ground. Larger meteorites can form craters.

meteoroid A rocky object in orbit around the sun that meets the International Astronomical Union (IAU) definition of a small solar system body (SSSB). Unlike an asteroid, a meteoroid is too small to be observed from Earth.

meteorology The scientific study of the atmosphere, including its composition, structure, weather, and alteration.

microgravity A condition such as a free fall or orbital motion in which an object appears weightless even in the presence of a strong gravitational force. Astronauts in orbit around the earth live and work in microgravity.

mid-ocean ridge A submerged mountain system uplifted by rising mantle rock at a divergent boundary between oceanic tectonic plates.

Milky Way The barred spiral galaxy in which the earth is located.

mineral A naturally occurring, inorganic, crystalline solid that has a definite chemical composition.

mineralogy The study of minerals including their formation, composition, structure, and alteration.

mixed-gas diving A diving system used for extremely deep dives that reduces the medical hazards of breathing normal air under high pressures. High-pressure oxygen is toxic, and high-pressure nitrogen produces a narcotic effect. Mixed-gas diving reduces the pressure of oxygen at great depths and replaces most of the nitrogen with less dangerous gases such as helium.

mixture A physical combination of two or more substances. Mixtures can be separated by physical means.

model A simple, useful, workable representation of something in the world.

Mohorovičić discontinuity (Moho) The boundary between the earth's crust and mantle.

Mohs scale A scale of 1 to 10 used for rating the relative hardness of a mineral. Using simple tests, mineralogists rate the hardness of an unknown mineral and compare it with the hardness of known minerals, ranging from talc (1) to diamond (10).

molecule Two or more chemically bonded atoms. Molecules can be made of just one kind of element or a combination of elements.

moon A natural satellite of a planet.

moraine A sedimentary ridge or surface formed by glacial till after a glacier melts or retreats. Moraines can form at the front of or to the side of a glacier (terminal or lateral moraine), in a glacial valley (ground moraine), or where two or more glaciers merge (medial moraine).

mountain A natural elevation of the earth's surface rising to a summit. Mountains are higher than hills as defined by local and historical traditions. There is no scientific distinction between hills and mountains.

mountain range A series of mountain peaks in the same geographic area.

mountain system A group of mountain ranges connected over a large area. Mountain systems are often produced by related tectonic processes.

mouth (river) The point at which a stream empties into another body of water; located at base level for that stream.

N

native mineral A mineral made up of only one element that naturally occurs in its pure state.

natural gas A gaseous fuel often found with petroleum; consists mostly of the organic gas methane.

natural resource Any raw material from the natural environment that we can use.

neap tide A tide that occurs when the sun's gravity works against the moon's gravity. Neap tides change sea level less than normal tides. Neap tides occur when the sun, earth, and moon form a right angle.

nebula An immense cloud of interstellar gas and dust. Nebulas may glow, reflect light, or form dark regions visible against a brighter background of stars.

nebular hypothesis The secular model for the origin of all stars and planetary systems throughout the universe. It suggests that a star system forms slowly from a cloud of gas and dust (called a *nebula*) as the matter clumps together under the influence of the gravity within the nebula. The densest part becomes the star, and planets form from the dusty disk surrounding the star.

neutron star A city-sized dense object that is the remnant of a supernova or collapsed white dwarf. Matter is so condensed that it is made of only neutrons. Some spinning neutron stars produce beams of radio waves and are detected as pulsars.

nitrogen cycle The flow of the element nitrogen and its compounds between the nitrogen reservoirs. These reservoirs include the atmosphere, ocean, and soil. The nitrogen cycle replenishes the nitrogen compounds required by living organisms.

nonclastic sedimentary rock A sedimentary rock composed of materials that are not clastic erosion sediments; nonclastic sedimentary rocks formed from chemical precipitation or from the deposition of marine fossils.

nonfoliated metamorphic rock A metamorphic rock lacking bands or layers; tends to break into sharp, angular pieces.

nonrenewable resource A natural resource that people cannot replenish. People can conserve nonrenewable resources by recycling or finding alternative materials to meet a need.

normal fault A dip-slip fault in which the upper body of rock drops relative to the lower body of rock.

nova A violent nuclear explosion that occurs when a white dwarf draws in hydrogen from a larger companion star in a stellar binary system. The hydrogen accumulates until it begins an unstable fusion reaction, which blows off the outer shell of matter.

nuclear change A change of one element into another element by changes occurring in the nucleus of the atom. Nuclear changes can occur when atoms emit or absorb rays or particles. It can also occur in nuclear reactions when atoms split apart through fission or join together through fusion.

nuclear fission A nuclear reaction in which one large nucleus splits into smaller nuclei, releasing large amounts of energy.

nuclear fusion A nuclear change in which small atomic nuclei are forcefully combined to make larger nuclei, releasing tremendous amounts of energy as nuclear mass is converted to energy; the process that powers stars.

occluded front A front where a cold front displaces a warm front. This happens when a cold air mass quickly catches up to a warm air mass, which was replacing another cold air mass. The warm air mass is lifted between the two cold air masses and loses contact with the ground.

ocean A vast body of salt water that separates the continents.

oceanic trench Deep, relatively narrow troughs in the ocean floor found in subduction zones adjacent to convergent plate boundaries.

oceanography The exploration and study of the oceans, including their formation, structures, composition, and biology.

operational science The study of presently occurring scientific events.

orbit The regular, elliptical path one astronomical body takes around another. It can also refer to the act of following such a path.

ore A rock containing valuable minerals.

orogeny The tectonic processes and landforms of mountain building.

orographic lifting The upward movement of air masses as they flow over mountains.

oxide A mineral family composed of minerals containing oxygen and a metal. Oxide ores are important sources of metals.

ozone layer A layer of concentrated ozone (O_3) in the stratosphere that shields the earth from harmful ultraviolet light.

paleontology The study of life in geologic time, as represented by the fossils of animals, plants, and other organisms found in the geologic column.

Pangaea The last of a series of supercontinents proposed by secular geologists that became the continents of today through continental drift.

Panthalassa The ocean that surrounded the last supercontinent according to secular geologists.

parallax The apparent change in position of a distant object as one's viewing position changes; useful for determining the distances to relatively nearby stars (within 100 pc).

parsec (pc) A unit of astronomical distance equal to 3.26 light-years.

pedology The study of soils, including their formation, composition, and alteration.

pegmatite A coarse-grained igneous rock with very large interlocking crystals. Grains are often larger than 2.54 cm across.

pelagic organisms Animals and plants that are active swimmers or that drift with ocean currents.

perennial stream A continuously flowing stream; a stream that flows year-round.

perigee The point in an orbit around the earth at which the orbiting body is nearest to the earth.

perihelion The point in an orbit around the sun at which an astronomical body is nearest to the sun.

permeability The property of a substance that allows other material to pass through it. In geology, the property of some rock to allow water to pass through it.

petroleum A liquid fossil fuel found underground as crude oil. Petroleum is thought to be made of decomposed fossil marine organisms.

petrology The field of geology that focuses on the identification, classification, and analysis of rocks.

phanerite An igneous rock composed of distinct mineral grains that are small and more or less uniform in size.

photosphere The visible surface of the sun.

physical change A change in matter that doesn't alter its chemical identity. Physical changes include changes of state, changing shape, polishing, dissolving, etc.

planet An astronomical body in orbit around the sun that has enough mass to form a spherical shape and enough gravity to clear the region of its orbit of smaller objects of any significant size other than its moons according to the International Astronomical Union (IAU).

planetesimal Any natural object other than planets that orbit the sun.

plasma A fluid state of matter made of extremely high-temperature ionized particles. Plasmas have both a changeable shape and volume. Since stars are made of plasma, it is the most common state of matter in the universe.

plate tectonics theory A model of earth's structure that describes the earth's crust as made of semirigid plates floating on plastic rock in the mantle. Geologists use plate tectonics to explain sea-floor spreading, subduction, mountain formation, earthquakes, volcanoes, and all major changes to the earth's surface.

pluton An intrusive igneous feature that formed when a large volume of magma collected, cooled, and solidified within a chamber in the earth's crust.

polar air mass polar (P) A cool or cold air mass originating over upper latitudes.

polar climate A climate zone located in the polar zone that receives varying amounts of frozen precipitation with consistently cold temperatures; includes tundra and ice cap subzones.

polar easterly A global wind flowing out of the high-pressure zones over the poles toward the sub-polar lows. The Coriolis effect turns these winds so that they flow mostly from the east.

pollution The addition of anything to a resource that reduces its usefulness. Pollution may be in the form of matter or energy.

porphyry An igneous rock containing a few large mineral grains embedded in a very finely textured matrix.

precipitation (geology) Any process that causes chemical substances dissolved in water to form a solid compound that settles out of solution.

precipitation (meteorology) Any form of water that falls to the earth's surface from the sky.

pressure gradient The rate of change of atmospheric pressure with horizontal distance.

presupposition A basic assumption about the world that a person assumes is true.

prevailing westerly A global wind of the mid-latitudes that flows from the subtropical high toward the sub-polar low. The Coriolis effect turns the winds so that they consistently flow from the southwest to northeast (Northern Hemisphere) or from the northwest to the southeast (Southern Hemisphere).

Prime Meridian The semicircle on the globe that connects the North and South Poles and passes through Greenwich, England. This is 0° longitude.

principle of cause and effect A scientific principle that tells us that any observable result of a process—an effect—must have an adequate cause.

principle of superposition The assumption that the layers of rock in any undisturbed sequence of strata contain the oldest stratum at the bottom with the younger or later layers placed in order above it. According to this principle, strata and the features they contain, such as faulting or intrusive formations, can be assigned ages.

principle of uniformity A scientific principle asserting that the same process will always produce the same results. It assumes that the world operates in a reliable and consistent way that makes it predictable.

probe An unmanned space vehicle that follows a one-way path from Earth for the purpose of exploring space or another celestial object.

prominence A stream of solar plasma that rises into the corona from the chromosphere and then gradually falls back. Many prominences arch or loop as they follow strong local magnetic fields that are produced by sunspots.

pure substance A material made of only one kind of element or compound; not a mixture.

P wave The first earthquake wave that reaches a seismic station from an earthquake. P waves are the fastest seismic waves. They pass through all parts of the earth's interior and tend to be the least destructive.

pyroclastic material Sharp, fragmented, solidified lava, including ash, cinders, bombs, and blocks.

Q

quasar An unusual celestial object that is as bright as a galaxy but appears as compact as a distant star in a telescope. Quasar stands for *quasi-stellar object*.

R

radiation The emission of electromagnetic waves, such as with sunlight or heat lamps. A method of transferring thermal energy even in a vacuum. It also refers to the rays and particles emitted by the nuclei of radioactive materials.

radiometric dating A scientific and mathematical process by which scientists estimate the ages of objects by using the relative amounts of different elements in a sample and the half-life of that element.

radio telescope A telescope that detects objects in space that emit radio waves.

ray A bright streak that radiates from some craters on the moon's surface; in general, a geometric description of something that moves in a straight line away from a source, such as an electromagnetic wave or a light ray.

recharge zone An area on the land's surface that resupplies groundwater.

red giant A large, luminous, reddish star that develops when a star with about the mass of the sun ages to its final state of existence as its hydrogen fuel is used up.

red shift (cosmological) The apparent shift in the spectrum of a star to longer wavelengths as a result of the distance to the star. This condition was predicted by the Big Bang theory and is caused by the stretching of space time.

red shift (Doppler) The apparent shift in the spectrum of a star to longer wavelengths at the red end of the visible spectrum. This effect can be explained by the object's movement away from the earth.

reflecting telescope A telescope that uses mirrors and reflection to gather light.

refracting telescope A telescope that uses only lenses and refraction to gather light.

regional metamorphism Changes in the composition, texture, and structure of rocks (metamorphism) that occurred over large areas by heat, the pressure of overlying rock, or widespread tectonic forces.

relative humidity The ratio of the amount of water vapor in the air to the maximum that could be vaporized at the current temperature.

relief The variation in elevation within a specific region.

remotely operated vehicles (ROV) An unmanned, tethered, robotic submersible used to work on the ocean bottom, especially on deep-sea oil wells and in shipwreck salvage operations.

renewable resource A natural resource that has an unlimited supply or that can be easily replenished.

resource management Efforts to wisely use natural resources.

retrograde motion The apparent motion of superior planets to slow down, stop, reverse their direction, and then resume their normal motion. Retrograde motion is caused by the earth's higher orbital speed.

reverse fault A dip-slip fault in which the upper body of rock slides upward relative to the lower body of rock.

revolution Orbital motion around a point, such as the earth revolving around the sun.

Richter scale A scale that rates the magnitude of the intensity of an earthquake.

rille A long, narrow channel on a moon or planet that is believed to have originated from tectonic processes.

rip current A narrow, fast current that sometimes forms as water piling up on a beach flows back toward the ocean. Rip currents are very dangerous to swimmers.

rock A naturally occurring, solid material made of various minerals that compose most of the earth's crust.

rock cycle An old-earth theory that rock is recycled in large geochemical cycles similar to other natural resources such as water, carbon, and nitrogen. According to this view, igneous rocks were the first to form, changing into other kinds of rocks by erosional and tectonic processes that continue today. According to this theory, all rocks eventually melt into magma again. One complete cycle requires millions of years according to current rates of these individual processes.

rocket An artificial object powered by a reaction engine that produces and expels hot gases to create the reaction force needed to propel the rocket. Rockets do not need an atmosphere to work and have been used in air, underwater, and in space.

rotation The spin of an object around its center of mass, such as the earth rotating about its axis.

rover A type of space probe that is a mobile, robotic vehicle delivered by a lander and is designed to explore a planet's surface by remote control from Earth.

S

salinity The amount of dissolved salts in water from any source.

satellite A small natural or artificial object that orbits a larger astronomical body. Natural satellites of planets are called *moons*.

science The collection of observations, explanations, and models produced through an organized study of nature and the processes found in nature, for enabling people to exercise good and wise dominion over God's world. The word *science* also refers to the organized methods that produce these observations, explanations, and models.

scientific process An orderly way of investigating a question in science by using measurable and repeatable observations to test a hypothesis.

scuba A diving system that permits a diver to carry a supply of air for extended periods underwater without any connection to the surface. Scuba is an acronym for *Self-Contained Underwater Breathing Apparatus*.

sea A large section of ocean that is mostly surrounded by land or islands.

sea-floor spreading The motion of oceanic tectonic plates away from the mid-ocean ridges where mantle rocks rise to form new ocean crust.

sea level The level of the sea's surface at a particular place and time. Sea level changes with rising and falling tides.

seamount A submerged mountain. Most seamounts appear to be extinct volcanoes rising from the abyssal plain.

season Cyclical natural changes in temperature, precipitation, and the amount of daylight produced by the earth's axial tilt and its revolution around the sun.

sediment Small particles of eroded material, such as rock, soil, sand, or clay.

sedimentary rock A rock made of sediments produced by erosion, deposition, compaction, and cementation or by the precipitation of chemicals from a solution.

seiche Small oscillations in water level, similar to sloshing in a bathtub, caused by changes in air pressure as weather systems move across a body of water. Seiches in large lakes can produce oscillations up to several meters that can last for many hours or even days.

seismic wave Waves moving though the earth originating at the focus of an earthquake. Seismic waves include P waves, S waves, and surface waves.

seismology The field of geology that studies forces and movement within the earth's crust. This includes the study of tectonics, earthquakes, and resulting effects.

seismometer An instrument that detects the movement of the earth due to seismic waves.

shear Forces acting in opposite directions on different parts of the same object or substance. Opposing sheer forces are not aligned so that matter within the object tends to slip in layers, like spreading a deck of cards.

shield volcano A broad, flattened, dome-shaped volcano built up by successive quiet eruptions of runny lava.

shore The strip of land bounded by the lowest and highest tides that separates the coastal region from the ocean.

silicate A mineral family composed of minerals that contain silicon and oxygen (or silica, SiO_2). Silicates make up about 25% of all known minerals and about 40% of the common ores. More than 90% of the earth's crust is made of silicates.

small solar system body (SSSB) Any astronomical body in the solar system that doesn't meet the International Astronomical Union (IAU) definition of a planet, a dwarf planet, or moons. Asteroids, comets, and meteoroids are SSSBs.

soil Layers of eroded earth materials and organic matter (humus) on the earth's surface.

soil conservation Any method used to prevent soil erosion; includes strip cropping, contour plowing, flood control dams, and windbreak construction.

solar constant The flow rate of radiant energy from the sun through space to reach Earth. At the top of the atmosphere, the solar constant is about 1.367 kW/m².

solar eclipse An astronomical event that occurs when the sun, earth, and moon are aligned so that the moon blocks a portion of the sunlight falling on the earth; can occur only at a new moon.

solar flare A sudden explosion of the plasma from the sun's surface that emits a burst of energetic rays and particles into space. Solar flares are very dangerous to humans and orbiting satellites. Large flares are called *coronal mass ejections (CME)*.

solar spectrum The entire range of electromagnetic wavelengths that the sun emits; includes radio waves, microwaves, infrared light, visible light, ultraviolet rays, x-rays, and gamma rays.

solar system The sun, planets, small solar system bodies, and other objects held in orbit around the sun by the sun's gravity.

solar wind The sun's output of protons, electrons, and small atomic nuclei into the solar system; extends as far as three times the radius of the solar system.

solid A rigid state of matter in which the particles align in a fixed arrangement. A solid substance has an unchanging shape and volume.

solstice The time during the year at which the sun reaches its greatest angular distance from the Equator; occurs twice within a year.

solute A substance that is dissolved in a solvent to form a solution.

solution A mixture with a uniform appearance throughout; made by dissolving a solute in a solvent.

solution cave A cave that appears to have formed when acidic ground water dissolved native rock.

solvent A substance, usually a liquid, that dissolves other substances to form a solution.

sonar A form of artificial echolocation using sound waves to measure distance; an acronym for *SOund Navigation And Ranging*.

sorting The orderly pattern in which sediments deposit as a stream slows down and turbulence decreases. This process results in sediment particles forming layers of sediments of similar sizes.

sounding A remotely measured physical quantity in the depths of the sea or in the heights of the atmosphere. In the past, ships measured the depth of the ocean by taking soundings with a lead line; in modern times, sonar is used.

source (stream) The origin and the highest elevation of a stream.

space station A manned Earth satellite designed to be a permanent human habitat and laboratory in space.

Space Transportation System (STS) The STS was intended to be an economical, reusable manned space vehicle designed to launch heavy satellites and space probes, perform experiments in low-earth orbit, and lift International Space Station (ISS) components and supplies into orbit. The STS was retired from service in 2011 and all remaining vehicles are now on display in various museums. Also known as a *space shuttle*.

specific gravity (sp gr) The ratio of a mineral's density to the density of water at 4 °C. A mineral's specific gravity is used for identification.

speleology The scientific study of caves, including their origin, formation, structure, and the changes they undergo.

speleothem A cave formation made of precipitated minerals deposited by dripping or flowing groundwater; also called a *secondary limestone deposit*.

spring tide A tide that occurs when the sun's gravity works with the moon's gravity. Spring tides change sea level more than normal tides. Spring tides occur when the sun, earth, and moon are in line with each other.

stalactite An icicle-like speleothem that hangs down from a cave's ceiling.

stalagmite A conical speleothem that grows from a cave's floor; usually deposited from water dripping from an overhead stalactite.

star A celestial body that generates electromagnetic energy. It consists of a mass of plasma held together by its own gravity.

star cluster A group of stars close enough to be held together by gravity. Stars in clusters have the same proper motion.

state of matter The physical form of matter described by its physical characteristics. The state of a pure substance largely depends on its temperature. Matter can be a solid, liquid, gas, or plasma. *Phase of matter* is another term for state of matter.

stationary front A front where a cold air mass and a warm air mass meet, neither displacing the other.

station model A symbolic representation of data from a weather station on a weather map. Each number and symbol and their locations in the model provide specific weather information.

stellar aging The change in stars over time.

storm A severe weather disturbance involving high winds, heavy precipitation, and other extreme conditions, such as lightning or low temperatures.

storm surge A higher-than-normal local sea level caused by seawater flowing toward the lower air pressure under a hurricane and pushed ahead of the associated high winds.

strain Any change in the shape of a solid due to stresses exerted on the material.

stratigraphy The field of geology that examines how rock layers form.

stratosphere A temperature layer in the earth's atmosphere above the troposphere and below the mesosphere extending from about 11 km to about 50 km. Temperature increases with altitude in this region.

stratovolcano A medium-sized volcano built up of deposits of lava from quiet eruptions alternating with layers of pyroclastic materials from explosive eruptions. The tall conical landform is what people usually associate with volcanoes. Also called a *composite cone volcano*.

stratum A rock layer in a sequence of strata. According to the principle of horizontality, sedimentary strata form nearly horizontally. Tectonic processes can tilt, fold, or fault strata.

streak test A color test used to identify a mineral. In this test, a mineral is rubbed on a ceramic plate called a *streak plate* to produce a powder with its characteristic color.

stream Flowing water that follows a distinct course over the land. Streams may be any size, ranging from trickles to broad rivers, and may be perennial or intermittent.

stream gradient A stream channel's change of elevation with horizontal distance; also referred to as the *steepness* or *slope* of the channel.

stream system A stream together with its tributaries.

stress Any force exerted on an object.

strike The compass direction of the horizontal plane of the surface of a fault face or rock stratum.

strike-slip fault A fault along which the main movement is horizontal or parallel to the fault's strike; also called a *transform fault*.

subduction The tectonic process by which relatively thin and denser oceanic crust slowly slides down and under more massive but less dense continental crust.

sublimation The change in state from a solid directly to a gas caused by adding energy. Sublimation is the opposite of deposition.

submarine canyon An underwater erosional feature often found in a continental slope that seems to have been formed by turbidity currents.

subsurface current Any current in deeper water that flows in directions other than those caused by surface winds. These currents can result from the Ekman spiral, downwelling, upwelling, or density currents.

sulfide A mineral family composed of minerals containing one or more metals and sulfur. Sulfide ores are important sources of metals.

summer solstice The moment at which the sun's angular distance from the Equator is at a maximum. This is the longest day of the year and marks the start of summer in the hemisphere tilted toward the sun.

sunspot A small, cool area on the sun's photosphere. Though very hot, it appears dark next to the full brightness of the photosphere. Sunspots are associated with intense magnetic field disturbances at the sun's surface.

supergiant An immense star with a mass up to 70 solar masses. Supergiants can be anywhere from 30 to 500 solar diameters and are located in the top of the Hertzsprung-Russell (H-R) diagram.

superior planet A planet whose orbit lies at a distance from the sun greater than the earth's orbital distance. This group includes Mars, Jupiter, Saturn, Uranus, and Neptune.

supernova An astronomical object that is the violent end of stars of greater than about 8 solar masses. The brilliant explosion that forms the supernova leaves behind a less energetic astronomical object, such as a white dwarf, neutron star, or black hole.

surface current The near continuous flow of seawater on the surface of a body of water set in motion by wind. The Coriolis effect and continental coastlines both affect surface currents.

surface wave (geology) One of several kinds of earthquake waves that travel only along the surface of the earth. Surface waves are far more destructive because they occur mainly where humans live and build their structures.

surface wave (hydrology) A wave that occurs at the surface of the ocean, in comparison to an internal wave. Surface waves are formed by the friction of wind blowing over the water.

S wave The second earthquake wave that reaches a seismic station from an earthquake. S waves don't pass directly through the earth's core and are much more destructive than P waves.

synoptic weather map A weather map that presents a summary or synopsis of weather data for a given time frame. There are four synoptic weather maps: the Surface Weather Map, the Highest and Lowest Temperatures Map, the Precipitation Areas and Amounts Map, and the 500 Millibar Height Contours Map.

T

tectonic mountain A mountain landform created by tectonic processes.

tectonic plate A section of the earth's crust that moves as a unit compared to other regions of the crust. A tectonic plate may be mainly thick continental crust, thin oceanic crust, or a combination of both. Tectonic plates meet at tectonic boundaries.

tectonics The forces, energy, and processes that formed and reshaped the earth's continents, mountains, and ocean basins. Plate tectonics is the major unifying concept that accounts for nearly all geologic processes known today.

temperate continental climate A climate zone located in the temperate zone that has large temperature swings and moderate amounts of precipitation year-round, including humid continental and subarctic climate subzones.

temperate marine climate A climate zone located in the temperate zone with moderate temperatures and either seasonal or continual moderate precipitation, including Mediterranean and chaparral climate subzones.

temperature The hotness or coldness of a substance, stated in degrees; a measure of the average kinetic energy of the particles of a substance.

tension Forces that act to pull an object apart. Tensile forces act in line with each other.

tephra Rock fragments and particles ejected by a volcanic eruption.

terminator The line separating the light and dark sides of any astronomical body.

terrestrial planet A planet similar in size and composition to the earth, including the inner planets (Mercury, Venus, Earth, and Mars).

texture (rock) How a rock looks and feels according to the sizes of the mineral grains and other particles that compose the rock. Large grains give a rock a coarse texture; smaller grains give a rock a fine texture.

theory A model that attempts to explain a set of observations. A theory *explains*.

thermal turnover Seasonal exchange of surface and bottom water in lakes. It is caused by variations in density due to temperature differences. This process moves nutrients and oxygen throughout lakes.

thermohaline current A density current flowing along the deep ocean basin floor caused by gravity acting on large masses of cold, salty ocean water. Flow rates are far slower than surface currents.

thermosphere A temperature layer in the earth's atmosphere above the mesosphere extending from about 80 km to about 500 km. It has the hottest temperatures in the atmosphere.

thrust fault A reverse fault with a dip of less than or equal to 45°.

thunderstorm A rainstorm that produces lightning. May include hail, bursts of strong winds, microbursts, and even tornadoes.

tide A change in local sea level caused by the gravitational pull of the moon and sun.

topography The features of the earth's surface in a particular region, including different landforms with varying elevation.

tornado A destructive, localized, rapidly rotating cyclonic windstorm forming a funnel; usually associated with a special cumulonimbus cloud called a *supercell*.

trade wind A global wind of the low latitudes that flows from the subtropical high toward low-pressure centers at the Equator. The Coriolis effect deflects these winds toward the west, causing them to blow consistently northeast to southwest in the Northern Hemisphere or southeast to northwest in the Southern Hemisphere.

transform boundary The margin between two tectonic plates that are sliding in opposite directions parallel to the margin. Transform boundaries occur near faults that usually form across divergent boundaries in oceanic crust. On land, seismologists refer to them as *strike-slip faults*.

transit The act of passing over, across, or through; the motion of inferior planets as they cross through the earth's view of the sun.

transitional form An organism that scientists consider a step in the evolution from one kind of organism to another higher up on the evolutionary tree of life. Also known as a *missing link*.

tributary A smaller stream that feeds into a larger one.

tropical air mass (T) A warm air mass originating in the tropics.

tropical rainy climate A climate zone located in the tropical zone that is both very warm and very wet year-round; includes tropical wet zones and tropical rainforests.

tropical storm A strong cyclonic storm with winds of at least 63 kph. Strengthening tropical storms can develop into hurricanes.

troposphere The lowest temperature layer of the earth's atmosphere extending from ground level to about 11 km. Air temperature generally decreases steadily with altitude (called the *lapse rate*). This region is the location of most of the earth's weather.

tsunami A far-reaching, devastating water wave caused by seismic activity.

turbidity current A fast, subsurface density current similar to an underwater mudslide that is caused by large amounts of suspended sediments.

U

underwater habitat An underwater outpost that allows people to live and work on the ocean bottom for long periods without diving equipment. Two types of habitats are possible: a habitat at normal air pressure requiring an air lock and a habitat at sea pressure permitting a direct ocean entrance through a moon pool.

uniformitarianism The belief that all geologic processes are natural, have always been the same, and have always happened at the same gradual rate. It can be summed up by the statement, "The present is the key to the past." This view was once held as a fundamental principle in geology. Most secular geologists today do not strongly hold to this view, though they are still committed to naturalism.

unmanned underwater vehicle (UUV) Any type of unmanned submersible vehicle.

upwelling Upward movement of dense, cold, nutrient-rich water from deep in the ocean to the surface. Upwelling occurs when deeper seawater flows toward places where sea level is lower than average because surface winds have blown surface water elsewhere.

V

vaporization The change in state from a liquid to a gas caused by adding energy. Vaporization is the opposite of condensation.

vent An opening in the earth's crust from which lava, ash, and gases can escape.

vernal equinox The moment at which the sun is directly over the Equator, marking the start of spring. *Equinox* means "equal night."

viscous Having a high resistance to flow.

Volcanic Explosivity Index (VEI) An intensity scale that rates a volcano's destructiveness and explosive power on the basis of the amount of matter that it ejects during an eruption.

volcano An opening in the crust of the earth through which lava, ash, and gases flow; a depositional mountain built from the accumulation of the material that flows from such an opening.

volume The amount of space that an object occupies.

W

warm front A front where a warm air mass displaces a cold air mass.

water cycle The continuous movement of water from the oceans to the atmosphere by evaporation, then to the continents by condensation and precipitation, and returned to the oceans by streams. The water cycle is responsible for naturally purifying and replenishing fresh water supplies. Also called the *hydrologic cycle*.

water table The water surface of a groundwater reservoir; the boundary between the zone of aeration and the zone of saturation.

wave In general, a disruption in a material that carries energy through the material. In water waves, disruptions occur in the layers of water due to the interactions of forces on and between those layers.

wave base The lowest point of an ocean wave in which the water is affected by the wave's passage. The wave base of a deep-water wave is at a depth equal to about half its wavelength.

wave height The vertical distance between a wave's crest and its trough. A wave's *amplitude* is half this distance.

wavelength The distance between identical points on two successive waves (e.g., the distance between the crests of two successive waves).

wave period The time for one entire cycle of a wave to pass a fixed location.

wave speed The speed of a wave crest in the direction of movement. Wave speed depends on water depth and the wavelength of the wave.

weather The condition of the atmosphere at a given time and place.

weathering The process by which factors or conditions in the environment break rocks down into smaller pieces; includes biological, chemical, and physical weathering.

weight The force of gravitational attraction between an object's mass and its resident planet.

white dwarf The hot, dense core of a dead star. No nuclear fusion exists in a white dwarf, so there is no continuing source of heat. A white dwarf continues to slowly cool and eventually becomes a black dwarf.

wind Moving air traveling from regions of high pressure to those of low pressure.

wind-chill factor The perceived lowering air temperature as wind blows over unprotected skin. Wind-chill applies only to people and warm-blooded animals, not to inanimate objects.

winter solstice The moment at which the sun's angular distance from the Equator is at a maximum. This is the shortest day of the year and marks the start of winter in the hemisphere tilted away from the sun.

winter storm A storm that brings heavy snow, sleet or freezing rain, high winds, and cold weather.

work Work is done when a force acts on a moving object in the same or opposite direction of its motion or when energy is transferred from one object to another.

worldview The overall perspective that is based on presuppositions that a person uses to view and interpret the world.

Z

zone of aeration Porous rock and soil above the water table. Spaces between rock and soil particles contain air or a mixture of air and water.

zone of saturation Porous rock and soil below the water table. Spaces between rock and soil particles are filled with groundwater.

INDEX

Boldface page numbers denote the location of definitions of key terms.

A

3D map, 69, 71, 72
'a'a, 178
abrasion, **285**
absolute humidity, 457
absolute magnitude, **593**
absorption spectrum, 533
abyssal plain, 310, **311**
acceleration, gravitational, 30, 34
accretion, 207
acid test (mineral), **210**
acoustic sensing, 69
actual height, **155**
adaptive optics, 621
aeration, zone of, **400**
Aerophore, 370
aerospace engineer, 634
aftershock, 136, 145
age
    absolute, 102
    of the earth, 102, 110–14
    relative, 103
Age of Exploration, 358
AGILE, 623
air mass, 456, **475**–80
    arctic, **476**
    cold, 476, 477, 478
    continental, **476**
    maritime, **476**
    polar, **476**
    source region, 475–76
    tropical, **476**
    warm, 476, 478
albedo, 574
Aldrin, Edwin, 639
Aleutian Islands, 313
algal bloom, 326
Algol, 602
Alkaid, 591
alluvial fan, 284
alpha particle, 46
Alpine-Himalayan volcano belt, 181
aluminum, 202, 203
Alvin (DSV-2), 365

Amazon River, 380, 382–83, 384
ammonia, 438
amplitude, 346
Analogous Days theory, 403
Anders, Will, 76, 98
Andromeda, 603, 604
anemometer, 456
Angel Falls, 384
Annapolis Tidal Power Plant, 336
Antarctic Circle, 505
anthracite, 266–267
anticline, 160
anticyclone, 459–60
aphelion, 566
apogee, 547
Apollo 1, 638, 639
Apollo 6, 630
Apollo 8, 76–77, 98, 644
Apollo 13, 522, 634
Apollo 15 rover, 543
Apollo 16, 645
Apollo 17, 645
apparent magnitude, **593**
apparent motion, 546
appearing stream, 421
aquifer, **403**, 404, 405
Archaeopteryx, 113, 260
Arches National Park, 294
arctic air mass, **476**
Arctic Circle, 505
Arecibo Radio Telescope, 420, 622
arête, 289
Argyle Diamond mine, 216–17
Aristotle, 43, 561
Ark, the, 116–17, 121, 263
Armstrong, Neil, 639
arroyo, 386
artesian well, 403
asterism, 589
asteroid, 541, **578**–79
asteroid belt, 578–79
asteroid strike, 580
asthenosphere, 91
astrobiology, 605

astrogeologist, 584
astronomer, 19, 529
astronomical unit (AU), **531**
astronomical year, 549
astronomy, 19
Atacama Desert, 164, 341, 479
Atafu Atoll, 315
Atlantis, 361, 642
atmosphere, 80–81, **429**–49
    and carbon cycle, the, 437
    and Creation, 430
    and energy, 445–49
    and Flood, the, 430
    and nitrogen cycle, the, 438
    and temperature, 433–36
    composition of, 431–36
    origin of, 429–30
atmosphere (atm), 322
atmospheric pressure, **455**–56
atoll, **315**
atom, 9, **41**–46
atomic number, 42
aurora, 435
aurora australis, 444, 538
aurora borealis, 39, 444
autonomous underwater vehicles, **374**–75
autumnal equinox, 550
avalanche, 281, 282
average sea level. *See* mean sea level
axiom, 1
axis of symmetry, 211

B

Badwater alluvial fan, 284
Badwater Basin, 390
bar-headed goose, 463
Barnard's star, 596
barometer, 455
    aneroid, 455
barrier island, 351
barrier reef, 315
Barringer meteorite crater, 583
Barton, Otis, 372

basalt, 177
basalt trap, 183
base, 155
base level, **381**, 382
basin, 163
batholith, 193
bathyscaphe, **372**–73
Bathysphere, **372**
Baumgardner, John, 119
Baumgartner, Felix, 431
bauxite, 203
bay barrier, 351
Bay of Fundy, 334, 335
Bayer, Johann, 590
beach, **309**, 310
bedrock, 295
Beebe, William, 372
benchmark, 67
bends, the, 370
benthic organism, **324**
Bentley, Wilson "Snowflake", 469
berm, 309
Betelgeuse, 596, 597
Big Bang theory, **608**–11
Big Dipper, 591, 602
Bikini Atoll, 315
binary planet, 578
biological soil crust, 294
bioluminescence, 325
biome, 170
Biosphere 2, 523
bioturbation, 247
bituminous coal, **266**–67
black hole, **598**, 603, 606
Black Rock Salt Flats, 391, 392
Black Sea, 386
black smoker, 180
blowout, 285
blue shift, 595
bog, 392
boiling, 28, 45
boiling point, 28
bolide, 582
bond, chemical, 39
Bonneville Salt Flats, 391, 392
booster rocket, 629
Borman, Frank, 76, 98
bottom profiler, 366
boundary
    convergent, **131**, 140, 158

divergent, **131**, 140, 159
transform, **131**, 135, 140
Brachiosaurus, 263
brackish water, 319
Brahe, Tycho, 564
breaker, 309, **348**, 349
breccia, 232
brine, 319
brown coal, 266–67
Bryce Canyon National Park, 166
buoy, 366, 367
    weather, 367, 471
buoyant force, **131**
butane, 269
butte, 166

C

caldera, 176
calendar year, 549
Callisto, 574
calving, 517
Canadian Shield, 183
canopy theory, 443, 507
capillary action, 401
capillary fringe, 401
carat, 214
carbon cycle
    in the atmosphere, 437
    marine, **325**–26
carbon dioxide, 448
    atmosphere levels, 288
    increase in, 522
carbon-14, 270
carbonate, **216**
    in water, 407
carbonic acid, 277
cardinal direction, **54**
Carolina bay, 390, 392
cartographer, **53**, 54, 61–63, 70, 157, 358
Cascadia Subduction Zone, 151
CASIS PCG5, 35
Caspian Sea, 386
Cassegrainian reflector, 621
Cassini, 630
Cassini, Giovanni, 531
Cassini-Huygens, 631, 632
catastrophic plate tectonics, 119
catastrophism, 85
cause and effect, principle of, **16**
cave, **414**–21

glacier, 415
ice, 415
solution, **415**–16
Cave of the Crystals, 220, 238
caver, 421, 425
caves
    and the Flood, 416, 417
    and worldview, 416
celestial sphere, 546
Celsius, Anders, 454
Celsius, degree, 26
Celsius scale, 454
cementation, 231
Cenozoic era, 262
center pivot irrigation, 409
Centralia, PA, 268
century leap year, 549
Cepheid variable, **594**
Ceres, 578
CFC, 441
Challenger, 642, 644
    disaster, 638
Challenger Deep, 312, 373
Challenger expedition, **359**, 362
Chandra X-ray Observatory, 644
Chandrayaan-1, 629
change
    chemical, **46**
    nuclear, **46**–47
    physical, **45**
change of state, 27–29
chart, 52
chemical change, **46**
Chesapeake Bay, 319
Chicxulub, 263
Chief Mountain, 142–43
chinook, 462
chlorofluorocarbon (CFC), 441
cholera, 51
chromosphere, **535**
circuit, hydrothermal, 195
Circum-Pacific volcano belt, 181
cirque, 289
Civilian Conservation Corps (CCC), 297
Clarey, Dr. Timothy, 201
Clark, William, 387
clast, **232**
cleavage (mineral), **207**
climate, 502–25 (**504**)
    dry, 511, **513**

highlands, 511, **513**
history of change, 516
polar, 511–**12**
temperate continental, 511–**12**
temperate marine, 511–**12**
tropical rainy, 511, **513**
climate change, 288, 502, 503 515–25
and glaciers, 288
and volcanism, 519
and worldview, 516, 524
climate graph, 511
Climate Prediction Center, 498
climate subzone, 512
climate zone, **511**–13
climatologist, 107, 504, 510
climatology, 19, 504, 510
cloud, **464**–67
altocumulus, 466
altostratus, 467
cirrocumulus, 466
cirrostratus, 467
cirrus, 466–67
cumulonimbus, 467, 472
cumulus, 466–67
nimbostratus, 467
nimbus, 467
noctilucent, 436
stratocumulus, 466
stratus, 466–67
cloud formation, 464–65
cloud types, 466–67
CMBR. *See* cosmic microwave background radiation
coal, 93, 265, **266**–67, 270
anthracite, 266–**267**, 270
bituminous, **266**–67, 270
coking, 267
lignite, **266**–67, 270
coalescence, **468**
coastal region, 309
coccolith, 234
coelacanth, 261
coking coal, 267
cold front, 477, 478, 484
coliform bacteria, 412
Collins, Eileen, 644
Collins, Michael, 639
Columbia, 642
accident, 638
comet, **579**–81

Tempel 1, 631
Wild 2, 631
command module, 639
compaction, 231
compound, 42, **44**, 46
compression, 33, **130**
condensation, 27–**29**, 45, , 457, 469
condensation nucleus, **465**
conduction, **447**
cone of depression, **404**
conglomerate, 232
conservation
of energy, 40
of matter, 223
of resources, **503**
of soil, **297**
of water, 410–11
conservation of matter, law of, 47
constellation, **589**–91
continental air mass, **476**
continental crust, 154
Continental Divide, 383
continental drift, 107, 110
Continental Drift theory, **104**–5, 107, 110
continental glacier, 112, 120
continental plate, 105, 140, 244
continental rise, **310**
continental shelf, **309**, 310
continental slope, **310**
contour interval, 156
contour line, 63, **64**, 156, 183
contour plowing, 297
controlled experiment, 15
convection, **447**, 479, 484
convection current, 447
convergence, **480**, 484
convergent boundary, **131**, 140, 158
convergent landform, 159–60
Copernican Revolution, 17, 561–63
Copernicus, Nicholas, 17, 563
coprolite, 250
coral polyp, 316
coral reef, **314**–15, 316
core (Earth's), **92**
Coriolis effect, **339**, 340, 341, 342, 458, 460, 489, 491
corona, **535**
coronagraph, 536
coronal mass ejection, 538
cosmic microwave background radiation, 607–**8**, 609, 611, 613

cosmogony, **606**–614
and young-earth view, 612–13
cosmological red shift, 600, **607**, 609
cosmology, **606**
Cousteau, Jacques-Yves, 369, 370, 371
crater (lunar), 541–42
Bailly, 541
Giordano Bruno, 541
Tycho, 541
crater, meteorite, 582
crater, volcanic, **176**
crater chain, 542
Crater Lake, 176, 186, 388, 391
Crowfoot Mountain, 279
Creation, 7, 114, 115, 307
theories of cosmogony, 612–13
Creation Mandate, **3**, 4, 12, 117, 288, 298, 399, 503
creep, 281
Cretaceous-Paleogene extinction, 263
critternaut, 645
crude oil. *See* petroleum
crust, 78, **90**–91, 104–5, 108, 117, 119, 120, 194
crystal, 45, 220
crystalline, 45
Curiosity, 633
current, 331, 338–45
convection, 447
deep-water, 344
density, **344**–45
longshore, **349**
ocean, 337
rip, **349**
subsurface, **343**
surface, **338**–43
thermohaline, **344**
turbidity, 310, **345**
currents, and climate, 341
cyanobacteria, 326
cyclone, 459–**60**, 489

D

dark energy, 610
dark matter, 603, **610**
Darwin, Charles, 85, 260
Dasha solution, 613
data, **11**–12, 15
derived, 11
descriptive, **11**

measured, **11**
Davy Crater Chain, 542
Day Age theory, 124, 403
Dead Sea, 318, 319, 380, 390
Death Valley, 155, 284
declination (DEC), 592
decompression sickness, 370
Deep Impact, 570, 631
deep mining, 217
deep submergence vehicle, **373**–74
Deepsea Challenger, 373
deep-time view, **86**, 582
deep-water current, 344
deferent, 562
deflation, **285**
deforestation, 521
Deimos, 573
delta, 284, 383
demand regulator, 370
Democritus, 43
Denali, 154, 155, 157
dendritic pattern, 382
density, 30
density current, **344**–45
deposition
    change in state, 28, **29**, 470
    sediment, 165, 231, **277**, 349, 350–51
depositional landform, 164
desalination, 304, 320
desert pavement, 285
desiccant, 217
design, evidence for. *See* evidence for design
dew, 469–70
dew point, 29, **464**, 469
diamond pipe, 213
diatom, 319
dichlorodifluoromethane, 441
diffraction grating, 600
dike, 177, 192, 237
diluvial geologic time scale, **119**
dinosaur, 263
disappearing stream, 419, 421
Discovery, 642, 644
divergent boundary, **131**, 140, 159
divergent landform, 161–62
divide (basin), **383**
diving bell, **368**
DMSP Block-5D2, 625, 626
DNA, 256, 259

doldrums, 461
dome, 163
dominion, 3, 5, 12, 14, 20, 52, 69, 83–84, 95–96, 99, 297, 298, 301, 305, 355, 378–79, 553, 584, 619
    and climate, 503, 510
    and fossil fuels, 272
    and fresh water, 388
    and ocean motion, 331
    and oceanography, 360
    and paleontology, 262
    and resources, 95–96
    and soils, 297, 298
    and space exploration, 634
    and speleology, 422
    and water use, 399, 408, 410–11
    and weather, 498
Dominion Mandate. *See* Creation Mandate
Don Juan Pond, 390
Doppler effect, 481, 595, 600, 607
Doppler radar, 481
downdraft, 484
downwelling, **342**–43, 395
drag, 459
drainage basin, 380, 382–**83**
Drake, Frank, 605
Dreyer, Johan, 601
dripstone, 418
driverless car, 72, 73
drizzle, 468
drumlin, 169, 290
ductility, 132
dune, **286**
Dust Bowl, 297
dust storm, 285
dwarf planet, **567**, 577–78

E

E. coli (*Escherichia coli*), 412
Earth, 572
    age of, 110–14, 247
    core, **92**
    and dominion, 83–84
    magnetic field, 81
    rotation, 79
    structure of, 78, 90–92
    tilt, 79, 505, 520
    See also sun, moon, and earth system
Earth history, 110–11

Earth motion, 546–47
earth tide, 553
earthquake, 64, 90, 128–48 (**130**), 151, 276
    early warning system, 129, 147
    hazard, 146
earthquakes
    Bam, Iran (2003), 146
    Great Chilean (1960), 147
    Haiti (2010), 67, 145, 146
    Indian Ocean Sumatra-Andaman (2004), 145, 147
    Liaoning Province, China (1975), 148
    New Madrid, MO (1811), 140
    Northridge, CA (1994), 145
    Parkfield, CA (2004), 148
    San Francisco (1906), 147
East African Rift zone, 161
Easter Island, 182
Ebola, 52
echolocation, 323
    artificial, 366
eclipse, 550–52
    annular, 551
    lunar, 54, **552**
    partial solar, 551
    solar, 536, **551**–52, 536, 556
    total solar, 551
ecliptic, **546**, 547–48, 590
ecliptic plane, 546, 551, 552
eddy, 339
Eduardo Avaroa Andean Fauna National Reserve, 285
ejecta, 541
Ekman effect, 342, 343
Ekman grab sampler, 364
Ekman spiral, **341**, 342, 343
Ekman, V. Walfrid, 341
El Niño Southern Oscillation (ENSO), **518**–19
elasticity, 132
electric force, 34
electromagnetic wave, 445, 532
electron, 41–42, 43
electroplating, 203
element, **42**–44, 655
elevation, **155**
ellipse, **565**
elliptical crater, 542
Elpistostege, 261
emission spectrum, 533

Emperor Seamounts, 182, 312
Enceladus, 570
Endeavor, 642, 647
energy, **36**–40
    chemical, 39
    conservation of, 40
    dark, 610
    elastic potential, 136
    electrical, 38
    geothermal, **196**–97
    hydroelectric, 379
    kinetic, 36–37
    light, 39
    magnetic, 38
    mechanical, 37
    nuclear, 39
    potential, 37
    sound, 38
    sun, of the, 531, 532–33
    thermal, 37
engineer, 19
Enhanced Fujita-Pearson scale, 488, 663
Eniwetok, 315
Enterprise, 642
epicenter, **139**–40
epicycle, 562
epidemiology, 52
Equator, **55**–57, 61–62
equatorial orbit, 626
equinox, 550
erosion, 165, 230–31, 276, **277**, 281–91, 297, 298, 349, 350–51, 353
    and glaciers, 281, 287–89
    and gravity, 281–82
    and water, 281, 283–84
    and wind, 281, 285–86
esker, 169, 290
ethane, 269
Eurasian tectonic plate, 153, 181
Europa, 574
eutrophication, **393**
evaporation, **28**, 45, 457
evidence for design, 77–83, 121, 227, 407, 430, 431, 439–44, 570, 582, 615
exfoliation, **279**
exoplanet, 112, 570
exosphere, 435, **436**
Explorer 1, 442, 625
extinction, 99, **261**
extravehicular activity, 638
extremophile, 180, 417

extrusive volcanism, **175**
eye (hurricane), 489, **490**
eyepiece, 620
Eyjafjallajökull, 174, 175, 182

F

facula, 535, 537
Fahrenheit, Daniel Gabriel, 454
Fahrenheit, degree, 26
Fahrenheit scale, 454
fathom, 362
Faulkner, Danny, 611, 613
fault, 132, **133**–36, 140, 162
    dip, **134**
    dip-slip, **135**
    locked, 135
    normal, **135**, 161, 162
    reverse, **135**, 142
    strike, 133–34
    strike slip, 126, **135**
    thrust, 135
    transform, 135
fault line, 134
fault plane, 133
felsic, 177
fetch, 346–**347**
fin, 166, 167
fission, nuclear, 47
fjord, 289
flame test (mineral), 209
flash flood, 386
flood basalt, **183**
Flood geology, 154, 182–83, 307
Flood, the, 115, 116–20, 121, 143, 230, 263, 264, 293, 303, 307, 310, 312, 317–19, 357, 416, 417, 430, 443, 516, 521
floodplain, 284, 382, **384**
flowstone, 418
focus, **139**
foehn, 462
fog, **466**
foraminifera, 234
force, 31–35
    contact, 32–33
    electric, 34
    field, 34
    gravitational, 34
    magnetic, 34
    tectonic, 129–32, 158–59, 264
Forel, François-Alphonse, 394

fossil, 103, 104–5, 107, 113, 118, 123, 246, 248–72 (**249**)
    age of, 256
    carbon print, 250
    cast, 250, 254
    compression, 250
    index, 103, 256, **260**
    microfossil, 250
    mold, 250, 254
    original material, 250
    petrified, 250
    trace, 250
fossil fuel, **265**–72, 452
fossil record, 111, 113, 118–19
fossil trap, 255
fossilization, 249–57
Foulke, William, 263
Fourpeaked Volcano, 179
fracture, 132
    conchoidal, 208
fracture (mineral), **208**
fracture zone, 104, 131, 311
Framework hypothesis, 509
freezing, 27, 28, 45
freezing nucleus, **465**
freezing point, 27
freezing rain, 468
Freon, 441
frequency. *See* wave
friction, 33, 459, 460
Friendship 7, 637
fringing reef, 314
front, **477**–80
    cold, 477, 478, 484
    occluded, 478
    stationary, 477, 478
    warm, 477, 478
frontal wedging, **480**, 484
frost, 469–70
frost heaving, 279
frost wedging, **278**
Fukushima Daiichi nuclear power plant, 144
fumarole, 179
fusion, nuclear, 47, **534**

G

Gagarin, Yuri, 644
galactic wind, 560
galaxy, 546, **603**–4

barred spiral, 604
classification of, 603–04
elliptical, 604
irregular, 604
spiral, 604
GALEX, 623
Galileo, 541, 564, 619
Galileo (probe), 631
gamma ray, 39
Ganymede, 574
Gap theory, 7
gas, 26–29
gas giant, 569
Geiger, Rudolf, 511
geocentric theory, 17, **561**–62, 563, 564
geographic coordinate, 56, 57, 67
geographic information system. *See* GIS
geoid, 307, 308
geologic column, **102**–3, 111–12, 118–19, 260
geologic time scale, 110–11, 118–19, 659
diluvial, **119**
geologist, 18, 85–89, 90–92
geology, **85**–89
historical, 87–88
operational, 86–87
structural, 18
geometer, 54
geostationary orbit, 626, 627
geosynchronous orbit, 626, 627
geothermal energy, **196**–97
geothermal gradient, **194**
geothermal process, 28
geyser, **196**
Giardia lamblia, 413
GIS, **65**–69, 70, 137, 147, 361
and weather, 66, 69, 498
glacial drift, **290**
glacial erratic, 108
glacial period, 111, 120
glacial till, 168–69, **290**
glacier, 33, 106, 108, 111–12, 113, 120, **287**–91
continental, 112, 120, 288
terminus, 288
valley, 287
Glacier Bay National Park, 289, 294
Glacier National Park, 142
glaciers, and climate change, 288
glass eel, 337
Glenn, John, 637, 644

Global Positioning System. *See* GPS
global warming, **515**, 521, 522, 526, 527
and sea level, 308
globe, 52
globular cluster, 603
globular cluster M80, 603
glowing avalanche, 179
Gobi Desert, 168, 508
God's glory, declaring, 4, 20, 582
Goddard, Robert H., 625
gospel, 5
Gospel for Asia, 398, 399
GPS, 58–59, 67, 73, 367, 626
grab sampler, 364
graben, 161
gradient, 194
geothermal, **194**
stream, **380**–81
Grand Canyon, 8–9, 381
Grand Prismatic Springs, 191
Grand Tetons, 162
granite, 118, 226
granule, 535
gravitational acceleration, 30, 34
gravitational force, 34
gravitational lensing, 599
gravity, 29, 30, 34–35, 131
gravity assist, 630
Great Barrier Reef, 315
Great Basin and Range Province, 162
great circle, 55–56, 62
Great Depression, 297
Great Lakes, 388
Great Pacific Garbage Patch, 330, 331
Great Salt Lake, 319, 390
Great White Spot, 575
greenhouse effect, 443, **448**–49, 507
greenhouse gas, 272, **448**, 502, 521
greenhouse hypothesis, **521**
Greenwich, England, 56
Gregorian year, 549
Grenville Dome, 163
groin, 350
groundwater, 28, **399**–421
groundwater chemistry, 406–8
groundwater landform, 414–16, 418–21
Gruithuisen Dome, 543
Gulf Stream, 337, 341
Guilin province of China, 420
gust front, 484

guyot, 311, 312
gyre, 338, **339**, 340

H

Haber-Bosch Process, 438
habitability zone, 81–82
Hadley Rille, 543
hail, 469, 484, 485
Half Dome, 192–93
halide, **217**
Halley, Edmund, 581
hanging valley, 289
hard-hat diving, 368, 370
hardness (mineral), **208**
hardness (water), **407**
hard water, 407
Harrison, John, 357
Hawaiian Islands, 182
hazard, 146
Heart Mountain, 143
heat, 37
heliocentric theory, 17, **563**
heliopause, 631
Helix Nebula, NGC 7293, 616
HEO zone, 626
Herschel, Sir William, 575
Herschel Space Observatory, 623
Hertzsprung-Russell (H-R) diagram, **597**, 598
heterosphere, 432–**33**
hill, **154**
Hillary, Sir Edmund, 152
Himalaya Mountains, 149, 152, 156, 508
Hipparchus, 593
history, of earth, 108–23
old-earth view, 108–14
young-earth view, 115–23
HMS *Challenger*, 359, 372
HMS *Endurance*, 517
homosphere, 432–**33**, 434
hook (landform), 351
hook echo (weather), 481
Hope Diamond, 214
horizon (soil), **295**
horn, 289
Horsehead Nebula, 617
horse latitudes, 461
Horseshoe Falls, 396
hot spring, 28, 191, 195
H-R diagram, **597**, 598

Hubble, Edwin, 603
Hubble Space Telescope, 576, 587, 588, 604, 623, 627
Hubble's law, 607
humic acid, 277
humidity, **457**, 464
    absolute, 457
    relative, **457**, 468
humus, **295**
hurricane, **489**–94
    categories, 490, 663
    extratropical, 489
    formation of, 489
    hazards, 492, 493–94
    naming, 490
    storm surge, **494**
    storm swell, 493
    tropical, 489
Hurricane Dennis, 494
Hurricane Isabel, 489
Hurricane Katrina, 491
hurricane hunter, 491
Hutton, James, 85, 89
Huygens lander, 632
hydrocarbon, 269
hydrologist, 18, 380
hydropower, 94, 379
hydrothermal circuit, 195
hydrothermal cycle, 403
hydrothermal feature, 194–95
hydrothermal fluid, 237, 238, 303
hydrothermal vent, 365
hygrometer, 457
hypothesis, **9**–10, 14, 15

I

ice age, 109, 111–12, 113, 520
Ice Age, the, 114, 120, 123, 263, 286, 288, 516, 521
iceberg, 517, 526
ice cap, 111, 121, 288, 512
ice core, 111
ice floe, 517, 525
ice giant, 575
ice shelf, 517
Ichthyosaurus, 251
icing, 465
iDino Project, 256
igneous, **175**
    rock, 223, **226**–29, 242, 243, 254

impermeable, 269, 402
index fossil, 103, 256, **260**
Indian plate, 153
inference, 12
infrared, 39, 446, 532
inlet stream, 387
Institute for Creation Research, 443
intensity, **145**
International Astronomical Union, 567
International Date Line, 56
International Space Station, 35, 435, 449, 538, 573, 627, **641**, 646, 647
International System (SI) standard, 212
intrusion, 192
intrusive volcanism, 175
Io, 574
ion, **42**
ionosphere, **442**
irrigation, **409**, 410
Isa Lake, 390
island arc, 105, **313**
isobar, 497
isohyet, 497
isostasy, **154**
isotherm, 497
I Zwicky 18, 604

J

James Webb Space Telescope, 587, 588, 623
Jason (ROV), 375
Jefferson Memorial, 221
Jesus Well project, 398
jet stream, **434**
joint, 132, **133**
Jordan River, 380, 381
Juan de Fuca plate, 151
Jupiter, 574

K

kame, 169, 290
Kant, Immanuel, 108
karst topography, 397, **419**–21
Keck Observatory, 621
kelvin, 26
Kennecott open pit copper mine, 217
Kennedy, John F., 635, 638
Kepler, Johannes, 17, 564–65
    laws of planetary motion, 17, 564–65

kettle, 290
kettle lake, 389
Khongoryn Els dunes, 168
Kilauea, 222, 244
Köppen, Wladimir, 511
Krafft, Maurice and Katia, 190
Kras Plateau, 419
Kuiper Belt, 559, 560, 578, 581
Kuril Islands, 313
Kuroshio Current, 341
Kutiah Glacier, 289

L

La Brea Tar Pits, 248, 249, 255
La Niña, 518
laccolith, 193
Laguna Colorada, 391
Laguna Verde, 164
lahar, 179
lake, **386**–95
    crater, 391
    life phases of, 393–94
    origin of, 388–89
Lake Baikal, 387, 388
Lake Champlain, 254–55
lake effect snow, 482
Lake Erie, 395
Lake Manicouagan, 392
Lake Powell, 282–83
Lake Sørvágsvatn, 377
Lake Superior, 386
Lake Tanganyika, 161, 388
Lake Toba, 391, 392
Lake Victoria, 161
lake, unusual, 390–92
land breeze, 462
lander, **632**
landform, **154**, 158–59
    convergent, 159–60
    depositional, 164
    divergent, 161–62
    groundwater, 414–16, 418–21
    subsidence, 159, 163
    uplift, 159, 162–63
landslide, 146, **282**
Langjökull Glacier, 287
lapse rate, **433**–34
large cane toad, 121
Large Hadron Collider, 23–24, 34, 36, 42, 43, 610

latitude, **55**–58
    and climate, 504
lava, **176**–78, 222
lava tube, 222
law, 9–**10**
    of universal gravitation, 10, 566, 576
laws of planetary motion, 17, 564–65
leach field, 413
lead line, 362
leap year, 549
Lemaître, Georges, 608
lenticular cloud, 500
LEO, 627, 636
LEO zone, 627
Lewis, Merriweather, 387
LHC. *See* Large Hadron Collider
LIDAR, 68
life shear cutter, 618, 619
light
    created in transit, 612
    infrared, 39
    ultraviolet, 39, 532–33
    visible, 39, 532
lightning, 484–**85**
light-year (ly), 82, **594**
lignite, **266**–67
limnologist, 18, 393
limnology, 393, **394**–95
Lincoln Memorial, 419
line spectrum, 533
Linnaeus, Carolus, 259, 454
liquid, **26**–29
lithosphere, **91**, 131, 153
Little Ice Age, 520
loam, **295**
Local Galactic Group, 614
local noon, 55, 57–58
loess, **286**
logic, 1
Lonar Lake, 392
longitude, **55**–58, 357
Longitude Act, 357
longshore current, **349**
longshore sandbar, 309
longshore trough, 309
Lovell, Jim, 76, 98
low earth orbit, 627, 636
Lowell, Percival, 577
Lower Falls, 380
Luna 9, 629

lunar eclipse. *See* eclipse, lunar
lunar excursion module, 639
lunar month, 83
lunar phase, **547**–48
lunar probe, 628–29
luster (mineral), **207**
Lyell, Charles, 85, 89

M

macaw, 219
Mackenzie Large Igneous Province, 183
mackerel sky, 466
Magellan, Ferdinand, 358
Magellanic Clouds, 603
magma, **104**–5, 177–78, 242
    andesitic, 177, 228
    basaltic, 177–78, 228
    mafic, 177
    rhyolitic, 177–78, 228
magma chamber, 176, 177
magnetic force, 34
magnetism, 210
magnetosphere, **442**–44
magnitude, **141**
main sequence, **597**
mammatus cloud, 501
Mammoth Hot Springs, 195
mantle, **91**–92, 104–5, 131, 153
map, **52**–69
    3D, 69, 71, 72
    and dominion, 69
    features, 58
    geographic, 63
    geologic, 134
    legend, 58
    oblique view, 52
    political, 63
    relief, 64, 156
    scale, **53**, 58
    thematic, **64**, 148
    topographic, 63–64, 153, 656–58
map projection, 60–63 (**61**)
    conic, 61, 62
    cylindrical, 61, 62
    equatorial, 61, 62
    gnomic, 62
    oblique, 61, 62
    polar, 61
mare, **542**
Mare Imbrium, 542, 543

Mare Tranquillitatis, 542
mare's tail, 466
Mariana Trench, 312
marine organism, 249
marine scientist, 19, 305
marine snow, 325, 327
maritime air mass, **476**
Mars, 572–73
Mars Polar Lander, 632
Mars Reconnaissance Orbiter, 631
mass, 25, 29–30, 36–37
mass wasting, **281**
matter, 24–47 (**25**)
    change in, 27–29, 45–48
    classification of, 25
    composition of, 25, 41–48
    dark, 24, 34
    state of, 25–29
    structure of, 44–45
Matterhorn, 171, 289
Mauna Kea, 155, 186
Mauna Loa, 185, 186
mean sea level (MSL), 155, **308**
mean solar day, 549
meander, **384**, 385
Mediterranean Sea, 318
melting, **27**, 28, 45
melting point, 27
Mercury, 561, 562, 571, 630, 631
meridian. *See* longitude
mesa, 166
mesopause, 434, 436
mesosphere, **434**, 435
Messenger, 631
Messier, Charles, 601
metamorphic rock, 223, **237**–41, 242, 243, 254, 266
metamorphism, 223, 237, 243
    chemical, 238
    contact, 237, **239**
    dynamic, 237, **239**, 240
    regional, **237**
meteor, 81, **582**
meteorite, **582**
meteoroid, 436, 541, **582**
meteorologist, 19, 107, 454, 475–98
meteorology, 426, 443, **454**
methane, 269
Michelangelo, 240
microclimate, 504

microfossil, 250
microgravity, 35, 636, 645
microwave, 39, 442
Mid-Atlantic Ridge, 156, 182
mid-ocean ridge, 104–5, 119, 131, 135, 158, **311**
Midway Island, 315
migration, 463
    ocean, 337
Milky Way, 82, 546, 560, 603, 604, 610
mineral, **203**–18
    acid test, **210**
    cleavage, **207**
    compound, **204**
    crystal, 207, 211
    facet, 207
    flame test, 209
    fracture, **208**
    hardness, **208**
    impurity, 206
    luster, **207**
    native, **204**
    streak, **206**
    streak test, 206
mineral spa, 407
mineralogist, 204, **205**
mineralogy, 18, 205
mining, 217
    mine fire, 268
Mir, 627, 640
missing link, **260**
Mississippi River, 284, 380, 381, 384, 387
Missouri River, 380, 387
mixed-gas diving, **369**, 370
mixture, **25**, 44
MMI scale, 145, 660
model, **9**–13, 15
model, solar system, 17
modeling, 523
Modified Mercalli Intensity scale, 145, 660
Moho (Mohorovičić discontinuity), **91**
Mohorovičić, Andrija, 91
Mohs scale, **208**
Mohs, Friedrich, 208
molecule, **44**
moment magnitude scale, 141
monadnock, 167
monarch butterfly, 463
monocline, 160
monoclonal antibody, 35

monsoon, 508
Monterey Submarine Canyon, 310
Montes Alpes, 542
Montes Apenninus, 543
Montreal Protocol, 441
Monument Valley, 166
moon, **567**
moon pool, 371
moon, the, 83, 539–44
    and tides, 553
    motion of, 547
    phases of, **547**
    race to, 635, 637–39
    structure of, 539–41
    unique in solar system, 83
    viewing, 547–48, 557
    See also sun, moon, and earth system
moraine, 169, 288, **290**
    ground, 290
    lateral, 290, 91
    medial, 290, 91
    terminal, 290
Morris, Henry, 86, 201, 443
Mount Edziza Plateau, 185
Mount Etna, 173, 185
Mount Everest, 152, 153, 155, 463
Mount Fairweather, 157
Mount Fuji, 185
Mount Kilimanjaro, 170, 506
Mount Krakatoa, 197, 294
Mount Mayon, 185
Mount Mazama, 186
Mount McKinley. *See* Denali
Mount Monadnock, 167
Mount Pico, 156
Mount Pinatubo, 519
Mount Rainier, 185, 186
Mount Rushmore, 226–27
Mount Shasta, 186
Mount St. Helens, 121, 137, 176, 181, 185, 186, 187, 191, 276, 277, 294, 296
    eruption (1980), 187, 189, 276, 277, 296
Mount Stromboli, 186
Mount Tambora, 519
Mount Tambora, eruption (1815), 188–89, 191
Mount Vesuvius, 185, 186, 187, 189, 254
    eruption (AD 79), 179

Mount Washington Observatory, 450, 451, 473
Mount Whitney, 154, 155, 162
mountain, 153–70 (**154**)
    depositional, **168**–69
    erosional, **166**–67
    fault-block, 162
    fold, **159**
    residual, 166
    tectonic, 158–64
mountain breeze, 462
mountain range, 134, **156**
mountain system, **156**
mudflow, 282
mud pot, 195
multistage rocket, 629
Murrah Federal Building, 618

N

Namib Desert, 341
Nansen bottle, 362–63
NASA, 367, 371, 522, 559, 570, 573, 580, 587, 588, 605, 611, 619, 623, 625, 633, 635, 637–39, 643
NASA Jet Propulsion Lab, 584
National Aviation and Space Administration. *See* NASA
National Hurricane Center, 491, 499
National Oceanographic and Atmospheric Administration, 491
National Weather Service, 475, 487, 498, 499
native species, 412
natural bridge, 419, 420
Natural Bridge, VA, 420
natural gas, 265, **269**, 271
natural resource, **93**–96
natural selection, 260
nautical mile, 56
NEAR Shoemaker, 632
near-earth asteroid, 579
Near-Earth Asteroid Tracking program, 580, 632
nebula, 108–9, **601**, 602
nebular hypothesis, 108–10 (**109**), 112, 113
neocatastrophism, 87
Neptune, 576, 578
neutron, 41–42, 43, 47
neutron star, **598**
Nevada del Ruiz eruption (1985), 179

New General Catalogue, 591, 601
New Horizons, 559, 560, 577, 582, 630, 631
newton (N), 29
Newton, Isaac, 566, 620, 624
    law of universal gravitation, 10, 566, 576
NEXRAD, 481, 498
NGC 1316, 604
Niagara Falls, 381, 384, 396
Nile River, 284, 384
Niskin bottle, 363
nitrogen, in Earth's design, 81
nitrogen cycle, in the atmosphere, 438
    marine, **326**–27
nitrogen fixing, 326, 438
nitrogen narcosis, 370
noctilucent cloud, 436
Nor'easter, 482
Norgay, Tenzing, 152
North American plate, 151
North Atlantic Drift Current, 337
North Atlantic Gyre, 337
North Pole, 55, 57
North Star, 591
northern lights, 39
Northwest Passage, 387
nova, **598**
nuclear fission, 47
nuclear fusion, 47, **534**
nuclear change, **46**–47
nucleus, 41–43
NWS. *See* National Weather Service

O

Oard, Michael, 120, 426
Oberth, Hermann, 625
objective lens, 620
oblique, view of map, 52
occluded front, 478
ocean, **305**–27
    density, 321, 322
    open, 324–25
    origin of, 306–7, 317
    temperature, 321, 322
ocean basin, 305–15
ocean current, 337
ocean exploration, 355–75
ocean sampling, 361–67
ocean zone
    abyssal, 325
    aphotic, 325
    Hadal, 325
    intertidal, 324, 325
    littoral, 324, 325
    neritic, 324, 325
    photic, 324, 325
oceanic crust, 153
oceanic plate, 243, 244
oceanic trench, **312**
oceanographer, 305, **356**, 360
oceanography, 360–67
Oklahoma City bombing, 618, 619
Old Faithful geyser, 196
old-earth story, 108–14
old-earth view, 86, 108–14, 242, 256, 261–62, 270–72, 292, 306–7, 416, 429–30, 516
Olympus Mons, 573
On the Origin of Species, 260
Oort cloud, 581
open cluster, 602
operational definition, 25, 31, 36
Opportunity, 633
orbit, **565**
orbital period, 565
orbits, of planets, 568–69
ore, **203**
organism
    benthic, 364
    pelagic, 364
Orion, 589, 590
Orion Nebula (M42), 601
orogeny, **154**
orographic lifting, **479**, 484
oscillation, 346, 348
Ötzi, 255, 257
outlet stream, 379, 387
outwash plain, 290
overthrust, 142–43
oxbow lake, 383, 384
oxide, **215**
ozone, 439–41
ozone hole, 440–41
ozone layer, 439–41 (**440**), 442

P

Pacific Gyre, 330
Pacific leatherback turtle, 337
Pacific plate, 140, 182
pahoehoe, 178
paleontologist, 111, 259–64
paleontology, 258–64 (**259**)
Panderichthys, 260, 261
Pangaea, 107, 109, **110**, 112, 119, 183, 608
Panthalassa, 110, 307
parallax, **595**
parallel. *See* latitude
parasitic vent, 176
Parícutin, 185
parsec (pc), **594**
pascal, 455
peat, 270
pedologist, 293, 301
pedology, 301
pegmatite, 227
pelagic organism, **324**
perc test, 404
percolation, 400
perigee, 547
perihelion, 566
periodic comet, 581
permeability, **402**
Permian-Triassic extinction event, 261
permineralization, 250
petroleum, 265, **269**, 271
petrologist, 18, 223
petrology, 18, 223
phanerite, 227
phase (of matter), 25
Philippines Islands, 313
Phobos, 573
photocell, 363
photon, 609
photosphere, **535**
physical change, **45**
physical property, 25
phytoplankton, 271, 448
Piccard, Auguste, 372–73
Pietà, 240
pillow lava, 178
pinnacle, 166, 285
placer deposit, 212
plage, 535, 537
planet, **561**
    albedo, 574
    classification of, 567, 569–71
    defined, 567–76
    dwarf, **567**, 577–78
    gas giant, 569

ice giant, 575
inferior, 570
Jovian, **569**, 574–76
retrograde motion, 562, 571
superior, 571, 572
terrestrial, **569**, 571–73
planetary motion, laws of, 17, 564–65
planetary nebula, 598
planetary probe, 629–31
planetesimal, 109
plankton net, 364
plasma, **26**, 42
sun, 534, 537
plate
continental, 105, 131, 140
oceanic, 105, 131, 140
tectonic, 78, **105**, 119–20, 131, 135, 140
plate tectonics
catastrophic, **119**
theory, 86
plateau, 162–63
Plato, 561
Pliny, 207
plucking, 289
Pluto, 559, 560, 567, 577, 578
pluton, **193**
polar air mass, **476**
polar high, 461
polar orbit, 626
polar zone, 79, 505
Polaris, 591
pollutant, 411–12
pollution, 95–96, 330, 331, 428, 429, 448
air, 452
biological, 412
chemical, 411
thermal, 412
water, **411**–12
Pompeii worm, 180
porosity, 402
porphyry, 227
position fix, 57–58
potable water, 305, 407
pound (lb), 29
practical salinity scale, 318
precipitation, 283, **457**, 468–69, 479, 480
and climate, 508–9
orographic, 508
precipitation (chemical), **233**

pressure gradient, **458**, 460
pressure gradient force, 458, 460
presupposition, **8**
prevailing wind, 508
prime hour circle, 592
Prime Meridian, **56**, 58
principle of cause and effect, **16**
principle of superposition, 89, **103**, 142, 260
principle of uniformity, 8, **16**
Principles of Geology, 85, 89
probe, **628**–31
lunar, 628–29
planetary, 629–31
Progressive Creationism, 253
Project Apollo, 639
Project Gemini, 638
Project Mercury, 637
prominence, solar, 26, **535**, 537
propane, 269
proper motion, 595, 596
property
of a substance, 26–27
physical, 25
proton, 41–42, 43, 46
protostar, 610
province, 162
Proxima Centauri, 594
psychrometer, 457
Ptolemy, 562, 563
Puakatike Volcano, 182
pure substance, **25**, 44, 46
P wave, 91
pyroclastic flow, 179, 232, 276
pyroclastic material, **180**–81

Q

Quake Catcher Network, 129
quasar, **606**
quasar HE 1013-2136, 606
quasar ULAS J1120+0641, 606
quasi-stellar object, **606**

R

radar, 68, 69, 481
radial motion, 595, 596
radiation, **447**
infrared, 446
solar, 436
ultraviolet, 448

radio wave, 39
radioactivity, 47
radiometric dating, 102–3, 113
rain gauge, 457
rain shadow, 479, 508
rapids, 382, 383
ray (crater), **542**
reaction, chemical, 46
reaction engine, 624
real motion, 595, 596
recharge zone, **404**
recycling, 96
red shift, 595, 600, 613
redemption, 5–6
reef
barrier, 315
coral, **314**–15, 316
fringing, 314
refraction, 620
relative humidity, **457**, 468
relief, **155**
relief map, 156
remote sensing, 68
remotely operated vehicle, **374**, 375
resolution, 620
resource
actual, 93
and dominion, 95–96
natural, **93**–96
nonrenewable, 93–**94**
potential, 93
renewable, 93–**94**
water as a, 408–13
resource management, **95**–96
retrograde motion, 562, 571
revolution, **546**
Richter scale, **141**
Ride, Sally, 644
rift, 104–5
rift valley, 161
right ascension (RA), 592
rille, **543**
Ring of Fire, 181
rip current, **349**
risk, 146
Robinson world map, 62
rock, 204, 222, **223**–44, 247
chemical sedimentary, **233**
classification of, 223–25
clastic sedimentary, **232**

extrusive, 228
foliated metamorphic, **239**
grain, **225**, 226–29
igneous, 223, **226**–29, 242, 243, 254
intrusive, 228
metamorphic, 223, **237**–41, 242, 243, 254, 266
nonclastic sedimentary, **233**
nonfoliated metamorphic, 239–**40**
sedimentary, 159, 223, **230**–34, 236, 242, 243, 250, 254, 262, 263, 266, 271, 303
texture, **225**, 226–29
rock cycle, **242**–44
rock flour, 287
rock grain, 247
Rock of Gibraltar, 352
rocket, **624**–25
rockslide, 281, 282
Ross Ice Shelf, 517
rotation (Earth), 79, **547**
rover, **633**
runaway subduction, 119
Rutherford, Earnest, 43
Ryukyu Islands, 313

S

Saffir-Simpson scale, 490, 663
Salar de Uyuni, 390–91
salinity, **317**
salt dome, 233
salt flat, 390, 391
saltation, 283
Salyut 1, 640
Samec, Dr. Ron, 529
San Andreas Fault, 135, 136, 148, 151
San Jose copper and gold mine, 218
sand dune, 168
sandstorm, 285
Sarcosuchus imperator, 263
Sarfati, Dr. Jonathan, 1
satellite, 539, **625**–28
applications, 628
artificial, 625–27
technology, 367
saturation, zone of, **401**
Saturn, 575, 585
Saturn V, 639
scale (water), 407
Schmidt-Cassegrainian telescope, 621

science, **12**
science, earth, 9, **13**
historical, 16–**17**
operational, **16**
scientific process, **14**–15
scientist
atmospheric, 443
marine, 19, 305
secular, 8–9
space, 18
scuba diving, **369**, 370
sea, **306**
sea arch, 351
sea breeze, 462
sea cave, 351
sea clock, 357
sea-floor spreading, **105**, 119, 131
sea level, **307**–8, 331, 332, 333
sea stack, 351
sea surface salinity, 367
sea watch, 357
seamount, 164, 310, **312**
Search for Extraterrestrial Intelligence, 599, 605
season, **504**, 550
seasons, and climate change, 518
seawater, 317–23
salinity, 317–19
Secchi disk, 363
secondary limestone formation, 418
secular scientist, 8–9
sediment, 8, 159, 242, 303
sorting, 284
suspended, 283
sediment core, 311
sediment corer, 364–65
sedimentary rock, 159, 223, **230**–34, 236, 242, 243, 250, 254, 262, 263, 266, 271, 303
sedimentation, 165
sedimentologist, 235
seed crystal, 227
seiche, **395**
seismic, 90
seismic gap, 147
seismic station, 136, 137, 139–40
seismic wave, 90, 91, 92, 133–40, 145
seismogram, 137
seismograph, 136–37
seismologist, 18, 138
seismology, 18, 136–40

seismometer, **129**, 136–37
service module, 639
SETI, 599, 605
Shackleton, Earnest, 517
shadow zone, 139
Shandong Province, China, 258
shear, 33, **130**, 131, 135
Shepard, Alan, 637
shooting star, 582
shore, **309**
shoreline, 309
sidereal month, 549
sidereal year, 548
Sierra Nevada Mountains, 162
silica, 177, 215
silicate, **215**
sill, 177, 192, 237
Silvestru, Dr. Emil, 303
Singing Sand Dunes, 168
sinkhole, 405, 419
sinkhole pond, 420
Skylab, 640
sleet, 468
slump, 281
small solar system body, **567**, 578–83
Smith, Tilly, 2, 20
Snelling, Dr. Andrew, 75
snow, 469
Snow, Dr. John, 51, 52, 64
snowball Earth, 112
soft coal. *See* bituminous coal
soft water, 407
soil, 277, **293**–98
acidic, 296
composition, 295
conservation, **297**
and Creation, 293
and dominion, 297, 298
formation, 294
horizon, **295**
living, 294
native, 295
transported, 295
soil classification graph, 295
solar constant, **445**–46
solar day, 549
solar eclipse. *See* eclipse, solar
solar energy, 532
solar flare, 444, 535, 536–37, **538**
solar mass, 598

solar minimum, 536–37
solar prominence, 26, **535**, 537
solar radiation, 436
solar spectrum, **532**
solar system, 82, 101–2, 112, 559–83 (**560**)
solar wind, 81, 442, 443, 444, 445, **536**, 560, 631
solid, **26**–29
Solomon Islands, 313
solstice, 504–5, 530, 550
solute, 406
solution, 25, 231, 406
solution cave, **415**–16
    origin of, 416
solvent, **406**
    universal, 406
Sombrero galaxy, 594
sonar, 69, 323, 362, **366**
sorting, **284**
Soufriere Hills eruption (1997), 179
sound
    audible, 38
    infrasonic, 38
    ultrasonic, 38
    See also wave
sound velocity profile, 323
sounding, **362**
sounding rocket, 628
South Pole, 55, 57
Soyuz, 641
Soyuz 11, 645
Space Launch System, 625
space probe, 559
space race, 625, 628, 635
space scientist, 18
space shuttle, 619, **642**, 647
space station, **640**–41
Space Transportation System, 642
SpaceX, 643
speciation, 121
specific gravity (sp gr), **209**
specimen, 206
spectrograph, 533
spectroscope, 533, 600
spectroscopy, 600
spectrum
    absorption, 533, 600
    bright-line, 600
    continuous, 600
    dark-line, 600
    emission, 533, 600
    line, 533, 600
    visible, 39, 532
speed of sound, 322–23
speleologist, 303, **416**, 417, 421, 422
speleology, 303
speleothem, **418**–19
    cave bacon, 423
    cloud, 418
    column, 418
    drapery, 418
    dripstone, 418
    flowstone, 418
    rapid formation, 419
    shelf, 418
    shelfstone, 424
    splash cup, 418
    stalactite, **418**–19
    stalagmite, **418**–19
spelunker, 421
spicule, 537
spiral galaxy M51, 604
Spirit, 633
spit, 351
Spitzkoppe, 167
spray toad, 121
spring, 379, 401, 402
    hot, 191, 195
spring scale, 31
Sputnik 1, 625, 627
Sputnik 2, 645
squall line, 484
stalactite, **418**–19
stalagmite, **418**–19
star, 588–98, 600
    binary, **602**
    black dwarf, 598
    brightness, 593–94
    classification, 596–97
    cluster, **602**–3
    double, 602
    dwarf, 597
    eclipsing binary, 602
    giant, 597
    luminosity, 596–97
    naming, 589–91
    primary, 602
    properties of, 592–95
    red giant, **598**
    supergiant, **597**
    white dwarf, 597, **598**
Stardust, 631
starlight travel, 612
state of matter, **25**–29
station model, **495**–96
stationary front, 477, 478
Statue of Liberty, 213
Stegosaurus, 263
stellar aging, 597–98, 608, 610, 613
stellar disk, 109
stellar evolution, **610**
stellar magnitude, 593
Steno, Nicholas, 89, 110
stock, 193
stone arch, 167
Stone Mountain, 167
Stonehenge, 530, 531
storm, 481–95
    tropical, **490**
    winter, **482**–83
storm surge, **494**
storm swell, 493
strain, **132**
Strait of Gibraltar, 344
stratigraphy, 18, 89
stratopause, 434
stratosphere, **434**, 435, 440, 441
stratovolcano, **185**
stratum, **102**–3, 111, 123
streak (mineral), **206**
    plate, 206
    test, 206
stream, 283, **379**–86
    base level, **381**, 382
    cross-section, 380, **381**–82
    elevation profile, **380**
    gradient, **380**–81, 458
    headwaters, 380
    intermittent, **385**–86
    load, 283
    mouth, **380**, 383
    order, 382
    perennial, **385**
    source, **380**
    system, **382**
streambed, 283
streamless valley, 419, 421
stress, **130**, 132, 133, 135
strike (fault), **133**–34
strip cropping, 297, 298

strip mining, 217
Subaru telescope, 621
subduction, **105**, 119–20, 131, 135, 243
  runaway, 119
  zone, 243
sublimation, 28, **29**
submarine canyon, **310**
submersible, 372–75
suborbital flight, 637
subpolar low, 461
subsidence, 405
subsidence landform, 159, 163
subsurface current, **343**
sulfide, **216**
summer solstice, 550
summit, 154
sun, the, 531–38
  and Earth's design, 81–82
  composition of, 533
  convective zone, 534
  core, 534
  radiative zone, 534
  structure of, 534–36
  surface, 534
sun, moon, and earth system, 545–53
  and seasons, 550
  and time, 548–49
sunspot, 520, 535, **536**–37
sunspot activity, and climate change, 520
sunspot cycle, 536
sun-synchronous satellite, 627
supercell, 486
supercontinent, 104–5, 109, **110**, 119
supercooled water, 465
supernova, **598**
supernova 1994D, 613
superposition, principle of, 89, **103**, 142, 260
surface current, **338**–43
surface mining, 217
surface water, 378–95
surface wind, 460–62
Surtsey, 184, 185
surveying, 67
sustainable yield, 95
S wave, 91
symmetry, 211
syncline, 160
synodic month, 549
synoptic weather map, **497**

T

talus, 166, 279
tarn, 289
Tau Tona gold mine, 217
technology, 6, 19
tectonic boundary
  convergent, 158
  divergent, 159
  transform, 131, 135, 140
tectonic force, 129–32, 158–59, 264
tectonic mountain, 158–64
tectonic plate, 78, **105**, 119–20, 140
tectonics, 104–5 (**105**), 107, 109, 110–11, 119–20
telescope, 619, 620–23
  composite, 621
  optical, 620–21
  radio, **622**
  reflecting, **620**–21
  refracting, **620**
  space, 623
Tempel 1, 570
temperate zone, 79, 505
temperature, **26**, 27, 28, 29, 454
  and climate, 504–6
tension, 32, **130**
tephra, **181**, 190
terminator, **548**
terrace, 195
terrace farming, 298, 99
terrain, 155
tetrapod, 261
The Genesis Flood, 86, 201
The Twelve Apostles sea stacks, 300
Theistic Evolution, 615
thematic map, **64**, 148
theory, 9–**10**
  Analogous Days, 403
  Big Bang, **608**–11
  canopy, 507
  Continental Drift, **104**–5, 107, 110
  Day Age, 124, 403
  Gap, 7
  geocentric, 17, **561**–62, 563, 564
  heliocentric, 17, **563**
  plate tectonics, 86
thermal (wind), 434
thermal turnover, **394**–95
thermodynamics, 40
  first law of, **40**, 47

thermograph, 454
thermohaline current, **344**
thermometer, 454
thermopause, 436
thermosphere, 435, **436**
Thompson, Charles Wyville, 359
Thompson, J. J., 43
Thompson, Lonnie, 510
Three Gorges Dam, 378, 379, 385
thunder, 484–85
thunderstorm, **484**–85
  formation, 484
  hazards, 485
Tibesti Mountains, 382
Tibetan Plateau, 479
tidal bore, 11
tidal bulge, 332–33, 334
tidal force, 553
tidal power generation, 336
tidal wave, 347
tide, **331**–36
  earth, 553
  high and low, 331, 332–33
  lunar, **332**–33
  neap, **334**–35
  solar, 334
  spring, **334**
tide gauge, 307, 331
tide table, 333
tilt (of Earth), 79, 505, 520
Tombaugh, Clyde, 577
tombolo, 351
Topex Poseidon, 367
topographic map, 153
topography, **153**–54
tornado, **486**–88
  categories, 488, 663
  formation, 486
  hazards, 488
Tornado Alley, 486
tornado chaser, 487
tornado warning, 487
tower karst topography, 420
transform boundary, **131**, 135, 140
transit, 570
transitional form, **260**
transneptunian object, 578
transpiration, 400
transportation (particle), 231
tributary, **382**

Triceratops, 256
Triceratops horridus, 259
Trieste, **373**
Trieste II (DSV-1), 373
trilobite, 265
trimix, 370
Triple Divide Peak, 383
Tropic of Cancer, 505
Tropic of Capricorn, 505
tropical air mass, **476**
tropical cyclone basin, 489
tropical storm, **490**
tropics, 505
tropopause, 433, 434
troposphere, **433**–34, 435, 436, 440
trough. *See* wave
Tsiolkovsky, Konstantin, 625
tsunami, 2, 142, 144, **147**, 151, 346, 347
tundra, 512
Tunguska Event, 580
turbidity, 393
turbidity current, 310, **345**
turbulence, 283–84
typhoon, 489
Tyrannosaurus rex, 256, 258

U

ultraviolet, 39, 448, 532
underground reservoir, 399–404
underwater habitat, **371**
uniformitarianism, **85**
uniformity, principle of, 8, **16**
United States Geological Survey, 138
universal gravitation, law of, 10, 566, 576
universe, origin of, 606–15
unmanned underwater vehicle, **375**
updraft, 484, 486
uplift landform, 159, 162–63
upwelling, **342**, 395
Uranus, 575–76
urban heat island effect, 504
Urey-Miller experiment, 430
Ursa Major, 591
USGS, 138
U-shaped valley, 289
USS *Thresher*, 373

V

V-2 rocket, 625, 645
valley breeze, 462

Van Allen belts, 442–43
Van Allen, James, 442
vapor, 27–29
vaporization, **27**, 28
Vardiman, Larry , 443
variable star, 594
varve, 292
VEI. *See* Volcanic Explosivity Index
vein, 212
vent, **176**
ventifact, 285
Venus, 571–72
vernal equinox, 550
Very Large Array, 622
Viking lander, 632
Villumsen, Rasmus, 107
Virgin Galactic, 643
viscous, 177
visible light, 39, 532
volcanic ash, 180
volcanic block, 181
volcanic bomb, 180–81
volcanic cinder, 180–81
volcanic cone, 176
volcanic eruptions
    Mount St. Helens (1980), 187, 189, 296, 276, 277
    Mount Tambora (1815), 188–89, 191
    Mount Vesuvius (AD 79), 179
    Nevada del Ruiz (1985), 179
    Soufriere Hills eruption (1997), 179
Volcanic Explosivity Index, **189**–91
volcanic neck, 167
volcanism, and climate change, 519
volcano, 164, 174–97 (**175**), 243
    active, 181, 182, **186**, 187
    cinder cone, **185**
    composite cone, 185
    continuous, 186
    dormant, 181, **186**
    extinct, 181, **186**, 187
    hot spot, 182
    shield, 184, **185**
    stratovolcano, **185**
    types, 184–91
volcanoes, and island arcs, 313
volcanologist, 190
volume, 25–26, **31**
von Braun, Wernher, 625
Voskhod 1, 638

Voyager 1, 631
Voyager 2, 576, 631
Vulcano Island, 175

W

wadi, 386
Wake Island, 315
Walker, Tasman, 118–19
wall cloud, 472
waning, 548
warm front, 477, 478
Wasatch Mountains, 237
Washburn, Brad, 152, 153, 157
Washington National Cathedral, 234
water, 80
    as a resource, 408–13
    fresh, 388
    hardness, **407**
    potable, 305, 407
water (hydrologic) cycle, **400**
water conservation, 410–11
water molecule, 406
water pollution, **411**–12
water softening, 408
water table, **400**
    perched, 402
water treatment, 413
waterfall, 382, 384
wave, 38, 345–51 (**346**)
    amplitude, 346
    body, 138
    compression, 38
    crest, 346, 347, 348, 349
    deep-water, 346
    electromagnetic, 445, 532
    height, **346**
    infrared, 532
    internal, 347
    Love, 139
    P, **138**–40
    period, **347**–47
    Rayleigh, 139
    S, **139**–40
    seismic, 90, 91, 92, 133–40, 145
    shallow-water, 348–49
    surface, **139**–40
    tidal, 347
    transitional, 348
    trough, 346
    tsunami, 346, 347

ultraviolet, 532
wave base, **346**, 347, 348
wave diffraction, 349
wave fetch, 346–**347**
wave refraction, 348
wave speed, **347**–47
wavelength, **346**–47, 348
waxing, 548
weather, 453–70 (**454**), 504
weather data, 454–57
weather forecast, 495–99
weather front, 462
weather map, 491, 496–97
weather map symbol, 478
weather prediction, 475
Weather Prediction Center, 496, 498
weather satellite, 491
weather vane, 456
weathering, **277**–91
    biological, **280**
    chemical, **277**–78, 280
    mechanical, 277, **278**–79
Wegener, Alfred, 86, 104, 107, 110, 119
weight, 25, **29**, 30, 31, 34, 35
weightlessness, 636
Werner, Abraham, 110
Whitcomb, John, 86, 201

White Hole cosmology, 613
White Nose Syndrome, 425
Wild 2, 631
wildfire, 527
Wilkerson Microwave Anisotropy Probe, 611
wind, **456**, 458–463
    polar easterly, 338, **461**
    prevailing, 338
    prevailing westerly, 338, **461**
    seasonal, 508
    trade, 338, 460–**61**
wind direction, 456
wind farm, 453
wind power, 453
wind shear, 486
wind turbine, 453
wind vane, 456
windbreak, 285, 297, 298
wind-chill, 456
wind-chill factor, **456**
winter solstice, 550
winter storm, **482**–83
woolly mammoth, 100–1, 112, 120
work, **36**
worldview, 1, 6, 7–10 (**8**), 17, 121, 123, 143, 429–30, 612, 613

and caves, 416
and climate change, 516, 524
and exploration, 560
and space exploration, 643

x-ray, 39

Yangtze River, 284, 385
Yellowstone Caldera, 190
Yellowstone National Park, 191
Yerkes Observatory, 620
Yeti crab, 365
young-earth story, 115–23
young-earth view, 86, 115–23, 244, 262, 264, 272, 293, 307, 317, 416, 430, 516, 521, 582
    and cosmogony, 612–13

zone
    of aeration, **400**
    of saturation, **401**
    recharge, **404**
Zuma Rock, 245

PHOTO CREDITS

Key: (t) top; (c) center; (b) bottom; (l) left; (r) right, (i) inset, (bg) background

COVER
front © Terry Davis/123RF; **spine** alice-photo/Shutterstock.com; **back** GalapagosPhoto/Shutterstock.com

FRONT MATTER
iiit leungchopan/Shutterstock.com; **iiib** David Herraez Calzada/Shutterstock.com; **ivt** Sergey Lavrentev/Shutterstock.com; **iv–vb** © Likrista82 | Dreamstime.com; **vt** Dja65/Shutterstock.com; **vc** kaluginsergey/Shutterstock.com; **vit** Daria Rosen/Shutterstock.com; **vib** frantic00/Shutterstock.com; **vii** Delpixel/Shutterstock.com; **viii** © iStock.com/Chelnok; **ix** "ISS-43 Earth view from the cupola onboard the ISS"/Wikimedia Commons/Public Domain; **x** © iStock.com/Valérie Koch; **xi** Getty Images/Hemera/Thinkstock; **xiitl** © iStock.com/ivandan; **xiitr** Martin Fowler/Shutterstock.com; **xiibl** "Mount St. Helens, one day before the devastating eruption" by Harry Glicken/USGS/Wikimedia Commons/Public Domain; **xiibr** "MSH82 st helens plume from harrys ridge 05-19-82" by Lyn Topinka/USGS/Wikimedia Commons/Public Domain; **xiiitl** sl_photo/Shutterstock.com; **xiiitr** Billion Photos/Shutterstock.com; **xiiicr** iStockphoto/Thinkstock; **xiiibr** Data SIO, NOAA, U.S. Navy, NGA, GEBCO/Image IBCAO/Image © 2011 DigitalGlobe/Image © 2011 TerraMetrics

UNIT OPENERS
xiv–1 © iStock.com/Becart; **1** Creation Ministries International; **74–75** Solcan Sergiu/Shutterstock.com; **75** Answers in Genesis; **200–201** vvoennyy/123RF; **201** Image provided by the Institute for Creation Research, ICR.org. Used by permission.; **426** Kichigin/Shutterstock.com; **426–27** © iStock.com/brainmaster; **427** Mike Oard; **528** Myrleen Pearson/Alamy Stock Photo; **529** Nathan Hawkins

CHAPTER 1
2 © iStock.com/shannonstent; **3** US Coast Guard Photo/Alamy Stock Photo; **4** © iStock.com/JazzIRT; **5** Courtesy of Samaritan's Purse; **6** mark higgins/Shutterstock.com; **8t** Billion Photos/Shutterstock.com; **8b** iStockphoto/Thinkstock; **10–11** Scott Dickerson/Perspectives/Getty Images; **12, 13** philippe giraud/Corbis Historical/Getty Images; **14–15bg** Stefan Christmann/Corbis Documentary/Getty Images; **15l** Geology magazine cover/CRSQ magazine cover/Answers in Genesis Research Journal Page/Used under Fair Use; **15tr** ANT Photo Library/Science Source; **15br** British Antarctic Survey/Science Source; **16t** NASA; **16b** Image by V.L. Sharpton, Courtesy: Lunar & Planetary Institute; **17** CICLOPS, JPL, ESA, NASA; **18** Jon Wilson/Science Source; **19t** Cultura Creative (RF)/Alamy Stock Photo; **19c** "Astronaut-EVA" by NASA/Wikimeda Commons/Public Domain; **19b** © 2010 HERRENKNECHT Tunnelling; **20** PA Images/Alamy Stock Photo; **21** aaltair/Shutterstock.com; **22** Wlad74/Shutterstock.com

CHAPTER 2
23 NASA, ESA, and the Hubble SM4 ERO Team; **24** "1e0657 scale"/NASA/CXC/M. Weiss/Wikimedia Commons/Public Domain; **25** Dudarev Mikhail/Shutterstock.com; **26** SOHO (ESA & NASA); **27** Alexander Kazantsev/Shutterstock.com; **28** Media Bakery/MediaBakery; **29t** jaroslava V/Shutterstock.com; **29c** Science Source; **29b** BJU Photo Services; **31t** design56/Shutterstock.com; **31b** PRILL/Shutterstock.com; **32–33** Dragon Images/Shutterstock.com; **33t** © iStock.com/Dave Logan; **33ti** geogphotos/Alamy Stock Photo; **33b** Agencja Fotograficzna Caro/Alamy Stock Photo; **34t** © iStock.com/t_kimura; **34b** iodrakon/Shutterstock.com; **35** ESA/NASA; **37t** Libor Píška/Shutterstock.com; **37b** © iStock.com/Justin Reznick; **38t, 39t** NASA; **38b** ©2005 JupiterImages/Photos.com. All rights resesrved.; **39b** bogdan ionescu/Shutterstock.com; **40** © iStock.com/troutnut; **43** "Ernest Rutherford 1905"/Wikimedia Commons/CC BY 4.0; **45t** "Bismuth crystals 2" by Maxim Bilovitskiy/Wikimedia Commons/CC BY-SA 4.0; **45bl** danymages/Shutterstock.com; **45br** Peter Dazeley/Photographer's Choice/Getty Images; **46** Roger Ressmeyer/Science Faction/Getty Images; **47** © iStock.com/CUTWORLD; **48** Carlos Chavez/Los Angeles Times/Getty Images

CHAPTER 3
51 Classix/E+/Getty Images; **52** CDC; **53tl** Encyclopaedia Britannica/Universal Images Group/Getty Images; **53tr, 68,** NASA; **54** "Lunar eclipse March 2007" by Navy/Wikimedia Commons/Public Domain; **57** © iStockphoto.com/Arne Bramsen; **59t** © iStock.com/BlackJack3D; **59bl** "GPS-IIRM"/US Government/Wikimedia Commons/Public Domain; **59bli** "LAAS Architecture" by LAAS_Architecture.png/Wikimedia Commons/Public Domain; **59br** "Osmand auf Samsung Galaxy S Advance (Hamburg)"/Alexrk2/Wikimedia Commons/CC BY-SA 3.0; **63** Bardocz Peter/Shutterstock.com/ekler/Shutterstock.com/AridOcean/Shutterstock.com; **64t** "Hawaii-Big-Island-TF"/USGS/Wikimedia Commons/Public Domain; **64c** "United States Soil Moisture Regimes"/USDA/Wikimedia Commons/Public Domain; **64b** "2004US election map"/MetaBohemian/Wikimedia Commons/CC BY-SA 4.0, 3.0, 2.0, 1.0; **66, 68i;** NOAA; **67t** Nicolesa/Shutterstock.com; **67b** 2010 Google Earth/Pacific Disaster Center; **69** ©Google Earth Pro/Data MBARI/Data SIO, NOAA, U.S. Navy, NGA, GEBCO/©2010 Google/Image AMBAG; **70** Stock Connection/SuperStock; **71** Vitaliy Redko/Shutterstock.com; **72** HERE; **73t** © iStock.com/BlueRingMedia/BlueRingMedia/Shutterstock.com; **73b** © iStock.com/chombosan

CHAPTER 4
76, 81, 82 NASA; **77** Bruce Dale/National Geographic/Getty Images; **78** Pasieka/Science Source; **80** "WOA09 sea-surf TMP AYool" by Plumbago/Wikimedia Commons/CC BY-SA 3.0; **82bg** European Space Agency & NASA; **83** Claudio Divizia/Shutterstock.com; **85** King's College London/Science Source;

86l Keith Douglas/All Canada Photos/Getty Images; 86r The Genesis Flood : 50th anniversary edition ISBN 978-1-59638-395-1 by Henry M Morris and John C Whitcomb Pub date 8/23/2011 Publisher P&R Publishing P O Box 817, Phillipsburg, N J 08865 www.prpbooks.com; 87 John Cancalosi/age fotostock/Getty Images; 88–89bg sumikophoto/Shutterstock.com; 89t Paul Maguire/Alamy Stock Photo; 89b BarryTuck/Shuttterstock.com; 90t REUTERS/Alamy Stock Photo; 90b USGS; 93 Sandstone-coal-tonsteins (Ericson Sandstone over Rock Springs Formation, Upper Cretaceous; hairpin curve roadcut along Superior Cutoff Road, east of Superior, Wyoming, USA) 2/James St. John/Flickr/CC BY 2.0; 94 © iStock.com/KeithSzafranski; 95t Artush/Shutterstock.com; 95b Steve Proehl/Corbis Documentary/Getty Images; 96–97 Jat306/Shutterstock.com; 98 "Apollo 8 Crewmembers-GPN-2000-001125"/NASA/Wilimedia Commons/Public Domain; 99t Encyclopaedia Britannica/Universal Images Group/Getty Images; 99b vvoe/Shutterstock.com

CHAPTER 5

100 © iStock.com/ivandan; 101 ITAR-TASS Photo Agency/Alamy Stock Photo; 102 © Media Bakery/MediaBakery; 102i NASA/JPL/STScI; 103 Laura C. Walthers/Shutterstock.com; 106–7bg Dara J/Shutterstock.com; 107 "Wegener Expedition-1930 026"/Archive of Alfred Wegener Institute/Wikimedia Commons/Public Domain; 108 Martin Fowler/Shutterstock.com; 109, 119 NASA; 111 Jason Edwards/National Geographic Creative/Getty Images; 113 Education Images/Universal Images Group/Getty Images; 116–17bg airn/Shutterstock.com; 121t Joel Sartore/National Geographic/Getty Images; 121b Aleksey Stemmer/Shutterstock.com; 122–23 © iStock.com/wisanuboonrawd; 124 Wellford Tiller/Shutterstock.com; 125 Sheila Fitzgerald/Shutterstock.com; 126 B.A.E. Inc./Alamy Stock Photo

CHAPTER 6

128 Design Pics/Reynold Mainse/Perspectives/Getty Images; 129 Photo: Colourbox.com; 130 BJU Photo Services; 131 © 2010 Google/© 2010 A riGis (Fly) Ltd./© 2010 LeadDog Consulting/Data SIO, NOAA, U.S. Navy, NGA, GEBCO; 133tr MehmetO/Shutterstock.com; 133l © iStock.com/pepmiba; 133br © iStock.com/hippostudio; 136 Kevin Schafer/Photolibrary/Getty Images; 137t USGS; 137b Checubus/Shutterstock.com; 138 © Roger Ressmeyer/Corbis/VCG/Corbis Documentary/Getty Images; 142l JIJI PRESS/AFP/Getty Images; 142r Athit Perawongmetha/Getty Images News/Getty Images; 143 BGSmith/Shutterstock.com; 144–45bg Michael S. Yamashita/National Geographic Magazines/Getty Images; 145 Visual China Group/Getty Images; 146t Tatsiana Volskaya/Moment Open/Getty Images; 146b E+/Getty Images; 147t US Air Force Photo/Alamy Stock Photo; 147b epa european pressphoto agency b.v./Alamy Stock Photo; 148 Map data ©2016 Google; 148i USGS; 149 © iStock.com/luxcreative; 150 Coprid/Shutterstock.com

CHAPTER 7

152 Vixit/Shutterstock.com; 153 Boston Globe/Getty Images; 156 iStock/Getty Images; 157 Robert Lackenbach/The LIFE Images Collection/Getty Images; 158–59bg fotoVoyager/Vetta/Getty Images; 158 Photographer's Choice/Getty Images; 160t Marco Simoni/robertharding/Getty Images; 160b Ken Barber/Alamy Stock Photo; 161t Auscape/Universal Images Group/Getty Images; 161b, bi © Clipart Of, LLC; 162t AlexTIZANO/Moment/Getty Images; 162b Migel/Shutterstock.com; 163 NASA; 164–65 Adhemar Duro/Moment Open/Getty Images; 166t Hiroya Minakuchi/Minden Pictures/Getty Images; 166b LordRunar/E+/Getty Images; 167t Christophe Lehenaff/Photononstop/Getty Images; 167b Mieka Sawatzki/EyeEm/Getty Images; 168t David Tipling Photo Library/Alamy Stock Photo; 168c Image courtesy of Serge Andrefouet, University of South Florida/USGS National Center for EROS and NASA Landsat Project Science Office; 168–69b paranyu pithayarungsarit/Moment/Getty Images; 169t © Tom Bean/Alamy; 169c Georg Gerster/Science Source; 170 Kyslynskyy/iStock/Getty Images; 171 vencavolrab/iStock/Getty Images; 173 Giuseppe Torre/EyeEm/Getty Images

CHAPTER 8

174 © Pall Gudonsson; pallgudjonsson.zenfolio.com/Moment/Getty Images; 175 NordicPhotos/Getty Images; 176–77bg The Asahi Shimbun/Getty Images; 177t Science Stock Photography/Science Source; 177c Kicky_princess/Shutterstock.com; 177b © iStock.com/Andreas_Wass; 178l Grant Kaye/Aurora/Getty Images; 178tr G. Brad Lewis/Science Faction/Getty Images; 178br NOAA Okeanos Explorer Program, Galapagos Rift Expedition 2011; 179t AP-PHOTO/hk/SUB-YOMIURI; 179b Fourpeaked-fumaroles-cyrus-read1/Cyrus Read, USGS/Wikimedia Commons/Public Domain; 180 "Sully Vent"/NOAA/Wikimedia Commons/Public Domain; 182 © iStock.com/anharris; 183t USGS; 183b "Canada geological map" by Qyd/Wikimedia Commons/Public Domain; 184 Georg Gerster/Science Source; 185t judytally/Shutterstock.com; 185b Gary Fiegehen/All Canada Photos/SuperStock; 186t © iStock.com/luiginifosi; 186b Greg Vaughn/Design Pics/Perspectives/Getty Images; 187l "Mount St. Helens, one day before the devastating eruption" by Harry Glicken/USGS/Wikimedia Commons/Public Domain; 187r "MSH82 st helens plume from harrys ridge 05-19-82" by Lyn Topinka/USGS/Wikimedia Commons/Public Domain; 188 John Crux Photography/Moment/Getty Images; 189 "Sumbawa Topography" by Sadalmelik/Wikimedia Commons/CC BY-SA 2.5, 2.0, 1.0; 190t "Yellowstone Caldera map2"/USGS/Wikimedia Commons/Public Domain; 190c Explorer/Krafft/Science Source; 190b Explorer/Krafft/Science Source; 191 Smith Collection/Gado/Archive Photos/Getty Images; 192–93 Lynn Y/Shutterstock.com; 195 John and Tina Reid/Moment/Getty Images; 196l Marco Simoni/Cultura/Getty Images; 196r © iStock.com/Rhoberazzi; 197 tom pfeiffer/Alamy Stock Photo; 198 Kletr/Shutterstock.com; 199t Vladislav S/Shutterstock.com; 199b © iStock.com/VvoeVale

CHAPTER 9

202, 208tr © iStock.com/VvoeVale; 203 © iStock.com/BruceBlock; 204l © iStock.com/iluziaa; 204c Bjoern Wylezich/Shutterstock.com; 204r dipressionist/123RF; 205 Oleksandr Lysenko/Shutterstock.com; 206t Brian C. Weed/Shutterstock.com; 206c Ksyutoken/Shutterstock.com; 206b Mark A. Schneider/Science Source; 207tl PNSJ88/Shutterstock.com; 207tr, 208tl, 215b all, 217 vvoe/Shutterstock.com; 207c GC Minerals/Alamy Stock Photo; 207b Albert Russ/Shutterstock.com; 208ctl Only Fabrizio/Shutterstock.com; 208cbl Gozzoli/Shutterstock.com; 208bl "Topaz-tz03a" by Rob Lavinsky

/iRocks.com/Wikimedia Commons/CC BY-SA 3.0; **208**br © iStock.com/JKristoffersson; **209** GIPhotoStock/Science Source; **209**i Clive Streeter/Dorling Kindersley/Science Museum, London/Science Source; **210**t SPL/Science Source; **210**b Roy Palmer/Shutterstock.com; **211**tr Harry Taylor/Dorling Kindersley/Science Source; **211**ctr Fokin Oleg/Shutterstock.com; **211**tl Albert Russ/Shutterstock.com; **211**cl WILDLIFE GmbH/Alamy Stock Photo; **211**bl Katarzyna Białasiewicz/123R; **211**cbr Harry Taylor/Dorling Kindersley/Getty Images; **211**br Cagla Acikgoz/Shutterstock.com; **212**t National Physical Laboratory/Crown Copyright/Science Source; **212**b NikoNomad/Shutterstock.com; **213** © iStock.com/Kaellman; **213**i © iStock.com/dibrova; **214**l Colin Keates/Dorling Kindersley/Natural History Museum, London/Science Source; **214**c ©2017 National Gem Collection, National Museum of Natural History, Smithsonian.; **214**r Shawn Hempel/Alamy Stock Photo; **214**bg Rigamondis/Shutterstock.com; **215**t Science Stock Photography/Science Source; **216**t Albert Russ/Shutterstock.com; **216**b aregfly/Shutterstock.com; **216–17**bg robertharding/Alamy Stock Photo; **218** "Chilean Mine Rescue (507302505)" by Hugo Infante/Government of Chile/Wikimedia Commons/CC BY 2.0; **219** NHPA/SuperStock; **220** Javier Trueba/MSF/Science Source; **221** Steve Heap/Canopy/Getty Images

CHAPTER 10

222 Reggie David/Perspectives/Getty Images; **223**t paleontologist natural/Shutterstock.com; **223**b Jiri Vaclavek/Shutterstock.com; **223**c, **226**b, **228**cr, **239**b vvoe/Shutterstock.com; **225**t Zelenskaya/Shutterstock.com; **225**cl, **228**tr www.sandatlas.org/Shutterstock.com; **225**cr, **227**, **242**cr © iStock.com/VvoeVale; **225**b VvoeVale/iStock/Getty Images; **226–27**bg © iStock.com/crsdsgn; **226**t © iStock.com/lucentius; **228**tl Nastya22/Shutterstock.com; **228**cl © iStock.com/CribbVisuals; **228**c Valery Voennyy/Alamy Stock Photo; **228**bl Science Stock Photography/Science Source; **228**bc Kicky_princess/Shutterstock.com; **228**br © iStock.com/voljurij; **229**t Bjoern Wylezich/Shutterstock.com; **229**b Vladislav S/Shutterstock.com; **229**bg © iStock.com/OlegAlbinsky; **230** Universal Images Group via Getty Images; **232**t © iStock.com/sonsam; **232**c Scientifica/Corbis Documentary/Getty Images; **232–33**bg © iStock.com/moorhen; **232**bi © iStock.com/NLink; **233**l, r BJU Photo Services; **234**t Andreas von Einsiedel/Dorling Kindersley/Science Source; **234**c © iStock.com/malerapaso; **234**b © iStock.com/bmcent1; **235** Sumit buranaroththrakul/Shutterstock.com; **236–37**bg Aneta Waberska/Shutterstock.com; **237** "WasatchMtns ISS011-E-13889"/NASA/Wikimedia Commons/Public Domain; **238**t MarcelClemens/Shutterstock.com; **238**b Javier Trueba/MSF/Science Source; **239**t © Marli Bryant Miller; **240**t Dirk Wiersma/Science Source; **240**c © iStock.com/prmustafa; **240**b Laurie Chamberlain/Corbis Documentary/Getty Images; **241** © iStock.com/cheri131; **242**t slobo/iStockphoto/Getty Images; **242**c Dirk Wiersma/Science Source; **242**cl paleontologist natural/Shutterstock.com; **242**b Grant Kaye/Aurora/Getty Images; **245** Compass Odyssey/Gallo Images/Getty Images; **246** Pascal Goetgheluck/Science Source

CHAPTER 11

248 © iStock.com/CoreyFord; **249** "La Brea Tar Pits" by John Fladd/Flickr/CC BY-SA 2.0; **250**t Colin Keates/Dorling Kindersley/Science Source; **250**c © iStock.com/phodo; **250**b © iStock.com/clu; **250**bg © iStock.com/fotocelia; **251**tr MarcelClemens/Shutterstock.com; **251**tl "Ichthyosaurus sp 2" by Ballista/Wikimedia Commons/CC BY-SA 2.0, GFDL 1.2; **251**bl © iStock.com/tacojim; **251**br Dominique Braud/Dembinsky Photo Associates/Alamy/Alamy Stock Photo; **251**cr Gary Ombler/Dorling Kindersley/Science Source; **251**cl © iStock.com/MarcelC; **253** "Wooly Rhino at the Natural History Museum" by Paul Hudson/Wikimedia Commons/CC BY 2.0; **254**t © iStock.com/JFsPic; **254**b "Cast of a Dog Killed by the Eruption of Mount Vesuvius, Pompeii" by Giorgio Sommer/Wikimedia Commons/Public Domain; **255**t Photo courtesy of UVM Geology Department/Bill DiLillo UVM Photo; **255**b Bjoern Wylezich/Shutterstock.com; **256**t, b Republished with permission of the American Association for the Advancement of Science, from "Soft-Tissue Vessels and Cellular Preservation in Tyrannosaurus rex," Science, 25 March 2005, vol. 307, Mary H. Schweitzer, et al. Copyright © 2005. Permission conveyed through Copyright Clearance Center, Inc.; **257** Andrea Solero/AFP/Getty Images; **258–59**bg Lou Linwei/Alamy Stock Photo; **259** © iStock.com/Syldavia; **260** "Panderichthys BW" by Nobu Tamura/Wikimedia Commons/CC BY-SA 3.0, GFDL 1.2; **262** Image by V. L. Sharpton, Courtesy: Lunar & Planet Institute; **263** © iStock.com/solarseven; **264** Layne Kennedy/Corbis Documentary/Getty Images; **265** © iStock.com/bobainsworth; **266**, **267**t Science Stock Photography/Science Source; **266–67** U.S. Geological Survey; **267**b Coprid/Shutterstock.com; **268** DON EMMERT/AFP/Getty Images; **269** © iStock.com/luoman; **270** Lee Prince/Shutterstock.com; **270**i Unusual Films/Dr. Hensen; **271** JOHN CLEGG/SCIENCE PHOTO LIBRARY/Getty Images; **272–73** Matt Moyer/National Geographic Magazines/Getty Images; **274** "Ice age fauna of northern Spain" by Mauricio Antón/© 2008 Public Library of Science./Wikimedia Commons/CC BY 2.5; **275** sarkao/Shutterstock.com

CHAPTER 12

276 tusharkoley/Shutterstock.com; **278**t "Tafoni by Endico" by Dawn Endico/Wikimedia Commons/CC BY-SA 2.0; **278**c "Lateritic weathering shells around rock core. C 020" by Werner Schellmann/Wikimedia Commons/CC BY-SA 2.5; **278**b bjul/Shutterstock.com; **278–79**bg Kichigin/Shutterstock.com; **279** Krishna.Wu/Shutterstock.com; **279**i Elena Dijour/Shutterstock.com; **280**tl © iStock.com/filipe_lopes; **280**tr Spencer Platt/Getty Images; **280**b KPG Payless2/Shutterstock.com; **281** Ed Darack/SuperStock; **281**i Nick Hawkes/Shutterstock.com; **282–83** Videowokart/Shutterstock.com; **284**l, **285**t © Marli Bryant Miller; **284**r USGS; **285**b Byelikova Oksana/Shutterstock.com; **286** Fedor Selivanov/Shutterstock.com; **287** Matthias Breiter/National Geographic Magazines/Getty Images; **289**t U.S. Geological Survey/Photograph by Bruce F. Molnia/Public Domain; **289**b Beth Davidow/Visuals Unlimited/Getty Images; **290–91**bg Benny Marty/Shutterstock.com; **290**t E. R. Degginger/Science Source; **290**c Grambo Photography/All Canada Photos/Getty Images; **290**b Doug Houghton/Alamy Stock Photo; **292**t Visuals Unlimited, Inc./Gerald & Buff Corsi/Getty Images; **292**b Avalon/Photoshot License/Alamy Stock Photo; **293**t LZ Image/Shutterstock.com; **293**b Lukas Gojda/Shutterstock.com; **294**t Courtesy: National Science Foundation/Peter West/Public Domain; **294**b J. Norman Reid/Shutterstock.com; **295** Noppharat4569/Shutterstock.com; **296**t vagabond54/Shutterstock.com;

296c, b David A. Anderson; **297**t NOAA - edited version © Science Faction/Getty Images; **297**c Photo Researchers/Science Source/Getty Images; **297**b Georg Gerster/Science Source; **298**t slobo/iStockphoto/Getty Images; **298**bl Harrison Shull/Aurora Creative/Getty Images; **298**br iSiripong/Shutterstock.com; **299** © iStock.com/NanoStockk; **300** © iStock.com/robynmac; **301** Photodiem/Shutterstock.com

CHAPTER 13

302–3 Jag_cz/Shutterstock.com; **303** Answers in Genesis (AIG) Canada; **304** Marcin Sylwia Ciesielski/Shutterstock.com; **305** Cultura Creative (RF)/Alamy Stock Photo; **306–7**bg mexrix /Shutterstock.com; **307** European Space Agency/Science Source; **308** NASA/JPL/Caltech; **308–9**bg ixpert/Shutterstock .com; **310**t ©2011 Google/Data MBARI/Data SIO, NOAA, U.S. Navy, GEBCO; **310**b Sequeiros, O. E., H. Naruse, N.; **311** "Romanche Trench"/NOAA/Wikimedia Commons/Public Domain; **312**t © 2011 Google/© 2011 Europa Technologies /Data SIO, NOAA, U.S. Navy, NGA, GEBCO/Image USDA Farm Service Agency; **312**b "Atlantic-trench"/USGS /Wikimedia Commons/Public Domain; **314–15**bg atiger /Shutterstock.com; **315** NASA; **316**t stockphoto-graf /Shutterstock.com; **316**b AMatveev/Shutterstock.com; **317** trubavin/Shutterstock.com; **318** Sean Pavone/Shutterstock .com; **319**i NOAA/Public Domain; **319** Tony Camacho/Science Source; **320** Andrea Izzotti/Shutterstock.com; **321** NASA /Science Source; **322** Catmando/Shutterstock.com; **324**t Douglas Klug/Moment/Getty Images; **324**b © iStock.com /fmajor; **325**t © iStock.com/nutthaphol; **325**b Solvin Zankl /Alamy Stock Photo; **327** MAR-ECO/MIR-dives; **328** sl_photo /Shutterstock.com; **329** © iStock.com/shannonstent

CHAPTER 14

330, 351cbl NASA/Goddard Space Flight Center Scientific Visualization Studio; **331**t Fotos593/Shutterstock.com; **331**b JL-Pfeifer/Shutterstock.com; **332, 333**b Gordon Bell/Shutterstock .com; **333**t George Collins; **335**r Vlad G/Shutterstock.com; **335**l Daniel Hussey/Shutterstock.com; **337**b © 2011 Transnavicom,Ltd /Data SIO, NOAA, U.S. Navy, NGA, GEBCO/Image IBCAO /Image © 2011 DigitalGlobe/Image © 2011 TerraMetrics; **337**t IrinaK/Shutterstock.com; **343** Nature Picture Library/Alamy Stock Photo; **348** "Waves in pacifica 1" by Brocken Inaglory /Wikimedia Commons/CC BY-SA 4.0, GFDL 1.2; **349** "Rip currents at La Arean beach, Soto del Barco. Asturias. 02" by Geologist15/Wikimedia Commons/CC BY-SA 4.0; **350**t Stephen Rose/Rainbow/SuperStock; **350**b Frank Fennema/Shutterstock .com; **351**t Bildgigant/Shutterstock.com; **351**ti Chintla /Shutterstock.com; **351**ctl Pixeljoy/Shutterstock.com; **351**ctr World Pictures/Alamy Stock Photo; **351**cbr NASA; **351**b "Tombolo Paximadhi Eboea" by Tim Bekaert/Wikimedia Commons/CC BY-SA 3.0, GFDL 1.2; **352** Alex Tihonovs /Shutterstock.com; **353** Michail Makarov/Shutterstock.com

CHAPTER 15

355 Frank Pey/Alamy Stock Photo; **356** kldy/shutterstock.com; **357**t Ancient and Modern Ships/Public Domain; **357**b Photo Researchers, Inc/Alamy Stock Photo; **358–59**bg World History Archive/Alamy Stock Photo; **358** "Detail from a map of Ortelius - Magellan's ship Victoria" by Ortelius/Wikimedia Commons/Public Domain; **359** Time Life Pictures/Mansell /The LIFE Picture Collection/Getty Images; **360** Photo courtesy of Michal Koblizek, Czech Academy of Sciences; **361** Photo by Chris Linder, Woods Hole Oceanographic Institution; **362** 1928 and 1931 Hydrographic Manual/NOAA; **363**t "Ctd hg" by Hannes Grobe, Alfred Wegener Institute/Wikimedia Commoms/CC BY-SA 2.5; **363**b AP Photo/Douglas Healey; **364**t Gavin Parsons/Oxford Scientific/Getty Images; **364**b Courtesy of Wildlife Supply Company, Part #3-196-B15; **365**tl "PS1920-1 0-750 sediment-core hg" by Hannes Grobe /Wikimedia Commons/CC BY 3.0; **365**tr Photograph by Curt Stager; **365**b Ifremer/A. Fifis", 2006; **366** Image courtesy of Lewis and Clark 2001, NOAA/OER; **367**t NOAA/Public Domain; **367**bl Courtesy NASA/JPL-Caltech.; **367**br NASA; **368–69**bg littlesam/Shutterstock.com; **368**t © iStock.com /serg269; **368**c "Alexander in a submarine - British Library Royal MS 15 E vi f20v (detail)"/British Library/Wikimedia Commons/Public Domain; **368**b WaterFrame/Alamy Stock Photo; **369**t Marka/UIG via Getty Images; **369**b Poelzer Wolfgang/Alamy Stock Photo; **370**t "Dykeri, fig 6, Nordisk familjebok"/Wikimedia Commons/Public Domain; **370**c Sergiy Zavgorodny/Shutterstock.com; **370**b U.S. Navy photo by Mass Communication Specialist 2nd Class Byron C. Linder; **371**t International Ocean Discovery Program (IODP), JOIDES Resolution Science Operator (JRSO), Photographer: Bill Crawford/Public Domain; **371**b NASA/Karl Shreeves; **372** "WCS Beebe Barton 600" by NOAA/Wikimedia Commons /Public Domain; **373**t OAR/National Undersea Research Program (NURP); **373**b Jason LaVeris/FilmMagic/Getty Images; **374–75**bg aquapix/Shutterstock.com; **374** Iakov Filimonov/Shutterstock.com; **375**t NOAA Ocean Explorer; **375**b LEUT Kelli Lunt/Australia Department of Defence via Getty Images; **376** "La Chalupa research laboratory"/NOAA /Wikimedia Commons/Public Domain; **377** Oscar Bjarnason /Cultura/Getty Images

CHAPTER 16

378 © iStock.com/Tamer_Desouky; **379** © Yk32 | Dreamstime .com; **380**t Warren Price Photography/Shutterstock.com; **380**b "Little-river-townsend-y" by Brian Stansberry/Wikimedia Commons/CC BY 3.0, GFDL 1.2; **381** © iStock.com /JoeChristensen; **382** "Tibesti tributaries" by Juan Ramon Rodriguez Sosa/Wikimedia Commons/CC BY-SA 2.0; **383** Google Earth; **384**bg Jacques Jangoux/MediaBakery; **384**t wigwam press/The Image Bank/Getty Images; **384**b Voran /Shutterstock.com; **385** Image © 2012 DigitalGlobe/Google Earth; **387** "Lewis and Clark Expidition Map" by Meriwether Lewis, William Clark, Nicholas Biddle, and Paul Allen /Wikimedia Commons/Public Domain; **387**i Bettmann/Getty Images; **388–89**bg topseller/Shutterstock.com; **389** Source: Natural Resources Canada http://open.canada.ca/en/open-government-licence-canada; **390** Olesya Baron/Shutterstock .com; **391**t Hulton Archive/Getty Images; **391**c Byelikova Oksana/Shutterstock.com; **391**b © peteleclerc – Fotolia; **392**t scaners3d/Shutterstock.com; **392**b Andia/UIG via Getty Images; **393** Pyty/Shutterstock.com; **396** Olesya Baron /Shutterstock.com; **397** xbrchx/Shutterstock.com

CHAPTER 17

398 Kjell Sandved/Alamy Stock Photo; **399** Louise Gubb /Corbis Historical/Getty Images; **402**t "Quartz-arenite" by Michael C. Rygel/Wikimedia Commons/CC BY-SA 3.0; **402**b Kurt Hollocher, Union College Geology Department;

404 niti_h/Shutterstock.com; **405t** AP Photo/The Ocala Star-Banner, Alan Youngblood; **405b** "Generalized altitude for the top of the Floridan aquifer system" courtesy of U.S. Geological Survey; **406** Mariyana M/Shutterstock.com; **407t** © iStock.com/DonNichols; **407b** Accidental Photographer/Bigstock.com; **408** Africa Studio/Shutterstock.com; **409bg** Jim Wark/Passage/Getty Images; **409** © iStock.com/tothemoonphoto; **410–11bg** nipastock/Shutterstock.com; **410** AJCespedes/Shutterstock.com; **411** Ken Graham/The Image Bank/Getty Images; **412t** Raymond Gehman/National Geographic/Getty Images; **412b** Fabien Monteil/Shutterstock.com; **412bi** Eye of Science/Science Source; **414–15** Avalon/Photoshot License/Alamy Stock Photo; **417t** Patrick Landmann/Science Source; **417c** Barry Mansell/SuperStock; **417b** "Heterophrynus 02" by Graham Wise/Wikimedia Commons/CC BY 2.0; **418t, 418bi, 423, 424, 425t** © Dave Bunnell; **418b** Santi Rodriguez/Shutterstock.com; **419t** BJU Photo Services; **419b** Christopher Selin; **420t** © iStock.com/1001slide; **420–21bg** © iStock.com/zhangguifu; **420b** PhotoDisc, Inc.; **421** Dmytro Gilitukha/Shutterstock.com; **422** Luis Javier Sandoval Alvarado/Science Faction/Getty Images; **425b** "Little Brown Bat with White Nose Syndrome (Greeley Mine, cropped)" by Marvin Moriarty/USFWS/Wikimedia Commons/Public Domain

CHAPTER 18

428 Stripped Pixel/Shutterstock.com; **429** Gyuszko-Photo/Shutterstock.com; **430** Getty Images/Hemera/Thinkstock; **431t** "Space Elevator - Green Mars" by Bruce Irving/Flickr/CC BY 2.0; **431b** RIA Novosti/Science Source; **436** Juhku/Shutterstock.com; **440–41bg** beboy/Shutterstock.com; **441** NASA's Goddard Space Flight Center; **443** Image provided by the Institute for Creation Research, ICR.org. Used by permission.; **444** Russ McElroy/123RF; **445** SOHO (ESA & NASA); **446–47bg** Dmitry Pichugin/Shutterstock.com; **447t, ti** USGS/Hawaiian Volcano Observatory; **447c** © Dabldy | Dreamstime.com; **447b** "Etang Chalais reflet" by Myrabella/Wikimedia Commons/CC BY-SA 4.0; **448** Coralie.T/Shutterstock.com; **449** NASA/JSC Gateway to Astronaut Photography of Earth; **450** littleny/Shutterstock.com; **451** Christopher Morris/Corbis via Getty Images

CHAPTER 19

452 Stockr/Shutterstock.com; **453** Mimadeo/Shutterstock.com; **454** David Hay Jones/Science Source; **455t** Ruslan Gusev/Shutterstock.com; **455b** Rob kemp/Shutterstock.com; **456t** Cico/Shutterstock.com; **456b** Ramil Gibadullin/Shutterstock.com; **457t** Garry McMichael/Science Source/Getty Images; **457b** Guy J. Sagi/Shutterstock.com; **459** Laurence Dutton/Stone/Getty Images; **462t** Grzejnik/Shutterstock.com; **462c** © iStock.com/narvikk; **462bl** © iStock.com/john davies; **462br** Lee Yiu Tung/Shutterstock.com; **463** Noradoa/Shutterstock.com; **464–65bg** © iStock.com/Jasmina007; **465** Kelly/Mooney Photography/Corbis Documentary/Getty Images; **466tl** CE Photography/Shutterstock.com; **466tr** SanchaiRat/Shutterstock.com; **466cl** Korionov/Shutterstock.com; **466cr** "Altocumulus se 2" by Kr-val/Wikimedia Commons/Public Domain; **466bl** © iStock.com/YavorKalev; **466br** Kent Wood/Science Source; **467tl, cl** alybaba/Shutterstock.com; **467bl** "Nimbostratus virga grey with hills" by Simon A. Eugster/CC BY-SA 3.0, GFDL 1.2; **466–67bg** elbud/Shutterstock.com; **467r** Greg Dale/MediaBakery; **468** Tatiana Popova/Shutterstock.com; **469t** Warren Faidley/Corbis Documentary/Getty Images; **469cl, cr** Dover Publications/Public Domain; **469b** Hassan Akkas Reine – Fotolia; **470** Kyreichenko Anastasiia/Shutterstock.com; **471** Aneese/Shutterstock.com; **472** NebraskaSC - Dale Kaminski/Moment/Getty Images; **473** Mount Washington Observatory

CHAPTER 20

474 GryT/Shutterstock.com; **476** lexuss – Fotolia; **477** Ryan McGinnis/Moment/Getty Images; **479t** Alison Eckett/Alamy Stock Photo; **479b** AridOcean/Shutterstock.com; **479bti** elisalocci/iStock/Getty Images Plus; **479bbi** By/Shutterstock.com; **481** Signature Exposures Photography by Shannon Bileski/Moment/Getty Images; **481i, 491t, br, 496, 497** NOAA; **482** "March 2014 nor'easter 2014-03-26"/NOAA/Wikimedia Commons/Public Domain; **482–83bg** © iStock.com/BanksPhotos; **483t** nd700 – Fotolia; **483b** Steve Heap/Shutterstock.com; **484** Martin Larcher/Shutterstock.com; **485r** Jim Reed/Corbis Documentary/Getty Images; **485l** NASA; **486–87** Cultura RM Exclusive/Jason Persoff Stormdoctor/Getty Images; **488** Niccolò Ubalducci Photographer - Stormchaser/Moment/Getty Images; **489** PhotoDisc, Inc.; **491bl** NOAA/NWS/Public Domain; **492** Mario Tama/Getty Images; **493** AP Photo/Eric Gay; **494** Jim Reed/SuperStock; **496–97bg** jessicahyde/Shutterstock.com; **498** Bennett Barthelemy/Aurora/Getty Images; **499** Martin Barraud/MediaBakery; **500** © iStock.com/CampPhoto; **501** © iStock.com/Bracker58

CHAPTER 21

502 coolbiere photograph/Moment/Getty Images; **503** © iStock.com/Pgiam; **503i** Mark Van Scyoc/Shutterstock.com; **506** John Warburton Lee/SuperStock; **507** Public Domain; **508t** AridOcean/Shutterstock.com; **508b** NASA; **510** Courtesy of Lonnie Thompson; **512–13bg** "World Koppen Map" by Peel, M. C., Finlayson, B. L., and McMahon, T. A.(University of Melbourne)/Wikimedia Commons/CC BY-SA 3.0; **512l** Auhustsinovich/Shutterstock.com; **512tr** leoks/Shutterstock.com; **512br** kkaplin/Shutterstock.com; **513t** MOROZ NATALIYA/Shutterstock.com; **513c** Quang Vu/Shutterstock.com; **513b** pisaphotography/Shutterstock.com; **514–15** Wolfgang Kaehler/SuperStock; **516t** NASA image courtesy of the LANCE/EOSDIS MODIS Rapid Response Team at NASA GSFC; **516b** Matt Champlin/Moment/Getty Images; **517t** blickwinkel/Alamy Stock Photo; **517b** Photo by Josh Landis/National Science Foundation/Public Domain; **518l, r** NOAA/NESDIS; **519t** Sergieiev/Shutterstock.com; **519b** Philippe Bourseiller/The Image Bank/Getty Images; **520** NOAA+RDC (D.V. Hoyt)+CNRS/INSU (J.-P. Legrand) + Ondrejov Obs. (K. Krivsky); **522t** Chart first published by Dr Euan Mearns at http://euanmearns.com; **522b** Carbon Engineering; **523** Joseph Sohm/Shutterstock.com; **524** Pingpao/Shutterstock.com; **525** Danita Delimont/Gallo Images/Getty Images; **526** Paul Souders/The Image Bank/Getty Images; **527t** "20130817-FS-UNK-0034" by USDA/Flickr/CC BY 2.0; **527b** "GHCN Temperature Stations" by Robert A. Rohde/Wikimedia Commons/GFDL, CC BY-SA 3.0

CHAPTER 22

530 David Nunuk/Science Source; **531** NASA/SDO; **534–35** "The Sun by the Atmospheric Imaging Assembly of NASA's Solar Dynamics Observatory - 20100819-02" by NASA/SDO

(AIA)/Wikimedia Commons/Public Domain; **535**t Vacuum Tower Telescope, NSO, NOAO; **535**b NASA/JAXA/Hinode; **536**t © 2001 by Fred Espenak, www.Mr.Eclipse.com; **536**b Sacramento Peak Observatory Association of Universities for Research in Astronomy, Inc.; **537** SOHO - EIT Consortium, ESA, NASA; **537**li SOHO/NASA; **537**ci NASA/Goddard Space Flight Center; **537**ri TRACE Project, NASA; **538** "ISS-44 pass through Aurora australis" by NASA/Scott Kelly/Wikimedia Commons/Public Domain; **539, 540**tr, **540–41, 547**b Claudio Divizia/Shutterstock.com; **540**tl "The Earth seen from Apollo 17 with white background" by NASA/Apollo 17 crew /Wikimedia Commons/Public Domain; **540**b USGS/NASA /JPL; **541**t "Apollo 13 - bright-rayed crater" by NASA /Wikimedia Commons/Public Domain; **541**b Peter van de Haar; **542**t "Mare Tranquillitatis (LRO)" by NASA/Wikimedia Commons/Public Domain; **542**b NASA/GSFC/Arizona State Univ./Lunar Reconnaissance Orbiter; **543**t "Montes Apenninus (LRO)" by NASA/Wikimedia Commons/Public Domain; **543**bg, b, **544, 546** NASA; **545** © iStock.com/aryos; **547**t Herman Eisenbeiss/Science Source/Getty Images; **548**l © 2004 Hemera Technologies, Inc.; **548**r JPL/NASA; **551** Detlev Van Ravenswaay/Science Source; **552** Derek Croucher /Photographer's Choice/Getty Images; **553** "Tideforcenw" by Rmo13/Wikipedia/GFDL, CC BY-SA 3.0; **554–55** Vadim .Petrov/Shutterstock.com; **556** "Partial solar eclipse Oct 23 2014 Minneapolis 5-36pm Ruen1" by Tomruen/Wikimedia Commons/CC BY-SA 4.0; **557** "Planetary Conjunction over Paranal" by G.Hüdepohl (atacamaphoto.com)/ESO/Wikimedia Commons/CC BY 4.0

CHAPTER 23

559 Universal History Archive/UIG via Getty Images; **560–61**t Detlev van Ravenswaay/Science Source; **561**b Public Domain; **563** Copernicus (colour litho), Huens, Jean-Leon (1921-82) /National Geographic Creative/Bridgeman Images; **564**t "Tycho-Brahe-Mural-Quadrant"/Wikimedia Commons/Public Domain; **564**c Public Domain work by Giuseppe Bertini (1858), of Galileo showing the Doge of Venice how to use the telescope; **564**b "Johannes Kepler 1610"/Wikimedia Commons/Public Domain; **566**t ullstein bild/ullstein bild via Getty Images; **566**b Public Domain; **568** (Sun) "The Sun by the Atmospheric Imaging Assembly of NASA's Solar Dynamics Observatory - 20100819-02" by NASA/SDO (AIA)/Wikimedia Commons/Public Domain; **568, 569** (Mercury, Saturn), **571**b, **575**t Vadim Sadovski/Shutterstock.com; **568, 569** (Venus, Mars, Jupiter), **569** (Uranus), **572**t, b, **574**t, **575**b, **577** NASA images/Shutterstock.com; **568, 569** (Earth) "The Earth seen from Apollo 17 with white background" by NASA/Wikimedia Commons/Public Domain; **569** (Neptune) © iStock. com/3quarks; **570** NASA/JPL/Space Science Institute; **571**t NASA/SDO, HMI; **572**c "The Earth seen from Apollo 17 with white background" by NASA/Wikimedia Commons/Public Domain; **573**t NASA; **573**b NASA/JPL/Cornell; **574**b Tristan3D/Shutterstock.com; **576** © iStock.com/3quarks; **580** Universal History Archive/Getty Images; **581** Photo Researchers/Science Source/Getty Images; **582**i © iStock.com /Africanway; **582–83** Zack Frank/Shutterstock.com; **584** NASA/JPL; **585** NASA/CampoAlto/V. Robles; **586** NASA Ames

CHAPTER 24

587 NASA/JPL-Caltech/STScI; **588** Charles Mostoller/Barcroft USA/Getty Images; **589** Image Work/amanaimagesRF/Getty Images; **589**i © iStock.com/Hollygraphic; **591** Dorling Kindersley/Getty Images; **594** "Sombrero Galaxy M104" by Hubble Heritage/Flickr/CC BY-SA 2.0; **596**t Sunti /Shutterstock.com; **596**b John Sanford/Science Source; **597** ESO/CC BY-SA 3.0; **598**t "Size IK Peg" by RJHall, chris (vector)/Wikimedia Commons/GFDL 1.2, CC BY-SA 3.0; **598**b "Keplers supernova" by NASA/ESA/JHU/R.Sankrit & W.Blair /Wikimedia Commons/Public Domain; **599** "Black Hole Milkyway" by Ute Kraus, Physics education group Kraus, Universität Hildesheim, Space Time Travel, (background image of the milky way: Axel Mellinger)/Wikimedia Commons/CC BY-SA 2.0 Germany; **600** NASA, ESA, H. Bond (STScI) and M; **601**l "Orion Belt 2009-01-29" by Astrowicht/Wikimedia Commons/GFDL 1.2, CC BY-SA 3.0; **601**r John R Foster /Science Source/Getty Images; **602** Luciano Corbella/Getty Images; **603**t "A Swarm of Ancient Stars - GPN-2000-000930" /NASA, The Hubble Heritage Team, STScI, AURA/Wikimedia Commons/Public Domain; **603**b "Milky Way 2005" by NASA /JPL/R. Hurt/Wikimedia Commons/Public Domain; **604–5**bg NASA, ESA, and the Hubble Heritage Team (STScI/AURA); Acknowledgment: M. Sun (University of Alabama, Huntsville); **604**tl X-ray: NASA/CXC/SAO; Optical: Detlef Hartmann; Infrared: NASA/JPL-Caltech; **604**bl NASA, ESA, and The Hubble Heritage Team (STScI/AURA); **604**r NASA, ESA, and A. Aloisi (Space Telescope Science Institute and European Space Agency, Baltimore, Md.); **605** NASA/JPL-Caltech/MSSS; **606** "Artist's rendering ULAS J1120+0641" by ESO/M. Kornmesser/Wikimedia Commons/CC BY 4.0; **606**i "Quasar HE 1013-2136 with Tidal Tails" by ESO/Wikimedia Commons /CC BY 4.0; **608** "Horn Antenna-in Holmdel, New Jersey" by NASA/Wikimedia Commons/Public Domain; **610**bg NASA images/Shutterstock.com; **610**t NASA/UH88/Nedachi et al.; **610**b NASA/DOE/Fermi LAT Collaboration; **611**t "Ilc 9yr moll4096" by NASA/WMAP Science Team/Wikimedia Commons/Public Domain; **611**b Answers in Genesis; **612–13**bg Basti Hansen/Shutterstock.com; **613** NASA/ESA, The Hubble Key Project Team and The High-Z Supernova Search Team; **614** "5 Local Galactic Group (ELitU)" by Andrew Z. Colvin/Wikimedia Commons/CC BY-SA 3.0, GFDL; **615** © iStock.com/den-belitsky; **616** NASA/JPL-Caltech/Univ.of Ariz.; **617** Stocktrek Images/Getty Images

CHAPTER 25

618 NASA/JPL-Caltech/Malin Space Science Systems; **619**l © Fotolia/modustollens; **619**r NICHOLAS KAMM/AFP/Getty Images; **619**ri Steve Patlan/CC BY-NC 2.0; **620**t "Yerkes 40 inch Refractor Telescope-2006" by user Kb9vrg/Wikimedia Commons/Public Domain; **620–21**b Richard Wainscoat /Alamy Stock Photo; **621**t NAOJ; **622** Arecibo Radio Observatory National Astronomy and Ionosphere Center; **622**i © iStock.com/strayarts; **623** STS-82 Crew, STScI, NASA; **624** Nerthuz/Shutterstock.com; **625**t NASA/MSFC; **625**b antonsav /Shutterstock.com; **626** David Ducros/Science Source; **628** Kittipong Jirasukhanont/123RF; **629**t Novosti Photo Library /Science Source; **629**b NASA/JPL; **629**bbg Claudio Divizia /Shutterstock.com; **630**t "Ap6-68-HC-191" by NASA/Public Domain; **630**bl NASA/Jet Propulsion Laboratory – Caltech; **630**br, **631, 632–33**bg NASA/JPL; **632** NASA/JPL/Corby

Waste; **633** NASA/JPL-Caltech/MSSS; **634** NASA/Goddard/Chris Gunn; **635** "President John F. Kennedy speaks at Rice University" by NASA/Wikimedia Commons/Public Domain; **636**t, **637**t, b, bi, **639**, **639**i, **641**t, **645**b, **646**, **647**, **648** NASA; **636**b SPL/Science Source; **638** Everett Historical/Shutterstock.com; **640**t NASA/MSFC; **640–41**b xtock/Shutterstock.com; **642** NASA - digital version copyright Science Faction/Getty Images; **643** NASA Photo/Alamy Stock Photo; **644** Irina Kuzmina/Shutterstock.com; **645**t OFF/AFP/Getty Images

BACK MATTER

649 USGS; **659** © 2012 The Geological Society of America; **661** USGS, Gail P. Thelin and Richard J. Pike

Maps from Map Resources